THE OXFORD HANDBOOK OF

INNOVATION MANAGEMENT

D1264751

THE OXFORD HANDBOOK OF

INNOVATION

MANAGEMENT

Edited by

MARK DODGSON,

DAVID M. GANN,

and

NELSON PHILLIPS

OXFORD

UNIVERSITY PRESS

OXFORD
UNIVERSITY PRESS

Great Clarendon Street, Oxford, OX2 6DP,
United Kingdom

Oxford University Press is a department of the University of Oxford.
It furthers the University's objective of excellence in research, scholarship,
and education by publishing worldwide. Oxford is a registered trade mark of
Oxford University Press in the UK and in certain other countries

© Oxford University Press 2014

The moral rights of the authors have been asserted

First published 2014
First published in paperback 2015

Published in the United States of America by Oxford University Press
198 Madison Avenue, New York, NY 10016, United States of America

British Library Cataloguing in Publication Data
Data available

Library of Congress Cataloging in Publication Data
Data available

ISBN 978–0–19–969494–5 (Hbk.)
ISBN 978–0–19–874649–2 (Pbk.)

For Sheridan, Anne, and Cristina

ACKNOWLEDGEMENTS

Our first and greatest thank you should be to the contributors to this Handbook. Writing authoritative chapters that concisely summarize substantial literatures and identify their most important themes is a challenging task, and the authors continually surpassed our high expectations. We are grateful to them all.

We would also like to acknowledge the funds supporting the production of this volume from the University of Queensland Business School, Australian Research Council, and the Innovation Studies Centre at Imperial College, supported by the UK Engineering and Physical Sciences Research Council.

Chapter authors were prevailed upon to review other chapters, for which many thanks are due. We also wish to thank Andreas Eisingerich, Anna Plodowski, and Ileana Stigliani for their comments on individual chapters.

Contents

LIST OF FIGURES

List of Tables

LIST OF CONTRIBUTORS

Pamela Adams, Professor of International Management, Franklin College, Switzerland.

Gautam Ahuja, Harvey C. Fruehauf Professor of Business Administration, Ross School of Business, University of Michigan.

Oliver Alexy, Professor of Strategic Entrepreneurship, School of Management, Technical University of Munich.

Erkko Autio, Professor of Technology Venturing and Entrepreneurship, Imperial College Business School, Imperial College.

Michelle Barton, Assistant Professor in Organizational Behaviour, School of Management, Boston University.

Frans Berkhout, Professor of Innovation and Sustainability, and Director of the Amsterdam Global Change Institute, the VU University Amsterdam.

Michael A. Cusumano, Sloan Management Review Distinguished Professor of Management, Sloan School of Management, Massachusetts Institute of Technology.

Linus Dahlander, Associate Professor, European School of Management and Technology.

Andrew Davies, Professor in the Management of Projects, Bartlett Faculty of the Built Environment, University College London.

Claudio Dell'Era, Assistant Professor, Department of Management, Economics and Industrial Engineering, Politecnico di Milano.

Mark Dodgson, Professor and Director of the Technology and Innovation Management Centre, University of Queensland Business School.

Graham Dover, Mindset Foundation.

Nicolai J. Foss, Professor, Department of Strategic Management and Globalization, the Copenhagen Business School, and Department of Strategy and Management and Center for Service Innovation, Norwegian School of Economics.

Nik Franke, Professor and Founder, Institute for Entrepreneurship and Innovation, Vienna University of Economics and Business.

Sascha Friesike, Researcher, Alexander von Humboldt Institute for Internet and Society, Humboldt University of Berlin.

Takahiro Fujimoto, Professor, Faculty of Economics, University of Tokyo.

Bryan Gallagher, Researcher, Beedie School of Business, Simon Fraser University.

Alfonso Gambardella, Professor of Corporate Management, Bocconi University.

David M. Gann, Vice-President (Development and Innovation), Imperial College, and Professor, Imperial College Business School.

Oliver Gassmann, Professor and Director, Institute for Technology Management, University of St. Gallen.

Annabelle Gawer, Assistant Professor in Strategy and Innovation, Imperial College Business School.

Paola Giuri, Associate Professor, Department of Management, University of Bologna.

Andrew Hargadon, Charles J. Soderquist Chair in Entrepreneurship, Professor of Technology Management, Graduate School of Management, University of California, Davis.

Lars Håkanson, Professor of International Business, Department of International Economics and Management, Copenhagen Business School.

Alan Hughes, Margaret Thatcher Professor of Enterprise Studies and Director, Centre for Business Research, University of Cambridge and Director, UK Innovation Research Centre.

Tim Kastelle, Senior Lecturer, University of Queensland Business School.

Jerry Kim, Assistant Professor, Columbia Business School.

Keld Laursen, Professor, Department of Organizational Economics and Innovation, Copenhagen Business School and Centre for Service Innovation, Department of Strategy and Management, Norwegian School of Economics.

Thomas B. Lawrence, Weyerhaeuser Professor of Change Management, Faculty of Business Administration, Simon Fraser University.

Aija Leiponen, Associate Professor, Charles H. Dyson School of Applied Economics and Management, Cornell University and Imperial College Business School, Imperial College London.

Dorothy Leonard, William J. Abernathy Professor of Business Administration Emerita, Harvard Business School.

Franco Malerba, Professor of Applied Economics and Director, Centre for Knowledge, Internationalization, and Technology Studies, Bocconi University.

Lorenzo Massa, Researcher, University of Bologna and Vienna University of Economics and Business.

Rita Gunther McGrath, Associate Professor, Columbia Business School.

Maureen McKelvey, Professor, Institute of Innovation and Entrepreneurship, University of Gothenburg.

Elena Novelli, Lecturer in Management, Cass Business School, City University London.

Ritsuko Ozaki, Senior Research Fellow, Imperial College Business School.

Nelson Phillips, Professor of Strategy and Organizational Behaviour, Imperial College Business School.

Jaideep Prabhu, Jawaharlal Nehru Professor of Indian Business and Enterprise and Director of the Centre for India and Global Business at Judge Business School, University of Cambridge.

Ammon Salter, Professor of Technology and Innovation Management, Imperial College Business School.

John Steen, Associate Professor, University of Queensland Business School.

Bruce S. Tether, Professor of Innovation Management and Strategy, Manchester Business School.

Llewellyn D. W. Thomas, Lecturer, Imperial College Business School.

Salvatore Torrisi, Professor of Strategic Management, University of Bologna.

Christopher L. Tucci, Professor of Management of Technology, Ecole Polytechnique Fédérale de Lausanne.

Roberto Verganti, Professor of Management of Innovation, Politecnico di Milano.

Maximilian von Zedtwitz, Professor and Director of Research Centre for Global R&D Management, University of St. Gallen and Tongji University.

Marina Yue Zhang, Associate Professor, School of Economics and Management, Tsinghua University.

PART I

INTRODUCTION

CHAPTER 1

PERSPECTIVES ON INNOVATION MANAGEMENT

MARK DODGSON, DAVID M. GANN, AND NELSON PHILLIPS

INTRODUCTION

INNOVATION is an essential means by which organizations survive and thrive. As a result, innovation must be managed, but before it can be managed it needs to be understood. This Handbook addresses the wide range of management processes and structures supporting innovation. It is concerned with understanding the nature and dynamics of innovation and the contextual influences affecting innovation choices: historical, social, economic, cultural, legal, and technological. These shape the strategies and practices decision-makers use to improve organizational benefits from innovation. It encompasses the choices managers make regarding what innovations to pursue, and how they develop, introduce, and gain value from their endeavours.

Innovation management is an important area of study because the differing abilities of organizations to obtain benefits from innovation depend upon how well it is managed. Innovation contributes centrally to economic performance, corporate competitiveness, environmental sustainability, levels and nature of employment, and, in the final analysis, overall quality of life. There are widespread social and economic benefits from innovation, but the organizational returns from it are skewed towards those better at managing its risks and complexities.

The immense contributions innovation has made to economic welfare and social well-being have depended on innovation managers successfully overcoming its many challenges. The risks, costs, and timescales of innovation often conflict with the financial objectives, operational routines, and managerial incentives found in most organizations. The best returns to innovation may be accrued not by the innovator, but by those that emulate and copy. Innovation disrupts markets, technologies, and workplaces. It requires levels of collaboration across professional and organizational boundaries, and tolerance of failure, that organizations find difficult to coordinate and sanction. In many instances it involves efforts to manage activities and events that are beyond the control of even their most influential contributors. At the same time, and despite these difficulties, innovation can be the most stimulating and rewarding of all organizational activities.

The study of innovation management builds upon understanding of the sources, nature, and outcomes of innovation and the economic, technological, and social context in which it occurs. There is a long tradition of research in this broader area of innovation studies, ably portrayed in *The Oxford Handbook of Innovation* (Fagerberg et al., 2005), but innovation management is a more specific, recent, and emerging area of study. This Handbook will take careful account of knowledge about innovation in general, but its interests lie particularly with how innovation is managed and the broader contextual factors that influence its management. Its concern lies with innovation within the organization and factors that affect its occurrence: its sources, strategies, and practices. It will also address the dramatic changes that have occurred over recent years in innovation resulting from new strategies and practices in companies, for example around business models, design and innovation ecosystems, and the opportunities provided by new digital technologies. There has been a recent paradigm shift in our understanding of innovation that significantly expands its scope, and this is captured in this Handbook.

This chapter offers a number of perspectives on innovation management as a developing field of study, on explanatory theories, recurrent challenges, and on its application to innovation processes. The chapter briefly introduces the rich contributions, made on a wide range of issues of innovation management, from the leading scholars whose efforts have produced this Handbook.

Before these explorations into innovation management begin, it is useful to summarize some general features of innovation. In the following chapter by Salter and Alexy, contemporary thinking about innovation in general is captured by a number of 'stylized facts', which help establish the basis for the discussions of innovation management in the rest of the book. It shows how innovation creates growth, takes different forms, is pervasive, and is based on relationships and new combinations. It discusses the patterns, speed, geography, and routines underlying innovation. Along with this chapter on perspectives on innovation management, the chapter by Salter and Alexy provides context for what follows in the rest of the Handbook.

THREE CHALLENGES IN THE STUDY OF INNOVATION MANAGEMENT

1. Defining the Scope of Innovation Management

The term 'innovation' is used widely and promiscuously. As a result there is an unhelpfully extensive range of activities included under the rubric of innovation management.

If innovation management is said to include breakthroughs at the cutting edge of science, or revolutionary new business models, on the one hand, and providing new colour options for products, or forms for reporting, on the other, then its scope is too broad to develop coherent and meaningful analysis.

We are content with the widely accepted definition of innovation as the successful application of new ideas, but believe that for analytical and practical purposes the definition of innovation management has to be more nuanced. Clearly ascertaining the specific aspects, levels, and types of innovation to be managed is crucial for improving understanding.

Innovation is both an outcome and a process, a fact and an act. An innovative outcome involves the successful application of new ideas, which results from organizational processes that combine various resources to that end. Its objectives are to produce positive results for organizations and their employees, customers, clients, and partners—such as growth, profit, sustainability, and job security—with better and cheaper products and services for consumers, and personal satisfaction for its contributors. Achieving these requires a process that creates, delivers, and captures innovative outcomes by combining and coordinating resources—including people and knowledge, finance, technology, physical spaces, and networks—and their capabilities—that is, their bundles of skills.

The innovative outcomes that have received the most attention by management researchers in the past have been in new and improved products, followed by operational processes, with services lagging a long way behind. These all remain important, even as the boundaries between them become blurred (smartphones, for example, can represent all three), but innovation is also found in new markets, ways of organizing, and constructing means of producing value in business models. Innovation management addresses all these types of innovation.

Innovation has always been driven by new market and technological opportunities, but innovation emerges from many potential sources and has a multiplicity of influences. The stimulus to innovate, for example, may derive from new regulations or technical standards, competition forcing firms to develop new solutions, new funding prospects, collaborative partners, small entrepreneurial firms, or the ideas of employees across the organization. These combine to produce a complex and interrelated array of contributors to the innovation process.

Innovation extends well beyond the mechanisms that drive it—such as invention, creativity, and the imaginative recombination of existing ideas and technologies—or the processes that encourage its implementation, such as change management. Creativity contributes to the origination of ideas and invention entails showing how ideas work in practice. The classic Schumpeterian notion of innovation as the recombination and reconstitution of resources highlights the importance of merging existing ideas and artefacts in new ways. Innovation management requires knowledge of all these sources and of how ideas can be successfully applied. The application of ideas may involve learning and re-skilling, and change management that transitions people and organizations along pre-determined and well-charted paths, but is also often characterized by experimentation, risk, and uncertainty. As pointed out in chapter 20 on innovation strategy by McGrath and Kim, change management is less of an issue for innovative organizations that continually adjust and renew their capabilities as a matter of course.

The extent of risk and uncertainty associated with innovation depends upon its ambition and amplitude. Incremental innovations occur in established markets, technologies, and ways of doing things close to an organization's existing activities. Radical innovations involve breakthroughs in markets, technologies, and ways of doing things very different from those supported by an organization's established resources and capabilities. Between these two levels on the innovation continuum are those substantial innovations that build upon existing activities, extending and diversifying them into new areas. Incremental innovations involve the renovation of existing products and processes and are the most common form of innovation. Radical innovations are rare, but can be highly consequential. Individual chapters in this book address the management of incremental and radical innovation, but the vast majority of chapters are concerned with those intermediate levels of innovation that require significant changes in resources and capabilities. They reflect the way the major concern of innovation management lies less with doing everyday things better or engaging in highly uncertain projects, and more with the controlled ambition and risk of doing challenging new things.

2. The Changing Nature of Innovation Management

Time is a crucial issue in understanding and managing innovation. The costs of investing in innovation are immediate, while the returns can be long term. The long-term benefits may create value unappreciated at the time of investment. The investment in underground railways and sewers in Victorian London produced billions of pounds of value 150 years later. Changes occur over time: today's incremental innovations may be based on yesterday's radical innovations, and these can occur quickly. One of the difficulties in studying innovation management is that all types of innovation can occur with remarkable speed: substantial new businesses and technologies can emerge in a very short time. Researchers studying the latest innovations may discover that their findings have been superseded by the time they publish. The innovation process itself, furthermore, also changes as a result of the application of new organizational approaches and technologies that speed up the manipulation of information and ideas, for example, by the Internet and social media. Research into innovation management has evolved as innovation processes change over time.

Joseph Schumpeter, the doyen of innovation economists, began his analysis of innovation in the early twentieth century predominantly focusing on the actions of individual entrepreneurs. The growth of formally organized research and development (R&D) departments in the 1920s and 1930s occurred during his lifetime, and his later works on the economics of innovation in the 1940s focused on the role of corporations. The transformational impact of research and analysis can also be seen historically. Adam Smith wrote of the advantages of the division of labour—the specialization of replicable tasks—and his observations were shortly thereafter put to productive use by pioneers of the Industrial Revolution such as Matthew Boulton and Josiah Wedgwood.

The consequence of such research is seen in the history of the automotive industry. The development of mass production techniques for automobiles, epitomized by Henry Ford, led to research on improving productivity in industry more widely through further specialization using the 'time and motion' studies associated with F. W. Taylor and Gilbreth. In contrast to this approach, which generally led to the de-skilling of workers, the Quality of Working Life movement emerged in the 1970s, allied to experiments with multitasking in the Volvo Company in Sweden. Studies of Japanese car making and the Toyota production system—described as 'lean production'—in the 1980s inspired the replication of its practices, such as 'just in time' delivery of components and certain quality management techniques, around the world. Innovation and innovation management research continue to co-evolve, and they necessarily have to be studied in a dynamic and interrelated manner.

There are robust lessons for innovation management in past experiences, but as innovation outcomes and processes are continually evolving, understanding contemporary practices is crucial. Here the study of innovation management not only faces the problem of the uncertain progress of businesses and technologies, but also that of particular management fads, to which the field is especially vulnerable. The complexity of organizational problems is often in inverse proportion to the enthusiasm for finding simple or all-encompassing solutions to them. Innovation management has seen a plethora of supportive tools and techniques emerge, mostly originating from academic research into a few organizations and generalized into consulting offerings. Some of these, which will be described later in this book, have retained value, but most have at one time or another been oversold and used inappropriately. The challenge for innovation researchers is to determine and retain the value of the tried and tested, while maintaining interest in the new and emerging with sufficient degrees of circumspection and caution.

3. Merging Disciplines, Levels of Analysis, and Research Methods

As revealed by the diverse backgrounds of the contributors to this book, the study of innovation management draws on a wide range of academic disciplines. Authors in this volume are scientists, engineers, economists, historians, geographers, psychologists, sociologists, and students of management and organizations. This plurality is inevitable because innovation management has wide-ranging concerns. A major challenge for innovation management scholarship generally, and more particularly for this book, is to build synergies between its different aspects being studied.

There is considerable value in connecting practice and context. Although innovation management can be highly idiosyncratic, reflecting differences in an individual organization's markets, technologies, resources, and capabilities, it is broadly affected by the wider context in which it occurs. Chapter 9 by Hargadon, on brokerage, shows how the genesis and impact of innovation are affected by the interrelationships between institutions, organizations, small teams, and individuals. Other research shows how the position of the organization in the industry and product life cycle affects the kind of innovation sought (Abernathy and Utterback, 1975). Whether or not organizations are part of particular technological trajectories (Dosi, 1982), or how their circumstances depend on the accumulation of particular assets, can be influential. Reaping returns to innovation depends on the extent to which organizations rely on the provision of complementary assets—the related resources needed to gain value from an innovation— by other organizations, and the method by which returns are appropriated (what Teece (1986) calls appropriability regimes).

Innovation also occurs in the context of various collections or systems of institutions and the character of connections within them. National innovation systems include the institutions of research, education, finance, and law, and the quality of relationships amongst their various contributors. These importantly include the nature of the relationships between users and suppliers, and those within geographical or industrial clusters. National and pan-national regulations are highly influential. History also matters. As shown by Fujimoto in Chapter 17 on Japanese innovation management, its practices owe much to the legacy of labour shortages after the Second World War.

Many studies of innovation management have addressed particular sectors or technologies. Much research in the 1980s, for example, focused particularly on the automotive industry, and there continues to be special interest in 'high-tech' sectors such as advanced engineering, information and communication technologies (ICT), and biotechnology. This has been balanced to some extent by the study of more traditional, but not necessarily less innovative sectors, such as construction. There remains a paucity of good studies of innovation management in service sectors, such as banking and insurance. Malerba and Adams in Chapter 10 discuss the important influence of sectoral differences on innovation management. Sectoral systems of innovation in ICT, for example, are in many ways unlike those in textiles. Using examples of pharmaceuticals, machine tools, and services, Malerba and Adams provide a framework that links knowledge and sources of innovation with the actors and institutions involved to explain the dynamics of innovative activity within and across sectoral boundaries. This framework is a valuable addition to the innovation manager's toolbox in helping analyse the context in which their organizations innovate.

Further analysis of services is provided by Tether in Chapter 30, which highlights the specific characteristics of services, including their intangible and perishable nature. Tether shows how services innovation differs from innovation in manufacturing, in that it is typified by frequent involvement by users and providers of complementary services, is less reliant on specific departments, such as R&D, and is more distributed with many diverse contributors. Many service innovations, he argues, involve business model innovation, and he offers a framework of stages and associated tools for services design.

The extent to which innovation management strategies and practices are transferable across sectors and technologies remains a germane question for researchers.

How do lessons about private, for profit innovation, for example, translate into areas of social innovation? Chapter 16 by Lawrence, Dover, and Gallagher argues that interest in managing social innovation has been growing, but there has not been a corresponding increase in research in the area. They review the existing literature around four themes that characterize understanding of social innovation: starting with social problems, focusing on novel solutions, varying potential organizing models, and benefiting beyond the innovators. They argue future research should recognize the construction of social problems and their historical and social embeddedness, and how the need for political and ethical considerations has to be taken into account.

Most research in innovation management has, furthermore, focused on US, European, and Japanese firms, and this needs to change (Dodgson et al., 2008). As Chapter 18 by Zhang shows, there has been remarkable growth of innovative capabilities in China. The Chinese model of innovation management is strongly influenced by government policies and China's culture, but through learning from multinational companies and developing their own practices, Chinese approaches to innovation provide an important future direction for the study of innovation management. One of the intriguing insights in the chapter by Zhang is the distinction between efficiency-led business models in the West and effectiveness-led models in China.

Decisions about innovation inevitably involve issues of finance. Whether it is concerned with levels and quality of venture capital, or the capacity of firms to raise capital in markets or invest retained earnings, the availability of finance is essential for innovation. Chapter 13 by Hughes places the issue of innovation funding within the broader context of national governance of capital markets and financial systems. Hughes shows the considerable variation in the balance of public and private funding of R&D, and draws on analysis of varieties of capitalism and systems of innovation to identify trends in financing. He also considers the impact of the 2008/9 financial crisis for the financing of innovation.

Case studies reveal a great deal about innovation management. There are rich case histories of both large multinational companies such as DuPont (Hounshell and Smith, 1988), Toyota (Fujimoto, 1999), Microsoft (Cusumano and Selby, 1995), and Corning (Graham and Shuldiner, 2001). The best of these illuminate how innovation complemented overarching corporate strategies, and provide insights and examples of the management practices used to innovate efficiently. Scope remains for many more case studies of innovation management in small firms, the especial challenges they face due to relative shortages of resources, and the advantages they possess in flexibility and responsiveness. Chapter 4 by McKelvey on science, technology, and business research discusses the importance of small firms as vehicles for transferring science into innovation. Studies of particular innovations, from the hovercraft (Rothwell and Gardiner, 1985) to the Internet (Tuomi, 2002), also throw light on effective management strategies and practices. These case studies do not have to be contemporary to be valuable, with many insights provided by great innovators such as Thomas Edison (Hargadon and Douglas, 2001) and Josiah Wedgwood (Dodgson, 2011). One of the most illustrative studies of organizational opposition to innovation is that of continuous aim gunfire by the British and US navies, developed at the end of the nineteenth century (Morison, 1988). Case studies are also useful in examining the inevitable organizational and interpersonal tensions involved (Webb, 1992).

Surveys of numbers of innovations, R&D expenditures, and patents, produced by organizations such as the OECD and European Union, are useful from an innovation management perspective when they highlight the different contexts in which organizations innovate. One of the earliest and most original empirical studies of innovation—Project SAPPHO (Rothwell et al., 1974)—showed how the challenges of innovation differed between sectors. One of the problems with many studies of innovation management based on patent data is the frequent inappropriate association of patenting with innovation. Patenting is at best a proxy measure of an element of innovation that is important in some sectors and irrelevant in others. Innovation studies are fortunate nowadays to have access to the power of Social Networking Analysis (SNA) as a new method for studying innovation management. As discussed in Chapter 6 by Kastelle and Steen, by mapping connections between people, groups, and organizations, SNA provides one of the best tools for innovation managers. Kastelle and Steen show how new statistical methods that examine large networks and test hypotheses about network structures and dynamics have dramatically changed the theories and techniques of network analysis, and they provide a guide on how to conduct an analysis. They outline some of the benefits of SNA for innovation managers, which include the identification of organizational silos, finding hubs and key actors, locating isolated people and groups, and identifying bottlenecks.

The challenge for the study of innovation management lies in integrating qualitative findings from rich, idiosyncratic case studies examining the history, structure, strategy, and environment of particular organizations, with testable and generalizable findings from quantitative research.

INNOVATION MANAGEMENT THEORY

The study of innovation management is driven by its practice. It is an applied field. There is no unified theory of innovation management, just as there is no unified theory of innovation. There are, however, diverse theories that can help explain various aspects of innovation management as a social and economic process. Elements of psychology, for example, explain the motivations of innovative individuals, while sociology explains the power relationships between and within groups and organizations that affect innovation as a social endeavour, and political science enlightens us about the influences institutions can exert. Organization theory tells us about how new fields of knowledge and effort are formed and institutionalized, and how practices are negotiated and become embedded. Our focus on the management of innovation as a purposive, instrumental activity leads us more towards theories in economics and strategic management, with a common concern to explain how resources and capabilities are deployed and value is created through the introduction of new ideas. That is not to underestimate the value of other theories and the explanations and insights they offer, but it does reflect the value of three approaches—evolutionary economics, dynamic capabilities, and innovation management—that emphasize the connections between context, strategy, and practice. It also suggests their value compared to alternative, often deeply embedded, theories in the same field, such as neo-classical economics (Foster and Metcalfe, 2004) or strategy based on industrial structure analysis (see chapter 20 by McGrath and Kim).

By identifying the three analytical lenses—evolutionary economics, dynamic capabilities theory, and innovation management—it is possible to recognize several strands or connections that help frame understanding. These are shown in a highly simplified and stylized manner in Figure 1.1.

Evolutionary economics is concerned with the dynamic processes by which economies develop and change, and the transformational influences of entrepreneurship,

Evolutionary Economics	Dynamic Capabilities	Innovation Management
Create variety	**Search** for new market opportunities	**Create options**
New firms, technologies, and business models	Create, access, and mobilize resources needed to engage in new activity to exploit selected opportunities	Search for innovation opportunities: internally and externally
	Absorptive capacity	
Select and eliminate	**Create and capture value**	**Select innovations to pursue**
Decisions by investors, customers, regulators, partners	Devise business models to produce outcomes that deliver value	Strategic/risk assessment and choice
	Develop specific capabilities and generate revenue to sustain returns	**Configure and deploy**
	Complement activity of co-evolving organizations	Resources and capabilities
	Complementary assets	
	Protect patents, institute customer switching costs	**Capture value**
		Produce distinct advantages Create IPRs and standards
Propagate	**Adapt**	**Build capabilities**
Selected innovations Reinvest to create more variety	Capabilities to changing business environment	Across innovation portfolio
Learn	**Learn**	**Learn**
Dynamic improvements in economy through creative destruction	Organizational learning in routines	Evaluate returns and review performance

FIGURE 1.1 Analytical lenses

technological change, and recombinations of organizational routines (Nelson and Winter, 1982; Foster and Metcalfe, 2004). The historical periods of transformational changes associated with this pattern of development see massive economic and technological shifts and also profound changes in organizational structures, industrial relations, and skill patterns (Freeman and Soete, 1997). The virtues of capitalism, in the evolutionary economics approach, lie in the continual creation of variety in response to turbulence and uncertainty, from which markets and other mechanisms make selections, the most successful of which are propagated and re-innovated to create the resources for investing in new variety creation. Notable in this formulation is the preponderance of failure. Alongside the creation of new innovations, firms and technologies fail continually in a Schumpeterian process of creative destruction.

Dynamic capabilities theory is concerned with the capacity of organizations to reconstruct their resources (Teece, 2009) to fit with changing and uncertain environments. Various dynamic capabilities are analysed, including the capacity to search for new ideas, choose between them, and then create and capture value. A key aspect of these capabilities is their ability to adapt as business opportunities change. Notable in this formulation are the recognition of the importance of integrating with co-evolving institutions, such as collaborative partners, and the value capturing strategies of intellectual property protection and creation of high customer switching costs.

The innovation management lens is much more applied, yet it draws on a number of analytical frameworks, such as complementary assets (Teece, 1986) and absorptive capacity (Cohen and Levinthal, 1990), which also inform strategic capabilities theory. The development of a pharmaceutical, for example, requires access to the complementary assets of production expertise, knowledge of regulatory approval processes, and distribution networks, before it reaches the market. Absorptive capacity is the organizational equivalent of radio communication needing receivers as well as transmitters. Knowledge flows only when there is the capacity to receive it, and investments in R&D aid the capacity of organizations to absorb externally sourced knowledge. Notable in this formulation, and in contrast to the previous two lenses that are primarily concerned with outcomes and performance, this lens also includes analysis with an internal focus into the processes of configuring and deploying resources and capabilities within the organization.

All these lenses are dynamic, responding to contextual change and disruption, and involve the search for and creation of variety and options; selection from within that variety from which to deliver and capture value; and propagation of successful choices creating resources and learning with which to re-invest into the cycle. Each involves learning as a core process and outcome: at the level of the economy, in the capabilities and routines that organizations possess, and in improving the management of innovation. These ways of theorizing support definitions of innovation management that move beyond the continuous improvements that lead to reduction of variety and increases in predictability, and include those approaches that involve risk and experiment.

INNOVATION MANAGEMENT PRACTICE

Organizations manage innovation, rather than leaving it to chance, by creating supportive structures, practices, and processes. Although the nomenclature varies, organizations define roles such as Chief Innovation Officer and Innovation Manager, and establish advisory bodies such as Innovation Boards. They have innovation strategies and plans, and offer incentives and rewards for innovators. Funds for internal venturing encourage entrepreneurship, intellectual property is protected by official policies, and prescribed project management processes guide decision-making. Resource allocation processes see budgets for innovation assigned in portfolio approaches with various time horizons, and R&D centres provide support for business units and options for the future.

These examples of management structures and practices are examined throughout this Handbook. They help explain the success of innovating organizations. But they only succeed when they accord with the contextual conditions in which organizations operate. Successful innovators furthermore manage in a way that balances the need to produce value through existing business that generate the resources that allow them at the same time to create opportunities to develop new ways of creating, delivering, and capturing value.

We now turn to five interrelated recurrent and enduring challenges of innovation management: dealing with disruption; balancing portfolios; integrating organizationally, technologically, and commercially; building advantage in intangible assets and activities; and encouraging creativity and playfulness. These represent a different order of challenges to the more general, day-to-day, management of budgets, projects, and personnel, and are essential to obtaining more long-term and sustainable advantage from innovation.

RECURRENT CHALLENGES

Dealing with Disruption

Disruption has many causes. The world is unfortunately not immune from extreme events—political, economic, environmental, geological, biological—that continually introduce new kinds of turbulence for organizations. Innovation is itself a major source of disruption for organizations, as competitors find ways of doing things better, cheaper, and faster. Competitors can increasingly benefit from global access to ideas, production capacities, and deregulated markets and from cheap and ubiquitous digital technologies. Essentially, as economic systems become ever more complex, interdependent and rapidly changing, the level of disruption that confronts organizations increases.

Disruption in this sense is not unexpected—it is inevitable—but its unpredictable manifestations mean it is something that organizations may not have planned for, have no ready response to, and cannot easily adjust their resources and capabilities to deal with.

Disruption occurs in business models and cost structures, such as the effect on telecommunications companies of providers of voice over the Internet services, or the consequences for high street stores of online shopping. Rapid market change can be disruptive, such as when competitors introduce superior offers, or products lose their appeal. Examples would be the smartphone replacing personal digital assistants, or the growing distaste for cigarettes in Europe. Changes in regulation can be disruptive, such as environmental controls on automobiles, or restrictions on the ability of banks to offer both retail and investment services. Technological change is a major source of disruption, especially when new platforms emerge as in the case of hard drive devices replacing CDs, or new methods of drug discovery by means of genetic engineering. The largest challenges emerge when different forms of disruption combine, such as the newspaper industry being confronted by electronic news sources. The consequences of disruption can sometimes be very painful because skills that were previously highly valued are no longer needed. Innovation management can involve making people redundant.

Balancing Portfolios

The bulk of most organizations' investments in innovation address small improvements. Archetypically, these continuous, incremental improvements mainly apply to day-to-day operations and improve performance with relatively low risk. Organizations that focus entirely on doing what they currently do slightly better are often exposed to innovative new entrants, therefore part of the portfolio of innovation investments and projects should aim to help the company diversify and grow new business by building upon and developing beyond existing capabilities. This involves taking risks. To create value through possessing options for an organization in a changing world, a relatively small proportion of most organizations' portfolios should be speculative, with high risks and potentially high rewards. Being capable of initiating or rapidly responding to radical, breakthrough technologies future-proofs organizations by having options and balance in the portfolio, and although risky, these risks are smaller than not having prospective possibilities for change. These more adventurous investments not only produce new knowledge, but also allow engagement with other innovation leaders around the world. Some highly innovative, often science-based, firms operate fully in this radical and unknown section of the portfolio, searching for breakthroughs with which they can trade. A simple representation of an innovation portfolio is shown in Figure 1.2.

A normal innovation management challenge is to balance the portfolio across the 45-degree axis on the diagram. The enduring challenge for most innovative organizations, however, is to invest in upper right hand areas of the diagram when organizational and managerial attention is inevitably directed towards existing activities in the bottom left that provide the core of the organization and deliver crucial objectives such as

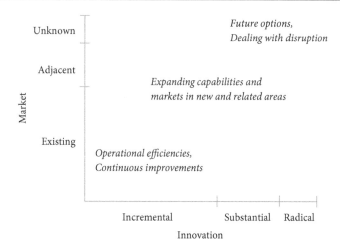

FIGURE 1.2 A simple innovation portfolio

income generation. There are, however, a wide variety of innovation management challenges and opportunities, such that the priority for some firms may be to concentrate their investments ever more deeply in existing capabilities when they provide a source of distinctive advantages.

Integrating the Innovation Process

In practice, ideas for innovation emerge from multiple sources and it often requires the collision and blending of many diverse insights into possibilities and opportunities. Encapsulating and focusing that diversity requires high levels of organizational, technological, and commercial integration.

Because innovation can derive from and involve many contributors, organizational integration within and between organizations, and between different professions, occupations, and skill sets, is a prerequisite for supportive and effective innovation processes. Many innovations occur within technological systems or architectures that require integration between modular components. They may also require connections between different vintages of technology, and integration between physical and digital technologies, for example in augmented reality, that supplements real world observations with computerized sounds, images, and senses. The capacity to unify diverse technical inputs is the key to technological integration, and there are supportive technologies and design tools to assist. The technology that produces a common digital platform for the integration of computer-aided design and manufacture has existed for decades, and new technologies assist the integration of all aspects of the innovation process. These require the management of tools for analysing 'big data': the vast amounts of data produced from scientific research, the

'Internet of things' created by ubiquitous sensors in devices such as Radio Frequency Identification and smartphones, and social networking technologies. As discussed in Chapter 19 by Dodgson and Gann, these 'innovation technologies' also rely on the management of simulation and modelling techniques and virtualization and visualization technologies that improve the speed and efficiency of developing and testing complicated systems.

Commercial integration ensures that innovations meet the requirements of customer and clients by delivering value to them in price, performance, or utility, and utilizes channels to market that the organization already has or can access. Supply chain integration ensures security of supply, and complementarity of components and standards.

Hargadon's chapter on brokerage argues innovation occurs through the process of recombining and integrating past knowledge and practices in new ways. Innovation occurs when individual practices and the organizational strategies to support them are integrated with the larger social structure. Using historical and contemporary cases, Hargadon identifies the central role of brokerage in explaining the generation and success of innovation, addressing key management questions such as continuing success in innovation and the virtues and challenges of diversity.

Fujimoto's chapter on innovation management in Japan combines insights from trade theory, architectural thinking in design theory, and an evolutionary framework of capability building. His contention is that, for a variety of historical reasons, Japan developed rich endowments in coordinative capabilities, such as teamwork of multi-skilled engineers, applied especially to coordination-intensive products, such as automobiles. These integrative abilities provide sources of great strength for relatively high-value, highly engineered products, but Fujimoto shows the shortcomings of this approach in modular, digital, and relatively cheaper products. He outlines a range of future strategic options for Japanese innovation management.

Managing Intangibles

In advanced economies, investment in intangible assets—knowledge and intellectual property, for example—exceeds that of tangibles, such as factories and equipment (Haskel and Wallis, 2013). Intangibles are commonly defined as things that cannot be seen or touched, and their management is often different from that of physical assets.

Broadly defined, a company's reputation, mindset, and culture for innovation are intangible assets. Other intangibles could include design and business models. When we consider that an organization that creates and delivers services—which comprise more than 70 per cent of gross domestic product in most developed economies—is creating and delivering an intangible, then the overall significance of managing intangibles is clear. These issues are examined by Verganti and Dell'Era in Chapter 8, Massa and

Tucci in Chapter 21, and Tether in Chapter 30. Services are also addressed in Chapter 10 by Malerba and Adams.

One of the biggest issues confronting the management of intangibles is the difficulty of measuring them. Progress in development cannot be observed in improved proto-types as, for example, in manufactured products. Innovation in services occurs as they are used in the market—innovation starts at the point of consumption rather than invention—so measuring inputs is less exact. Expenditures on R&D are readily account-able; customer engagement in new service development is less so. For these reasons the use of project management and marketing techniques in industrial innovation is lim-ited. It explains why software companies often release their services in beta form to be developed and tested by use.

Encouraging Creativity and Play

Creativity is commonly seen as the origination of ideas, insights, and innovation as their successful application. Much of the management literature on creativity has tended to focus on individuals or the role of teams, addressing techniques to extract the best per-formance from them. These are crucial contributors, but the connection between crea-tivity and innovation is so important that it is core to the strategic development of the firm. Creativity therefore needs to be considered within the strategies and practices that shape its manifestation as innovation. Chapter 7 by Leonard and Barton argues how creativity and innovation have a paradoxical relationship with knowledge. Whether at an individual, group, or organizational level, knowledge can both stimulate innova-tive ideas and prevent their fruition. Using concepts such as core rigidities and deep smarts, Leonard and Barton provide insights and guidance on ways to counter the downsides of knowledge and use its power to stimulate creativity and inspire people and organizations.

One way of connecting creativity and innovation is the notion of play (Dodgson et al., 2005). Play at work is important for individual and organizational performance. It encompasses those activities where people explore, template, model, prototype, rehearse, and tinker with new ideas, often in combination with others with different skills in stimulating environments where work rules are relaxed. Play, in this sense, is an antidote to the procedures and bureaucracy that inevitably develop in organizations over time and are anathema to innovation (Dodgson et al., 2013).

Jazz improvisation is a common metaphor used in the organization and management literature to reflect this appreciation of the nature of play (Meyer et al., 1998). Jazz pro-vides an idiom for understanding the balance in the relationship between individuals as they collectively explore the unexpected within the confines of accepted styles and structures. It reflects the way that effective improvisation, seen as spontaneous experi-ment, actually reflects depth of experience and degrees of discipline by its players.

The notion of play also introduces the challenging high incidence of failure in innova-tion, which happens constantly around any ambitious ideas. Only a small proportion of innovations being explored at any one time will ever succeed in the market. For many in organizations, especially those whose job it is to control expenditure and whose remuneration package depends on short-term performance, the remainder of these investments are often construed as failures. Such failures are, however, inevitable and provide valuable learning experiences.

Different Types of Innovation Process

Managing the many challenges of innovation requires the combination of resources in different business and organizational processes. To help frame analysis, six broad processes used to coordinate resources to create, deliver, and capture innovation are determined, each requiring different underlying management capabilities. Innovative organizations use most if not all of these types of process in different combinations.

Type 1—Research and technology led. These processes support the use of science, research, and technology as the stimulus to innovation in an organization. The key management capabilities required are selecting, conducting, and applying R&D and technology projects. A number of chapters in the Handbook inform us about this type of innovation process.

McKelvey in Chapter 4 on science, technology, and business innovation discusses the differences in the types of knowledge underlying each, the role of public financing, and the interactions between universities and business, including science in entrepreneurial firms and academic spin-offs. She discusses the motivations for public investments in science, and the specific demands of science-based industries and other sectors that rely on scientific research. Critical of a restricted 'technology transfer' model of university-business interactions, McKelvey argues for a more broad-based 'engagement' model. Amongst the key challenges for innovation management she identifies is the manner in which scientific advances are by definition unknown before research occurs, and while firms may wish to invest in the creation and use of knowledge, they are uncertain about the value of that knowledge.

Von Zedtwitz, Friesike, and Gassmann in Chapter 26, on managing R&D and new product development, explain the contributions these activities make, and describe their central elements. These include the product development funnel, R&D portfolio management, and the organization of R&D. The chapter discusses concepts such as the 'fuzzy front end' that provide valuable analytical and practical tools for innovation management.

Research and development is an increasingly globalized activity, and this is the focus of Chapter 27, by Håkanson, who outlines trends in the internationalization of R&D in multinational companies. Håkanson discusses the motivations behind decisions to perform R&D overseas, and the managerial issues that result. Firms internationalize their R&D for reasons varying from the adaptation of products to local market requirements to linking with global centres of basic science. The chapter discusses the managerial implications of these different objectives in the systems, processes, and practices used by multinational firms.

The results of R&D and new product development feed into organizations' goods, processes, and services and can also be traded in markets for technology. Chapter 12, by Gambardella, Giuri, and Torrisi, outlines the size and characteristics of markets for technology, as organizations exploit their technology or outsource it from third parties using methods such as licensing, cross-licensing, and the sale of patents. This chapter examines the incentives firms have to participate in markets for technology, including the differences between those in large and small firms, and considers the barriers to technology trade. Gambardella, Giuri, and Torrisi argue that markets for technology are an important strategic consideration, increasing in size and range, and as a result are a significant issue for innovation management.

The strategic significance of intellectual property is a theme developed in Chapter 28 by Leiponen. She examines legal and competitive strategies to control and benefit from intellectual property and from technical standards that are crucial for the interoperability of many product and service systems. By examining the ICT industries in particular, Leiponen argues the need for business models that respond to weakening appropriability regimes. She shows how innovation strategies that encompass intellectual property are crucial as negotiations and litigation can determine the success or failure of innovations.

Type 2—Market-facing. These processes begin with understanding of the nature of market demand, and the organization of resources in response to market opportunities. Key management capabilities are collecting, analysing, and responding to information about markets, users, and consumers, and the capacity to make decisions on when to create and lead markets ahead of demand.

In Chapter 3, on marketing and innovation, Prabhu examines how marketing influences innovation both as a source of and location for innovation. As an innovation in marketing itself, Prabhu examines the who, what, and how questions that marketing helps answer for innovators. As a source of innovation, he explains how a firm's orientation towards its market affects the ways it innovates, and how marketing is a crucial element of the cross-functional coordination needed for successful innovation.

It has long been appreciated that innovations are enhanced by engagement with their users during the process of their development. Chapter 5 by Franke, on user innovation, argues that this practice is gaining momentum as the Internet provides information relevant for innovation ever more quickly and cheaply. Franke explains why users innovate, how they organize, and their motivations for sharing their innovations with other users. He describes three methods by which companies can benefit from user innovativeness, including ways of identifying lead users, toolkits for self-design, and crowdsourcing for solutions to particular problems.

Chapter 14 by Ozaki and Dodgson argues that innovation managers have to dig deeper than simply understanding why customers buy innovations, such as their functionality, utility, and price, and consider also how those innovations are consumed. This, they argue, requires understanding of consumers' underlying values, and more emotional and socially contextual factors. Using a historical example, and modern cases of hybrid vehicles and green electricity tariffs, Ozaki and Dodgson discuss the complexities of the decision to consume innovation, and how better appreciation of these complexities improves the management of innovation.

Type 3—Internal coupling. To avail itself of market and/or technological opportunities, an organization needs the internal communications and connections between all its various contributors to aid the realization of an innovative outcome. The most important management capabilities in this type of process are communications and the capacity for feedback and iterations on projects. Also valuable are the abilities of people to combine their deep expertise in particular areas with a capacity to work effectively across different aspects of an organization's activities. Apart from R&D, marketing, and sales, other contributing domains of activity might include: intellectual property protection, prototyping and testing, and operations and servicing. Such coupling may involve cross-departmental coordination and budgets, and can draw on the different perspectives and skills of multidisciplinary and inter-departmental teams. A new product development project, for example, commonly includes representatives from marketing, R&D, and production and operations (see Chapter 3 by Prabhu and Chapter 26 Von Zedtwitz, Friesike, and Gassmann). These internal links can also be facilitated by the use of computer-assisted integration between design, development, and operations (see Chapter 19 by Dodgson and Gann).

In a related chapter, Chapter 24, Phillips discusses how different aspects of organizations affect innovation and argues for the importance of managing the organizational context when managing innovation. He identifies a number of mechanisms that can enhance integration and enable innovation, including leadership, culture, organizational structure, networks, and teams. The organizational context can either enhance innovation or impede it, depending on how well these aspects of organization are managed. In addition to aspects of organization that have been explored in the existing literature, he also speculates about the role of organizational identity, institutional context, and the organization's willingness and ability to adopt new practices in innovation.

Laursen and Foss in Chapter 25 emphasize the importance of extensive lateral and vertical organizational communications in their chapter on Human Resource Management (HRM) and innovation. They argue the value in combining these patterns of communications with high levels of delegated decision-making and use of particular reward systems in 'new' or 'modern' HRM practices. As the innovation process changes, for example, by being more distributed and inclusive, Laursen and Foss argue the need for HRM practices to change as well. They develop a model for considering the moderators and mediators of the relationship between innovation and HRM.

Chapter 31 by Davies argues that projects provide an important organizational form for innovation. Projects are a temporary organization and process established to create a novel or unique outcome. Davies argues how project management tools and techniques were developed to help select, plan, manage, and reduce the uncertainties associated with innovation. Distinguishing between optimal and adaptive models of project management, he argues the latter are emerging as a new paradigm for understanding the relationship between project-based innovation and uncertainty. A fundamental means for internal organization, projects also provide key mechanisms of engagement with clients, customers, partners, and suppliers.

Type 4—External collaboration. These processes connect organizations with external parties as they search for, choose, and implement innovations. They may involve research links with universities and research institutes (see Chapter 4 by McKelvey), and collaboration with companies working in similar markets and technologies in various forms of consortia. Connections with customers and suppliers are important, and often the ability to work with demanding 'lead' customers is a stimulus to innovation (see Chapter 5 by Franke). The capacities to select partners within established value chains and work effectively with them are key management capabilities. The management of innovation in such processes additionally involves the ability to search widely for ideas within wider innovation ecosystems, select from them judiciously, manage the potentially increased contest over intellectual property rights, and ensure good information flow and cooperation within the broad ecology.

Autio and Thomas in Chapter 11, on innovation ecosystems, review how the concept has evolved and how it can be applied to the analysis, design, and implementation of innovation strategy. Their chapter provides insights into the boundaries, structures, and dynamics of innovation ecosystems and offers three theoretical lenses through which to examine them. Autio and Thomas argue that while a large body of research has addressed innovation ecosystems, study of their implications for innovation management remains in its infancy, providing rich future opportunities for scholars.

Firms belong in innovation ecosystems because they cannot innovate by themselves, and some of the connections firms make in order to innovate are especially intimate and involve mutual commitment of resources to agreed objectives. This is the definition of collaboration in Chapter 23, by Dodgson. Collaboration, he argues, contributes to an organization's ability to attain complementarities, encourage learning, develop capabilities, and deal with uncertainty and complexity. It is often a challenging process, Dodgson contends, and managing the inherent instabilities and tensions in collaboration requires careful partner selection and effective structuring and organization.

Type 5—Strategic integration. These processes provide the strategic overview for all other innovation processes, as they involve decisions about how innovation supports overall organizational objectives and what innovations to pursue. They encourage high levels of internal and external organizational integration in support of overall corporate objectives, rather than individual projects, and this may involve investments in coordinating technological infrastructure and platforms. It is strategic oversight that prevents organizations falling prey to the dangers of research and technology push when there is no market for its outcomes, and over-reliance on demand-pull processes where customers can be conservative and stifle potentially disruptive innovation. The ability to formulate and implement innovation strategy and encourage highly coordinated internal and external organizational support for innovation is a key management capability.

Chapter 20 by McGrath and Kim reveals the considerable shortcomings of mainstream theories of strategy when addressing innovation. The industrial organization and resource-based views of strategy, they argue, fail to account for the turbulence and dynamics in the 'hypercompetition' confronting contemporary firms. McGrath and Kim argue that in a world where competitive advantage is transient, and competitive threats can emerge from diverse and unexpected sources, new metrics of performance are required, greater account should be made of networks of people and organizations, and more attention should be paid to the role of the general manager.

Strategic integration within and across organizations is a theme in a number of other chapters, and is especially relevant in the chapters on business model innovation, platforms, design, and open innovation.

The relationships between business models and innovation have been an increasing focus of research attention. In Chapter 21, Massa and Tucci define business models as the rationale of how an organization creates, delivers, and captures value in relationship with a network of exchange partners. They argue business models represent both an important vehicle for innovation and a source of innovation in and of itself. Massa and Tucci examine business model innovation in three contexts: their design in newly formed organizations, reconfiguration in incumbents, and as a means of encouraging sustainability. They provide a synthesizing meta-framework and identify tools and perspectives of business models to assist innovation management.

The notion of design, as discussed in Chapter 8 by Verganti and Dell'Era, is fluid and slippery, and is commonly considered in a very restricted way. Recently, however, design has become better understood as a fundamentally integrative contributor to and source of innovation. Verganti and Dell'Era consider design as the form of things, as creative problem solving, and as the innovation of meaning, and focus on the latter as a means of understanding why people use things (a question also pursued with a different perspective by Ozaki and Dodgson). They develop the idea of design-driven innovation and its relationships with technology push and market pull innovation, and show how it offers a vital new paradigm for managing innovation through its power to interpret and envision meaning.

The capacity of a number of technologies to integrate the innovation process within and across organizations leads Dodgson and Gann in Chapter 19 to describe them in combination, as innovation technology. Another integrative concept is that of the platform that uses new technology and helps organize markets for innovation around them, thereby adding value. Using a number of cases of ICT companies, Gawer and Cusumano in Chapter 32 examine the implications of platforms for innovation management. They distinguish between internal, supply chain, and industry platforms, and consider their strategic implications. Gawer and Cusumano use examples of how platforms compete and evolve, and draw out lessons for where they can encourage and discourage innovation.

The concept of open innovation promises to leverage internal R&D and gain benefits from access to externally sourced innovation. As Alexy and Dahlander argue in Chapter 22, however, the level of interest of researchers in the subject is not matched by the ease with which organizations develop successful open innovation strategies. They distinguish between four aspects of openness: acquiring, sourcing, selling, and revealing, and consider the conditions in which their combination is beneficial. By highlighting a number of influential contingencies on openness, Alexy and Dahlander guide understanding of its benefits and limitations.

Type 6—Future ready. These processes prepare organizations for the future by building their awareness of, and responsiveness to, changing business models and disruption in technologies, markets, regulations, demands for sustainability, and in general business circumstances. The continuity and success of organizations when confronted with disruption depends significantly on the ways in which they manage innovation; a theme developed in many chapters in this Handbook. Early sensing of potential disruption is extremely valuable, and may involve high degrees of openness, including deep immersion in the research community, keen observation of peripheral developments in start-up companies and competitors, and active engagement in the policy-making process in areas such as regulation. It involves understanding the nature of the innovation ecosystem in which the organization operates, what points of influence and control it possesses, and what levers organizations possess, such as mergers and acquisitions (M&A), to respond to opportunities.

The ability of M&A to increase innovation and enhance an organization's innovative capacity are some of the potential benefits discussed in Chapter 29, by Ahuja and Novelli. One of the key conclusions of this chapter is the need for more research into the subject of the impact on innovation of M&A in, for example, their capacity to improve future readiness. Ahuja and Novelli's theoretical and empirical review encompasses the managerial challenges of M&A, the diversity of views on their consequence for innovation, and their potential value.

Future ready innovation processes provide organizations with the adaptive capacity to continually deal with, and profit from, uncertainty and disruption.

Nowhere is the challenge of being future ready more important than in the area of sustainability. Berkhout in Chapter 15, on sustainable innovation management, outlines why environmental sustainability has become such a crucial innovation management concern. He identifies three main influences of technology on business environmental performance: sensing and providing information, improving efficiencies, and transforming resource-use and environmental impacts. He argues that because of the systemic complexities of environmental challenges, responses need to be transformative, requiring a mixture of old and new innovation capabilities, new business models, and linkages.

As Phillips discusses in his chapter, being future ready is also deeply dependent on leadership and culture. Culture shapes the degree to which an organization looks forward or focuses on the past, and also determines the rate of change and innovation that organizational members are comfortable dealing with. Leadership plays a similarly central role in the degree to which an organization is future ready. If leaders are forward-looking and provide the sort of transformational leadership that makes organizational members feel secure and empowered, then awareness of the need to change, ideas for innovation, and a willingness to change to meet future challenges will all increase.

Responses to disruption are unlikely to be completely autarkic, and inevitably involve external collaboration with customers and occasionally with government, for example through technical standards bodies. Preparedness for disruption depends on an organization's strategic appetite for risk and its taking early bets on potential developments. It rests upon openness, ability to use supportive technology, and experimenting with low cost and more 'inclusive' innovation that involves wider community involvement. Internally, the capacity of organizations to employ people whose energies are directed towards sensing external threats and opportunities, and then responding flexibly and quickly to them, is crucial for their future readiness.

Key management capabilities here are the management of less observable and measurable intangibles, such as organizational culture and mindset, service orientation and entrepreneurial spirit, and the encouragement of creativity and playfulness. Tolerance of failure is important for attracting people to work in adventurous ways, as is the provision of tools and techniques such as innovation technologies that allow quick and cheap failure, and effectively learning lessons when things do not go to plan. As intangibles are less readily measured, there is greater reliance for decision-making on judgement, expertise, experience, and intuition.

Conclusions

As this chapter suggests, the perspectives that are needed to understand innovation management are broad and diverse. To reflect this diversity the book is divided into four sections. The first part, including this and the following chapter, is offered by way of introduction to the subject. The second section addresses various sources of innovation. The third section analyses contextual influences on innovation management. The fourth and final section considers issues of strategy, management, and organization. Many chapters traverse sources, context, and practice, but are allocated according to their primary contributions.

The Handbook offers a rich collection of insights and cases on innovation management that not only capture what we know about this subject but what we do not know and need to know, and it offers an extensive range of suggestions on future research agendas. Innovation management is a field of research rich in significance and ripe for better understanding. It provides fertile ground for further exploration.

References

Abernathy, W. J., and Utterback, J. M. (1975). 'A Dynamic Model of Process and Product Innovation', *Omega*, 3(6): 639–56.

Cohen, W., and Levinthal, D. (1990). 'Absorptive Capacity: A New Perspective on Learning and Innovation', *Administrative Science Quarterly*, 35: 128–52.

Cusumano, M., and Selby, R. W. (1995). *Microsoft Secrets: How the World's Most Powerful Software Company Creates Technology, Shapes Markets and Manages People*. New York: The Free Press.

Dodgson, M. (2011). 'Exploring New Combinations in Innovation and Entrepreneurship: Social Networks, Schumpeter, and the Case of Josiah Wedgwood (1730–1795)', *Industrial and Corporate Change*, 20(4): 1119–51.

Dodgson, M., Gann, D., and Phillips, N. (2013). 'Organizational Learning and the Technology of Foolishness: The Case of Virtual Worlds in IBM', *Organization Science*, 24(5): 1358–76.

Dodgson, M., Gann, D., and Salter, A. (2005), *Think, Play, Do: Technology, Innovation and Organization*. Oxford: Oxford University Press.

Dodgson, M., Gann, D., and Salter, A. (2008). *The Management of Technological Innovation: Strategy and Practice*. Oxford: Oxford University Press.

Dosi, G. (1982). 'Technological Paradigms and Technological Trajectories: A Suggested Interpretation of the Determinants and Directions of Technical Change', *Research Policy*, 2 (3): 147–62.

Fagerberg, J., Mowery, D., and Nelson, R. (eds) (2005). *The Oxford Handbook of Innovation*. Oxford: Oxford University Press.

Foster, J., and Metcalfe, J. S. (2001). *Frontiers of Evolutionary Economics: Competition, Self-Organization and Innovation Policy*. Cheltenham: Edward Elgar.

Foster, J., and Metcalfe, J. S. (eds) (2004). *Evolution and Economic Complexity*. Cheltenham: Edward Elgar.

Freeman, C., and Soete, L. L. G. (1997). *The Economics of Industrial Innovation*. London: Pinter.

Fujimoto, T. (1999). *The Evolution of a Manufacturing System at Toyota*. Oxford: Oxford University Press.

Graham, M., and Shuldiner, A. (2001). *Corning and the Craft of Innovation*. New York: Oxford University Press.

Hargadon, A., and Douglas, Y. (2001). 'When Innovations Meet Institutions: Edison and the Design of the Electric Light', *Administrative Science Quarterly*, 46: 476–501.

Haskel, J., and Wallis, G. (2013). 'Public Support for Innovation, Intangible Investment and Productivity Growth in the UK Market Sector', *Economics Letters*. Amsterdam: Elsevier, 195–8.

Hounshell, D., and Smith, J. (1988). *Science and Corporate Strategy: Du Pont R&D, 1902–1980*. Cambridge: Cambridge University Press.

Meyer, A., Frost, P. J., and Weick, K. J. (1998). 'The Organization Science Jazz Festival: Improvisation as a Metaphor for Organizing (Overture)', *Organization Science*, 9(5): 540–2.

Morison, E. (1988). 'Gunfire at Sea: A Case Study of Innovation', in M. Tushman and W. Moore, *Readings in the Management of Innovation*. New York: Harper Business.

Nelson, R., and Winter, S. (1982). *An Evolutionary Theory of Economic Change*. Cambridge, MA: Belknap Press, 437.

Rothwell, R., et al. (1974). 'SAPPHO Updated: Project SAPPHO, Phase II', *Research Policy*, 3: 258–91.

Rothwell, R., and Gardiner, P. (1985). 'Invention, Innovation, Re-innovation and the Role of the User: A Case Study of the British Hovercraft Development', *Technovation*, 3: 167–86.

Teece, D. (1986). 'Profiting from Technological Innovation: Implications for Integration, Collaboration, Licensing and Public Policy', *Research Policy*, 15: 285–305.

Teece, D. (2009). *Dynamic Capabilities and Strategic Management: Organizing for Innovation and Growth*. Oxford: Oxford University Press.

Tuomi, I. (2002). *Networks of Innovation: Change and Meaning in the Age of the Internet*. New York: Oxford University Press.

Webb, J. (1992). 'The Mismanagement of Innovation', *Sociology*, 26(3): 471–92.

....

CHAPTER 2

THE NATURE OF INNOVATION

....

AMMON SALTER AND OLIVER ALEXY

INTRODUCTION

....

THE past fifty years has seen the growth of efforts by academics and practitioners around the world to better understand the nature, sources, and determinants of innovation. Research involves attempts to map, measure, and refine our understanding of how novelty is introduced into the economic system. Inspired originally by Schumpeter (1911, 1942), this field has moved beyond a narrow group of researchers at the margins of economics and sociology to become one of the major topics of interest across management, economics, sociology, and social psychology. Over this time, our understanding of innovation has become richer, more detailed, and refined (Martin, 2012). The goal of this chapter is to briefly review some of the lessons from this research programme.

Our approach to this review is to bring to the surface some 'stylized facts' that have emerged from the study of innovation. The concept of a 'stylized fact' was first proposed by Nicholas Kaldor (1957) to capture some of the main lessons of research on the economics of growth. A stylized fact is a simplified representation of a set of empirical findings. It should be essentially true, but may not fully apply to all settings. Looking at stylized facts helps to survey the broad area, without becoming lost in the small print. Or, as Kaldor put it, it allows one to 'concentrate on broad tendencies, ignoring individual detail' (Kaldor, 1961: 178).

A related way to understand a research discipline originates from Lakatos' description of the 'hard core' of a 'progressive' field of research (Lakatos, 1970). For Lakatos, a progressive area of research is open to stunning new facts, novel experiments, new sources of data and methods, and more precise predictions. At the centre of any field of research is a set of ideas that are widely held by members of the field. They may be captured by a set of stylized facts, a set of generalizations that most members of the field would subscribe to about the nature of knowledge within the field. These statements represent the 'hard core' of a discipline, a focused set of ideas and understandings shared by members of a community of researchers. Of course, the hard core of ideas is not static and is open to change through new discoveries. As Lakatos suggested, a progressive field of research has a limited 'protective belt', an openness to new ideas or discoveries that may change the 'stylized facts' within the hard core of the field.

In this chapter, we build on these two conceptual tools to characterize our current understanding of the nature of innovation. We focus on the stylized facts at the hard core of innovation studies. We develop a set of statements that are based on consistently reoccurring results of decades of empirical research. Although these statements are not always true, they are liable to be true in most cases. In doing so, we try to quickly and effectively summarize what has been learnt about innovation over the past fifty years. Or, as a student commented to their history professor, they wanted 'more years with fewer words' (Gaddis, 2005: viii).

We would suggest that the stylized facts below are held in broad agreement by innovation researchers and are largely uncontroversial. Going beyond this safe ground, we highlight several occurrences of new ideas that have put in doubt existing 'facts' or even created changing perceptions of the nature of innovation, such as the shifting attention away from 'the firm' as the central actor in the innovation process towards distributed or community-based models of innovation. We suggest newly emergent ways of innovating present challenges to previous conceptions of innovation and in turn open a wide range of different research topics.

Innovation in Historical Perspective

The interest in the nature of innovation is not new. In 1772, Samuel Johnson complained to Sir William Scott that 'the age is running mad after innovation; and all the business of the world is to be done in a new way; men are to be hanged in a new way; Tyburn itself is not safe from the fury of innovation' (Boswell, 1791).[1] And shortly thereafter, the events of the French Revolution only further confirmed Johnson's insight through a significant innovation introduced in the early 1790s: the 'guillotine'. Previously, France had used the 'breaking wheel' for executions, which inflicted immense pain before causing death. Similarly agonizing were other methods in use at the time, such as hanging or beheading by the sword. For its inventor, the guillotine had several advantages over the past methods of execution: it was efficient, instantaneous, and pain-free. For the French Revolutionaries, it offered a new, more humane way of ensuring justice. Like most successful innovations, it had a long life. The guillotine stayed in use in France until the late twentieth century, when the last person (a convicted murderer) was executed in 1977, almost 200 years after its first development.

Over the past thirty years, the interest in innovation in the popular press, governments, and business firms has accelerated, creating a crescendo of concern and enthusiasm for innovation. In 1999, *The Economist* described innovation as 'the industrial religion of the late 20th century'.[2] Nowadays, innovation features as a prominent buz-zword amongst heads of states—take the following example from US president Barack Obama:[3]

> Now, history should be our guide. The United States led the world's economies in the 20th century because we led the world in innovation. Today, the competition is keener; the challenge is tougher; and that's why innovation is more important than ever. That's the key to good, new jobs in the 21st century. That's how we will ensure a high quality of life for this generation and future generations.

At the same time, underneath all this enthusiasm for innovation, a greater understand-ing of how innovation happens and what impact it has on economic development has emerged. A recent summary by Martin (2012) provides an overview of some of the most cited papers and books on the topic.

The Hard Core of Innovation Studies

Innovation and Growth

By its capacity to increase the rate of productivity growth in the economy, innovation is one of the main driving forces of economic growth. Estimates of the contribution of innovation originally focused on the 'residual'—that share of economic growth that could not be accounted for by capturing the increasing quantity and quality of labour and capital inputs in the economic system. Solow's (1957) early estimates placed 87.5 per cent of economic growth in the residual, which he referred to as 'technical change'. This approach was widely criticized, however, for treating innovation as a leftover, something that was unmeasured. New growth theory, developed in the early 1980s, sought to more fully incorporate the effects of innovation in growth accounting (Romer, 1986, 1990). This approach involved the development of new models that reflect the informational properties of ideas, their non-rivalry,[4] and potential for reuse. These models showed that innovations influence growth primarily by generating spillovers: the transfer of an idea from one place to another at little or no economic cost to the actor receiving the idea.

Although new growth theory helped to more effectively model the contribution of innovation to economic development, it still left its measurement relatively unat-tended. Some scholars, however, have been trying to measure more effectively the con-tributions of major technological changes, such as the Internet or the Information and Communications Technology (ICT) revolution on productivity (e.g. Brynjolfsson, 1993). This work showed that much of the surge of productivity growth in the USA in the 1990s was driven by the adoption of ICT by downstream sectors, such as retail-ing. In effect, it was the use of ICT by Wal-Mart and other large retailers that induced significant productivity improvements. Moreover, the major change in the nature of productivity in the semiconductor sector, the centre of the technological revolution of the 1980s and 1990s, was itself responsible for a significant portion of the productivity gains of this period.

In the 2000s, scholars have sought to capture business firms' investment in intangibles, such as R&D, organizational change, and marketing, and link these investments to economic growth. Using surveys of firm expenditures on these intangibles, it was possible to estimate what share of the growth of productivity was accounted for by investments in innovation. This research showed that almost two-thirds of productivity growth between 1999 and 2006 could be accounted for by investments in intangibles or innovation (Haskel et al., 2010). This evidence provided strong support for the idea that innovation plays an important role in shaping economic development when measured directly alongside changes in the levels and quality of capital and labour. It also helped to renew interest in the measurement of the contribution of innovation to economic development, leading to an increase in new growth accounting approaches that measure and map the contribution of innovation (Acemoglu et al., 2012; Marrano et al., 2009).

Stylized fact 1: Innovation plays a major role in productivity growth.

Combinatorial Power of Innovation

Schumpeter, the father of the study of innovation, suggested that innovation should be defined as 'new combinations' (Schumpeter, 1911, 1942). His idea was that most innovations are not novel in themselves; they are novel combinations of elements that already exist. The main challenge for the innovator in this context is not to think of something new, but to find a new combination of existing things. This is not to suggest that novelty does not enter the system through the development of new technologies, processes, or ways of organizing, but that such novelty is primarily a process of recombining existing elements in new ways. A clear example of this is the case of the development of the assembly line and Model T Ford, widely acknowledged as one of the most significant innovations of the twentieth century. Specifically, Ford's innovation involved a new combination of four elements—the electric motor, continuous flow production, assembly line, and interchangeable parts (Hargadon, 2003). But as Henry Ford himself commented,

> I invented nothing new. I simply assembled into a car the discoveries of other men behind whom were centuries of work...Had I worked fifty or ten or even five years before, I would have failed. So it is with every new thing. Progress happens when all the factors that make for it are ready, and then it is inevitable. To teach that a comparatively few men are responsible for the greatest forward steps of mankind is the worst sort of nonsense. (Greenleaf, 1961, citing an article from the *New Outlook*, 1934)

Thus, for (almost) every innovation, it is possible to look to its pre-history, and the series of ideas, attempts, and failures which are similar in nature and scope. This means that a single innovation is not an isolated event; it springs from the body of materials, experiments, and ideas of previous innovative efforts (Edgerton, 2008). Changes in a single component or module may allow individuals or organizations to create a new way of integrating systems or increasing systems performance dramatically by rearranging the relationships between their different elements.

This combinatorial perspective suggests that the rate and direction of innovation in an industry or market is largely influenced by the potential for the development of new combinations. When the scope for novelty through recombination is exhausted, the speed and pace of innovation slows. Yet, the scope for novelty through recombination is considerable, and much greater than the space for creation of novelty through the introduction of new discoveries. This is because the scope for recombination is almost infinite, as distinct elements can be endlessly combinable to yield new and valuable products, processes, and services (Kogut and Zander, 1992). Although it is often lamented that much of the space for innovation is exhausted, it may reflect perceptions of the opportunity for recombination. A major combinatorial breakthrough may spur a series of related combinations, which, in turn, can unleash a cluster of further innovations. The iPhone, for example, helped to create a new market for thousands of small, innovative software applications. This suggests that one should be wary about claims of the saturation of innovation in a market, as such slowing down of innovation may create the potential for subsequent effort and opportunity to introduce novelty through recombination.

> *Stylized fact 2: Most innovation involves new combinations of existing elements, bodies of knowledge or technology.*

Pervasiveness of Innovation

It is a common perception that innovation is highly concentrated in a few, leading high technology sectors. Research shows that the pace and direction of innovation differs across sectors, with some sectors moving more quickly to introduce new products, processes, and services than other sectors. In part, the pace of innovation can be captured by measures of investment, such as investments in R&D (Griliches, 1981; Griliches et al., 1991) or skilled labour (Cohen and Levinthal, 1989; Leiponen, 2005), or in the churn of the industry, such as the number of new firms and exits of old firms (Abernathy and Utterback, 1978; Klepper, 1997). It also reflects the potential for recombination. Yet, research on the innovation process highlights that innovation is pervasive across all parts of the economic system. Although it is conventional to assign industries into buckets with labels, such as 'high tech' or 'innovative', it is clear that many sectors are home to significant processes of what Schumpeter described as the creative destruction associated with innovation.

Differences in the pace of innovation should not lead away from looking for innovation in all parts of the economic system. Even traditional, slow-moving sectors can be home to important innovations and have been transformed by the development of new products, processes, and ways of working. For example, the use of CATIA, a software system originally created to design fighter airplanes, provided the key tool to allow Frank Gehry to create the Guggenheim Museum in Bilbao (Boland et al., 2007; Dodgson et al., 2005). This technology created a 'wake of innovation' across different parts of the process of constructing a building, including changes in manufacturing, design, and fire and safety. By using CATIA, Gehry Partners could visualize complex structures in a comprehensible form and communicate clearly with customers, collaborators, and subcontractors about interfaces, materials, and eventual cost estimates for their designs. Another example is the mass diffusion of mobile phones across the world, providing a wide range of opportunities to create financial services for millions of people in sub-Saharan Africa and Latin America with no formal bank account (Dodgson et al., 2013).

Stylized fact 3: Innovation is pervasive throughout the economic system.

The Pace of Innovation

Although the pace of innovation differs across sectors and time, a fundamental fact of innovation is that most changes in our knowledge and technology are evolutionary in nature. These changes come about through the introduction of incremental or modest improvements in existing products, processes, and services. Radical innovations generate attention and excitement, capturing the interest of the popular press and consumers. Yet, the vast bulk of corporate investment and management effort is directed towards incremental innovation, looking for opportunities to make small improvements in existing products, processes, and services. Since developing entirely new products, processes, and services is costly and uncertain, major innovators usually put most of their effort into improving what they already have. Organizations, such as Procter & Gamble and Unilever, relentlessly seek incremental ways of making their products more attractive by changing their colour and smell, packaging and positioning on the shelf. These firms have major investments in brands, such as Ariel or Tide, large manufacturing facilities, dedicated R&D teams, and strong supplier and distribution channels. Building up these assets is expensive and many large firms are loath to change their routines around them unless they are forced to by competitive pressure or they have an opportunity which is simply too great to pass up. As a result, they tend to focus on the tried and tested, directing innovative efforts to short-term, near market innovations that will help leverage past investments and offer little risk (Leonard-Barton, 1992).

  Incremental innovative efforts can have significant effects. Leading automotive production facilities, for example, can aim for a yearly target of 10 per cent improvement in operational efficiency per year (Womack et al., 1990). Much of these efficiency gains are achieved thanks to Kaizen, a relentless pursuit of small improvements in production systems. Such a factory would double its productivity every seven years and triple productivity every eleven years. Another case of the power of incremental innovation is the ability of major airports in Europe to dramatically expand their ability to handle passengers without an increase in runway capacity. London Heathrow, for example, increased its passengers per year from 5 million to over 50 million, even though it has used only two runways for the past thirty years. The increase in air traffic and capacity was brought about by a persistent pursuit of minor improvements in airport operational efficiency, including changing the queuing system for planes, developing spill-off runways to move planes off the runway more quickly, and training pilots to land and quickly exit the main runway (Tether and Metcalfe, 2003). This suggests that, in the medium term, even small changes can have a major impact on economic output.

The importance of incremental innovation does not limit the economic impact of more radical innovation. Radical innovation—which is often defined by a shift in the performance-price ratio by a factor of five or even ten—can spur the generation of new industries and lead to a long progression of incremental innovations. Attempts to measure the frequency of radical innovation have suggested that in most industries radical innovations are infrequent, occurring every thirty years (Anderson and Tushman, 1990; Tushman and Anderson, 1986). In part, this is reflected in the fact that in the USA only 20 per cent of industries underwent a major shake-up in market share of large incumbents (McGahan, 2004). This means workers entering the average industry may work their entire lives without ever experiencing a radical innovation happening in their industry. In turn, it seems the most critical radical innovations to the economic system are those that have a wide range of applications across *different* industries, what are sometimes referred to as 'general purpose technologies' (Helpman, 1998). A clear example of such an innovation was Fritz Haber and Carl Bosch's development of synthetic nitrogen, creating new, more powerful weapons and fertilizers. These fertilizers enabled a dramatic increase in food production worldwide, helping to feed the world's population as it soared from 1.6 billion in 1911 to almost 7 billion in 2011.

The challenge for organizations is that sources and timing of the emergence of a radical innovation are unpredictable and even unknowable. Industry experts often fail to see radical innovations within their sectors and the history of innovation is littered with estimates of the future state of the world that are almost always badly wrong. The difficulty of anticipating a radical innovation is that their effects are not simply quantitative in nature; they are usually qualitative in character. They do not modify the way something is done; they often totally transform it. As a result, radical innovations are hard to anticipate and prepare for. Even though large firms account for a significant share of radical innovation (Christensen, 1997; Tellis et al., 2009), they often fail to reap the advantages of these breakthroughs. For example, in 1992, IBM developed the world's first smartphone, called Simon. It had a touchscreen, and email and organized diary functionality. But it was others, such as RIM, Apple, and Samsung, that were able to commercially exploit this idea. Moreover, since large firms have heavy investments in incremental efforts, they often struggle to respond to radical changes in their markets. They may listen too closely to their current customers (Christensen, 1997), be unwilling to cannibalize their existing assets (Tellis et al., 2009), or unwilling to change in their current business models and routines (Tripsas and Gavetti, 2000).

Stylized fact 4: Most change brought about through innovation is evolutionary, incremental adaptations of existing elements, products, and technologies.

Stylized fact 5: Radical and revolutionary changes are rare and largely unknowable.

Innovation is Relational

The early years of the study of innovation were dominated by stories of heroic inventors, such as Thomas Edison or Alfred Nobel. Inventors, such as Nobel created whole new industries on the basis of their discoveries, such as nitro-glycerine and later dynamite. Alfred Nobel himself clearly fits the image of a lone inventor. He worked almost entirely by himself in his lab at the bottom of his Paris mansion. He was a difficult and lonely character, with few friends and passions outside of his work. He carefully guarded his inventions and the practices he used to arrive at them. When he shared his ideas with others, it usually ended badly in protracted legal disputes over priority, including a lengthy court case in England with English chemist Frederick Abel over the invention of cordite. His inventions were radical and had applications across a range of industries, including mining and railways, and in warfare. They also allowed Nobel to create a global industrial empire spanning seventy countries, and provided resources upon his death for the launching of the richest prize for science, medicine, and peace (Brown, 2005).

Even modern accounts of innovation tend to privilege the exploits of individuals. Apple's success in the early twentieth century is commonly seen as the direct result of Steve Jobs and his passion for design and relentless pursuit of innovation.

Although all new ideas emerge from the inspirational efforts of individuals, innovation is primarily a relational activity, in that it requires interaction between different people, teams, and functions to be successfully achieved. Individuals may provide the spark and direction to allow great innovation to emerge, but it is usually teams that do the hard graft of turning ideas into innovations. In the case of Apple, for example, Jobs was able to draw on the operational skills of Tim Cook and the design flair of Jonathan Ive, along with the rest of the Apple team. Indeed, research has shown that an individual's ability to generate good, innovative ideas is profoundly shaped by their social capital: the goodwill and resources they can draw upon from their personal contacts (Burt 2005). This effect is also strong for teams and organizations. Organizations that can draw upon ideas, resources, and support from other organizations have greater potential for developing innovations and they also have greater opportunities to capture value from these innovative efforts (Ahuja, 2000; Burt, 2009; Powell et al., 1996).

The relational character of innovation is already reflected in the fact that its value is based on customers' and users' reactions to it. An innovation by itself has no value; it is only the consumer or business demand that innovation creates that leads to value creation and later value capture. In this respect, early customers and lead users provide the seeds to enable the spread and development of innovations and engagement of these users is a critical first step in building up interest in an innovation (Rogers, 2003; von Hippel, 1988, 2005).

Innovations also typically require close coordination with suppliers to design and create critical components. Apple's first iPod, for example, relied closely on Toshiba providing a high storage memory chip to enable it to hold many more songs than competing music players. It also needed Sony and other music rights holders to agree to sell their copyrighted music through the iTunes music store. New products and processes also have to be aligned to regulations and standards of performance and health and safety, which may be subject to lengthy and critical reviews. Obtaining approval for a new drug requires years of patient and careful preparation to convince government agencies, such as the US Food and Drug Administration or the UK National Institute for Clinical Excellence, of the efficacy and value of the new medicine compared to alternatives. Innovations may also require firms to work with competitors and universities to help sustain a new area of development. When the engineering design company Arup, for example, sought to expand the market for fire engineering services, it found considerable hostility by insurers, builders, and regulators to its innovations, such as using elevators for egress in extreme situations. Only by sharing its technology with competitors and universities did it create a wider community of fire engineering practitioners to judge and validate its own work (Dodgson et al., 2007). In addition, innovators may need to reach out to external communities to sustain and develop their products. Propellerhead Software, a Sweden-based computer music program, has created a vibrant community of musicians who rely on its software. It allows its users to develop modifications to its main program and incorporates these modifications in subsequent generations of the software (Jeppesen and Frederiksen, 2006).

Stylized fact 6: Innovation is relational and usually involves collaboration between two or more parties.

Unpacking Creativity, Invention, and Innovation

In an attempt to explain the nature of innovation, many researchers have focused on the source of creativity and novelty arising from individuals (see Chapter 7 by Leonard and Barton). Much of this research focuses on the creativity of individuals, with creativity seen as the ability to develop 'novel' and 'useful' ideas. In turn, creativity related to innovation comes from a person's innate skills and abilities and the human mind is a wonderful instrument for creative endeavours. Everyone has the potential for creative outputs (Boden, 2004), but some individuals possess greater likelihood of achieving an innovation than others. Some of this ability may be innate, based on an individual's genetic make-up partly inherited from their biology (Nicolaou et al., 2008), but much of it is based on personal experience, training, and effort.

To explore the sources of creativity, researchers have sought to probe the character of inventors and innovators. To this end, research has sought to understand how individuals' psychology and perceptions of their environment shape their likelihood of developing creative ideas. This research shows that individuals with a strong self-determination or intrinsic motivation are able to generate greater creative output (Amabile, 1983; Deci and Ryan, 1985). Moreover, an individual's perception of whether the organizational climate supports creative output strongly influences the likelihood they will generate new and useful ideas (Scott and Bruce, 1994). In addition, individuals need to work in teams with empathic leaders, who tolerate failure and provide them with a degree of safety to undertake activities that break away from the norm (Edmondson, 1999).

Creative ideas may provide the wellspring of inventions, but creativity is not always directed towards invention and later innovation. Many creative ideas have no practical application and although useful may fail to lead to innovation. In this sense, creativity is input to the development of an invention, a novel idea that has practical application. Inventions may be of sufficient novelty that they can be used to apply for a patent, granting the inventor a period of exclusivity of the use of this idea. However, even if patented, most inventions do not succeed in being translated into innovations. Of the total share of inventions eventually patented, for example, only a few will be of significant financial value. The patent system only requires that an invention have a potential usefulness and therefore it is up to the inventor or holder of the patent to make the additional effort to turn a practical idea into a commercially useful product, process, or service. In this sense, innovations are the rarest of ideas, those ideas that can be commercialized or implemented to allow the developer of the ideas to capture value from their efforts.

> *Stylized fact 7: Creativity is as critical to invention, as invention is to innovation, but these concepts are separate and distinct elements of the innovation process.*

Capturing Returns from Innovation

The literature on innovation studies can further tell us why capturing the returns from innovation is not easy (see Chapter 12 by Gambardella et al., and Chapter 28 by Leiponen). Partly, this is because these returns are highly skewed. For many classes of innovative activities, one finds that a very small number of activities, projects, or events account for the lion's share of the total returns (Scherer et al., 2000). It is not uncommon in an R&D portfolio, for example, for 10 per cent of projects to account for 90 per cent of all the total returns. In part, this skewness is a result of the uncertainty—the 'unknown unknowns'—of investments in innovative efforts. It also reflects the cumulative advantages that small differences in the early stages of development of an innovation can make to its eventual success. In this sense, the field of innovation is concerned with finding 'black swans', rare events that capture significant returns (Fleming, 2007). An example of such skewness can be seen in Microsoft's long-term investment in R&D over a period of ten years, which was widely perceived in the consumer electronics industry to be a costly failure. Except, of course, that in 2010 Microsoft launched Kinect, which sold 8 million units in its first sixty days on the market.

Innovation research has further taught us that the skills required to generate innovation differ significantly from the skills required to capture their returns. In fact, many innovators find that the returns to their innovative efforts are captured by others. This pattern can be seen for a range of different industries, as the organizations that originate an innovation lose out to skilled competitors. Examples abound. Royal Crown Cola first developed a diet soft drink, yet saw Pepsi and Coca-Cola—its bitter rivals— profit from this innovation rather than itself. The first Magnetic Resonance Imaging product was developed by EMI, yet GE and Siemens become the dominant players in the market. Xerox in its PARC lab developed the first graphical user interface for the personal computer, but saw first Apple and then later Microsoft exploit this idea for commercial gain.

A key explanation of these patterns comes from the work of David Teece, captured in his profiting from innovation framework (Teece, 1986). Teece suggested that the ability of firms to gain from an innovation is a function of their ability to capture the value of their intellectual property as well as the nature of knowledge in their industry. Although intellectual property protection such as patents can be effective mechanisms to stop other firms from copying an innovation, patents are imperfect and many skilled rivals can invent around them. Moreover, small, new firms who lack bargaining power may find that their ideas are simply taken from them because they lack the legal resources or firepower to enforce their intellectual property. Teece cites the example of Robert Kearns, later captured in the film *Flash of Genius*, who developed the first intermittent windshield wiper. Kearns later found that the idea was copied by Ford and, subsequently, by Chrysler. He was unable to find legal support to challenge this infringement, as few lawyers were willing to take on the mighty Ford. It was only by learning the law and representing himself that he was able to secure an eventual victory in a lengthy and costly court battle. Although he was eventually successful, the ordeal damaged his health and family relationships.

Teece's approach points to the industrial environment—what he called the 'appropriability regime' of an industry—that shapes the links between innovators and the returns to innovation. The appropriability regime covers the nature of knowledge in an industry as industries with complex, cumulative patterns of knowledge development are hard to penetrate by new entrants as opposed to industries that rely on new knowledge and/or new combinations of existing knowledge (Levin et al., 1987; Pisano and Teece, 2007; Winter, 1987). It also reflects the availability and effectiveness of intellectual property protection mechanisms (Cohen et al., 2000). In addition, Teece highlights the importance of complementary assets, tangible or intangible items required to enable a successful commercialization of an invention, such as marketing or sales forces or manufacturing capability. He points out that, in many cases, ownership of complementary assets determines who eventually gets to benefit from innovation. In particular, he shows that complementary assets often reside with large firms, with whom innovators will need to partner in order to have any hope of commercial success.

Stylized fact 8: Most innovators fail to capture returns from their innovative efforts and capturing returns from innovation requires different skills from creating innovations.

Varieties of Innovation

Innovations come in many forms and types. Scholars have sought to characterize both the degree and type of innovation, and it sometimes seems there is a whole industry of academics and consultants putting new words in front of the word innovation. The most critical distinction lies in the end points in the continuum between incremental and radical innovations, which speaks to the degree of change introduced by an innovation into the economic system.

Types of innovation also differ. A classic distinction in innovation studies is between product and process innovations. Product innovations are easy to identify, as they involve the creation and launch of new goods and services. In contrast, process innovations are often silent, hidden from public view as they involve changes in operations, tasks, and ways of working in organizations. Process innovations do not require changes in the nature of the product. For example, the development of float glass manufacturing revolutionized the productivity of glass making, but the product—glass—remained largely the same. In this sense, process innovations are largely cost-reducing, as they involve ways of producing a given good or service with lower levels of inputs (Utterback, 1994).

Alongside the distinction between product and process innovation, scholars have suggested that innovations may be architectural or modular (Henderson and Clark, 1990). Architectural innovation involves changes in the interfaces between different components or aspects of knowledge. They do not themselves require the development of new products or processes, but may lead to significantly new ways of bringing together elements of a product or system. An example of architectural innovation is the movement from tricycles towards two-wheel 'safety' cycles. In contrast, modular innovation involves significant changes in a single component of a product, such as a bicycle light, but these changes do not affect the way a component works with other components. Here, efficiency-driven organizations specializing in advancing the components of their systems have a clear advantage in driving forward a technology along its trajectory. A simple example would be the use of new rechargeable battery in a 'phone.

Christensen (1997) describes the situation in which incumbents fail in the face of seemingly easy-to-handle innovation. His concept of disruptive innovation describes how companies that continuously improve their products to satisfy customers may eventually end up providing products that are over-performing for the needs of the markets in which they are offered. In this situation, firms may be vulnerable to be attacked from below by other companies offering inferior products, which are, however, 'good enough' for consumers and which beat incumbents on price or a previously irrelevant performance dimension. Christensen illustrates how listening too closely to current customers led industry leaders who were actually often the inventors of disruptive innovations to choose *not* to bring them to the market. This was because of concerns current customers would not like them, or fears over lower margins and cannibalization of their other products. He shows how disruptive innovations led to a repeated change in industry leadership in the hard-disk drive industry over several generations of products and further uses it to explain the competitive dynamics caused by the introduction of hydraulics in the excavator industry.

Finally, the last decade has seen the emergence and acceptance of several new categories of innovation—such as for example open innovation (Chesbrough, 2003, and see Chapter 22 by Alexy and Dahlander). Given the prominence associated with the 'discovery' of a new type of innovation, we would not hesitate to predict a further increase in 'new types' of innovation over the years to come. We also see increased research into important new frameworks for analysing innovation, such as platforms (see Chapter 32 by Gawer and Cusumano) and ecosystems (see Chapter 11 by Autio and Thomas).

Stylised fact 9: There is a vast array of different types of innovation.

Patterns of Innovative Activity

Since the earliest studies of innovation, scholars have tried to explore patterns of activity that reoccur periodically. Schumpeter, for example, picked up Kondratieff's concept of the long wave, which illustrated that economic growth, based on innovative activity, would proceed in waves of about fifty years of length. Great inventions, such as the steam engine, steel, electrical engineering, the automobile, computers, and biotechnology, have been suggested to represent transformative underlying technologies.

Patterns have also been found with individual technologies. Here, Abernathy and Utterback's concept of the product life cycle (PLC) features most prominently (Abernathy and Utterback, 1978). The PLC argues that over the lifetime of a technology, firms place varying levels of emphasis on product and process innovation. In the initial 'fluid' stage, firms propose an array of different products and designs incorporating the new technology. In the 'transitional' stage, a dominant product design emerges, and while not necessarily the highest performing product configuration, this design becomes a commonly accepted standard by producers and consumers. Accordingly, the rate of product innovation decreases and efforts begin to focus on variants of the design, while at the same time the first significant investments into process innovations are made. Finally, in the 'specific' stage, the product has moved on to become a commodity, and concerns about production cost are dominant. Hereafter, product innovation activities are limited, and innovative activity mainly revolves around the optimization of process technologies. Whereas the concept of the PLC has been widely confirmed for a variety of industries, several extensions and criticisms exist. For example, Klepper's work argued when the marginal advantages of additional investments in process versus product R&D is reached, there is a rush for scale and an industry shake-out (e.g. Klepper, 1997). In addition, Barras (1986, 1990) points out that for services innovation, especially financial services, the PLC sometimes applies in reverse: first, process technologies need to be established and standardized that then facilitate the generation of new services upon them. Barras argues that this is due to fundamental differences between product and service innovation, which reside in their co-terminality, intangibility, and low capital intensity. Other authors, however, have pointed out a wide range of examples for the standard PLC also applying to services industries.

Finally, there are strong complementarities between different types of innovation. Service innovation can create opportunities for product innovation, processes for new products, and new products for new processes. Evidence from innovation surveys finds strong complementarities between different types of innovation. They are both often present simultaneously and the creation of multiple complementary forms of innovation at the same time can help stimulate greater firm performance (Damanpour and Gopalakrishnan, 2001).

Stylised fact 10: There are 'regular' patterns in innovative activity over time, and strong complementarities between different types of innovation.

The Geography of Innovation

Despite the fact that innovative activity is becoming an increasingly global and inter-connected phenomenon (see below as well as Chapter 27 by Håkanson), innovation tends to remain 'sticky' to particular places. Innovation investments and outputs tend to be concentrated in global centres, where leading actors congregate, mingle, and compete.

Within organizations, co-location of individuals is still a crucial mechanism to enable the effective flow of knowledge between people who have to work together to produce innovation. As a result, organizations give careful attention to the design of their R&D and development facilities to create 'spaces for innovation', hoping to maximize exchange and cross-fertilization. When BMW constructed a new R&D facility, for example, it sought to ensure that engineers working on related problems were no more than 25 metres from one another. One reason for this need to be close is that knowledge itself can often be characteristic as 'sticky' (von Hippel, 1994); difficult to express and transfer, and contextually dependent. Anticipated users of a planned product may, for example, be unable to articulate their own needs plausibly when asked by marketers, and only explore them by actually using or modifying a product themselves. von Hippel goes on to argue how this sticky nature of knowledge might even lead to predict where innovation comes from and who profits from it, in particular emphasizing the role of users as the actual source of innovation when sticky information resides with them (see Chapter 5 by Franke).

Beyond the level of the organization, the sticky nature of knowledge implies that certain types of knowledge will not travel far. In particular, valuable knowledge spillovers have a clear tendency to only bridge small geographic distances. Also, collaborative activity between firms benefits from face-to-face interaction to facilitate knowledge exchange and transfer. And finally, even investments in innovative activity, for example venture capital, have a clear local bias (Sorenson and Stuart, 2001). All in all, it is clear that to a certain degree, innovation remains a face-to-face business in which geography plays a crucial role (Storper and Venables, 2004).

In turn, this also implies that regional and national differences in how innovation is fostered and supported may matter significantly, and specifically innovation is affected by the variance in institutional set-ups that govern interactions between firms and individuals (Lundvall, 1992). The sheer existence of a patent system and different configurations thereof may shape domestic and foreign investment in R&D. Moreover, varying institutions may give rise to different inputs into the innovation process, whether they are supportive or less helpful. The German manufacturing sector, for example, is famed for powerful work councils and long-term employment and a highly skilled labour force, whereas UK manufacturing firms are more strongly controlled by management, feature shorter-term employment, and have tended to employ more workers in lower-skilled jobs.

Certain national and regional innovation systems are notorious for their ability to become successful launch pads for innovation. At the national level, the famous example of nationally organized catch-up strategies in Korea and Taiwan highlight the potential of these efforts (Hobday, 1995; Kim, 1997). At the regional level, clusters may emerge into hotspots of innovative activity. Silicon Valley or the Boston region in the USA are powerhouses for ground-breaking, first-to-world products and services (Saxenian, 1994). In contrast, regions in Italy and Chile are famed for having sustained excellence in the traditional shoe-making or wine industry, respectively, for decades (Boschma and Frenken, 2007; Giuliani and Bell, 2005). Other prominent examples include the USA's Hollywood, India's Bollywood, or Nigeria's Nollywood in the movie industry. These examples also highlight the need to balance tight local links with global pipelines to ensure diversity in both knowledge inputs as well as pathways to markets (Bathelt et al., 2004; Powell et al., 1996).

Stylized Fact 11: Innovation is a 'sticky' activity in which location matters.

The Organizational Routines of Innovation

Research has sought to understand what types of organizational routines support innovation. Originally, this work started with an attempt to understand what makes a 'technologically progressive firm', highlighting the importance of organic, fluid organizational structures to support innovation (Burns and Stalker, 1961). At the same time, this work highlighted that mechanisms or formal structures could help regulate and regularize innovative efforts, ensuring that processes and products could be replicated and scaled up. This tension has long been at the heart of innovation management, with attempts to develop organizational routines that support the creation and development of new ideas as well as enabling their execution and delivery. One solution to this challenge is to create separate organizational structures to support different forms of innovation; one unit for exploring creative, radical new ideas and another for the exploiting and developing of incremental improvements in existing ideas (Tushman and O'Reilly, 1996). In doing so, an organization would be ambidextrous, taking advantage of both models of innovative support. Indeed, this organizational model has been widely adopted by firms, which often create separate organizational units for different innovative tasks with contrasting work and human resources practices.

Organizations have sought to develop routines to support creativity. These include providing autonomy for innovators, possibly by providing a share of time for individuals to work outside their official project plans. They have also nurtured a tolerance of failure and culture of forgiveness for those individuals or teams that attempt to achieve innovation but fail. They seek to create fluid and dynamic teams that bring together disciplines and functions. In particular, innovative organizations have adopted integrated product development teams, including representatives of different departments to help work together on a R&D project. They also seek to refresh team membership, and ensure that they display an openness to outsiders (Leonard and Swap, 1999).

Another critical routine for innovative organizations has been the development of tools to manage and select R&D projects. The alternative is that efforts to generate ideas by letting a thousand flowers bloom may lead to a garden of weeds (Kanter et al., 1997). Since resources are always limited and the costs of scaling up any idea are liable to be high, organizations need to think very carefully about how they choose and manage ideas. In R&D management, the development of stage-gate systems that create a series of stop-go decision gates provides an opportunity for organizations to reflect on R&D projects at different stages of their maturity (Cooper, 1990, 2001). Moreover, these projects can be judged against one another and a range of criteria, such as potential market value, costs of development, and so on. Using multi-criteria assessment to judge the quality of projects helps to avoid the tendency within organizations to rely on 'gut-feel' about the value of projects. These tools also help to avoid the danger than firms overcommit to single projects and help to ensure a good allocation of resources between products at different stages and with different degrees of radicalness.

Attempts to assess the value of R&D projects before they are completed are useful, but often problematic. Good projects may be killed by internal stage-gate process as they go against established ways of working, leading to a tendency to short-term, incremental efforts. Even techniques that offer statements about the financial value of a project are based on expected returns and costs, estimates that are liable to suffer from dangers of bias and misstatement.

Stylized fact 12: There is a clear set of organizational routines that can help organizations to better manage the innovation process.

THE PROTECTIVE BELT OF INNOVATION STUDIES

Early work on the nature of innovation focused mostly on innovation driven by technical change, usually in the manufacturing sector. There were some obvious reasons for this. The industrial revolution is often seen to be driven by technological changes—embodied in inventions such as the steam engine or the spinning jenny. Yet, the industrial revolution was also the result of social, political, and economic changes, and it does not lend itself to simplistic and often misleading explanations of economic development based on technological determinism (Mokyr, 2004).

In addition, part of the focus on technical change in the study of innovation was driven by measurement. The main measurement instruments of innovation studies—R&D surveys, patents, and academic publications—all tend to focus on the generation and use of new scientific and technological knowledge. Since the measurement tools concentrate on this topic, researchers and governments have done likewise, tending to focus on innovation in the 'measured' sectors where there is considerable R&D, patenting, and publications. Indeed, much of the modern focus of research on innovation has focused on pharmaceuticals, semiconductors, and biotechnology industries, all sectors whose innovative efforts are captured by the current toolkit of innovation studies. This approach has created numerous blind spots for the research tradition, and created opportunities for researchers to develop new ideas in areas that are distant from the conventional focus on the generation and use of scientific and technological knowledge.

One idea that has begun to penetrate the protective belt of innovation studies is that managers and researchers should not give primacy to technological innovation over other types of innovation. It is clear that many innovations are not primarily 'technological' in nature. For instance, most service innovations are largely organizational, involving new ways of bringing together information and creative routines (Gallouj and Weinstein, 1997). These services require deep knowledge of a range of systems, and the ability to integrate diverse sets of activities in new and productive ways. Dell's success in the 1990s and early 2000s, for example, was driven by its strong electronic commerce systems, such as websites and telephone ordering. These systems allowed it to bypass conventional sales channels to directly interface with customers. In addition, over the past ten years, Xerox has transformed itself from an organization focused on the development of hardware and technology to a solution provider, with more than half of its sales arising from services. Much of the sales of Xerox now arise from activities that have little or nothing to do with photocopiers. As the Dell and Xerox cases attest, building a successful business model can be a tremendous stimulus to innovation (Chesbrough, 2011, and see Chapter 21 by Massa and Tucci).

A second area of activity that has been open to a major shift in thinking is the role of R&D in the innovation process. In the early stages of the study of innovation, capturing investment in R&D by governments, firms, and universities represented a major breakthrough in our understanding and, since 1965, this information has been collected systematically across the developed world (OECD 2002). Yet, it was clear early on that R&D only captures a modest share of total economic and social investment in innovation, and that R&D investment and resulting inventive outputs such as patents were at best incomplete predictors of innovation and growth (Griliches, 1981; Griliches et al., 1991). This is true for countries as well as companies. And with increasing levels of connectedness due to phenomena such as the Internet or globalization, many corporations have shifted away from an R&D-led model of innovation, focusing on more open and distributed models (Chesbrough, 2003; von Hippel, 2005). Although it is clear that R&D is a still a critical resource for firms to develop new products, services, and processes and to learn about the efforts of others (Cohen and Levinthal, 1989, 1990), there are other mechanisms that support innovation and learning that operate within the firm. Accordingly, companies have also sought to capture information on their expenditures on different types of intangibles, such as customer goodwill, networks, and brands. They have also moved away from focusing on the level of R&D expenditure as a measure of corporate vitality and growth. It is also clear that there is no direct link between expenditures on R&D and corporate performance, as many firms compete effectively against large R&D spenders although they spend much less on R&D. Over a period of five years, for example, Apple spent less than a third on R&D than Nokia, yet was able to overcome Nokia's dominance of the mobile 'phone market.

These shifts in corporate perspectives towards the salience of R&D have not always been reflected in government thinking. Many governments remained focused on expenditures on R&D as a key measure of national innovative effort. It is common for major countries to adopt targets for R&D spending as a share of the economy. The European Union, for example, has a target of spending 3 per cent of Gross Domestic Product on R&D in 2020, which is the same target it sought to achieve in 2010, but failed to reach.

A problem with R&D is that it misses out on much of the investment in services innovation, especially in critical and growing sectors such as professional services. Generally, the notion of R&D is often based on the idea of corporate research labs coalescing the efforts of scientists and engineers to develop new knowledge. Yet, many other people and functions are involved in knowledge production and the creation of novelty in the economic system, including management consultants, designers, or software programmers. Many of the activities of these individuals are not captured by conventional R&D measurements and therefore by focusing on R&D we are looking at the tip of the iceberg when it comes to social and economic investment on innovation. Software development, for example, is a hugely important part of the banking industry, but it cannot be easily accounted for by traditional R&D reporting categories. The scale of this measurement problem was recently demonstrated in the UK, where attempts were made to capture the total level of expenditures on intangibles in the economy. These estimates showed that R&D accounted for only 9 per cent of total intangible investment, dwarfed by expenditures on software development, organizational development, and training (Haskel et al., 2010). As result, attention has been newly focused on the measuring and refining of understanding of other firm investments that shape innovative outcomes, shifting attention away from R&D as the central mechanism that supports innovation in the economic system.

The third break from the past in the study of innovation concerns the role of the firm in the innovation process. Traditionally, the study of innovation saw the firm as the central actor in the process of innovation, as it was assumed that firms provided the means to create, diffuse, and capture value from innovative efforts. Yet, with the advent of more collaborative and distributed models of innovation (Chesbrough, 2003; von Hippel, 1988, 2005), it is not clear that the firm is always the most critical actor in the innovation process. Innovations are increasingly the product of collaborations between a range of actors, including users, universities, firms, and governments. The decline of the salience of the firm in creating and capturing value from innovation reflects in part the way that innovations increasingly rely on complex knowledge, sourced from a range of actors. As a result, firms rely on greater levels of collaboration to generate and commercialize their ideas. ARM, a UK-based design-based semiconductor firm, for example, relies on a network of collaborators that includes over 300 different chipmakers, designers, and chip users. This ecosystem supports a range of developments outside the direct control of ARM, but provides a rich pool of resources to facilitate the development of ARM chip designs (Garnsey et al., 2008). Moreover, it is now easier for organizations that seek to profit from an innovation to find partners to help them in the manufacturing, delivery, service, and support of their products. This deepening of the innovative division of labour allows organizations to become more specialized at those parts of the value chain that they are best able to contribute to (Arora et al., 2001; Gambardella et al., 2007). In addition, organizations are increasingly utilizing third parties to help them innovate, including investing in crowdsourcing, innovation intermediaries, and co-creation with customers. All of these changes at the heart of the innovation process suggest that the firm of today rarely controls its own destiny when it comes to innovation, and that its innovative potential is largely determined not by the assets and knowledge it holds, but by its ability to draw upon resources, knowledge, and skills from others.

A fourth area of change to our understanding of innovation has been about the nature of public and private knowledge. Traditionally, studies of innovation have assumed that firms develop private knowledge, whereas universities develop public knowledge. Yet, the past twenty years have altered this perspective. On one side, universities are increasingly seeking to patent their discoveries and profit from them by licensing them to established firms or establishing university spinouts. In part, this commercial effort has been driven by government pressures on university finances, but also expectations that universities were missing out on significant sources of potential funding. Although the returns to university patenting have been relatively modest, the movement of universities to create private knowledge has altered the division of labour between universities and firms in the innovation process (Mowery et al., 2001; Nelson and Nelson, 2002). No longer can it be assumed that new knowledge created at universities is freely available for firms to use. Instead, this knowledge is increasingly accessible only by signing collaborative agreements or through direct licensing.

At the same time, firms have become more and more active in creating public knowledge. They publish in the scientific literature (Hicks, 1995). They may donate patents to support open source software and devote resources to helping build, sustain, and develop these communities (Alexy and Reitzig, 2013). They also join forces with their competitors to help develop public repositories of knowledge, such as Merck, Eli Lilly, and GlaxoSmithKline (GSK)'s support for the Structural Genomics Consortium (Perkmann, 2009). As a result, the openness of new knowledge cannot be determined just by looking at whether it was created by public or private organizations and, as a result, the landscape of the knowledge for innovation has become complex and layered.

As the level of research and interest in innovation has increased, we could expect more breakthroughs and changes to our understanding of its nature. As a progressive science, the study of innovation is open to the 'creative destruction' of its hard core of stylized facts. With the advent of new, richer, and more powerful information sources on the nature of innovative efforts by public and private actors, there is a significant opportunity to transform what is known about innovation to assist its management.

NOTES

1. A notorious site in London where hangings took place.
2. From 18 February 1999. See also <http://www.economist.com/node/186620>.
3. Speech given in August 2009 in Elkhart County, Indiana. See also <http://www.whitehouse.gov/blog/Spurring-Innovation-Creating-Jobs/>.
4. That is, their ownership by one party does not preclude access to them by another.

REFERENCES

Abernathy, W. J., and Utterback, J. M. (1978). 'Patterns of Industrial Innovation', *Technology Review*, 80: 41–7.

Acemoglu, D., Gancia, G., and Zilibotti, F. (2012). 'Competing Engines of Growth: Innovation and Standardization', *Journal of Economic Theory*, 147(2): 570–601.

Ahuja, G. (2000). 'Collaboration Networks, Structural Holes, and Innovation: A Longitudinal Study', *Administrative Science Quarterly*, 45(3): 425–55.

Alexy, O., and Reitzig, M. (2013). 'Private-Collective Innovation, Competition, and Firms' Counterintuitive Appropriation Strategies', *Research Policy*, 42(4): 895–913.

Amabile, T. M. (1983). *The Social Psychology of Creativity*. New York: Springer.

Anderson, P., and Tushman, M. L. (1990). 'Technological Discontinuities and Dominant Designs: A Cyclical Model of Technological Change', *Administrative Science Quarterly*, 35(4): 604–33.

Arora, A., Fosfuri, A., and Gambardella, A. (2001). *Markets for Technology*. Cambridge, MA: MIT Press.

Barras, R. (1986). 'Towards a Theory of Innovation in Services', *Research Policy*, 15(4): 161–73.

Barras, R. (1990). 'Interactive Innovation in Financial and Business Services: The Vanguard of the Service Revolution', *Research Policy*, 19(3): 215–37.

Bathelt, H., Malmberg, A., and Maskell, P. (2004). 'Clusters and Knowledge: Local Buzz, Global Pipelines and the Process of Knowledge Creation', *Progress in Human Geography*, 28(1): 31–56.

Boden, M. A. (2004). *The Creative Mind: Myths and Mechanisms*, 2nd edn. London: Routledge.

Boland, R. J., Lyytinen, K., and Yoo, Y. (2007). 'Wakes of Innovation in Project Networks: The Case of Digital 3-d Representations in Architecture, Engineering, and Construction', *Organization Science*, 18(4): 631–47.

Boschma, R. A., and Frenken, K. (2007). 'Applications of Evolutionary Economic Geography', in K. Frenken (ed.), *Applied Evolutionary Economics and Economic Geography*. Cheltenham: Edward Elgar.

Boswell, J. (1791). *The Life of Samuel Johnson, LL.D.* London: Henry Baldwin for Charles Dilly.

Brown, S. R. (2005). *A Most Damnable Invention*. Toronto: Viking Canada.

Brynjolfsson, E. (1993). 'The Productivity Paradox of Information Technology', *Communications of the ACM*, 36(12): 66–77.

Burns, T., and Stalker, G. M. (1961). *The Management of Innovation*. London: Tavistock.

Burt, R. S. (2005). *Brokerage and Closure: An Introduction to Social Capital*. New York: Oxford University Press.

Burt, R. S. (2009). *Neighbor Networks: Competitive Advantage Local and Personal*. New York: Oxford University Press.

Chesbrough, H. (2003). *Open Innovation: The New Imperative for Creating and Profiting from Technology*. Boston: Harvard Business School Press.

Chesbrough, H. (2011). *Open Services Innovation: Rethinking your Business to Grow and Compete in a New Era*. San Francisco: John Wiley & Sons.

Christensen, C. M. (1997). *The Innovator's Dilemma*. Boston: Harvard Business School Press.

Cohen, W. M., and Levinthal, D. A. (1989). 'Innovation and Learning: The Two Faces of R & D', *Economic Journal*, 99(397): 569–96.

Cohen, W. M., and Levinthal, D. A. (1990). 'Absorptive Capacity: A New Perspective on Learning and Innovation', *Administrative Science Quarterly*, 35(1): 128–52.

Cohen, W. M., Nelson, R. R., and Walsh, J. P. (2000) (February). Protecting their Intellectual Assets: Appropriability Conditions and Why U.S. Manufacturing Firms Patent (or not)'. Available at <http://www.nber.org/papers/w7552> (accessed 7 April 2006).

Cooper, R. G. (1990). 'Stage-Gate Systems: A New Tool for Managing New Products', *Business Horizons*, 33(3): 44–54.

Cooper, R. G. (2001). *Winning at New Products: Accelerating the Process from Idea to Launch*, 3rd edn. New York: Basic Books.

Damanpour, F., and Gopalakrishnan, S. (2001). 'The Dynamics of the Adoption of Product and Process Innovations in Organizations', *Journal of Management Studies*, 38(1): 45–65.

Deci, E. L., and Ryan, R. M. (1985). *Intrinsic Motivation and Self-Determination in Human Behaviour*. New York: Plenum.

Dodgson, M., Gann, D., and Salter, A. (2005). *Think, Play, Do: Technology, Innovation, and Organization*. Oxford: Oxford University Press.

Dodgson, M., Gann, D., and Salter, A. (2007). ' "In case of fire, please use the elevator": Simulation Technology and Organization in Fire Engineering', *Organization Science*, 18(5): 849–64.

Dodgson, M., Gann, D., Wladawsky-Berger, I., and George, G. (2013). 'From the Digital Divide to Inclusive Innovation: The Case of Digital Money', RSA Pamphlet, Royal Society of Arts, London.

Edgerton, D. (2008). 'The Charge of Technology', *Nature*, 455(7216): 1030–1.

Edmondson, A. (1999). 'Psychological Safety and Learning Behavior in Work Teams', *Administrative Science Quarterly*, 44(2): 350–83.

Fleming, L. (2007). 'Breakthroughs and the "Long Tail" of Innovation', *MIT Sloan Management Review*, 49(1): 69–74.

Gaddis, J. L. (2005). *The Cold War*. London: Penguin Books.

Gallouj, F., and Weinstein, O. (1997). 'Innovation in Services', *Research Policy*, 26: 537–56.

Gambardella, A., Giuri, P., and Luzzi, A. (2007). 'The Market for Patents in Europe', *Research Policy*, 36(8): 1163–83.

Garnsey, E., Lorenzoni, G., and Ferriani, S. (2008). 'Speciation through Entrepreneurial Spin-off: The Acorn-ARM Story', *Research Policy*, 37(2): 210–24.

Giuliani, E., and Bell, M. (2005). 'The Micro-Determinants of Meso-Level Learning and Innovation: Evidence from a Chilean Wine Cluster', *Research Policy*, 34(1): 47–68.

Greenleaf, W. (1961). *Monopoly on Wheels*. Detroit: Wayne State University Press.

Griliches, Z. (1981). 'Market Value, R&D, and Patents', *Economics Letters*, 7(2): 183–7.

Griliches, Z., Hall, B. H., and Pakes, A. (1991). 'R&D, Patents, and Market Value Revisited: Is There a Second (Technological Opportunity) Factor?', *Economics of Innovation and New Technology*, 1(3): 183–201.

Hargadon, A. B. (2003). *How Breakthroughs Happen: The Surprising Truth About How Companies Innovate*. Cambridge, MA: Harvard Business School Press.

Haskel, J., Clayton, T., Goodridge, P., Pesole, A., Barnett, D., Chamberlain, G., Jones, R., Khan, K., and Turvey, A. (2010) (February). 'Innovation, Knowledge Spending and Productivity Growth in the UK: Interim Report for Nesta Innovation Index Project. Available at <http://spiral.imperial.ac.uk/bitstream/10044/1/5279/1/Haskel%202010-02.pdf> (accessed 26 November 2011).

Helpman, E. (1998). *General Purpose Technologies and Economic Growth*. Cambridge, MA: MIT Press.

Henderson, R. M., and Clark, K. B. (1990). 'Architectural Innovation: The Reconfiguration of Existing Product Technologies and the Failure of Established Firms', *Administrative Science Quarterly*, 35(1): 9–30.

Hicks, D. (1995). 'Published Papers, Tacit Competencies and Corporate Management of the Public/Private Character of Knowledge', *Industrial and Corporate Change*, 4(2): 401–24.

Hobday, M. (1995). *Innovation in East Asia: The Challenge to Japan*. London: Elgar.

Jeppesen, L. B., and Frederiksen, L. (2006). 'Why Do Users Contribute to Firm-Hosted User Communities? The Case of Computer-Controlled Music Instruments', *Organization Science*, 17(1): 45–63.

Kaldor, N. (1957). 'A Model of Economic Growth', *Economic Journal*, 67(268): 591–624.

Kaldor, N. (1961). 'Capital Accumulation and Economic Growth', in F. A. Lutz and D. C. Hague (eds), *The Theory of Capital*. London: MacMillan, 177–222.

Kanter, R. M., Kao, J., and Wiersema, F. (1997). *Innovation: Breakthrough Thinking at 3M, DuPont, GE, Pfizer, and Rubbermaid*. New York: HarperCollins.

Kim, L. (1997). *Imitation to Innovation: The Dynamics of Korea's Technological Learning*. Boston: Harvard Business School Press:

Klepper, S. (1997). 'Industry Life Cycles', *Industrial and Corporate Change*, 6(1): 145–81.

Kogut, B., and Zander, U. (1992). 'Knowledge of the Firm, Combinative Capabilities, and the Replication of Technology', *Organization Science*, 3(3): 383–97.

Lakatos, I. (1970). 'Falsification and the Methodology of Scientific Research Programmes', in I. Lakatos and A. Musgrave (eds), *Criticism and the Growth of Knowledge*. Cambridge: Cambridge University Press, 91–196.

Leiponen, A. (2005). 'Skills and Innovation', *International Journal of Industrial Organization*, 23(5–6): 303–23.

Leonard, D., and Swap, W. (1999). *When Sparks Fly: Igniting Creativity in Groups*. Boston: Harvard Business School Press.

Leonard-Barton, D. (1992). 'Core Capabilities and Core Rigidities: A Paradox in Managing New Product Development', *Strategic Management Journal*, 13: 111–25.

Levin, R. C., Klevorick, A. K., Nelson, R. R., and Winter, S. G. (1987). 'Appropriating the Returns from Industrial-Research and Development', *Brookings Papers on Economic Activity*, 3: 783–831.

Lundvall, B. (1992). *National Systems of Innovation: Towards a Theory of Innovation and Interaction Learning*. London: Pinter.

McGahan, A. M. (2004). *How Industries Evolve: Principles for Achieving and Sustaining Superior Performance*. Boston: Harvard Business School Press.

Marrano, M. G., Haskel, J., and Wallis, G. (2009). 'What Happened to the Knowledge Economy? ICT, Intangible Investment, and Britain's Productivity Record Revisited', *Review of Income and Wealth*, 55(3): 686–716.

Martin, B. (2012). 'The Evolution of Science Policy and Innovation Studies', *Research Policy*, 41(7): 1219–39.

Mokyr, J. (2004). *The Gifts of Athena: Historical Origins of the Knowledge Economy*. Princeton: Princeton University Press.

Mowery, D. C., Nelson, R. R., Sampat, B. N., and Ziedonis, A. A. (2001). 'The Growth of Patenting and Licensing by U.S. Universities: An Assessment of the Effects of the Bayh-Dole Act of 1980', *Research Policy*, 30(1): 99–119.

Nelson, R. R., and Nelson, K. (2002). 'Technology, Institutions, and Innovation Systems', *Research Policy*, 31(2): 265–72.

Nicolaou, N., Shane, S., Cherkas, L., Hunkin, J., and Spector, T. D. (2008). 'Is the Tendency to Engage in Entrepreneurship Genetic?', *Management Science*, 54(1): 167–79.

OECD (2002). *Frascati Manual: Proposed Standard Practice for Surveys on Research and Experimental Development*, 6th edn. Paris: OECD.

Perkmann, M. (2009). 'Trading off Revealing and Appropriating in Drug Discovery: The Role of Trusted Intermediaries', *Academy of Management Proceedings*: 1–6.

Pisano, G. P., and Teece, D. J. (2007). 'How to Capture Value from Innovation: Shaping Intellectual Property and Industry Architecture', *California Management Review*, 50(1): 278.

Powell, W. W., Koput, K. W., and Smith-Doerr, L. (1996). 'Interorganizational Collaboration and the Locus of Innovation: Networks of Learning in Biotechnology', *Administrative Science Quarterly*, 41(1): 116–45.

Rogers, E. M. (2003). *Diffusion of Innovations*, 5th edn. New York: Free Press.

Romer, P. M. (1986). 'Increasing Returns and Long-Run Growth', *Journal of Political Economy*, 94(5): 1002–37.

Romer, P. M. (1990). 'Endogenous Technological Change', *Journal of Political Economy*, 98(5): S71–S102.

Saxenian, A. L. (1994). *Regional Advantage: Culture and Communication in Silicon Valley and Route 128*. Cambridge, MA: Harvard University Press.

Scherer, F. M., Harhoff, D., and Kukies, J. (2000). 'Uncertainty and the Size Distribution of Rewards from Innovation', *Journal of Evolutionary Economics*, 10: 175–200.

Schumpeter, J. A. (1911). *Theorie der wirtschaftlichen Entwicklung: Eine Untersuchung über Unternehmergewinn, Kapital, Kredit, Zins und den Konjunkturzyklus*. Berlin: Dunkler & Humblot.

Schumpeter, J. A. (1942). *Capitalism, Socialism, and Democracy*. New York: Harper and Brothers.

Scott, S. G., and Bruce, R. A. (1994). 'Determinants of Innovative Behavior: A Path Model of Individual Innovation in the Workplace', *Academy of Management Journal*, 37(3): 580–607.

Solow, R. M. (1957). 'Technical Change and the Aggregate Production Function', *Review of Economics and Statistics*, 39(3): 312–20.

Sorenson, O., and Stuart, T. E. (2001). 'Syndication Networks and the Spatial Distribution of Venture Capital Investments', *American Journal of Sociology*, 106(6): 1546–88.

Storper, M., and Venables, A. J. (2004). 'Buzz: Face-to-Face Contact and the Urban Economy', *Journal of Economic Geography*, 4(4): 351–70.

Teece, D. J. (1986). 'Profiting from Technological Innovation: Implications for Integration, Collaboration, Licensing and Public Policy', *Research Policy*, 15(6): 285–305.

Tellis, G. J., Prabhu, J. C., and Chandy, R. K. (2009). 'Radical Innovation across Nations: The Preeminence of Corporate Culture', *Journal of Marketing*, 73(1): 3–23.

Tether, B. S., and Metcalfe, J. S. (2003). 'Horndal at Heathrow? Capacity Creation through Co-operation and System Evolution', *Industrial and Corporate Change*, 12(3): 437–76.

Tripsas, M., and Gavetti, G. (2000). 'Capabilities, Cognition, and Inertia: Evidence from Digital Imaging', *Strategic Management Journal*, 21(10–11): 1147–61.

Tushman, M. L., and Anderson, P. (1986). 'Technological Discontinuities and Organizational Environments', *Administrative Science Quarterly*, 31(3): 439.

Tushman, M. L., and O'Reilly III, C. A. (1996). 'Ambidextrous Organizations: Managing Evolutionary and Revolutionary Change', *California Management Review*, 38(4): 8–30.

Utterback, J. M. (1994). *Mastering the Dynamics of Innovation*. Boston: Harvard Business School Press.

von Hippel, E. (1988). *The Sources of Innovation*. New York: Oxford University Press.

von Hippel, E. (1994). ' "Sticky Information" and the Locus of Problem Solving: Implications for Innovation', *Management Science*, 40(4): 429–39.

von Hippel, E. (2005). *Democratizing Innovation*. Cambridge, MA: MIT Press.

Winter, S. G. (1987). 'Knowledge and Competence as Strategic Assets', in D. J. Teece (ed.), *The Competitive Challenge Strategies for Industrial Innovation and Renewal*. Cambridge, MA: Ballinger, 159–84.

Womack, J. P., Jones, D. T., and Roos, D. (1990). *The Machine that Changed the World*. New York: Maxwell Macmillan International.

PART II

THE SOURCES OF

INNOVATION

CHAPTER 3

MARKETING AND INNOVATION

JAIDEEP PRABHU

INTRODUCTION

MARKETING is a process by which companies create value for customers and build strong relationships to capture value from customers in return (Kotler et al., 2008). As such, marketing is both a business philosophy, that is, a way to succeed at business, as well as a business function, that is, a set of activities that (marketing) managers perform on a day-to-day basis.

Innovation, on the other hand, is the successful commercial exploitation of new ideas (Schumpeter, 1942; Rosenberg, 1982; von Hippel, 1988; Dodgson et al., 2008; Tellis et al., 2009). As such, innovation (and hence the management of innovation) involves identifying, developing, and exploiting new ideas to generate value.

Marketing and innovation are therefore closely intertwined. As Peter Drucker put it many years ago: 'Because the purpose of business is to create a customer, the business enterprise has two—and only two—basic functions: marketing and innovation...Marketing and innovation produce results; all the rest are costs', (see Drucker, 2003). Specifically, marketing attempts to create and keep customers. To do so, it must identify and satisfy customers' evolving needs. The main way to do this, in turn, is to develop new offerings for customers and find new ways to develop and deliver those offerings, namely to innovate. Firms that are good at innovation are likely to be good at marketing, and vice versa.

There are at least two ways in which marketing influences innovation in firms. First, marketing is a *location for* innovation within firms. Because marketing is a business function, the marketing department is itself a place where innovation occurs within the firm. Second, marketing acts as a *source of* innovation within firms. Because marketing offers the firm a philosophy of how to succeed (namely, meeting the evolving needs of customers), it informs innovation in other parts of the company as well.

Moreover, the process of developing new offerings and delivering them in new ways involves other functional areas within the firm such as R&D and operations. Again, the inter-functional and systemic nature of innovation implies a close relationship between marketing and other areas of the firms.

This chapter will explore both of these ways in which marketing influences innovation in firms. Regarding marketing as a location for innovation, the chapter will examine three major issues. First, it will examine how firms innovate in terms of *who* to market to, specifically, how they identify new customer segments, choose new segments to target, and identify new market spaces. Second, it will examine how firms innovate in terms of *what* to market to target consumers, that is, how they create new value propositions and innovate around product, price, promotion, and distribution. Third, it will examine *how* to market to target consumers, how to create new ways to deliver the value proposition by reducing fixed costs, achieving economies of scale, and exploiting experience (learning) curve effects.

Regarding marketing as a source of innovation, the chapter will examine the following issues: (a) how a firm's market orientation, namely its unrelenting focus on meeting customers' evolving needs better than its competitors, impacts how it innovates, and (b) how, through the inter-functional coordination needed to identify and develop new offering as well as new ways to develop and deliver them, marketing informs innovation in other parts of the firm such as R&D, HR, and finance.

Innovation in Marketing Itself

Successful innovation involves addressing three questions well: *who* to market to (target consumers), *what* to market (the value proposition), and *how* to market (how to deliver this value proposition). As it turns out, these are key marketing questions as well. Indeed, marketing managers within firms, whether they be product or brand managers, are typically involved with performing precisely these functions on a day-to-day basis. This section will examine in detail each of these three questions and how marketing acts as a source of innovation in each case. In doing so, the section addresses the profound links between marketing and what is increasingly referred to as business model innovation, namely the simultaneous adoption of new ways of offering a new value proposition with new ways of delivering it (see IBM Global CEO Study, 2006; Zott and Amit, 2008; Johnson et al., 2008; Gambardella and McGahan, 2010; Velu et al., 2010).

Innovation in *who* to Market to

Innovation is the commercialization of new ideas. It therefore presupposes the existence of a group of customers to commercialize new ideas for. The process of identifying such a group of customers typically falls to the marketing function within the firm and, by extension, to the marketing manager. Identifying such a group, in turn, involves processes such as segmenting markets, identifying consumer needs, choosing segments to target, and identifying new ways of creating/finding a market space. This subsection looks in detail at innovation in the context of each of these issues.

Segmenting Markets and Identifying Consumer Needs

Segmentation is the process of dividing the market into distinct groups that: (a) have common needs and (b) respond in a similar way to marketing actions (e.g. price, promotion, etc.). Firms segment markets by using a variety of data and approaches including demographics (age, gender, incomes, education, geography, ethnicity, etc.), psychographics (attitudes, lifestyles, and value), usage (heavy versus light users), and benefits sought (convenience, affordability, availability, etc.). A significant way in which marketing contributes to innovation is in how it goes about segmenting markets. Thus the move from a purely demographic approach to using psychographic and behavioural data is a major example of this.

A key aspect of segmentation is identifying the needs of consumers and tying this to demographic data. There are two broad ways in which marketers go about identifying consumer needs: (a) qualitatively, though ethnographic and projective approaches, and (b) quantitatively, through surveys and secondary, behavioural data. While qualitative approaches help answer 'why' questions and provide insights and tentative hypotheses, quantitative approaches help answer 'how much' questions and test hypotheses. Marketing can also act as a source of innovation by developing both quantitative and qualitative approaches to identifying consumer needs. For example, many approaches such as focus groups, interviews, and surveys rely on consumers being able to articulate their needs. However, in really new categories (e.g. mobile phones when they first became available) consumers lack experience and therefore are either unaware of their needs or are unable to articulate them. In such cases, a major innovation has been the use of ethnographic approaches which involve deep, longitudinal observation of consumers in their habitat to generate inferences about their behaviour, motivations, and need. For instance, Nokia uses large number of ethnographers it calls 'global nomads' who travel the world, live with prospective consumers, and identify their needs based on this experience. These insights then go into new phones that Nokia makes or into the re-design of existing phones.

Selecting Segments to Target

Once marketers have segmented markets, they need to decide on which segment or segments to target with potential market offerings. This process is called targeting and involves two related questions: (a) How many segments to target and (b) Which segments to target? Addressing these questions, in turn, requires the marketer to not only know a great deal about consumers in each segment (their purchasing power, their lifetime customer value, etc.) but also to be able to assess their firm's ability to attract and retain customers in each segment relative to competitors.

Finding and Creating New Market Spaces

A major way for firms to innovate is by finding new market spaces. These new spaces can either be in gaps within existing markets that have been hitherto undiscovered or in entirely new markets that have been hitherto ignored.

Finding a new space within existing markets. When markets grow and mature, the competitive space becomes increasingly crowded, even saturated. It is hard, in such a situation, for any players, new or old, to successfully differentiate themselves from others and provide a new value proposition to consumers within the rules of the existing game. In the language of strategy: one is now playing a red ocean. In order to move to a blue ocean of new possibilities and relatively little competition, firms need to innovate in how they think about the market: they need to create new market spaces. There are at least six possible ways in which they can do so (Chan and Mauborgne, 1999). First, they can identify a space between existing substitute industries. For example, Home Depot created the market space of DIY in the space between existing substitute industries of hardware stores and building contractors. Similarly, Southwest Airlines created a market space for low-cost airlines between the existing substitute industries of long-distance flying and driving/car rentals. Second, firms can create a market space between strategic groups within industries. For example, Ralph Lauren Polo created a market space between designer labels and high-volume classics. Third, firms can create a market space between chains of buyers. For example, Bloomberg created a new market space by focusing on the needs of traders and analysts within broker firms rather than the needs of the IT managers within these firms (as had been the case previously). Fourth, firms can create a market space between complementary products. For example, Borders innovated by creating a market space between the book and leisure industries. Fifth, firms can create a new market space by switching from an emotional appeal to a rational appeal, or vice versa. Thus, Starbucks innovated by making coffee a lifestyle choice rather than an off-the-shelf choice about price. And the Body Shop created a market space by making cosmetics about functionality rather than emotionality. Finally, firms can innovate by looking across time and creating a market space in advance of a new trend. Thus, Cisco Systems created a market space by innovating for the time when the Internet would be so pervasive that high speed data transfer would be a major market need.

Finding a new space within new markets. While the above typology focuses on finding a new space within an *existing* saturated market, firms can also find new market spaces in hitherto untapped markets. These can be either new geographical markets or segments of an existing market that have been ignored.

Perhaps the most significant such market is the Bottom of the Pyramid (BoP) market, or the so-called next 4 billion consumers who live on less than $3000 purchasing power parity and make up more than half the world's population (Hammond et al., 2007). These consumers have typically been ignored by for-profit firms, but this is changing as firms increasingly recognize that such consumers (a) are increasingly aspirational and empowered; (b) are very large in number; and (c) are a relative blue ocean with little competition targeted at their needs. Of course, significant innovation is needed in order to successfully reach and attract such consumers. And marketing has a substantial role to play in making this innovation happen. First, because the needs of these consumers are very different from relatively affluent consumers, and because little market research has traditionally been done on these segments, marketers have much to contribute by studying such consumers' needs. Second, once their needs have been understood, firms have to develop and deliver appropriate solutions for these needs. Products have to be simple and affordable, distribution has to be extensive and economical, and promotion has to be suited to the culture and background of these consumers. Again, marketing has an important role to play in ensuring this happens within firms targeting the BoP.

Innovation in *what* to Market to Target Consumers: the Value Proposition

At the core of marketing as a business function sits the marketing mix or the 4 Ps of product (or service), price, promotion, and place (or distribution). The 4 Ps together form the value proposition that firms offer to their customers. Thus, at the heart of marketing managers' role on a day-to-day basis is innovation around these tools to ensure a better fit between the firm's offerings and customers' preferences. This section examines innovation in each of the 4 Ps in turn.

Product/Service Innovation

A product is anything that can be offered to a market that might satisfy a want or need (Kotler et al., 2006). Thus, products can be tangible goods like mobile phones (e.g. the iPhone), MP3 players (e.g. the iPod), or tablet computers (e.g. the iPad). Or they can be intangible services like overnight delivery (e.g. UPS), financial services (e.g. Internet banking), or leisure spaces (e.g. Starbucks). Moreover, they can be business to business (e.g. enterprise software like SAP) or business to consumer (e.g. video games).

Innovation to products includes not only changes to their design and features (e.g. the user interface of the iPod) but also their packaging (e.g. milk and juice in tetrapack containers) and branding (extending a brand, e.g. when Bic, which used to make disposable ball point pens, also began offering disposable lighters under the same brand). Frequently the boundaries between tangible products and intangible services break down. And increasingly, innovation to tangible products involves adding intangible service elements such as convenience or after-sales maintenance (e.g. Amazon doesn't just sell books and e-readers, it also sells convenience and information), while innovation to intangible services involves adding tangible technology that underpins the delivery or use of the service (e.g. Apple doesn't just sell software, it also sells the products that run the software).

Price Innovation

In developing a value proposition suited to consumers, marketing managers not only innovate around the product features, design, packaging, or brand, they also frequently innovate around the price. Pricing innovation can take many forms including the pricing of individual products, groups of products (such as bundling of substitute or complementary products), or whole portfolios of products (such as with retailers).

Pricing innovations around *individual* products include: *reference* pricing (a strategy in which a product is sold at a price just below its main competing brand); *psychological* pricing (pricing designed to have a positive psychological impact, e.g. selling a product at £1.95 or £1.99, rather than £2.00); *dynamic* pricing (a flexible pricing mechanism which allows companies to adjust the prices of identical goods to correspond to a customer's willingness to pay, e.g. when airlines charge different rates depending on when customers make their booking); *pay what you want* pricing (where buyers pay any desired amount for a given commodity, sometimes including zero); and '*freemium*' pricing (where the basic product or service is free but a premium is charged for advanced features, functionality, or related products and services).

Pricing innovations around *grouped* products include: *bundling* (a strategy that involves offering several products for sale as one combined product with the combined price less than the sum of the individual products); *loss leader* or *leader* pricing (when a product is sold at a low price to stimulate other profitable sales, e.g. when firms sell razors cheap to make a profit on the blades); and *premium decoy* pricing (where the price of one product is set very high in order to boost the sales of a lower priced product).

Finally, pricing innovations around *portfolios* of products (as in retailing) include: *high–low* pricing (where the goods or services offered by the organization are regularly priced higher than competitors, but through promotions, advertisements, and or coupons, lower prices are offered on key items); and *everyday low pricing* (a pricing strategy where consumers can always expect a low price without the need to wait for sale price events or comparison shop, e.g. Walmart).

Promotion Innovation

Promotion is a complex and powerful tool at the disposal of marketing innovators. As such, promotion takes many different forms including advertising, sales promotions, public relations, contests, sponsorships, and so on. The recent rapid diffusion of technological innovations such as the Internet and mobile phones (along with changes in demographics and lifestyles) has led to a revolution in how firms do promotion.

The twentieth century was mostly characterized by the dominance of mass communication vehicles such as radio and television and the 60 second spot. The dominant model of promotion, therefore, was the 'interruption model', one in which the advertiser interrupted consumers as they were going about their lives to send them a company sponsored message.

In the twenty-first century this model has given way to a 'permission model' in which the advertiser asks consumers for permission to send them a company-sponsored mission through a medium of their choice and at a time of their preference. Indeed, given the democratization of social media, consumers increasingly actively seek out information about firms and products themselves or even create their own sources of information either individually (for instance through blogs) or collectively (by setting up brand communities on Facebook and the like). With this revolution and the shift of power from producers (and hence advertisers) to consumers, innovation in promotion could even be said to have shifted from firms to its consumers. Nevertheless, creative firms are innovating how they promote their products by using social media to monitor and engage consumer-led promotional initiatives. Firms increasingly maintain their own Facebook sites, systematically use Twitter to send messages and engage relevant stakeholders, and employ company-sponsored bloggers to write about the firm's products and activities. Even television advertising has undergone a change with the advent of devices, such as TiVo, which enable consumers to take a more proactive role in what ads they watch and when. Again, creative firms and advertisers have found ways to use the technology to their advantage, for instance by targeting their ads more accurately to only those consumers who have an active interest in their products and services. And again, the Internet is a medium which firms increasingly exploit to ensure their advertising is targeted at only high-value consumers with an active interest in what the firm has to offer.

Place or Distribution Innovation

Distribution involves the process of getting products and services from the firm to the consumer. Prior to the rise of the Internet, even digitizable products such as music, books, and news required some form of physical distribution chain. Hence the dominance in these industries of music retailers, bricks and mortar booksellers, and newsagents. In recent years, however, thanks to the Internet (and latterly mobile telephony and high-speed broadband) firms can go direct to consumers with music or news or books, leading to the rise of virtual booksellers such as Amazon. com, e-book readers such as the Kindle, and consumer electronic devices (and ecosystems) such as the iPod and iTunes. Even in physical distribution systems, mobile telephony, the Internet, and broadband have enabled process innovations in supply chain management and logistics.

Increasingly, exchange takes place not only between firms and consumers, but consumers themselves. Such P2P or person to person exchange may, however, be mediated by firms (such as in the case of eBay for goods and services or Zopa in the case of peer to peer lending). With the inevitable advent of mobile-based payments solutions (such as already exist in African and Asian countries), P2P innovations are likely to be a significant aspect of how marketers innovate the way products and services are distributed in the twenty-first century.

Innovation in *how* to Market to Target Consumers: Delivering the Value Proposition

Marketing managers not only innovate around who to market to or what to market, they also innovate around how to market to consumers, namely in how to deliver the value proposition in an efficient and effective way. Much of this sort of innovation is to do with how to reduce the cost of marketing operations. There are three broad ways in which firms can do this: they can innovate to (a) reduce fixed costs, (b) achieve economies of scale, or (c) reduce variable costs. This section discusses each of these broad strategies in turn.

Reducing Fixed Costs

There are three major ways in which firms can innovate to reduce the fixed costs associated with making, distributing, and promoting products and services: outsourcing (to reduce manufacturing costs), franchising (to reduce distribution costs), and the use of generics (to reduce advertising/promotion costs).

By outsourcing the manufacturing of elements of products or entire products, firms avoid incurring the major fixed costs of setting up and maintaining factories and a salaried workforce. This strategy is especially effective in commodity businesses where the market is more efficient than any individual firm can expect to be. But the strategy doesn't just help the firm reduce its costs and become efficient. It also frees up management time and helps the firm to focus on other marketing activities such as branding and differentiation. Thus, Nike does very little of its own manufacturing, preferring instead to focus on design and branding activities that enable it to remain globally effective against stiff competition in a fast moving sector. Similarly, the Indian telecoms service provider Bharti Airtel chose to outsource key functions like setting up and maintaining network equipment (to Ericsson) and customer billing (to IBM), thus reducing fixed costs as well as freeing up time to focus on rapidly reaching more consumers affordably.

Franchising, pioneered by the American sewing machine company Singer in the nineteenth century, has come to be a dominant means by which firms from various sectors, such as fast food and retail, achieve global distribution reach in a short period of time without incurring the huge fixed costs of setting up and maintaining a physical distribution system themselves. Besides the costs and time efficiency benefits, franchising also reduces risk for the franchising firm.

Finally, many firms avoid the fixed costs involved in advertising and promoting products by selling generic products that are unbranded. This strategy is particularly popular with retail chains which offer consumers who visit their stores the unbranded versions of products such as cereals, beer, milk, eggs, and bread at lower prices (and often equivalent quality) to branded products in these categories. This strategy is also seen in the pharmaceutical industry where generic firms rush to produce a non-branded version of a branded drug as soon as the drug goes off patent. Such generics have the advantage of not having to incur the huge fixed costs of R&D that go into drug discovery and testing.

Achieving Economies of Scale

A powerful way in which firms can reduce their fixed costs indirectly is by achieving economies of scale. For instance, by pursuing market share, firms can spread the fixed costs of manufacturing, distribution, and promotion out across a larger number of consumers thus reducing per unit costs overall. This strategy is particularly popular in sectors which are commodities and where the fixed costs are so high that a business is only viable if it is assured of a large part of the market. Industries such as telecommunications, electricity, postal services, airlines, and defence were once so capital intensive that they were designated as natural monopolies that were typically owned and maintained by the state. With technological progress and the reduction of minimum efficient scale, these sectors have been increasingly deregulated, but high fixed costs still mean that only a few firms, each with a large market share, tend to dominate. The strategy is also popular with large fast-moving consumer goods firms such as Unilever and P&G that pursue large market shares of commodity sectors such as soap and shampoo as a way of gaining economies of scale.

Reducing Variable Costs

As firms gain more experience with manufacturing products, their unit costs of manufacturing typically decline exponentially. Actively pursuing these 'experience curve' effects can be a powerful way in which firms can reduce their variable costs over time. In sectors where such costs can be particularly large, the promise of such learning effects could induce firms to attempt to enter early and thus erect barriers to entry for competitors, or to do R&D to identify new process technologies that might move the firm onto an entirely different learning curve with substantially lower costs from the start. The drawback of pursuing such experience curve effects is that it doesn't necessarily provide firms with a long-term competitive advantage as other firms can also learn over time.

HOW MARKETING INFORMS INNOVATION IN OTHER PARTS OF THE FIRM

Innovation involves at least three phases: the detection or identification of ideas, the development of these ideas into processes or products, and the commercialization of these processes or products (Chandy et al., 2006; Yadav et al., 2007). As discussed in the section above, the marketing function within firms has an important role to play in each of these phases. Marketing helps identify unmet or latest consumer needs and hence aids detection. It helps design and test prototypes to ensure they meet customer specifications, and hence aids development. And marketing helps with hastening the adoption and use of offerings in the marketplace, and hence aids commercialization. However, other functional areas within the firms, such as R&D, operations, and finance also have an important role to play in each of these phases. As such, marketing informs and works with other functional areas within the firm to make successful innovation happen.

This section elaborates on how marketing informs innovation activities in other parts of the firm. Specifically, it addresses the following issues. First, it looks at how marketing informs the strategic orientation of the firm and the push towards inter-functional coordination in the firm's innovation activities. A key focus here is on the notion of market orientation, namely the need for firms, in order to survive and prosper, to systematically collect, analyse, and respond to information on customers and competitors over time. Second, the section considers the role that top managers, including C-level marketing executives, play in driving innovation within the firm. Third, and related to the issue of leadership, is the role that the human resources department (in relation to marketing) plays in creating and fostering a culture of innovation with the firm. Fourth, the section addresses the relationship between marketing and R&D in the conversion of ideas into products and processes. Fifth, the role interplay between marketing and finance in the commercialization of innovations is examined.

Market Orientation and Inter-Functional Coordination

At least since Kohli and Jaworski (1990), a major stream of research has focused on how marketing informs the firm's approach towards its survival and growth (see also Narver and Slater, 1990; Jaworski and Kohli, 1993, 1996; Kohli et al., 1993). A major assertion and finding of this research is that firms that are more market oriented tend to be both more innovative (Athuene-Gima, 1995, 1996; Ottum and Moore, 1997; Han et al., 1998; Hurley and Hult, 1998; Narver et al., 2000; Frambach et al., 2003) and to have higher profitability in the long run. Market orientation in turn involves the 'organization-wide generation of market intelligence, dissemination of the intelligence across departments and organization-wide responsiveness to it' (Kohli and Jaworski, 1990).

By their very nature, therefore, market-oriented firms have proactive marketing departments that work closely with other functional areas within the firm such as R&D and operations, to not only identify new market opportunities but also devise new products and services in response to these opportunities which the firm as a whole then seeks to commercialize ahead of the competition. In market-oriented firms, therefore, the marketing function plays a central role driving and coordinating innovation in other parts of the firm.

CEOs, CMOs, and Innovation

Top managers have an important role to play in driving innovation within their firms. A survey by Boston Consulting Group (BCG) found that 45 per cent of managers believed that their firms' CEO was 'the biggest force driving innovation' in their company (Boston Consulting Group, 2006). And the business press is full of stories of the legendary exploits of CEOs like Steve Jobs (Apple), and Andy Grove and Gordon Moore (Intel) and their role in driving innovation in their firms.

Top managers play a crucial role in driving innovation in their firms in at least four ways (Tellis et al., 2009; Boyd et al., 2010). First, top managers help identify new market opportunities and direct the attention of others in the firm towards these opportunities. Second, top managers decide the level and type of innovation-related investments the firm makes. Third, top managers determine the firm's relationships with its main innovation-related stakeholders such as major customers, investors, alliance partners, and employees. Finally, top managers drive the attitudes and practices within the firm that determine its innovation culture.

Given the important role that top managers play in driving innovation in the firm, various approaches have been used to examine what types of managers are more likely to play such a role well. Accordingly, one approach examines top managers' education and experience as a means to predict and explain their focus on innovation. For instance, some research in this stream suggests that top managers with experience in 'output'-oriented functions, such as marketing, R&D, and sales, focus more on product innovation than top managers with experience in 'throughput'-oriented functions, such as accounting/finance, production, administration, and legal (Hambrick and Mason, 1984; Finkelstein and Hambrick, 1996). Still other research suggests that CEOs who focus on the future and on entities external to the firm are more innovative than others (Yadav et al., 2007). Taken together, a great deal of this research suggests that marketing (and related areas such as sales) drive innovation in the firm from the top through CEOs and other C-level executives with a marketing and sales background.

Marketing and Human Resources: Creating a Culture of Innovation

A key challenge for firms is not merely the creation of value, but its capture. Namely, even if firms manage to develop the next big thing, they are not always successful at being able to commercialize it successfully. Indeed, firms that have been particularly successful at innovation in the past are particularly susceptible to failing at commercializing new products in the future. This is because their commitment to existing markets and technologies makes it hard for them to focus on new markets and technologies (see Christensen, 1997; Chandy and Tellis, 1998). Thus, Kodak, which for over nearly a century created and dominated the photographic film industry, failed to repeat the same trick with digital photography, despite actually having been the first to invent the digital camera (see Munir, 2005). Similarly, Xerox, the pioneer in copy machines, invented in its Palo Alto Research Center (PARC) all the key elements of the paperless office of the future (including the desktop PC, mouse, email, Ethernet, printer, etc.) but failed to commercialize any of these inventions.

A major impediment to commercialization therefore is organizational (as opposed to technological). The solution to this problem is in turn one of culture: namely, the management of attitudes and practices within the firm. It is here that marketing, innovation, and human resources intersect. Marketing's forte is commercialization and markets; human resources' forte is the creation of culture; and innovation is the link between organization, technologies, and markets. Firms that possess key attitudes and practices deep within their organizational DNA are consistently more successful at innovation than those that do not. There are three key attitudes: a focus on future markets and technologies (not just current markets and technologies), a willingness to cannibalize current products and services in favour of new ones, and a willingness to take on the risks that doing so entails. Supporting such attitudes entails three key practices: (a) the use of product champions (i.e. employees at all levels of the firm empowered to identify and take forward new ideas); (b) the use of asymmetric incentives (rewarding success but tolerating failure up to a point); and (c) using internal markets and competition to break organizational monopolies and guard against complacence or inertia (see Tellis et al., 2009). Creating such a culture of innovation requires close cooperation between the firm's marketing and human resources departments and managers.

Marketing and R&D: Commercialization and Complementary Assets

In many firms, especially large firms in industries like automobiles, pharmaceuticals, consumer electronics, food, and fast-moving consumer goods, the division within the firm most central to innovation is R&D. Such firms have huge R&D departments with large budgets, massive numbers of science and technology employees, and deep patent pools. The R&D department in these firms develops new technologies, tests them, and—in many cases—identifies routes to market for them. Even in such R&D-driven firms, marketing has an important role to play. For instance, marketing, because of its close relationship with customers, can direct R&D towards new market opportunities which in turn determine the projects that R&D works on. However, marketing's key role in the innovation process is typically in the commercialization of new products and services. Even in R&D-intensive and technically driven industries like pharmaceuticals, the firm's marketing resources, such as the salesforce, as well as its marketing assets such as brands, can play a vital role in capturing the value that R&D creates. Specifically, complementary assets such as advertising and salesforce (namely product support) enhance the effectiveness of the firm's new products by convincing more people to adopt more of these products more quickly (see Sorescu et al., 2003).

Another role that marketing increasingly plays even in traditional R&D-driven firms is through its traditional links with other external stakeholders such as customers, suppliers, competitors, and industrial partners (Prabhu et al., 2005; Rao et al., 2008). Given that even large R&D-driven firms such as P&G are moving towards an open innovation model in which upwards of 50 per cent of the firm's new ideas may come from outside the firm, marketing has an increasingly important role to play in making this model successful (see Rigby and Zook, 2002; Chesbrough, 2004; Huston and Sakkab, 2006). Marketing's relationship with lead users can bring crucial feedback on new ideas back into the firm as well as take new ideas from within the firm out into the market for early and relatively risk-free feedback (von Hippel, 1986). More generally, marketing's relationship with consumers enables the firm to tap into the growing participation of customers in the innovation process (see Prahalad and Ramaswamy, 2000; O'Hern and Rindfleisch, 2009; Hoffman et al., 2010; Hoyer et al., 2010) Marketing's links with suppliers can also help identify solutions for new product challenges that the firm lacks or would find time consuming or expensive to develop.

Marketing and Finance

The role of the finance function within the firm is to maximize cash flow and shareholder value. This role suggests a crucial link between marketing, finance, and innovation. Given marketing's role in helping firms capture the value inherent in their new products, marketing can help firms better realize their obligations to shareholders (Chaney et al., 1991; Sood and Tellis, 2008). Marketing assets such as brands and marketing tools such as salesforce and advertising can enable the firm to convince more consumers to adopt its products and services sooner, thus increasing not only the cash flows that the firm accrues but also their net present value (Sorescu et al., 2003; Sorescu et al., 2007).

Firms increasingly manage relationships not only with their consumers but also with sources of ideas and knowledge such as universities. Managing these relationships often falls to marketing executives within the firm. Research shows that universities not only provide a vital source of new ideas and solutions for the problems the firm faces, they also provide legitimacy to firms that then increase their access to key resources including funding (see Zucker et al., 1998). For instance, new ventures in biotechnology attract more venture capital and gain more from the launch of their new products if they have star scientists on their board than firms that do not (see Rao et al., 2008). Marketing is therefore involved with more than simply managing relationships with consumers. By managing relationships with other key stakeholders such as scientists and investors, marketing can help firms achieve their innovation objectives through accessing ideas, gaining legitimacy, and attracting investment.

CONCLUSION

This chapter has explored two major ways in which marketing influences innovation in firms. First, it has examined how marketing is a *location for* innovation within firms. Second, it has examined how marketing acts as a *source of* innovation within firms.

Regarding marketing as a location for innovation, the chapter examined three major issues: (a) how firms innovate in *who* to market to; (b) how firms innovate in terms of *what* to market to target consumers; and (c) how firms innovate in *how* to market to target consumers.

Regarding marketing as a source of innovation, the chapter considered: (a) how a firm's market orientation impacts how it innovates; (b) the role of top managers, especially top marketing managers, in driving innovation in the firm; and (c) how, through the inter-functional coordination needed to identify and develop new offerings as well as new ways to develop and deliver them, marketing informs innovation in other parts of the firm such as R&D, HR, and finance.

By highlighting these issues, the chapter has attempted to show the profound and various ways in which marketing contributes to innovation in firms. This influence, while it has been gathering momentum during the last few decades, is only likely to grow in the years to come. As consumer power grows and as technologies break down the barriers between firms and consumers, the role of marketing in innovation is likely to become more important than it has been before. As such, academics and managers involved in studying and developing these links are likely to find themselves in much demand in a brave new world in which marketing and innovation, as Peter Drucker (1974: 54) once put it, form the 'two...basic functions or the firm...[that] produce results...[and] all the rest are costs.'

References

Athuene-Gima, K. (1995). 'An Exploratory Analysis of the Impact of Market Orientation on New Product Performance', *Journal of Product Innovation Management*, 12: 275–93.

Athuene-Gima, K. (1996). 'Market Orientation and Innovation', *Journal of Business Research*, 25(2): 93–103.

Boston Consulting Group (2006). *Innovation 2006*. Boston: Boston Consulting Group.

Boyd, J. E., Chandy, R. K., and Cunha, M. Jr. (2010). 'When Do Chief Marketing Officers Affect Firm Value? A Customer Power Explanation', *Journal of Marketing Research*, 47: 1162–76.

Chan, W., and Mauborgne, R. (1999). 'Creating New Market Space', *Harvard Business Review*, January.

Chandy, R. K., and Tellis, G. J. (1998). 'Organizing for Radical Product Innovation: The Overlooked Role of Willingness to Cannibalize', *Journal of Marketing Research*, 35(4): 474–487.

Chandy, R., Hopstaken, B., Narasimhan, O., and Prabhu, J. (2006). 'From Invention to Innovation: Conversion Ability in Product Development', *Journal of Marketing Research*, 43(3): 494–508.

Chaney, P. K., Devinney, T. M., and Winer, R. S. (1991). 'The Impact of New Product Introductions on the Market Value of Firms', *Journal of Business*, 64(4): 573–610.

Chesbrough, H. (2004). *Open Innovation*. Cambridge, MA: Harvard Business School Press.

Christensen, C. M. (1997). *The Innovator's Dilemma: When New Technologies Cause Great Firms to Fail*. Boston: Harvard Business School Press.

Dodgson, M., Gann, D., and Salter, A. (2008). *The Management of Technological Innovation: Strategy and Practice*. Oxford: Oxford University Press.

Drucker, P. (1974). *Management: Tasks, Responsibilities, Practices*. Oxford: Butterworth Heinemann.

Drucker, P. (2003). *The Essential Drucker: The Best of Sixty Years of Peter Drucker's Essential Writings on Management*. London: Harper Collins.

Finkelstein, S., and Hambrick, D. (1996). *Strategic Leadership: Top Executives and Their Effects on Organizations*. St. Paul, MN: West Publishing Co.

Frambach, R. T., Prabhu, J., and Verhallen, T. (2003). 'The Influence of Business Strategy on New Product Activity: The Role of Market Orientation', *International Journal of Research in Marketing*, 20(4): 377–97.

Gambardella, A., and McGahan, A. (2010). 'Business-Model Innovation: General Purpose Technologies and their Implications for Industry Structure', *Long Range Planning*, 43(2–3): 262–71.

Hambrick, D., and Mason, P. A. (1984). 'Upper Echelons: The Organization as a Reflection of its Top Managers', *Academy of Management Review*, 9(2): 193–206.

Hammond, A., Kramer, W., Katz, R., Tran, J., and Walker, C. (2007). 'The Next Four Billion: Market Size and Business Strategy at the Base of the Pyramid', *World Resources Institute*.

Han, J. K., Kim, N., and Srivastava, R. K. (1998). Market Orientation and Organizational Performance: Is Innovation a Missing Link? *Journal of Marketing*, 62: 30–45.

Hoffman, D. L., Kopalle, P. K., and Novak, Th. P. (2010). 'The "Right" Consumers for Better Concepts: Identifying Consumers High in Emergent Nature to Develop New Product Concepts', *Journal of Marketing Research*, 47: 854–65.

Hoyer, W., Chandy, R., Dorotic, M, Krafft, M, and Singh, S. (2010). 'Customer Participation in Value Creation', *Journal of Service Research*, 13(3): 283–96.

Hurley, R. F., and Hult, G. T. M. (1998). 'Innovation, Market Orientation, and Organizational Learning: An Integration and Empirical Examination', *Journal of Marketing*, 62: 42–54.

Huston, L., and Sakkab, N. (2006). 'Connect and Develop: Inside Procter & Gamble's New Model for Innovation', *Harvard Business Review*, 84(3): 58–66.

IBM Global CEO Study (2006), *Expanding the Innovation Horizon*. IBM Global Business Services. Available at <http://www-07.ibm.com/sg/pdf/global_ceo_study.pdf> (accessed 7 February 2011).

Jaworski, B. J., and Kohli, A. K. (1993). 'Market Orientation: Antecedents and Consequences', *Journal of Marketing*, 57: 53–70.

Jaworski, B. J., and Kohli, A. K. (1996). 'Market Orientation: Review, Refinement, and Roadmap', *Journal of Market Focused Management*, 1(2): 119–35.

Johnson, W., Christensen, C., and Kagermann, H. (2008). 'Reinventing Your Business Model', *Harvard Business Review*, 86(12): 50–9.

Kohli, A. K., and Jaworski, B. J. (1990). 'Market Orientation: The Construct, Research Propositions, and Managerial Implications', *Journal of Marketing*, 54: 1–18.

Kohli, A. K., Jaworski, B. J., and Kumar, A. (1993). 'MARKOR: A Measure of Market Orientation', *Journal of Marketing Research*, 30: 467–77.

Kotler, P., Armstrong, G., Brown, L., and Adam, S. (2006). *Marketing*, 7th edn. Harlow: Pearson Education Australia/Prentice Hall.

Kotler, P., Wong, V., Saunders, J., Armstrong, G., and Wood, M. B. (2008). *Principles of Marketing: Enhanced Media European Edition*. London: Prentice Hall.

Munir, K. A. (2005). 'The Social Construction of Events: A Study of Institutional Change in the Photographic Field', *Organization Studies*, 26(1): 93–112.

Narver, J. C., and Slater, S. F. (1990). 'The Effect of a Market Orientation on Business Profitability', *Journal of Marketing*, 54(4): 20–35.

Narver, J.C., Slater, S. F., and MacLachlan, D. (2000). 'Total Market Orientation, Business Performance and Innovation', Marketing Science Institute, Working Paper Series, Report No. 00-116.

O'Hern, M. S., and Rindfleisch, A. (2009). 'Customer Co-Creation: A Typology and Research Agenda', in N. K. Malholtra (ed.), *Review of Marketing Research*, Vol. 6. Armonk, NY: M.E. Sharpe, 84–106.

Ottum, B. D., and Moore, W. L. (1997). 'The Role of Market Information in New Product Success/Failure', *Journal of Product Innovation Management*, 14: 258–73.

Prabhu, J. C., Chandy, R. K., and Ellis, M. E. (2005). 'Acquisition and Innovation in High-tech Firms: Poison Pill, Placebo, or Tonic?', *Journal of Marketing*, 69(1): 114–30.

Prahalad, C. K., and Ramaswamy, V. (2000). 'Co-Opting Customer Competence', *Harvard Business Review*, 78: 79–87.

Rao, R. S., Chandy, R. K., and Prabhu, J. C. (2008). 'The Fruits of Legitimacy: Why Some New Ventures Gain More from Innovation than Others', *Journal of Marketing*, 72(4): 58–75.

Rigby, D., and Zook, C. (2002). 'Open-Market Innovation', *Harvard Business Review*, 80(10): 5–12.

Rosenberg, N. (1982). *Inside the Black Box: Technology and Economics*. Cambridge: Cambridge University Press.

Schumpeter, J. (1942). *Capitalism, Socialism, and Democracy*. New York: Harper.

Sood, A. and Tellis, G. J. (2008). 'Do Innovations Pay Off? Total Stock Market Returns to Innovation', *Marketing Science*, 28(3): 442–56.

Sorescu, A. B., Chandy, R. K., and Prabhu, J. C. (2003). 'Sources and Financial Consequences of Radical Innovation: Insights from Pharmaceuticals', *Journal of Marketing*, 67: 82–102.

Sorescu, A. B., Chandy, R. K., and Prabhu, J. C. (2007). 'Why Some Acquisitions do Better than Others: Product Capital as a Driver of Long-term Stock Returns', *Journal of Marketing Research*, 44(1): 57–72.

Tellis, G. J., Prabhu, J. C., and Chandy, R. K. (2009). 'Radical Innovation across Nations: The Preeminence of Corporate Culture', *Journal of Marketing*, 73(1): 3–23.

Velu, C., Prabhu, J. C., and Chandy, R. K. (2010). 'Evolution or Revolution: Business Model Innovation in Network Markets', Working Paper, Judge Business School, University of Cambridge.

von Hippel, E. (1986). 'Lead Users: A Source of Novel Product Concepts', *Management Science*, 32(7): 791–805.

von Hippel, E. (1988). *The Sources of Innovation*. New York: Oxford University Press.

Yadav, M. S., Prabhu, J. C., and Chandy, R. K. (2007). 'Managing the Future: CEO Attention and Innovation Outcomes', *Journal of Marketing*, 71: 84–101.

Zott, C., and Amit, R. (2008). 'Exploring the Fit between Business Strategy and Business Model: Implications for Firm Performance', *Strategic Management Journal*, 29(1): 1–26.

Zucker, L. G., Darby, M. R., and Brewer, M. B. (1998). 'Intellectual Human Capital and the Birth of US Biotechnology Enterprises', *American Economic Review*, 88(1):290–306.

CHAPTER 4

....

SCIENCE, TECHNOLOGY, AND BUSINESS INNOVATION

....

MAUREEN MCKELVEY

INTRODUCTION

....

THIS chapter discusses the interactions between science, technology, and business innovation. Science and technology are broad concepts used to describe new types of knowledge and artefacts that are invented, developed, and used. The focus is particularly upon science and its relationship to innovation management.

This chapter will review the types of knowledge developed, and the different organizations involved, as well as the key issues of managing business relationships with science in different industries. Managers have to consider how to cope with uncertainty about the directions in which science and technology will be developed, as well as with which types of advances may be useful to the future of their company. They must also consider the diverse types of organizations involved, and especially how to manage the interaction between universities and industry. This chapter will therefore explore the ways in which science and technology are 'manageable', and in which industries this way of thinking must especially be taken into consideration.

One peculiarity of the interactions between science, technology, and business innovation is the important role of governments and public financing of research and education. Much investment occurs through public, rather than private, financing, although both sources are important for stimulating later business innovations as products, processes, and services. Science, as well as business research and development activities, helps renew the pool of technological opportunities available in society (Sherer, 1965). Science and technology are key sources of knowledge that generate business opportunities, for many different types of firms. Their broader impacts on society more generally provide governments with a rationale to invest in the science and education system.

Thus, discussing the relationship between science and innovation requires focus, not only upon firms, but also on the public-sector aspects of this type of knowledge. Some organizations in society, such as universities, and related bodies such as public research institutes and organizations, specialize in developing science. Public financing—or a mix between public and private financing—of the science and education system usually supports them. Managing the relationships with these types of organizations is thus important for certain types of firms, with variation across different types of industrial and service sectors.

So as to incorporate scientific and technological knowledge into business innovation, firms need to manage their networks with universities. Science may also help stimulate new organizational forms such as start-up companies and academic entrepreneurship.

Section 2 of this chapter discusses the nature of science and technology and Section 3 addresses the particular role of public policy. Section 4 discusses science-based industries and Section 5 considers the interactions between universities and industries, and the different motivations and channels for collaboration. Section 6 addresses science in entrepreneurial firms and academic spin-offs. The chapter concludes by considering future challenges.

Scientific Research and Technology

There are many ways to define and distinguish between science and technology, and between different types of research. Defining the parameters of science and technology helps establish principles for understanding how businesses manage innovation, and the role of the public sector.

Science refers to accumulated and systematic bodies of knowledge as well as research results emanating from research organizations. Science is often contrasted with technology. They are considered different types of knowledge, and have different types of impacts on new business opportunities.

The Organization for Economic Co-operation and Development (OECD) has long been a champion of defining, measuring and gathering data about science, technology, and innovation (<www.oecd.org>). Its definitions are relevant for understanding scientific debates and the development of cross-country data. In 1963, the OECD's so-called Frascati manual defined the following three types of research:

- *Basic research* is experimental or theoretical work undertaken primarily to acquire new knowledge of the underlying foundation of phenomena and observable facts, without any particular application or use in view.

- *Applied research* is also original investigation undertaken in order to acquire new knowledge. It is, however, directed primarily towards a specific practical aim or objective.
- *Experimental development* is systematic work, drawing on existing knowledge gained from research and/or practical experience, which is directed to producing new materials, products, or devices, to installing new processes, systems, and services, or to improving substantially those already produced or installed.

These three types of research later became standard definitions, which have been further expanded and operationalized into national statistics across the globe. They are useful in differentiating more basic science from technical development work.

Taxonomies related to the degree of novelty are also relevant to understanding science and its impact on business innovation. Radical innovations are generally based upon large advances of knowledge, which sometimes lead to new disciplines and 'paradigms', and these radical changes usually require new knowledge in several fields of science and technology before the ideas can be implemented in practice in a company. Incremental innovations, such as improvements to a wind turbine, are generally based upon many smaller and cumulative improvements within a dominant design or technological paradigm (Kuhn, 1974; Utterback and Abernathy, 1978; Dosi, 1982).

In the economics of science tradition, key starting points distinguishing science and technology are the incentives and institutional structures that stimulate individuals to either focus upon scientific or technological and practical results (Dasgupta and David, 1994). In contrast, Nightingale (1998) examines the cognitive foundation of the different types of problem solving. He argues that there are key differences, in that science starts with known conditions and tries to discover unknown results, whereas technology and innovation are driven by an idea of the final result and the researcher then tries to solve all the necessary conditions to produce the desired results.

Several streams of literature—such as economic history, history of science and technology, economics of science, and innovation studies—have developed more nuanced definitions of science and technology and understanding of the relationships between them. Contributions by researchers such as Rosenberg and Birdzell (1986), Price (1965), Mokyr (2002), McCloskey (1994), and Von Tunzelmann (1995) have produced deep and detailed studies of the dynamics of developing science and technology, and of their subsequent impacts on economic growth.

These conceptual and empirical studies provide ideas for how to differentiate science and technology in relation to business innovation and economic growth. Yet, the historical case studies also demonstrate the difficulties of defining and establishing clear distinctions between science and technology based upon the purpose and intent of research. In reality individuals and organizations undertaking scientific and technical research often have multiple objectives and draw upon different types of knowledge.

In some cases, solving specific technical issues in a company may require basic research (Rosenberg, 1994). In the early use of biotechnology in pharmaceuticals, for example, biotech start-up firms needed to be engaged in basic science to demonstrate the feasibility of the idea and signal the quality of their scientific results, but they also had to develop many complementary assets and develop network relationships with established pharmaceutical firms (McKelvey, 1996).

In other cases, having very detailed and systemic practical knowledge about the application area around a technology leads to the more basic research being carried out in the engineering fields, along a design trajectory (Vincenti, 1990). Stokes (1997) introduces the idea of 'use-inspired basic research' as a key element of science to solve particular problems (or use-inspired problems), calling this 'Pasteur's Quadrant'. Thus, the actual details of how knowledge develops in society usually involve the blurring of the lines of these types of research.

SCIENCE AND PUBLIC POLICY

This section discusses the organizations carrying out science, theoretical justifications for government funding, and the trade-offs between private and public financing and control of research. Governments and public investment play a central role in financing the science and education system and thereby in stimulating new scientific outcomes and producing educated individuals. The innovation management issues related to science can thus be framed by relating public policy and public goods arguments to business strategy.

The science and university system expanded significantly after the Second World War. Science is largely—but not exclusively—carried out in universities, colleges, and public research organizations, based upon public monies. Science is also carried out in some companies, industrial institutes, and non-governmental organizations (NGOs), and its funding may also derive from private sources. The balance between public and private financing of research tends to change over time. In recent years, there has been a clear shift in the financing of universities and science from the public purse to private sources and competitive research grants, particularly in Europe (Geuna, 1999) and the US (Mowery and Sampat, 2005). At the same time, companies tend to reduce their research and development (R&D) spending during downturns and crises, and to rely more upon public financing.

The particular characteristics of research and information as public goods provide one reason why governments around the world invest so significantly in research and education in universities and public research organizations. The OECD countries spent approximately $968 billion US on R&D and had 4.2 million researchers (full-time equivalents) in 2010 (OECD 2012). The rationale for investing public monies in research, and especially in universities, colleges, and public research organizations, is related to the particular qualities of knowledge. Scientific knowledge is a public good in the sense that it can be used by many users at the same time, and it is difficult to stop others from using information once it is discovered.

Theoretical support for funding basic science initially relied upon a type of market failure argument, related to the particular characteristics of information and knowledge. Scholarship by Arrow (1962) and Nelson (1959) established the economic rationale for a public investment in science, set in relation to the incentives for private actors to invest in R&D.

Information and knowledge are types of public goods, or public-private goods, with particular characteristics. One characteristic is 'non-rivalry', implying that one person's use does not prevent (or diminish) another's use of the same knowledge. This enables spillovers from one party to another and increasing scales of production and benefits to all parties, in the long run. Another characteristic is non-excludability, which means that once the information is created, others can gain access to it, and knowledge becomes diffused.

The characteristics of non-rivalry and non-excludability help explain elements of firm strategies. Firms may want control of their research results in the short run, in order to use them to develop their specific business innovations, using secrecy and first-mover advantage as well as patenting strategies (Cohen et al., 2004; see Chapter 28 by Leiponen and Chapter 12 by Gambardella et al.). Yet, firms also benefit from broad access to scientific information and knowledge. In the long run, there are positive effects on the economy as a whole due to these types of diffusion effects, or what are known as spillovers.

The innovation manager's need to direct the firm's search for innovation and technology raises the question about how much firms should invest in R&D, and how they can benefit from their investment into R&D and skilled personnel. The 'appropriability problem', reflects the degree to which the inventor realizes value, or obtains returns, from an investment in R&D. The modern understanding of appropriability relates value to many characteristics of the firm and industry (Winter, 2006). In his analysis of the fundamental issues of appropriability problems, Winter argues that a key element of the classic contribution by Teece (1986) is the emphasis upon the innovator's access to complementary assets as part of the appropriability problem, as well as the contracting issues beyond the inventor per se.

An implication of this view is that scientific and technical knowledge should simultaneously be seen as having characteristics of a private good, in that certain types of knowledge are difficult to transfer without deep know-how, and can remain secret or protected through complementary assets. Patents, for example, are useful tools to protect products in the pharmaceutical industry, but secrecy is more important for process innovations in this industry. Surveys of firm managers provide empirical results about the very different ways of protecting—or appropriating—the returns to the investment in knowledge creation and diffusion (Levin et al., 1987).

The nature of knowledge able to be developed and applied in companies will affect their decision about whether to invest in R&D and skilled employees internally, or whether to use networks to obtain information and knowledge externally. From the firm's perspective, the types of appropriability regime, as well as the key role of complementary assets, are crucial in realizing value from an idea (Teece, 1986).

SCIENCE-BASED INDUSTRIES

Some industrial sectors and firms are more directly dependent than others upon science for innovation and growth. A classic sectoral taxonomy proposed by Keith Pavitt (Pavitt, 1984) considers the special characteristics of science-based industries.

Pavitt contrasts the science-based industries with those that can be predominately classified as supplier-dominated, scale- and production-intensive, and specialized suppliers. These categories of industries are differentiated based upon their source of technology, requirements of users, appropriability regime, and the typical size of the firm. These variables are used to differentiate the key characteristics of how business innovation occurs within these sectors.

For the science-based industries, public science and R&D play key roles as sources of technology, and these firms actively use external relations with universities and scientists in the development of innovations. Moreover, firms in these industries tend to work on both process and product innovations and are price-sensitive, but they also target high performance. Pavitt particularly focuses upon large firms, such as chemical, pharmaceutical, and electronics as examples of science-based firms.

Through the years, the Pavitt taxonomy has been revised and extended by many authors, and one discussion has been how to understand services, particularly business services dependent upon knowledge (Archibugi, 2001). Services firms are difficult to classify according to the OECD definition of different types of research, but they clearly innovate through other mechanisms (see Chapter 30 by Tether). A report by the Royal Society shows how dependent service companies are upon scientific knowledge and their employment of science and engineering graduates (Royal Society, 2009).

In more recent years, the role of skilled employees has become a key indicator of science intensity, as a complement to the traditional one of R&D, or R&D intensity relative to sales. The average R&D intensity of industrial sectors has been used for many years, leading to rather imprecise classifications of high-tech, medium-tech and low-tech industries. One reason for the post-Second World War focus upon the management of R&D in large firms lay with the development of dedicated departments in those firms, leading to technical advances (Dodgson et al., 2008). In more recent years, the understanding of business innovation being broader than R&D has thus been mirrored by new types of indicators to capture what 'science' may be within the firm or industry, and also includes measures such as the number or percentage of employees with advanced degrees.

Having an understanding of science is important within many different kinds of firm. Firms in varied industrial and service sectors need a basic understanding of technology and science, obtained through hiring skilled graduates and conducting their own R&D. One reason has to do with the economic value of knowledge in stimulating business innovations in many different types of firms and sectors (Dodgson et al., 2008). Another reason is that the firm may need internal competencies, or absorptive capacities (Cohen and Levinthal, 1990), in order to benefit from the spillovers discussed above, including being able to understand the results of research carried out in a university. Without these absorptive capacities, the firm that wants to innovate will have difficulties in assessing external sources of research results, and in particular, they will have difficulties in appreciating and applying advances in knowledge made elsewhere to their products, processes, and services. To use the economists' terminology, the individual firm will have difficulties benefitting fully from knowledge spillovers if it does not have appropriate absorptive capacities and complementary assets.

Finally, an understanding of how to manage science and technology is closely linked to the methods and conceptualizations of the whole range of innovation management issues discussed in this Handbook, related to R&D, new product development, collaboration and networks, intellectual property, strategy, and organization.

University–Industry Interactions

One particular aspect of managing science and technology for innovation is firms' direct contacts with universities. University–industry interactions, and especially their impact upon innovation, matters at the societal level by providing and extending potential network partners for industry development, and it matters for firms in how they choose to interact with specific universities, and how they gain value from those networks.

At the societal level, many simplistic prescriptions, in theoretical and empirical terms, have emerged about how to tackle the complexity of university–industry interactions. These simplifications have been necessary first steps, in order to try to 'measure' the efficiency and efficacy of public policies to support the science and education system. The role of the university in stimulating regional and economic growth may be demonstrated through, for example, measureable outcomes such as academic patents and start-up companies at a specific university. Another way of framing the consideration of the role of universities is to argue that society should have an overarching goal of stimulating an 'entrepreneurial university' (Etzkowitz, 1998).

These prescriptions and measureable outcomes are based upon the expectations that public investment in science will lead directly to economic benefits for society. This so-called 'linear model' suggests that investments in basic science lead to applied research and technical development, which in turn leads to economic growth in the form of commercialization of product innovations and new companies.

Despite rich debates on alternative models of innovation policy (Mytelka and Smith, 2001; Borrás, 2003), the linear model implicitly guides much public policy. Perhaps the strength of the linear model lies in the fact that the model is logically coherent and appears to offer easy answers, and there are considerable policy difficulties in adopting alternative approaches (Dodgson et al., 2012). Even though alternative models of feedback loops have been suggested, and empirical studies show that the impact of science is usually not as straightforward as this linear model suggests (Kline and Rosenberg, 1986), no unified and clear alternative model nor quantitative indicators currently exist in a convincing manner to measure university–industry interactions.

The problem lies in the way the justifiable need to prove the efficient and transparent use of public policy investments can lead to an emphasis of 'measureable outcomes' rather than the real drivers of innovation. Investing in basic science and universities may well stimulate economic growth and job creation in the long run, not least because knowledge and education are quasi-public goods that help drive economic growth (Fagerberg and Verspagen, 1996). In the short run, however, these effects from public investment can be difficult to measure, and the existing proxies used can be misleading on the real impacts on business innovation.

From a societal perspective, universities have much more complex roles in the knowledge economy than the linear model assumes (Deiaco, Hughes, and McKelvey, 2012; Hughes and Kitson, 2012). The problem with simple indicators such as academic patents and start-up companies is that they rarely capture actual impacts and contributions to economic growth, or the role of knowledge in the economy or importance to a specific firm. Numbers of patents and start-up companies provide very limited knowledge about the real interactions between university and industry. Indeed, Salter and Martin (2001) reviewed the literature and found that the main impacts of science on society are broad impacts, such as knowledge flows and the education of students.

At the firm level, empirical and theoretical studies demonstrate that the interactions between science and innovating practices in companies are complex, with different motivations, channels for communication, and expected results (Bercovitz and Feldman, 2007; Perkmann and Walsh, 2008).

From the firm's perspective, the questions are what exactly is expected from interactions with universities, and whether it wants directly useful results or whether it is looking for more long-term ideas. Managers should also consider the incentives and rationale for the broader-based and more strategic concerns of continuing search for improvements and new knowledge, techniques, fields of application, instruments, and research results. Surveys of firm managers about the sources of new ideas often rank universities much lower than suppliers and customers, who they use for more directly commercializable results.

Perkmann et al. (2013) propose that 'academic engagement with industry' differs from the direct commercialization of university–industry interactions through patents and start-up companies. Academic engagement with industry is a broader concept about the knowledge base of interaction, including multiple channels and rationales for creating and diffusing knowledge. A broad range of indicators can therefore be developed, including various advances in science, but also recruitment of students, interactions in long-term projects, developing competencies in firms, and consultancy work.

A key challenge for managing business innovation in this context relates to the fact that interacting with universities in order to develop science and technology will, by definition, provide advances to knowledge in different domains and disciplines. Moreover, there are many dimensions to these scientific and technical advances, such as new ways of thinking, new instruments and measurement techniques, and new research results. These advances, per definition, are not known before the research, development, or creative process occurs. This creates a particular challenge for innovation management, given that the firm and manager want to invest in the creation and use of new knowledge, but are at the same time uncertain about the rate of returns or even the potential 'usefulness' of that knowledge. This has to do with intrinsic properties of the advancement of knowledge (Campbell, 1987).

SCIENCE IN ENTREPRENEURIAL FIRMS AND ACADEMIC SPIN-OFFS

Although the earlier work on science-based industries stressed the management of R&D-intensive large firms, research in more recent decades has focused upon innovation management in entrepreneurial firms, and especially academic spin-offs.

Science in entrepreneurial firms has primarily been studied in high tech industries, especially in emerging sectors such as biotechnology, IT, material sciences, and more recently, nanotechnology (Faulkner et al., 1995; McKelvey, 1996; Powell et al., 2005; Robinson et al., 2007). Biotechnology was of particular interest because firms in pharmaceuticals—a science-based industry according to the Pavitt taxonomy—were suddenly challenged by small science-based firms. These academic spin-offs were usually started by PhD scientists. Moreover, this is an industry where knowledge spillovers and networks have transformed the relationships needed for innovation, by redefining what was 'basic science' and what was 'commercialization' (see McKelvey and Orsenigo, 2006). Emerging sectors (or technologies) pose the question of whether large firms or small firms will have a long-term advantage in innovating. This was a classical issue raised by Schumpeter as to whether industries dominated by large firms or small firms would be more innovative (Malerba and Orsenigo, 1996; Nelson and Winter, 1982).

These science-based entrepreneurial firms are commonly studied through the channel of academic spin-offs. Shane (2004), for example, studies the interactions between technologies, individuals, industrial sectors, and the university environment in promoting academic entrepreneurship. Academic spin-offs are ventures that are started by employees of universities, public research institutes, or by students. These spin-offs may retain formal agreements with the university—through licensing, employment, contact with technology transfer offices, IPR, contracts, and so forth, or they may be based upon informal agreements and the casual movement of resources and people between the organizations.

This type of science-based entrepreneurship may also be seen as the result of long-term interactions between the founder and the external environment—hence helping to confirm the earlier point of why public funding of science may stimulate broader economic growth. A key issue has been how resources are moved, and re-configured, around a university or region between many firms that are academic spin-offs.

A long line of scholarship has examined the high tech start-ups and clustering of firms around Cambridge University, in what has been called the 'Cambridge Phenomenon' (Druilhe and Garnsey, 2000; Garnsey, Lorenzoni, and Ferriani, 2008). The university is clearly one driver of this science and technology cluster, but there are also the relationships between the firms themselves—or between founders, some of who create serial start-ups—and these forces also stimulate the regional clustering in particular industries. Theoretically, Garnsey argues that the dynamics of the entrepreneurial process are a series of relationships and decisions which affect the configuration of resources and the exploitation of opportunities at the regional level (Garnsey and Heffernan, 2005; Garnsey, Stam, and Heffernan, 2006). In economic geography, important issues about the relationships between universities and regional clusters include the study of academic spin-offs. These approaches explain why science-based firms help stimulate the creation of clusters as well as why they disappear or fail to take off as agglomerations of activities and firms (Feldman, 2007).

These types of firm are generally small, and work within a resource-constrained environment. For that reason, the networks and social capital of the founders has been found to strongly influence the later performance of the firm (Jones et al., 1997). Moreover, the types of networks that matter to the firm may shift over time and are dependent upon the types of resources acquired. For biotechnology, Bagchi-Sen (2007) demonstrates that these firms need networks with university scientists in order to develop their reputation, but they also need networks with other firms for product development.

A recent conceptualization of knowledge intensive entrepreneurship in a European project—AEGIS—has demonstrated that these entrepreneurial firms based upon science may emerge from low-tech and services industries, as well as from the high-tech sectors (Malerba et al., 2013; McKelvey and Lassen, 2013a; McKelvey and Lassen, 2013b). This work suggests that small, new firms may derive innovative opportunities through science, even in sectors such as machine tools or digital advertising that do not immediately fit into the classification of science-based industries. They often require direct contacts with universities and highly skilled employees, despite their small size. These types of firms are dependent upon scientific, technological, or creative knowledge, and their main challenge is to use that knowledge to develop innovations. Hence, understanding how science affects entrepreneurial firms and academic spin-offs needs also to be studied in 'low-tech' industries and creative industries.

FUTURE ISSUES

Science and technology open up new opportunities for firms as well as for society and economies through the development of new knowledge, which is why they are important for innovation management.

One of the most basic challenges for managers is how much science to invest in and perform within the firm, and how much to try to gain through external linkages by means of spillovers, networks, and partnerships. The answer to these questions depends partly upon the firm's access to complementary assets and the appropriability regime. It also depends upon the extent to which research results and business innovations are within the radical or incremental continuum, and whether related changes are needed in technologies and markets.

Another challenge of continuing interest is the extent of science's long-run positive or negative impacts upon society, and what this means for innovation managers. The positive impact of science upon business innovation and society is widely recognized. It is clear that with a long-term perspective, advances in scientific and technological knowledge—and applications through innovations—matter profoundly for society. One impact is the development of products and services that were unthinkable in earlier generations. More importantly, the resulting innovations help solve societal problems, such as providing treatment for diseases and for medical conditions such as diabetes. Science, technology, and business innovation will help solve societal challenges such as ageing, global warming, and public health where both public and private actors will be involved in providing solutions.

At the same time, scientific results and business innovations may also have mixed, or negative, consequences in some dimensions. When science-based innovations are widely diffused they can shift society into new ways of living and consuming. Recent debates about low carbon emissions and an environmentally sustainable society have focused upon the negative long-term impacts of contemporary ways of living and working, and upon the need for alternatives. Emerging negative consequences thus create new problems, which generally lead to search for innovative solutions, such as public transportation instead of the car, or the need to reduce projected future energy consumption and increase alternative renewable sources around the world.

The negative effects that society must find a way to address may lead to additional public and private investments in science, technology, and innovation, opening up new business opportunities, such as the race for green technologies.

REFERENCES

Archibugi, D. (2001). 'Pavitt's Taxonomy Sixteen Years On: A Review Article', *Economics of Innovation and New Technology*, 10(5): 415–25.

Arrow, K. (1962). 'Economic Welfare and the Allocation of Resources for Invention'. In R. Nelson (ed.), *Economic Welfare and the Allocation of Resources for Invention: The Rate and Direction of Inventive Activity*. Princeton, NJ: Princeton University Press, 609–25.

Bagchi-Sen, S. (2007). 'Strategic Considerations for Innovation and Commercialization in the US Biotechnology Sector', *European Planning Studies*, 15(6): 753–66.

Bercovitz, J., and Feldman, M. (2007). 'Fishing Upstream: Firm Innovation Strategy and University Research Alliances', *Research Policy*, 36(7): 930–48.

Borrás, S. (2003). *The Innovation Policy of the European Union: From Government to Governance*. Cheltenham: Edward Elgar Publishers.

Campbell, D. T. (1987). 'Blind Variation and Selective Retention in Creative Thought as in Other Knowledge Processes', in Radnitzky, G., and Bartley, W. W. (eds), *Evolutionary Epistemology, Rationality, and the Sociology of Knowledge*. La Salle, IL: Open Court.

Cohen, W. M., and Levinthal, D. A. (1990). 'Absorptive Capacity: A New Perspective on Learning and Innovation', *Administrative Science Quarterly*, 35(1): 128–52.

Cohen, W. M., Nelson, R. R., and Walsh, J. P. (2004). *Protecting their Intellectual Assets: Appropriability Conditions and Why US Manufacturing Firms Patent (Or Not)*. Cambridge, MA: National Bureau of Economic Research.

Dasgupta, P., and David, P. A. (1994). 'Towards a New Economics of Science', *Research Policy*, 23(55): 487–521.

Deiaco, E., Hughes, A., and McKelvey, M. (2012). 'Universities as Strategic Actors in the Knowledge Economy', *Cambridge Journal of Economics*, 36(3): 525–41.

Dodgson, M., Gann, D., and Salter, A. (2008). *The Management of Technology Innovation: Strategy and Practice*. Oxford: Oxford University Press.

Dodgson, M., Hughes, A., Foster, J., and Metcalfe, S. (2011). 'Systems Thinking, Market Failure and the Development of Innovation Policy: The Case of Australia', *Research Policy*, 40(9): 1145–56.

Dosi, G. (1982). 'Technological Paradigms and Technological Trajectories: A Suggested Interpretation of the Determinants and Directions of Technical Change', *Research Policy*, 11(3): 147–62.

Druilhe, C., and Garnsey, E. W. (2000). 'Emergence and Growth of High-Tech Activity in Cambridge and Grenoble', *Entrepreneurship and Regional Development*, 12: 163–71.

Etzkowitz, H. (1998). 'The Norms of Entrepreneurial Science: Cognitive Effects of the New University-Industry Linkages', *Reseach Policy*, 27(8): 823–33.

Fagerberg, J., and Verspagen, B. (1996). 'Heading for Divergence? Regional Growth in Europe Reconsidered', *Journal of Common Market Studies*, 34(3): 431–48.

Faulkner, W., Senker, S., and Velho, L. (1995). *Knowledge Frontiers: Public Sector Research and Industrial Innovation in Biotechnology, Engineering Ceramics, and Parallel Computing*. Oxford and New York: Clarendon Press.

Feldman, M. (2007). *The Geography of Innovation*. Dordrecht Kluwer.

Garnsey, E. W., and Heffernan, P. (2005). 'High-Technology Clustering through Spin-out and Attraction: The Cambridge Case', *Regional Studies*, 39: 1127–44.

Garnsey, E., Lorenzoni, G., and Ferriani, S. (2008). 'Speciation through Entrepreneurial Spin-off: The Acorn-ARM Story', *Research Policy*, 37: 210–24.

Garnsey, E., Stam, E., and Heffernan, P. (2006) 'New Firm Growth: Exploring Processes and Paths', *Industry and Innovation*, 13: 1–20.

Geuna, A. (1999). *The Economics of Knowledge Production: Funding and the Structure of University Research*. Cheltenham: Edward Elgar.

Hughes, A., and Kitson, M. (2012). 'Pathways to Impact and the Strategic Role of Universities: New Evidence on the Breadth and Depth of University Knowledge Exchange in the UK and the Factors Constraining its Development', *Cambridge Journal of Economics*, 36(3): 723–50.

Jones, C., Hesterley, W. W., and Borgatti, S. P. (1997). 'A General Theory of Network Governance: Exchange Conditions and Social Mechanisms', *Academy of Management Review*, 22(4): 911–45.

Kline, S. J., and Rosenberg, N. (1986). 'An Overview of Innovation', in R. Landau and N. Rosenberg (eds), *The Positive Sum Strategy: Harnessing Technology for Economic Growth*. Washington DC: National Academy Press, 275–305.

Kuhn, T. S. (1974). 'Second Thoughts on Paradigms', in F. Suppe (ed.), *The Structure of Scientific Theories*. Urbana: University of Illinois Press, 459–82.

Levin, R., Klevorick, A., Nelson, R. R., Winter, S. G., Gilbert, R., and Griliches, Z. (1987). 'Appropriating the Returns from Industrial Research and Development', *Brookings Papers on Economic Activity* 1987 (3): 783–831.

McCloskey, D. N. (1994). *Knowledge and Persuasion in Economics*. Cambridge: Cambridge University Press.

McKelvey, M. (1996). *Evolutionary Innovations: The Business of Biotechnology*. Oxford: Oxford University Press.

McKelvey, M., and Lassen, A. H. (2013a). *Managing Knowledge Intensive Entrepreneurship*. Cheltenham: Edward Elgar Publishers.

McKelvey, M., and Lassen, A. H. (2013b). *How Entrepreneurs Do What They Do: Case Studies of Knowledge Intensive Entrepreneurship*. Cheltenham: Edward Elgar Publishers.

McKelvey, M., and Orsenigo, L. (2006). *The Economics of Biotechnology: Volume I and II*. Cheltenham: Edward Elgar Publishers.

Malerba, F., Caloghirou, Y., McKelvey, M., and Radosevic, S. (2013). *The Dynamics of Knowledge Intensive Entrepreneurship in Europe*. London and New York: Routledge.

Malerba, F., and Orsenigo, L. (1996). 'Schumpeterian Patterns of Innovation are Technology-Specific', *Research Policy*, 25(3): 451–78.

Mokyr, J. (2002). *The Gifts of Athena: Historical Origins of the Knowledge Economy*. Princeton, NJ: Princeton University Press.

Mowery, D., and Sampat, B. (2005). 'Universities in National Innovation Systems', in *The Oxford Handbook of Innovation*. Oxford: Oxford University Press, 209–39.

Mytelka, L., and Smith, K. (2001). 'Policy Learning and Innovation Theory: An Interactive and Coevolutionary Process', *Research Policy*, 31(8–9): 1467–79.

Nelson, R. R. (1959). 'The Simple Economics of Basic Scientific Research', *Journal of Political Economy*, 67(3): 297–306.

Nelson, R. R. (1993). *National Innovation Systems: A Comparative Analysis*. Oxford: Oxford University Press.

Nelson, R. R. (2011). 'The Moon and the Ghetto Revisited', *Science and Public Policy* 38(9): 681–90.

Nelson, R. R., and Winter, S. G. (1982). 'The Schumpeterian Trade-Off Revisited', *American Economic Review*. 72: 114–32.

Nightingale, P. (1998). 'A Cognitive Model of Innovation', *Research Policy* 27(7): 689–709.

OECD (2012). 'Key Figures', in *Main Science and Technology Indicators 2011* (2), OECD Publishing. Available at <www.oecd.org> (accessed 23 July 2013).

Pavitt, K. (1984). 'Sectoral Patterns of Technical Change: Towards a Taxonomy and a Theory', *Research Policy*, 13(6): 343–73.

Perkman, M., Tartari, V., McKelvey, M., Autio, A., Broström, E., D'Este, P., Fini, R., Geuna, A., Grimaldi, R., Hughes, A., Krabel, S., Kitson, M., Llerena, P., Lissoni, F., Salter, A., and Sobrero, M. (2013). 'Academic Engagement and Commercialisation: A Review of the Literature on University Industry Relations', *Research Policy*, 42(2): 423–42.

Perkmann, M., and Walsh, K. (2008). 'Engaging the Scholar: Three Types of Academic Consulting and their Impact on Universities and Industry', *Research Policy*, 37(10): 1884–91.

Powell, W. W., White, D. R., Koput, K. W., and Owen-Smith, J. (2005). 'Network Dynamics and Field Evolution: The Growth of Interorganizational Collaboration in the Life Sciences', *American Journal of Sociology*, 110(4): 1132–1205.

Price, D. J. de Solla (1965). 'Is Technology Historically Independent of Science? A Study in Statistical Historiography', *Technology and Culture* 6(4): 553–68.

Rosenberg, N., and Birdzell, J. L. E. (1986). *How the West Grew Rich: The Economic Transformation of the Industrial World*. New York: Basic Books.

Robinson, D., Rip, A., and Mangematin, V. (2007). 'Technological Agglomeration and the Emergence of Clusters and Networks in Nanotechnology', *Research Policy*, 36(6): 871–79.

Rosenberg, N. (1994). *Exploring the Black Box: Technology, Economics, and History*. Cambridge: Cambridge University Press.

Royal Society (2009). 'Hidden Wealth: The Contribution of Science to Service Sector Innovation'. Available at <http://royalsociety.org/uploadedFiles/Royal_Society_Content/policy/publications/2009/7863.pdf> (accessed 23 July 2013).

Salter, A., and Martin, B. (2001). 'The Economic Benefits of Publicly Funded Basic Research: A Critical Review', *Research Policy*, 30(3): 509–32.

Shane, S. (2004). *Academi Entrepreneurship: University Spin-offs and Wealth Creation*. Cheltenham: Edward Elgar Publishing.

Sherer, F. M. (1965). 'Firm Size, Market Structure, Opportunity, and the Output of Patented Inventions', *American Economic Review*, 55(5): 1097–1125.

Stokes, D. (1997). *Pasteur's Quadrant: Basic Science and Technological Innovation*. Washington, DC: The Brookings Institution.

Teece, D. J. (1986). 'Profiting from Technological Innovation', *Research Policy*, 15(6): 285–305.

Utterback, J. M., and Abernathy, W. J. (1978). 'Patterns of Industrial Innovation', *Technology Review*, 80(7): 40–47.

Vincenti, W. G. (1990). *What Engineers Know and How They Know It: Analytical Studies from Aeronautical History*. Baltimore, MA: Johns Hopkins University Press.

Von Tunzelmann, G. N. (1995). *Technology and Industrial Progress: The Foundations of Economic Growth*, Cheltenham: Edward Elgar Publishers.

Winter, S. (2006). 'The Logic of Appropriability: From Schumpeter to Arrow to Teece', *Research Policy*, 35(8): 1100–06.

CHAPTER 5

USER-DRIVEN INNOVATION

NIK FRANKE

WHAT IS USER-DRIVEN INNOVATION?

WHEN we term a new product or service as being a *user innovation*, we mean that it was invented and prototyped by an institution that aims to benefit from the innovation by *using* it, not by *selling* it (producer innovation). User innovations thus have a very direct benefit for the innovator, such as making things easier, more practical, or safer. Of course, users exist both in consumer markets and in industrial markets. In the former case, users are individual consumers such as, for example, those sports aficionados who were bored by ordinary skiing and hence invented the snowboard for their own use (Shah, 2000). In the latter case, users are often firms or professionals like the surgeon who needed more precise equipment for conducting brain surgery and therefore invented a medical robot system for neurosurgery (Lettl et al., 2006).

A PARADIGM SHIFT

Innovations by users are an ancient mode of innovation. For a long time, user innovations might have been the only mode—we can hardly imagine prehistoric man inventing the use of fire for commercial purposes. The division of labour, industrialization, and increasing complexity of technology and production processes, however, led to firms with specialized R&D departments and professional innovation functions. Scholarly research on innovation management followed this and since Schumpeter's (1911) models of innovation assumed that the dominant mode is producer innovation (Baldwin and von Hippel, 2011). The rationale for this is rooted in scale effects. For an institution that is able to sell the outcome of the innovation process—a new standard product—to many customers, the benefit is n^*u (with n being the number of users and u being the benefit for the individual user). For the individual user it is only u. Given that markets are large and firms often have thousands or millions of customers, it follows that the incentive to innovate is much higher for a producer than for a user (Baldwin and von Hippel, 2011). Thus, theoretically, user innovations should hardly exist, at least not in industrialized and developed economies.

However, they do. A number of empirical studies have demonstrated that user innovation is both a frequent and important phenomenon. The first type of study draws samples of users and analyses what proportion reports having innovated. Studies cover different industries, such as printed circuit Computer Aided Design (CAD) software, pipe hanger hardware, library information systems, surgical equipment, and several types of consumer products, and find that user innovation is a relatively frequent behaviour (von Hippel, 2005). Recently, some studies have provided particularly systematic evidence. von Hippel, de Jong, and Flowers (2012), drawing on a nationally representative sample of 1173 UK household residents aged 18 and over based on standardized telephone interviews, find that 6.2 per cent of them have engaged in creating or modifying consumer products they use during the prior three years.[1] This represents 2.9 million people—about two orders of magnitude more than the number of product developers employed by all the consumer goods producing firms in the UK. Consumer product innovation spans a wide range of fields, from toys, to tools, to sporting equipment, and to personal solutions for medical problems. Replications of this study in Japan and the USA show comparable numbers (von Hippel, Ogawa, and de Jong, 2011). There are also recent studies confirming that firms are frequent user innovators. de Jong and von Hippel (2008) analysed a nationally representative sample of 2416 small and medium-sized enterprises (SMEs) in the Netherlands and found that 21 per cent engage in user innovation, that is, they develop and/or significantly modify existing techniques, equipment or software to satisfy their own process-related needs.

The finding that user innovation is a frequent behaviour does not necessarily mean that it bears importance. Thus, a second strand of study investigated the economic value of user innovations. Here, scholars started by interviewing experts and based on this determined a list of the most important innovations in a given industry. Then they studied trade publications and conducted a series of interviews in order to find out who actually built the first prototype of each of these innovations. The stable finding in industries such as petroleum processing, pultrusion, scientific instruments, windsurfing, or skateboarding is that the roots of many of the most important innovations are users (see von Hippel, 2005 for an overview).

Two questions immediately arise from this surprisingly strong pattern. First, one wonders why the significance of users as innovators has been underestimated for such a long time. Why do we think that these figures are counterintuitive? Why is the first reaction of many students and managers straight disbelief? Why did older textbooks of innovation management often ignore this source of innovations completely or, in other cases, treat it as a somehow 'exotic' exception? Typical quotes are: 'Customers should not be trusted to come up with solutions; they aren't expert or informed enough for that part of the innovation process' (Ulwick, 2002) or 'The truth is, customers don't know what they want. They never have. They never will. The wretches don't even know what they don't want' (Brown, 2001).

The reason is both a neglect of user innovation in most public statistics and innovation surveys (de Jong and von Hippel, 2008) and a general perceptual bias resulting from communication patterns. Producer innovations are being advertised and marketed to as many potential users as possible—after all, the producer wants to sell them. User innovations by contrast are being developed for the users' own use, thus it is often the case that not very many other people come into contact with them. Users simply have a lower incentive for popularizing their achievements among the mass of users. Also, when a producing firm picks up a user innovation and commercializes it, the producer has no interest in revealing the original source to the public. Claiming authorship might also appear subjectively justified, as producers usually further develop and design user innovations before introducing them to the mass market.

The second question arising from studies into the sources of innovation is whether there is a trend. Is the significance of user innovation increasing? There is no immediately obvious systematic longitudinal study and any such analysis would be difficult to attain. However, a number of arguments make it plausible that the frequency and importance of user innovation has been increasing dramatically in the past years.

First, the Internet and social network media have enabled individuals to exchange information much more easily than when contact was more or less restricted to friends, relatives, colleagues in the working environments, and neighbours. Geographical and social impediments decreased, and users have now easy access to likeminded people around the globe. For individual innovators, this means that it is quite easy to complement their creativity, knowledge, and technical capabilities, and thereby establish the critical mass that is often necessary for major innovative developments. Good examples of user innovations enabled by the Internet are open-source software projects such as Linux, Apache, or Firefox, and digital products based on user-generated content such as Wikipedia or YouTube.

Developments in ICT also resulted in the evolution of many tools that assist and support the individual user to actively convert an idea into a product. Personal computers and both general purpose and specialized software for writing texts, doing calculations, creating designs, and assembling machinery are cheap and easy to handle today and thus greatly decrease the costs of user innovation. Only a few years ago, for example, creating a pop song recording required not only musical talent but also access to an extremely expensive studio, a professional producer, and a large amount of money for the physical manufacturing. Today it can be done with a PC at home and it can be distributed via the Internet.

The attention scholars pay to the phenomenon may also be indicator of an increase in the significance of user innovation. A simple count analysis reveals that the number of academic articles devoted to the topic simply exploded in recent years (see Figure 5.1). While in the 1980s only two articles appeared in peer-reviewed journals, this number jumped to sixty in 2006–2010, and in 2011 alone there were twenty-one. Several scholars interpret this increase in importance as a paradigm change (e.g. Baldwin and von Hippel, 2011; von Hippel et al., 2011, Dahlander and Frederiksen, 2012).

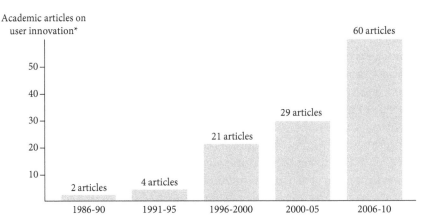

Academic articles on
user innovation*

* Hits in the data base EBSCO, search string "user innovation" or "lead user" or "user-driven innovation", accessed 2011

FIGURE 5.1 Importance of user innovation in scholarly research

THEORETICAL EXPLANATIONS

Which Users Innovate and Why?

User innovation is not an activity randomly distributed among users. It follows relatively clear patterns. von Hippel (1986) uses the term 'lead user' to capture the nature of those users who are most likely to come up with valuable new products and services. The original theoretical thinking that led to the definition of 'lead users' as having (a) high expected benefits from an innovation and (b) a position ahead of an important market trend was built on findings from two different streams of literature (von Hippel, 1986, 2005).

The 'high expected benefits' component of the lead user definition was derived from research on the economics of innovation. Studies of industrial product and process innovations have shown that the greater the benefit an entity expects to obtain from a required innovation, the greater that entity's investment in obtaining a solution (e.g. Schmookler, 1966). The benefits a user expects can be higher than those expected by a manufacturer, for example if the market is new and uncertain, if customer preferences are heterogeneous and change quickly in the market, or if the costs of innovation are lower for users than for manufacturers due to the 'stickiness' of preference information (von Hippel, 1994). In all these cases, a firm expecting benefits from selling the innovation will hesitate, while a user with a strong need for a solution might have hardly any choice. Component (a) of the lead user definition was therefore intended to serve as an indicator of innovation likelihood.

The second component of the lead user definition, namely being 'ahead of an important marketplace trend', was included because of its expected impact on the commercial attractiveness of innovations developed by users residing at that location in a market-place (von Hippel, 1986). Studies on the diffusion of innovations regularly show that some adopt innovations before others (Rogers, 1994). Furthermore, classic research on problem solving reveals that subjects are heavily constrained by their real-world experience through an effect known as 'functional fixedness': for example, those who use an object or see it used in a familiar way find it difficult to conceive of novel uses (e.g. Allen and Marquis, 1964). Taken in combination, these findings led to the hypothesis that users at the leading edge would be best positioned to understand what many others will need later. After all, their present day reality represents aspects of the future from the viewpoint of those with mainstream market needs. Component (b) of the lead user definition therefore indicates the commercial attractiveness of an innovation created by such a user.

Note that these two components of the lead user definition are conceptually independent. They stem from different areas of literature, and they serve different functions in lead user theory. Although they may be correlated in some cases and to some degree—especially as a position ahead of the trend—may well be accompanied by a high need for innovative solutions, this is not necessarily always the case. Therefore, the lead user construct can be described as consisting of two (formative) dimensions. These theoretical considerations were tested and confirmed in a study of 456 kite surfers and their user innovations (Franke et al., 2006). Figure 5.2 shows the main findings in graphic form: first of all, we see that both components are indeed relatively independent. Users (represented by small crosses or bubbles) are broadly distributed, and a considerable number of users are far ahead of the trend but would hardly expect any benefit from innovating. At the same time, many users would derive high benefits from an innovation but are not at all ahead of the trend. Second, moving from left to right (i.e. from low to high benefit), we can see that the proportion of innovators (represented by the grey bubbles) rises relative to the number of non-innovating users (represented by the small dots). Third, moving upwards (i.e. from a position behind the trend to a position ahead of the trend), we can see that the attractiveness of innovations (represented by the size of the grey bubbles) increases. Hence, both the proportion of users with innovative ideas and the commercial attractiveness of the innovations they develop are highest in the lead user quadrant (top right) of Figure 5.2.

The graph also illustrates what Morrison et al. (2004) discovered, namely that the lead user construct is distributed over a continuum: users are not simply 'lead users' or 'non-lead users' (as some literature appears to suggest); instead, they may have more or fewer lead user characteristics. Recent studies have found that an individual's 'lead user-ness' with respect to a specific market is correlated with other characteristics, such as innovativeness, adoption behaviour, or opinion leadership (Schreier and Prügl, 2008). This suggests that lead users may very well constitute valuable informational resources in other phases of the innovation process, such as new product forecasting, product and concept testing, product design, and the diffusion of innovations (e.g. Ozer, 2009).

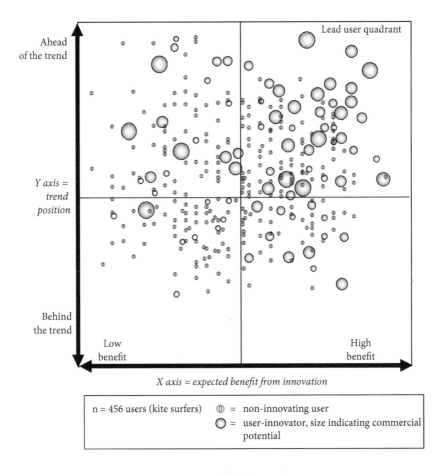

Low
benefit

High
benefit

X axis = expected benefit from innovation

n = 456 users (kite surfers)   ⓞ = non-innovating user
                                ◯ = user-innovator, size indicating commercial
                                    potential

FIGURE 5.2 A test of lead user theory

How do Users Innovate?

If lead user innovators are individuals, the question arises as to how they tackle the often complex task of product development. An individual may easily develop an idea, but developing the idea into a functioning prototype often requires diverse and specific knowledge which a lone individual is unlikely to possess. Franke and Shah (2003) found that from fifty-three user innovation cases they investigated in the area of extreme sports, not a single user innovator developed his or her product without assistance. This appears to be a general pattern. Innovative individual users purposefully seek other users in order to complement their capabilities, that is, exchange ideas and experiences, discuss possible solutions, and further refine intermediate solutions. As a result, user innovators often organize into communities. These communities can be off-line such as clubs or informal groups and organizations. Increasingly, they are online, such as open source project groups, social networks, or brand communities (von Hippel,

2007). Within these communities, lead users usually have a central position and often bridge different clusters or segments of users, a feature termed 'betweenness centrality' (Kratzer and Lettl, 2009).

Why do Users Often Freely Reveal their Innovations?

Another interesting pattern of user innovation is that it is often freely revealed. This means that the innovator gives proprietary information on the functioning and construction of the innovation openly to others at no cost. This behaviour has been documented by von Hippel and Finkelstein (1979) for medical equipment, by Lim (2000) for semiconductor process equipment, by Morrison et al. (2004) for library information systems, and by Franke and Shah (2003) for sporting equipment. In open source software projects, the General Public License (GPL) enforces the free revealing of new software code and thus makes it the norm (Stallman, 1999). When it is freely revealed, the innovation becomes a public good (Harhoff et al., 2003). At first this appears strange—after all, the individual invested resources at private expense and the outcome is potentially valuable. Why give it away freely? The answer is not that user innovators are necessarily altruists. Rather, they benefit in a different way than they would from selling it:

- Being known for a helpful innovation increases a user's reputation. Reputational gains could in turn increase profits for an innovating firm (Allen, 1983) or the career prospects or salaries of individual innovators (Lerner and Tirole, 2002).
- By freely giving away the innovation, the innovator usually increases its diffusion relative to what would occur if they charged for it. If the innovation allows for network effects, the innovator directly profits from a broader user base. But even if not, other users may give valuable feedback or even further refine and develop the innovation—and also freely reveal that to the innovator (Raymond, 1999).
- Sometimes, an adopting manufacturer may be able to produce the innovation and sell it at a price lower than users' in-house production costs (Harhoff et al., 2003).
- In innovation communities in particular, individual participants may derive valuable private benefits from the fun and learning they gain from participation (Lakhani and Wolf, 2005).
- Communities may also have established a 'norm of sharing' with regard to giving assistance to and freely revealing the outcome of innovation projects (Franke and Shah, 2003). Ignoring such norms would result in social costs.

In sum, it may be beneficial for user innovators to freely reveal their accomplishments—to the benefit of other users. von Hippel and von Krogh (2003) summarize this fascinating pattern in the so-called 'private–collective innovation model'.

However, it is important to note that, sometimes, it appears beneficial for user innovators to change their functional role and become producers themselves, that is, to start benefitting from selling their innovation (Shah and Tripsas, 2007). This is particularly the case when their opportunity costs are low and they have direct access to peers with an interest in the innovation (Baldwin et al., 2006).

METHODS FOR EXPLOITING USER INNOVATIVENESS

For producing firms, it is advisable to proactively respond to this continuing paradigm change and (at least) complement their traditional mode of producer-centred innovation engagement with user-driven approaches. In the following, three different ways are proposed in which firms can benefit from user innovativeness. They are, of course, not mutually exclusive. As the field of user innovation is quite dynamic and many firms experiment with methods, the list is also not exhaustive.

The Lead User Method

The lead user method proposed by Urban and von Hippel (1988) is a managerial heuristic that enables companies to search for particularly attractive user innovations and identify radically new business opportunities. Usually, this method is described as comprising four phases (e.g. Lüthje and Herstatt, 2004).

- Phase 1: In the beginning, during the starting phase, objectives are defined (e.g. 'finding an innovative solution to problem X' or 'identifying an innovative product concept in market Y') and a cross-functional team is set up (usually comprising employees from different functions such as R&D, marketing, production, etc.). The latter is important to ensure that solutions found have a sufficient fit with regard to strategy, R&D, production capabilities, and objectives. Also, broad anchorage reduces the risk of 'not invented here' problems arising from the fact that solutions external to the company are being sought.
- Phase 2: Identification of major needs and trends. In the second phase, the three to five most important trends are identified. 'Trends' are those dimensions in which lead users are far ahead of the mass market. The function of the trends in the lead user method is to narrow the problem and to allow a systematic search for lead users. They can be based on technology (e.g. a trend towards modularization or new materials) market information (e.g. a trend towards security or wireless solutions). The selection is usually based on interviews with experts, information from online forums, and literature research.
- Phase 3: Identification of lead users. The third phase involves a broad search for individuals who are far ahead with regard to the trends identified and have

high personal benefits from innovations. Earlier studies usually employed a mass screening approach in which a large sample of users (typically from customer databases) was systematically filtered in order to identify those users who score highest on both lead user dimensions. More recently, lead user studies have increasingly turned to the pyramiding method for the purpose of lead user identification (Lilien et al., 2002). In the latter approach, researchers start with a few users and ask them who has especially high needs and is leading the trend. Those individuals are then contacted and asked the same questions, and the process continues until a sufficient level of 'lead user-ness' is achieved (which is usually the case after two or three steps). Recently, experiments have demonstrated the superior efficiency of the pyramiding search strategy compared to screening (von Hippel et al., 2009). Another advantage of pyramiding is the possibility of identifying individuals outside of a pre-defined population or sample (Poetz and Prügl, 2010). Particularly analogous markets, that is, markets which are different from the target market but characterized by the same trends, are valuable sources in the search for lead users (Franke et al., 2013). Consider the example of a lead user study which aims to find methods of preventing infections in clinical surgery. For this purpose, one important trend would be 'methods for increased air purity'. Outside of leading hospitals, experts from the analogous field of chip production or CD production may also be able to provide valuable creative input. There are two reasons why it might make sense to ask such people. First, they might possess solution-related knowledge which is worth transferring from the analogous field to the target field, and second, they are less likely to be cognitively 'blocked' by existing solutions in the target field. Naturally, selected individuals must also be open, creative, willing to jointly work in a team with other lead users, possess sufficient verbal skills, and so on (Hoffman et al., 2010).

- Phase 4: The lead user workshop. Once the lead users have been identified they are invited to a two- or three-day workshop in which company members from different functional areas also participate (Lüthje and Herstatt, 2004). At these workshops, techniques such as brainstorming, group discussions, and so on are used to capitalize on the creativity of the participants. It is important for the company to address the issue of intellectual property rights prior to the workshop and to ensure that the ideas and concepts generated can be commercialized without the risk of legal infringement. Often this is unproblematic as in many cases it is economically profitable for users to reveal their innovations freely (e.g. because they expect to profit from the use of the resulting product).

A number of detailed studies document that the lead user method can effectively and systematically generate ideas for commercially attractive new products (Urban and von Hippel 1988; Herstatt and von Hippel, 1992; Olson and Bakke, 2001). The most systematic analysis is Lilien et al.'s (2002) study in which the authors studied forty-seven new product development projects conducted at 3M and found that those projects that used the lead user method were much more successful than those in which more traditional ways of idea-generation were employed. Most importantly, projected sales of the lead user projects were $146 million, while sales of the projects which did not use the lead user method were only an eighth of that ($18 million). Despite the increasing interest among firms, there is not yet very much scholarly research on the lead user method as such. For example, we lack knowledge about which phases are most critical, why lead users actually participate (or refrain from doing so) and hence how they should be incentivized, and what role different appropriation regimes play for revealing creative ideas and concepts, and how lead users and lead experts differ in their motivations and contributions.

Toolkits for User Innovation and Design

A different method building on user creativity is to outsource individual product design to customers. If people are creative and dissatisfied with standard offerings, why not provide them with equipment that decreases the individual users' costs and allows the producer to do what it is best at, namely *producing*? von Hippel conceptualized 'toolkits for user innovation and design' as coordinated sets of design tools that allow individual users to self-design their own individual product according to their individual preferences and give visual and informational feedback on (virtual) interim solutions (von Hippel, 1998, 2001; von Hippel and Katz, 2002). If the customers like what they design, they can order their product and the toolkit provider will produce it according to their individual design specifications.

   Many companies have started offering toolkits which enable users to create their own individual computer chips, machines, flavours, custom food, software, plastic polymers, industrial refrigerators, security systems, climate control and air conditioning systems, windows, electronic equipment, t-shirts, watches, breakfast cereals, cars, kitchens, sofas, skis, jewellery, laptops, pens, sneakers, and so forth online which the manufacturer can then produce to order. Albeit their differences with regard to the product arena, toolkits share two common principles: first, they all contain some form of design tools that enable the user to create and modify a design. Some are quite restricted, coming in the form of lists to choose from. Others are of the drag-and-drop-style, for example when users can choose graphic symbols, place them on skis, define their size, and shift them around in order to find the position where the skis look best. There are also toolkits that allow users to combine product components modular-wise, similar to building an artefact with Lego® bricks. Still others allow free design like a graphic computer program. They comprise functional aspects of the product (the material, size, shape, or functions included), the product's aesthetics (such as colour, graphics incorporated, or other forms of style), and the possibility of personalization (e.g. by adding one's name or other symbols). Toolkits can of course be applied in service industries, too. There are examples of websites that allow the individual user to customize events such as wedding feasts, trips, electronic newspapers, financial investments and insurances, music, ring tones, mobile phone contracts, and so on, and there are many other services where toolkits do not yet exist but would make a lot of sense. In service industries, toolkits also provide design tools that allow the user to determine the design of the service, and give some sort of feedback. Very often toolkits are incorporated in PC games and allow the user to extend, modify, and create new game characters, maps, surroundings, and so forth. An interesting feature of software is that in this case the design simulation and the product often go together: the simulation is the product. A new map designed for a game no longer needs to be produced by a producer—the outcome of the self-design process is already the product, and the user is also the producer.

A second feature that these heterogeneous toolkits have in common is that they give some sort of feedback information during the design process. In consumer goods settings, the most common form of feedback is a virtual, simulated visual representation of the current product design that is updated in real time with every design change users make. If the toolkit allows for functional product manipulation, feedback, of course, should also be functional. For example, a gardening toolkit exists that allows the creation of one's own garden and provides an alarm when a garden pond is positioned too close to a broadleaf tree: 'in fall leaves might fall on the pond—they might quickly silt up your pond, in summer this might offer too much shade and so stopping aquatic plants from growing'. Others inform about weight, price, or technical performance. In sum, a good toolkit provides the user with information about the anticipated consequences of design decisions, just as a capable salesperson would do. This enables the user to conduct what is called trial-and-error learning (von Hippel, 1998). Few of us have the imagination to come up with a precise, detailed, and definitive product specification. Most cannot design a product just in their mind—they need to play around, try different things, and iteratively find out what they like best. The toolkit supports this way of problem solving.

Consumers can derive considerable value from using toolkits. Franke and Piller (2004) found that consumers' willingness-to-pay is twice as high for a self-designed watch than for the bestselling standard watches of the same objective quality. This dramatic value increase has been confirmed in a number of studies in the product areas of breakfast cereals, carving skis, mobile phone covers, fountain pens, kitchens, newspapers, scarves, and t-shirts (Schreier 2006; Franke and Schreier 2008; Franke et al., 2009; Franke et al., 2010). A number of factors have been identified that cause this value increment: customized products better fit consumers' preferences (Dellaert and Stremersch, 2005; Franke et al., 2009), are perceived as more unique (Franke and Schreier, 2008), and provide the consumer with a 'I designed it myself' feeling, that is, a sense of accomplishment (Franke et al., 2010). Also the enjoyment of the design process augments the value consumers derive from self-designing a product (Franke and Schreier, 2010).

Research into situational moderators has found that consumers' level of category knowledge can influence their satisfaction with the toolkit itself as well as with the resulting product (Randall et al., 2007). Consumers' insights into their own preferences (Bharadwaj et al., 2009; Franke et al., 2009) and consumers' social comparisons to other designers can also influence evaluations of these custom products (Moreau and Herd, 2010). Finally, the intended recipient of the product also matters: products designed as gifts for others are more highly valued by the self-designer than those designed for one's own use (Moreau et al., 2011).

Most studies are experiments and use existing mass customization toolkits within the area of low-price consumer goods. Our knowledge with regard to motives, use patterns, and success factors in industrial goods settings is still quite limited, although externalizing product design to customers plays an important role in these markets, too.

Crowdsourcing

A third form of profiting from user creativity is to 'crowdsource' the task. This method is also known as 'broadcast search' (Jeppesen and Lakhani, 2010), 'innovation contest' (Terwiesch and Xu, 2008), 'virtual co-creation' (Füller, 2010), 'innovation tournament' (Terwiesch and Ulrich, 2009) or 'virtual customer environment' (Nambisan, 2002). The underlying idea is straightforward: a company poses a question to 'crowds' by way of open online calls for solutions. The sponsor then evaluates the solutions submitted, and rewards those who submit the best solutions (Nambisan, 2002; Ogawa and Piller, 2006; Dahlander and Magnusson, 2008; Terwiesch and Xu, 2008). The value of crowdsourcing comes from the fact that 'crowds' usually consist of many contributors with a wider range of skill sets, perspectives on the problem, and solution heuristics than company insiders possess (Jeppesen and Lakhani, 2010). Such crowds often exhibit 'wisdom' superior to the far smaller number of in-house specialists, particularly in cases where problems are new, complex, and ill-defined, and require a substantial amount of creativity or a transfer of analogous knowledge (Surowiecki, 2004; von Hippel, 2005; Hoyer et al., 2010; Jeppesen and Lakhani, 2010). In such situations, having many heterogeneous problem solvers who compete for a solution (Terwiesch and Ulrich, 2009) and in some instances also collaborate and assist each other (Bullinger et al., 2010) can constitute a major advantage. Raymond (1999: 41) integrates this argument into what he terms 'Linus' law', namely 'given enough eyeballs, all bugs are shallow'. Crowdsourcing principles therefore can be quite powerful and result in surprisingly innovative solutions (Nambisan and Baron, 2009, 2010; Terwiesch and Ulrich, 2009; Bullinger et al., 2010; Harhoff and Mayrhofer, 2010). Evidence of crowdsourcing's advantages is found in a study by Poetz and Schreier (2011) in which consumers out-performed a firm's professionals in generating new product ideas in the consumer goods market for baby products. In their real-world study, consumers produced ideas that scored significantly higher in terms of both novelty and consumer benefits than those produced by the firm's new product developers.

Given these potential benefits, it is not surprising that firms are capitalizing on crowdsourcing principles not only in the apparel and clothing industry, but also in areas such as consumer electronics, software, music, jewellery, photography, household goods, cars, mobile phones, autonomous robotic vehicles, TV casting shows, biotechnology, pharmaceuticals, high-tech R&D problems, and many others (Füller, 2006; Ogawa and Piller, 2006; Sawhney et al., 2005; Humphreys and Grayson, 2008; Terwiesch and

Xu, 2008; Leimeister et al., 2009; O'Hern and Rindfleisch, 2009; Bullinger et al., 2010; Nambisan and Baron, 2010).

A number of studies have helped us understand why users contribute to commercial crowdsourcing business models in which the firm benefits directly from contributor input. The dominant theoretical basis of this research is social exchange theory (Blau, 1964), which posits that voluntary exchange relationships are initiated and maintained when benefits exceed costs. Accordingly, researchers have identified a number of participation motives characterized by self-interest. These motives include seeking monetary rewards (e.g. Hall and Graham, 2004; Füller, 2006, 2010; Ebner et al., 2009; Leimeister et al., 2009; Brabham, 2010; Nambisan and Baron, 2010), showing ideas and getting into contact with the company (e.g. Füller, 2006, 2010; Shah, 2006), getting appreciation from the company (e.g. Jeppesen and Frederiksen, 2006; Ebner et al., 2009; Leimeister et al., 2009; Füller et al., 2010) and from peers (e.g. Wu and Sukoco, 2010), having the opportunity to learn and develop one's skills (e.g. Nambisan and Baron, 2009; Brabham, 2010; Füller, 2010), and intrinsic factors such as enjoyment (e.g. Füller, 2006, 2010; Füller et al., 2006; Füller et al., 2008, Füller et al., 2009; Nambisan and Baron, 2009). Additionally, it was found that potential participants not only calculate whether participation will pay off; they also form a subjective evaluation of fairness in the crowdsourcing business model (Franke et al., 2013). They consider whether they get a 'fair share', that is, whether the benefits and costs are divided evenly between them and the organizing company (distributive fairness), whether they have a 'voice' in decisions, and whether processes are consistent and transparent (procedural fairness). These fairness perceptions have clear behavioural consequences, as they inform the individuals' propensity to submit a design to a crowdsourcing firm even accounting for self-interest.

Knowing the motivation of potential contributors to crowdsourcing projects is one thing; knowing which factors impact this motivation is another. We still have little knowledge of how the concrete organizational features of crowdsourcing systems influence the individuals' motivation to contribute (or not). As Hoyer et al. (2010: 290) put it: 'Currently, we know little about the actions and approaches that firms can take to stimulate participation. . . . Research on the most effective means of stimulating cocreation will therefore have great value'.

Combinations

We have modelled the three methods of lead user method, toolkits, and crowdsourcing as independent approaches. It is clear, however, that they overlap and can be combined in many ways. For example, it is possible to combine a toolkit with a crowdsourcing approach (Piller and Walcher, 2006). In a way this is what Apple has done with its iPhone: Apple openly called for innovative applications and provided a specific software allowing them to be created and simulated. This toolkit also ensures that any app programmed actually runs on a smartphone. It is difficult to imagine that any in-house R&D department would ever come up with so many and such heterogeneous apps. Crowds may also help individual toolkit users to develop an initial idea, handle the toolkit (e.g. how certain toolkit functions work) and evaluate preliminary design solutions (Jeppesen and Molin, 2003; Jeppesen, 2005; Jeppesen and Frederiksen, 2006; Franke et al., 2008). Finally, lead user search in applications of the lead user method is often done via crowdsourcing search techniques in which the company posts open calls for solutions in expert communities.

Future Developments

Currently, we can observe an increase in user innovation. We also observe that many firms react to this trend and frequently use and experiment with ways of benefiting from user innovativeness. Scholarly research also focuses on these new possibilities. Thus, it is quite probable that our understanding of the most effective methods will increase in the future. It is also quite likely that the factors facilitating user innovation will develop further and users will be increasingly able to innovate for and communicate among themselves without a producing firm, as is already the case with open source software projects. Technology is now available that enables consumers to both design and produce new products independently. Google Sketchup is an example of a general-purpose design toolkit for consumers. It is a user-friendly computer-aided design (CAD) toolkit that enables users to create—collaboratively or singly—and share designs for manufacturable 3D products of any type. In the future, 3D printers will become increasingly more affordable, and consumers may gain the ability to prototype rapidly as well. As the costs to design products and to communicate with other consumers decline, the viability of consumer innovation steadily increases. Consumers can increasingly choose to 'go it alone', independent of producers (Baldwin and von Hippel, 2011). Over time, we predict, producers will increasingly cede design to consumers and seek to establish advantageous relationships with them via community support and/or the provision of toolkits to assist consumer innovators.

References

Allen, R. C. (1983). 'Collective Invention', *Journal of Economic Behavior and Organization*, 4(1): 1–24.

Allen, T. J., and Marquis, D. G. (1964). 'Positive and Negative Biasing Sets: The Effects of Prior Experience on Research Performance', *IEEE Transactions on Engineering Management*, EM-11 (4): 158–61.

Baldwin, C., and von Hippel, E. (2011). 'Modeling a Paradigm Shift: From Producer Innovation to User and Open Collaborative Innovation', *Organization Science*, 22(6): 1399–1417.

Baldwin, C., Hienerth, C., and von Hippel, E. (2006). 'How User Innovations Become Commercial Products: A Theoretical Investigation and Case Study', *Research Policy*, 35(9): 1291–313.

Bharadwaj, N., Naylor, R., and ter Hofstede, F. (2009). 'Consumer Response to and Choice of Customized Versus Standardized Systems', *International Journal of Research in Marketing*, 26(3): 216–27.

Blau, P. (1964). *Exchange and Power in Social Life*. New York: Wiley.

Brabham, D. C. (2010). 'Moving the Crowd at Threadless', *Information, Communication & Society*, 13(1): 1122–45.

Brown, S. (2001). 'Torment your Customers (They'll Love It)', *Harvard Business Review*, 83: 82–88.

Bryla, A., Kardinal, A., Schirg, F., and Franke, N. (2012). 'How Many End-users Actually Innovate? Results Depend on Measurement', Working Paper WU Wien.

Bullinger, A. C., Neyer, A.-N., Rass, M., and Moeslein, K. (2010). 'Community-based Innovation Contests: Where Competition Meets Cooperation', *Creativity and Innovation Management*, 19(3): 290–303.

Dahlander, L., and Frederiksen, L. (2012). 'The Core and Cosmopolitans: A Relational View of Innovation in User Communities', *Organization Science*, 23(4): 988–1007.

Dahlander, L., and Magnusson, M. (2008). 'How Do Firms Make Use of Open Source Communities?', *Long Range Planning*, 41(6): 629–49.

de Jong, J. P. J., and von Hippel, E. (2008). 'User Innovation in SMEs: Incidence and transfer to producers', Zoetermeer Working Paper.

Dellaert, B. G. C., and Stremersch, S. (2005). 'Marketing Mass-Customized Products: Striking a Balance Between Utility and Complexity', *Journal of Marketing Research*, 42(2): 219–27.

Ebner, W., Leimeister, J. M., and Krcmar, H. (2009). 'Community Engineering for Innovations: The Idea Competition as a Method to Nurture a Virtual Community for Innovations', *R&D Management*, 39(4): 342–56.

Franke, N., and Piller, F. (2004). 'Value Creation by Toolkits for User Innovation and Design: The Case of the Watch Market', *Journal of Product Innovation Management*, 21(6): 401–15.

Franke, N., Poetz, M., and Schreier M. (2013). 'Integrating Problem Solvers from Analogous Markets in New Product Ideation', *Management Science*, forthcoming.

Franke, N., and Schreier, M. (2008). 'Product Uniqueness as a Driver of Customer Utility in Mass Customization', *Marketing Letters*, 19(1): 93–107.

Franke, N., and Schreier, M. (2010). 'Why Customers Value Mass-Customized Products: The Importance of Process Enjoyment', *Journal of Product Innovation Management*, 27: 1020–31.

Franke, N., and Shah, S. (2003). 'How Communities Support Innovative Activities: An Exploration of Assistance and Sharing among End-users', *Research Policy*, 32: 157–78.

Franke, N., Keinz, P., and Schreier, M. (2008). 'Complementing Mass Customization Toolkits with User Communities: How Peer Input Improves Customer Self-design', *Journal of Product Innovation Management*, 25(6): 546–59.

Franke, N., Keinz, P., and Steger, C. (2009). 'Testing the Value of Customization: When do Customers Really Prefer Products Tailored to their Preferences', *Journal of Marketing*, 73(5): 103–21.

Franke, N., Klausberger, K., and Keinz, P. (2013). 'Does This Sound Like a Fair Deal? The Role of Fairness Perceptions in the Individual's Decision to Participate in Firm Innovation', *Organization Science*, forthcoming.

Franke, N., Schreier, M., and Kaiser, U. (2010). 'The "I Designed it Myself" Effect in Mass Customization', *Management Science*, 56: 125–40.

Franke, N., von Hippel, E., and Schreier, M. (2006). 'Finding Commercially Attractive User Innovations: A Test of Lead User Theory', *Journal of Product Innovation Management*, 23(4): 301–15.

Füller, J. (2006). 'Why Consumers Engage in Virtual New Product Developments Initiated by Producers', *Advances in Consumer Research*, 33(1): 639–46.

Füller, J. (2010). 'Refining Virtual Co-creation from a Consumer Perspective', *California Management Review*, 52(2): 98–122.

Füller, J., Bartl, M., Ernst, H., and Mühlbacher, H. (2006). 'Community Based Innovation: How to Integrate Members of Virtual Communities into New Product Development', *Electronic Commerce Research*, 6: 57–73.

Füller, J., Matzler, K., and Hoppe, M. (2008). 'Brand Community Members as a Source of Innovation', *Journal of Product Innovation Management*, 25(6): 608–19.

Füller, J., Mühlbacher, H., Matzler, K., and Jawecki, G. (2009). 'Consumer Empowerment through Internet-Based Co-Creation', *Journal of Management Information Systems*, 26(3): 71–102.

Füller, J., Faullant, R., and Matzler, K. (2010). 'Triggers for Virtual Customer Integration in the Development of Medical Equipment: From a Manufacturer and a User's Perspective', *Industrial Marketing Management*, 39(8): 1376–83.

Hall, H., and Graham, D. (2004). 'Creation and Recreation: Motivating Collaboration to Generate Knowledge Capital in Online Communities', *International Journal of Information Management*, 24(3): 235–46.

Harhoff, D., and Mayrhofer, P. (2010). 'Managing User Communities and Hybrid Innovation Processes: Concepts and Design Implications', *Organizational Dynamics*, 39(2): 137–44.

Harhoff, D., Henkel, J., and von Hippel, E. (2003). 'Profiting from Voluntary Information Spillovers: How Users Benefit by Freely Revealing their Innovations', *Research Policy*, 32(10): 1753–69.

Herstatt, C., and von Hippel, E. (1992). 'From Experience: Developing New Product Concepts via the Lead User Method: A Case Study in a "Low tech" Field', *Journal of Product Innovation Management*, 9(3): 213–22.

Hoffman, D. L., Kopalle, P. K., and Novak, T. P. (2010). 'The "Right" Consumers for Better Concepts: Identifying Consumers High in Emergent Nature to Develop New Product Concepts', *Journal of Marketing Research*, 47: 854–65.

Hoyer, W. D., Chandy, R., Dorotic, M., Krafft, M., and Singh, S. S. (2010). 'Consumer Cocreation in New Product Development', *Journal of Service Research*, 13(3): 283–96.

Humphreys, A., and Grayson, K. (2008). 'The Intersecting Roles of Consumer and Producer: A Critical Perspective on Co-production, Co-creation and Prosumption', *Sociology Compass*, 2(3): 963–80.

Jeppesen, L. B. (2005). 'User Toolkits for Innovation: Consumers Support Each Other', *Journal of Product Innovation Management*, 22: 347–62.

Jeppesen, L. B., and Frederiksen, L. (2006). 'Why do Users Contribute to Firm-Hosted User Communities? The Case of Computer-controlled Music Instruments', *Organization Science*, 17(1): 45–63.

Jeppesen, L. B., and Lakhani, K. (2010). 'Marginality and Problem Solving Effectiveness in Broadcast Search', *Organization Science*, 21(5): 1016–33.

Jeppesen, L. B., and Molin, M. J. (2003). 'Consumers as Co-developers: Learning and Innovation Outside the Firm', *Technology Analysis & Strategic Management*, 15(3): 363–84.

Kratzer, J., and Lettl, C. (2009). 'Distinctive Roles of Lead Users and Opinion Leaders in the Social Networks of Schoolchildren', *Journal of Consumer Research*, 36(4): 646–59.

Lakhani, K. R., and Wolf, B. (2005). 'Why Hackers Do What They Do: Understanding Motivation and Effort in Free/open Source Software Projects', in J. Feller, B. Fitzgerald, S. Hissam, and K. R. Lakhani (eds), *Perspectives on Free and Open Source Software*. Cambridge, MA: MIT Press.

Leimeister, J. M., Huber, M., Bretschneider, U., and Krcmar, H. (2009). 'Leveraging Crowd-sourcing: Activation-supporting Components for IT-based Ideas', *Journal of Management Information Systems*, 26(1): 197–224.

Lerner, J., and Tirole, J. (2002). 'Some Simple Economics of Open Source', *Journal of Industrial Economics*, 50(2): 197–234.

Lettl, C., Herstatt, C., and Gemuenden, H. G. (2006). 'Users' Contributions to Radical Innovation: Evidence from Four Cases in the Field of Medical Equipment Technology', *R&D Management*, 36: 251–72.

Lilien, G. L., Morrison, P. D., Searls, K., Sonnack, M., and von Hippel, E. (2002). 'Performance Assessment of the Lead User Idea Generation Process', *Management Science*, 48(8): 1042–59.

Lim, K. (2000). 'The Many Faces of Absorptive Capacity: Spillovers of Copper Interconnect Technology for Semiconductor Chips', MIT Sloan School of Management. Working paper No. 4110.

Lüthje, C., and Herstatt, C. (2004). 'The Lead User Method: An Outline of Empirical Findings and Issues for Future Research', *R&D Management*, 34(5): 553–68.

Moreau, C. P., and Herd, K. B. (2010). 'To Each His Own? How Comparisons with Others Influence Consumers' Evaluations of their Self-designed Products', *Journal of Consumer Research*, 36: 806–19.

Moreau, C. P., Bonney, L., and Herd, K. B. (2011). 'It's the Thought (and the Effort) That Counts: How Customizing for Others Differs from Customizing for Oneself', *Journal of Marketing*, 75: 120–33.

Morrison, P. D., Roberts, J. H., and von Hippel, E. (2004). 'The Nature of Lead Users and Measurement of Leading Edge Status', *Research Policy*, 33(2): 351–362.

Nambisan, S. (2002). 'Designing Virtual Customer Environments for New Product Development', *Academy of Management Review*, 27(3): 392–413.

Nambisan, S. and Baron, R. A. (2009). 'Virtual Customer Environments: Testing a Model of Voluntary Participation in Value Co-creation Activities', *Journal of Product Innovation Management*, 26(4): 388–406.

Nambisan, S. and Baron, R. A. (2010). 'Different Roles, Different Strokes: Organizing Virtual Customer Environments to Promote Two Types of Customer Contributions', *Organization Science*, 21(2): 554–72.

Ogawa, S., and Piller, F. (2006). 'Reducing the Risks of New Product Development', *Sloan Management Review*, 47(2): 65–71.

O'Hern, M., and Rindfleisch, A. (2009). 'Customer Co-creation: A Typology and Research Agenda', *Review of Marketing Research*, 6: 84–106.

Olson, E. L., and Bakke, G. (2001). 'Implementing the Lead User Method in a High Technology Firm: A Longitudinal Study of Intentions Versus Actions', *Journal of Product Innovation Management*, 18(2): 388–95.

Ozer, M. (2009). 'The Roles of Product Lead-Users and Product Experts in New Product Evaluation', *Research Policy*, 38(8): 1340–49.

Piller, F. T., and Walcher, D. (2006). 'Toolkits for Idea Competitions: A Novel Method to Integrate Users in New Product Development', *R&D Management*, 36(3): 307–18.

Poetz, M. K., and Prügl, R. (2010). 'Crossing Domain-Specific Boundaries in Search of Innovation: Exploring the Potential of 'Pyramiding', *Journal of Product Innovation Management*, 27(6): 897–914.

Poetz, M. K., and Schreier, M. (2011). 'The Value of Crowdsourcing: Can Users Really Compete with Professionals in Generating New Product Ideas', *Journal of Product Innovation Management*, forthcoming.

Randall, T., Terwiesch, C., and Ulrich, K. T. (2007). 'User Design of Customized Products', *Marketing Science*, 26(2): 268–80.

Raymond, E. S. (1999). 'The Cathedral and the Bazaar', *Knowledge, Technology & Policy*, 12(3): 23–49.

Raymond, E. S. (2001). *The Cathedral and the Bazaar*, 2nd edn. Sebastopol, CA: O'Reilly Media.

Rogers, E. M. (1994). *Diffusion of Innovation*, 4th edn. New York: The Free Press.

Sawhney, M., Verona, G. M., and Prandelli, E. (2005). 'Collaborating to Create: The Internet as Platform for Customer Engagement in Product Innovation', *Journal of Interactive Marketing*, 19(4): 4–17.

Schmookler, J. (1966). *Invention and Economic Growth*. Cambridge, MA: Harvard University Press.

Schreier, M. (2006). 'The Value Increment of Mass-Customized Products: An Empirical Assessment', *Journal of Consumer Behaviour*, 5(4): 317–27.

Schreier, M., and Prügl, R. (2008). 'Extending Lead User Theory: Antecedents and Consequences of Consumers' Lead Userness', *Journal of Product Innovation Management*, 25(4): 331–46.

Schumpeter, J. (1911). *Theorie der wirtschaftlichen Entwicklung*. Berlin: Duncker & Humblodt.

Shah, S. (2000). 'Sources and Patterns of Innovation in a Consumer Products Field: Innovations in Sporting Equipment', MIT Sloan School of Management. Working Paper No. 4105.

Shah, S. (2006). 'Motivation, Governance, and the Viability of Hybrid Forms in Open Source Software Development', *Management Science*, 52(7): 1000–14.

Shah, S., and Tripsas, M. (2007). 'The Accidental Entrepreneur: The Emergent and Collective Process of User Entrepreneurship', *Strategic Entrepreneurship Journal*, 1(1–2): 123–40.

Stallman, R., (1999). 'The GNU Operating System and the Free Software Movement', in C. DiBona, and S. Ockman, S. (eds), *Open Sources: Voices from the Open Source Revolution*. Sebastopol, CA: O'Reilly, 53–70.

Surowiecki, J. (2004). *Why the Many Are Smarter Than the Few and How Collective Wisdom Shapes Business, Economies, Societies and Nations*. New York, NY: Doubleday.

Terwiesch, C., and Ulrich, K. T. (2009). *Innovation Tournaments*. Cambridge, MA: Harvard University Press.

Terwiesch, C., and Xu, Y. (2008). 'Innovation Contests, Open Innovation, and Multiagent Problem Solving', *Management Science*, 54(9): 1529–43.

Ulwick, A. W. (2002). 'Turn Customer Input into Innovation', *Harvard Business Review*, January: 91–7.

Urban, G. L., and von Hippel E. (1988). 'Lead User Analyses for the Development of New Industrial Products', *Management Science*, 34(5): 569–82.

von Hippel, E. (1986). 'Lead Users: A Source of Novel Product Concepts', *Management Science*, 32: 791–806.

von Hippel, E. (1994). 'Sticky Information and the Locus of Problem Solving: Implications for Innovation', *Management Science*, 40: 429–40.

von Hippel, E. (1998). 'Economics of Product Development by Users: The Impact of "Sticky" Local Information', *Management Science*, 44(5): 629–44.

von Hippel, E. (2001). 'Perspective: User Toolkits for Innovation', *Journal of Product Innovation Management*, 18(4): 247–57.

von Hippel, E. (2005). *Democratizing Innovation.* Cambridge, MA: MIT Press.

von Hippel, E. (2007). 'Horizontal Innovation Networks: By and for Users', *Industrial & Corporate Change,* 16(2): 293–315.

von Hippel, E., de Jong, J., and Flowers, S. (2012). 'Comparing Business and Household Sector Innovation in Consumer Products: Findings from a Representative Study in the United Kingdom', *Management Science,* 58(9): 1669–81.

von Hippel, E. and Finkelstein, S. N. (1979). 'Analysis of Innovation in Automated Clinical Chemistry Analyzers', *Science & Public Policy,* 6(1): 24–37.

von Hippel, E. and Katz, R. (2002). 'Shifting Innovation to Users via Toolkits', *Management Science,* 48(7): 821–34.

von Hippel, E. and von Krogh, G. (2003). 'Open Source Software and the "Private-Collective" Innovation Model: Issues for Organization Science', *Organization Science,* 14(2): 209–23.

von Hippel, E., Franke, N., and Prügl, R. (2009). 'Pyramiding: Efficient Search for Rare Subjects', *Research Policy,* 38: 1397–406.

von Hippel, E., Ogawa, S., and de Jong, J. P. J. (2011). 'The Age of the Consumer-Innovator', *MIT Sloan Managemnt Review,* 53(1): 27–35.

Wu, W.Y., and Sukoco, B. M. (2010). 'Why Should I Share? Examining Consumer Motives and Trust on Knowledge Sharing', *Journal of Computer Information Systems,* 50(4): 11–19.

CHAPTER 6

NETWORKS OF INNOVATION

TIM KASTELLE AND JOHN STEEN

INTRODUCTION

NOVEL ideas start out as networks. At the most basic level, an innovative idea represents a new set of connections between neurons within the brain. Steven Johnson
(2010:99) explains it thus:

> Like any other thought, a hunch is simply a network of cells firing inside your
> brain in an organized pattern. But for that hunch to blossom into something more
> substantial, it has to connect with other ideas. The hunch requires an environment
> where surprising new connections can be forged: the neurons and synapses of the
> brain itself, and the larger cultural environment that the brain occupies.

In other words, networks are an essential element of innovation. Johnson's idea raises a bigger issue as well: these networks function at multiple levels. An individual thought is built
on a network of cells, a new innovation is built out of a network of connected ideas, and it
comes to life through networks of people. And, of course, networks of people scale up into
networks within a firm, networks of firms, networks of clusters, and networks of regions.

The idea that networks are central to innovation goes back at least to Schumpeter
(1911/1983), who talked about innovation arising from new combinations of ideas. For
a long period of time this concept has been operationalized more as a metaphor than
as a rigorous analytical approach to the management of innovation. However, the rapid
and recent development of new quantitative approaches to network analysis has allowed
us to move beyond using connections as a proxy for networks to actually being able to
measure and evaluate the structures of innovation networks.

In this chapter we will investigate the potential for using network analysis to advance
our understanding of the innovation process, and use it to illustrate how innovation in

research methods raises new opportunities for understanding innovation management. When this methodology originated it was referred to as Social Network Analysis (SNA), since nearly all of the early work looked at interpersonal networks. Recent increases in computing power and software applications have led to a surge in new techniques and approaches that are able to analyse very large data sets. These have sometimes been called 'Complex Network Analysis'. We will use the older term, SNA, but when we do so it will refer to all of the available analytical techniques, including those developed in Complex Network Analysis. We will focus on why network analysis is such a promising approach, particularly in examining innovation in large systems such as organizations and industries.

WHAT IS NETWORK ANALYSIS AND WHY IS IT IMPORTANT?

A network is any system that can be described by a set of things or actors (people, firms, regions, computers, and so on), and the connections between them. Network analysis is the set of techniques used to statistically describe these systems (Wasserman and Faust, 1994). Analytically, the things or actors in a network are referred to as nodes, and the connections, links. Examples of networks include computer networks, where the nodes are pieces of hardware and the links are wired or wireless connections; social networks where the nodes are people and the links are social connections (for example—friendship, or interaction through a medium such as LinkedIn); citation networks where the nodes are academic papers and the links are citations; or financial exchange networks where the nodes are people or firms, and the links are financial transactions. Network analysis is concerned with measuring the characteristics of these networks.

Network data is gathered in several different ways. Individuals can be asked about who they interact with, and the characteristics of these individual networks can be compared. This type of network is called an 'ego-network', because the focal point is the individual. It is more common now to collect data for all of the members in a particular network, so that the characteristics of the full network can be analysed. One difficulty with this approach is defining the boundary of the network and dealing with or ignoring significant links that cross the boundary. However, the main benefit of full networks is that they enable system-level analysis, which can capture emergent behaviours (Dopfer and Potts, 2006).

Many innovation scholars might ask what is new about network analysis. Social network analytical techniques have been in use in management studies since the 1960s, and they have also been used in innovation research (Allen and Cohen, 1969; Crane, 1969). In this chapter we will not discuss these older techniques and theories that have been based on the analysis of ego-networks. While there are some clear advantages in restricting data collection and analysis to the immediate contacts of an individual, there has been more significant development in the analysis of entire networks. Unlike studies based on ego-networks, analysis of complete networks acknowledges that even dyadic relationships are affected by a much larger system. While ego-networks are more amenable to regression models, they do miss the multilevel interactions between individuals, local networks, and the wider network (Newman, 2010).

The techniques and theories of network analysis have changed dramatically in response to new statistical methods that can examine large networks and also test hypotheses about network structure and dynamics. For example, Watts and Strogatz's (1998) work on small-world networks, where actors are separated by a small number of links in the network, has led to a proliferation of new techniques designed to measure the characteristics and dynamics of large complex networks. This work has led to two important advances in the theoretical underpinnings of network studies (Newman, Barabasi, and Watts, 2006). The first is that there is now an emphasis on studying network dynamics and evolution. In the past, many network studies focused only on data developed at a single point in time, rather than longitudinal data. Recent work has found that the history of a network often has an impact on its future development (Newman, Barabasi, and Watts, 2006). Therefore, it is crucial that the dynamics of a network be properly understood (Barabasi, 2002). The second advance is the development of a perspective that drives researchers not just to investigate the topology of a network, but also to try to understand the agent-level behaviours that create its structure (Watts, 1999). Complex network analysis is an excellent technique for finding the standardized patterns that arise from the idiosyncratic behaviour of heterogeneous agents. This links network analysis with complexity theory, where the macro-structures of networks are emergent properties of agent-level action (Holland, 1995). These new perspectives have combined with the increase in computing power available now to facilitate the analysis of the dynamics of many extremely large and very complex networks. At the same time, development of analytical software such as SIENA, discussed in the section called 'Network Analysis in Innovation Research: The Future', has enabled theory building through the possibility of testing hypotheses on the evolutionary dynamics of particular networks.

There are at least three good reasons to use network analysis for managing innovation. The first is that there is a strong theoretical justification for doing so from the perspective of evolutionary economics, with its focus on innovation. The researchers that have built on Schumpeter's groundbreaking work now view the economy as an evolving complex network (Saviotti, 1996; Potts, 2000; Foster, 2005; Dopfer and Potts, 2006). In this view, economic growth arises through an evolutionary process. The key elements in this process are variation, selection, and retention (Metcalfe, 2005). Innovation leads to variation (Dodgson, Gann, and Salter, 2005), while network connections are the key drivers of the selection and retention processes (Dopfer, Foster, and Potts, 2004). Network analysis is a key tool in evolutionary econometrics. As this view develops, the benefits of modelling the economy as a sparsely connected network are becoming increasingly apparent (Kirman, 1997).

These benefits are based on the premise that the economy and its subsystems are complex adaptive systems. This leads to the second reason to use network analysis in in novation studies. Complex adaptive systems are best analysed as whole systems, rather than as individual parts interacting at arm's length. There are several characteristics that identify complex adaptive systems, but two key ones are that they include large numbers of interacting elements, and that these interactions lead to important emergent properties that cannot be predicted through the study of individual elements (Mitchell, 2009). People within firms, collaboration and exchange networks within firms, and trade between different industries of geographic regions all have these characteristics. Consequently, research that only investigates the nature of individual actors within these systems, whether people or firms, is likely to miss important factors that drive change. Studying these systems as networks is one of the best ways to get around this problem (Barabasi, 2012). The structure of networks is an emergent property of the behaviours of the actors within the system. As Vonartas (2009: 27) says:

> It has become almost a cliché to argue that the behavior and performance of firms can only be understood fully by examining their social, technological, and exchange relationships with other economics agents. The image of atomistic agents competing for profits in impersonal markets has become increasingly inadequate in view of the explosion of inter-firm collaboration the past two to three decades, as well as the growing empirical evidence formally substantiating the influence of the social context in which firms are embedded on their conduct and performance.

The final reason to study innovation through the network lens is more practical: managing the network is often the quickest and most effective way to enact change (Cross, Liedtka, and Weiss, 2004). This makes network knowledge an important managerial tool. At an intuitive level, this makes sense. It is widely acknowledged that communication is of central importance in managing organizational change, and communication, of course, happens through networks. Mohrman, Tenkasi, and Mohrman (2003) show that successful organizational change efforts were primarily distinguished by the activation of informal networks, while those that were less successful relied primarily on formal communication through the normal hierarchy. It is also much easier to intercede to change less formal network structures than it is to reconfigure the official organizational chart (Johnson, 2009).

Network Analysis Basics

The first step in undertaking network analysis is to gather network data. This requires defining the boundaries of the network. A network can be closed or open. Closed networks are those with a clearly definable boundary, such as everyone in one firm, or all of the firms in one industry. An open network does not have pre-defined boundaries. Ego-networks are a common example, where focal actors are asked to name all of their connections of a particular type (e.g. friends or collaborators). Closed networks are easier to analyse, in that you know in advance who the members are. That said, even though the boundaries are defined as closed, in reality nearly all networks are open, so the boundaries can be viewed as somewhat arbitrary. One example of a closed network would be 'all of the people in firm X working on project Y'. If you ask all of the members of the network a question such as, 'Who do you go to for help solving innovation problems?' it is very reasonable to believe that they could go to people outside of this particular group for help, such as ex-colleagues from their previous firms, friends in other industries, or career mentors. Consequently, it is important to think through the boundaries that will be used for any particular network study, and just as important to document the choices and assumptions that underlie the location of these boundaries.

Data on network connections can be gathered through primary or secondary sources. Primary data is usually gathered either through interviews or surveys. Secondary data comes from databases or other large data pools. The most common source of secondary data for network studies in innovation is patent data. These studies use co-authorship on patents as the links between either individual researchers or the firms that file the patents. The main advantage to using secondary data is that you can get larger amounts of data than you can get through interviews or surveys. The drawback is that few of the available data sets actually include data on connections, and without this, there is no network to analyse. In both forms of data collection, it is normal to also collect as much data about the characteristics of the actors in the network. In network analysis terms, these characteristics are referred to as attributes.

Once the data are gathered, they are organized for analysis. The majority of the networks studied in relation to innovation are single-mode. This means that all of the actors in the network can link directly to each other using whatever form of connection is specified. This is in contrast to two-mode, or affiliation networks. These are networks where the focal actors are connected via membership within groups or events. In single-mode networks with N nodes, the data is organized in an NxN matrix. Connections between two actors are indicated by a 1 in the matrix, while those that are unconnected are indicated by a 0. In two-mode networks with N nodes and M groups or events, the data is organized in an NxM matrix, with membership by a particular person in a particular group or event indicated by a 1, and non-membership by a 0. The analysis of the network then usually requires specialized software.

There are a number of tools available for the analysis of networks. In order to utilize these tools, it is necessary to become familiar with some of the terminology that is used in network analysis. A clearly written and comprehensive review is available in Newman (2003), and there are less mathematical treatments in Robins, Pattison, and Woolcock (2005), Barabasi (2002), and Watts (2003).

Network Analysis in Innovation Research: A Brief Review

The network concept has been used in a number of ways in innovation research. At the most basic level, two main questions have been interrogated using network ideas. The first is to look at the number of connections that actors have, based on the premise that more connections are generally good. Bergenholtz and Waldstrom (2011) characterize this perspective as 'metaphor' because the idea of networks is used conceptually. These studies are often highly quantitative, as variables measuring network degree, or the number of connections held by individuals or firms, are easily incorporated into multi-variate analytical approaches. They contrast this with a second perspective, 'analytical', which looks at the overall structure of an innovation network, based on the premise that these collaborative structures will strongly influence innovation outcomes. Both of these approaches occur across multiple levels of analysis. They can look at interpersonal networks between individuals. These people can be within one firm, or across multiple firms. Inter-firm collaboration is another common level of analysis, where a network of firms is analysed. These can be firms within one region, one supply chain, or one indus-try sector. Networks of collaboration between geographic regions can be mapped—these are essentially networks of networks since the actors within a region are clusters or firms that also have a network structure.

Two special issues on the topic of network analysis in innovation studies contain literature reviews (Kastelle and Steen, 2010a; Colombo, Laursen, Magnusson, and Rossi-Lamastra, 2011). The first, by van der Valk and Gijbers (2010), looked at the use of network analysis in innovation studies. They reviewed forty-nine papers from the top ten innovation journals which used the analytical approach. In other words, all of them used social network analysis to measure network properties related to innovation performance. They discovered that the most common applications were to study inter-personal and inter-organizational collaboration networks, and communication net-works, both within and between organizations, and technological and sectoral network structures.

In the second review, Bergenholtz and Waldstrom (2011) start with over 1000 arti-cles that address both networks and innovation. However, the majority of these do not include any actual network data. This narrows down the number of articles that they review to 306. Of these, SNA is only applied in fifty of the articles. The rest use the meta-phor perspective to study the topic. They also find that the most common unit of analysis is the industry network—for example, collaboration networks within the biotech indus-try (see Owen-Smith and Powell (2004) for an example). This is important because one of the key questions to look at in designing networks studies is how to draw the bound-ary of the network. In essence, about two-thirds of the papers reviewed here draw only a loose boundary around the network of study. The rest identify a focal network of study. The papers are split evenly between those that use data from a snapshot in time versus those that look at the evolution of networks over time. About 38 per cent of the papers look at relationships that occur across multiple levels. Very few of the papers, only 6 per cent, compare different networks to each other.

Both literature reviews, as well as both editorials for the special issues conclude that this is a very young field of study. But there are efforts to help shape the field. Mitzenmacher's (2005) proposal for a new research programme in complex network research contains five levels of analysis: *observing* or measuring a complex network; *interpreting* the importance of the network characteristics to the behaviour of the system; developing a *model* that underlies the development of the network; *validating* the model with empirical data or observations; and using the model to *control* or improve the behaviour of the system. He further states that, while the steps of validation and control are probably the most important, the majority of complex network studies focus only on the first three levels of observing, interpreting, and modelling.

The vast majority of innovation research to date has only looked at the first two levels: observing and interpreting. In many fields, it has been sufficient to identify that a particular network has a small world structure, or a power law degree distribution (where most of the network is dominated by a relatively small number of actors). The move to interpretation occurred relatively recently, not just in innovation research, but in most of the fields that utilize SNA. Given how new many of the analytical techniques are, it is not surprising that research has not yet moved much beyond these basic objectives. However, it is time now to move into the more challenging and sophisticated uses of this methodology to assist innovation management.

Before outlining how this might be done, we evaluate the current state of the use of networks in innovation studies: what do we currently know? Network analysis has shown that network structures are a significant determinant of innovation outcomes, and this has been summarized by Steen and Macaulay (2012) (Table 6.1).

Thematically, there is general support for the proposition that networks that enable the connection of different knowledge sets result in more innovation. This can be seen in studies that find evidence for small world networks being more innovative. For example, Uzzi and Spiro's (2005) research into small worlds in the Broadway musical industry showed that blockbuster musicals followed periods where the collaboration network of writers, choreographers, and librettists could be statistically described as a small world. Also, there is evidence from ego-network studies that shows how actors bridging gaps in the network, or 'structural holes', have an improved innovation performance compared to their peers in other parts of the network. Similarly, studies of the role of weak ties, where actors are in relatively infrequent contact with each other, show how bridging disparate groups can support innovation.

It has been noted that not all studies have found a relationship between weak ties, structural holes, and innovation. However, a recent study of networks in a US executive recruitment business may offer an explanation for these disparate results (Aral and Van Alstyne, 2011). If we accept the premise that connecting different skills and ideas is a precondition for innovation, then spanning structural holes isn't the only way that this can happen. When there is a lot of change in the network environment, this diversity of new connections is happening anyway and in this case a well-connected network will result in more innovation. Similarly, when there is a lot of change and uncertainty, strong ties will outperform weak ties because keeping track of the changes in the network and knowing where knowledge resides is more important than maintaining diverse pockets of knowledge.

Given that we know so much about the relationship between network structures and innovation performance, the practical question is how we shape the networks to create these optimal configurations. This remains a challenge, but one guiding principle

Table 6.1 Key network measures and links to innovation performance (Steen and Macaulay, 2012)

Concepts	Common Network Measures	Relevant Authors	Link to Innovation
Informal power	Centrality	Freeman (1979); Bonacich (1987)	Power provides actors with better access to and control over resources. Actors with higher centrality can leverage these advantages to improve innovation performance.
Strength of ties	Frequency of interaction. Frequent interaction produces strong ties	Granovetter (1973)	Strong ties are likely to communicate redundant information whereas weak ties convey novel information. Strong ties are thus contexts for exploitation, with weak ties being sources of exploration (March, 1991).
Social capital: Structural holes	Constraint measures: degree to which an actor's ties are non-redundant.	Burt (1992)	An individual who spans multiple social worlds is able to benefit from transferring information and insights between these contexts. A structural hole describes the situation where an actor not only spans these social worlds, but is spanning otherwise poorly connected worlds. The diversity of information resulting from this structural position puts these actors at a distinct advantage. Improved innovation performance is one of many outcomes that result.
Social capital: Closure	Density	Coleman (1990)	The density of relations within a social network improves coordination and reduces exchange risk. Organizations (e.g. project teams; firms) with these structural features are more likely to succeed when engaging in innovation.

for building networks is known as 'the law of propinquity' (e.g. Reagans, 2011). This is the principle that the probability of tie formation is related to physical proximity and has been demonstrated in numerous studies. Building networks requires people to be in contact with each other, and this has important implications for work environments and job design.

One structure that has often been identified in innovation networks is a power law degree distribution, particularly in those networks using large secondary datasets. One of the attractions of identifying a power law distribution is that many believe that this indicates something about the generative mechanisms of the network. This is because the first power law networks identified showed strong evidence of following a 'rich-get-richer' pattern in forming new connections, referred to as 'preferential attachment' (Albert and Barabasi, 2002). This form of distribution occurs when the chances of a new connection within the network being formed with a particular agent are proportional to the number of connections that the agent already has. In other words, new connections are more likely to go to actors that are already well connected. The story is very clean then: if a degree distribution shows a power law pattern, then the driver of network evolution must be preferential attachment.

There are, however, a few problems with this story. The first is that most of these distributions are not actually power law distributions. When power law degree distributions are graphed on a log-log scale, they show up as a straight line. The techniques used to fit empirically observed distributions against a best-fit trend line are fraught with difficulty. Gallegati, Keen, Lux, and Ormerod (2006:3) describe the problem:

> ... there is no reason to believe that one should find simple power laws in all types of socio-economic data that are universal over countries and time horizons. Although their interpretation of power laws as signatures of complex, possibly self-organizing systems makes them a much wanted object, one should be careful in not seeing a power-law decline in each and every collection of data points with a negative slope.

Clauset, Shalizi, and Newman (2007) performed a meta-analysis of twenty-four papers which purported to find power law distributions in their data, drawn from a number of disciplines. Most of the papers that fit the degree distribution to a power law curve use linear regression. This is an error, as the strong interdependency of the variables violates the assumptions of the method (Gallegati et al., 2006). Instead of this, Clauset, Shalizi, and Newman use the more appropriate Kolmogorov–Smirnov test and the Hill Estimator to fit the empirical data to theoretical curves. The outcome of this is that the actual distributions in all but three are conclusively not power laws, and the three that might be still do not show a great fit. The majority of the data sets are better described by a log-normal distribution.

The second problem with the story is that, in reality, there are many ways that a power law distribution can be generated. Even if one is correctly identified, this does not guarantee that preferential attachment is the generative mechanism of the distribution. There are many network growth mechanisms that can lead to power law, log-normal or other fat-tailed distributions (Andriani and McKelvey, 2007).

Two points arise from this discussion. The first is cautionary: the casual application of methodologies from other disciplines must be undertaken with care. The physics-based network studies have placed a great deal of emphasis on finding a small number of mechanisms that drive the evolution of networks. As is the case with preferential attachment, finding such mechanisms can lead to simple, clean stories about the evolution of the system being studied. However, it is often not the case that human systems that demonstrate these complex structures are acting in the same way as other systems such as the World Wide Web do as they evolve (Gallegati et al., 2006).

The second point helps find a way forward: the time has indeed come to start looking for the micro-level behaviours that drive the evolution of the macro-structures that we see in networks. Doing this will lead to the higher levels in Mitzenmacher's hierarchy: the development and verification of models, and the normative guidelines for effectively managing innovation networks. We now turn to the approaches that can be used to achieve these ends.

NETWORK ANALYSIS IN INNOVATION RESEARCH: THE FUTURE

The shortcomings that are identified in the review by Bergenholtz and Waldstrom (2011) reflect the current state of knowledge. There are four analytical approaches that are relatively new, which have not been widely used in innovation studies, and which all have the potential to overcome these shortcomings and help connect individual actions and collective structures. These approaches are Exponential Random Graph Models (ERGM), and longitudinal, multilevel, and weighted network analysis.

All four methods are on the cutting edge of network analysis methodologies. Consequently, they open up new opportunities for developing insights from network data, which may lead to more effective management of innovation networks. The drawback with these techniques is that because they are new, there is not yet a consensus on how best to use them, or even on what some of the basic network measures mean.

ERGM (exponential random graph models) is the most refined of these approaches, and consequently the one that is most able to be used more widely in innovation studies. The phrase sounds complex, but the idea behind it is relatively simple. It starts by measuring the network that you are studying, taking some of the basic network statistics, such as density, average degree, and clustering. If you have a network with 300 nodes, and an average degree of 10, the number of possible networks that can be built with those basic characteristics is enormous. ERGM answers the question: out of all of those possible networks, how likely is it that the structures observed in our actual network have arisen by chance?

ERGM is used to test hypotheses about the relative importance of various network measures within the network. Typically, the ones that are identified as most important will be split evenly between those based on network position (e.g. reciprocity, or centrality) and those based on the attributes of the actors in the network. To date, the technique has been used most frequently on secondary data sets. For example, several studies have been done using the extensive banking and social records collected on the Medici in Florence during the Renaissance. These studies have been useful in identifying where and how ERGM work best, and developing an understanding of which network structures are important and why.

One study of virtual collaboration, for example, has shown that reciprocity is a significant mechanism in explaining network structure where the most active contributors to the community also received more help in return. Contrary to previous assumptions, preferential attachment, where popular members received more responses, was not a significant driver of network dynamics (Faraj and Johnson, 2010). In another example, Lomi and Pattison (2006) examine the relationships between manufacturing businesses in southern Italian industry and find that the high degree of relational embeddedness that supports technology transfer is revealed in the network being based upon triangle structures between actors. These triadic structures have been show to support higher levels of trust in the network through governance of exchange relationships via third parties (Jones, Hesterly, and Borgatti, 1997; Molm, Schaefer, and Collett, 2009).

Bergenholtz and Waldstrom (2011) show that nearly half of the network studies they evaluated have a longitudinal component, although it should be remembered, however, that nearly all of these use the network as metaphor perspective. Such studies measure the evolution of some network variables (typically the number of connections of the actors), and then use regression analysis to map the importance of these variables in innovation outcomes of some sort. An excellent example of this approach is shown in Laursen and Salter's (2006) investigation of the determinants of open innovation success. The key finding in the paper is that when the number of open innovation sources (network connections) is graphed against innovation performance, there is an inverse-U shape. Increasing network connections improve performance up to a point, after which performance declines, with further increases in connections. The longitudinal aspect comes from used lagged performance variables.

While this approach leads to important insights into the innovation process, new longitudinal network analysis techniques, also using ERGM-based tools, enable the investigation of a different set of questions. The most common and best-developed methodology in this category is Simulation Investigation for Empirical Network Analysis (SIENA) (Snijders, 2001). The premise behind SIENA is that network structure and actor behaviour co-evolve. SIENA is an analytical programme designed to measure the factors that influence the evolution of networks over time (Snijders, Steglich, Schweinberger, and Huisman, 2007). The method is based on stochastic modelling of evolution of tie structure within a network over time. It measures the factors that influence the formation of new ties (and the breaking of existing ones), including both those that are based on network structure (e.g. clustering, degree) and also the attributes of the actors in the network.

This is an actor-oriented model, which means that, in deciding to change its outgoing tie variables $(X_{i1}, \ldots, X_{ij})$, the actor tries to make changes that result in the network configuration x that provides the highest expected utility. In the case of a collaboration network, this means that an actor makes a new connection with the expectation that this will help improve their collaboration outcomes and probably also their economic performance.

These analyses start with sampling the same network at multiple points in time. The technique allows for actors to enter and leave the network, and works best with turnover of 20 per cent or less in between samples. As in ERGM, the evolution of the network is then simulated over thousands of iterations. This again provides statistically significant results, which have the added benefit of showing the relative importance of the different network statistics that are included in the analysis.

The third new methodology is multilevel network analysis. There is a need for theories of innovation that connect individuals to groups and to higher levels of analysis including organizations and industries. Both ERGM and SIENA are able to include actor-level variables in their models to examine the effect of individuals on the network. Ohly, Kase, and Skerlevaj (2010), for example, showed that seniority significantly affected the structure of idea realization networks in a software development company. When employees needed to act on new ideas, supervisors became involved in the network. While this result is not surprising, senior managers were not significant in determining the structure of the idea generation network.

Finally, there is weighted network analysis. All of the methods that we have described so far are based on binary networks. A connection between two actors is either present or absent. In situations where the networks have never been mapped at all, there is a great deal of insight that can be gained using binary approaches. However, one weakness in these is that it is very hard to capture the quality of connections between actors. One of the landmark network analysis studies is Granovetter's (1973) investigation of how people found new jobs. He measured ego-networks, and he asked people how well they knew the others in their networks. These connections were then sorted into strong and weak ties. The surprising result was that people were much more likely to get a lead on a new job from weak ties in their network than from strong ties. The reason for this is that because the weak ties had fewer duplicate connections with the focal actors, they were able to bring in information from a wider variety of people.

This is an example of the kind of insights that can be gained by using weighted measures of connections between actors. In these studies, data must be gathered on the weight of connections between actors. These connections can have multiple discrete values, as in the Granovetter study (connections were absent, weak, or strong). They can also be continuous variables, such as measures of monetary exchange.

At the present time, the big issue with both multilevel and weighted network analysis is that the meaning of the different variables is still not clear. This makes it much harder to build models, and to make normative prescriptions. Here is an example from weighted network analysis. What constitutes reciprocity? If I get twenty new ideas from you, and you get no new ideas from me, then the relationship is clearly not reciprocal. But what if the ratio is 20:1 instead? The issues become even more complex when considering triangles, such as clustering measures.

The opportunity with these methods is that when they are applied well to a new problem, occasions arise to make a significant methodological contribution in addition to the work that is at the core of the research. Granovetter's study is so well known, not because the insight was so novel (although it was certainly interesting), but because he introduced a new way to conceptualize network problems.

Managing Innovation Networks

There are several important management implications that arise from this. To some extent, these can be acted upon even if the network within a particular innovation system (firm, industry, region) has not been formally measured. Viewing management challenges through a network lens can provide useful insights. However, the most effective way to manage networks is to measure them, design an intervention, and measure the outcomes.

Consider the network shown in Figure 6.1, described in detail in Kastelle and Steen (2010b). It is a network from an engineering company. The data was collected as part of a study investigating the impact that knowledge-sharing network structures have on innovation in project-based firms. The network includes 134 people working on a processing plant design project. The team was split across two locations, the darker circles were people based in Brisbane, and the lighter in Perth. The size of the circle indicates the person's rank in the firm, with top-level managers the largest circles, followed by

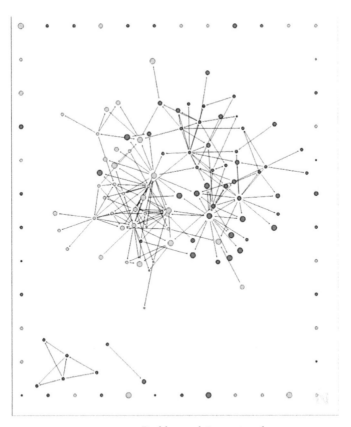

FIGURE 6.1 Problem-solving network

middle managers, engineers, and administrative staff the smallest. This network was drawn based on the team's responses to the question, 'Who gives you help in solving work problems?'

Some of the easily observable characteristics of this network illustrate how network analysis can be used to address important innovation issues.

- *Identification of silos*: one of the most obvious features is the geographical split within the network. This reflects the impact of propinquity—people were much more likely to be connected to others in the same office than to those in other locations. While this may not be surprising in a project working across a distance of 3600 kilometres, similar structures often arise based on differences in which floor of the building people work on, areas of expertise or specialization, educational background, and so on. Network analysis is a powerful tool for identifying silos. This is significant because of the importance of diversity in idea generation and innovation.
- *Finding hubs and other key actors*: network analysis can find people who are highly connected (hubs), those that span two groups (bridges), and those that carry a high information load. This is particularly useful in that the importance of a person in the network does not necessarily correlate with their status within the organizational hierarchy. In the engineering sample shown, however, most of the key players actually are high-level managers. This raises questions about whether or not problem solving is the best use of their time, and also about how engineers are able to effectively communicate with each other, since these conversations appear to be mediated by the organizational hierarchy.
- *Locating isolates*: there are two types of isolates in this network. The people around the edge apparently are not involved in problem solving at all, while the two groups in the bottom left corner only work amongst themselves. While most analysis is done at the level of the network, it is possible to evaluate the role of individuals and groups as well. In this case, the isolated groups were based on unique functional responsibility within the project.
- *Identifying bottlenecks*: the highly connected actors within a sparsely connected network such as this play several roles. This network has a skewed degree distribution, indicating that the most connected people are genuine hubs. This means that they have a disproportionately high influence on the overall functioning of the flow of knowledge through the network. One disadvantage that arises from this type of structure, particularly when the hubs are mostly people who are relatively high in the hierarchy, is that they actually become bottlenecks. When this happens, the key actors can inhibit the flow of information rather than facilitate it.

Those are just the issues that are obvious from a visual inspection of the network map. Utilizing some of the more sophisticated techniques that we have discussed can yield further insights.

The final point to consider, then, is how to effectively intervene in response to the findings from a network analysis. This is moving to the highest level in Mitenmacher's hierarchy—using analysis to facilitate control. The first issue to consider is the trade-offs involved in adding connections within a network. In most cases, when managers decide that their network structures are ineffective, the logical intervention is to add connections. This is the way to span structural holes, for example. However, network ties are expensive to maintain. Personal ties require the investment of time, effort, and emotional investment. Even ties based strictly on information sharing require time and attention: two resources that are scarce within most organizations. It is impossible to connect everyone with everyone. The data load would be overwhelming. This is why networks structure is so important.

When adding connections it is important to keep the trade-offs discussed in mind. Optimal network structures will differ depending on the task being undertaken by the network. The network above is a reasonable structure for an idea execution network, but idea generation networks tend to need less hierarchical structures, with a more evenly balanced distribution of connections among actors. These considerations provide guidance to those managing innovation networks.

Conclusions

Ideas start as networks in the brain, and then develop and come to life within networks of people and firms. The new network analysis methodologies discussed here provide significant opportunities to gain new insights into this process. They can be used to evaluate how the networks that are central to innovation development and diffusion evolve.

In Chapter 1 by Dodgson, Gann, and Phillips three current issues in innovation management are identified: defining its scope, managing the changing nature of the innovation process, and merging disciplines, levels of analysis, and research methods. The network perspective can help address all three areas.

In scope, innovation is best viewed at a systemic level. The linear models of innovation tend to miss important interactive factors, such as feedback loops, co-creation, and the non-linear nature of idea diffusion. Analysing innovation networks can provide important insight into all of these issues. Those responsible for managing innovation within organizations cannot afford to deal only with internal processes. Their systems and approaches must allow for interacting with the complex environment in which they operate. This involves dealing with multiple actors all the way through the innovation process: from idea generation to execution and through to diffusion. Each of these activities can be analysed at a network level. Furthermore, gaining a network understanding is critical to managing the process, because different network structures are required to be effective in each activity. The scope of innovation management needs to include network thinking, even if it does not include formal network analysis. One major benefit of formal analysis is that it can help turn network capital from an intangible asset into a more tangible one.

The changing nature of the innovation process can be read as a network story as well. The different combinations of resources and actors reflect different network configurations. An organization using a research and technology push strategy will be structured differently from one using internal coupling. The more recent types of configuration, external collaboration, strategic integration, and future ready, all imply different network configurations again. In considering issues such as whether or not R&D should be centralized or decentralized, or how to best manage distributed open innovation, managers are grappling with network questions. Again, thinking about networks metaphorically can be a significant aid in building the organizational structures required to support answers to these questions. In order to understand which structures work best in which circumstances, researchers will need to use network analysis. We can then use the results from such research to develop normative guidelines for managers.

With respect to Dodgson, Gann, and Phillips' final point, network analysis is by its very nature cross-disciplinary and multi-method. It is an excellent tool also for analysing multilevel systems. Because the approach is derived from complexity science, it is a method that is well suited to analysing complex adaptive systems such as a firm, or an alliance network—anything that is based on interactions between people. Innovation drives growth in such systems, and network analysis is an important tool for developing an understanding of precisely how this happens.

Innovation happens in networks—networks of the brain, networks of people, and networks of firms. The management of innovation can be greatly enhanced through gaining an understanding of how networks work, how they are built, and how they are best managed. Network analysis is not a panacea, but network understanding and network concepts are important tools, both for those studying innovation, and for those managing innovation.

References

Albert, R., and Barabasi, A.-L. (2002). 'Statistical Mechanics of Complex Networks', *Review of Modern Physics*, 74: 48–97.

Allen, T. J., and Cohen, S. I. (1969). 'Information Flow in Research and Development Laboratories', *Administrative Science Quarterly*, 14(1): 12–19.

Andriani, P., and McKelvey, B. (2007). 'Beyond Gaussian Averages: Redirecting Management Research Towards Extreme Events and Power Laws', *Journal of International Business Studies*, 38(7): 1212–30.

Aral, S., and Van Alstyne, M. (2011). 'The Diversity-Bandwidth Tradeoff', *American Journal of Sociology*, 117(1): 90–171.

Barabasi, A.-L. (2002). *Linked*. Plume/Penguin: New York.

Barabasi, A.-L. (2012). 'The Network Takeover', *Nature Physics*, 8(1): 14–16.

Bergenholtz, C., and Waldstrom, C. (2011). 'Inter-Organizational Network Studies: A Literature Review', *Industry and Innovation*, 18(6): 539–62.

Bonacich, P. (1987). 'Power and Centrality: A Family of Measures', *American Journal of Sociology*, 92(5): 1170–82.

Burt, R. S. (1992). *Structural Holes: The Social Structure of Competition*. Cambridge, MA: Harvard University Press.

Clauset, A., Shalizi, C. H., and Newman, M. E. J. (2007). 'Power-Law Distributions in Empirical Data', arxiv:0706.1062v1.

Coleman, J. S. (1990). *Foundations of Social Theory*. Cambridge, MA: Belknap Press of Harvard University Press.

Colombo, M. G., Laursen, K., Magnusson, M., and Rossi-Lamastra, C. (2011). 'Organizing Inter- and Intra-Firm Networks: What is the Impact on Innovation Performance?', *Industry and Innovation*, 18(6): 531–8.

Crane, D. (1969). 'Social Structure in a Group of Scientists: A Test of the "Invisible College" Hypothesis', *American Sociological Review*, 34(3): 335–52.

Cross, R., Liedtka, L., and Weiss, L. (2004). 'A Practical Guide to Social Networks', *Harvard Business Review*, March: 124–32.

Dodgson, M., Gann, D., and Salter, A. (2005). *Think, Play, Do: Technology, Innovation and Organization*. Oxford: Oxford University Press.

Dopfer, K., Foster, J., and Potts, J. (2004). 'Micro Meso Macro'. *Journal of Evolutionary Economics*, 14: 263–79.

Dopfer, K., and Potts, J. (2006). *The General Theory of Economic Evolution*. New York, NY.: Routledge.

Faraj, S., and Johnson, S. L. (2010). 'Network Exchange Patterns in Online Communities', *Organization Science*, published online before print, 29 December: 1–17. See <http://orgsci.journal.informs.org/content/early/2010/12/29/orsc.1100.0600.abstract>

Foster, J. (2005). 'The Self-Organizational Perspective on Economic Evolution: A Unifying Paradigm', in Dopfer, K. (ed.), *The Evolutionary Foundations of Economics*. Cambridge: Cambridge University Press, 367–90.

Freeman, L. C. (1979). 'Centrality in Social Networks: Conceptual Clarification', *Social Networks*, 1(3): 215–39.

Gallegati, M., Keen, S., Lux, T., and Ormerod, P. (2006). 'Worrying Trends in Econophysics', *Physica A*, 370: 1–6.

Granovetter, M. (1973). 'The Strength of Weak Ties', *American Journal of Sociology*, 78(6): 1360–80.

Holland, J. H. (1995). *Hidden Order: How Adaptation Builds Complexity*. Cambridge: Perseus.

Johnson, J. D. (2009). *Managing Knowledge Network*. Cambridge: Cambridge University Press.

Johnson, S. (2010). *Where Good Ideas Come From: The Natural History of Innovation*. New York, NY: Riverhead Books.

Jones, C., Hesterly, W. S., and Borgatti, S. P. (1997). 'A General Theory of Network Governance: Exchange Conditions and Social Mechanisms', *Academy of Management Review*, 22(4): 911–45.

Kastelle. T., and Steen, J. (2010a). 'Using Network Analysis to Understand Innovation', *Innovation: Management, Policy & Practice*, 12(1): 2–4.

Kastelle. T., and Steen, J. (2010b). 'Are Small World Networks Always Best for Innovation?', *Innovation: Management, Policy & Practice*, 12(1): 75–87.

Kirman, A. (1997). 'The Economy as an Evolving Network', *Journal of Evolutionary Economics*, 7: 339–53.

Laursen, K., and Salter, A. (2006). 'Open for Innovation: The Role of Openness in Explaining Innovation Performance among U.K. Manufacturing Firms', *Strategic Management Journal*, 27: 131–50.

Lomi, A., and Pattison, P. (2006). 'Manufacturing Relationships: An Empirical Study of the Organization of Production across Multiple Networks', *Organization Science*, 17(3): 313–32.

Metcalfe, J. S. (2005). 'Evolutionary Concepts in Relation to Evolutionary Economics', in Dopfer, K. (ed.), *The Evolutionary Foundations of Economics*. Cambridge: Cambridge University Press, 391–430.

Mitchell, M. (2009). *Complexity: A Guided Tour*. Oxford: Oxford University Press.

Mitzenmacher, M. (2005). 'The Future of Power Law Research'. *Internet Mathematics*, 2(4): 525–34.

Molm, L. D., Schaefer, D. R., and Collett, J. L. (2009). 'Fragile and Resilient Trust: Risk and Uncertainty in Negotiated and Reciprocal Exchange', *Sociological Theory*, 27(1): 1–32.

Mohrman, S. A., Tenkasi, R. V., and Mohrman Jr., A. M. (2003). 'The Role of Networks in Fundamental Organizational Change: A Grounded Analysis', *The Journal of Applied Behavioral Science*, 39(3): 324–36.

Newman, M. E. J. (2003). 'The Structure and Function of Complex Networks', *SIAM Review*, 45: 167–256.

Newman, M. E. J. (2010). *Networks*. Oxford: Oxford University Press.

Newman, M. E. J., Barabasi, A.-L., and Watts, D. J. (2006). *The Structure and Dynamics of Networks*. Princeton: Princeton University Press.

Ohly, S., Kase, R., and Skerlevaj, M. (2010). 'Networks for Generating and Validating Ideas: The Social Side of Creativity', *Innovation: Management, Policy & Practice*, 12(1): 41–52.

Owen-Smith, J., and Powell, W. W. (2004). 'Knowledge Networks as Channels and Conduits: The Effects of Spillovers in the Boston Biotechnology Community', *Organization Science*, 15(1): 5–21.

Potts J. (2000). *The New Evolutionary Microeconomics: Complexity, Competence and Adaptive Behaviour*. Cheltenham: Edward Elgar.

Reagans, R. (2011). 'Close Encounters: Analyzing How Social Similarity and Propinquity Contribute to Strong Network Connections', *Organization Science*, 22(4): 835–49.

Robins, G., Pattison, P., and Woolcock, J. (2005). 'Small and Other Worlds: Global Networks Structures from Local Processes', *American Journal of Sociology*, 110(4): 894–936.

Saviotti, P. P. (1996). *Technological Evolution, Variety and the Economy*. Cheltenham: Edgar Elgar.

Schumpeter, J. A. (1911/1983). *The Theory of Economic Development*. New Brunswick: Transaction Publishers.

Snijders, T. A. B. (2001). 'The Statistical Evaluation of Social Network Dynamics', in M. E. Sobel and M. P. Becker (eds.), *Sociological Methodology—2001*, London: Blackwell, 361–95.

Snijders, T. A. B., Steglich, C. E. G., Schweinberger, M., and Huisman, M. (2007). *Manual for SIENA version 3.2*, Oxford: University of Oxford Department of Statistics.

Steen, J., and MacAulay, S. (2012). 'The Past, Present and Future of Social Network Analysis in the Study of Innovation', in D. Rooney, G. Hearn, and T. Kastelle (eds), *Handbook on the Knowledge Economy, Volume Two*, Cheltenham: Edward Elgar Publishing, 216–37.

Uzzi, B., and Spiro, J. (2005). 'Collaboration and Creativity: The Small World Problem', *American Journal of Sociology*, 111(2): 447–504.

van der Valk, T., and Gijbers, G. (2010). 'The Use of Social Network Analysis in Innovation Studies: Mapping Actors and Technologies'. *Innovation: Management, Policy & Practice*, 12(1): 5–17.

Vonartas, N. S. (2009). 'Innovation Networks in Industry'. In F. Malerba and N. S. Vonartas (eds), *Innovation Networks in Industry*. Cheltenham: Edward Elgar Publishing, 27–44.

Wasserman, S., and Faust, K. (1994). *Social Network Analysis: Methods and Applications.* Cambridge: Cambridge University Press.

Watts, D. J. (1999). *Small Worlds: The Dynamics of Networks between Order and Randomness.* Princeton: Princeton University Press.

Watts, D. J. (2003). *Six Degrees: The Science of a Connected Age.* London: William Heinemann Ltd.

Watts, D. J., and Strogatz, S. H. (1998). 'Collective Dynamics of Small-World Networks', *Nature*, 393: 440–2.

CHAPTER 7

KNOWLEDGE AND THE MANAGEMENT OF CREATIVITY AND INNOVATION

DOROTHY LEONARD AND MICHELLE BARTON

INTRODUCTION

KNOWLEDGE has a paradoxical but critical relationship with both creativity and innovation: it is essential to creativity and innovation and yet (under certain conditions) also inimical to them. That is, knowledge both gives birth to innovative ideas and can also kill them. This chapter explores the management of that paradox at three levels: organizational, group, and individual. Before turning to that discussion, we need to provide some definitions.

Definitions

Knowledge

The term knowledge is often confused with either *data*, which Davenport and Prusak (1998: 3) define as 'discrete, objective facts about events' or, even more commonly, *information*, which the same authors define as 'a message...meant to change the way the receiver perceives something....data that makes a difference' (1998: 4). In contrast, *knowledge* is information that is 'relevant, actionable, and at least partially based on experience. It implies an understanding of processes, situations and interactions, and includes both skills and values. Knowledge may derive from science, history, structured education and vicarious as well as personal experience' (Leonard, 2011: xiv). Because it has an experiential component, knowledge has tacit dimensions, that is, unarticulated and sometimes unconscious aspects (Polanyi, 1966). Those tacit dimensions are particular to the individual, making them both potentially more competitively valuable (Spender, 1996) but also more difficult to convey to others.

These distinctions and definitions are important. One of the reasons that knowledge management systems have often failed to spur innovation is that they have focused more on delivering data or information than on providing access to knowledge. Moreover, as the above description implies, knowledge may be embodied in products and services, in equipment and organizational processes in employees' heads—even in norms of behaviour.

While both explicit and tacit knowledge may originate outside organizational boundaries as well as inside, we will focus in this chapter mostly on the knowledge inside corporations, leaving issues of outside sourcing and transfer across organizational boundaries to other essays.

Creativity

It is also important at the outset to specify what is meant by creativity. Usually conceived of as a personal, innate characteristic (e.g. Simonton, 1999), creativity can also characterize groups or teams (e.g. Kurtzberg and Amabile, 2000), and even organizations (e.g. Woodman et al., 1993). Writing about it in a business context, Leonard and Swap (1999: 6) define creativity as 'a process of developing and expressing novel ideas that are likely to be useful'. Managers have more influence—both positive and negative—over the expression of creativity around them than they might think. The stimulation of creativity gives rise to innovation in business, that is, the 'embodiment, combination, and/or synthesis of knowledge in novel, relevant, valued new products, processes or services' (Leonard and Swap, 1999: 7).

In the following pages, we discuss the interaction of knowledge with creativity and innovation at the previously mentioned three levels in the organization, while fully realizing that these are artificial bounds. Our contention is that organizations operate as fractals:[1] that is, the culture, norms, behaviours, and processes that are encouraged or discouraged at the organizational level are often replicated at the group and even individual level. Moreover, as discussed below, the attitudes and behaviours of individuals influence the creativity of the collective—especially when those individuals are in positions of leadership. Thus, the larger unit influences the smaller, and vice versa. In the sections that follow, we explore the positive and negative effects of knowledge at each of the three levels.

ORGANIZATIONAL LEVEL

Core Capabilities Underlie Innovation

Knowledge as the basis for organizational competitiveness and innovation has a venerable history. In 1964 Peter Drucker wrote, 'It is only in respect to knowledge that a business can be distinct, can therefore produce something that has a value in the marketplace'

(Drucker, 1964: 5). Researchers and managers alike recognized that the gap between the book value of a company and its market value could be explained largely in terms of the underlying deep knowledge that produced streams of innovations (Rumelt, 1986). Such knowledge was researched under various names, for example, distinctive (Snow and Hrebiniak, 1980) or organizational competencies (Hayes, Wheelwright, and Clark, 1988), invisible assets (Itami and Roehl, 1987), firm-specific competence (Pavitt, 1991), dynamic capabilities (Teece, Pisano, and Shuen, 1997), and core capabilities (Leonard-Barton, 1992). In a detailed study of paired successful and failed projects in five very different companies (Ford Motor, Chaparral Steel, Hewlett-Packard, and two others identified simply as Chemicals and Electronics), four dimensions of core capabilities were identified: (1) employee knowledge and skills; (2) technical systems that embodied proprietary knowledge; (3) managerial systems, that is, processes of knowledge creation and control; and (4) values and norms associated with the knowledge and its creation. Several early books on knowledge and innovation explored at depth both the nature of such knowledge and also how it was created (Leonard-Barton, 1995; Nonaka and Takeuchi, 1995). Corporations began making concerted efforts to manage their knowledge better, for example by appointing 'knowledge officers'. The growth in such practices has been paralleled by continued research interest (Eisenhardt and Martin, 2000). There are obvious benefits to be gained from exploiting existing knowledge assets. Some corporations have a well-deserved reputation for exploiting their in-house capabilities in the service of innovation. Their culture and managerial practices encourage both the creation and reuse of knowledge. A close examination of their practices reveals that inventions, that is, knowledge originally new to the world, has been converted in successive projects into a stream of derivative products—usually over a long period of years, and often in products far removed from the original market.

Consider, for example, a 3M technology platform that traces its roots back to an invention in the 1930s. Researcher Al Boese ran late-night experiments on alternative uses for a machine that was used during the day to heat and knead rubber used in the manufacture of adhesive tape. The machine itself exemplified knowledge the company already had about kneading and flattening substances, but Boese wanted to see if the machine could also bind fibres. He fed clumps of cellulose acetate fibres through the machine rollers. The technique produced a novel, non-woven fibre-based material, but then it took almost ten years before this knowledge was applied to produce a marketable application. After almost killing the project twice, management gave Boese three months to identify a viable new product. A bit more tinkering, and Sasheen decorative ribbon was launched. 3M sold a quarter of a million yards in the first year. More importantly, however, the ability to produce the non-woven material led to other product streams: floppy disk liners, insulation tape, Scotch-Brite scrubbing pads, mesh that could be used to clean up oil spills and after, another researcher, Dave Braun, created a way to widen the web produced, a very big product line: respirator masks. Constant refinement led to many derivatives. When particles were incorporated into the microfibres, the masks had better filtering properties. Next, 3M engineers devised ways to electrify the fibres so that they attracted dust, producing lighter, more comfortable masks that worked six times better. The masks were used in medical sites and manufacturing as well as in residences (Gundling, 2000). When the US economy declined steeply in 2008–09, 3M CEO George Buckley challenged the teams to come up with cost-reducing innovation. Another derivative product resulted: the ultra-low-cost respirator mask. This huge stream of products all flowed from the original invention, with many individual contributions of knowledge along the way—including knowledge of the markets.

Some research using patent data suggests that the flow of knowledge across divisional boundaries in diversified firms results in even more innovation than such flows within divisions (Miller, Fern, and Cardinal, 2007). Unfortunately, however, this potential advantage for multidivisional firms is often unrealized because knowledge is 'sticky'—difficult to move across organizational boundaries (Szulanski, 1996). However, knowledge can also be 'leaky'—moving across boundaries unintentionally. There are many examples of the latter in the innovation literature. One of the most famous is Steve Jobs' 1979 visit to Xerox's Palo Alto Research Center (PARC). Immediately recognizing the value of the graphical user interface and the mouse in PARC's prototype Alto computer, he set about replicating this technology and incorporating it into the Macintosh computer. Xerox reaped no tangible benefits from this knowledge transfer—mostly because the larger organization found it impossible to integrate PARC's innovations. There was no clear fit between a lot of what PARC scientists produced and Xerox product lines at the time. Today graphical user interfaces are common for everything from duplicating machines to cars—but at the time, there was no fit with Xerox core capabilities.

Core capabilities of organizations are part of the institution's taken-for-granted reality—an accretion of decisions made over time, accumulated behaviours and beliefs that reflect the history of the organization and its founders. If the competitive, social, and technological environment in which organizations function remained stable, these core capabilities would sustain success indefinitely, and path dependency (the strong influence of prior decisions and actions on current ones) would matter much less. However, as conditions change—alterations in market, technology, demography, social, and political environments—these core capabilities function as core rigidities (Leonard-Barton, 1992, 1995). Excellence and expertise in one domain inevitably mean less attention to, and knowledge about, other markets, technologies, and processes.

Core Rigidities Inhibit Innovation

The knowledge that has led to an organization's success makes two important innovative activities difficult: exploration and renewal (Henderson and Cockburn, 1994; Doz, 1997). In fact, Christensen and Raynor (2003: 177) write: 'most often the very skills that propel an organization to succeed in sustaining circumstances systematically bungle the best ideas for disruptive growth'. Because of their superior fit in nascent markets, 'disruptive' innovations favour new entrants over incumbent leaders in an industry. Thus, the old guard's failure to compete arises less from a lack of technical knowledge than from strong ties to a set of customers using the old technology and inability to comprehend the new market needs.

In such cases, several forms of knowledge stand in the way of innovation besides familiarity with the old market. Sets of skills that have been highly valued in the firm and whose practitioners have risen in power within the organization because of those skills become relatively unimportant (e.g. polymer chemistry in film companies such as Kodak, Fuji Film, Polaroid). It is not easy to swap out entire departments based on the knowledge that has made the company successful, for new skills. How does a manager even know how to set up criteria for hiring?

Less visible, but at least as potent a deterrent to innovation are managerial assumptions that masquerade as certain knowledge about the determinants of success. When an electronic imaging group was created within Polaroid, they started development on a digital camera. As Tripsas and Gavetti (2000) recount, although the company was well positioned technically by 1986 to deliver the camera, they did not do so for a decade. What occasioned the delay? In no small part, it was due to the challenge such a camera posed to ingrained beliefs about how to make money. Film was making 70 per cent profits, and managers were wedded to the 'razor blade model' of making business from what went into the hardware—not the hardware itself. They scoffed at the 38 per cent margins the new hires in the Electronic Imaging Division offered on cameras. Moreover, the old guard 'knew' that customers wanted a physical print, so video camcorders would not be competitive. By 1996, only about fifty employees were devoted to digital imaging research, from a high of about 300. On 12 October 2001, the company filed for bankruptcy. It is very difficult for innovations to take hold when the environment is so hostile and opinion leaders are so sure of their own beliefs, their knowledge.

Outsourcing Ideation to the 'Crowd'

One new tool that organizations have to challenge assumptions and address problems creatively is 'crowdsourcing' either within the organization or globally (see Chapter 5 by Franke). Companies such as Procter & Gamble or IBM, government agencies, and medical facilities now routinely post problems needing creative solutions on the World Wide Web. Google conducts global contests or 'jam fests', to create code. The reward for being a finalist in such contests is a three-month internship at Google—a singularly clever way to capture the tacit knowledge of the contest winner by co-locating him or her with current Google employees. Corporate intranets also offer the opportunity to tap into knowledge and creativity anywhere in their organization. Electronic suggestion boxes enable any person in the organization to suggest ideas easily. For example, Novartis, the huge pharmaceutical company with a highly respected R&D capability, nevertheless finds it useful to solicit ideas from their more than 100 000 employees through a system dubbed IdeaPharm. Contributors to IdeaPharm offer diverse perspectives.

Organizational Stories about Innovation Help or Hinder

Another powerful form of knowledge that can either help or hinder innovation is embodied in stories that are accessible and meaningful to the listener or reader. Several types of stories are relevant to creativity. First, there are stories about past innovation in an organization: how ideas were discovered, how managers rewarded or punished creative thinking, what happened to the teams that produced successful or unsuccessful novel products and services—such stories can influence the expectations and actions of employees. Among the seven types of archetypal stories identified by Martin et al. (1983) are two that particularly affect creativity: rule-breaking and managerial reactions to mistakes. In companies where everyone has heard of successful rule-breaking and managerial forgiveness, creativity has a chance to thrive. However, the obverse is also true: stories about would-be innovators who were penalized dampen the appetite for creativity.

Second are stories that serve as 'springboards' to drive organizational innovation because they provide 'the kind of plausibility, coherence, and reasonableness that enables people to make sense of immensely complex changes' (Denning, 2001: 37). Stories can aid creative endeavours in ways that pleas for innovation or analytical presentations of the need for change cannot. Stories are powerful because they incorporate details and emotion that human brains tend to remember. More than other forms of presentations, they communicate some of the tacit dimensions of knowledge possessed by the story-teller because they chronicle behaviour rather than sterile bullet points in an argument. They linger long after the particular event they chronicle; therefore managers wishing to enhance organizational creativity need to be aware of their power (Swap et al., 2001).

GROUP LEVEL

For groups, as for whole organizations, knowledge both helps and hinders creativity. A danger in groups charged with innovating is that innovative thought will be relegated to 'the creatives', that is, those individuals whose job titles or history mark them as the ordained sources of new ideas. Yet in practice, even individuals valued for their personal creativity cite the importance of interaction with others (Csikszentmihalyi and Sawyer, 1995). In fact, as Hargadon and Bechky (2006) argue, creative problem solving often arises in the interactions of a group, and cannot be traced to any one individual. Creative ideas in groups arise from the members. Group composition, cohesion, norms, processes, and leadership all determine the level of creativity a group expresses.

Intellectual Diversity in Group Composition Helps Innovation

One of the most powerful ways to enhance the probability that a group will be creative is to compose the group with individuals who access very different knowledge sources.

Creativity blossoms at the crossroads of diverse knowledge domains. Therefore, a certain amount of 'requisite variety' is highly desirable (Nonaka and Takeuchi, 1995). Laboratory research and the practice of highly creative companies suggest that composing groups of individuals with different life experiences and knowledge bases sets the stage for creative interactions (Leonard and Swap 1999, 2000; Rodan and Galunic, 2004). When the resulting inevitable disagreements among group members are substantive and impersonal (Eisenhardt, Kahwajy, and Bourgeois, 1997) the result is positive 'creative abrasion' (Hirshberg, 1998). In such groups, the collision of perspectives leads to exploration of different problem framing and solution identification. Of course the individualism of such groups has to be valued to be useful. That is, if the group values conformity over individual contributions, creativity may suffer (Goncalo and Staw, 2006).

Group Cohesion Can Hurt Creativity

The well-known research on 'groupthink', that is, the tendency for cohesive groups to seek premature consensus (Janis, 1972), suggests that highly cohesive groups may have difficulty initiating or accepting innovation. Janis' (1972) research, based as it was on case studies on decision-making, has been challenged on a number of grounds, including that it was studied more in crisis situations than in normal innovation activities. However, his basic finding about the social pressures on group members to agree with one another rather than to seek alternative solutions is supported by studies in engineering management on the relative creative performance of research teams.

Katz (1982) found that research groups with tenure longer than five years lost creativity. After testing extensively for alternative explanations (such as individual competence or the relative visibility of the fifty projects studied), Katz concluded that the higher performing groups did so because they communicated more outside their project membership. He explains that as project members work together over a long period, they 'reinforce their common views, commitments and solution strategies. Such shared perceptions created through group processes, act as powerful constraints on individual attitudes and behaviors' (Katz, 1982: 101). Similarly, Pelz and Andrews (1966) found that groups with more frequent turnover in membership were more creative than ones with stable membership—even if they were interdisciplinary.

While Katz considered his findings an explanation of the renowned 'Not Invented Here' syndrome, decried in the engineering management literature as responsible for the rejection of innovative ideas originating outside the organization or team, the phenomenon sounds very similar to Janis' 'groupthink'. Some researchers have noted that only a premature concurrence-seeking tendency before consideration of critical options is really detrimental to performance (Longley and Pruitt, 1980). This nuanced perspective on groupthink resonates with observations about the process of group creativity, which Leonard and Swap (1999) describe as moving through periods of divergence (option seeking) and convergence (agreement on a course of action). One may argue, therefore, that groupthink does not threaten convergence, that is, knowledge consolidation and agreement, but is inimical to the divergent processes of knowledge creation and ideation.

Group routines can inhibit creativity by framing problems as so familiar as to not require creative thinking. In his analysis of the Mann Gulch fire that claimed the lives of several experienced firefighters, Weick (1993) points out that the team's assumption that they were fighting a 'ten o'clock fire' (a fire that is expected to be easily controlled and extinguished by ten in the morning), prevented firefighters from making sense of what was actually a very different situation. More importantly, their assumptions constrained most of the team members to the tools and processes normally employed for a routine fire (e.g. fighting it head-on, trying to out-run it). Only one team member was able to recognize in time the need for a different approach and successfully improvise a novel, life-saving solution.

Minority Opinions Challenge Undesirable Group Cohesion

One way to avoid groupthink and challenge group assumptions is to introduce minority opinions. As we know from the famous experiments by Asch (1955), people in groups often succumb to peer pressure and accede to judgements that they know are incorrect—but even one dissenting voice motivates others to express their real opinion. Nemeth and Wachtler (1983) found that majority influence caused convergence on group judgements, whether correct or not, whereas minority influence caused laboratory subjects to give novel, correct judgements. They concluded that the expression of minority views influenced others in the group 'to reanalyse a problem and, in the process, perhaps function more creatively and accurately' (Nemeth and Wachtler, 1983: 54).

Group Norms Cut Both Ways

'Group norms are the informal rules that groups adopt to regulate and regularize group members' behaviors' (Feldman, 1984: 47). Such norms can greatly aid creativity but again, can likewise inhibit the behaviours that lead to innovation. Those governing the accessing and sharing of knowledge in a group are particularly influential.

Risk-taking and Psychological Safety

One norm especially important to creativity is the degree to which failure is tolerated or even encouraged. *Intelligent* failures are different from mistakes (Leonard and Swap, 1999). Scientific and technical research depends upon failing forward, taking acceptable and anticipated risks in the hopes of an innovative pay-off. But even in less risk-oriented environments, a norm of psychological safety, that is, 'a shared belief that the team is safe for interpersonal risk-taking…a sense of confidence that the team will not embarrass, reject or punish someone for speaking up', enables the honesty about mistakes and near-mistakes that leads to improvement and often creative solutions (Edmondson, 1999). If all failure is treated the same—as a mistake that dooms one's career, group members will not take risks. And of course, innovation requires risk, as by definition it involves action that is novel to some degree.

Hierarchy and Expertise

As further discussed in the section on Group Leadership, a norm that hierarchy and expertise are not to be questioned is antithetical to creativity and can stop novel ideas before they are even expressed. Organizations that require improvisation and creativity in response to crises push responsibility for action down to the lowest level at which the relevant knowledge resides. The US Coast Guard, although a military organization, is renowned for its ability to respond creatively to crises such as the hurricane that devastated the coastal city of New Orleans in the USA, or the infamous '9/11' terrorist attack destroying two skyscrapers in New York in 2001. One of the primary reasons for the organization's ability to improvise is freedom to act without waiting for approval from the organizational hierarchy. Even the lowest ranking officer is considered to have full authority over the vessel she commands—even if an admiral were to come on board. Consequently the person closest to the situation is the decision-maker, and can ignore standard operating procedures when warranted.

Group Processes that Guide Innovation

Norms enable the processes that support creative endeavours. Leonard and Swap (1999) treat creativity within businesses as a five-stage process of preparation, innovation opportunity identification, divergent thinking (creating options), incubation, and convergence (selection among options). This sequential model is what Fisher and Amabile (2009) call a 'componential' creativity process and is characteristic of new product and process development. They describe a contrasting model of 'improvisational creativity', which they define as 'actions responsive to temporally proximate stimuli, where the actions contain both a high degree of novelty and a low temporal separation of problem presentation, idea generation, and idea execution' (2009: 19). The major difference between the two models is the tight time connection between the creative idea and its execution. The Coast Guard example mentioned in the previous paragraph would be considered improvisional creativity. However, similar to Leonard and Swap, Fisher and Amabile note that improvisational creativity can be embedded within the stages in a componential model.

Brainstorming

Brainstorming is one such group process that can facilitate creativity. This technique was scoffed at as a knowledge-creating mechanism when years of laboratory research showed that individuals working alone came up with more and better ideas than they did when working in a group, and the larger the group, the greater the disparity. However, Sutton and Hargadon (1996), among others, found that brainstorming conducted according to process rules (e.g. famous design company IDEO's 'build on the ideas of others'), focused on a real world problem instead of the kind of irrelevant challenge often used in laboratory research (e.g. what if people had two thumbs on each hand?), and time limited, had excellent results in stimulating divergent thinking. Again, one of the principal benefits was the opportunity to tap into very different knowledge bases and thus stimulate creative abrasion.

Incubation

Incubation is an underrated group process. As individuals we may be aware that it helps sometimes to step away from a problem and let our unconscious hum along while we go about other business. But deflecting attention from a problem is often anathema to group managers. Who can take time out from an intensive product development project, for example? Yet there are examples of the benefit of doing just that, for groups as well as for individuals. Jerry Hirshberg (1998) attributes overcoming a block in the design of the Nissan Pathfinder to a break the entire design organization took to see a newly opening movie: *The Silence of the Lambs*! Given that the organization was already behind in their schedule, Nissan Design president Kengo Ishida understandably questioned why the designers would take a break at such a crucial juncture: 'We're going now, Kengo-san', Hirshberg explained, '*because* we are behind' (1998: 89).

An increasing body of research concludes there is a strong link between incubation and creativity (Amabile and Kramer, 2011). Psychologists believe that the reason incubation helps creativity may be that the unconscious mind, freed from the constraints of logic, convention, and habit, begins to associate ideas freely, and creative insights result. Whatever the underlying mechanism, 'it's clear that time away from the task is crucial for the creative process' (Leonard and Swap, 1999: 99). A creative group may therefore establish a norm that it is permissible to 'sleep on' an intractable problem in order to generate creative ideas.

Dysfunctional Momentum

Taking a break from ongoing processes may have another effect on group thought processes as well—breaks may disrupt the momentum of a project in beneficial ways. Once individuals or groups are engaged in an ongoing course of action, they rarely take time to re-evaluate or reconsider their processes or assumptions. Norms to 'get the job done' or 'keep working' prevent any reconsideration of the current approach. This can lead to what Barton and Sutcliffe (2009) call dysfunctional momentum— when groups, caught up in the momentum of existing processes, fail to reconsider or re-evaluate ineffective actions. In a study of wild land firefighters, the researchers found that when group members or leaders deliberately introduced interruptions into the flow of thought and action, groups were more likely to re-evaluate their assumptions and actions and take a new and different approach. So not only do breaks from the process allow ideas to incubate in individuals' heads, breaks may also trigger a tendency to re-evaluate the situation—often introducing new frameworks or new assumptions—so that individuals and groups come at the problem from a new direction.

Empathic Design

One process engaged in by many creative teams is empathic design, that is, anthropological expeditions into potential customers' environments to identify unarticulated needs (Leonard and Swap, 1999; Leonard-Barton, 1995). The knowledge obtained from empathic design is likely to be very different from that acquired in more traditional market research such as surveys or focus groups. One of the reasons is that, as noted below, we humans are frequently poor at knowing or explaining why we make the choices we do—the reasons are confounded by emotion and/or are inaccessible to our conscious mind. Even asking customers about their preferences for familiar product lines (e.g. automobiles), is fraught with difficulty. Market researchers have become very creative at tapping into the tacit dimensions of peoples' opinions and beliefs through the use of metaphors and visualization (see Zaltman, 2003). But seeking opinions about emerging, truly novel products, is even trickier, venturing into speculation. As potential customers, we can react to prototypes—even very crude ones—but it is hard for us to suggest solutions to problems when we don't know what technical knowledge is available to address the issues and may not even realize we have a problem or need.

No one asked Kimberly-Clark for pull-up diapers—but observations by design firm GVO suggested that both parents and toddlers desired a step between diapers and 'big-boy' (but leaky) underpants. Huggies pull-up diapers were an immediate and long-lasting hit on the market because they addressed both the practical issue of keeping the toddlers and their environments dry, and also a psychological ego need for parents and children to feel progress out of babyhood. By observing customers in their own natural environment, designers can originate solutions that neither they nor the customers would have otherwise envisioned. Sometimes they solve ergonomic, usability, comprehension problems; sometimes they suggest totally new ideas. Many design firms and product development divisions in consumer product companies routinely incorporate empathic design into their processes so as to tap into users' tacit knowledge and unarticulated needs.

Group Leadership

Group leaders can encourage creativity—or kill it. The foregoing discussion suggests ways to do both, although not all fall completely within the control of such leaders. Composing groups for creative abrasion, introducing minority opinions from many sources, encouraging norms supporting risk-taking, incubation, and empathic design— all of these stimulate creativity. In contrast, group creativity is damaged by leaders who support the tyranny of hierarchy or technical expertise at the expense of other knowledge flows. We turn now to a few specific leadership behaviours that strongly influence group creativity and which are the responsibility of the individual leader more than of the collective.

The Progress Principle: Motivating Creativity

Leaders are responsible for what Amabile and Kramer (2011) call the 'progress princi-ple'. In a far-reaching study of twenty-six project teams in seven firms, the researchers examined the 'inner work life' of 238 people—daily. Their research identified a power-ful force motivating not only hard work but also creativity, namely the extent to which those individuals were able to make meaningful progress on their work that day. The researchers identified seven major catalysts (and their mirror opposites, major inhibi-tors) affecting that progress. Among the catalysts were two already suggested in this chapter as influencing creativity positively: a free flow of ideas (including debate), and learning from both success and failure. Both of these, it should be noted, are related to knowledge creation. The researchers also found that this principle affected employees' moods—which in turn influenced creativity. Interestingly, there was a 'carry-over effect', that is, the more positive an individual's mood on a given day, the more creative she was on the next day, and even to some degree, on the day after that—regardless of her mood on those subsequent days. Amabile and Kramer attribute that effect to incubation, citing prior research by Isen (1999) that showed a link between pleasant moods and creativity. The argument is that pleasant moods 'stimulate greater breadth in thinking—greater cognitive variation…' (Amabile and Kramer, 2011: 52).

Time Pressure

Managers sometimes believe that time pressures will motivate their teams to perform more creatively. Research findings about the effects of time pressure on creativity are not totally consistent or simple, but the predominant result is that such pressure hurts more than it helps. For example, in the same study referenced above, Amabile and Kramer (2011) discovered that while people may *feel* more creative when they cope with very high time pressure, except for rare exceptions, they actually think more creatively when they have time to explore, collaborate with others, and come up with options. (See also George and Zhou, 2007, who posit a positive effect of time pressures only in combina-tion with positive moods and a supportive organization.)

Insisting on Hierarchy

As suggested in the sections above, creativity can bloom low in an organization, but lead-ers greatly influence the likelihood of its expression. In 1927, Harry Warner of Warner Brothers picture studios in Hollywood, notoriously proclaimed his opposition to the innovation of talking movies: 'Who the hell wants to hear actors *talk*?' All of us, it turns out. And a cautionary tale from the history books: in October 1707, twenty miles south-west of England, four of five British warships struck the rocks around the Scilly Isles and sank. Two thousand troops under the command of Admiral Sir Cloudesley Shovell per-ished in the fog. At the time, there was no way to reckon longitude reliably. Navigators relied upon elapsed time from leaving shore to reckon their position. However, pendulum-swinging clocks were notoriously subject to temperature, moisture, and the motion of the ships. All navigators in the British fleet agreed: the ships were safely west of Ile d'Ouessant. They were tragically wrong, and Admiral Shovell's flagship was the first to sink, with three more following. Only two men washed ashore alive. The tragedy could have been averted, had knowledge been allowed to flow up the hierarchy. One seaman on the flagship, who claimed to have kept his own reckoning, believed the danger of hitting the islands so severe that he risked his life to approach an officer to voice his concerns. Admiral Shovell had him hanged for mutiny on the spot. Hours later, his prophesy proved correct, but there was no one alive to say 'I told you so' (Sobel, 1995). Leaders need not be as tyrannical as Admiral Shovell to kill creative solutions. Simply by offering an opinion of her own before opening up discussion, a leader can shut off ideas from the group (Leonard and Swap, 1999) and the more technical expertise the leader has, the more likely she is to shut down a potential dissent this way.

Situated Humility

An attitude (and consequent behaviour) that helps leaders to listen is 'situated humility' (Barton and Sutcliffe, 2009). Despite their own high level of competence or expertise, leaders with this characteristic recognize that this *particular* situation is so dynamic or complex that any assessment of it must be considered malleable, subject to change in response to new information or different perspectives. They realize that no matter how expert they are, their assessment of the situation is only partial; other perspectives or set of events that they haven't seen yet are likely to emerge. This openness and expectation of more in-coming knowledge is especially helpful to the improvisational creativity mentioned earlier. Leaders who exhibit situated humility are more likely to solicit other expertise and listen to different viewpoints, even from low in the group hierarchy.

However, even such attempts may not bear fruit if the knowledge sought has a high proportion of tacit dimensions. As explained further in the section on individual deep smarts, people do not always know what they know until a question or problem is framed in a context, and even then, they may not be able to explain (Reber, 1989). Moreover, as individuals we are often unaware of the way our brains function. We turn now to the benefits and hazards for creativity of individual knowledge.

INDIVIDUAL LEVEL

Deep Smarts Help

It is because of the highly individual content in people's brains that intellectual diversity is so critical to creativity. Individuals become experts in a given knowledge domain, the research suggests, only after seven to ten years of diligent practice (Ericsson, 1996). However, once they achieve such expertise, they behave very differently from novices (Leonard and Swap, 2005). They make decisions more rapidly, recognize context (i.e. when and how their knowledge applies in particular situations), and make fine distinctions that are invisible to someone with less knowledge. Most important, from prior experience, they recognize patterns in situations and behaviours, and they have a lot of tacit knowledge (Klein, 1998).

All of this expertise adds up to a significant benefit both to the individuals and the organizations that hire them. We owe a great deal of invention to people with prepared minds, that is, enough smarts to pursue knowledge to its recognized edges and then make educated leaps of imagination. The microwave originated with such an individual. In 1946, Percy Spencer was touring a laboratory at Raytheon Corporation. He paused in front of a magnetron, the power tube that drives a radar set. To his surprise, he noticed a candy bar in his pocket began to melt. A less deeply smart individual might have thought his body heat was the cause. Instead, Spencer quickly made the connection to the magnetron. He sent out for popcorn, to see what would happen if he placed the kernels near the magnetron. When they popped, he then brought in a tea kettle, cut a hole inside, inserted a raw egg and was delighted when it exploded (Flatow, 1992). (It is somewhat ironic that today one of the principal uses of the microwave is popping popcorn!) Such stories abound about inventions that occurred to prepared minds.

Cognitive Biases Can Hurt Creativity

However, experts can be subject to arrogance and over-confidence. They are also just as apt as the rest of us to succumb to any number of cognitive biases that can destroy creative insights (see Bazerman, 1998).

Mental Set and Functional Fixedness

For decades, psychologists have known about a general human tendency towards *mental set*, that is, to continue diagnosing problems with a familiar approach that has worked in the past—even if it is not optimal in a given situation (Luchins, 1942). NASA's 2003 disaster, in which the Columbia space craft disintegrated upon re-entry, has been attributed to a piece of insulating foam that broke free and struck the wing during take off. During re-entry, superheated air melted a hole in the unprotected wing. The reason that no one anticipated this result was that losses of foam had never been a problem in the past. Even after the explosion, NASA personnel continued to discount the breach of thermal protection as a cause of the accident, until the physical evidence showed that it was (Columbia Accident Investigation Board, 2003). A closely related bias is functional fixedness, that is, the tendency to think of an object or process in terms of its usual function, without considering alternative, perhaps more creative uses. A coin is valued for its buying power—but it can be used to turn a screw head.

Confirmation Bias

We all also can fall prey to confirmation bias, that is, seeking information that confirms what we already believe to be true and a reluctance even to consider disconfirming information and sources. In the past couple of decades, this bias has been reinforced by the splintering of media sources into very specialized niches, appealing to tightly targeted audiences, and the capability that the Internet provides for us all to select just those sources that we trust to give us confirming information and to shield us from uncomfortable refutations. Scientists' ability to look into the brain has provided some fascinating data on the biological manifestation of this bias. During the 2004 national elections in the USA, researchers tested a number of hypotheses about how Republicans and Democrats would react when confronted with contradictory statements from their candidates (Westen et al., 2006). Not surprisingly, while judging harshly the contradiction presented by the opposition, each partisan rationalized away conflicting statements by their favoured candidate. However, the researchers made a discovery they had not anticipated. Once the cognitive dissonance was resolved, the brains lit up the circuitry involved in positive emotion. Biased reasoning was rewarded! Clearly the confirmation bias works against the open-mindedness that is so critical to creative thinking.

Conclusion

The manager's solutions to the paradoxical relationship of knowledge with creativity can be simply stated. (Operationalizing them is obviously more difficult!) The best defence against the perils and inhibitions of knowledge is to challenge assumptions at all levels, including one's own. The best positive contribution towards enhancing creativity at all levels is to recognize the value and power of disparate sources of knowledge.

Note

1. For a similar argument, see Rothaermel and Hess (2007). Seeking the foundations of dynamic capabilities, the authors look at individual, firm-level, and network-level variables and note that 'the antecedents to innovation capabilities clearly lie across different levels of analysis' (2007: 916).

References

Amabile, T. M., and Kramer, S. J. (2011). *The Progress Principle*. Boston, MA: Harvard Business Review Press.

Asch, S. E. (1955). 'Opinions and Social Pressure', *Scientific American*, 193: 31–5.

Barton, M., and Sutcliffe, K. (2009). 'Overcoming Dysfunctional Momentum: Organizational safety as a social achievement', *Human Relations*, 62(9): 1327–56.

Bazerman, M. (1998). *Judgment in Managerial Decision Making*. New York: John Wiley & Sons.

Christensen, C. M., and Raynor, M. E. (2003). *The Innovator's Solution*. Boston, MA: Harvard Business School Press.

Columbia Accident Investigation Board (2003). *Final Report*. Washington DC: U.S. Government Printing Office.

Csikszentmihalyi, M. (1997). *Creativity: Flow and the Psychology of Discovery and Invention*. New York: Harper Perennial.

Csikszentmihalyi, M., and Sawyer, K. (1995). 'Creative Insight: The Social Dimension of a Solitary Moment', in R. J. Sternberg and J. E. Davidson (eds), *The Nature of Insight*. Cambridge, MA: The MIT Press.

Davenport, T. H., and Prusak, L. (1998). *Working Knowledge: How Organizations Manage What They Know*. Boston MA: Harvard Business School Press.

Denning, S. (2001). *The Springboard: How Storytelling Ignites Action in Knowledge-Era Organizations*. Boston, MA: Butterworth-Heinemann.

Doz, Y. (1997), 'Managing Core Competency for Corporate Renewal: Towards a Managerial Theory of Core Competencies', in A. Campbell and K. S. Luchs (eds), *Core Competency-Based Strategy*. London: International Thomson Business Press.

Drucker, P. F. (1964). *Managing for Results: Economic Tasks and Risk-Taking Decisions*. New York: Harper & Row.

Edmondson, A. (1999). 'Psychological Safety and Learning Behavior in Work Teams', *Administrative Science Quarterly*, 44(2): 350–83.

Eisenhardt, K. M., and Martin, J. A. (2000). 'Dynamic Capabilities: What are they?' *Strategic Management Journal*, Special Issue 21: 1105–21.

Eisenhardt, K. M., Kahwajy, J. L., and Bourgeois, L. J. III (1997). 'Conflict and Strategic Choice: How Top Management Teams Disagree', *California Management Review*, 39: 42–62.

Ericsson, K. A. (1996). 'The Acquisition of Expert Performance: An Introduction to Some of the Issues', in K. A. Ericsson (ed.), *The Road to Excellence: The Acquisition of Expert Performance in the Arts and Sciences, Sports, and Games*. Mahwah, NJ: Lawrence Erlbaum.

Feldman, D. (1984). 'The Development and Enforcement of Group Norms', *The Academy of Management Review*, 9 (1): 47–53.

Fisher, C. M., and Amabile, T. M. (2009). 'Creativity, Improvisation and Organizations', in T. Rickards, M. A. Runco, and S. Moger (eds), *The Routledge Companion to Creativity*. Routledge: New York, 13–24.

Flatow, I. (1992). *They All Laughed*. New York: HarperCollins.

George, J. M., and Zhou, J. (2007). 'Dual Tuning in Supportive Context: Joint Contributions of Positive Mood, Negative Mood, and Supervisory Behaviors to Employee Creativity', *Academy of Management Journal*, 50: 605–22.

Goncalo, J. A., and Staw, B. M. (2006). 'Individualism—Collectivism and Group Creativity', *Organizational Behavior and Human Decision Processes*, 100: 96–109.

Gundling, E. (2000). *The 3M Way to Innovation: Balancing People and Profit*. Tokyo: Kodansha International.

Hargadon, A. B., and Bechky, B. A. (2006). 'When Collections of Creatives Become Creative Collectives: A Field Study of Problem Solving at Work', *Organization Science*, 17(4): 484–500.

Hayes, R. H., Wheelwright, S. C., and Clark, K. B. (1988). *Dynamic Manufacturing: Creating the Learning Organization*. New York: Free Press.

Helfat, C. E. (1997). 'Know-how and Asset Complementarity and Dynamic Capability Accumulation: The Case of R&D', *Strategic Management Journal*, 18: 339–60.

Henderson, R. M., and Clark, K. B. (1990). 'Architectural Innovation: The Reconfiguration of Existing Product Technologies and the Failure of Established Firms', *Administrative Science Quarterly*, 35: 9–30.

Henderson, R. M., and Cockburn, I. (1994). 'Measuring Competence? Exploring Firm Effects in Pharmaceutical Research', *Strategic Management Journal*, 15: 63–84.

Hirshberg, J. (1998). *The Creative Priority: Driving Innovation Business in the Real World.* New York: HarperBusiness.

Isen, A. (1999). 'Positive Affect', in T. Dagleish and M. Power (eds), *Handbook of Cognition and Emotion.* New York: Wiley, 521–39.

Itami, H., and Roehl, T. W. (1987). *Mobilizing Invisible Assets.* Boston, MA: Harvard University Press.

Janis, I. (1972). *Victims of Groupthink.* Boston: Houghton Mifflin.

Katz, R. (1982). 'The Effects of Group Longevity on Project Communication and Performance', *Administrative Science*, 27: 81–104.

Klein, G. (1998). *Sources of Power.* Cambridge, MA: MIT Press.

Kurtzberg, T. R., and Amabile, T. M. (2000). 'From Guilford to Creative Synergy: Opening the Black Box of Team-level Creativity', *Creativity Research Journal*, 13 (3–4): 285–94.

Leonard, D. (2011). *Managing Knowledge Assets, Creativity and Innovation.* London: World Scientific.

Leonard, D., and Swap, W. (1999; 2000). *When Sparks Fly: Igniting Creativity in Groups.* Boston, MA: Harvard Business School Press.

Leonard, D., and Swap, W. (2005). *Deep Smarts: How to Cultivate and Transfer Enduring Business Wisdom.* Boston, MA: Harvard Business School Press.

Leonard-Barton, D. (1992). 'Core Capabilities and Core Rigidities: A Paradox in Managing New Product Development', *Strategic Management Journal*, 13 (Summer Special Issue): 111–25.

Leonard-Barton, D. (1995). *Wellsprings of Knowledge: Building and Sustaining the Sources of Innovation.* Boston, MA: Harvard Business School Press.

Longley, J., and Pruitt, D. G. (1980). 'Groupthink: A Critique of Janis' Theory', in L. Wheeler (ed.), *Review of Personality and Social Psychology.* Newbury Park, CA: Sage, 507–13.

Luchins, A. S. (1942). 'Mechanizations in Problem Solving', *Psychological Monographs*, 54: 248.

Martin, J., Feldman, M., Hatch, M. J., and Sitkin, S. (1983). 'The Uniqueness Paradox in Organizational Stories', *Administrative Science Quarterly* 28(3): 438–53.

Miller, D. J., Fern, M. J., and Cardinal, L. (2007). 'The Use of Knowledge for Technological Innovation within Diversified Firms', *Academy of Management Journal*, 50(2): 308–26.

Nemeth, C. J., and Wachtler, J. (1983). 'Creative Problem Solving as a Result of Majority vs. Minority Influence', *European Journal of Social Psychology*, 13(1): 45–55.

Nonaka, I., and Takeuchi, H. (1995). *The Knowledge-Creating Company: How Japanese Companies Create the Dynamics of Innovation.* London: Oxford University Press.

Pavitt, K. (1991). 'Key Characteristics of the Large Innovating Firm', *British Journal of Management*, 2: 41–50.

Pelz, A., and Andrews, F. (1966). *Scientists in Organizations.* New York: John Wiley.

Pettigrew, A. (1979). 'On Studying Organizational Cultures', *Administrative Quarterly*, 24: 570–81.

Polanyi, M. (1966). *The Tacit Dimension.* New York: Doubleday.

Reber, A. S. (1989). 'Implicit Learning and Tacit Knowledge', *Journal of Experimental Psychology*, 118: 219–35.

Rodan, S. and Galunic, D. C. (2004). 'More than Network Structure: How Knowledge Heterogeneity Influences Managerial Performance and Innovativeness', *Strategic Management Journal*, 25(6): 541–62.

Rothaermel, F. T., and Hess, A. M., (2007). 'Building Dynamic Capabilities: Innovation Driven by Individual-, Firm-, and Network-level Effects', *Organization Science*, 18(6): 898–921.

Rumelt, R. P. (1986). *Strategy, Structure and Economic Performance*. Boston, MA: Harvard Business School Press.

Simonton, D. K. (1999). *Origins of Genius: Darwinian Perspectives on Creativity*. New York: Oxford University Press.

Snow, C. C., and Hrebiniak, L. G. (1980). 'Strategy, Distinctive Competence, And Organizational Performance', *Administrative Science Quarterly*, 25: 317–35.

Sobel, D. (1995). *Longitude: The True Story of a Lone Genius Who Solved the Greatest Scientific Problem of His Time*. New York: Walker and Company.

Spender, J. (1996). 'Competitive Advantage from Tacit Knowledge? Unpacking the Concept and its Strategic Implications', in B. Mosingeon and A. Edmondson (eds), *Organizational Learning and Competitive Advantage*. London: Sage Publications, 56–73.

Sutton, R., and Hargadon, A. (1996). 'Brainstorming Groups in Context: Effectiveness in a Product Design Team', *Administrative Science Quarterly*, 41(4): 685–718.

Swap, W., Leonard, D., Shields, M., and Abrams, L. (2001). 'Using Mentoring and Storytelling to Transfer Knowledge in the Workplace', *Journal of Management Information Systems*, 18 (1): 95–114.

Szulanski, G. (1996). 'Exploring Internal Stickiness: Impediments to the Transfer of Best Practice Within the Firm', *Strategic Management Journal*, 17 (Winter Special Issue): 27–43.

Teece, D. J., Pisano, G., and Shuen, A. (1997). 'Dynamic Capabilities and Strategic Management', *Strategic Management Journal*, 18(7): 509–533.

Tripsas, M., and Gavetti, G. (2000). 'Capabilities, Cognition, And Inertia: Evidence from Digital Imaging', *Strategic Management Journal*, 21: 1147–61.

Weick, K. E. (1993). 'The Collapse of Sensemaking in Organizations: The Mann Gulch disaster', *Administrative Science Quarterly*, 38(4): 628.

Westen, D., Blagov, P. S., Harenski, K., Kilts, C., and Hamann, S. (2006). 'Neural Bases of Motivated Reasoning: An fMRI Study of Emotional Constraints on Partisan Political Judgment in the 2004 U.S. Presidential Election', *Journal of Cognitive Neuroscience*, 18(11): 1947–58.

Woodman, R. W., Sawyer, J. E., and Griffin, R. W. (1993). 'Toward a Theory of Organizational Creativity', *Academy of Management Review*, 18(2): 293–321.

Zaltman, G. (2003). *How Customers Think: Essential Insights into the Mind of the Market*. Boston: Harvard Business School Press.

....

CHAPTER 8

DESIGN-DRIVEN INNOVATION

Meaning as a Source of Innovation

....

ROBERTO VERGANTI AND CLAUDIO DELL'ERA

INTRODUCTION

....

STUDIES of innovation management have often focused their investigations on two domains: technologies and markets (for an extensive review see Garcia and Calantone, 2002; Calantone et al., 2010). Technological innovation has captured most attention, especially as far as radical technological change is concerned. Indeed, in the past decades a rich stream of studies has explored the antecedents of technological breakthroughs (Abernathy and Clark, 1985; Henderson and Clark, 1990; Utterback, 1994; Christensen and Bower, 1996; Christensen, 1997). Other investigations have focused more on the applications of existing or new technologies and/or products to penetrate new market domains (Kim and Mauborgne, 2005; McGrath and MacMillan, 2009). Within this mix, design has recently gained more attention among practitioners and scholars as a source of innovation. Firms are increasingly investing in design and involving design firms in their innovation processes (Nussbaum, 2005). Academic journals are publishing articles that explore the contribution of design to product development and business performance (Gemser and Leenders, 2001; Platt et al., 2001; Hertenstein et al., 2005; Journal of Product Innovation Management 2005a, 2005b). And the practitioners' press has embraced the subject extensively (Verganti, 2006, 2009, 2011; Brown, 2008, 2009; Martin, 2009). Still, the role of design in innovation and competition remains a rather young (pre-paradigmatic) area, with blurred boundaries and often unclear or contrasting perspectives. In this chapter we aim to provide a theoretically solid and empirically grounded view on design from a very specific angle, one which is dictated from this book: design as a source of innovation. We first define innovation driven by design and how it stands apart from other approaches to innovation. We will see that design is related to the innovation of the meaning of products and services: and innovation that concerns the purpose, the 'why' people use things, rather than the functionality and performance of products (i.e. the 'what' and 'how'). This type of innovation aims to introduce new meaningful experiences to people, usually with significant implications for their cultural, symbolic, and emotional dimensions. Second, we discuss two different strategies that design contributes to innovation in the meaning of things: user-driven innovation, which induces mainly incremental innovation of meanings, and design-driven innovation, that drives radical changes in meaning. We focus our attention on the latter, by first discussing the value of radical innovation of meaning and then illustrating how design-driven innovation is managed.

DESIGN AND INNOVATION

One of the reasons why the scientific investigation of design is a hard challenge for scholars of innovation management is that the definition of 'design' is fluid and slippery (for a comprehensive analysis see Love, 2000). With a high level of aggregation we can cluster those definitions around three areas: design as the form of things, design as a creative approach to problem solving, and design as the 'making sense of things'.

Design as the Form of Things

The first approach looks at design from a narrow perspective: design is associated with the form of products, often in juxtaposition to the product function. Indeed, when things come down to the real essence of the concept, many people believe that design is basically dealing with form. If engineers make products function, by using technology, then designers make things beautiful. Indeed, the history of design has been punctuated by the debate about the prevalence and dominance of form versus function as a driver of innovation. Modernists, especially, questioned the predominance of form, from the claim of American architect Louis Sullivan that 'form follows function' at the dawn of the twentieth century, to the aphorism 'less is more' of Ludwig Mies van der Rohe, one of the directors of the Bauhaus (the art and architecture school active in Germany between 1919 and 1933). Yet, if we look at the most frequent applications of design in business in the past century, they have been infused by a prevailing attention to form, and have complied more to the dogma 'ugliness does not sell' coined by Raymond Loewy, the French-American designer who is a founding father of 'styling'. And most business people still today associate design with the beauty of products. We will not explore this concept further here. Indeed, beauty has little in common with innovation. Instead, they are sometimes even in contrast. People associate beauty with aesthetical standards they already have in their mind. But novel things, especially when they are radically innovative, do not conform with existing standards, and in fact they often try to impose new ones (Eco, 2004).

Design as Creative Problem Solving

If the concept of design as form is too narrow, the reaction of many experts recently has been to enrich and stretch its definition to a point where it embraces any kind of creative activity. Indeed, as Herbert Simon states, 'everyone designs who devise courses of action aimed at changing existing situations into preferred ones' (Simon, 1982). An interpretation that clarifies that, if we consider design in its broader meaning, concerns all major creative activities and professions that produce a modification in the environment: 'Engineering, medicine, business, architecture and painting are concerned not with the necessary but with the contingent—not with how things are but with how they might be—in short, with design' (Simon, 1996). We can therefore talk of product design, engineering design, software design, organization design, business model design, market design. All these activities have their own disciplines and field of investigation which we will not discuss here. Recently, however, there has been an attempt to associate design to a 'better way of thinking' (usually referred to as 'design thinking'), which can be used to address any kind of problem (Boland and Collopy, 2004; Brown, 2008, 2009; Martin, 2009). And since this acceptation is often promoted by design consultants, design thinking is assumed to express a better capability of designers (as a profession) to address and solve problems. Unfortunately, there is no clear convergence about what this slippery concept of design thinking is: for some, it is an abductive or intuitive as opposed to analytical approach to thinking, for others, a mix of experimentation, visualization, and user understanding, which may sound reasonable, apart from the fact that is not clear why those characteristics are peculiar to design and not to other creative activities. This perspective lacks theoretical foundation, and, in addition, no one has yet empirically demonstrated (and probably never will) that 'designers think better...'. We prefer to stay with Jonathan Ive (VP of Design for Apple) who states that what is better is not design, but 'good design'. And we prefer to stand by Simon, with his indication that design is simply an activity, in this case to create new things, which is proper to any profession. And finally, we support the warning of Tomas Maldonado, a well respected theorist of design, who claims '[There is] a progressive desemanticization of the word design. As it is applied to respond to the programmatic (and promotional) needs of all kinds of activities ... the word winds up losing its specific meaning... This indeterminacy appears today as the main obstacle to a definition of design as a discipline' and slows down scientific progress in the field (Maldonado, 2000).

Design as the Innovation of Meanings: 'Making Sense of Things'

Neither of these two extremes, one narrowly focused on the form of objects, and the other considering design as basically everything, can help an understanding of how design contributes to innovation and therefore to competitive advantage. What really is peculiar about design? What makes it different from other forms of innovation that have been widely investigated, such as technological innovation? We will refer here to Klaus Krippendorf and John Heskett, two major theorists of design, to capture its peculiarity:

> The etymology of design goes back to the latin de + signare and means making something, distinguishing it by a sign, giving it significance, designating its relation to other things, owners, users or gods. Based on this original meaning, one could say: design is making sense (of things).
>
> (Krippendorff, 1989).

> Design, can be defined as the human capacity to shape and make our environment in ways without precedent in nature, to serve our needs and give meaning to our lives.
>
> (Heskett, 2002)

Both scholars clearly point to the peculiar characteristic of design: it is concerned with making things more meaningful. Design is the activity through which we innovate the meaning of things. If technological innovation is often driven by the question, 'can we make this product work better?', innovation driven by design is triggered by the question, 'does this improvement in performance make sense?', can we instead create a new experience that, even if it does not have optimal technical performance, is more meaningful to customers? When we use design as a driver of innovation we therefore move from the 'what' of a product (its features) to the 'why': we innovate the reason why and the purpose for which people buy and use things. And inherently we act not only on the utilitarian dimension of use ('I buy a car because I need to move from A to B, safely and rapidly'), but also on the emotional/symbolic meaning (see Chapter 14 by Ozaki and Dodgson). Emotional meaning is connected to individual motivation, something that makes a person feel intimately gratified (sensory experiences, e.g. 'I buy this car because of the feeling of its leather steering wheel', or psychological experiences, e.g. 'I buy this car because my parents always bought this brand'). Symbolic meaning is linked to social motivation, what the product says about myself and about the others (e.g. 'I buy this car because it tells others that I'm wealthy and have a sporting attitude'). The product style (considered as its mere aesthetic appearance) is but one of many ways a product may bring symbolic and emotional messages to customers. Everything, from products to services, from processes to business models, has meaning.

The meaningful dimension of design has been recognized and underlined by several design scholars and theorists (Cooper and Press, 1995; Margolin and Buchanen, 1995; Petroski, 1996; Friedman, 2003; Karjalainen, 2003; Lloyd and Snelders, 2003; Bayazit, 2004; Norman, 2004; Redstrom, 2006). Research in marketing, consumer behaviour and anthropology of consumption has also demonstrated that the affective/emotional and symbolic/socio-cultural dimension of consumption is as important as the utilitarian perspective of classic economic models, even for industrial clients (Douglas and Isherwood, 1980; Csikszentmihalyi and Rochberg-Halton, 1981; Fournier, 1991; Sheth et al., 1991; Kleine et al., 1993; Mano and Oliver, 1993; Brown, 1995; Du Gay, 1997; Holt,

1997, 2003; Bhat and Reddy, 1998; Schmitt, 1999; Pham et al., 2001; Oppenheimer, 2005; Shu-pei, 2005).

What matters, however, and what moves us from marketing to the realm of design, is that meanings are not simply given (and therefore we can only try to understand them through market research), but meanings can be innovated, even radically, because of the evolution of the socio-cultural context and of the emergence of new technologies. Of course, meanings cannot be imposed (they depend on the interaction between a customer and a product), but firms can design several elements to encounter and stimulate meaningful interpretations by users: from product functionality to its design language (that is the set of signs, symbols, and icons associated with a product—of which style is just one instance—that includes materials, sensory features such as sound, the user interface, etc.).

DESIGN AND INNOVATION STRATEGIES

The above definition allows us to link design more precisely with other theories of innovation and to point out its peculiar nature.

The Dimensions of Innovation

Consider in particular the diagram in Figure 8.1. Building on the above discussion we may say that innovation may be driven by technology, by meaning or both. And similarly technological innovation may imply a position along the continuum of incremental or radical change; in the same way, innovation of meaning may be more or less radical. In particular, innovation of meanings is incremental when a product has a purpose that is in line with the current evolution of socio-cultural models: it supports the existing experience of users, and fits even better with what customers are looking for. However, innovation of meaning may also be radical, which is what happens when a product enables an experience whose purpose is significantly different than previous offerings.

An (extreme) example of radical innovation of meaning is the well-known Alessi product line called 'Family Follows Fiction'. In 1991 Alessi, a leading Italian design-intensive company that operates in the kitchenware industry, created playful, colourful, and metaphoric kitchenware, with corkscrews shaped like dancing women or parrots and orange squeezers shaped like Chinese mandarins. Although today these types of symbolic object are frequently imitated, before the 1990s no one would ever have thought that people would love to have 'dancing' corkscrews. This was a breakthrough change in what kitchenware meant for people: from simple kitchen tools to 'transitional objects', that is, objects of affection that talk directly to the child that is still living inside each adult.[1] Thanks to this breakthrough in meaning Alessi's turnover in the three years after market release increased by 70 per cent (the industry average increase in turnover

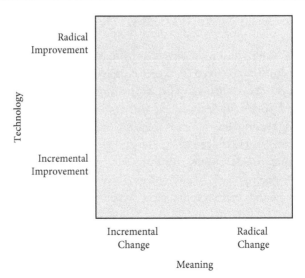

FIGURE 8.1 Technologies and meanings as dimensions of innovation

(Source: Verganti; 2009)

was 4 per cent) and its return on equity grew by 40 per cent (the industry average was an increase of 5 per cent). Nowadays Alessi releases new products within this family of toys every year. These can be considered incremental innovations of meaning: the meaning of kitchenware as affective playful objects is still there, even reinforced and updated as aesthetic taste and cultural interpretation evolve in society.

The area on the top right-hand corner of the framework in Figure 8.1. is of particular interest. Here radical change in meaning is enabled by the emergence of new technologies. Examples of this combined use of technologies and design include Nintendo, Apple, and Swatch. All three have used technologies to radically change the meaning of offerings in a category—why customers buy or how they use the product. Nintendo's clever application of MEMS (micro-electro-mechanical systems) accelerometers transformed the experience of playing with game consoles from passive immersion in a virtual world into active physical entertainment. Apple's creation of the iPod and the iTunes Store made it easier for people to discover and buy new music and organize it into personal playlists, and provided a solution to the piracy that was threatening to destroy the music industry. And Swatch used inexpensive quartz technology to change watches from timekeeping tools into fashion accessories. These companies weren't necessarily the first to introduce a new technology in the product category (the iPod was released in 2001, four years after the first MP3 player), but they unveiled its most meaningful and profitable form.

User-Driven Design

There are two strategies for innovating the meaning of things: user-driven and design-driven. User-driven design has been popular in the last decade. It has been in the spotlight thanks to the successes of major design firms such as IDEO (Kelley, 2001) or Continuum (Lojacono and Zaccai, 2004). It implies that product development should start from a deep analysis of user needs. Its assumption is that a firm may infer unique insights to inform product innovation by asking users about their needs or, more effectively, by observing them as they use existing products and by tracking their behaviour in consumption processes. The growth of interest on applied ethnographic research—that is, the practice of observing users in the context of use—is a signal and a direct consequence of this approach. Investigation of user-driven design and analysis of successful instances has helped to provide a deeper and more valuable interpretation of design as an organizational process, a process to get closer to users and their actual needs. And indeed, models of user-centred design processes, with proper steps and tools, have been proposed (Patnaik and Becker, 1999; Whitney and Kumar, 2003). Such models effectively combine, on the one hand, methods to better understand customer needs (such as ethnographic research and its variations, see for example Rosenthal and Capper, 2006) and, on the other hand, guidelines on how to improve creative skills (Sutton, 2001).

User-driven design proves to be effective for incremental innovation. As it brings a firm closer to existing behaviours, it enables a better understanding of how people *currently* give meanings to things. The purpose of the design process here is not to change that meaning, but to design products that better satisfy it. The assumption is that there is a mismatch between how people currently give meaning to things and existing products. By using ethnographic methods and observation, and therefore by getting closer to users, firms may better understand those meanings, and then, through creative problem-solving sections, may address the mismatch between existing meanings and existing products.

Design-Driven Innovation

Radical innovation of meaning, however, clearly requires a different process. People definitely didn't ask for Mandarin-like orange squeezers before 1991. But they loved Alessi's products after they saw them. Similarly, teenagers (and especially the most adept game players) did not ask for an experience of moving while playing with a console (they could not even imagine that this was possible, as they were unaware of the existence of MEMS accelerometers). They were interested in having more powerful consoles that could enable a more precise virtual experience, which led Microsoft and Sony to invest in expensive development of chips. Indeed, Nintendo envisioned the new Wii experience by stepping back from existing users' behaviour and working with game developers and the manufacturers of the accelerometers.

In fact, customers rarely help in anticipating possible radical changes in product meanings. The socio-cultural context in which they are currently immersed make them inclined to interpretations that are in line with what is happening today. Radical changes in meaning instead call for radically new interpretations of what a product is meant for, and this is something that might be understood (and affected) only by looking at things from a broader perspective. Design-driven innovation is therefore *pushed* by a firm's vision about possible breakthrough meanings that people could love. As this vision cannot be developed solely by looking at current user behaviour, the process of these firms has little in common with user-centred approaches.

In other words, in a similar way to how radical technological innovations ask for profound changes in technological regimes (Latour, 1987; Callon, 1991; Bijker and Law, 1994; Geels, 2004), radical innovations of meaning ask for profound changes in *socio-cultural regimes*. We are not talking of 'fashionable' or stylish products here, but rather of products that may contribute to the definition of new aesthetic standards, maybe something that could become an icon in the future, definitely something that plays a major role in changing socio-cultural models. In other words design-driven innovation may be considered as a manifestation of a 're-constructionist' (Kim and Mauborgne, 2004, 2005) or 'social-constructionist' (Prahalad and Ramaswamy, 2000) view of the market, where the market is not 'given' a priori (such as from the structural perspective in, e.g., Porter, 1980) but is the result of interaction between consumers and firms: new radical meanings are therefore co-generated. Design-driven innovation is not an answer to, but a dialogue with and a modification of, the market.

Tracing back design-driven innovation to theories of innovation management, we acknowledge that a similar perspective is shared by scholars of technology management. There was an intense debate in the 1970s about the direction of innovation processes (technology push versus market pull), culminating in the understanding the contributions of both technologies and markets, and that changes in technological paradigms (i.e. radical technological innovations) are mainly technology push, whereas incremental innovations within existing technological paradigms are mainly market pull (Dosi, 1982). This is an approach shared by more recent research on the relationship between disruptive innovations and user needs (Christensen and Rosenbloom, 1995; Christensen and Bower, 1996; Christensen, 1997; Dahlin and Behrens, 2005). These considerations are mapped in the diagram on the dimensions of innovation, highlighting the major areas of action of three modes of innovation (see Figure 8.2):

- *Design-driven innovation*, where innovation starts from the comprehension of subtle and unspoken dynamics in socio-cultural models and results in proposing radically new meanings and languages that often implies a change in socio-cultural regimes.
- *Market-pull innovation*, where innovation starts from the analysis of user needs, and subsequently searches for the technologies and languages that can actually satisfy them. We include user-centred innovation as a declination of market-pull innovation, as they both start from users to directly or indirectly identify directions for innovation. Although the user-centred approach is more advanced and sophisticated as its methodologies allow a better understanding of why and how people give meaning to existing things—which can lead to more innovative concepts compared to traditional market pull processes—it still operates within existing socio-cultural regimes.

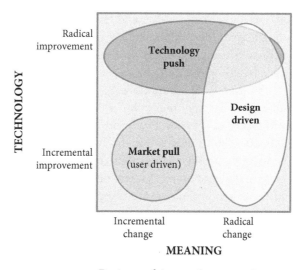

FIGURE 8.2 Design and innovation strategies

- *Technology push innovation*, that is, the result of the dynamics of technological research. The overlap between technology push and design-driven innovation in the upper right corner of the diagram highlights that breakthrough technological changes are often associated with radical changes in product meanings, that is to say, shifts in technological paradigms are often coupled with shifts in socio-cultural regimes (see Geels, 2004). The introduction of quartz watches in the 1970s, for example, was both a breakthrough change in technologies (the introduction of semiconductors) and in meanings (watches moved from being jewels to being instruments—some even had a small calculator as an additional feature!). And vice versa, radical innovations of meanings are often prompted by the availability or exploration of new technologies.

THE RELEVANCE OF DESIGN-DRIVEN INNOVATION

In today's business and academic arenas, design is increasingly viewed as an important strategic asset. The first decade of the twenty-first century has seen a rising interest in design in management and organization studies. Initially management scholars contributed to this debate through research in specialized journals (e.g. Dumas and Mintzberg, 1989, 1991; Verganti, 2003); more recent works have been published in journals of a more general orientation (e.g. Hargadon and Sutton, 1997; Boland and Collopy, 2004; Rindova and Petkova, 2007; Verganti, 2006, 2009, 2011; Michlewski, 2008). This growing attention to design has led scholars and executives to investigate and under-stand the links among design, innovation, and competitive advantage. Recent studies have specifically analysed the impacts of design management and design practices on company performance (Gemser and Leenders, 2001; Hertenstein et al., 2005; Veryzer, 2005). As demonstrated by several studies (e.g. Schmitt and Simonson, 1997; Bloch et al., 2003), consumers increasingly make choices on the basis of the aesthetic and symbolic value of products and services. Examining the significance of 'look and feel' in many industries, Postrel (2003) claims that the aesthetic and symbolic dimensions of a product are increasingly pertinent to a company's success. Consequently several companies that operate in different industries have been found to invest more and more resources in order to make their products more meaningful rather than more functional (Pesendorfer, 1995; Cappetta et al., 2006). If in the luxury and fashion industries the aes-thetic and symbolic dimensions represent critical success factors, they are also becom-ing increasingly relevant in those industries traditionally regulated by straight-forward technological evolution (Trueman and Jobber, 1998; Ravasi and Lojacono, 2005; Rindova and Petkova, 2007; Verganti, 2011). Indeed, despite the fact that companies such as Apple, Nokia, Nintendo, or Bang & Olufsen operate in industries that are usually shaped by the emergence of new technologies, the success of their products has been strongly connected to the prominent role played by aesthetic and symbolic dimensions (Cappetta et al., 2006; Cillo and Verona, 2008).

Diffusion Models in Design-Intensive Industries: Convergence and Re-interpretation

Cappetta et al. (2006) construct and test a conceptual framework for the creation and evolution of stylistic innovation in the high fashion industry. Unlike technology-based industries, for the fashion-based industries it is difficult to identify and recognize a dominant design; instead, very often groups of styles share several regularities. Cappetta et al. (2006) define *convergent design* as a style that most companies use as a reference point over a particular period of time. They explain the convergent design by idiosyn-cratic features of the context, such as the emergence of snob effects, consumers' need for differentiation, and the signalling power of style for companies.

As demonstrated by several studies on cultural anthropology and cultural branding (e.g. Holt, 2003, 2004), meanings that people associate with products very often con-verge around archetypes and icons able to survive longer than normal competitors. The rich literature on technology management proposes similar concepts when referring to dominant designs. Abernathy and Utterback (1978) define the dominant design as a particular configuration of technological parameters resulting from the successful synthesis of individual technological innovations introduced independently in prior products. Utterback (1994) elaborates this concept suggesting that a dominant design is the design that wins the allegiance of the marketplace, the one that competitors and innovators must adhere to if they hope to command significant market following. This theory of technology management illustrates that the emergence of a dominant design significantly shapes industry dynamics: competition moves from product innovation to process innovation and efficiency and the number of competitors significantly decreases (Utterback and Abernathy, 1975, Utterback and Suarez, 1993). Comparing the literature addressing the radical innovation of meanings and technologies, Verganti (2008) suggests differences and similarities between the concepts of dominant design and dominant language. Explorations of large datasets show that industry dynamics are less affected by the emergence of dominant languages (Cappetta et al., 2006; Dell'Era and Verganti, 2007).

Research conducted in the Italian furniture industry (Dell'Era and Verganti, 2007) demonstrates that several product meanings coexist over a given period of time, and they are often new interpretations of existing product languages. In other words, innovations within this industry do not necessarily require the use of new materials or new colours. *Re-interpretations* of old styles can certainly make a comeback if the meanings associated with them also become relevant again within society. Dell'Era and Verganti (2007) identify a link between design strategy and the research attitude demonstrated by a company, showing how the ability to interpret socio-cultural and aesthetic trends allows innovators to focus on a specific dominant product language, rather than experimenting with a variety of them. In other words, unlike dominant designs emerging in technological fields (Abernathy and Utterback, 1978; Utterback, 1994), several dominant languages coexist in the same industry; rapid evolutions over time and the contemporary presence of several dominant languages on the market make the identification of trends of selected languages in the market very difficult. Companies operating in this kind of industry (and especially imitators) are not able to clearly identify dominant languages, increasing the variety of proposals launched on the market.

Diffusion of New Product Meanings: Determinants and Dynamics

According to the previous reflections, innovations of product meanings can lead to articulated processes of diffusion and re-diffusion. Leveraging the rich literature on the diffusion of innovation, which is mainly dedicated to technological innovations (Fourt and Woodlock, 1960; Mansfield, 1961, 1963; Floyd, 1962; Chow, 1967; Bass, 1969; Mahajan et al., 1990, 1995, 2000; Geroski, 2000; Ruttan, 2000; Rice et al., 2002; Chakravorti, 2003, 2004; Rogers, 2003; Mukoyama, 2004; Alexander et al., 2008), Dell'Era and Verganti (2011) identify the determinants that impact the diffusion dynamics of product meanings. Specifically they analyse the impacts of marketing strategies adopted by innovators and their organizational characteristics on diffusion dynamics of new product meanings: *diffusion speed* represents the rapidity in which new product meanings spread in the market while *diffusion contagion* describes the ability to influence several adopters.

Frambach (1993) states that cooperation with other suppliers through shared technology or educating a target audience (including other producers) can increase the speed of innovation adoption. Consistent with the literature about technology-based innovations (Easingwood and Beard, 1989; Frambach, 1993), *collaboration* with other manufacturers in design-intensive industries through the adoption of the same product meaning increases *diffusion speed* (Dell'Era and Verganti, 2011). Diffusion processes activated by several companies are able to spread product meaning very quickly. The activation of diffusion processes by numerous groups of companies that share the same product meanings helps spread it very quickly (see Table 8.1). According to Frambach (1993), sharing the technology with other companies increases total demand and can allow new standards to be set. The early adoption of emerging technologies allows the introduction of 'lock-in' actions: proposing new solutions, companies that act fast and first can set new standards for competitors and partners alike. Furthermore, the number of companies proposing similar innovations can increase their capacity to influence the entire market and thus to set new standards. As demonstrated by Dell'Era and Verganti (2007) innovative Italian furniture companies establish informal collaborations with other manufacturers (even if some are competitors) and are part of the same design discourse. Supporting a continuous dialogue about socio-cultural models and patterns of consumption, companies develop collective and networked research processes on new product meanings and identify those solutions that can significantly influence the market. In other words, the more companies contemporaneously propose the same product meaning, the greater the ability to influence competitors in the adoption of the same product meaning. A *collaboration* among many companies in the launch of the same product meaning increases the capacity to influence the rest of the market as well as the possibility to facilitate *diffusion contagion* over time (see Table 8.1).

Each company can show different degrees of collaboration during the diffusion of new product meanings: in fact each company can focus its portfolio of new products on a few meanings or distribute its creativity across different styles. According to Karjalainen and Warell (2005), the recognition of similar connotations across multiple products allows for the development of consistent messages and the enhancement of identity. In design-intensive industries where several styles coexist (Cappetta et al., 2006; Dell'Era and Verganti, 2007), only those product meanings that have been contemporaneously proposed by several companies with precise identities represent signals that can be quickly perceived and followed by the rest of the market (Dell'Era and Verganti, 2011). In other words, the *focalization* positively impacts on the *diffusion speed* of new product meanings.

As demonstrated by several streams of research, innovation capability is significantly affected by the diversity of direct contacts developed by companies; the number of such contacts is relevant, to the degree that it increases the probability of network diversity (see the discussion in Chapter 6 by Kastelle and Steen). Several studies about network organizations suggest that the portfolios of partners that a firm maintains can be just

Table 8.1 Determinants of diffusion dynamics of new product meanings

	Diffusion speed	Diffusion contagion
Collaboration between companies launching new product meanings	+	+
Focalization of companies launching new product meanings	+	
System openness adopted by companies launching new product meanings		+

as influential as the dyadic characteristics of those alliances (Gulati, 1998). Different approaches and organizational backgrounds among partners can increase the number of information sources, making an organization more likely to become aware of an innovation (Zaltman et al., 1973). Rogers (2003) argues that system openness, measured as the number of links that members of an organization are able to establish with other people who are external to the organization, is positively related to innovation adoption. In design-intensive industries, collaboration with several partners increases the ability to intercept and interpret the weak signals that have the potential to become future trends. Leveraging rich networks that provide knowledge diversity, innovators are able to influence large segments of the market, pushing their competitors to adopt their product meanings (Dell'Era and Verganti, 2010). Diffusion processes activated by companies characterized by high levels of *system openness* are able to influence many other competitors and consequently positively impact *diffusion contagion* (see Table 8.1). In other words, companies immersed in dense and rich networks of designers are able to activate and anticipate diffusion processes around new product meanings adopted by the majority of the market. This means that innovativeness does not depend solely on the capacity of a single designer, but also on the amount of knowledge provided by an entire portfolio of designers characterized by significant diversity. The value of a single collaboration lies in the externalities generated by other collaborations (Dell'Era and Verganti, 2011).

DESIGN-DRIVEN INNOVATION: MAJOR THEMES, RESEARCH FINDINGS, AND THEIR IMPLICATIONS

If design-driven innovation is the creation of a radically new meaning, rather than resulting from a process of problem solving, it derives from a process of 'interpretation' (or better, re-interpretation) of the reason why people buy and use products. Indeed, *meanings*, by definition, are the result of an interpretative process. Therefore firms that develop design-driven innovation step back from users and take a much broader perspective. They explore how the context in which people live is evolving, both in socio-cultural terms (how the reason why people buy things is changing) and in technical terms (how technologies, products, and services are shaping that context). Most of all, they envision how this context of life *could* change for the better. The word 'could' is not incidental. These firms are not simply following existing trends. They are making *proposals*, with which they will *modify* the context. They build scenarios that would perhaps never occur (or occur more slowly) if they do not deliver their unsolicited proposal. It is a process of *generative* interpretation. Their question, therefore, is, 'how could people give meaning to things in their evolving context of life?'

Design-Driven Innovation as a Process of Interpretation

When a company takes this broader perspective, it discovers that it is not alone in asking that question. Every company is surrounded by several agents (firms in other industries that target its same users, suppliers of new technologies, researchers, designers, artists, etc.) who share its same interest. Consider, for example, a food company that, instead of closely looking with a lens at how a person cuts cheese, asks, 'what meanings could family members search for when they are at home and are going to have dinner?' This same question is investigated by other actors: kitchen manufacturers, white-goods manufacturers, TV broadcasters, architects who design home interiors, food journalists, and so on. Although they have different users, they all look at the same person in the same context of life. And they conduct research about how that person could give meaning to things. They are, in other words, *interpreters*.

Companies that produce design-driven innovations value highly their interaction with this network of interpreters. They exchange information with them on scenarios, test the robustness of their assumptions, and discuss their own visions. What these companies have understood is that knowledge about meanings is diffused within their external environment; that they are immersed into a collective *research laboratory*, where interpreters make their own investigations and are engaged in a continuous mutual dialogue (see Figure 8.3). The process of design-driven innovation therefore gets close to interpreters. It leverages their ability to understand and influence how people could give meaning to things. This process consists of three *actions*.

The first one is *Listening*. It is the action of *accessing* knowledge about new possible product meanings by interacting with interpreters. Firms that 'listen' well are those that develop privileged relationships with a distinguished group of key interpreters. These are not necessarily the best known in the industry. Rather, successful firms are those that first identify unspotted interpreters, usually in fields where competitors are not searching. Key interpreters are forward-looking researchers that are developing, often for their own purposes, unique visions and explorations about how meaning could evolve in the context of life we want to investigate. Firms that realize design-driven innovations are able to detect, attract, and interact with key interpreters better than their competitors.

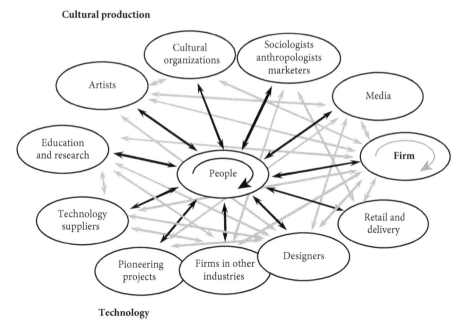

Cultural production

Technology

FIGURE 8.3 The interpreters in design-driven innovation

The second action is *Interpreting*. Its purpose is to develop a firm's unique proposal. It is the internal process in which knowledge accessed by interacting with interpreters is recombined and integrated with a firm's proprietary insights, technologies, and assets. This process implies a sharing of knowledge through exploratory experiments rather than extemporaneous creativity. In many ways it is more like the process of engineering research (although working on meanings instead of technologies) than of a creative agency. Its outcome is the development of a breakthrough meaning for a product family.

The third action is *Addressing*. Radical innovations of meanings, being unexpected, sometimes initially confuse people. To prepare the ground for groundbreaking proposals, firms leverage the *seductive power* of interpreters. By having interpreters discuss and internalize a firm's novel vision, they will inevitably change the broader context (through the technologies they will develop, the product and services they will design, the artwork they will create) in a way that will make the company proposal more meaningful and attractive once people see it.

The Role of Designers as Interpreters

A key role in this network of external interpreters is played, of course, by designers. Designers can support companies in the identification and interpretation of how people give meaning to things, and, most of all (which makes them different than anthropologists, or sociologists), they can *envision* new possible meanings, new experiences that do not exist, mostly because of their understanding of technology; their ability to investigate user needs and the evolution of socio-cultural models can support the scenario-building activities and consequently the development of radical innovation of product meanings. Specifically, leveraging their knowledge about technologies and processes, designers embed new meanings and insights about the changing culture of consumers into new products. Acting as cultural gatekeepers, designers can support companies in the interpretation of different cultures and complex social phenomena. As demonstrated by several studies, collaboration with external consultants rather than the exploitation of internal capabilities can provide several advantages. Internal designers, though familiar with the company's approach and products, tend to become complacent and, in turn, less innovative (Bruce and Morris, 1998). By contrast, external design consultants tend to provide fresh and more innovative concepts. The opportunity to collaborate with companies that operate in different industries allows designers to transfer languages from one sector to another (Capaldo, 2007). As demonstrated by Verganti (2003), languages and meanings can easily move from one industry to another. From a managerial perspective, this property implies great innovative and creative stimulus. By capturing, recombining, and integrating knowledge about socio-cultural models and product semantics in several different social and industry settings, designers act as brokers of design language and creators of breakthrough product meanings. Similar to technology brokers, designers are able to transfer product languages and meanings across industries, exploiting their connections and networks (Hargadon and Sutton, 1997). The culture in which people live influences the connection between product languages and meanings (Lloyd and Snelders, 2003). This means that designers of different nationalities can provide different viewpoints and support companies in the interpretation of product meanings to match the social and cultural needs of the people who live in different countries. Collaboration with external designers in fact represents a diffuse practice in several industries and aims at sourcing fresh insights, creativity, and new knowledge (Verganti, 2003; Cillo and Verona, 2008). Considering that in today's economy firms recognize that most of the valuable knowledge for innovation resides outside their boundaries, collaboration with external designers is in line with a general tendency towards open innovation and the development of business ecosystems (Rigby and Zook, 2002; Chesbrough, 2003; Soh and Roberts, 2003; Iansiti and Levien, 2004; Sorenson and Waguespack, 2005; Huston and Sakkab, 2006; Lakhani and Panetta, 2007; Pisano and Verganti, 2008; see also Chapter 22 by Alexy and Dahlander and Chapter 11 by Autio and Thomas).

Studies have underlined the importance of external designers in the innovation process to the point that some of them are considered 'superstars': Jacob Jensen and David Lewis for Bang & Olufsen, Michael Graves for Target, Philippe Starck for several furniture companies, as well as for Nike and Puma (Gierke et al., 2002; Ravasi and Lojacono, 2005; Durgee, 2006). Despite the rhetoric of design magazines, however, the success of firms seems not necessarily related to the choice of a specific designer, but rather to the capability to identify and manage an articulated *portfolio* of designers. Alessi, for example, leverages on a network of more than 200 external designers rather than collaborate with few of them (i.e. the more famous and successful). This indicates that Alessi's innovativeness cannot be explained by reference to the talent of few external designers, but to the firm's ability to build a complex portfolio. Similarly, single designers who work with Alessi do not seem to have provided analogous value when working with other firms (Heimeriks et al., 2009; Holmberg and Cummings, 2009; Verganti, 2009). Rather than an individual spark of creativity, the value of the contribution of each designer is hard to identify unless it is seen within the context of the knowledge sourced from the entire array of external collaborators. And from the opposite perspective, knowledge developed through collaboration with a specific designer can be exploited in a number of projects (eventually developed with other designers). In other words, the value of a single collaboration benefits from externalities generated by other collaborations. In this light, Dell'Era and Verganti (2010) demonstrate that innovators and their competitors build completely different designer portfolios. By analysing the educational backgrounds of designers, for example, it emerges that innovators mix different approaches and mindsets in order to design unique and original products. They build a network of multiple interpreters with different backgrounds and from several countries. The knowledge diversity developed through the collaboration with designers of different backgrounds and experiences has to be interpreted as a cumulative asset. The identification, selection, and attraction of key creative collaborators requires significant investment. These activities are crucial for companies in design-intensive industries. They allow a creative asset to grow over time and this can be exploited by the company in several ways: the knowledge developed through a specific collaboration typically remains 'stuck' to the company and can consequently be used in other projects. In other words, knowledge collected through collaboration with French cross-industrial architects can be re-interpreted and used in projects developed by Italian in-house furniture engineers. Relationships developed with creative talents represent a distinctive asset that provides the opportunity to enter into a new business. The ability required to identify and select key designers needs a deep knowledge of the domain of innovation. Innovativeness does not depend on the diversity brought by an individual designer, but on the diversity brought by the entire portfolio of designers of a firm. The bottom line implication is that firms should not focus only on the characteristics of single external parties when developing a collaborative innovation strategy, but, rather, manage carefully a balanced portfolio of collaborators.

A New Perspective on Innovation

Innovation managers have traditionally considered that their main challenge is to find and develop new technologies before their competitors. But as companies open their innovation process to external parties, new technological opportunities become more easily accessible. Thanks to intensifying cooperation between research institutions and corporations, to innovation marketplaces such as Innocentive or NineSigma, and to the growth of creative communities, there is today a proliferation of technologies, ideas, and solutions. In this context, the main challenge for innovation managers is therefore shifting from technology *development* to technology *interpretation*: 'what can we do with this wealth of technological opportunities?' It is not just a matter of screening options but, rather, of *envisioning* the meaningful application of a combination of new (and old) technologies that better suit the market. Being first in *launching* a new technology is therefore less relevant than being first in finding the right *meaning* of that technology.

Unfortunately, the subject of design as an innovation of product meanings has largely been neglected in management studies. Whereas literature on the management of innovation has explored the antecedents of radical change of technologies, we still miss a deep investigation of the dynamics of radical change in meaning. A cause of this lack of investigation is that the nature of the innovation of meaning is peculiar: it involves symbolic, emotional, and intangible factors. It is therefore problematic to investigate this type of innovation with the lenses of classical theories of innovation, conceived mainly for tangible factors, such as technology, utility, performance, and function. We therefore need new approaches and frameworks.

What is more interesting is that design, in this light, helps us to address innovation from a totally new perspective. Unlike established theories that consider innovation as stemming from a process of *problem solving* (i.e. the search for a *optimal* solution of a *given* problem), or from a process of *ideation* (where the assumption is that once one has a good idea, its value can be easily recognized and captured), design-driven innovation tells us that some of the most intriguing and yet valuable events in innovation come from a process of interpretation and envisioning. Innovation through design therefore becomes a hermeneutic process: a process to search for new meaning (Öberg and Verganti, 2011). It is a breakthrough in paradigm from the mantras of many innovation studies and approaches that have populated the innovation rhetoric in the past decade (focused on user-driven processes, where creativity based on several divergent ideas is considered to be a key factor). Here we talk of a vision-driven process in which instead of thousands of ideas, posits it is one where strong re-interpretation that matters. It is a new avenue for the world of innovation. And it is an intriguing one.

Although existing approaches to innovation management (innovation as problem solving and innovation as ideation) are still useful for investigating the dynamics of innovation of meaning, there is a need for a richer perspective to fully capture the real nature of this type of innovation. Our preliminary investigations of case studies on the radical innovation of meanings indicate that the major challenge is not to generate ideas, nor to solve problems, but to recognize the value of these ideas by envisioning them into a new context. The focus, therefore, moves from product development (the field of investigation of the 1990s) and idea generation (the focus of investigations of the 2000s) to how firms reframe the way they make sense of opportunities, which is a matter of innovation strategy and vision creation. Studies of Actor Network Theory (Latour, 1987; Bijker and Law, 1994), diffusion of innovation (Rogers, 2003) and sensemaking in organizations (Weick, 1995), adopt a similar stance. Whereas they introduce a sociological dimension to innovation, their approach considers meanings as a contextual factor of innovation: something that explains how innovation (in technology or strategy) occurs, through interactions in society, markets and within organizations. We suggest companies focus on meanings as the output of the innovation process: the target that a company wants to achieve.

NOTE

1. When creating the 'Family Follows Fiction' product line, Alessi was inspired by the theories of paediatrician and psychoanalyst David Winnicott on transitional objects, of psychoanalyst Franco Fornari on affective codes, and of Jean Baudrilliard on object systems. David Winnicott in particular focused on the role that objects have in the psychological development of children, who associate feelings and meanings to their daily objects. He investigated the role of transitional incremental innovations of meaning: the meaning of kitchenware as affective playful objects is still there, even reinforced and updated as aesthetic taste and cultural interpretation evolve in society.

REFERENCES

Abernathy, W., and Clark, K. (1985). 'Innovation: Mapping the Winds of Creative Destruction', *Research Policy*, 14: 3–22.

Abernathy, W. J., and Utterback, J. M. (1978). 'Patterns of Industrial Innovation', *Technology Review*, June–July: 40–7.

Alexander, D. L., Lynch, J. G., and Wang, Q. (2008). 'As Times Go By: Do Cold Feet Follow Warm Intention for Really-new vs. Incrementally-new Products?', *Journal of Marketing Research*, 45: 307–19.

Bass, F. M. (1969). 'A New Product Growth Model for Consumer Durables', *Management Science*, 15: 215–27.

Bayazit, N. (2004). 'Investigating Design: A Review of Forty Years of Design Research', *Design Issues*, 20(1) Winter.

Bhat, S., and Reddy, S. K. (1998). 'Symbolic and Functional Positioning of Brands', *Journal of Consumer Marketing*, 15(1): 32–47.

Bijker, W. and Law, J. (eds), (1994). *Shaping Technology / Building Society: Studies in Sociotechnical Change*. Cambridge, MA: MIT Press.

Bloch, P. H., Frederic, F. B., and Todd, J. A. (2003). 'Individual Differences in the Centrality of Visual Product Aesthetics: Concept and Measurement', *Journal of Consumer Research*, 29: 551–65.

Boland, R. J., and Collopy, F. (2004). *Managing as Designing*. Stanford, CA: Stanford University Press.

Brown, S. (1995). *Postmodern marketing*. London: Routledge.

Brown, T. (2008). 'Design Thinking', *Harvard Business Review*, 84–92.

Brown, T. (2009). *Change by Design: How Design Thinking Transforms Organizations and Inspires Innovation*. London: HarperCollins.

Bruce, M., and Morris, B. (1998). 'In-house, Outsourced, or a Mixed Approach to Design', In Bruce M. and Jevnaker, B. H. (1998), *Management of Design Alliances: Sustaining Competitive Advantage*. Chichester: Wiley, 39–64.

Calantone, R. J., Harmancioglu, N., and Dröge, C. (2010). 'Inconclusive Innovation "Returns": A Meta-Analysis of Research on Innovation in New Product Development', *Journal of Product Innovation Management*, 27: 1065–81.

Callon, M. (1991). 'Techno-economic Networks and Irreversibility', in J. Law (ed.), *A Sociology of Monsters: Essays on power, technology and domination*. London: Routledge, 132–61.

Capaldo, A. (2007). 'Network Structure and Innovation: The Leveraging of a Dual Network as a Distinctive Relational Capability', *Strategic Management Journal*, 28: 585–608.

Cappetta, R., Cillo, P., and Ponti, A. (2006). 'Convergent designs in fine fashion: An evolutionary model for stylistic innovation', *Research Policy*, 35: 1273–90.

Chakravorti, B. (2003). *The Slow Pace of Fast Change: Bringing Innovation to Market in a Connected World*. Boston, MA: Harvard Business Press.

Chakravorti, B. (2004). 'The New Rules for Bringing Innovations to Market', *Harvard Business Review*, 82(3): 58–67.

Chesbrough, H. W. (2003). *Open Innovation: The New Imperative for Creating and Profiting from Technology*. Boston, MA: Harvard Business School Press.

Chow, G. C. (1967). 'Technological change and demand for consumers', *American Economic Review*, 57: 1117–30.

Christensen, C. M. (1997). *The Innovator's Dilemma: When New Technologies Cause Great Firms to Fail*. Boston, MA: Harvard Business School Press.

Christensen, C., and Bower, J. (1996). 'Customer Power, Strategic Investment, and the Failure of Leading Firms', *Strategic Management Journal*, 17: 197–218.

Christensen, C., and Rosenbloom, R. (1995). 'Explaining the Attacker's Advantage: Technological Paradigms, Organizational Dynamics, and the Value Network', *Research Policy*, 24: 233–57.

Cillo, P., and Verona, G. (2008). 'Search Styles in Style Searching: Exploring Innovation Strategies in Fashion Firms', *Long Range Planning*, 41(6): 650–71.

Cooper, R., and Press, M. (1995). *The Design Agenda*. Chichester: John Wiley and Sons.

Csikszentmihalyi, M., and Rochberg-Halton, E. (1981). *The Meaning of Things: Domestic Symbols and the Self*. Cambridge: Cambridge University Press.

Dahlin, K. B., and Behrens, D. M. (2005). 'When is an Invention Really Radical? Defining and Measuring Technological Radicalness', *Research Policy* 34: 717–37.

Dell'Era, C., and Verganti, R. (2007). 'Strategies of Innovation and Imitation of Product Languages', *Journal of Product Innovation Management*, 24: 580–99.

Dell'Era, C., and Verganti, R. (2010). 'Collaborative Strategies in Design-intensive Industries: Knowledge Diversity and Innovation', *Long Range Planning*, 43: 123–41.

Dell'Era, C., and Verganti, R. (2011). 'Diffusion Processes of Product Meanings in Design-intensive Industries: Determinants and Dynamics', *Journal of Product Innovation Management*, 28: 881–95.

Dosi, G. (1982). 'Technological Paradigms and Technological Trajectories: A Suggested Interpretation of the Determinants and Directions of Technical Change', *Research Policy*, 11: 147–62.

Douglas, M., and Isherwood, B. (1980). *The World of Goods: Towards an Anthropology of Consumption*. Harmondsworth: Penguin.

Du Gay, P. (ed.) (1997). *Production of Culture: Cultures of Production*. London: Sage.

Dumas, A., and Mintzberg, H. (1989). 'Managing Design/Designing Management', *Design Management Journal*, 1(1): 37–43.

Dumas, A., and Mintzberg, H. (1991). 'Managing the Form, Function and Fit of Design', *Design Management Journal*, 2: 26–31.

Durgee, J. F. (2006). 'Freedom of Superstar Designers? Lessons from Art History', *Design Management Review*, 17(3): 29–34.

Easingwood, C., and Beard, C. (1989). 'High Technology Launch Strategies in the UK', *Industrial Marketing Management*, 18: 125–38.

Eco, U. (ed.) (2004). *History of Beauty*. Milan: Rizzoli International Publications.

Floyd, A. (1962). 'Trend Forecasting: A Methodology for Figure of Merit', in J. British (ed.), *Technological Forecasting for Industry and Government*. New Jersey: Prentice Hall, 95–105.

Fournier, S. (1991). 'Meaning-based Framework for the Study of Consumer/object Relations', *Advances in Consumer Research*, 18: 736–42.

Fourt, L. A., and Woodlock, J. W. (1960). 'Early Prediction of Early Success of New Grocery Products', *Journal of Marketing*, 25: 31–8.

Frambach, R. T. (1993). 'An Integrated Model of Organizational Adoption and Diffusion of Innovations', *European Journal of Marketing*, 27(5): 22–41.

Friedman, K. (2003). 'Theory Construction in Design Research: Criteria: Approaches, and Methods', *Design Studies*, 24: 507–22.

Garcia, R., and Calantone, R. (2002). 'A Critical Look at Technological Innovation Typology and Innovativeness Terminology: A Literature Review', *Journal of Product Innovation Management*, 19: 110–32.

Geels, F. W. (2004). 'From Sectoral Systems of Innovation to Socio-technical Systems: Insights about Dynamics and Change from Sociology and Institutional Theory', *Research Policy*, 33: 897–920.

Gemser, G., and Leenders, M. (2001). 'How Integrating Industrial Design in the Product Development Process Impacts on Company Performance', *Journal of Product Innovation Management*, 18: 28–38.

Geroski, P. A. (2000). 'Models of Technology Diffusion', *Research Policy*, 29: 603–25.

Gierke, M., Hansen, J. G., and Turner, R. (2002). 'Wise Counsel: A Trinity of Perspectives on the Business Value of Design', *Design Management Journal*, 13(1): 10.

Gulati, R. (1998). 'Alliances and Networks', *Strategic Management Journal*, 19(4): 293–317.

Hargadon, A., and Sutton, R. I. (1997). 'Technology Brokering and Innovation in a Product Development Firm', *Administrative Science Quarterly*, 42: 716–49.

Heimeriks, K. H., Klijn, E., and Reuer, J. J. (2009). 'Building Capabilities for Alliance Portfolios', *Long Range Planning*, 42(2): 96–114.

Henderson, R. M., and Clark, K. M. (1990). 'Architectural Innovation: The Reconfiguration of Existing Product Technologies and the Failure of Established Firms', *Administrative Science Quarterly*, 35: 9–30.

Hertenstein, J. H., Platt, M. B., and Veryzer, R. W. (2005). 'The Impact of Industrial Design Effectiveness on Corporate Financial Performance', *Journal of Product Innovation Management*, 22: 3–21.

Heskett, J. (2002). *Design, a very short introduction*. Oxford: Oxford University Press.

Holmberg, S. R., and Cummings, J. L. (2009). 'Building Successful Strategic Alliances: Strategic Process and Analytical Tool for Selecting Partner Industries and Firms', *Long Range Planning*, 42(1): 164–93.

Holt, D. B. (1997). 'A Poststructuralist Lifestyle Analysis: Conceptualizing the Social Patterning of Consumption in Postmodernity', *Journal of Consumer Research*, 23 (March): 326–50.

Holt, D. B. (2003). 'What Becomes an Icon Most?', *Harvard Business Review* (March): 3–8.

Holt, D. B. (2004). *How Brands become Icons: The Principle of Cultural Branding*. Boston, MA: Harvard Business Press.

Huston, L., and Sakkab, N. (2006). 'Connect and Develop; Inside Procter & Gamble's New Model for Innovation', *Harvard Business Review*, March 2006.

Iansiti, M., and Levien, R. (2004). *The Keystone Advantage: What the New Dynamics of Business Ecosystems Mean for Strategy, Innovation and Sustainability*. Boston, MA: Harvard Business School Press.

JPIM, (2005a). *Journal of Product Innovation Management*, 22: 1.

JPIM, (2005b). *Journal of Product Innovation Management*, 21: 2.

Karjalainen, T. M. (2003). 'Strategic Design Language: Transforming Brand Identity into Product Design Elements', *10th International Product Development Management Conference*, Brussels, June 10–11.

Karjalainen, T. M., and Warell, A. (2005). 'Do you Recognise this Tea Flask?: Transformation of Brand-specific Product Identity Through Visual Design Cues', *Proceedings of International Design Congress—IASDR 2005*, Taiwan, 31 October–4 November 2005.

Kelley, T. (2001). *The Art of Innovation*. New York: Curreny.

Kim, W. C., and Mauborgne, R. (2004). 'Blue Ocean Strategy', *Harvard Business Review* (October): 1–9.

Kim, W. C., and Mauborgne, R. (2005). 'Blue Ocean Strategy: From Theory to Practice', *California Management Review*, 47(3): 105–21.

Kleine, III R. E., Kleine, S. S., and Kernan, J. B. (1993). 'Mundane Consumption and the Self: A Social-identity Perspective', *Journal of Consumer Psychology*, 2(3): 209–35.

Krippendorff, K. (1989). 'On the Essential Contexts of Artifacts or on the Proposition that Design is Making Sense (of Things)', *Design Issues*, 5(2): 9–38.

Lakhani, K. R., and Panetta, J. A. (2007). 'The Principles of Distributed Innovation', *Innovations* (Summer): 97–112.

Latour, B. (1987). *Science in Action: How to follow scientists and engineers through society*. Cambridge, MA: Harvard University Press.

Lloyd, P., and Snelders, D. (2003). 'What Was Philippe Starck Thinking Of?', *Design Studies*, 24: 237–53.

Lojacono, G., and Zaccai, G. (2004). 'The Evolution of the Design-Inspired Enterprise', *Sloan Management Review* (Spring): 75–79.

Love, T. (2000). 'Philosophy of Design: A Metatheoretical Structure for Design Theory', *Design Studies*, 21: 293–313.

McGrath, R., and MacMillan, J. (2009). *Discovery Driven Innovation*. Boston, MA: Harvard Business Press.

Mahajan, V., Muller, E., and Bass, F. M. (1990). 'New Product Diffusion Models in Marketing: A Review and Directions for Research', *The Journal of Marketing*, 54(1): 1–26.

Mahajan, V., Muller, E., and Bass, F. M. (1995). 'Diffusion of New Products: Empirical Generalizations and Managerial Uses', *Marketing Science*, 14(3): 79–88.

Mahajan, V., Muller, E., and Wind, Y. (eds) (2000). *New Product Diffusion Models*. London: Kluwer Academic Publishers.

Maldonado, T. (2000). 'Opening lecture', Design + Research Conference, Milano, 18–20 May.

Mano, H., and Oliver, R. L. (1993). 'Assessing the Dimensionality and Structure of the Consumption Experience: Evaluation, Feeling, and Satisfaction', *Journal of Consumer Research*, 20: 451–66.

Mansfield, E. (1961). 'Technical Change and Rate of Imitation', *Econometrica*, 29: 741–66.

Mansfield, E. (1963). 'The Speed of Response of Firms to New Techniques', *Quarterly Journal of Economics*, 77: 290–311.

Margolin, V., and Buchanen, R. (eds) (1995). *The Idea of Design: A Design Issues Reader*. Cambridge: MIT Press.

Martin, R. (2009). *The Design of Business: Why Design Thinking is the Next Competitive Advantage*. Boston, MA: Harvard Business Press.

Michlewski, K. (2008). 'Uncovering Design Attitude: Inside the Culture of Designers', *Organization Studies*, 29: 373–92.

Mukoyama, T. (2004). 'Diffusion and Innovation of New Technologies under Skill Heterogeneity', *Journal of Economic Growth*, 9(4): 451–79.

Norman, D. A. (2004). *Emotional Design. Why We Love (or Hate) Everyday Things* (New York: Basic Books).

Nussbaum, B. (2005). 'The Power of Design', *BusinessWeek*, Cover Story, 17 May.

Öberg, Å., and Verganti, R. (2011). 'Vision and Innovation of Meaning: Hermeneutics and the Search for Technology Epiphanies', *18th EIASM International Product Development Management Conference*, 6–7 June 2011, Delft, the Netherlands.

Oppenheimer, A. (2005). 'Products Talking to People: Conversation Closes the Gap Between Products and Consumers', *Journal of Product Innovation Management*, 22: 82–91.

Patnaik, D., and Becker, R. (1999). 'Needfinding: The Way and How of Uncovering People's Needs', *Design Management Journal*, 2: 37–43.

Pesendorfer, W. (1995). 'Design Innovation and Fashion Cycles', *The American Economic Review*, 85: 771–92.

Petroski, H. (1996). *Invention by Design*. Cambridge, MA: Harvard University Press.

Pham, M. T., Cohen, J. B., Pracejus, J. W., and Hughes, G. D. (2001). 'Affect Monitoring and the Primacy of Feelings in Judgment', *Journal of Consumer Research*, 28: 167–88.

Platt, M. B., Hertenstein, J. N., and David, R. B. (2001). 'Valuing Design: Enhancing Corporate Performance through Design Effectiveness', *Design Management Journal*, 12(3): 10–19.

Pisano, G., and Verganti, R. (2008). 'Which Kind of Collaboration is Right for You?', *Harvard Business Review*, December.

Porter, M. (1980). *Competitive Strategy: Techniques for Analysing Industries and Competitors*. New York: Free Press.

Postrel, V. (2003). *The Substance of Style*. New York: Harper Collins Publishers.

Prahalad, C. K., and Ramaswamy, V. (2000). 'Co-opting Customer Competence', *Harvard Business Review* (January–February): 79–87.

Ravasi, D., and Lojacono, G. (2005). 'Managing Design and Designers for Strategic Renewal', *Long Range Planning*, 38: 51–77.

Redstrom, J. (2006). 'Towards User Design? On the Shift from Object to User as the Subject of Design', *Design Studies*, 27: 123–39.

Rice, M., Leifer, R., and Colarelli O'Connor, G. (2002). 'Commercializing Discontinuous Innovations: Bridging the Gap from Discontinuous Innovation Project to Operations', *IEEE Transactions on Engineering Management*, 49(4): 330–40.

Rigby, D., and Zook, C. (2002). 'Open-market Innovation', *Harvard Business Review*, 80(10): 80–9.

Rindova, V. P., and Petkova, A. P. (2007). 'When is a New Thing a Good Thing? Technological Change, Product Form Design, and Perceptions of Value for Product Innovations', *Organization Science*, 18(2): 217–32.

Rogers, E. M. (2003). *Diffusion of Innovations*, 6th edn. New York: The Free Press.

Rosenthal, S. R., and Capper, M. (2006). 'Ethnographies in the Front End: Designing for Enhanced Customer Experiences', *Journal of Product Innovation Management*, 23(3): 215–37.

Ruttan, V. W. (2000). 'Technology Adoption, Diffusion, and Transfer'. In V. W. Ruttan, *Technology, Growth, and Development: An Induced Innovation Perspective*. New York: Oxford University Press.

Sanderson, M., and Uzumeri, M. (1995). 'Managing Product Families: The Case of the Sony Walkman', *Research Policy*, 24(5): 761–82.

Schmitt, B. (1999). *Experiential Marketing: How to get Customers to Sense, Feel, Think, ACT, and Relate to your Company and Brands*. New York: The Free Press.

Schmitt, B., and Simonson, A. (1997). *Marketing Aesthetics: The Strategic Management of Brands Identity and Image*. New York: Free Press.

Sheth, J. N., Newman, B. I., and Gross, B. L. (1991). 'Why we Buy What we Buy: A Theory of Consumption Values', *Journal of Business Research*, 22: 159–70.

Shu-pei, T. (2005). 'Utility, Cultural Symbolism and Emotion: A Comprehensive Model of Brand Purchase Value', *International Journal of Research in Marketing*, 22: 277–91.

Simon, H. (1982). *The Sciences of the Artificial*, 2nd edn. Cambridge, MA: The MIT Press.

Simon, H. (1996). *The Sciences of the Artificial*, 3rd edn. Cambridge, MA: The MIT Press.

Soh, P. H., and Roberts, E. B. (2003). 'Networks of Innovators: A Longitudinal Perspective', *Research Policy*, 32: 1569–88.

Sorenson, O., and Waguespack, D. M. (2005). 'Research on Social Networks and the Organization of Research and Development: An Introductory Essay', *Journal of Engineering and Technology Management*, 22: 1–7.

Sutton, R. I. (2001). 'The Weird Rules of Creativity', *Harvard Business Review* (September): 95–103.

Trueman, M., and Jobber, D. (1998). 'Competing Through Design', *Long Range Planning*, 31: 594–605.

Utterback, J. M. (1994). *Mastering the Dynamics of Innovation*. Boston, MA: Harvard Business Press.

Utterback, J. M., and Suarez, F. F. (1993). 'Innovation, Competition, and Industry Structure', *Research Policy*, 22: 1–21.

Utterback, J. M., and Abernathy, W. J. (1975). 'A Dynamic Model of Process and Product Innovation', *OMEGA, The International Journal of Management Science* 3(6): 639–56.

Valente, T. M., and Davis, R. L. (1999). 'Accelerating the Diffusion of Innovations using Opinion Leaders', *The Annals of the American Academy* (November): 566.

Verganti, R. (2003). 'Design as Brokering of Languages: The Role of Designers in the Innovation Strategy of Italian Firms', *Design Management Journal*, 3: 34–42.

Verganti, R. (2006). 'Innovating Through Design', *Harvard Business Review*, December: 1–8.

Verganti, R. (2008). 'Design, Meanings, and Radical Innovation: A Metamodel and a Research Agenda', *Journal of Product Innovation Management*, 25: 436–56.

Verganti, R. (2009). *Design-Driven Innovation: Changing the Rules of Competition by Radically Innovating What Things Mean*. Boston, MA: Harvard Business Press.

Verganti, R. (2011). 'Designing Breakthrough Products', *Harvard Business Review*, October.

Veryzer, R. W. (2005). 'The Roles of Marketing and Industrial Design in Discontinuous New Product Development', *Journal of Product Innovation Management*, 22(1): 22–41.

Weick, K. (1995). *Sensemaking in Organizations*. Thousand Oaks, CA: Sage.

Whitney, P., and Kumar, V. (2003). 'Faster, Cheaper, Deeper User Research', *Design Management Journal*, 14(2): 50–57.

Wind, J., and Mahajan, V. (1987). 'Marketing Hype: A New Perspective for New Product Reseach and Introduction', *Journal of Product Innovation Management*, 4(1): 43–9.

Zaltman, G., Duncan, R., and Holbek, J. M. A. (1973). *Innovations and Organizations*. New York: Wiley.

CHAPTER 9

....

BROKERAGE AND
INNOVATION

....

ANDREW HARGADON

BROKERAGE models of innovation explain how individuals and organizations innovate by recombining past knowledge and practices in new ways. Common to these theories is an emphasis on connecting events, conditions, and actions at multiple levels of analysis, from the institutional to the organizational to the individual. Indeed, these theories draw dependencies between typically disparate approaches to innovation found in the literature: the larger social and structural conditions that shape the potential for (and record the impact of) innovation; the management strategies and work practices of organizations in pursuit of innovation; and the individual and collective action and cognition that make up the innovation process.

Such brokerage theories begin by seeing the process as one of recombining existing ideas, artefacts, and people. This is a well-recognized fact of innovation. In 1922, sociologist William Ogburn defined it as 'combining existing and known elements of culture in order to form a new element' (1922, in Basalla, 1988: 21). In 1929, historian Abbot Payton Usher (1929: 11) noted that 'Invention finds its distinctive feature in the constructive assimilation of pre-existing elements into new syntheses, new patterns, or new configurations of behaviour.' And in 1934, economist Joseph Schumpeter wrote that '[Innovation], to produce other things, or the same things by a different method, means to combine these materials and forces differently' (1934: 656). Yet, research into the management of innovation has more often emphasized the pursuit of novel ideas than the constructive assimilation of old ones.

So where traditional models focus on the generation of novel solutions, brokerage theories focus instead on how managers recognize and recombine existing resources. Thomas Edison, for example, remains the icon of invention yet his work can alternatively be seen as the recombination of pre-existing elements—ideas, artefacts, and—people—beginning in the telegraph industry and moving into new applications for other markets (Conot, 1979; Friedel and Israel, 1986). In Edison's case, his most prolific period of invention took place in the five years from 1876 to 1881, when his company operated as a consulting firm spanning multiple industries and often adapting the electro-mechanical technologies first developed for the telegraph industry for uses elsewhere. And where traditional models often implicitly equate novel ideas (e.g. radical) with revolutionary impact (e.g. disruptive), brokerage models recognize how continuity rather than novelty drives the resulting impact. Edison's success in building the first large-scale, reliable, and profitable system of incandescent lighting, for example, depended on the advances in generators, wiring, and bulbs developed by others in the decade prior to his own work (Hughes, 1989).

By embedding innovation within a larger context, brokerage theories focus on the potential created by the varieties of knowledge and practice that exist across (relatively) disconnected domains and their possible combination as applications for any one domain. Rather than view Edison as a particularly inventive individual, brokerage theories place his work practices and organizational strategy within the context of a fragmented industrial landscape in which he could profit by moving the knowledge and practice of solutions from where they were known (the telegraph industry) to where they were not. IDEO, a modern exemplar of innovation, similarly works as a consultant across a range of disparate industries and its success hinges on its capacity to constructively assimilate the ideas, artefacts, and people it learned of through its past engagements (Hargadon and Sutton, 1997).

In this way, brokerage theories recognize how individual practices and the management strategies intended to support them are both enabled and constrained by the larger social structures in which they take place. Innovative organizations, such as Edison and IDEO, should be studied within the context of their larger institutional landscapes and, in particular, as operating within and across multiple domains. And in this way, they can be distinguished from organizations that operate within one or a few domains. Similarly, practices like group brainstorming can be considered in the larger institutional context from which participants are drawn, and seen as means not for generating novel ideas but rather for eliciting and recombining existing ideas from across individual experiences drawn from multiple domains (Sutton and Hargadon, 1996; Sawyer 2007). In other words, the effectiveness of brainstorming may depend less on the specifics of its practice than on the variety of relevant but disparate experiences that participants bring.

Brokerage theories have been used to explain a range of phenomena in the study of innovation. For example,

- how some organizations are able to pursue strategies of continuous innovation, by finding and exploiting brokerage positions spanning multiple disconnected domains (Hargadon and Sutton, 1997; Hargadon, 1998);
- how particular innovations emerge from diverse communities, through the work of individuals and organizations that bridge these larger communities and build new solutions from their diversity (Burt, 2004; Lingo and O'Mahony, 2010);

- why the recombinant nature of such innovations shapes their impact, built as they are from extant institutional resources (Hargadon, 2003; Chen and O'Mahony, 2007); and
- how individual and group cognition is both constrained and enabled by exposure to the schemas, scripts, and identities of multiple domains rather than immersion in one (DiMaggio, 1997; Hargadon and Douglas, 2001; Hargadon and Fanelli, 2002).

ORIGINS IN LITERATURE

Before discussing the scope of research within brokerage models, it is useful to briefly recognize its origins. In addition to extant research on the management of innovation, brokerage models draw significantly from several other streams: social network theory; actor network theory; cognitive psychology and problem solving; and neo-institutional theory. The most overt dependence is on social network theory, and particularly the work on brokerage and structural holes, which recognizes the role of individuals and organizations and the opportunities presented them by their surrounding network structures (Granovetter, 1973; Gould and Fernandez, 1989; Burt, 1992). Social network theory, with its conceptions of weak ties (Granovetter, 1973), structural holes (Burt, 1992), and brokers, conceptualizes larger networks as populations of smaller, tightly coupled subgroups that are connected by a few well-connected individuals. Brokers occupy positions linking otherwise disconnected subgroups and profit by enabling or preventing the flow of resources between them (Gould and Fernandez, 1989; Burt, 1992; DiMaggio, 1992; Padgett and Ansell, 1993).

However, beyond this structural conception, brokerage models gain considerable insight from also building on Actor Network Theory (ANT), which recognizes that individuals, artefacts, and ideas all have agency and thus all represent 'nodes' in larger networks (Callon, 1989; Law and Hassard, 1999). Further, ANT conceptualizes technological systems as networks of individuals, artefacts, and ideas that come together into larger assemblages that ultimately, in turn, become institutions. Thus, ANT recognizes that institutions represent assemblages, or networks, of individual elements—ideas, people, and artefacts—that, when disassembled and re-assembled, provide the raw materials for new innovation. Similarly, while neo-institutional theory focuses on the similarities between organizations and logics *within* institutional fields, it also provides a valuable perspective in highlighting the differences *across* fields. Finally, cognitive psychology and, particularly, the problem-solving literature captures the individual and small-group cognitive processes by which people recognize opportunities for innovation by recombining their past knowledge in new ways, and also captures the institutional forces acting against innovation for those individuals and small groups acting within only a single domain.

COMMON THREADS

Three characteristics underlie brokerage theories and their approach to managing innovation. First, that the process of innovation plays out across three levels: the institutional (or field), the organizational, and the individual and small group. The alignments of events and conditions, behaviours, and context, across all three levels explain the genesis and the impact of any particular innovation. The main question is how innovation is shaped by the interactions across levels, across boundaries, and across time. Perhaps most simply stated, it is concerned with how innovation reflects both the existing social structures and the individual and collective actions that manage to change them. Figure 9.1 presents a schematic of the interactions that take place across multiple levels of analysis: from institutional structures and logics to organizational strategies and work practices to individual and collective actions and cognitions. The innovation process reflects the dynamic interaction between these three levels. In other words, it is impossible to understand innovation without understanding how the social, organizational, and individual elements of any given time and place shape the process.[1]

Second, brokerage theories of innovation are inherently process theories. Process theories, or models, tell a story of how a particular outcome comes about. In this case, how innovations emerge and are valued within an organizational context. These models comprise discrete events and states rather than variables and the outcome of the model, the creation of innovations, is an event outcome (Mohr, 1982). By contrast, variance theories are concerned with the degree to which changes in a predictor variable

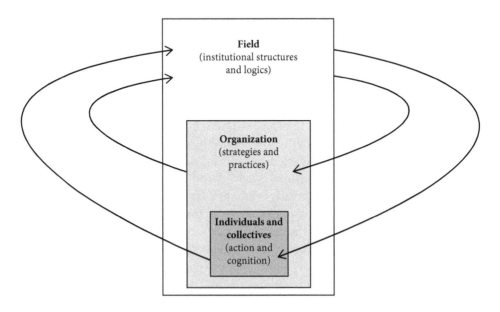

FIGURE 9.1 Three levels of innovation process

effects changes in a dependent variable.[2] Variance theories inherently view innovation as a continuing process continually affected by a range of variables (e.g. levels of organizational slack, supportiveness of managers, intrinsic motivation) and, as a result, remove the process from the particular conditions and timing of each effort. The pursuit of innovation becomes represented as a process of tweaking the dials of managerial control up or down. And yet no amount of organizational slack or intrinsic motivation would make the development of a high-voltage, low-current incandescent bulb innovative ten years after Edison's lighting system was introduced, or ten years before.[3]

Finally, in brokerage theories, the unit of analysis—the outcome of interest—is the innovation and so we must take care to clarify what that means. A working definition should recognize both novelty and continuity because, while all innovations are novel in some ways, they are old in others. Second, it should be consistent in recognizing when something becomes an innovation: with the *conception* of a 'new idea', with its *demonstration*, or with its *acceptance and institutionalization* by the larger organization or society. The difference is non-trivial. As Mathematics professor Larry Shepp once said, when told that he was credited with discovering a theorem previously developed by another mathematician: 'Yes, but when I discovered it, it stayed discovered' (Kolata, 2006). Here, the only innovations that will be considered are those novel, valuable, and non-obvious ideas that are adopted by a larger group and ultimately institutionalized— however small and inconsequential or large and disruptive.

Combining these three threads, brokerage theories recognize that innovation is a process that unfolds at the individual and small group level, often embedded in organizational context and, in turn, embedded within and across institutional fields. These social structures constitute and constrain the process of innovation which, in turn, changes the social structure. From the top down, structure shapes innovation through the institutional preconditions that create both the barriers to innovation (i.e. the institutional structures that maintain stable behaviour within fields) and the potential for innovation as those same elements can be disassembled and reassembled to form new ones (O'Reilly and Chatman, 1996; Hargadon and Fanelli, 2002). Organizations both act within these structural conditions and represent the arena in which individuals and groups pursue innovation. Finally, at the individual and group level, cognition and action reflect processes embedded within these larger contexts—where the raw materials for innovation (the ideas, artefacts, and people) are drawn. From this perspective, we can see how those firms that move easily across a range of industries or markets are in a better position to manage how the elements of one domain can be used in new ways (and in new combinations with other elements) to solve problems elsewhere. This brokerage position provides two critical advantages; it enables innovators to build on the decades of development behind the elements they find and it enables them to avoid many of the entangling cognitive, political, and technical dependencies that make change so difficult for individuals and organizations entrenched in single domains. Conversely, from the bottom up, innovation brings structural change as individual and group actions reshape organizations and ultimately the larger institutional fields in which they occur. When viewing innovation as the creation of enduring changes in organizations and fields, rather than novel ideas or transient demonstrations, the process entails building new and lasting relationships between ideas, people, and artefacts. In this way, brokerage theories of innovation focus not only on the conditions that lead to *conceptions* of novel combinations, but also on the *construction* of new networks—new social orders—around those conceptions.

The balance of this chapter addresses the three levels of analysis that brokerage theories of innovation span: the institutional, organizational, and the individual.

THE INSTITUTIONAL LEVEL

To understand brokerage models of innovation it is necessary to account for the field-level events and conditions that enable (and constrain) innovation. The landscape at the field level is fragmented into multiple, loosely connected domains. Domains are defined by and constructed from the habitualized actions, interactions, and beliefs of the inhabitants (Giddens, 1979, 1984; DiMaggio, 1992). Over time, they develop what Friedland and Alford (1991: 248–9) describe as 'institutional logics' that are 'symbolically grounded, organizationally structured, politically defined and technically and materially constrained'. The repeated interactions that arise from the dense interrelations within domains create socially shared but individually held schemas and scripts (DiMaggio, 1997; Hargadon and Fanelli, 2002). Domain-specific schemas and scripts constitute an individual's portfolio of available understandings and appropriate actions in that context. These portfolios constrain actors to only appropriate actions for their perceived roles but, equally, enable and legitimize the actions of those roles. Neo-institutional theory has recognized that, within a single field, institutional dynamics drive individuals and organizations towards isomorphism in structure, strategy, and logics, reducing diversity within fields. Yet, while these dynamics constrain innovation, they also create the conditions that enable it.

The dynamic towards isomorphism isolates each field from others, driving conformity in knowledge and practice within individual fields and reducing the interactions between them (Swidler, 1986; DiMaggio, 1997). Such fields are defined as sets of 'organizations that, in the aggregate, constitute a recognized area of institutional life; key suppliers, resource and product consumers, regulatory agencies, and other organizations that produce similar services or products' (DiMaggio and Powell, 1983: 148). Within neo-institutional theory, the emphasis in defining field boundaries is cognitive and cultural: 'the notion of field connotes the existence of a community of organizations that partake of a common meaning system and his participants interact more frequently and faithfully with one another and with actors outside the field' (Scott, 1995: 56). These fields can be delineated as markets, industries, or other domains in which the resident knowledge and practice is (relatively) uniformly shared and, indeed, is maintained and constrained by institutional pressures. They can be seen as relatively tightly coupled networks, loosely connected to one another. Indeed, similar language in social network theory describes what distinguishes subgroups in larger networks. Burt (1983: 180), for example, describes how actors residing within particular subgroups 'know one another, are aware of the same kinds of opportunities, have access to the same kind of resources, and share the same kinds of perceptions'.

At the same time, however, the phenomenon of isomorphism within fields creates different stocks of well-developed knowledge and practice within each field. This variation, in turn, creates the potential for innovations that recombine these existing elements—ideas, artefacts, and people—in novel and valuable ways. Within fields, institutions are constituted by particular and enduring arrangements of ideas, artefacts, and people. These elements become so closely associated as to be taken for granted and yet, to outsiders who do not hew to the native logics, these elements represent raw materials that can be disassembled and re-assembled in new arrangements and for new applications.

The recombinant nature of innovation can perhaps best be illustrated with the origins of Henry Ford's mass production 1907–14. Henry Ford did not invent the automobile. Rather, he developed a manufacturing process, later named 'mass production', that dramatically reduced the cost of production. Ford's challenge and accomplishment, in other words, lay in building a car good enough for the mass market, at the scale of the mass market, while profitable for the company and its investors. In 1906, Ford's engineers developed the basic design of the Model T and, based on the specifications and an initial price of $850, sales agents pre-ordered 15,000 cars. In 1907, the Ford Motor Company built close to 1,600 cars; in 1908, 6,000; and, by 1914, Ford was producing 265,000 cars a year. During that time, the price dropped from $850 to $490 to, ultimately, $260 and Ford held 55 per cent of the US automobile market.

Henry Ford didn't invent mass production so much as combine elements of technologies in use, some for almost a century, in other industries. In armory, bicycle, and sewing machine production he found the technologies of interchangeable parts; in canneries, granaries, and breweries he found the technologies of continuous flow production; in the meatpacking plants of Chicago, the assembly line (Hounshell, 1984; Hargadon, 2003). To many within the automobile industry, for example, interchangeable parts was a radical idea. But to Max Wollering, one of the engineers Ford hired to design the car, the tools, and the factory around interchangeable parts,

> There was nothing new [about interchangeability] to me, but it might have been new to the Ford Motor Company because they were not in a position to have much experience along that line. (Hounshell, 1984: 221)

Similarly, Ford credits the 'disassembly' lines of the Chicago meatpacking plants for the original idea of the assembly line. Ford's engineers visited these plants and took copious notes. William Klann, head of the engine department at Ford, recalled touring Swift's Chicago plant thinking, 'If they can kill pigs and cows that way, we can build cars that way'. Indeed, Ford was well aware he neither invented the automobile nor the technologies of mass production, once testifying,

> I invented nothing new. I simply assembled into a car the discoveries of other men behind whom were centuries of work…Had I worked fifty or ten or even five years before, I would have failed. So it is with every new thing. Progress happens when all the factors that make for it are ready, and then it is inevitable. To teach that a comparatively few men are responsible for the greatest forward steps of mankind is the worst sort of nonsense. (Gordon, 2001: 103)

Ford's mass production had an irreversible impact on the automobile industry, on manufacturing, and on society—not because the technology was novel but just the opposite. The existing elements that Ford exploited in constructing this system of mass production enabled the company to build on sixty years of learning-in-use in the machine tool industry and on the foundries, breweries, granaries, and meatpacking plants where new equipment and methods had already been developed (Rosenberg, 1963; Hounshell, 1984; Hughes, 1989). Indeed, the time spent moving down the learning curve was considerably shortened by recombining known elements in new ways. Ford's first experiment with the assembly line, for example, took place in the Magneto Assembly room and the first day brought an almost 40 per cent improvement in productivity. By the end of the first year, productivity had increased over 400 per cent. Rather than attributing success to the novelty of his ideas, Ford's mass production succeeded precisely because it hinged on the recombination of existing ideas, artefacts, and people.

Conversely, the more an innovation is built from scratch the more effort is required to develop and market it. In 1891, Whitcomb Judson first applied for a patent on the ideas of the zipper (Friedel, 1994). His idea for 'A Clasp Locker or Unlocker for Shoes' had few precedents to draw from. Unfortunately for Judson and his investors, this meant Judson and his inventors not only had to design all the elements of the new device, but also the machinery to manufacture it. It took two more decades of continuous development before Gideon Sundback, working for Judson's investors, developed a model that worked smoothly and could be manufactured cheaply. And, once the manufacturing process was finally developed, it took another two decades before designers, retailers, and finally the public would fully embrace the product.

ORGANIZATIONS AS ACTORS AND ARENAS

Brokerage theories of innovation recognize the role of organizations in the innovation process as both actor within larger organizational fields and the arena for action by individuals and groups. In this way, brokering helped explain two aspects of innovation: first, how some organizations are able to routinely innovate and, second, how the organizational context shapes individual and group level innovation.

Brokering describes how firms spanning multiple domains may innovate by moving ideas from where they are known to where they are not, in the process creating new combinations of existing ideas. Edison's Menlo Park Research Laboratory, the Ford Motor Company, and modern counterparts like 3M, Apple, or IDEO are remembered, for example, because they proved capable of creating one breakthrough after another. These firms organized to pursue continuous innovation through strategies for recombining existing technologies rather than inventing new ones. Such strategies place firms in positions spanning multiple domains to routinely see how the problems of one field might be solved by combining ideas, artefacts, and people of other fields.

The structural advantages that shape innovation come into sharper focus when looking at firms that engage in this process regularly. Thomas Edison and his Menlo Park, New Jersey Laboratory exploited the process of recombinant innovation to generate a stream of patented innovations (involving over 400 patents within a six-year period) (Hargadon and Sutton, 1997; Hargadon, 2003). By combining ideas they learned while working in the telegraph industries, Edison's lab produced innovations in stock tickers, fire alarms, mimeographs, and telephones. The mimeograph pen, for example, was Edison's first commercial success and simply repurposed a component pulled from an automatic telegraph machine designed to rapidly puncture paper to record incoming messages (Conot, 1979). While the mimeograph is now obsolete, this design remains in use today with only minor modifications as a tattooing needle (Morton, 2002). Similarly, lessons learned in building an Atlantic cable gave Edison's team insights into designing a microphone for the telephone. Millard (1990) described how any one development project often offered valuable spillovers that Edison would exploit in other projects. Should any insights emerge from one project, Millard (1990: 48) explained, 'If it provided the key to another problem in a totally different project, [Edison] was prepared to quickly exploit it. The new lab was built with this kind of flexible innovation in mind.'

At Menlo Park, Edison had created the ideal conditions for pursuing innovation through technology brokering. Working for a range of clients and in a range of industries, the company moved easily through these different industries—enabling it to be the first to see how ideas developed in one industry might be useful in another. Indeed, to exploit these opportunities, Edison modelled the laboratory after the machine shops from which he and many of the others emerged, where mechanics and independent entrepreneurs would work side by side, sharing machines, telling stories, and passing along promising ideas or opportunities. The group at Menlo Park numbered approximately fourteen. Edison worked most closely with Charles Batchelor, whose training as both a mechanic and a draftsman so complemented (and grounded) Edison's more flighty visions that the two split all patent royalties 50–50. Many of the lab's breakthroughs were attributed to Batchelor or one of the others who worked on the projects while Edison dealt with clients or investors. As one such assistant, Francis Jehl, said: 'Edison is in reality a collective noun and means the work of many men' (Conot, 1979: 469). The laboratory's range of clients from many different industries meant that any one development project offered valuable information that Edison might exploit in other projects.

Modern firms pursuing brokerage strategies include IDEO and Design Continuum in product development and firms like McKinsey and Accenture in management consulting. Design Continuum, for example, works with over a hundred companies from a range of industries including computers, office equipment, communication systems, consumer electronics, appliances, consumer non-durables (e.g. diapers), sporting goods, industrial products (e.g. robotics), medical instrumentation, and health care products. Gian Zaccai, CEO of Design Continuum, believed that working in different industries 'frees you from the dogma of any one industry and their firm belief in the links between problems and solutions'. This is not to say only firms occupying 'structural holes' can innovate by combining the knowledge and practice of other fields, but rather that such a position supports a sustained level of innovation. Ford Motor Company, for example, pursued a strategy of becoming central, and dominant, in just one field while taking advantage of ideas from a range of fields. It should be noted that Ford's demise, however, a decade later, came because they were unwilling to consider ideas and opportunities that emerged outside their own organization.

Whether firms pursue brokerage positions across larger networks or central positions within one field, their structure and strategy create the conditions—the arena—for individual and group innovation within. For example, the same fragmented conditions that exist at the field level are often replicated within organizations. Organizational structures create separations between divisions, and between groups within divisions to match the divisions among markets and industries they face. These internal divisions take on institutional structures of their own. As Salancik (1995: 346–7) points out, 'all interactions occur in a context of institutions, including rules and roles. Organizational policies impose some of these: units are explicitly directed to interact with one unit but not others or are instructed to report to one unit rather than another.' Informal norms, such as competition for scarce resources or market opportunities, also reduce interactions between groups, as do the development of specialized languages and perspectives within groups (Dougherty and Hardy, 1996; Bechky, 1999). As a result, domains emerge within organizations that reflect and reinforce the relatively few interactions between them. Under these conditions, ideas about organizing, valuable technical artefacts, and skilled individuals in one domain go unknown and untapped in others. In other words, large organizations often become fragmented by their very structure—the presence of multiple business units, through physical separation and competition, creates local domains and inhibits the flow of people, ideas, artefacts across them but also creates the conditions for brokers to innovate by moving ideas from one corner of the organization to another.

When 3M researchers developed a new technology for creating tiny geometric prisms on a plastic surface—now called microreplication—they came to realize it could have broader applications. 'That was a profound realization', scientist Roger Appledorn explained, 'Because all 3M products have surfaces, microreplication had the potential to add something to all of them.'[4] But for two decades, that technology languished in a small business unit with responsibility for manufacturing and selling lighting systems. In 1983, 3M created the Optical Technology Center, a group dedicated to finding new uses for microreplication across the rest of 3M's business units. Rather than focus on inventing novel technologies, they focused on combining their microreplication technology with the existing products 3M was already producing. Thus, 3M created an internal group that could profit by pursuing a brokerage position between the otherwise disconnected business units, moving an idea from where it was known to where it was not and, in the process, combining it with existing elements in those domains. Today, the technology is embedded in industrial grinders, surface coatings for boat hulls, compact disks, computer screens, street lights, and a range of other products.

Brokerage models of innovation highlight how organizations are embedded within larger field landscapes and how managerial strategies, structures, and actions aimed at innovation will be enabled or constrained by the institutional dynamics unfolding at these larger levels. The innovation lessons of an Edison or Design Continuum, for example, should be tempered by appreciating the opportunities they experienced by virtue of their network position. Similarly, the creativity and innovation of individuals and groups within organizations can and should be seen as embedded in the larger landscapes of the organization and field.

Individuals and Groups

Ultimately the focus of brokerage theories sits at the level of the individual and small group. In other words, the balkanization of institutional fields shapes the opportunities for (recombinant) innovation when resources in one domain are valuable yet unknown in others. In addition, the organization amplifies or buffers these opportunities by creating the local context in which individuals and groups pursue innovation. Understanding innovation at the individual level requires seeing it embedded within these larger organizational and institutional context. Such an approach hews closely to the microsociological perspective, which is concerned with how individual cognition and action is shaped by, and in turn shapes, the social structures in which they take place (Fine, 1991; DiMaggio, 1997). Individuals construct novel interpretations and actions through an intellectual bricolage, dis-assembling and re-assembling their past experiences in ways that enable them to understand and respond to new situations (Lévi-Strauss, 1966; Swidler, 1986; DiMaggio, 1997). From this perspective, the process of innovation at the level of individuals and small groups involves the *acquisition* of past experiences, the *conception* of novel combinations of institutional elements (artefacts, ideas, and people) that make up those past experiences, and the *construction* of real innovations-as-outcomes that combine those elements as durable solutions in new situations.

Acquisition entails individuals and groups gaining exposure to and learning about the elements that constitute the knowledge and practice of multiple domains—in the form of both enacted solutions and the defined problems they address. The more and more varied experiences that individuals and groups acquire, the wider the range of possible permutations they may generate from their shared experiences. A social network perspective suggests that, by bridging otherwise disconnected domains, brokers can innovate by transferring solutions across domains (Burt, 2004; Hargadon and Sutton, 1997). However, such a purely structural approach neglects the social and cognitive effects on individuals and groups acting across multiple domains. Mere exposure to the ideas, artefacts, and people of disparate domains exerts a complex influence on the innovation process. Learning, for example, represents the process by which experience becomes encoded into rules, routines, and technologies that become, in turn, subject to (relatively) mindless application (Cohen et al., 1972; March, 1972; Weick, 1979). And in stable environments, such learning processes are effective at improving performance (Wright, 1936; Epple et al., 1991; Tyre and Orlikowski, 1994). However, not all learning is the same. Individuals and groups that learn within a single domain will differ from those that learn through experiences across a range of domains. Brokers have neither the goal of codifying knowledge for repetition in similar circumstances, nor the promise of status and power that accrue to those who use it to gain advantage within a single domain. Instead, they tend to acquire knowledge with an eye towards its usefulness in often very different (and unforeseen) situations. In Hargadon and Sutton's (1997) study of the design firm IDEO, for example, an engineer referred to this type of learning as throwing a bunch of tools into his tool box. He 'remembered the tools, but not where they came from'. So individuals can learn whether they inhabit a single domain or move through many, and what (and why) they learn can differ considerably.

Conception describes the process of recognizing how and when old resources can be combined in new ways to address new and problematic situations. Acquisition of the knowledge and practice of multiple domains shapes the potential for innovation through brokerage, but it is not sufficient. All knowledge is contextual. Even when learned, it remains entangled in its original situation in meaning (Berger and Luckman, 1967; Nonaka, 1994). Conception thus remains difficult because it requires disentangling and recombining institutional elements previously embedded in one or more domains (DiMaggio, 1997; Hargadon and Fanelli, 2002). Indeed, failures and problem solving are often the result not of ignorance (the lack of knowledge) but rather the inability to see when old knowledge fits new situations (Lave, 1988; Reeves and Weisberg, 1993; Thompson et al., 2000).

To disentangle what individuals and groups learned in the context of one domain and see its value in another relies on what cognitive psychologists refer to as analogical reasoning (Gick and Holyoak, 1980, 1983; Gentner and Gentner, 1983). Analogies highlight non-obvious similarities between two things that appear to be dissimilar, and individuals solve novel problems by making analogic connections to other problems they have faced in the past and by adopting and adapting their existing solutions to fit the new problem (Reeves and Weisberg, 1993, 1994). This process also occurs in groups in a form of collective cognition, where analogic connections emerge in the interactions between individuals with diverse experiences (Hargadon and Bechky, 2006). Again, a purely structural approach to brokerage neglects the social and cognitive effects on individuals and groups of acting within multiple domains. As DiMaggio (1997: 280) points out, 'When persons or groups switch from one domain to another, their perspectives, attitudes, preferences, and dispositions may change radically.' In other words, by moving between domains, brokers are better prepared to overcome the cognitive constraints that exist in the local beliefs and actions of the domains from which knowledge comes and to which it is applied, and thus they are better prepared to see connections that others might miss.

Finally, *Construction* describes bringing to reality the novel conceptions of new combinatorial possibilities. Innovation can be measured in multiple ways: as ideas, as demonstrations, and as enacted solutions. Historically, we tend to study those innovations that have been enacted—that survived the test of time to become themselves institutionalized through acceptance and use. While recognizing potential new combinations is challenging, innovators and entrepreneurs must also attend to building new and enduring ties that link previously disparate elements to one another (see Walker et al., 1997) for a discussion of the differential advantages of bridging and building network ties). As new ties emerge around an innovation, they bring coordinated effort, legitimacy, and social capital that turn innovations into institutions of their own (Baker and Obstfeld, 1999). Martin and Eisenhardt (2001), for example, found that successful new product development ventures that bridged multiple groups within the firm did so by building strong ties where previously only weak or no ties existed.

Consider how these three activities at the individual and group level conspired to enable one innovation in Design Continuum.[5] In 1988, Reebok approached Design Continuum to develop a competitive response to Nike's new Air™ technology, a heel cushion and 'active energy return system'. Instead, Design Continuum created the Pump™. The Pump™ concept produced a form-fitting shoe by incorporating an inflatable air bladder into the sides of an athletic shoe. This solution first emerged when one of the designers, who had previously designed an inflatable splint, recognized how such splints might prevent injuries by building ankle support into a basketball shoe. Another on the team had worked on medical equipment and, familiar with IV bags, saw how these small sealed bags could be modified to provide the oddly shaped air bladders necessary to make this 'splint-in-a-shoe' concept work. Still, the problem remained of how customers could easily inflate and deflate the shoe. Others in the firm became involved, several having worked with diagnostic instruments and the little pumps, tubing, and valve components that made up those products. When they recognized how such components could be adapted to fit in the tongue of the shoe, the major ideas were in place. Within six months of these ideas coming together, the engineering design was complete and, six months later, the Reebok Pump™ shoe was introduced to the market. In the following year it accounted for over $1 billion in revenue in the highly competitive athletic shoe market and gained wide praise in the business press for its creativity.

The series of events and conditions that created to the Reebok Pump™ shoe idea at Design Continuum illustrate the explanatory value of brokerage theories of innovation.

Innovation involved first the conditions that created and isolated the knowledge and practice of the athletic shoe industry from those of inflatable splints, IV bags (and their manufacture), and the pumps and valves of other fields. Within the firm Design Continuum, it also involved exposing designers to this diverse knowledge and the opportunity and encouragement to recognize and connect that knowledge in new ways. Finally, innovation entailed building a real solution that connected these elements materially and socially: Reebok, IV bags, and IV bag manufacturer (now a supplier of inflatable bladders for basketball shoes), valve manufacturers, equipment, and people. The ultimate success of the Reebok Pump™ sneaker built on both Design Continuum's brokerage position, to develop the innovative design, and Reebok's central position in the athletic shoe market to rapidly manufacture, distribute, and promote the shoe.

CONCLUSION

The value of a brokerage strategy for innovation, as compared to pursuing success within a single field, was summed up by Elmer Sperry, a contemporary of Edison whose major contributions included pioneering work in feedback control mechanisms, particularly using electric motors (but also of autopilot for ships and planes). Like Edison, he acted as a broker spanning multiple domains and exploited the advantage of moving between many small worlds, explaining once,

> If I spend a lifetime on a dynamo I can probably make my little contribution toward increasing the efficiency of that machine six or seven percent. Now then, there are a whole lot of [industries] that need electricity, about four- or five hundred percent. Let me tackle one of those. (Hughes, 1989)

Sperry's strategy was to introduce his knowledge of electro-magnetic technologies to new domains rather than to specialize in just one. And it was effective, as Sperry's work coincided with the rapid diffusion of electricity and electro-mechanical solutions. In 1899, electric motors accounted for 5 per cent of the total installed horsepower in American manufacturing; in 1909, the number was 25 per cent; in 1919, 55 per cent; in 1929 over 80 per cent (Rosenberg, 1982).

Future research in this stream could pursue several fertile directions. First, by continuing to develop historical accounts of the social and diachronic nature of the innovation process, we could better understand why ideas and entrepreneurs are successful in moments of time and not before or after. This approach would inform scholars of innovation and strategy. Second, further development of the microsocial nature of creativity would help us understand how individual and collective creative acts are products of individual and group level histories as much as their immediate surroundings. Such investigations would help to contextualize psychological studies of individual and group creativity. Finally, attention to the individual and organizational capabilities and behaviours that enable particular ones to accomplish innovation would better inform models of the innovation process that currently focus predominantly on idea-generation and novelty.

From Edison, Ford, and Sperry to IDEO, Design Continuum, and 3M, the actions of the individuals and groups involved in the pursuit if innovation can be understood by their embeddedness in their organization and in the larger fields in and across which they worked. Traditional approaches to innovation might focus on the organizational strategies or management practices that foster such innovation: for example, organizational slack, matrixed or horizontal organizational structures, risk tolerance in decision-making, open sourcing of innovative ideas or, at the level of individual and group practices, design thinking, brainstorming, and other practices designed to generate novel insights. However, the effectiveness of these tools for managing innovation hinge on the organizational and field-level landscapes in which they are employed. These larger landscapes provide the raw materials and opportunities for innovation as well as the means to measure the resulting impacts. To dis-embed the practice of innovation from these larger landscapes is to obscure how, when, and where the innovation process unfolds. Brokerage theories of innovation represent one attempt to bring clarity to this embeddedness and, in the process, generate practical and actionable strategies for managing innovation.

Notes

1. For example, organizational innovation has narrowly focused on the management of innovation as an process unaffected by changing institutional conditions in which it occurs (e.g. Tushman and Anderson, Era of Ferment; E. Rogers and diffusion). With the notable exceptions of Foster on S-Curves and G. Moore on crossing the chasm. Similarly, field-level studies of innovation have often neglected the causal significance of individual and organizational actions to shape the rate and direction of large-scale innovation. For example, the role of design, of standards setting, and of entrepreneurial initiatives in shaping the particular nature and timing of emerging markets and technologies. And finally, research that focuses on individual and group creativity in organizations, like brainstorming, has narrowly focused on the actions of individuals while neglecting the larger organizational and institutional context.

2. In a process theory the order of events is essential while variance theories are often less, if at all, concerned with the causal sequence of variables. Finally, in a process theory, each event or state is a necessary but not sufficient condition for the outcome to occur. A variance theory, on the other hand, views each variable (or interaction between variables) as an independent cause of changes in the dependent variable. Process theories are then subject to disconfirmability when the outcomes they study can be shown to occur without the presence of the events or states described by the model, or by passing through those states in a different sequence than is described by the model.

3. Early process theories of innovation, known as stage model research, attempted to outline the stages of the innovation process within organizations. Largely theoretical, this research produced a broad range of stage models that roughly followed a common pattern (i.e.

from concept development to product planning to engineering design to detailed design to pre-production to production), but such stage models suffered from the very simplicity and linearity that they implied but that organizations did not exhibit. Often, in empirical work, innovation is found to be a complex, iterative process with many feedback and feed-forward loops.

4. from <http://www.3M.com/about3m/pioneers/appeldorn.jhtml><http://www.3M.com/about3m/pioneers/appeldorn.jhtml> (accessed 28 March 2002).

5. This example draws from Hargadon (2003).

REFERENCES

Baker, W. E., and Obstfeld, D. (1999). 'Social Capital by Design: Structures, Strategies, and Institutional Context', in R. T. A. J. Leenders and S. M. Gabbay (eds), *Corporate Social Capital and Liability*. Boston: Kluwer Academic Publishers, 88–105.

Basalla, G. (1988). *The Evolution of Technology*. New York: Cambridge University Press.

Bechky, B. A. (1999). 'Crossing Occupational Boundaries: Communication and Learning on a Production Floor', Ph.D. Dissertation, Stanford University.

Berger, P. L., and Luckman, T. (1967). *The Social Construction of Reality*. New York: Doubleday.

Burt, R. S. (1983). 'Range', in R. S. Burt and M. J. Minor (eds), *Applied Network Analysis: A Methodological Approach*. Thousand Oaks, CA: Sage Publications, 176–194.

Burt, R. S. (1992). *Structural Holes: The Social Social Structure of Competition*. Cambridge, MA: Harvard University Press.

Burt, R. S. (2004). 'Structural Holes and Good Ideas', *American Journal of Sociology*, 110(2): 349–99.

Callon, M. (1989). 'Society in the Making: The Study of Technology as a Tool for Sociological Analysis', in W. E. Bijker, T. P. Hughes, and T. J. Pinch (eds), *The Social Construction of Technological Systems*. Boston: MIT Press, 83–103.

Chen, K., and O'Mahony, S. (2007). *The Selective Synthesis of Competing Logics*. Wayne, NJ: William Paterson University, 67.

Cohen, M. D., et al. (1972). 'A Garbage Can Model of Organizational Choice', *Administrative Sciences Quarterly*, 17(1): 1–25.

Conot, R. E. (1979). *A Streak of Luck*. New York: Seaview Books.

DiMaggio, P. (1992). 'Nadel's Paradox Revisited: Relational and Cultural Aspects of Organizational Structure', in N. Nohria and R. G. Eccles (eds), *Networks and Organizations: Structure, Form and Action*. Boston: Harvard Business School Press, 118–42.

DiMaggio, P. (1997). 'Culture and Cognition', *Annual Review of Sociology*, 23: 263–87.

DiMaggio, P. J., and Powell, W. W. (1983). 'The Iron Cage Revisited: Institutional Isomorphism and Collective Rationality in Organizational Fields', *American Sociological Review*, 48(April): 148–60.

Dougherty, D., and Hardy, C. (1996). 'Sustained Product Innovation in Large, Mature Organizations: Overcoming Innovation-to-Organization Problems', *Academy of Management Journal*, 39(5): 1120–53.

Epple, D., et al. (1991). 'Organizational Learning Curves: A Method for Investigating Intra-Plant Transfer of Knowledge Acquired through Learning by Doing', *Organization Science*, 2(1): 58–70.

Fine, G. A. (1991). 'On the Macrofoundations of Microsociology: Constraint and the Exterior Reality of Structure', *Sociological Quarterly* 32(2): 161–77.

Friedel, R. D. (1994). *Zipper: An Exploration in Novelty*. New York: W. W. Norton.

Friedel, R., and Israel, P. (1986). *Edison's Electric Light: Biography of an Invention*. New Brunswick, NJ: Rutgers Unversity Press.

Friedland, R.,and Alford, R. (1991). 'Bringing Society back in: Symbols, Practices, and Institutional Contradictions', in W. W. Powell and P. DiMaggio (eds), *The New Institutionalism in Organizational Analysis*. Chicago: University of Chicago Press, 232–63.

Gentner, D., and Gentner, D. R. (1983). 'Flowing Waters or Teeming Crowds: Mental Models of Electricity', in D. Gentner and A. Stevens (eds), *Mental Models*. Hillsdale, NJ: Lawrence Erlbaum Associates, Inc., 99–129.

Gick, M. L., and Holyoak, K. J. (1980). 'Analogic Problem Solving', *Cognitive Psychology* 12: 306–55.

Gick, M. L. and Holyoak, K. J. (1983). 'Schema Induction and Analogic Transfer', *Cognitive Psychology*, 15: 1–38.

Giddens, A. (1979). *Central Problems in Social Theory: Action, Structure and Contradiction in Social Analysis*. Berkeley and Los Angeles: University of California Press.

Giddens, A. (1984). *The Constitution of Society*. Berkeley and Los Angeles: University of California Press.

Gordon, J. S. (2001). *The Business of America: Tales from the Marketplace—American Enterprise from the Settling of New England to the Breakup of AT&T*. New York: Walker Publishing Company.

Gould, R. V., and Fernandez, R. M. (1989). 'Structures of Mediation: A Formal Approach to Brokerage in Transaction Networks', *Sociological Methodology*, 19: 89–126.

Granovetter, M. (1973). 'The Strength of Weak Ties', *American Journal of Sociology*, 6: 1360–80.

Hargadon, A. B. (1998). 'Firms as Knowledge Brokers: Lessons in Pursuing Continuous Innovation', *California Management Review*, 40(3): 209.

Hargadon, A. B. (2003). *How Breakthroughs Happen: The Surprising Truth about How Companies Innovate*. Cambridge, MA: Harvard Business School Press.

Hargadon, A. B., and Bechky, B. A. (2006). 'When Collections of Creatives become Creative Collectives: A Field Study of Problem Solving at Work', *Organization Science*, 17(4): 484–500.

Hargadon, A. B., and Douglas, Y. (2001). 'When Innovations Meet Institutions: Edison and the Design of the Electric Light', *Administrative Science Quarterly*, 46: 476–501.

Hargadon, A. B., and Fanelli, A. (2002). 'Action and Possibility: Reconciling Dual Perspectives of Knowledge in Organizations', *Organization Science* 13(3): 290–302.

Hargadon, A., and Sutton, R. (1997). 'Technology Brokering and Innovation in a Product Development Firm', *Administrative Science Quarterly*, 42(4): 716–49.

Hounshell, D. A. (1984). *From the American System to Mass Production*. Baltimore: Johns Hopkins University.

Hughes, T. P. (1989). *American Genesis: A Century of Invention and Technological Enthusiasm, 1870–1890*. New York: Viking.

Kolata, G. (2006). 'Pity the Scientist who Discovers the Discovered', *New York Times*, from <http://www.nytimes.com/2006/02/05/weekinreview/05kolata.html>

Lave, J. (1988). *Cognition in Practice: Mind, Mathematics, and Culture in Everyday Life*. Cambridge: Cambridge University Press.

Law, J., and Hassard, J. (1999). *Actor Network Theory and after*. Oxford: Blackwell Publishers.

Lévi-Strauss, C. (1966). *The Savage Mind*. Chicago: University of Chicago Press.

Lingo, E. L., and O'Mahony, S. (2010). 'Nexus Work: Brokerage on Creative Projects', *Administrative Science Quarterly*, 55(1): 47–81.

March, J. G. (1972). 'Model Bias in Social Action', *Review of Educational Research*, 44: 413–29.

Martin, J. A., and Eisenhardt, K. M. (2001). *Exploring Cross-Business Synergies*, Best Paper Proceedings, Academy of Management Conference, Washington, DC.

Millard, A. (1990). *Edison and the Business of Innovation*. Baltimore: Johns Hopkins University Press.

Mohr, L. B. (1982). *Explaining Organizational Behavior*. San Francisco: Jossey-Bass.

Morton, D. (2002). 'Tattoing', *Invention & Technology*, Winter 2002: 36–41.

Nonaka, I. (1994). 'A Dynamic Theory of Organizational Knowledge Creation', *Organization Science*, 5(1): 14–37.

Ogburn, W. F. (1922). *Social Change*. New York: Viking Press.

O'Reilly, C. A., and Chatman, J. A. (1996). 'Culture as Social Control: Corporations, Cults, and Commitment', *Research in Organizational Behavior: An Annual Series of Analytical Essays and Critical Reviews*, 18: 157–200

Padgett, J. F., and Ansell, C. K. (1993). 'Robust Action and the Rise of the Medici, 1400–1434', *American Journal of Sociology*, 98(6): 1259–319.

Reeves, L. M., and Weisberg, R. W. (1993). 'On the Concrete Nature of Human Thinking: Content and Context in Analogical Transfer', *Educational Psychology*, 13(3 and 4): 245–58.

Reeves, L. M., and Weisberg, R. W. (1994). 'The Role of Content and Abstract Information in Analogical Transfer', *Psychological Bulletin*, 115(3): 381–400.

Rosenberg, N. (1963). 'Technological Change in the Machine Tool Industry. 1840–1910', *Journal of Economic History*: 414–43.

Rosenberg, N. (1982). *Inside the Black Box*. New York: Cambridge University Press.

Salancik, G. R. (1995). 'WANTED: A Good Network Theory of Organization', *Administrative Science Quarterly*, 40: 345–9.

Sawyer, R. K. (2007). *Group Genius: The Creative Power of Collaboration*. New York: Basic Books.

Schumpeter, J. (1934). *The Theory of Economic Development*. Cambridge, MA: Harvard University Press.

Scott, W. R. (1995). *Institutions and Organizations*. Thousand Oaks, CA: Sage.

Sutton, R. I., and Hargadon, A. (1996). 'Brainstorming Groups in Context: Effectiveness in a Product Design Firm', *Administrative Science Quarterly*, 41(4): 685–718.

Swidler, A. (1986). 'Culture in Action: Symbols and Strategies', *American Sociological Review*, 51: 273–86.

Thompson, L., et al. (2000). 'Avoiding Missed Opportunities in Managerial Life: Analogical Training More Powerful than Case-Based Training', *Organizational Behavior and Human Decision Processes*, 82: 60–75.

Tyre, M. J., and Orlikowski, W. J. (1994). 'Windows of Opportunity: Temporal Patterns of Technological Adaptation in Organizations', *Organization Science*, 5(1): 98–118.

Usher, A. P. (1929). *History of Mechanical Invention*. Cambridge, MA: Harvard University Press.

Walker, G., et al. (1997). 'Social Capital, Structural Holes and the Formation of an Industry Network', *Organization Science*, 8: 109–25.

Weick, K. E. (1979). *The Social Psychology of Organizing*. Reading, MA: Addison-Wesley.

Wright, T. P. (1936). 'Factors Affecting the Cost of Airplanes', *Journal of the Aeronautical Sciences*, 3: 122–8.

PART III

··

THE CONTEXT FOR

INNOVATION

··

CHAPTER 10

SECTORAL SYSTEMS OF INNOVATION

FRANCO MALERBA AND PAMELA ADAMS

INTRODUCTION

THIS chapter examines sectoral differences in innovation and in the organization of innovative activities. Understanding differences across sectors in terms of the rate, characteristics, and sources of technological change, and in terms of the management of innovative activities, is extremely relevant for any analysis that aims to foster innovation in firms. Just think of how different such elements are in sectors such as ICT and textiles. This chapter shows that analyses of sectoral differences have progressed significantly over the past decades, from early studies based on simple distinctions between high-tech and low-tech industries (in terms of R&D expenditures), to studies focused on market structure and innovation (Schumpeter Mark I and II), to analyses of innovation sources (the Pavitt taxonomy) and competition (Porter's five forces). Over time, these different approaches have widened the attention of scholars and managers from a narrow focus on R&D intensity to a broader view of the multiple dimensions of innovation in and across sectors. This chapter takes these analyses one step further to propose a systemic and evolutionary framework for analysing sectoral differences. This framework links knowledge and the sources of innovation with the actors and the institutions involved, and with the structure of innovative activities. It attempts to broaden the scope of analysis for innovation management and firm strategies and to lay the basis for a more solid understanding of the dynamics of innovative activities both within and across sectoral boundaries. By using this framework, managers may better understand both the forces that drive innovative activities in their sectors and how these forces change over time.

Do Sectoral Differences Matter?

In his studies of economic development, Joseph Schumpeter clearly recognized the importance of sectoral differences in the innovation process. Schumpeter distinguished between what others have called Mark I sectors (Schumpeter, 1911) characterized by 'creative destruction' where entrepreneurs and new firms play a key role in innovative activities, and Mark II sectors (Schumpeter, 1942) characterized by cumulative technological advancement and the prevalence of large, established firms ('creative accumulation'). He also proposed that sectoral differences were relevant for understanding economic growth and long-term industrial transformation.

The importance of sectoral differences has also been noted in more recent work in both economics and management. In industrial economics, sectoral differences have been related to variations in the equilibrium structure of industries when examining technology and demand. Using different methodological approaches—from the structure-conduct-performance tradition, to sunk cost and econometric analyses—these studies differentiate sectors in terms of their degree of concentration, vertical integration, diversification, technical progress, and firm entry and growth (Bain, 1956; Tirole, 1988; Scherer, 1990; Sutton, 1991, 1998). In the management literature, the 'five forces' framework developed by Michael Porter led the way in applying a structural analysis of industries to an understanding of competitive advantage and firm strategy. His framework expanded the traditional boundaries of industries based on current competitors to include market interactions with buyers and suppliers and the potential for new entrants and substitute products (Porter, 1979). Later, Porter's approach to the value chain advanced the idea that differences in industrial structure may influence firm level choices concerning the primary and support activities performed in order to design, produce, and commercialize products (Porter, 1985).

While these contributions from both economics and management represent important advances in the understanding of the sectoral context of firm competitiveness and profitability, they have limitations for understanding innovation. First, because the focus of such studies is on firms and on market-based transactions between firms, little attention is given to the role of non-firm organizations such as finance, universities, research centres, industry associations, standards bodies, and government agencies. Yet, the contribution of such organizations to innovation is often critical. Second, these frameworks focus on the effects of firm size, firm strategy, and market relationships on innovation and prices and, hence, on profitability and growth. Non-market relations between firms remain outside of the scope of such studies. Innovation, by contrast, may involve relations that go beyond market transactions to include processes of communication and cooperation through which learning and knowledge creation occur. Third, these approaches provide little basis for analysing the links and interdependencies among related industries that often drive innovation and determine changes in the boundaries of sectors over time (Dahmen, 1989; Geroski, 1995). Even Porter, who recognized that the boundaries between industries should be viewed as a continuum rather than as rigid and bounded spaces, and who allowed for substitutability on both the supply and the demand side, focused his analysis on statically defined industry structures without considering dynamic processes that might work as feedback mechanisms to change such structures. Such an approach makes it difficult to look beyond current boundaries to understand the learning processes that underlie innovation both across industries and over time (Rosenberg, 1976, 1982; Granstrand, 1994; Grant, 1996). It also hinders an understanding of innovation in firms with activities in multiple markets and/or multiple technologies (Granstrand et al., 1997). Fourth, concepts such as the value chain are centred on the supply side and focus on the range of activities used by firms along the chain to bring a product/service to market. Such concepts leave little room for the role of the demand side in innovation processes or for the input of actors and institutions outside of the direct supply chains of a particular industry. Finally, while many of the frameworks proposed by industrial economists and management scholars emphasize the importance of market structure for performance, there is little evidence to date to support the existence of a clear, unidirectional link between market structure and innovation (Cohen and Levin, 1989; Sutton, 1998; Gilbert, 2006). Rather, research indicates that market structure and innovation co-evolve (Nelson and Winter, 1982; Nelson, 1994).

Reference to the importance of sectoral differences may also be found in the literature related to innovation studies. Here industrial economists who study innovation tend to complement case studies with quantitative analyses based on indicators such as R&D intensity, R&D alliances, patents, and patent intensity. Such research has also been used to develop classifications of different types of sectors based on a limited number of key variables, much along the lines of the Schumpeter Mark I/II sectors. One such classification used by international organizations such as the Organisation for Economic Cooperation and Development (OECD) and the European Union, defines sectors as either '*high R&D intensive*' (e.g. electronics and pharmaceuticals) or '*low R&D intensive*' (e.g. textiles and shoes). Scherer (1982), on the other hand, distinguishes between sectors that are '*net sources*' of R&D (e.g. computers and instrumentation) and sectors that are '*net users*' of technology (e.g. textiles and metallurgy). A similar analysis by Robson, Townsend, and Pavitt (1988) identifies '*core sectors*' (e.g. electronics, machinery, instruments, and chemicals) which generate most of the innovations in an economy and are net sources of technology, '*secondary sectors*' (e.g. auto and metallurgy) which are less innovative and may draw on technologies from the core, and '*user sectors*' (e.g. services) which mainly absorb technology. Perhaps the most widely used taxonomy, however, is the one developed by Keith Pavitt (1984). He distinguishes between four different types of sectors in terms of the sources of innovation and appropriability mechanisms. In '*supplier dominated*' sectors (e.g. textiles, services), new technologies are embodied in new components and equipment, and diffusion and learning take place through learning-by-doing and learning-by-using. In '*scale intensive*' sectors (e.g. auto, steel) process innovation is relevant and the sources of innovation may be both internal (R&D and learning-by-doing) and external (involving mainly suppliers of materials, components, and machinery). In such sectors, appropriability is obtained through secrecy and patents. In '*specialized suppliers*' sectors (e.g. equipment producers), innovation is focused on improvements in performance, reliability, and customization, while the sources of innovation are both internal (tacit knowledge and experience of skilled technicians) and external (user–producer interactions). In this case, appropriability comes mainly from the localized and interactive nature of knowledge. Finally, '*science-based*' sectors (e.g. pharmaceuticals, electronics) are characterized by high rates of product and process innovation, internal R&D, and scientific research done at universities and public research laboratories. In these sectors, science is a major source of knowledge and innovation, and appropriability is obtained through a variety of mechanisms, such as patents, lead times, learning curves, and secrecy.

The empirical research at the base of such classifications has enriched our understanding of sectoral patterns of innovation over the past decades. This research has also highlighted the important role of actors outside of the firm and of the various learning mechanisms that occur within sectors. Still, studies based on a limited number of variables such as R&D expenditures and patents, or on classifications focused on learning and appropriability, cannot provide a comprehensive understanding of sectoral differences in innovation. What is required is a broader framework that incorporates not only the sources of innovation, but also the wider context in which innovation occurs and the dynamics at play within such contexts.

SECTORAL SYSTEMS OF INNOVATION

The Foundations

The sectoral systems framework attempts to provide a more comprehensive approach to the analysis of sectoral differences in innovation and innovative activities. Sectoral systems is a multidimensional, integrated, and dynamic approach which draws on the work of three areas of research in economics and innovation studies.

The first area consists of the literature on change and transformation in industries. This research includes studies on industry life cycles (Utterback, 1994; Klepper, 1996) as well as broader analyses of the long-term evolution of industries as found in Schumpeter (1942), Kuznets (1930), and Clark (1940), and more recent work focused on the patterns of innovative activities and technological regimes (Malerba and Orsenigo, 1996; Dosi, 1997). Such studies underscore the importance of adopting a dynamic perspective in sectoral studies. Using a Schumpeterian classification, in fact, it is possible to identify situations in which a Schumpter Mark I pattern of innovative activity evolves into a Schumpter Mark II pattern. In the early stages of an industry, when knowledge is changing rapidly, uncertainty is high and barriers to entry are low. In such a context, new firms will most likely be the major sources of innovation and the key drivers of industrial dynamics. As the industry develops and eventually matures, however, and as technological change begins to follow well-defined trajectories, economies of scale, learning curves, barriers to entry, and financial resources become important elements in the competitive process. At this point, large firms with monopolistic power may come to the forefront of the innovation process (Gort and Klepper, 1982; Utterback, 1994; Klepper, 1996). Inversely, in cases in which major knowledge, technological, and market discontinuities exist, a Schumpeter Mark II pattern of innovative activities may, in turn, be replaced by a Schumpeter Mark I pattern. In this situation, a rather stable organization, characterized by incumbents with monopolistic power, will be displaced by a more turbulent one with new firms using new technologies or focusing on new segments of demand (Henderson and Clark, 1990; Christensen, 1997). Although rather archetypical, such dynamics underline the importance of examining sectoral differences over time in terms of knowledge and learning regimes.

The second area of research in which the sectoral systems approach is grounded is evolutionary economics. Evolutionary theory places an emphasis on dynamics, innovation processes, and economic transformation. Learning and knowledge are seen as key elements in the evolution of economic systems. In addition, evolutionary theory focuses on the cognitive aspects of learning processes, such as beliefs, expectations, and objectives, which are, in turn, affected by previous learning and experience and by the environments in which agents act. 'Boundedly rational' agents act, learn, and search in uncertain and changing environments. Agents know how to do different things or do the same thing in different ways. Thus, learning, knowledge, and behaviour entail agent heterogeneity in experience and organization. Distinctive competences result in persistent performance differences (Nelson, 1995; Dosi, 1997; Metcalfe, 1998). However, the heterogeneity of agents, their different learning processes and their distinctive competences are somewhat constrained and 'bounded' by the external sectoral context in which they operate. Firms within the same sector, in fact, will face the same set of technologies, will search within a similar knowledge base, will undertake similar production activities, and will be 'embedded' in the same institutional setting. As a result, these firms will develop a range of learning patterns, behaviours, and organizational forms that are specific to that sector.

The third area of research regards innovation systems. According to this literature, firms do not innovate in isolation. Rather, innovation is seen as an interactive process that involves a wide variety of actors including both firms and non-firm organizations such as universities, research centres, government agencies, regulatory bodies, and financial organizations (Lundvall, 1992; Carlsson, 1995; Edquist, 1997). This tradition draws from both interdisciplinary and historical approaches to the study of innovation and views learning as a key determinant of innovation (Edquist, 1997). While the point of departure for such studies is similar to 'environmental' analyses that examine external factors that influence firm decisions and performance (Narayanan and Fahey, 2001), the richness of the systems approach comes from its vision of the broader environment as a system of elements that work and interact together, rather than as single elements that work independently.

Much of the work on innovation systems has focused on national (Freeman, 1987; Lundvall, 1992; Nelson, 1993), regional or local systems of innovation (Cooke et al., 1998). The objective of such studies has been to understand the role of national or local institutions, government policies, regulations, and standards in influencing the innovative performance of firms in particular countries and regions. While this focus is useful for examining the national context or the local environment in which innovation occurs and the role of public agencies, universities and scientific research centres, and other national or regional organizations in promoting innovation, it hinders an understanding of how innovation within or across geographic boundaries is affected by sectoral characteristics. In a world increasingly defined by global competition, these defining characteristics need to be brought to the forefront in innovation management.

Recently, the systems approach has also been applied to the development of firm strategies under the label of business 'ecosystems' (Moore, 1996; Adner, 2006; and see Chapter 11 by Autio and Thomas). Ecosystems are normally defined as loose networks of intermediaries, suppliers, competitors, customers, and other stakeholders that affect, and are affected by, a firm's strategy (Iansiti and Levien, 2004). In business ecosystems, as in any ecosystem, survival and success depends on the ability of companies to manage their own strategies while, at the same time, promoting the overall health of other members of the system. For a single company or business, this concept is useful for underscoring the importance of interdependencies between different agents involved in the innovation process over time and across the value chain (Adner, 2012). It does not go far enough, however, to define the broader characteristics of the processes of interaction between these elements. The focus on the immediate business environment of a company, moreover, risks obscuring elements of the system that extend across multiple businesses and that define the wider sectoral environment in which individual companies operate. While it may be important for firms to analyse the state of the immediate business ecosystems of which they are a part, it is just as critical for firms to understand the characteristics of the broader sectoral context in which such businesses emerge and develop.

The Elements of Sectoral Systems

The sectoral systems approach draws on each of these three areas of research to propose a methodology for analysing the characteristics of sectors and for comparing the drivers of innovation across different sectors. According to this approach, a sector is seen as a set of activities which are associated with broad and related product groups, address similar existing or emerging demands, needs and uses, and share common knowledge bases (Malerba, 2002). The sectoral systems framework focuses on three main elements (see Figure 10.1):

   a. *Knowledge and technological domains:* A sector is characterized by a specific knowledge base and specific technologies and inputs needed for research, production, and distribution. Knowledge plays a central role in the sectoral

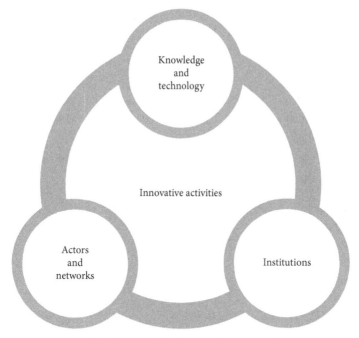

FIGURE 10.1 The framework of sectoral systems of innovation

systems approach. Knowledge is highly idiosyncratic at the firm level, does not diffuse automatically and freely among firms (Foray, 2004), and must be absorbed by firms through the capabilities which they have accumulated over time (Cohen and Levinthal, 1990). Knowledge, especially technological knowledge, involves varying degrees of specificity, tacitness, complexity, complementarity, and independence (Winter, 1987; Cowan et al., 2000). In a dynamic perspective, it is important to understand how knowledge and technologies are created, how they flow and are exchanged across firms, and how such processes might result in the redefinition of sectoral boundaries.

b. *Actors and networks:* A sector is composed of heterogeneous agents that include firms, non-firm organizations (e.g. universities, financial organizations, industry associations) and individuals (e.g. consumers, entrepreneurs, scientists). These heterogeneous agents are characterized by specific learning processes, competencies, beliefs, objectives, organizational structures, and behaviours. They interact through processes of communication, exchange, competition, control, and cooperation. Thus, within a sectoral systems framework, innovation is considered to be a process that involves systematic interactions among a wide variety of actors for the generation and exchange of knowledge relevant to innovation and its commercialization (networks).

c. *Institutions:* The cognitive frameworks, actions and interactions of agents are influenced by institutions, which include norms, common habits, established practices, rules, laws, and standards. Institutions may be created and imposed on agents from above, or may be generated through processes of interaction among agents themselves (such as contracts). They may be more or less binding and more or less formal (e.g. patent laws or specific regulations vs. traditions and conventions). Many institutions have national dimensions (e.g. patent laws or regulations), while others are specific to sectors and may cut across national boundaries (such as international conventions related to the exchange of specialized materials, or established practices concerning the types of knowledge that are freely exchanged within a sector).

Each of these elements has its own characteristics and its own set of dynamics which are important to understand. But each of these elements is also part of a broader system in which interaction between the parts drives change and innovation. Some of the ways in which these elements work both individually and as a system are discussed in the following section.

The Dynamics of Sectoral Systems

Knowledge is a major driving force of development and transformation in economic systems. Learning and knowledge by individuals and organizations are at the base of innovation and define the dynamics of firm and industrial growth. Knowledge works to redefine the boundaries of firms and industries in various ways. First, knowledge differs in terms of its sources (suppliers, universities, users, etc.), its domains (i.e. the specific scientific and technological fields at the base of innovative activities in a sector) and its applications (in terms of uses). Knowledge may be more or less cumulative (i.e. the degree to which the generation of new knowledge builds upon current knowledge) and may have different degrees of accessibility (i.e. opportunities to obtain knowledge that is external to firms). Knowledge may also spread more or less intentionally across individuals and organizations, as illustrated in the literature on knowledge spillovers and the work on the effects of the mobility of inventors, managers, and skilled labour (Malerba and Orsenigo, 2000; Malerba, 2002).

The recent focus on knowledge and sectoral systems has changed the traditional view of firms and industries by placing attention on four key drivers of innovation within and across sectors: learning and capabilities, networks, demand and uses, and institutions (Lundvall and Johnson, 1994; Kim, 1999; Dosi et al., 2000). In terms of learning, firms accumulate knowledge through internal processes as well as through processes that involve interaction with external actors that have heterogeneous knowledge and capabilities. Capabilities refer to the ability to absorb, develop, and integrate tacit and codified knowledge and to use it for specific functions, applications, and technological and productive transformations (Nelson, 1991; Dosi et al., 2000; Malerba and Orsenigo, 2000). They may be seen as sets of skills, routines, and complementary assets which are not necessarily formalized. They are based on procedural knowledge which is linked to specific applications or technological domains and may, therefore, not be easily transferable among firms. In other words, firms become repositories of knowledge (embodied in organizational routines which are largely tacit), that evolve though processes of learning, adaptation, and search (Nelson and Winter, 1982; Malerba, 1992).

Within this perspective, the 'knowledge' boundaries of a firm are related to the use and creation of knowledge for innovation and production and do not necessarily correspond to the legal boundaries of a firm (Brusoni et al., 2001). In their innovative activities, firms may access scientific knowledge in various ways, including R&D collaboration, informal networks, personnel mobility, and research contracts. The knowledge boundaries of a firm may encompass links with universities, research centres, or other firms involved in scientific activity. Knowledge boundaries may reach suppliers from whom firms may get specialized knowledge, or may tap into consumers and users who have knowledge concerning a specific context or a particular application. For understanding innovation, these networks of knowledge often redefine the boundaries of firms in a more meaningful way than the traditional legal boundaries.

In terms of networks, the variety of links and connections among agents greatly affects the dynamics of sectoral systems. As mentioned above, within sectoral systems, heterogeneous agents are connected in various ways through both market and non-market relationships. Early analyses of industrial organization examined agents as involved in processes of exchange, competition, and control (such as vertical integration). Later, the focus of analysis expanded to include processes such as collusion, hybrid governance forms, and formal R&D cooperation. The evolutionary approach and the innovation systems literature have broadened such analyses even further by considering networks as flows of both codified information and tacit knowledge which need to be absorbed and integrated by actors that are different in terms of capabilities. Therefore, according to this perspective, in uncertain and changing environments, networks emerge not because agents are similar (the reason at the base of more traditional perspectives that focus on spillovers and information asymmetry), but because they are different in terms of knowledge, capabilities, and specialization (see Teubal et al., 1991; Lundvall, 1992; Nelson, 1995; Edquist, 1997; Powell and Grodal, 2005). This approach has been used, in fact, to explain how relationships between firms and non-firm organizations (such as universities and public research centres) have been a source of innovation and change in sectoral systems such as pharmaceuticals and biotechnology, information technology, and telecommunications (Nelson and Rosenberg, 1993).

Third, changes in demand, needs, and uses have also been key drivers of transformation in many sectoral systems. The demand side and the actors that compose it (such as consumers, industrial and professional users, and public agencies) have been central to innovation in sectors as diverse as semiconductors, machine tools, and software (Mowery and Nelson, 1999). Demand is recognized as an important element in generating new knowledge and in defining and changing the knowledge boundaries of sectoral systems. Consumers, individuals, industrial users, and public agencies, each characterized by their own needs, learning processes, competencies, and goals, provide relevant knowledge for innovation and production (Lundvall, 1992). This happens in many different ways and with varying degrees of intensity across sectors (DiStefano et al., 2012). In some cases, demand may be a source of information, feedback or testing for new ideas. In other cases, users themselves may design product innovations to suit their specific needs which may be different from, or ahead of, the needs of the general mass market (von Hippel, 1988, 2005). In further cases, users with application specific knowledge may not only design innovative solutions for their needs, but also produce and commercialize these innovations through their own entrepreneurial ventures (Shah and Tripsas, 2007; Fontana and Malerba, 2010; Adams et al., 2013).

Finally, institutions play a major role in the dynamics of all sectoral systems by influencing the rate of technological change, the organization of innovative activity and the innovative performance of firms. The type and the nature of the institutions involved vary across different sectoral systems. For example, both the impact of intellectual property rights and the behaviour and strategies of firms in protecting their innovations through patents differ between pharmaceuticals and software (Levin et al., 1987). Similarly, the norms that guide investment decisions and labour markets vary significantly across sectors such as machine tools and software. Such differences underline the importance of including an analysis of institutions in studies of sectoral systems of innovation.

It should be noted, however, that institutions also have national level characteristics that may impact the innovative activities of firms differently from country to country (Nelson, 1993). For example, the well-known diversity between the first-to-invent and the first-to-file rules in the patent systems of the United States and Japan had a significant impact on the behaviour of firms in these two countries. In addition, the characteristics of national institutions may favour sectors that fit the specificities of the national institutions better than others. Thus, a sectoral system may become predominant in a country due to the fact that the existing institutions of that country provide an environment that supports firms in that sector. For example, in France, the behaviour of public agencies with regard to procurement and investment choices favoured sectors in which public demand was an important factor in stimulating innovation (Chesnais, 1993). In other cases, national institutions may constrain the development of, or innovation in, specific sectors, or may create mismatches between national and sectoral institutions. Examples of such mismatches and, more broadly, of the impact of differences across national institutions on the evolution of different sectors, are examined in Dosi and Malerba (1996) for advanced economies, and in Malerba and Nelson (2012) for developing economies. Of course, the interaction between national institutions and sectoral systems is not only unidirectional, going from national institutions to sectoral variables. Sometimes, the direction is reversed, such that developments in specific sectors work to influence national institutions. Specific policies designed to support new biotechnology firms or regulations made to protect consumers in banking, for example, have, in some instances, inspired broader changes in national institutions (i.e. to support start-ups in a broader range of sectors or to develop laws for privacy protection). Sometimes this has proved successful, and sometimes not (Dodgson et al., 2008).

Sectoral Systems of Innovation at Work

The elements of sectoral systems interact to create processes of exchange and learning from which innovation emerges. Different sectors will have distinct systems of innovation that involve different types of knowledge, different sets of actors, and different kinds of institutions. The sectoral systems framework helps us to understand these differences and their implications for innovation management. Depending on the purpose of the analysis, moreover, the sectoral systems framework may be used at varying levels of aggregation to understand how these processes work across broad sectoral boundaries or within more narrowly defined product groups or categories of use.

In this section, the framework will be applied to the analysis of three sectoral systems in order to show how it may be used to understand the processes underlying innovation in different settings. The section begins with an analysis of the pharmaceutical sector and then examines two very different sectors (machine tools and services) to compare and contrast the elements of their systems with that of pharmaceuticals. Only a few of the features of each system have been selected for analysis here in order to highlight differences across them and to show how these differences may affect the management of innovation. More in-depth analyses for each sector may be found in the references.

Pharmaceuticals

The actors in pharmaceuticals include both large and small pharmaceutical companies, new biotechnology firms, universities, research organizations and hospitals, the medical profession (physicians, nurses, etc.), financial organizations (i.e. venture capital), regulatory agencies, and consumers. In terms of the knowledge base and technology, the sector has witnessed a period of constant transformation and growth over the past four decades. Scientific advances in physiology, pharmacology, enzymology, and cell biology were followed by discoveries in molecular genetics and recombinant DNA technology and, subsequently, in genomics, gene sequencing, transgenic animal creation, molecular biology, and chemistry. These advances were accompanied by the introduction of a host of new techniques and equipment for drug design and testing, all of which led to major changes in the processes used to develop new drugs (from 'random search' to 'guided discovery' to biology and platform technologies) (Henderson et al., 1999). In terms of the institutional framework in pharmaceuticals, three main elements may be highlighted. The first is the impact of public policy on innovative activities. The approval process for new drugs, as well as the organizational structure of, and the policies adopted by, the national health systems have all influenced the path of innovation in pharmaceuticals. The second is the importance of patent legislation and intellectual property rights in the sector. Closely linked to this element are the norms and behaviours that have defined the role of financial organizations, especially venture capitalists, in pharmaceuticals. Strong patent rights, in fact, were needed to provide a guarantee to firms, research organizations, and their financing partners that they would be able to reap the returns of their investments (McKelvey et al., 2004).

Although these elements have displayed important local variations in different national settings, within the broader framework of the sector, the institutional and technological changes that characterized pharmaceuticals over this period resulted in the creation of both horizontal and vertical networks between firms. Horizontally, networks of exchange and transfer emerged which allowed incumbent firms to access and eventually absorb cutting-edge knowledge and techniques from younger and more dynamic firms. Such dynamics resulted in a division of labour between large, established drug companies and new entrants that offered specialized knowledge and innovation in a wide range of scientific fields. Alongside of these horizontal networks, new vertical links were created through processes of vertical integration that took place through successive mergers and acquisitions within the industry (Gambardella, 1995; Pisano, 1991). The learning regimes and selection processes that characterized this sector, therefore, had an impact not only on the intensity of the relationships among existing actors in the sector (both firms and other organizations), but also on the emergence of new actors and networks from expanding boundaries of knowledge.

Machine Tools

The sectoral system of innovation in machine tools varies in several significant ways from that of pharmaceuticals. In contrast to pharmaceuticals, the actors involved in machine tools have traditionally included a stable base of medium-sized and family-owned enterprises on the supply side and a large base of user firms on the demand side. Research organizations have played a much smaller role in machine tools, and external financial organizations, government bodies, and regulatory agencies have had almost no role. Innovation in machine tools has been characterized by incremental product development and has involved only limited investment in R&D (Wengel and Shapira, 2004). Moreover, while the demand side was not directly involved in product development processes in pharmaceuticals, innovation in machine tools has traditionally come from a close interaction between suppliers and user firms. In fact, innovation was driven by established firms that were able to develop strong relationships with leading user firms for the exchange of application-specific knowledge (Mazzoleni, 1999) rather than by the entrance of entrepreneurial firms with advanced scientific knowledge. Such interactions allowed supplier firms to absorb tacit knowledge about user applications and to generate learning processes inside their own organizations that would allow them to design customized solutions to meet user needs. Finally, it should be noted that institutions, such as government policies and patent legislation, were much less relevant in machine tools than in pharmaceuticals. Rather, firms considered their ability to create tailored solutions for their customers as a better guarantee for appropriating the returns from innovation than the use of patents. Given the low levels of R&D and outside finance in machine tools, moreover, these firms did not face the same pressures as pharmaceutical companies did to protect their investments through patents.

These characteristics also fostered the development of different types of networks within machine tools. Networks linking firms with external research organizations in leading edge technologies did not develop in this sector. Rather, strong regional clusters were established where firms could access skilled labour pools, remain close to large users and leading customers, and benefit from complementary linkages for services such as finance (local banks), maintenance, parts, and communications (Wengel and Shapira, 2004).

Within this context, it is interesting to note how the recent expansion of the knowledge bases relevant for machine tools has changed the dynamics in this sector. New technologies from fields such as lasers, new materials, software, and microelectronics, now play a significant role in innovation in machine tools and have introduced a layer of codified knowledge on top of the existing layers of tacit knowledge within supplier firms. The introduction of rapid prototyping (RP) technologies, for example, has facilitated the production of single objects and complex shapes at rapid speeds without the use of moulds. This has made it easier for companies to develop and test different models before final production. As a result, many organizations are shifting from specification-driven strategies based on the expressed needs of current users towards prototype-driven strategies which are able to draw on inputs from a broader range of disciplines and functions (Schrage, 1993). The development of new knowledge bases relevant for innovation, moreover, has resulted in the generation of new capabilities within firms, higher levels of investment in R&D, and new links with external organizations with advanced capabilities in specialized fields. While relations with users remain critical in machine tools, innovation is increasingly being driven by new entrepreneurial ventures with broader communities of stakeholders. Finally, as new customer bases grow around the globe, the traditional regional networks that characterized the sector are slowly being replaced by international partnerships and by collaboration with research and finance organizations.

Services

Services encompasses a wide variety of sectors that involve humans (e.g. medicine, insurance, tourism), physical artefacts (e.g. maintenance and repair, storage, transportation) or data, symbols, and information (e.g. telecommunications, financial services). This section will highlight only a few general characteristics that distinguish most innovation systems in services from those of the other sectors already examined. These characteristics will not apply to all services, but the following discussion may contribute to an understanding of how the framework may apply in such sectors.

Services differ from other manufacturing sectors in many ways. Services are intangible, heterogeneous, perishable, and produced and consumed simultaneously. More than in manufacturing sectors, innovation in services includes business models, organizational forms, and business processes. The actors involved range from service providers, to customers and client firms, suppliers, and government organizations concerned with public policies and public services. What is noteworthy, however, is that somewhat like machine tools, close interaction with users is a distinguishing feature of most services. Services, in fact, typically involve high levels of contact with clients in the design, production, delivery, and consumption of the service activity. As a result, innovation in services revolves much more around social and cultural norms than other sectoral systems whose focus is on tangible products (Miles, 2005).

Although service innovation involves a high degree of learning-by-doing with users, it would be misleading to characterize services as mere adopters of technologies developed by manufacturing sectors (Pavitt, 1984). Recent surveys, in fact, indicate that, along with the purchase of machinery and equipment, investment in technologies (including software), R&D, and training are significant in many service sectors. What characterizes services in terms of knowledge and learning is the importance of technique and procedure alongside of devices and artefacts. Innovation in equipment technology goes hand in hand with innovation in skills and in the procedures used to organize and offer services (Tether and Metcalfe, 2004). Learning through experience and cooperation occurs among networks that extend from service organizations to both ancillary services and technology suppliers and customers. These networks tend to encompass a vast array of small, independent providers and customers, often active at the local level, and are relatively fluid, allowing for entry and exit according to the service problem to be solved. In recent years, these networks have become increasingly national and international in scope and have been extended to include relations with sources of more formal and scientific knowledge.

As was the case in pharmaceuticals, institutions play a significant role in innovation in services. These institutions include a wide variety of policies and regulations covering issues such as market structure and the provision of public services, health and safety regulations, labour laws, and privacy rules. Yet, as underlined above, given the close interaction between suppliers and users in service innovation, cultural and social norms also play a crucial role in services (Tether, 2005). Such norms may determine how, how often, and with whom such interactions occur as well as the depth and the nature of the information that is be exchanged during such interactions. Finally, intellectual property protection in the form of patents is not used for many service innovations. Rather, service firms tend to rely on the customization of their offerings and on trademarks and copyrights to protect their ideas from imitation. It may be noted, however, that this may change as more attention is given to the development of mechanisms that would allow companies to protect both intangible assets and 'business models' from competition.

Although many of the observations regarding these three sectors are schematic and do not capture the richness of the systems of innovation at work in each context, this brief analysis allows us to highlight the usefulness of the sectoral systems framework for analysing the elements and the dynamics of innovation across different sectors.

Major Themes and Research Findings

In the last decade, research in sectoral systems has progressed along three broad lines of inquiry.

Expansion of the types of sectoral systems examined. The variety of sectors examined using a sectoral systems framework has expanded significantly in recent years and has been extended beyond the range of manufacturing industries that were the subject of most of the earlier work in the field (Mowery and Nelson, 1999; Malerba, 2002, 2004). Some basic conclusions emerge from an examination of these new studies. The first is that knowledge represents a major driving force behind the development and transformation of sectoral systems. Changes in knowledge and learning provoke modifications in the external links of sectors, pave the way for the emergence of new actors and new institutions, and redefine sectoral boundaries. Understanding knowledge, its continuous development and its impact on the characteristics of a sectoral system is thus of utmost importance for innovation management. Yet, knowledge is an element that is rarely operationalized in management studies. The sectoral systems framework provides a way to do that. Second, with the availability of new longitudinal data on firms and other actors (see, for example, the CIS survey now carried out regularly at the firm level, or patent and patent citation data at the firm and individual levels), it is now possible to develop quantitative analyses of sectoral systems both over time and across countries. Finally, these studies show that a sectoral systems approach to the study of innovation and production is not a straightjacket, but a broad, flexible, and adaptable tool. Having identified the key elements and characteristics of a sectoral system, researchers can chose the level of analysis according to the objective at hand. Within broadly defined sectoral systems such as services, for example, it is possible to use the framework to analyse highly innovative services (e.g. airport runway systems, logistics) or less innovative ones (e.g. hotels and restaurants). Similarly, the framework can be used to examine a broadly defined sectoral system such as ICT, where the goal is to understand broader processes of integration between different technologies, the convergence of previously separated industries and the redefinition of sectoral boundaries, or a more narrowly defined system within ICT (such as custom software) where the goal is to conduct a detailed analysis of the actors, knowledge bases, and networks involved in specific innovative activities and to identify distinctive co-evolutionary processes (Steinmueller, 2004).

Catching up in different sectoral systems. A second area of progress regards the application of the sectoral systems framework to processes of catching up in different national contexts. Studies on China, India, and Brazil provide many interesting examples. India has been quite successful in catching up in pharmaceuticals, but unsuccessful in telecom equipment; Brazil has been successful in catching up in agro-food and less so in pharmaceuticals; China is moving up the ladder in autos and telecommunications, but faces greater challenges in semiconductors. The existing literature on catching up recognizes the importance of learning and capabilities by firms for such processes (Lee and Lim, 2001). The sectoral system approach adds to these studies by showing how the learning processes, capabilities, organizations, and strategies of domestic firms need to be analysed in conjunction with the specificities of the sector involved. The analyses in Malerba and Nelson (2012) on different sectors (telecommunications, pharmaceuticals, software, semiconductors, automobiles, and agro-food) in five countries, in fact, show that innovation and catching up have been affected by a range of different elements. The key drivers of catching up varied in these cases from government policies to industrial structures (of either small or large firms, or both), financial organizations, users and demand, universities, public research laboratories, and standards and regulations. But more than just pointing to such diversity, these studies show that the role played by these elements is best understood if analysed within a systemic framework.

Simulation models. Finally, a great deal of progress has been made in recent years in the development of simulation models to analyse the causal mechanisms that affect innovation and the dynamics of sectoral systems. Simulations, in the form of agent-based models (Tesfatsion, 2006), provide a useful tool for examining the structures, multiple components, and feedback processes that characterize sectors. They also provide a way to represent complex environments, to develop a process view of innovation, and to place sectoral analyses in evolving, dynamic settings.

Within the family of agent-based models, history-friendly models have recently been used to analyse the relationship between innovation, the dynamics of market structure, and industry evolution. History-friendly models start from an empirical analysis of the main elements that characterize the configuration and evolution of an industry and from qualitative interpretations of the causal relationships that affect innovation and the dynamics of market structure. From these bases, models are constructed with the objective of explaining the specific empirical patterns observed. History-friendly models can play an important role in bridging different levels and styles of analysis by developing a dialogue between rich and articulated qualitative research and formal analyses of the role of specific elements within a system (Malerba et al., 2012).

Studies based on history-friendly models have been done on the computer industry (Malerba et al., 2012), the semiconductor DRAM industry (Kim and Lee, 2003), the pharmaceutical industry (Garavaglia et al., 2012), and the synthetic dye industry (Brenner and Murmann, 2003). These studies have used such models to examine the role of specific elements of sectoral systems such as firms (in terms of capabilities, organization, and strategies), demand and users, suppliers, governments, universities, and institutions (e.g. IPR). However, such analyses may easily be expanded to include additional elements of sectoral systems, such as the knowledge base, the role of venture capital, and the presence of collaborative networks among firms. They may also be used to give direction to firms in terms of strategy by examining factors such as the role of first mover advantage, imitation, entry, and user–supplier interaction (Malerba et al., 2012).

IMPLICATIONS FOR MANAGEMENT AND FUTURE DEVELOPMENTS

The development and application of a dynamic, comparative, and systemic framework for understanding innovation in different sectors has three major implications for innovation management. The first implication is that innovation strategies at the firm level need to be managed within a broader and systemic perspective. Innovation involves not only the firms in a sector and their decisions, but also knowledge flows and learning processes, the institutional setting in which such decisions are made, and other actors that are linked through networks and feedback processes.

The second implication is that the so-called 'statistical and official' boundaries of sectors may not represent the most useful definitions when discussing innovation management. Such boundaries, in fact, ignore the vertical and horizontal links within and across sectors in terms of knowledge, technologies, and products, as well as the interdependencies that shape innovation. Think of the convergence of ICT, the recent use and integration of new technologies in so-called traditional sectors, or the presence of biotechnology and nanotechnology in so many sectors. Static and rigid perspectives need to be replaced by frameworks and models that allow for change and dynamics.

The third implication regards our understanding of the basic co-evolutionary processes that characterize firms and other actors active in a sectoral system. In a broad sense, co-evolution entails feedback mechanisms across elements that lead them to change together. These co-evolutionary processes involve actors, knowledge, technology, demand, and institutions. They are often path-dependent (Nelson, 1994) and sector-specific. The challenge for innovation managers is to move from broad and unhelpful observations that all factors seem to be interconnected, to more useful analyses that increase their understanding of how specific processes of co-evolution occur, at what intensity they do so, and with what kinds of feedback mechanisms.

How can managers use the sectoral systems framework? This framework provides a systemic view that may help managers to see not only what elements are changing in an environment, but also what effects those changes may have on other elements in the system. It thus enriches traditional environmental scanning techniques used in many businesses by extending studies of single factors to include analyses of their interdependencies as well. This approach also allows managers to see how elements of a system work together to create value, often in excess of what any one of the single parts could have created on its own. Such observations are at the base of many new strategies for innovation management, including open innovation, innovation platforms, and value networks. Success in dynamic sectoral systems requires a deep understanding of how systems work and change, as well as of the challenges of being part of systems in which the success of any one actor may depend heavily on the characteristics and performance of the others actors within them.

The perspective of a sectoral system also provides a useful framework for managers to look beyond the direct upstream and downstream parts of their businesses for new ideas and new knowledge to stimulate innovation. The impetus for innovation, in fact, may come from elements that may be outside of the immediate business environment of a company. By developing a deeper understanding of the characteristics of their own sectoral system and its dynamics over time, managers will have a clearer sense of the position of their companies within the system and of which elements are working to support their strategies and objectives, and which are not.

References

Adams, P., Fontana, R., and Malerba, F. (2013). 'The Magnitude of Innovation by Demand in a Sectoral System: The Role of Industrial Users in Semiconductors', *Research Policy*, 42(1): 1–14.

Adner, R. (2006). 'Match your Innovation Strategy to your Innovation Ecosystem', *Harvard Business Review*, 84(4): 98–107.

Adner, R. (2012). *The Wide Lens*. New York: Penguin Portfolio.

Bain, J. (1956). *Barriers to New Competition*. Boston: Harvard University Press.

Brenner, T., and Murmann, P. (2003). *The Use of Simulations in Developing Robust Knowledge about Causal Processes: Methodological Considerations and an Application to Industrial Evolution*, Papers on Economics and Evolution #0303, Max Plank Institute, Jena.

Brusoni, S., Prencipe, A., and Pavitt, K. (2001). 'Knowledge Specialization, Organizational Coupling and the Boundaries of the Firm: Why do Firms Know More than they Make?' *Administrative Science Quarterly*, 46(4): 597–621.

Carlsson, B. (ed.) (1995). *Technological Systems and Economic Performance*. Dordrecht: Kluwer.

Chesnais, F. (1993). 'The French National System of Innovation', in R. Nelson (ed.), *National Innovation Systems*. Oxford: Oxford University Press, 192–229.

Christensen, C. M. (1997). *The Innovator's Dilemma*. Boston: Harvard Business School Press.

Clark, C. (1940). *The Conditions of Economic Progress*. London: Macmillan.

Cohen, W. M., and Levin, R. (1989). 'Empirical Studies of Innovation and Market Structure', in R. Schmalensee and R. Willig (eds), *Handbook of Industrial Organization*. New York: North Holland, 1060–1107.

Cohen, W. M., and Levinthal, D. (1990). 'Absorptive Capacity: A New Perspective on Learning and Innovation', *Administrative Science Quarterly*, 35(1): 128–52.

Cooke, P., Heidenreich, M., and Braczyk, H. J. (1998). *Regional Innovation Systems*. London: Routledge.

Cowan, R., David, P., and Foray, D. (2000). 'The Explicit Economics of Codification and the Diffusion of Knowledge', *Industrial and Corporate Change*, 9(2): 211–53.

Dahmen, E. (1989). 'Development Blocks in Industrial Economics', in B. Carlsson (ed.), *Industrial Dynamics*. Dordrecht: Kluwer Academic Press, 109–22.

DiStefano, G., Gambardella, A. and Verona, G. (2012). 'Technology Push and Demand Pull Perspectives in Innovation Studies: Current Findings and Future Research Directions', *Research Policy*, 41(8): 1283–95.

Dodgson, M., Mathews, J., Kastelle, T., and Hu, M. (2008). 'The Evolving Nature of Taiwan's National Innovation System: The Case of Biotechnology Innovation Networks', *Research Policy*, 37(3): 430–45.

Dosi, G. (1997). 'Opportunities, Incentives and the Collective Patterns of Technological Change', *Economic Journal*, 107(444): 1530–47.

Dosi, G., and Malerba, F. (1996). *Organization and Strategy in the Evolution of the Enterprise*. London: Macmillan.

Dosi, G., Nelson, R., and Winter, S. (2000). *The Nature and Dynamics of Organizational Capabilities*. Oxford: Oxford University Press.

Edquist, C. (ed.) (1997). *Systems of Innovation*. London: Frances Pinter.

Fontana, R., and Malerba, F. (2010). 'Entry, Demand and Survival in the Semiconductor Industry', *Industrial and Corporate Change*, 19(5): 1629–54.

Foray, D. (2004). *The Economics of Knowledge*. Cambridge, MA: MIT Press.

Freeman, C. (1987). *Technology Policy and Economic Performance: Lessons from Japan*. London: Frances Pinter.

Gambardella, A. (1995). *Science and Innovation: The U.S. Pharmaceutical Industry during the 1980s*. Cambridge: Cambridge University Press.

Garavaglia, C., Malerba F., Orsenigo L., and Pezzoni M. (2012). 'Technological Regimes and Demand Structure in the Evolution of the Pharmaceutical Industry', *Journal of Evolutionary Economics*, 22(4): 677–709.

Geroski, P. (1995). 'What Do We Know About Entry?', *International Journal of Industrial Organization*, 13(4): 421–40.

Gilbert, R. (2006). 'Looking for Mr. Schumpeter: Where Are We on the Competition—Innovation Debate?', in A. Jaffe, J. Lerner, and S. Stern (eds), *Innovation Policy and the Economy*. Cambridge, MA: MIT Press, 159–215.

Gort, M., and Klepper, S. (1982). 'Time Paths in the Diffusion of Product Innovations', *Economic Journal*, 92(367): 630–53.

Granstrand, O. (1994). *The Economics of Technology*. Amsterdam: Elsevier Science Publishers.

Granstrand, O., Patel, P., and Pavitt, K. (1997). 'Multi-technology Corporations: Why They Have Distributed rather than Distinctive Core Competencies', *California Management Review*, 39(4): 8–25.

Grant, R. (1996). 'Towards a Knowledge-based Theory of the Firm', *Strategic Management Journal*, 17: 109–22.

Henderson, R., and Clark, K. (1990). 'Architectural Innovation: The Reconfiguration of Existing Product Technologies and the Failure of Established Firms', *Administrative Science Quarterly*, 35(1): 9–30.

Henderson, R., Orsenigo, L., and Pisano, G. (1999). 'The Pharmaceutical Industry and the Revolution in Molecular Biology', in D. Mowery and R. Nelson (eds), *The Sources of Industrial Leadership*. Cambridge, MA: Cambridge University Press, 267–312.

Iansiti, M., and Levien, R. (2004). 'Strategy as Ecology', *Harvard Business Review*, 82(3): 68–78.

Kim, C., and Lee, K. (2003). 'Innovation, Technological Regimes and Organizational Selection in Industry Evolution: A "History Friendly Model" of the DRAM Industry', *Industrial and Corporate Change*, 12(6): 1195–221.

Kim, L. (1999). *Learning and Innovation in Economic Development*. London: Edward Elgar Publishing.

Klepper, S. (1996). 'Entry, Exit, Growth and Innovation over the Product Life Cycle', *American Economic Review*, 86(3): 562–83.

Kuznets, S. (1930). *Secular Movements in Production and Prices: Their Nature and their Bearing upon Cyclical Fluctuations*. Boston: Houghton Mifflin.

Lee, K., and Lim, C. (2001). 'Technological Regimes, Catching-up and Leapfrogging: Findings from the Korean Industries', *Research Policy*, 30(3): 459–83.

Levin, R., Klevorick, A., Nelson, R., and Winter, S. (1987). 'Appropriating the Returns from Industrial R & D', *Brookings Papers on Economic Activity*, 3, 783–831.

Lundvall, B. Å. (1992). *National Systems of Innovation*. London: Frances Pinter.

Lundvall, B. Å., and Johnson, B. (1994). 'The Learning Economy', *Journal of Industry Studies*, 1(2): 23–42.

McKelvey, M., Orsenigo, L., and Pammolli, F. (2004). 'Pharmaceuticals as a Sectoral Innovation System', in F. Malerba (ed.), *Sectoral Systems of Innovation: Concepts, Issues and Analyses of Six Major Sectors in Europe*. Cambridge: Cambridge University Press, 73–120.

Malerba, F. (1992). 'Learning by Firms and Incremental Technical Change', *Economic Journal*, 102(413): 845–59.

Malerba, F. (2002). 'Sectoral Systems of Innovation and Production', *Research Policy*, 31(2): 247–64.

Malerba, F. (ed.) (2004). *Sectoral Systems of Innovation: Concepts, Issues and Analyses of Six Major Sectors in Europe*. Cambridge: Cambridge University Press.

Malerba, F., and Nelson, R. (2011). 'Learning and Catching up in Different Sectoral Systems: Evidence from Six Industries', *Industrial and Corporate Change*, 20(6): 1645–75.

Malerba, F., and Nelson, R. (eds.) (2012). *Economic Development as a Learning Process: Variation across Sectoral Systems*. Cheltenham: Edward Elgar.

Malerba, F., and Orsenigo, L. (1996). 'Schumpeterian Patterns of Innovation', *Cambridge Journal of Economics*, 19(1): 47–65.

Malerba, F., and Orsenigo, L. (2000). 'Knowledge, Innovative Activities and Industry Evolution', *Industrial and Corporate Change*, 9(2): 289–314.

Malerba, F., Nelson, R., Orsenigo, L., and Winter, S. (forthcoming). *Innovation and the Evolution of Industries: History-friendly Models*. Cambridge: Cambridge University Press.

Mazzoleni, R., (1999). 'Innovation in the Machine Tool Industry: A Historical Perspective on the Dynamics of Comparative Advantage', in D. Mowery and R. Nelson (eds.), *Sources of Industrial Leadership*. Cambridge: Cambridge University Press, 169–216.

Metcalfe, S. (1998). *Evolutionary Economics and Creative Destruction*. London: Routledge.

Miles, I. (2005). 'Innovation in Services', in J. Fagerberg, D. Mowery, and R. Nelson (eds.), *The Oxford Handbook of Innovation*. Oxford: Oxford University Press, 433–58.

Moore, J. F. (1996). *The Death of Competition: Leadership and Strategy in the Age of Ecosystems*. New York: Harper Business.

Mowery, D., and Nelson, R. (eds.) (1999). *The Sources of Industrial Leadership*. Cambridge: Cambridge University Press.

Narayanan, V. K., and Fahey, L. (2001). 'Macro-environmental Analysis: Understanding the Environment Outside the Industry', in L. Fahey and R. Randall (eds.), *The Portable MBA in Strategy*. New York: John Wiley and Sons, 189–214.

Nelson, R. (1991). 'Diffusion of Development: Post-World War II Convergence among Advanced Industrial Nations', *American Economic Review*, 81(2): 271–5.

Nelson, R. (ed.) (1993). *National Innovation Systems: A Comparative Study*. Oxford: Oxford University Press.

Nelson, R. (1994). 'The Coevolution of Technology, Industrial Structure and Supporting Institutions' *Industrial and Corporate Change*, 3(1): 47–63.

Nelson, R. (1995). 'Recent Evolutionary Theorizing About Economic Change', *Journal of Economic Literature*, 33(1): 48–90.

Nelson, R., and Rosenberg, N. (1993). 'Technical Innovation and National Systems', in R. Nelson (ed.), *National Innovation Systems*. Oxford: Oxford University Press, 1–18.

Nelson, R., and Winter, S. (1982). *An Evolutionary Theory of Economic Change*. Cambridge, MA: The Belknapp Press of Harvard University.

Pavitt, K. (1984). 'Sectoral Patterns of Technical Change: Towards a Taxonomy and a Theory', *Research Policy*, 13(6): 343–73.

Pisano, G. (1991). 'The Governance of Innovation: Vertical Integration and Collaborative Arrangements in the Biotechnology Industry', *Research Policy*, 20(3): 237–49.

Porter, M. E. (1979). 'How Competitive Forces Shape Strategy', *Harvard Business Review*, 57: 86–93.

Porter, M. E. (1985). *Competitive Advantage*. New York: The Free Press.

Powell, W., and Grodal, S. (2005). 'Networks of Innovators', in J. Fagerberg, D. Mowery, and R. Nelson (eds.), *The Oxford Handbook of Innovation*. Oxford: Oxford University Press, 56–85.

Robson, M., Townsend, J., and Pavitt, K. (1988). 'Sectoral Patterns of Production and Use of Innovation in the U.K.: 1943–1983', *Research Policy*, 17(1): 1–14.

Rosenberg, N. (1976). *Perspectives on Technology*. Cambridge: Cambridge University Press.

Rosenberg, N. (1982). *Inside the Black Box: Technology and Economics*. Cambridge: Cambridge University Press.

Scherer, M. (1982). 'Inter-industry Technological Flows in the United States', *Research Policy*, 11(4): 227–46.

Scherer, M. (1990), *Industrial Market Structure and Economic Performance*. Boston: Houghton Mifflin.

Schrage, M. (1993). 'The Culture(s) of Prototyping', *Design Management Journal*, 4: 55–65.

Schumpeter, J. A. (1911). *The Theory of Economic Development*. Boston: Harvard Press.

Schumpeter, J. A. (1942). *Capitalism, Socialism and Democracy*. New York: Harper.

Shah, S. K., and Tripsas, M. (2007). 'The Accidental Entrepreneur: The Emergent and Collective Process of User Entrepreneurship', *Strategic Entrepreneurship Journal*, 1(1–2): 123–40.

Steinmueller, W. E. (2004). 'The Software Sectoral Innovation System', in F. Malerba (ed.), *Sectoral Systems of Innovation: Concepts, Issues and Analyses of Six Major Sectors in Europe*. Cambridge: Cambridge University Press, 193–242.

Sutton, J. (1991). *Sunk Costs and Market Structure*. Cambridge, MA: MIT Press.

Sutton, J. (1998). *Technology and Market Structure*. Cambridge, MA: MIT Press.

Tesfatsion L. (2006). 'Agent-based Computational Economics: A Constructive Approach to Economic Theory', in L. Tesfatsion and K. Judd (eds.), *Handbook of Computational Economics, Volume 2: Agent-Based Computational Economics*. Amsterdam: Elsevier/North-Holland, 831–80.

Tether, B. S. (2005). 'Do Services Innovate (Differently)?: Insights from the European Innobarometer Survey', *Industry and Innovation*, 12(2): 153–84.

Tether, B. S., and Metcalfe, J. S. (2004). 'Services and Systems of Innovation', in F. Malerba (ed.), *Sectoral Systems of Innovation*. Cambridge: Cambridge University Press, 287–324.

Teubal, M., Yinnon, T., and Zuscovitch, E. (1991). 'Networks and Market Creation', *Research Policy*, 20(5): 381–92.

Tirole, J. (1988), *The Theory of Industrial Organization*. Cambridge, MA: MIT Press.

Utterback, J. (1994). *Mastering the Dynamics of Innovation*. Boston: Harvard Business School Press.

von Hippel, E. (1988). *The Sources of Innovation*. Oxford: Oxford University Press.

von Hippel, E. (2005). *Democratizing Innovation*. Cambridge, MA: MIT Press.

von Tunzelman, N., and Acha, V. (2005). 'Innovation in Low-tech Industries', in J. Fagerberg, D. Mowery, and R. Nelson (eds), *The Oxford Handbook of Innovation*. Oxford: Oxford University Press, 407–32.

Wengel, J., and Shapira, P. (2004). 'Machine Tools: The Remaking of a Traditional Sectoral Innovation System?' in F. Malerba (ed.), *Sectoral Systems of Innovation: Concepts, Issues and Analyses of Six Major Sectors in Europe*. Cambridge: Cambridge University Press, 243–86.

Winter, S. (1987). 'Knowledge and Competencies as Strategic Assets', in D. Teece (ed.) *The Competitive Challenge: Strategies for industrial innovation and renewal*. Cambridge, UK: Ballinger Publishing Company, 159–84.

CHAPTER 11

INNOVATION ECOSYSTEMS

Implications for Innovation Management?

ERKKO AUTIO AND LLEWELLYN D. W. THOMAS

INTRODUCTION

THE notion of 'ecosystems' provides an attractive metaphor to describe a range of value creating interactions and relationships between sets of interconnected organizations. First introduced in practitioner literature in the mid-1990s (Moore, 1993, 1996), this metaphor has been increasingly adopted in research journals such as the Strategic Management Journal (Teece, 2007; Pierce, 2009; Adner and Kapoor, 2010; Gulati, Puranam, and Tushman, 2012). It has recently been argued that the focus of marketing and strategy must be on shaping the ecosystem in which the firm resides (Singer, 2006), and firms should increasingly move away from industry-focused strategic planning towards strategizing within and around ecosystems (Iansiti and Levien, 2004b). Arguably, the attractiveness of this rather loosely defined and versatile metaphor rests on its ability to evoke and highlight interdependencies between organizations and to provide a fresh way to think about specialization, co-evolution, and co-creation of value (Frels, Shervani, and Srivastava, 2003; Adner and Kapoor, 2010).

But, to use another often (ab?)used metaphor, do ecosystems have legs? Do notions of ecosystem add insight beyond existing constructs of broadly similar content, such as value chains and supply networks, or are we dealing simply with yet another convenient catchphrase that allows management consultants to substitute impression for substance? A cursory reading of the literature would certainly suggest ample reason to be sceptical. Most of the uses of the term are found in practitioner literature, with only few treatments in academic journals; definitions proliferate and are often difficult to reconcile; and the term has been used in a variety of contexts without much cross-fertilization between domains and levels of analysis. As we will demonstrate in this chapter, however, conceptual proliferation does not necessarily mean absence of progress. While the field is certainly fragmented, pockets of advance exist and, taken together, the literature does appear to provide insight into the management of innovation in evolving networks of interconnected actors organized around a focal firm or platform. In this chapter, we therefore review the proliferating ecosystem literature and extract insight for innovation management. We show that the ecosystem construct does have at least some legs and we try to illustrate what those legs might look like.

In this chapter, we first discuss definitions of the ecosystem concept. We then review received ecosystem literature from two perspectives. First, drawing on ecosystem research as carried out in different empirical contexts, we summarize insights with regard to ecosystem boundaries, structure, and coordination. Second, we review three theoretical perspectives that can be applied to ecosystem research: notably, the value creation perspective, the network embeddedness perspective, and the network management perspective. Drawing on these reviews, we then elaborate on the application of the ecosystem concept in innovation strategy analysis, design, and implementation.

Definitions

The 'ecosystem' term has been applied in a wide variety of contexts outside its original application in biological systems. In management research, the term 'ecosystem' usually refers to a network of interconnected organizations that are linked to or operate around a focal firm or a platform (Moore, 1993, 1996; Iansiti and Levien, 2004a, b; Teece, 2007). As a theoretical construct, the ecosystem differs from other network-centric constructs in management research by its inclusion of both production and use side participants, including complementary asset providers and customers. We formally define an innovation ecosystem as *a network of interconnected organizations, connected to a focal firm or a platform, that incorporates both production and use side participants and creates and appropriates new value through innovation.*

In our definition, ecosystems are organized around a shared focal point or asset. The production and use side participants of ecosystems can be connected to either a focal firm in the locality (Teece, 2007; Adner and Kapoor, 2010), a 'hub' firm (Moore, 1993, 1996; Iansiti and Levien, 2004a, b), or a shared technology platform (Cusumano and Gawer, 2002; Gawer and Cusumano, 2002; and see Chapter 32 by Gawer and Cusumano). The explicit inclusion of use side participants differentiates the ecosystem construct from other network-centric constructs in management literature, such as clusters, innovation networks, and industry networks, which tend to focus on the production side. User networks, on the other hand, focus exclusively on the use end of industrial value chains. In terms of the variety of stakeholders covered (i.e. producers, users, competitors, and complementors), thus, the ecosystem construct is perhaps the most broad-based of the different network-based constructs used in management research. It is distinguished by its broad-based coverage and by its focus on value co-creation and appropriation.

As a stream of management literature, ecosystems research constitutes part of a wider, heterogeneous body of network literature in management research (See Chapter 6 by

Kastelle and Steen). In this literature, the inclusion of use side participants is not a unique characteristic of ecosystems, as use side participants are also considered in the context of 'strategic networks' (Jarillo, 1988; Gulati, Nohria, and Zaheer, 2000), 'business networks' (Möller and Svahn, 2006; Anderson, Hakansson, and Johanson, 1994), 'value nets' (Nalebuff and Brandenburger, 1996), 'value networks' (Christensen and Rosenbloom, 1995; Stabell and Fjeldstad, 1998) and 'value constellations' (Normann and Ramirez, 1993). This broad body of work usually treats networks as a distinct mode of organization that is different from the market and from organizational hierarchy (Thorelli, 1986), emphasizes the social embeddedness of economic action (Granovetter, 1985; Uzzi, 1997), draws on the idea of the 'network organization' and 'virtual organization' (Miles and Snow, 1986), emphasizes value chain, market structure and value appropriation considerations (Porter, 1980, 1985; Teece, 1986), often builds on the notions of resource-based, relational and core competence-based advantage (Prahalad and Hamel, 1990; Barney, 1991; Dyer and Singh, 1998), and acknowledges that innovation is a complex process that may be driven by multiple different stakeholders (von Hippel, 1988). These theoretical underpinnings have provided a fertile soil for the ecosystems thinking to develop.

Although seldom explicitly defined, the various applications of the ecosystem construct exhibit distinctive characteristics that help set it apart from related constructs. We noted earlier that this is one of the few constructs that explicitly covers conceptually both upstream (production side) and downstream (user side) activities. This 'whole-system' view echoes the original biological meaning of the term. The ecosystem construct is distinguished from value chain and supply chain constructs by its non-linear aspect, as it includes both vertical and horizontal relationships between actors. The ecosystem construct is also distinguished from value creation-oriented constructs such as value networks and value constellations by its focus on value appropriation and use. A distinctive—although not universally applied—aspect associated with this construct relates to its focus on the evolution of networks of interconnected actors towards new states, rather than emphasizing the optimization of the output potential of the an existing and unchanging network configuration (Gustafsson and Autio, 2011).

Ecosystem Boundaries, Structure and Dynamics

The diversity in ecosystem concepts and definitions reflects the variety of contexts in which the concept has been applied. A sampling of studies across different contexts provides insight into the nature of innovation ecosystems. Three sets of insights are discussed here—relating to ecosystem boundaries, structure, and dynamics arising from relationships and interactions between ecosystem participants.

Since the introduction of the concept to management literature, there have been quite a few attempts to discuss innovation ecosystems in a substantive manner. Moore (1993) first used the term to describe a set of producers and users around a focal organization that contributed to its performance. Moore's insight was that a given company can be fruitfully viewed not as a member of a given industry, but rather, as part of a 'business ecosystem'. Moore was making the point that interactions between firms and collective value creation processes are often much more complex than what received strategy frameworks drawing on an industrial organization perspective had implied (Porter, 1980, 1985). This insight was echoed in the notion of value networks by Christensen and Rosenbloom (1995), which considered the value network as the context, or the nested commercial system, within which a given firm competes and solves customers' problems. Compared to Christensen and Rosenbloom's concept, the ecosystem concept is broader, as it covers the diverse community of organizations, institutions, and individuals that impact the fate of the focal firm and its customers and suppliers, including complementors, suppliers, regulatory authorities, standard-setting bodies, the judiciary, and educational and research institutions (Teece, 2007).

Moving beyond this basic contextual view, ecosystems have also been seen as dynamic and purposive networks in which participants co-create value (Adner and Kapoor, 2010; Lusch, Vargo, and Tanniru, 2010). In this perspective, ecosystem participants co-evolve capabilities around a shared set of technologies and cooperate and compete to support new products, satisfy customer needs, and eventually incorporate the next round of innovation (Moore, 1993, 1996). In this sense, ecosystems are collaborative arrangements through which firms combine their individual offerings into a coherent, customer-facing solution, and which allow firms to co-create value in ways that few individual firms could manage alone (Adner, 2006; Moore, 1996). Ecosystems hence extend the concept of a value chain to that of a system that includes any organization that contributes to the shared offering in some way (Iansiti and Levien, 2004a, b). This means that an ecosystem may include participants from outside the traditional value chain of suppliers and distributors, for example, outsourcing companies, financial institutions, technology providers, competitors, customers, and regulatory and coordinating bodies.

The variety of ecosystem participants makes it difficult to define the boundaries of ecosystems. Ecosystem boundary definition is made even more difficult by the wide spread treatment of ecosystem boundaries as open and permeable (Gulati et al., 2012). This difficulty is echoed in the variety of approaches proposed for the operationalization of ecosystem boundaries. As an example, in their value-based model, Adner and Kapoor (2010) defined an innovation ecosystem as consisting of only those participants (suppliers, complementors, customers) that were only one network link away from the focal firm or customer. Other operationalizations are not as clear-cut. Iansiti and Levien (2004b) emphasized participant identification with the ecosystem community and argued that ecosystem boundaries are focal firm specific and drawn through the identification of ecosystem participants with the wider ecosystem community. Santos and Eisenhardt (2005) pointed out that ecosystems have broader boundaries than those implied by market efficiency. They specifically highlighted organizational power and the specialization of organizational competencies as the central features upon which the definition of ecosystem boundaries should be based. Although different boundary definitions may be applicable to different purposes and perspectives, it is clear that ecosystem boundaries only rarely overlap with traditional industry boundaries, as defined by a given set of products and their producers. The defining element of innovation ecosystems is not a given product, but rather a coherent set of interrelated technologies and associated organizational competencies that glue a variety of participants together to co-produce a set of offerings for different user groups and uses. Instead of thinking about ecosystems as an industry, it is more useful to think about ecosystems as an evolving community that specializes in the development, discovery, delivery, and deployment of evolving applications that exploit a shared set of complementary technologies and skills.

The above implies that a defining characteristic of innovation ecosystems is their ability to adapt and evolve (Basole, 2009). A healthy ecosystem is *productive*, in that it consistently transforms technology and other inputs to innovation into lower costs and new markets; and *robust*, that is, capable of surviving disruptions such as unforeseen technological change and able to create niches to increase meaningful diversity (Iansiti and Levien, 2004a). Relationships among ecosystem participants are often symbiotic, as members co-evolve with the system (Moore, 1996; Iansiti and Levien, 2004a; Li, 2009). Each member of an innovation ecosystem ultimately shares in the fate of the system as whole (Li, 2009). To the extent that these characteristics are met, a given participant can be thought of as residing within the boundaries of a given ecosystem and not outside it.

The question of who belongs to an ecosystem and who doesn't evokes the corollary question of ecosystem structure. The multiplicity of participant types, roles, and interdependencies implies that challenges are not equally distributed across participants (Adner and Kapoor, 2010). The interdependence among ecosystem participants also raises the question of how ecosystems are coordinated and managed. In many contexts, a hub firm, or firms, exists that coordinates services to the system (Cusumano and Gawer, 2002; Iansiti and Levien, 2004a; Li, 2009; Pierce, 2009).[1] Such firms may control the technological architecture or the brand that drives value in the ecosystem, and coordination may be based, for example, on architectural control or perhaps on the regulation of access to a given shared platform, as in the case of eBay or Android. Indeed, a substantive subset of the literature proposes 'platforms' as the coordinating artefact that the hub firm uses, or the services, tools, and technologies that other members of the ecosystem can use to enhance their own performance (Cusumano and Gawer, 2002; Iansiti and Levien, 2004a, b; Li, 2009).

Whatever the coordination device, they are central to the health and stability of an ecosystem, as they drive collective performance by enabling and facilitating value creation and sharing (Gawer and Cusumano, 2002; Iansiti and Levien, 2004a; Evans, Hagiu, and Schmalensee, 2006;). Network research suggests that hubs naturally emerge in networks, regardless of the quality of the networked system, its participants, or the specific nature of their connections (Barabási and Albert, 1999; Newman, 2001; Barabási, 2002; Cohen, 2002). For instance, research on digital services has found that while the number of digital services grow in a linear fashion, the distribution of complementors to hub firms tends to follow a power law, implying that a small number of hub firms provided for a majority of complementors (Weiss and Gangadharan, 2010). The control of the coordination device may reside with a single company, a collection of firms, a consortium, or a not-for-profit organization (Chesbrough and Appleyard, 2007). This underlying architecture may be a 'platform', but it does not have to be (Cusumano and Gawer, 2002; Iansiti and Levien, 2004a, b; Jacobides, Knudsen, and Augier, 2006; Teece, 2007). Although a successful platform typically has an ecosystem surrounding it, an ecosystem does not necessarily have a platform at its core.

In addition to formal and ownership-based control devices, there are also informal coordination mechanisms that influence the evolution of innovation ecosystems. These include social and behavioural coordination devices embedded in ecosystem relations, such as trust, tact, professionalism, openness, transparency, and complementarity, all of which are seen as crucial in the development of ecosystem relationships (Agerfalk and Fitzgerald, 2008). Key enablers of trust building within ecosystem environments include complementarity of obligations over the product lifecycle, alignment of differing perceptions of obligation fulfilment, and balance between value creation and community values. In the context of software firms, Iyer, Lee, and Venkatraman (2006) found that the software sector operates as a small world ecosystem, which continued to be a small world during the emergence of the Internet and despite technological changes. Because of well-established informal control devices, the ecosystem was efficient at transferring information and diffusing innovative advances and resources throughout itself. Thus, informal mechanisms facilitate innovation by promoting information disclosure and sharing, and therefore innovation through knowledge combination, whereas formal mechanisms are instrumental in preventing dissipation of effort and in channelling attention to promising areas of development.

Some researchers have taken an institutional approach towards understanding the coordination and evolution of innovation ecosystems (Thomas and Autio, 2013; Gawer and Phillips, 2013). Innovation ecosystems can be portrayed as organizational fields, or 'those organizations that, in the aggregate, constitute a recognized area of institutional life: key suppliers, resource and product consumers, regulatory agencies, and other organizations that produce similar services or products' (DiMaggio and Powell, 1983:148). As a theoretical construct, an innovation ecosystem is analogous to an organizational field in that it has its own institutional actors, logics, and governance structures (Scott, 2007). In particular, an ecosystem can be considered an organizational field that has value co-creation as its recognized area of institutional life, as distinct to the more traditional considerations of a common industry, technology or issue as the recognized area of institutional life (Thomas and Autio, 2013). The introduction of institutional theory into innovation ecosystem analysis provides a useful theoretical lens for understanding the organizing principles and behavioural rules and norms that support and regulate informal coordination and coherence in the allocation of effort.

Ecosystem Behavioural Logics

Having reviewed received insights into the boundaries, structure, and governance mechanisms of innovation ecosystems, we next turn to the theoretical lenses that can be applied to understanding their behavioural logics. Our review of the literature uncovered three thematic streams, with broadly coherent and distinctive theoretical underpinnings, that enable the consideration of various aspects of operation of and within innovation ecosystems. We labelled these as value creation, network embeddedness, and network management streams. These are summarized in Table 11.1. In Table 11.2 we provide some examples of empirical research for each stream.

Table 11.1 Overview of ecosystem and related research

Stream	Ecosystem	Value Creation	Network Embeddedness	Network Management
Construct variants	Ecosystem; innovation ecosystem; business ecosystem	Value network; value constellation	Strategic network; business network; innovation network	Business network; value net
Description	Dynamic and purposive networks of production and consumption side participants	Non-linear value creating system	Long term, purposeful arrangements among distinct but related organizations	Intentional inter-organizational structures designed deliberately for specific purposes
Core Discipline	General Management	General Management; Marketing	General Management; Marketing	Marketing
Key Concepts	Value co-creation; participant symbiosis; ecosystem coordination	Value co-creation; interdependence	Co-specialization; complementarity	Network coordination;
Key Influences	Chandler (1962); Williamson (1975); Porter (1980; 1985); Wernerfelt (1984); Teece (1986); Brandenburger and Nalebuff (1996); Gawer and Cusumano (2002)	Porter (1980; 1985); Katz and Shapiro (1985); Normann (1993); Christensen (1997); Chesbrough and Rosenbloom (2002)	Williamson (1975); Granovetter (1985); Thorelli (1986); Von Hippel (1988); Womack (1990)	Anderson (1994); Uzzi (1997); Dyer and Singh (1998); Achrol (1999); Zollo (2002); Gulati et al (2000); Hakansson and Ford (2002)
Empirical contexts	Information and internet technologies; telecommunications; manufacturing	Information technologies; telecommunications	Cross industry; manufacturing; information technologies	Cross industry; manufacturing
Main Journals	Harvard Business Review; Sloan Management Review; Strategic Management Review	Telecommunications Policy; Long Range Planning; Journal of Academy of Marketing Science	Industrial Marketing Management; Sloan Management Review; Strategic Management Journal	Industrial Marketing Management
Main works	Moore (1993); Iansiti and Levien (2004); Teece (2007); Adner and Kapoor (2010)	Normann and Ramirez (1993); Christensen and Rosenbloom (1995); Stabell and Fjeldstad (1998)	Jarillo (1988); Anderson et al (1994); Gulati et al (2000); Afuah (2000)	Ritter et al (2004); Moller et al (2005); Moller and Svahn (2006)

Table 11.2 Examples of empirical research

Stream	Study	Context	Type	Sample	Key Results
Ecosystem	Iyer, Lee, and Venkatraman (2006)	Information Technologies	Quantitative	Packaged software industry	The software sector operates as a small world ecosystem, the structure of which continued during the emergence of the Internet and despite technological changes. This indicates that the network is very efficient at moving information, innovations, and resources through the ecosystem.
	Adner and Kapoor (2010)	Information Technologies	Quantitative	Semiconductor lithography equipment	Challenges in the external ecosystem can either enhance or erode a firm's competitive advantage from technology leadership. Specifically, the advantage from technology increases from component challenges, and decreases from complement challenges.
Value Creation	Christensen and Rosenbloom (1995)	Information Technologies	Case Study	Disk drive industry	In addition to the characteristics or magnitude of the technological change relative to capabilities of the incumbent, and the managerial processes and organizational dynamics through which entrant firms respond, there is a third factor, the value network, which affects whether incumbent or entrant firms will most successfully innovate.
	Huemer (2006)	Logistics	Case Study	Logistics service providers	The value network logic provides a fruitful alternative to the value chain, and reveals a number of complexities in supply relationships, such as differences in value creation logic, additional structural dimensions, and multiple interdependencies.
Network Embeddedness	Afuah (2000)	Information Technologies	Quantitative	RISC workstation manufacturers	A firm's performance decreases after a technological change with the extent to which the technological change renders co-opetitors' capabilities obsolete.

(Continued)

Table 11.2 (*Continued*)

Stream	Study	Context	Type	Sample	Key Results
	Rabinovich, Knemeyer, and Mayer (2007)	Information Technologies	Quantitative	Ecommerce sites	Low levels of asset specificity and uncertainty drive, Internet commerce firms to establish these relationships, as well as the fact that these relationships offer access to networks that bundle many complementary services.
Network Management	Hughes, Ireland, and Morgan (2007)	Cross Industry	Quantitative	Incubators in the UK	The network in incubators only dictates opportunities for value creation, and it is firm behaviour that dictates the extent to which these opportunities can be realized.
	Oberg, Hennenberg, and Mouzas (2007)	Cross Industry	Case Study	Truck manufacturer, electronic billing system integrator, IT sales	Following a merger or acquisition managers need to adapt their previous network pictures in a radical way; however not all managers adjust their network pictures and networking activities to adjust to new reality.

The value creation theme focuses on value creation processes within innovation eco-system contexts. It builds upon and extends industrial organization frameworks of stra-tegic management, notably those addressing industry structure, industry value chains, and value appropriation in industry contexts (Porter, 1980, 1985; Teece, 1986), by empha-sizing the non-linear, iterative, and non-sequential nature of value creation and appro-priation processes in industry networks. This stream contributes to ecosystem research by explicating the theoretical logic of value creation and appropriation in ecosystem contexts, as well as the mechanisms that drive value in network contexts, such as net-work externalities and complementary innovation. The network embeddedness stream emphasizes the structural and relational aspects of networks and considers prerequisites and constraints of operation within innovation ecosystems. This stream also extends the theoretical base of ecosystem research by introducing notions from social theory, such as trust, and legitimacy at the dyad level such as those proposed in virtual organizations (Miles and Snow, 1986). The network management stream emphasizes the management strategies and tactics for coordinating and managing within network contexts (Miles and Snow, 1986; Möller, Rajala, and Svahn, 2005). This stream extends the ecosystem research by explicitly including considerations of the hub firm and differing approaches to net-work management. Together, these three additional streams deepen the theoretical base of ecosystem research and in doing so, make it more amenable for empirical application.

VALUE CREATION THEME

The value creation thematic stream constitutes somewhat of a 'hodgepodge' theme. Value creation in networks has been considered from a number of perspectives, with slightly different theoretical framings (Lee, Lim, and Soriano, 2009). The general thrust, however, has centred around the idea of adding horizontal linkages between partici-pants of value networks in contrast to the predominantly vertical and sequential orien-tation of Porter's (1985) infamous value chain model.

In this thematic stream, an early approach focused on the notion of 'value constel-lations', as coined by Normann and Ramirez (1993). They observed that value is not built through sequential processes, as suggested by Porter (1985), but rather, as the result of a complicated set of economic transactions and institutional arrangements between suppliers, customers, specialized service providers, and other complemen-tors. This observation of the non-sequential and systemic nature of value creation was presented by the authors as a 'new logic' of value creation. In particular, Normann and Ramirez (1993) argued that the goal of business is to mobilize customers to co-create value; that the most attractive offerings involve networks of customers, suppliers, allies, and business partners in new combinations; and that the only true source of competi-tive advantage is the ability to conceive of an entire value-creating system and make it work. Such themes have been echoed in subsequent work on value creation in networks. For instance, Normann noted that innovative value constellations *'identify economic*

actors and link them together in new patterns which allow the creation of new busi-
nesses that did not exist previously, or change the way certain types of value are created'
(2001: 107). This value-creation logic has been emphasized in the context of non-linear
value-creating systems where the goal is to 'innovate customers' through changing the
way value is co-created in conjunction with suppliers, partners, allies, and customers
(Michel, Brown, and Gallan, 2008). Echoing meanings attached to the innovation ecosystem construct, value constellations have also been seen as a particular type of inter-
organizational network established to create value in situations where any individual
company would be unable to single-handedly launch the product and where each actor
within the constellation can capture a sufficient portion of the overall value to justify their
participation (Lin, Wang, and Yu, 2010). The value constellation construct has become
integral to many service-dominated business logics (Michel, Vargo, and Lusch, 2008).

Similar to the work on value constellations, Stabell and Fjeldstad (1998) asked how
the analytical concepts provided by value chain theory could be extended beyond traditional manufacturing contexts, thereby highlighting limitations of the hitherto
dominant value chain theory by Porter (1985). In addition to the value chain, Stabell
and Fjeldstad identified the 'value shop' and the 'value network' as alternative conceptual tools with which to study value co-creation. A firm in a value network creates
value through the facilitation of network relationships with customers using a mediating technology (Thompson, 1967). In this concept value is created through a managed
mediation service, where value is driven by positive demand-side externalities and by
the service opportunity and associated service delivery capacity. Competitive advantage
results from scale building, capacity utilization, and linkages between participants and
learning (Stabell and Fjeldstad, 1998). The value network concept has also been applied
usefully to supply chain contexts, where value creation processes exhibit a number of
complexities, such as differences in value creation logics across value chains, additional
structural dimensions, and multiple interdependencies (Huemer, 2006). Thus, the value
network construct by Stabell and Fjeldstad offers an alternative conceptual framework
for understanding value creation in mediated network contexts.

Another approach to understanding value creation in network contexts has been
to integrate value chain and network concepts. In an early approach, Weiner, Nohria,
and Hickeman (1997) proposed that value networks emerge where value chain assets
are disaggregated and no single organization controls them all. In this concept the
focal company connects and exploits the strengths of each complementary value
provider, by coordinating production and delivery across companies to deliver value
to a specific customer segment. The value network thus enables a coalition of players to exert greater market power and extend the scope of the markets they address
and offerings they produce. An alternative approach considers the value network as
a series of intertwined value chains where some nodes are simultaneously involved
in more than one value chain (Li and Whalley, 2002). In this concept a multitude
of market entry points exist where a diverse range of companies enter the market
through different routes, and the exit point—where the company interacts with its
chosen end customers—may differ significantly according to the business model of

the different players (Li and Whalley, 2002). Similarly, Funk (2009) considered value networks as connecting multiple buyers and sellers at a single node, where the node can be part of a value chain or of a larger value network, integrating notions of value constellations and value networks (Weiner et al., 1997). For Funk (2009), a value network implies increased complexity of firm interrelationships, network externalities, standards, critical mass, multi-sided markets, and policy considerations within the value network. In these value networks, each participant shares in the success or failure of the network (Pagani and Fine, 2008).

Another related concept for the study of value creating systems is that of the 'value net' (Parolini, 1999). The value net is a dynamic, collaborative network which combines each participant's core competences (Bovet and Martha, 2000). A value net is formed around customers, and it captures the customers' choices in real time in an effort to satisfy actual demand. Value nets are fluid and flexible, comprising a group of collaborators that unite to exploit a specific opportunity—and once the opportunity is met, the net often disbands. In the concept's purest form, each company that links up with others to create a virtual corporation contributes only what it regards as its core competencies (Christopher and Gaudenzi, 2009).

In summary, the value creation theme has explored various aspects of value creation dynamics in innovation ecosystems. This line of research was prompted as a reaction to the linear and sequential conception of value creation processes in the Porter (1985) value chain theory. In contrast with this perspective, the value creation theme has emphasized co-creation of value, the importance of collaboration between network participants, and value creation through the combination of each participant's specialized capabilities and core competences within the value network.

Network Embeddedness Theme

The network embeddedness theme emphasizes the structural and relational aspects of the social network in which ecosystem participants are embedded and considers the prerequisites and constraints of operation within innovation ecosystems from the perspective of the participating organizations (i.e. not necessarily that of hub firms). This thematic stream seeks to understand how individual ecosystem participants can best take advantage of the innovation ecosystem that surrounds them. As ecosystem relationships are characterized by intense interactions between complementary participants, much of the effort has gone into understanding how ecosystem participants can build relational assets into their ecosystem relationships so as to facilitate smooth transactions and collaborations (Dyer and Nobeoka, 2000); establish favourable initial network positions (Ozcan and Eisenhardt, 2009; Hallen and Eisenhardt, 2012); encourage knowledge sharing for innovation (Yli-Renko, Autio, and Sapienza, 2001); exploit favourable structural positions for performance (Wincent, Anokhin, Örtqvist, and Autio, 2010a); promote movement to desired directions (Wincent, Örtqvist, Eriksson, and Autio, 2010b); and to generate trust and norms and rules that facilitate the efficient operation of the network.

An early framing of the challenges of operating as part of a wider network was introduced by the concept of 'strategic networks', defined as long-term, purposeful arrangements among interconnected firms that seek to build competitive advantage relative to competitors outside the network (Jarillo, 1988). Jarillo argued that strategic networks exhibit some of the properties of both markets and hierarchies, as the activities necessary for the production of a given good or service can be carried out either by an integrated firm or by a network of firms. Within strategic networks, participating firms farm out some activities while specializing more fully on those in which they have an opportunity to build comparative advantage. This combination of outsourcing and specialization creates interdependency among the network participants, which grows stronger as a function of mutual co-specialization. In this approach, value is either appropriated fairly if sufficient trust is built into the relationships between network participants, or on the basis of power and control of critical assets if power relations are asymmetric and the scope of abuse is not mitigated by enlightened self-interest (Jarillo, 1988; Casciaro and Piskorski, 2005; Adner and Kapoor, 2010). Thus, for the participating firm, having a good understanding of the activities in which they have a comparative advantage, combined with a realistic understanding of the potential dangers inherent in co-dependent relationships, is key to successful operation in an innovation ecosystem.

Balancing the benefits of specialization with the hazards of dependence necessitates the building of relational assets that mitigate opportunism. Building upon the notion of strategic networks, Gulati, Nohria, and Zaheer (2000) took a broader relational view (Dyer and Singh, 1998) to understand participant network embeddedness, arguing that the constituting relationships are enduring and are of strategic significance for the firms entering them. As such, these relationships can be strategic alliances, joint ventures, long-term buyer-supplier partnerships, and so on, essentially encompassing the firm's set of relationships, both horizontal and vertical, be they with suppliers, customers, competitors, or other entities, including relationships across industries and countries. This network of relationships acts as a source of both opportunities and constraints to the participating firm. In particular, network embeddedness can be a source of opportunity as it potentially provides a firm with access to information, resources, markets, and technologies such as contractual power, increased innovation generation, improved technological transfer, and improved entry opportunities (Nosella and Petroni, 2007). In addition, it allows participating firms to achieve strategic objectives, such as sharing risks and outsourcing value chain stages and organizational functions, enabling learning, scale, and scope economies (Gulati et al., 2000; Rabinovich, Knemeyer, and Mayer, 2007). In these contexts, learning can usefully be regarded as occurring at the inter-organizational level as well as between groups of organizations (Knight, 2002). The network embeddedness of a participating firm can also be a constraint, as it may lock it into unproductive relationships or preclude partnering with other partners (Gulati et al., 2000). Firms need to balance between broadening the number of relationships and maintaining existing relationships, as these have an interlinked effect on firm performance (Wincent et al., 2010a). Similarly, being embedded in a network does not necessarily lead to value creation, only to opportunities to do so, and it is how the participating firm behaves and pursues opportunities that leads to their success (Hughes, Ireland, and Morgan, 2007).

Network participants can also move beyond dyad-specific relational assets to promote ecosystem-wide norms that reinforce predictability in mutual exchanges and mitigate opportunism (Bosse, Phillips, and Harrison, 2009; Wincent et al., 2010a). Strategic networks pursue shared goals through collective efforts by multiple participants, all of whom also have their own strategic interests that do not always align with those of the wider network (Gulati et al., 2000). Strategic networks and innovation ecosystems therefore face distinctive governance challenges. An important challenge is created by the lack of any immediate link between individual member's efforts and the collective benefits (Winkler, 2006). To materialize the benefits of the strategic network, participants need to commit resources towards shared goals. Because reciprocation is not immediate, however, opportunities arise for free-riding (Rosenfeld, 1996; Vanhaverbeke, Gilsing, Beerkens, and Duysters, 2009). Reciprocation is not even always direct; a firm's commitments may rather be reciprocated by a third party within the ecosystem. The delay and often indirect nature between resource commitment and reciprocation creates an incentive for free-riding and complicates the evaluation of partners' goodwill (Powell, Koput, and Smith-Doerr, 1996; Human and Provan, 1997). To overcome this governance challenge, firms participating in strategic networks need to establish and reinforce generalized reciprocity designed to mitigate the risks of opportunism and free-riding (Das and Teng, 2002; Das and Teng, 2003; Bercovitz, Jap, and Nickerson, 2006). Without such norms there is a significant risk that the efforts of one party are not reciprocated. Strong shared norms that encourage reciprocity and increase the social cost of free-riding can operate as a strong informal governance mechanism that reduces opportunism and promotes collaborative behaviour (Bercovitz et al., 2006). Therefore, generalized reciprocity norms provide a particularly potent alternative to contractual governance mechanisms especially in multi-stakeholder collaborations involving shared development efforts, such as those often prevailing in innovation ecosystems (Wincent et al., 2010a).

Network change can alter the balance of network relationships and thus create adaptation challenges (Halinen, Salmi, and Havila, 1999). Change events in one part of the dyad can have effects at the network level, and similarly network changes can reflect on the balance of each dyad. In order to understand sense-making by embedded participants during eras of network change, Oberg, Henneberg, and Mouzas (2007) proposed a cognitive approach, 'picturing', where the position of the participant within the network is visualized by integrating perceptions of customers' needs and developments. Change efforts can be understood as a negotiated process where overlapping network representations are re-negotiated to fit multiple actor constituencies (Kragh and Andersen, 2009). The cognitive framing of networks has been further developed to include 'network insight', which not only includes the pictures held by individual managers but is also grounded in the practice of inter-firm exchange (Mouzas, Henneberg, and Naudé, 2008).

To summarize, the network embeddedness stream emphasizes the structural and relational view of networks and introduces wider marketing considerations into ecosystem research. This stream extends the theoretical base of ecosystem research by introducing notions of social theory, trust, and legitimacy at the dyad level.

NETWORK MANAGEMENT THEME

The network management theme considers how organizations can proactively manage the innovation ecosystem, or the 'business network' (Möller and Svahn, 2003; Ritter, Wilkinson, and Johnston, 2004). This stream differs from the network embeddedness theme in that whereas the former theme seeks to understand how firms can best adapt to and take advantage of innovation ecosystems, the network management theme considers how firms can manage innovation ecosystems themselves and influence their operation. This stream builds on early observations that strategic networks can be managed (e.g., Dyer and Singh, 1998; Jarillo, 1988). In much of this thematic stream, the focus has been on the 'business network', the 'value net' or the 'strategic nets' (e.g., Nalebuff and Brandenburger, 1996; Ritter et al., 2004). These are inter-organizational structures set up deliberately for specific purposes and consisting of coalitions of autonomous but interdependent firms that are willing to coordinate some of their actions and sometimes even to submit part of their activities and decision domains to centralized control in order to achieve benefits that are greater than any single member of the net can create independently (Möller and Svahn, 2006). Others have extended the reach to include alliance partners (Afuah, 2000, 2004) and followers, imitators, universities, professional bodies and other institutions (Kang and Afuah, 2010; Möller and Svahn, 2009).

The earliest consideration of network management is that which considers 'co-opetition', a term introduced by Nalebuff and Brandenburger (1996). Here the focal firm is able to utilize games to coordinate and appropriate value from its network. This stream refers to the 'value net' as the network of customers, suppliers, competitors, and complementors. Key to understanding this stream of research is the relationship between the participants: customers and suppliers play symmetric roles and competitors and complementors play mirror-image roles (Nalebuff and Brandenburger, 1996). Hence Nalebuff and Brandenburger (1996) developed a game theoretic approach that focused on balancing competitive and cooperative challenges. The classic 'co-opetitive' strategies include imitation, combination, shut-out, entry, and holdup. Such game strategies enable firms to better position themselves to capture rent from innovations and enable further innovation (Kang and Afuah, 2010). This game theoretic approach has also been extended to include institutional interactions between industry and government, as efforts to influence government are often a form of business competition in disguise (Watkins, 2003). This extension includes two further types of games where the government can act as rule makers and as referees, value-net games and public interest games (Watkins, 2003).

Much of the work in the network management theme springs from Möller and Svahn (2003), who argued that the management of a network requires specific organizational capabilities. Building on this observation, Ritter, Wilkinson, and Johnston (2004) distinguished between 'managing in' and 'managing of' aspects of inter-organizational relationships. 'Managing in' refers to coping within a given network situation, whereas the 'managing of' refers to the management of those relationships themselves in terms of leading, determining, and organizing. Möller, Rajala, and Svahn (2005) distinguished between different levels of network operation and argued that management challenges differ for different levels. They suggested that network visioning and orchestration are most relevant when the focus is on the network as a whole. Network operation and coordination were relevant at the level of the hub firm when it managed an existing network. Tie portfolio management was relevant at the relationship portfolio level, and relationship management dominated at the level of individual dyads. They further identified three factors that acted as boundary conditions for the firm's ability to manage its network. First, the hub firm needed to be able to influence and control network value activities and other network participants. Such control could be achieved through the various coordination mechanisms and devices discussed above in the 'coordination' section. Second, sufficient commonality had to exist between the goals of the network as a whole and those of its constituent participants. Third, the structure of the network had to be amenable for coordination—for example, hub-and-spoke configurations lend themselves more readily for coordination than do distributed networks with no centrally positioned firms. Finally, these authors distinguished between three types of business networks, according to their maturity (Möller and Rajala, 2007; Möller et al., 2005; Möller and Svahn, 2006). Current (mature) business nets had a stable, well-defined value system consisting of well-known and specified value activities, well-known actors, technologies, and business processes, all of which enhanced the manageability and coordination of the network. Business renewal nets had an established value system with incremental improvements consisting of well-known value systems and change through local and incremental modifications within the existing value system. Finally, emerging business nets had an emergent value system with radical changes such as frequent entry by new actors, radical transformation in pre-established activities, constant creation of new value activities, uncertainty around both value activities and actors, and radical system-wide change. Such characteristics radically reduced the manageability and coordination of the network. More recently this framework has been developed to consider how network management capabilities can be utilized to influence the creation of new business fields. Here the central influencing mechanisms are cognitive—e.g. control of sense-making and agenda construction—as these influence the cognitive frames by the participants (Möller, 2010; Möller and Svahn, 2009). Research on ecosystem creation still remains in its infancy, however.

An interesting new stream has emerged that explores how entrepreneurial firms use behavioural strategies to create and shape innovation ecosystems (Hallen, 2008; Hallen and Eisenhardt, 2012; Ozcan and Eisenhardt, 2009; Zott and Huy, 2007). Ozcan and Eisenhardt (2009), for example, found that entrepreneurs who had a strategic vision of their industry were more likely to build high-performing alliance portfolios. In addition, they found that strategies to pre-emptively shape emerging industry structure with alliance relationships were likely to lead to better performance whereas structurally constrained tie-building strategies were less likely to do so. Hallen and Eisenhardt (2012) found that entrepreneurs could establish advantageous positions in innovation ecosystems by employing different catalysing strategies, such as casual dating and timing relationship activities around important milestones. Together, this emerging stream suggests that entrepreneurs can circumvent structural inertia in network creation by employing a range of relational, institutional, and coordination strategies, thereby establishing more advantageous initial network positions and promoting wider lock-in around a given ecosystem configuration.

Summarizing, the network management stream has explored management strategies and organizational capabilities that enable firms to proactively manage their innovation ecosystems. The bulk of the work has focused on game strategies and strategic plays that firms can employ to manoeuvre within ecosystems. Only recently have researchers started to consider whether and how firms could initiate and proactively shape innovation ecosystems to their own advantage. This work remains in very early stages, however, and little is known about the early stages of ecosystem development.

Discussion and Implications for Innovation Management

Although a major body of research has explored innovation ecosystems and closely related concepts, only recently has the literature begun exploring implications for innovation management. As a result, still too little is known about how firms can proactively create, steer, and leverage innovation ecosystems for enhanced innovation performance. In this concluding discussion, we offer our view of the most important gaps in the field and discuss implications for practicing managers. The most important gaps are insufficient understanding of value appropriation and ecosystem creation and insufficient elaboration of practitioner implications for strategic management.

Despite the importance of the logic of value for innovation ecosystems, the majority of the ecosystems literature to date has not explicitly considered value creation and appropriation. Although Adner and Kapoor (2010) empirically linked value creation and value capture within ecosystem contexts, this is perhaps the only paper to date to have done so. Given the importance of the value logic, and in particular the co-creation and appropriation of value in the ecosystem construct, a more coherent and detailed formulation similar to the value creation logics of Doz and Hamel (1998) for alliance contexts will aid both academic and practitioner understanding. We propose that strategic management practitioners should, when planning for value creation and value appropriation in innovation ecosystems, consider:

– **Control Mechanisms.** Which are the major control mechanisms that enable firms to influence ecosystem evolution and use as levers for value appropriation? Possible control mechanisms include:

  ◦ **Shared platforms** (Cusumano and Gawer, 2002): if the ecosystem has been formed around a focal platform (e.g. an operating system, hardware platform, or a cloud service) the control of such a platform usually constitutes a strong appropriation lever.
  ◦ **Critical assets** (Teece, 1986; Teece, 1998): critical assets are resources that are important for ecosystem operation, yet in scarce supply. Scarcity combined with criticality ensures strong appropriation ability. Examples of such assets include the Intel microprocessor architecture, exclusive distribution channels, and, for example, scarce and hard-to-substitute raw materials.
  ◦ **Pre-emptive alliances** (Ozcan and Eisenhardt, 2009). Sometimes pre-emptive alliances can become a strong control mechanism, particularly if they pre-empt access to critical assets by competitors. In their research, Ozcan and Eisenhardt (2009) showed how early entrants into the mobile gaming industry were able to secure valuable alliances, thereby locking themselves into a long-term positional advantage.

– **Value Creation Dynamics.** How is value to be created and delivered within the ecosystem? How much of it will be based on services, manufactured goods, or intangible assets? Are value processes sequential and distributed along value chains or are they parallel and horizontally distributed? How much of the value is co-produced at the point of use, and how much of it is stored into transferable goods and services? An understanding of the value creation dynamics is crucial for successful positioning within the ecosystem, and therefore, for successful appropriation of value.
– **Control Migration.** As ecosystems evolve, it is likely that the critical control mechanisms migrate elsewhere. If the firm fails to anticipate and proactively plan for ecosystem evolution, its position may be undermined by ecosystem developments. One classic example is provided by IBM's failure to anticipate the primacy of the user interface as a critical control device in the PC ecosystem. A more recent example is provided by Nokia's fall from grace, as it failed to anticipate the transformation of the mobile phone industry from a tightly controlled supply chain system towards a smartphone-dominated system, where applications developed by others constitute a major control lever.
– **Value Externalities.** An important aspect of value creation in innovation ecosystems is defined by the existence of value externalities—or direct and indirect

network effects that boost the overall value produced by the ecosystem. If the innovation ecosystem provides incentives and structures for complementary innovation, this may help encourage a superior value creation dynamic—e.g. in the case of the Android ecosystem.

As noted above, little is known about the processes by which innovation ecosystems are created. At present, the processes of ecosystem creation have been considered variously from both lifecycle and teleological perspectives (Van De Ven and Poole, 1995). For instance, a lifecycle approach has been proposed for ecosystem (Moore, 1993), network structure (Larson, 1992), and network management perspectives (Möller and Svahn, 2009). The lifecycle approach considers ecosystem creation as a series of path-dependent stages driven by a common underlying process. Conversely, in a teleological perspective, an end state is attained through a repetitive sequence of goal formulation, implementation, evaluation, and modification (Gawer and Cusumano, 2008). However, while the understanding of ecosystem *evolution* is quite substantial, less is known about the exercise of entrepreneurial agency in ecosystem *creation*. As noted above, an emerging stream of behavioural strategies has considered simultaneous and interlinked teleological and life cycle processes (Hallen, 2008; Hallen and Eisenhardt, 2012). However this literature is regarded, no model to date comprehensively considers how complementary markets themselves are initially created (with the exception of Santos and Eisenhardt (2009)), nor is there much systematic work exploring the underlying processes. We propose that a coherent understanding of ecosystem creation requires a multi-theoretic approach, as well as a careful consideration of three related architectures—the technological architecture, the activity architecture, and the value architecture (Thomas, Autio, and Gann, 2012):

- **Technological Architecture**, or the design principles of shared technological resources and platforms, will determine who will be able to connect to the innovation ecosystem and in which roles. Key design issues involve, for example, the modularity of the system; whether key interfaces are open or closed; questions of which design aspects to put into the open domain and which to keep concealed; and so on. By modifying such aspects of the technological architecture, the platform owner will influence who will be able to connect to the platform (i.e. activity architecture) and what the resulting value dynamic will be (i.e. value architecture).
- **Activity Architecture** defines the composition and structure of the innovation ecosystem that may emerge around the core platform. Aspects of activity architecture include not only who and in which roles, but also, co-specialization drivers and coordination mechanisms. The definition of participant roles defines the specialized competencies participants develop, and therefore, cements the long-term configuration of the activity network.
- **Value Architecture** describes the resulting value dynamic, as defined by the interplay between technological architecture and activity architecture. Key aspects of value architecture were already discussed above.

As such, although there has been little explicit discussion in the literature on ecosystem creation, we believe that the complex nature of innovation ecosystems requires the coordination of strategic activities on at least four levels (Autio and Thomas, 2012):

- **Technological Strategies** involve not only technology architectural decisions, as discussed above, but also standardization strategies, open-source strategies, and patenting and licensing strategies, to name a few.
- **Economic Strategies** involve the choice, access, and promotion of complementary assets and associated investment strategies. What will be the value chain functions included in the system, how will these be organized, and how are the necessary assets included within the system?
- **Behavioural Strategies** cover behavioural tactics in the creation of initial network ties and alliances and involve, for example, persuasion and influencing strategies, as briefly discussed previously.
- **Institutional Strategies** cover the creation of—and connectivity with—institutional structures (both formal and informal) necessary to provide for ecosystem coordination and establish an institutional and regulatory framework to ensure smooth coordination and operation of the ecosystem.

In conclusion, although an increasing literature argues that strategic networks and innovation ecosystems have become a new basis of competition (Moore, 1993, 1996; Normann and Ramirez, 1993; Gulati et al., 2000; Iansiti and Levien, 2004b; Iyer et al., 2006), the managerial implications of this insight remain insufficiently developed. An increasing number of researchers argue that in the 'information', 'knowledge', or 'digital' economy, the 'innovation ecosystem' provides the frame of reference for strategy design and implementation (Iyer et al., 2006). Specific tasks involved in innovation ecosystem strategies include ecosystem creation; ecosystem coordination; optimization of business models to take advantage of ecosystem externalities; and the creation of control strategies to ensure value appropriation. However, the basis for managerial insights remains fragmented, reflecting the general fragmentation of this important domain.

In this chapter we have summarized emerging empirical and conceptual insights regarding innovation ecosystems and outlined areas and tasks where they matter for managerial practice. We hope that the insights offered in this chapter will prompt further explorations into this important topic.

NOTES

1. Not all of the literature proposes the existence of a central coordinating firm, with much of the network embeddedness literature focusing on a focal firm that does not necessarily have a coordinating role. In that stream, the focal firm is not required for coordination purposes but instead is necessary for empirical and methodological reasons, as the research is more interested in the individual dyads and ties than the actual roles of the participants. In addition, empirical ecosystem research has identified network roles other than the hub

firm. For instance, in his study of the mobile technology ecosystem, Basole (2009) found that there was no central firm that played a coordinating role. Similarly, in their structural analysis of the software sector, in addition to identifying coordinating hub organizations, Iyer et al. (2006) also identified broker organizations that acted as liaisons, representatives, or gatekeepers, and bridging organizations. Thus, although there is empirical verification of hub firms in ecosystem contexts, other network roles are also present. Despite these reservations on the role of a coordinating firm in the network structure stream, and indications of other roles in the network from the ecosystem stream, the majority of the ecosystem and related streams explicitly discuss a central firm that coordinates the ecosystem.

References

Adner, R. (2006). 'Match your Innovation Strategy to your Innovation Ecosystem', *Harvard Business Review*, 84: 98.

Adner, R., and Kapoor, R. (2010). 'Value Creation in Innovation Ecosystems: How the Structure of Technological Interdependence Affects Firm Performance in New Technology Generations', *Strategic Management Journal*, 31: 306–33.

Afuah, A. (2000). 'How Much do your Co-opetitors' Capabilities Matter in the Face of Technological Change?' *Strategic Management Journal*, 21: 397–404.

Afuah, A. (2004). 'Does a Focal Firm's Technology Entry Timing Depend on the Impact of the Technology on Co-opetitors?' *Research Policy*, 33: 1231–46.

Agerfalk, P. J., and Fitzgerald, B. (2008). 'Outsourcing to an Unknown Workforce: Exploring Opensourcing as a Global Sourcing Strategy', *MIS Quarterly*, 32: 385–409.

Anderson, J. C., Hokansson, H., and Johanson, J. (1994). 'Dyadic Business Relationships within a Business Network Context', *Journal of Marketing*, 58: 1.

Autio, E., and Thomas, L. D. W. (2012). 'Tilting the Playing Field: Towards a Strategic Theory of Endogenous Action', *Innovation & Entrepreneurship Group Working Papers*: 1–46. London: Imperial College Business School.

Barabási, A. L. (2002). *Linked: The New Science of Networks*. New York: Perseus.

Barabási, A. L., and Albert, R. (1999). 'Emergence of Scaling in Random Networks', *Science*, 286: 509.

Barney, J. B. (1991). 'Firm Resources and Sustained Competitive Advantage', *Journal of Management*, 17: 99–120.

Basole, R. C. (2009). 'Visualisation of Interfirm Relations in a Converging Mobile Ecosystem', *Journal of Information Technology*, 24: 144–59.

Bercovitz, J., Jap, S. D., and Nickerson, J. A. (2006). 'The Antecedents and Performance Implications of Cooperative Exchange Norms', *Organization Science*, 17(6): 724–40.

Bosse, D. A., Phillips, R. A., and Harrison, J. S. (2009). 'Stakeholders, Reciprocity, and Firm Performance', *Strategic Management Journal*, 30(4): 447–56.

Bovet, D., and Martha, J. (2000). *Value Nets: Breaking the Supply Chain to Unlock Hidden Profits*. New York: Wiley and Sons.

Casciaro, T., and Piskorski, M. J. (2005). 'Power Imbalance, Mutual Dependence, and Constraint Absorption: A Closer Look at Resource-dependence Theory', *Administrative Science Quarterly*, 50: 167–99.

Chesbrough, H. W., and Appleyard, M. M. (2007). 'Open Innovation and Strategy', *California Management Review*, 50: 57.

Christensen, C. M., and Rosenbloom, R. S. (1995). 'Explaining the Attacker's Advantage: Technological Paradigms, Organizational Dynamics, and the Value Network', *Research Policy*, 24: 233–57.

Christopher, M., and Gaudenzi, B. (2009). 'Exploiting Knowledge Across Networks Through Reputation Management', *Industrial Marketing Management*, 38(2): 191–7.

Cohen, D. (2002). 'All the World's a Net', *New Scientist*, 174: 24–9.

Cusumano, M. A., and Gawer, A. (2002). 'The Elements of Platform Leadership', *MIT Sloan Management Review*, 43: 1–8.

Das, T. K., and Teng, B. (2002). 'A Social Exchange Theory of Strategic Alliances', in F. J. Contractor, and P. Lorange (eds), *Cooperative Strategies and Alliances*, Oxford: Elsevier Science, 429–60.

Das, T. K., and Teng, B. S. (2003). 'Partner Analysis and Alliance Performance', *Strategic Management Journal*, 19: 279–308.

DiMaggio, P. J., and Powell, W. W. (1983). 'The Iron Cage Revisited: Institutional Isomorphism and Collective Rationality in Organizational Fields', *American Sociological Review*, 48(2): 147–60.

Doz, Y. L., and Hamel, G. (1998). *Alliance Advantage: The Art of Creating Value through Partnering*. Boston, MA: Harvard Business School Press.

Dyer, J. H., and Nobeoka, K. (2000). 'Creating and Managing a High-performance Knowledge-sharing Network: The Toyota Case', *Strategic Management Journal*, 21(3): 345.

Dyer, J. H., and Singh, H. (1998). 'The Relational View: Cooperative Strategy and Sources of Interorganizational Competitive Advantage', *Academy of Management Review*, 23(4): 660–79.

Evans, D. S., Hagiu, A., and Schmalensee, R. (2006). *Invisible Engines: How Software Platforms Drive Innovation and Transform Industries*. Cambridge, MA: The MIT Press.

Frels, J. K., Shervani, T., and Srivastava, R. K. (2003). 'The Integrated Networks Model: Explaining Resource Allocations in Network Markets', *Journal of Marketing*, 67: 29–45.

Funk, J. L. (2009). 'The Emerging Value Network in the Mobile Phone Industry: The Case of Japan and its Implications for the Rest of the World', *Telecommunications Policy*, 33: 4–18.

Gawer, A., and Cusumano, M. A. (2002). *Platform Leadership: How Intel, Microsoft, and Cisco Drive Industry Innovation*. Boston, MA: Harvard Business School Press.

Gawer, A., and Cusumano, M. A. (2008). 'How Companies Become Platform Leaders', *MIT Sloan Management Review*, 49: 28.

Gawer, A., and Phillips, N. (2013). 'Institutional Work as Logics Shift: The Case of Intel's Transformation to Platform Leader', *Organization Studies*, 34(8): 1035–71.

Granovetter, M. (1985). 'Economic Action and Social Structure: The Problem of Embeddedness', *American Journal of Sociology*, 91(3): 481–510.

Gulati, R., Nohria, N., and Zaheer, A. (2000). 'Strategic Networks', *Strategic Management Journal*, 21(3): 203–215.

Gulati, R., Puranam, P., and Tushman, M. L. (2012). 'Meta-organization Design: Rethinking Design in Interorganizational and Community Contexts', *Strategic Management Journal*, 33(6): 571–86.

Gustafsson, R., and Autio, E. (2011). 'A Failure Trichotomy in Knowledge Exploration and Exploitation', *Research Policy*, 40(6): 819–31.

Halinen, A., Salmi, A., and Havila, V. (1999). 'From Dyadic Change to Changing Business Networks: An Analytical Framework', *Journal of Management Studies*, 36: 779–94.

Hallen, B. L. (2008). 'The Causes and Consequences of the Initial Network Positions of New Organizations: From Whom do Entrepreneurs Receive Investments.?', *Administrative Science Quarterly*, 53: 685–718.

Hallen, B. L., and Eisenhardt, K. M. (2012). 'Catalyzing Strategies and Efficient Network Tie Formation: How Entrepreneurs Obtain Venture Capital', *Academy of Management Journal*, 55(1): 35–70.

Huemer, L. (2006). 'Supply Management-Value Creation, Coordination and Positioning in Supply Relationships', *Long Range Planning*, 39: 133–53.

Hughes, M., Ireland, R. D., and Morgan, R. E. (2007). 'Stimulating Dynamic Value: Social Capital and Business Incubation as a Pathway to Competitive Success', *Long Range Planning*, 40(2): 154–177

Human, S. E., and Provan, K. G. (1997). 'An Emerging Theory of Structure and Outcomes in Small-firm Strategic Manufacturing Networks', *Academy of Management Journal*, 40(2): 368–403.

Iansiti, M., and Levien, R. (2004a). *The Keystone Advantage: What the New Dynamics of Business Ecosystems Mean for Strategy, Innovation, and Sustainability*. Cambridge, MA: Harvard Business School Press.

Iansiti, M., and Levien, R. (2004b). 'Strategy as Ecology', *Harvard Business Review*, 82: 68–78.

Iyer, B., Lee, C.-H., and Venkatraman, N. (2006). 'Managing in a Small World Ecosystem: Some Lessons from the Software Sector', *California Management Review* 48: 28–47.

Jacobides, M. G., Knudsen, T., and Augier, M. (2006). 'Benefiting from Innovation: Value Creation, Value Appropriation and the Role of Industry Architectures', *Research Policy*, 35: 1200–21.

Jarillo, J. C. (1988). 'On Strategic Networks', *Strategic Management Journal*, 9(1): 31–41.

Kang, J., and Afuah, A. 2010. 'Profiting from Innovations: The Role of New Game Strategies in the Case of Lipitor of the US Pharmaceutical Industry', *R and D Management*, 40: 124–37.

Knight, L. (2002). 'Network learning: Exploring learning by interorganizational networks', *Human Relations*, 55(4): 427–454.

Kragh, H., and Andersen, P. H. (2009). 'Picture This: Managed Change and Resistance in Business Network Settings', *Industrial Marketing Management*, 38: 641–53.

Larson, A. (1992). 'Network Dyads in Entrepreneurial Settings: A Study of the Governance of Exchange Relationships', *Administrative Science Quarterly*, 37(1): 76–104.

Lee, S. M., Lim, S.-B., and Soriano, D. R. (2009). 'Suppliers' Participation in a Single Buyer Electronic Market', *Group Decision and Negotiation*, 18: 449–65.

Li, F., and Whalley, (2002). 'Deconstruction of the Telecommunications Industry: From Value Chains to Value Networks', *Telecommunications Policy*, 26: 451–72.

Li, Y.-R. (2009). 'The Technological Roadmap of Cisco's Business Ecosystem', *Technovation*, 29: 379–86.

Lin, Y., Wang, Y., and Yu, C. (2010). 'Investigating the Drivers of the Innovation in Channel Integration and Supply Chain Performance: A Strategy Orientated Perspective', *International Journal of Production Economics*, 127: 320–32.

Lusch, R. F., Vargo, S. L., and Tanniru, M. (2010). 'Service, Value Networks and Learning', *Journal of the Academy of Marketing Science*, 38: 19–31.

Michel, S., Brown, S. W., and Gallan, A. S. (2008). 'Service-logic Innovations: How to Innovate Customers, not Products', *California Management Review*, 50: 49–66.

Michel, S., Vargo, S. L., and Lusch, R. F. (2008). 'Reconfiguration of the Conceptual Landscape: A Tribute to the Service Logic of Richard Normann', *Journal of the Academy of Marketing Science*, 36: 152–5.

Miles, R. E., and Snow, C. C. (1986). 'Network Organizations: New Concepts for New Forms', *California Management Review*, 28(2): 68–73.

Möller, K. (2010). 'Sense-making and Agenda Construction in Emerging Business Networks: How to Direct Radical Innovation', *Industrial Marketing Management*, 39: 361–71.

Möller, K., and Rajala, A. (2007). 'Rise of Strategic Nets: New Modes of Value Creation', *Industrial Marketing Management*, 36: 895–908.

Möller, K., Rajala, A., and Svahn, S. (2005). 'Strategic Business Nets: Their Type and Management', *Journal of Business Research*, 58: 1274–84.

Möller, K., and Svahn, S. (2003). 'Managing Strategic Nets: A Capability Perspective', *Marketing Theory*, 3(2): 201–26.

Möller, K., and Svahn, S. (2006). 'Role of Knowledge in Value Creation in Business Nets', *Journal of Management Studies*, 43: 985–1007.

Möller, K., and Svahn, S. (2009). 'How to Influence the Birth of New Business Fields: Network Perspective', *Industrial Marketing Management*, 38: 450–8.

Moore, J. F. (1993). 'Predators and Prey: A New Ecology of Competition', *Harvard Business Review*, 71: 75–86.

Moore, J. F. (1996). *The Death of Competition: Leadership and Strategy in the Age of Business Ecosystems*. New York, NY: HarperBusiness.

Mouzas, S., Henneberg, S., and Naudé, P. (2008). 'Developing Network Insight', *Industrial Marketing Management*, 37: 167–80.

Nalebuff, B., and Brandenburger, A. M. (1996). *Co-opetition*. Cambridge, MA: Harper CollinsBusiness.

Newman, M. E. J. (2001). *The Structure of Scientific Collaboration Networks*. Paper presented at the Proceedings of the National Academy of Sciences.

Normann, R. (2001). *Reframing Business: When the Map Changes the Landscape*. Winchester: Wiley.

Normann, R., and Ramirez, R. (1993). 'From Value Chain to Value Constellation: Designing Interactive Strategy', *Harvard Business Review*, 71: 65–77.

Nosella, A., and Petroni, G. (2007). 'Multiple Network Leadership as a Strategic Asset: The Carlo Gavazzi Space Case', *Long Range Planning*, 40(2): 178–201.

Oberg, C., Henneberg, S., and Mouzas, S. (2007). 'Changing Network Pictures: Evidence from Mergers and Acquisitions', *Industrial Marketing Management*, 36: 926–40.

Ozcan, P., and Eisenhardt, K. M. (2009). 'Origin of Alliance Portfolios: Entrepreneurs, Network Strategies, and Firm Performance', *Academy of Management Journal*, 52(2): 246–79.

Pagani, M., and Fine, C. (2008). 'Value Network Dynamics in 3G-4G Wireless Communications: A Systems Thinking Approach to Strategic Value Assessment', *Journal of Business Research*, 61: 1102–12.

Parolini, C. (1999). *The Value Net: A Tool for Competitive Strategy*. Chichester: Wiley and Sons.

Pierce, L. (2009). 'Big Losses in Ecosystem Niches: How Core Firm Decisions Drive Complementary Product Shakeouts', *Strategic Management Journal*, 30: 323–47.

Porter, M. E. (1980). *Competitive Strategy: Techniques for Analyzing Industries and Competitors*. New York: Free Press.

Porter, M. E. (1985). *Competitive Advantage: Creating and Sustaining Superior Performance*. New York: Free Press.

Powell, W., K., Koput, K. W., and Smith-Doerr, L. (1996). 'Interorganizational Collaboration and the Locus of Innovation: Networks of Learning in Biotechnology', *Administrative Science Quarterly*, 41(1): 116–45.

Prahalad, C. K., and Hamel, G. (1990). 'The Core Competence of the Corporation', *Harvard Business Review*, May/June: 275–92.

Rabinovich, E., Knemeyer, A. M., and Mayer, C. M. (2007). 'Why do Internet Commerce Firms Incorporate Logistics Service Providers in their Distribution Channels? The Role of Transaction Costs and Network Strength', *Journal of Operations Management*, 25(3): 661–81.

Ritter, T., Wilkinson, I. F., and Johnston, W. J. (2004). 'Managing in Complex Business Networks', *Industrial Marketing Management*, 33: 175–83.

Rosenfeld, S. A. (1996). 'Does Cooperation Enhance Competitiveness? Assessing the Impacts of Inter-firm Collaboration', *Research Policy*, 25: 247–63.

Santos, F. M., and Eisenhardt, K. M. (2005). 'Organizational Boundaries and Theories of Organization', *Organization Science*, 16: 491–508.

Santos, F. M., and Eisenhardt, K. M. (2009). 'Constructing Markets and Shaping Boundaries: Entrepreneurial Power in Nascent Fields', *Academy of Management Journal*, 52: 643–71.

Scott, W. R. (2007). *Institutions and Organizations: Ideas and Interests*, 3rd edn. London, UK: Sage Publications.

Singer, J. G. (2006). 'Systems Marketing for the Information Age', *MIT Sloan Management Review*, 48: 95.

Stabell, C. B., and Fjeldstad, O. D. (1998). 'Configuring Value for Competitive Advantage: On Chains, Shops, and Networks', *Strategic Management Journal*, 19: 413–37.

Teece, D. J. (1986). 'Profiting from Technological Innovation: Implications for Integration, Collaboration, Licensing', *Research Policy*, 15: 285–305.

Teece, D. J. (1998). 'Capturing Value from Knowledge Assets: The New Economy, Markets for Know-how, and Intangible Assets', *California Management Review*, 40(3): 55–79.

Teece, D. J. (2007). 'Explicating Dynamic Capabilities: The Nature and Microfoundations of (Sustainable) Enterprise Performance', *Strategic Management Journal*, 28: 1319–50.

Thomas, L. D. W., and Autio, E. (2013) 'The Fith Facet: The Ecosystem as Organizational Field', *Innovation and Entrepreneurship Group Working Papers*: 1–40. London: Imperial College Business School.

Thomas, L. D. W., Autio, E., and Gann, D. M. (2012). 'Value Creation and Appropriation in Ecosystem Contexts', *Innovation and Entrepreneurship Group Working Papers*: 1–35. London: Imperial College Business School.

Thompson, J. D. (1967). *Organizations in Action*. New York: McGraw-Hill.

Thorelli, H. B. (1986). 'Networks: Between Markets and Hierarchies', *Strategic Management Journal*, 7: 37–51.

Uzzi, B. (1997). 'Social Structure and Competition in Inter-firm Networks: The Paradox of Embeddedness', *Administrative Science Quarterly*, 42(1): 35–67.

van de Ven, A. H., and Poole, M. S. (1995). 'Explaining Development and Change in Organizations', *Academy of Management Review*, 20: 510.

Vanhaverbeke, W., Gilsing, V., Beerkens, B., and Duysters, G. (2009). 'The Role of Alliance Network Redundancy in the Creation of Core and Non-core Technologies', *Journal of Management Studies*, 46(2): 215–44.

von Hippel, E. (1988). *The Sources of Innovation*. New York: Oxford University Press.

Watkins, M. D. (2003). 'Government Games', *MIT Sloan Management Review*, 44: 91–6.

Weiner, M., Nohria, N., and Hickeman, A. (1997). 'Value Networks: The Future of the US Electric Utility Industry', *MIT Sloan Management Review*: 21–35.

Weiss, M., and Gangadharan, G. R. (2010). 'Modeling the Mashup Ecosystem: Structure and Growth', *R&D Management*, 40: 40–9.

Wincent, J., Anokhin, S., Örtqvist, D., and Autio, E. (2010a). 'Quality Meets Structure: Generalized Reciprocity and Firm-Level Advantage in Strategic Networks', *Journal of Management Studies*, 47(4).

Wincent, J., Örtqvist, D., Eriksson, J., and Autio, E. (2010b). 'The More the Merrier? The Effect of Group Size on Effectiveness in SME Funding Campaigns', *Strategic Organization*, 8(1): 43–68.

Winkler, I. (2006). 'Network Governance Between Individual and Collective Goals: Qualitative Evidence from Six Networks', *Journal of Leadership and Organizational Studies*, 12(3): 119–34.

Yli-Renko, H., Autio, E., and Sapienza, H. J. (2001). 'Social Capital, Knowledge Acquisition, and Knowledge Exploitation in Young Technology-Based Firms', *Strategic Management Journal*, 22: 587–613.

Zott, C., and Huy, Q. N. (2007). 'How Entrepreneurs Use Symbolic Management to Acquire Resources', *Administrative Science Quarterly*, 52: 70–105.

....

CHAPTER 12

MARKETS FOR TECHNOLOGY

....

ALFONSO GAMBARDELLA,
PAOLA GIURI, AND
SALVATORE TORRISI

INTRODUCTION

....

FIRMS typically try to profit from their technological innovations by selling them embedded in new processes, goods, and services. Less frequently, innovators rely on the market for technology for the exploitation of their technologies or for outsourcing technologies developed by third parties. Traditionally, external sources of knowledge were considered an important option for small and medium-sized firms, as they could not rely on in-house R&D laboratories. Also, small firms are typically more willing to offer their technology on the markets for technology because they lack the downstream complementary assets needed to reach the market for products. However, more recently large firms have also opened their innovative activities to external sources of knowledge and complementary assets (see Chapter 22 by Alexy and Dahlander).

As a result, markets for technology are becoming larger and more diffused, although their future growth depends on a more substantial involvement of large firms, many of which remain still reluctant to rely on technology trade.

To better understand the potential growth of the markets for technology we need to understand the peculiarities of these markets. Following Arora et al. (2001), markets for technology are characterized by transactions where the main focus of exchange is disembodied technology rather than a physical artefact. However, the distinction between pure technology and physical artefacts blurs when technology is embodied in computer programs or designs. The object of trade can be a bundle of intellectual property rights (IPR) (e.g. patents), know-how and services, or a particular technology. Strictly speaking, market transactions are arm's length, anonymous, and typically involve an exchange of a good for money. Market transactions for technology, however, often fail to meet at least one of these criteria. They frequently involve complex, non-anonymous contracts, and may be embedded in technological alliances between sellers and buyers and therefore are not arm's length. Technology can also be exchanged through joint ventures, M&A, or human capital mobility across firms. An important distinction is between ex-ante contracts (i.e. contract R&D or joint R&D alliances) and ex-post contracts (i.e. contracts for existing technology). From a transaction cost perspective, ex-ante deals give rise to greater potential contracting problems (e.g. moral hazard). Our focus will be mostly on ex-post contracts (licensing deals, technology sale, and co-development or alliances based on the exchange of existing technology). We do not consider acquisitions of technology-based firms and people mobility across firms as channels of knowledge transactions (see Chapter 29 by Ahuja and Novelli and Chapter 25 by Laursen and Foss).

Like other markets for 'ideas' or intangibles, the market for technology is relatively inefficient in terms of lack of thickness (few participants on both sides of the market relative to potential participants), high congestion (participants cannot easily compare alternative offers for lack of time or due to the long time taken by each transaction), and limited market safety (participants have incentives to manipulate information for strategic reasons or transact outside the marketplace) (Gans and Stern, 2003; Roth, 2008). These characteristics affect the firm decision to participate in the technology market, the characteristics of the transaction process (negotiation duration and costs) and the outcome of the transaction process (price level and structure, complexity of clauses, etc.).

Moreover, markets for technology do not always imply substantial transfer of knowledge. *License* agreements often result from litigation for the control of IPRs. This is the case of owners of large patent portfolios who sign *cross-licensing* agreements that give the parties the freedom to operate in their respective technological and market fields rather than allowing access to outside technology. Litigation sometimes is initiated by patent trolls, namely organizations whose main aim is to accumulate patent portfolios and force potential infringers to licensing-in their technology. Probably, these IPR agreements too do not imply any real technology transfer between the parties. Typically, large firms seek to quickly settle out of court a dispute with patent trolls or small patent owners to avoid the risk of an injunction to stop alleged patent infringement. An example is Research in Motion (RIM) of Canada, which was sued by NTP Software for alleged infringement of patents. After a four-year dispute and the risk of an injunction to stop alleged infringement by a US court, RIM accepted to pay NTP over $612 million to avoid the risk of shutting down its operations in the US market. Under the terms of the agreement, NTP licensed its products to RIM.

Other similar forms of technology transactions (patent licensing and cross-licensing) are generated by patent pooling like the MPEG-2 (digital image compression), the RFID (Radio Frequency Identification), and SNP (human genome), whose impact on the efficiency of the technology market and the market for products is under debate (e.g. Lerner and Tirole, 2004). Finally, licensing can be imposed by the public authority to limit market power (e.g. AT&T, IBM, and Microsoft).

These various avenues leading to licensing and other forms of technology transactions are characterized by different actors, incentives, costs, and benefits, of which innovation managers should be aware.

WHY ARE MARKETS FOR TECHNOLOGY IMPORTANT?

Trade in technology can yield important benefits. Firms can first benefit from the presence of well-functioning technology markets because they can rely on a wider set of options for the acquisition or exploitation of their technologies. Markets for technology can also be beneficial for the efficient organization of industries. Companies may specialize upstream or downstream according to their comparative advantages. For example, technology developers can specialize in the supply of technology to downstream producers who have a comparative advantage in the use of technology because of their better manufacturing and marketing assets (Arora and Gambardella, 1994).

An example of specialization and change in the organization of industries allowed by the market for technology is represented by the Specialized Engineering Firms (SEF) in the chemical processing industries. As discussed in Arora and Gambardella (1998), SEF focused on the design and engineering of chemical plants, rather than on producing chemicals. They typically sold process technology embedded in the design of the plant, accompanied by a variety of technical services. However, some SEFs, such as Universal Oil Product (UOP), specialized more narrowly in technology licensing. UOP was responsible for several major chemical processing innovations (Remsberg and Higdon, 1994). Another notable example of specialization favoured by the market for technology is the unbundling of software and service sales from hardware, a commercial practice pioneered by IBM in 1969 which spurred the growth of an independent software industry (Torrisi, 1998).

This division of labour leads to specialization, and to Smithian economies of scale and learning (Stigler, 1951). Industrial economics also emphasized the positive effects of markets for licensing on the use and diffusion of technology, reduced duplication of research efforts, and product market competition (e.g. Gallini, 1984; Shephard, 1987; Rockett, 1990).

The importance of technology markets is also attested by their substantial size and growth during the last decades. Empirical evidence also shows that licensing and cross-licensing are important motivations for patenting. However, the diffusion of licensing and cross-licensing activities, especially amongst large companies, is still not widespread.

The Size of Markets for Technology

Technology markets have grown significantly over the past twenty years. In the mid-1990s the US markets for technology amounted to about $25–35 billion, and the global market to about $35–50 billion according to estimates by Arora and Gambardella (2010). Payments for technology accounted for as much as 10 to 15 per cent of total civilian R&D in OECD countries in the 1980s (Arora et al., 2001: 43). Other studies provide similar estimates (e.g. Robbins, 2006; Athreye and Cantwell, 2007; Mendi, 2007). OECD data on disembodied technology trade in the G8 countries also indicate that technology royalty payments and receipts increased by an average annual factor of 10.7 per cent, from 1980 to 2003, a substantially higher rate than the world GDP growth rate in the same period (OECD, 2006).

Although a large share of international technology flows is accounted for by transactions among affiliated entities (Arora and Gambardella, 2010), data from different sources confirm that the market for technology is growing, albeit not uniformly across countries, industries, and firms.

The Importance of the Markets for Technology for Firms: Patent Licensing, Cross-Licensing, and Sale

To illustrate the differences in the importance of the market for technology across different countries, industries, and firms, it is useful to look at the available evidence on patent licensing, cross-licensing, and sale.

Licensing and Cross-Licensing as Reasons for Patenting

Although licensing and cross-licensing are less important than other reasons for patenting such as direct commercial exploitation, prevention of imitation, blocking patents, and prevention of infringement suits, their importance has increased between the 1990s and the 2000s.[1] Licensing as a reason for patenting is more frequent among North-American and Japanese than European firms (Sheehan et al., 2004). Moreover, licensing is considered a very important motivation for patenting in biotechnology, pharmaceuticals, organic chemicals, and nuclear technologies (InnoST, 2011). Cross-licensing is particularly important in Japan and in some technological fields (Telecommunications, Audiovisual technologies, Information technologies, and Semiconductors) (Grindley and Teece, 1997; Cohen et al., 2000; Cohen et al., 2002; InnoST, 2011).

Actual Licensing, Cross-Licensing, and Patent Sale

Based on PatVal 1 data, we find that about 8.3 per cent of patents are licensed to an independent party, and in 8.5 per cent of cases assignees are willing to license. The share of actual licensing is higher in the US than Europe and Japan. Only about 17 per cent of licensing deals are part of cross-licensing agreements (InnoST, 2011). These results are in line with previous studies (e.g. Zuniga and Guellec, 2009).

An additional indicator for assessing the extent of technology markets is the sale of patents. Unfortunately, the empirical evidence on this issue is very scarce (Lamoreaux and Sokoloff, 2001; Serrano, 2010). Serrano (2010) analyses patent transactions and finds that a large proportion of patents taken at the US patent and trademark office (USPTO) are traded at least once in their life. Individual inventors and small innovators (up to 5 patents granted per year) are more likely to sell their patents compared with large innovators (more than 100 patents granted per year). Computers, telecommunications, pharmaceuticals, and medical instruments are the technologies with the largest rates of patent trade. Citations received and technological generality increase the probability that a patent is traded.

Our calculations based on PatVal 2 data provide additional novel evidence on patent sales. We find that patent sale accounts for an important share of markets for technology as 5.47 per cent of patents are sold to independent owners, while 5.61 per cent of owners are willing to sell their patents. The share of patent sale is larger in the United States than Europe and Japan.

Though substantial in absolute value, patent licensing and sale are still not central to the innovation process, with some notable exceptions such as chemicals, bio-pharmaceuticals, audiovisual, and IT. In sum, while technology markets are growing, they are still limited to a relatively small number of sectors and countries.

Firms' Incentives to Participate in the Markets for Technology

In the past two decades there has been an increasing interest in the literature on firm strategies in technology markets. Most of the literature has focused on the supply side of technology markets, mainly looking at the determinants of licensing out by established companies and by small and new firms. Few recent studies aim at understanding the demand side of technology markets, and the barriers to technology trade.

Motivations of Technology Suppliers

The economics and management literature has identified several reasons why firms should license their technology to others, sometimes even to potential or actual rivals. We distinguish between revenue-driven motivations and strategic motivations.

Revenue-Driven Motivations

In most cases, the single most important motive for technology licensing is the revenue it generates, defined as the present value of the flow of payments accruing to the licensor. Such benefit must be compared with the costs of transferring the technology, which might be substantial. Finally, revenues from licensing net of transaction costs must be compared with the opportunity costs of giving up the benefits of direct exploitation of the technology in new products or processes. The importance of revenues as a motivation for licensing is determined by various factors.

Complementary assets. Innovators are not always the best users of their innovation. As Teece (1986) argued, the value of an innovation depends on the complementary downstream assets that are needed for its exploitation, including marketing, distribution, brand name, and finance. An innovator that lacks some of these downstream assets can profitably license the technology to another firm that is well endowed with these assets. However, complementary assets can be generic or co-specialized. In the latter case, licensing is problematic, as seen in the example of international licensing. It occurs when companies realize that they cannot enter a foreign market by a foreign direct investment, or that it would be extremely expensive or risky to do so because of geographical, cultural, and 'psychical' distance between countries (Hofstede, 1991). In contrast, unless there are significant contracting problems, the firm can reap the returns to its technology through a licensing deal or other collaborative agreements (Dunning, 1981), which implies a lower level of exposure to country-specific risk and more limited costs to deal with the local market (Hill et al., 1990).

Appropriability, information asymmetry, and transaction costs. A well-functioning system of IPR reduces transaction costs in technology trade because the seller of the technology is less concerned about the potential loss of property rights on the innovation (Teece, 1986).

In this respect, Caves et al. (1983) find that, due to imperfections in the licensing market, licensors capture only about one-third of the rents from their innovations. Property rights are far easier to define and enforce on tangible goods than on such intangible goods as designs, ideas, or technologies. It is difficult to realize the full value of a new piece of knowledge without fully disclosing it, but, once the knowledge is disclosed, the acquirer's incentive to pay for it evaporates. Because of incomplete appropriability then, the potential buyer will base his decision to purchase on less than optimal criteria (Arrow, 1962).

Competition and rent dissipation effect. This effect consists in the reduction in the licensor's profits (i.e. all profits other than payments from the licensing agreement) that may occur as a consequence of a licensing agreement permitting a newcomer to compete in the product market or an existing rival to increase its competitive advantage (Arora and Fosfuri, 2003; Fosfuri, 2006). The rent dissipation effect depends on several factors. Primary among these is the magnitude of the competitive pressure exerted by a new player in the product market. The rent dissipation effect is small when the licensor firm does not have a stake in the product market or its market share is tiny. Moreover, the rent dissipation effect is likely to be less binding when the licensees operate in distant markets, either geographically or in the product space, as we shall also discuss.

Technological generality. The combination of revenue and rent dissipation effects is particularly favourable to the use of the market in the case of general purpose technologies. Dedicated technologies address the needs of specific users or applications and therefore have a limited market potential. Instead, general-purpose technologies are characterized by high fixed costs and lower marginal costs for applying them to different uses (Bresnahan and Trajtenberg, 1995; Bresnahan and Gambardella, 1998). The diffusion of general-purpose technologies is driven by the possibility to spread high fixed costs over a large number of applications (breadth). These technologies then have a larger potential market compared with dedicated technologies which focus on a few large applications (depth). The application breadth of general-purpose technologies favours the division of labour between technology specialists and vertically integrated firms that can acquire a general technology on the market rather than developing a dedicated technology in-house. Clearly, larger firms have greater incentives to develop dedicated technologies in-house compared with smaller firms. However, when the price of external technologies declines because of exogenous changes that reduce the adaptation costs of the general-purpose technology to different applications (e.g. the introduction of platform technologies in the biotech field or application development tools in the software field), large firms have greater incentives to acquire a technological asset on the market rather than developing it in-house.

In recent years, some technology specialists have developed generic technologies that can be employed in different sub-markets, thus exploiting the opportunities of breadth. Gambardella and McGahan (2010) provide several examples of technology specialists—in biotech, software, and nanotech—that have bet on generic technologies. When generic technologies are supplied on the market, in-house development or acquisition of specialized suppliers will destroy value by restricting the application set of the technology. Technology specialists with no downstream operations can raise their profits by (a) making their technology more general in order to cover more applications; (b) searching for new uses, through alliances or other deals with downstream firms in different final markets to test and possibly co-develop applications.

Moreover, Gambardella and Giarratana (2012) note that markets for technology are more likely to arise when technologies are general purpose and product markets are fragmented with many different niches. This relates back to the earlier point that a fragmented market implies that licensing is less likely to dissipate the rents of the licensor in the product market. A dedicated technology cannot be profitably used in a distant sub-market, where the licensee would not compete with the licensor; a general-purpose technology, which can instead be used in a distant market, can only be licensed to a close competitor if the product market is homogenous. Thus, markets for technology likely thrive when technologies are general, product markets are fragmented, and the suppliers (and potential licensors) operate in different sub-markets. In this case, the licensee is willing to pay a royalty for the license because it is useful, and the licensor is willing to license it because the rent dissipation in its product market is minimal.

Strategic Motivations: 'Stick' Licensing and Cross-Licensing

There are situations in which the revenue generated by transactions is not the main reason why innovators offer their intellectual property for licensing. Several strategic incentives for licensing have been identified in the literature. First, Gallini (1984) has shown how an incumbent firm may license its production technology to reduce the incentive of a potential entrant to develop its own, possibly better, technology. Moreover, the incumbent firm might license to a weak rival in order to deter the entry of stronger competitors (Rockett, 1990). Second, licensing may be used to maintain market power. Even if collusion in the product market is prohibited, a firm can license its technology to rivals and set royalty rates that yield equilibrium profits identical to what a cartel would have yielded in the product market (Fershtman and Kamien, 1992). Third, competition in the market for technology may generate strategic incentives to license (Arora and Fosfuri, 2003): a technology holder is encouraged to license when the prospective licensee can obtain the technology from other potential licensors anyway and ultimately compete in the product market. Fourth, licensing can be used as a second sourcing strategy (Shepard, 1987; Farrell and Gallini, 1988), or it can result from technology or business focusing—for example, firms who aim to improve their innovativeness may decide to sell or license out non-core technologies (Corts, 2000). Fifth, licensing could be motivated by the attempt to create and control market standards (Khazam and Mowery, 1994).

Finally, in industries with cumulative technological change and complex products, firms are spurred to accept license offers by the owners of alleged infringed patents (stick licenses) or enter into cross-licensing agreements to have access to key technologies patented by other firms.

As discussed, cross-licensing is a strategy adopted especially by the owners of large patent portfolios operating in complex products industries like computers, semiconductors, and electronics (Grindley and Teece, 1997; Cohen et al., 2000). And a large proportion of patents owned by large firms such as IBM, TI, and HP is probably used as bargaining chips in litigation and cross-licensing deals (Rivette and Kline, 2000; Hall and Ziedonis, 2001).

Cross-licenses typically do not imply significant transfer of technology as both parties are often interested in gaining the freedom to design or manufacture. Cross-licensing is an important issue for technology and intellectual property management for several reasons (Giuri and Torrisi, 2011a).

First, it represents a coordination mechanism that allows the owners of overlapping patent portfolios to moderate the costs of litigation (Bessen and Meurer, 2008; Hall et al., 2009). The incentive to use cross-licensing (or similar arrangements) has increased over time with the explosion of patent applications since the 1980s and the associated rising patent litigation costs (Harhoff and Reitzig, 2004; Bessen and Meurer, 2008, and see Chapter 28 by Leiponen). The threat of litigation is particularly strong for firms with high capital-intensity because of their sunk costs in technology-related activities (see Hall and Ziedonis, 2001; Beard and Kaserman, 2002; Ziedonis, 2004). These firms then have more incentives to engage in cross-licensing and other forms of out of court agreements. Giuri and Torrisi (2010a) found that capital-intensity has a positive moderating effect on the association between product complexity and cross-licensing as a motivation for patenting. As Cohen et al. have noted, 'in complex product industries, firms rarely have proprietary control over all the essential complementary components of the technologies they are developing' (2002: 1356). For example, there are several complementary patents that are essential to implement technical standards such as GSM,

DVD6 video, and the MP3 patents. In the case of GSM, the owners of 'essential' patents (Nokia, Motorola, Ericsson, Siemens, and Alcatel) promoted the initial diffusion of the standard by signing a cross-licensing deal which was opened to other participants later on (Bekkers et al., 2002).

In industries such as semiconductors and biotechnology, firms often find it convenient to engage in cross-licensing or create a patent pool (where all blocking patents can be licensed on the basis of a package license agreement) to avoid mutually blocking patents, reduce 'multiple patent burdens' and the hold-up problem (i.e. the risk of infringement complaint by the holders of a patent the firm is not aware of), and moderate the risk of unintentional infringement (Hall and Ziedonis, 2001; Davis, 2008).

Besides providing assurance against the risk of accidental patent infringement and ensuring the freedom to design and manufacture, cross-licensing presents several advantages for firms engaged in R&D-intensive and patent-intensive industries. First, cross-licensing helps free internal resources that can be devoted to R&D activities that do not merely replicate earlier inventive efforts. Second, cross-licensing speeds-up the firms' development process and therefore allows a faster commercialization of innovation (Fershtman and Kamien, 1992).

Differences Between Large and Small Firms in the Market for Technology

Large firms (> 250 employees) account for about two-thirds of European Patent Office patents and are less likely to license their technologies compared with smaller firms. Large firms are also less willing to license their patents compared with small firms (Gambardella et al., 2007; InnoST, 2011).[2]

These findings are in line with theory, as smaller firms have limited complementary assets to benefit from technology and suffer less from the rent dissipation effect of additional rivals compared with larger firms (Arora et al., 2001; Arora and Fosfuri, 2003).

Beyond size, the organization of innovation management activities may strongly affect the incentive (and ability) to exploit the firm's technology externally. Arora et al. (2011) show that decentralizing the decision to 'license or produce' to business units makes licensing less likely to occur.

Case studies of IBM, Dow, Boeing, Motorola, Xerox, and Procter & Gamble suggest that in firms that license extensively, licensing is handled by a specialized business unit (often treated as an independent business) and it is incentivized in various ways (licensing revenues are typically shared with operating units). Arora et al. (2011) found that firms that decentralize the management of their patent portfolio experience faster sales growth. Since licensing cannibalizes sales in the product market, this finding is consistent with lower licensing propensity under decentralization. A survey of US patent holders by Jung and Walsh (2010) also found that patents where the inventor is from the manufacturing unit (rather than central R&D) are less likely to be licensed.

The lack of incentives to participate in the market for technology has major implications for the future growth of this market. A large number of deals in this market are about patents, and large firms account for a very large share of patents worldwide. Compared with smaller firms, a larger proportion of large firms' patents are never used, internally or externally. Large firms then represent a great unexploited repository of technology. The size of markets for technology would increase further if large firms increase the openness of their innovative activities.

The Demand Side of Markets for Technology

A large body of research has explained the incentives to license, primarily from the licensor's perspective. Few studies have adopted the licensee's perspective despite the increasing importance of technology outsourcing as an alternative to in-house R&D in various industries (e.g. Silverman, 1999; Arora et al., 2001; Fosfuri, 2006). Although scholars have started to acknowledge that '[b]oth the buying and the selling perspectives are necessary to improve the management of IP' (Chesbrough, 2003: 158), the theoretical and empirical investigation of technology sourcing strategy, especially through in-licensing, is limited to a few studies (e.g. Killing, 1978; Caves et al., 1983; Atuahene-Gima, 1993; Atuahene-Gima and Patterson, 1993; Lowe and Taylor, 1998). These studies on technology in-licensing have found that the acquired licenses were most often closely related to the focal firm's technological competencies. However, from these contributions it is not clear to what extent markets for technology allow innovators greater strategic flexibility and a larger number of feasible options as compared to in-house explorative search. Laursen et al. (2010) have compared the behaviour of in-licensing firms to the behaviour of comparable non-in-licensing firms and found that firms that rely on licensing-in explore more distantly from their existing technological portfolio compared to similar firms that do not rely on licensing-in. They also show that the use of licensing, as an alternative to internal R&D, reinforces the positive effect of the firm's monitoring ability (measured by the scale and diversification of its patents' backward citations) on the distance of technological exploration. This analysis points to the importance of markets for technology sourcing. It also suggests that gaining access to distant, unfamiliar, technologies through the market for technology requires prior investments in monitoring ability. This result is in line with Cohen and Levinthal's (1990) argument that knowledge is not a public good and requires specific investment to be absorbed. Markets for technology can reduce, but not eliminate, the costs of access to external knowledge.

BARRIERS TO TECHNOLOGY TRADE

Participation in the market for technology implies risks and transaction costs for both sides of the market. As discussed before, technology suppliers have to balance the positive 'revenue effects' with negative 'rent dissipation effects' generated by technology transfer to competitors. Moreover, firms face significant managerial difficulties in using the market for technology because of various sources of inefficiency.

Uncertainty, Information Asymmetry, and Contract Incompleteness

The seminal papers by Arrow (1962) and Nelson (1959) show how the nature of knowledge as a commodity is a fundamental reason why the market for inventions is inefficient. Subsequent studies have pointed out other characteristics of knowledge, such as its tacitness or stickiness, which can hamper technology transactions between different organizations (Winter, 1987; von Hippel, 1990; Kogut and Zander, 1992; Arora and Gambardella, 1994). Various more recent theoretical works rely on the theory of market imperfections to explain the incentives and obstacles to the growth of the licensing market (e.g. Arora et al., 2001; Anton and Yao 2002; Gans and Stern, 2003; Scotchmer, 2004). From a different perspective, the strategic management literature reaches similar conclusions by pointing out the difficulties encountered by firms who try to acquire intangible assets like R&D and customer loyalty in the 'strategic factor' markets (Barney, 1986; Dierickx and Cool, 1989). Dierickx and Cool have contended that strategic factor markets such as R&D do not exist because strategic assets (i.e. assets that yield abnormal profits to the user) are non-tradable, and firms have to develop these factors in-house.

Uncertainty and information asymmetry between buyers and suppliers of technology make it difficult to write enforceable, complete contracts. Scotchmer (1996) and Anton and Yao (2002) have pointed out the obstacles to optimal technology trade contracts that are particularly strong in the case of cumulative, sequential innovations. Moreover, trade partners may end up disclosing proprietary information to rivals, or exploiting it in future transactions (Mowery, 1983). More importantly, inventive activity and commercialization are often intertwined, and several feedbacks take place between different phases of the innovation chain. These interactions are difficult to anticipate and specify in advance (Kline and Rosenberg, 1986). This also hampers writing complete contracts and may also require that the two activities proceed in close proximity, with tight and frequent information exchange, typically best achieved within a firm rather than via market exchanges (Teece, 1986). These difficulties notwithstanding, it is possible to write efficient contracts for technology exchange under certain conditions: (a) the tacit and codified components of technology are complements and can be bundled together in the same contract package; (b) the codified component can be protected by IPR; (c) the payment structure is composed of two parts—an upfront fee that the licensee pays at the time the know-how is provided and a second fee to be paid afterwards. For an efficient contract to be written, it is important that the supplier can withdraw the license for the codified component if the licensee refuses to pay the second-stage fee. The threat of license withdrawal is made credible by complementarity (the value of the tacit component depends on the use of the codified component) and well-defined IPRs (Arora, 1995). A weak, uncertain protection of technological property rights is then an important source of uncertainty and inefficiency (Teece, 1986; 1998).

In theory, patent protection reduces transaction costs because it requires that the supplier reveals key information and, at the same time, is protected from knowledge misappropriation by the acquirer. This is shown by the fact that the growth of technology transactions has paralleled the explosion of patent applications since the 1980s (WIPO, 2011).

However, the patent system spurs excessive fragmentation of property rights, favours protection of low value inventions which are sought for purely strategic reasons, increases the risk of involuntary infringement on multiple patents, and increases transaction costs in the market for technology (Heller and Eisenberg, 1998; Harhoff et al., 1999; Hall and Ziedonis, 2001; Gambardella et al., 2008; Hall et al., 2009).

Cumulative Technologies, Strategic Patenting, and Unused Patents

In industries characterized by cumulative technical change and complex products, firms are induced to accumulate large patent portfolios. Portfolio patent races in which all competitors try to acquire as many patents as possible propagates a patent thicket (a dense set of overlapping patent rights), and does not necessarily favour technology transactions such as licensing and cross-licensing, especially when patent rights are very fragmented (Siebert and von Graevenitz, 2008).

Indeed, a large number of patents are not used. Drawing on PatVal 2 survey, we found that about 40 per cent of patents held by business enterprises are neither commercially used by the owner nor are sold, licensed, or used to found a new firm (Giuri and Torrisi, 2011b). This suggests that a significant share of unused patents could be used, and it raises the issue of innovation management practices that reduce the barriers to a more intensive use of patented technologies.

It is possible that patents are not used because patentees do not recognize the value of patents in their portfolios. Organizations that own large patent portfolios often do not carry out technology audits and therefore miss the opportunity to fully exploit the economic value of all their patents, either by internal use or by using the market for technology (Rivette and Kline, 2000).

An efficient market for technology would help patent owners to better evaluate and exploit their technology. However, a more efficient market for technology does not necessarily translate into a more intensive use of patents that are not used for purely strategic reasons. As discussed before, this is the case of blocking patents, that is, patents that are taken to avoid other parties patenting complementary or substitute inventions, and pure defence patents (Grindley and Teece, 1997; Hall and Ziedonis, 2001).

Additional Evidence on Barriers to Technology Trade

Besides strategic reasons and IPR fragmentation there are other obstacles to technology trade highlighted by empirical studies. Razgaitis (2004) shows that the reasons why many licensing deals are not concluded depend on subtler and harder-to-observe elements, such as the inability to find buyers, the difficulty in getting internal approval to conclude the deal, disagreements on exclusivity, or geographical restrictions. Other factors that hamper the growth of a market for licensing are represented by the investments needed to absorb knowledge from external sources, the costs of IP enforcement, institutional, and cultural barriers, and the difficulty of transferring tacit know-how across different organizations.

Based on data collected by face-to-face interviews with the IP and licensing managers of twenty-two European firms, Gambardella and Torrisi (2010) have explored further the barriers to participation in this market for both licensors and licensees. The most important obstacles to license are the threat of hold-up and opportunistic behaviour and cultural/organizational differences between partners. Another obstacle highlighted by several cases is represented by uncertainty and the difficulty of evaluating (and agreeing upon the value of) a technology, especially in early development stages. This difficulty is important in the pharmaceutical sector where a firm that licenses-in a new molecule can find out many unpredictable difficulties during clinical trials. In other sectors (e.g. automotive), a source of uncertainty is represented by the difficulty of predicting production cost when the technology evolves from the level of a patented invention or a sample product to production. Another source of uncertainty is evaluation which is made difficult by the fact that the market for licensing is a bilateral bargaining market, rather than a multilateral trading institution.

Another form of uncertainty from the licensor's viewpoint arises from the difficulty of evaluating the ability of the licensees to generate revenues, which does not entirely depend on the intrinsic value of the technology. This is an issue especially for small firms that do not easily get significant upfront fees and often offer an exclusive license because of their limited bargaining power. Since the real value for them comes later with milestones and royalties, it is extremely important to acquire information about the licensor's development and commercialization capabilities. Moreover, small technology firms are very concerned about opportunistic behaviour and the risk of hold-up. Large licensees have large project pipelines, can choose among several alternatives and can take some time to negotiate, whereas their smaller counterparts are more impatient because of much tighter financial constraints.[3]

Mechanisms for Reducing the Barriers to Technology Trade

Despite the barriers and managerial difficulties in technology transfer, the use of intermediary services that could facilitate the search for partners and the negotiation process is still very limited.

More recently, online marketplaces for technology trade have emerged such as yet-2and.com, ICAP, and Ocean Tomo. These markets have the advantage of reducing search and execution costs, but entail information asymmetry (and adverse selection) and a high risk of rent expropriation (Dushnitsky and Klueter, 2011). Dushnitsky and Klueter (2011) note that technology marketplaces use adverse-selection moderating mechanisms like upfront participation fees and disclosure requirements. Further analysis would be required to understand how adverse-selection mechanisms used by online marketplaces work as a substitute for social ties or geographical proximity and favour the growth of the markets for technology.

FUTURE DIRECTIONS OF RESEARCH

An issue for future research concerns the role of institutional and organizational barriers to the growth of the market for licensing. Several factors hampering technology transactions have been investigated by the literature, such as those related to knowledge characteristics (Arrow, 1962, von Hippel, 1990, Arora and Gambardella, 1994), contractual incompleteness (Mowery, 1983; Anton and Yao, 2002), market size (Bresnahan and Gambardella, 1998), and IPR protection (Lamoreaux and Sokoloff, 2001; Arora and Merges, 2004). However, the relative importance of these different barriers to technology licensing should be analysed more deeply to understand in particular the role of cultural, institutional, and organizational barriers.

As for the institutional barriers, Arora and Ceccagnoli (2006) have noted that patent protection affects licensing by influencing both the decision to patent and the decision to license, conditional on having patented. A strong patent regime is likely to encourage licensing among firms that lack complementary assets, but it may also reduce the incentive to licensing by the owners of such assets. Future research should take into account the interaction of IPR with complementary assets and its implications for the licensing decision.

As for cultural and organizational barriers, their importance for licensing may be moderated by geographical distance and the differences in the organization of R&D activities across firms (Furman, 2003). Alcacer et al. (2008) have shown that in the US biotech industry the probability of licensing-in is affected by co-location. Further, finer-grained research is also needed to explore the effectiveness of online marketplaces. To understand how these markets differ from traditional technology markets, future research should address the differences across technologies (e.g. IPR regime, generality, etc.) and participants (experience with the online or offline markets for technology, size, reputation, etc.). Moreover, the quality of technologies traded on these markets could be compared with those of traditional technology transactions.

A further research topic is about the association between licensing and firms' technological exploration. Only a small number of studies focus on the demand side of the licensing market (e.g. Laursen et al., 2010). It is important to understand more thoroughly the determinants of licensing-in and the link between the patterns of licensing-in and the overall R&D strategy of the firm. For example, it is important to understand better when a firm decides whether to acquire or license-in a patent or to acquire a technological start-up in its exploration of the technology landscape.

There is also a limited understanding of how licensing activities are organized at the firm level and of the complementarity between different strategies of external and internal exploitation of technologies. The implications of centralization of IP and licensing decision at the corporate level for the incentive to license are still not well understood (see Arora et al., 2011) and deserve further exploration.

Finally, there is limited research on the interactions between markets for technology and financial markets. Besides the revenue streams discussed before, licensing could have a signalling value, and it is important then to see how the market reacts to licensing announcements. While there is a quite large body of studies on the market reaction to alliances and M&A announcements, the analysis of market reaction to licensing is less understood (e.g. Anand and Khanna, 2000).

Notes

1. Throughout this chapter we will draw on data from two surveys conducted by the authors in two European projects. For the sake of simplicity, we will refer to these data sets as PatVal 1 (Giuri et al., 2007) and PatVal 2 (InnoST, 2011) respectively.
2. In line with previous studies (Zuniga and Guellec, 2009), our calculations based on PatVal 2 survey show a U-shaped relation between licensing and firm size.
3. We do not consider the case of IP firms and patent trolls whose core business consists in amassing patent portfolios and offering licenses to third parties under the threat of litigation. In these conditions, being a large, publicly listed firm could be a weakness because a large firm facing a court is likely to suffer larger financial and reputation losses compared to a smaller firm in case of an injunction to stop violation.

References

Alcacer, J., Cantwell, J., and Gittelman, M. (2008). 'Are Licensing Markets Local? An Analysis of the Geography of Vertical Licensing Agreements in Bio-pharmaceuticals', Draft. Harvard Business School.

Anand, B., and Khanna, T. (2000). 'Do Firms Learn to Create Value? The Case of Alliances', *Strategic Management Journal* 21(3): 295–315.

Anton, J. J., and Yao, D. A. (2002). 'The Sale of Ideas: Strategic Disclosure, Property Rights, and Contracting', *Review of Economic Studies*, 67: 585–607.

Arora, A. (1995). 'Licensing Tacit Knowledge: Intellectual Property Rights and the Market for Know-how', *Economics of Innovation and New Technology*, 4: 41–59.

Arora, A. Belenzon S., and Rios L. A. (2011). 'The Organization of R&D in American Corporations: The Determinants and Consequences of Decentralization', NBER Working Paper No. 17013.

Arora, A. and Ceccagnoli, M. (2006). 'Patent Protection, Complementary Assets, and Firms' Incentives for Technology Licensing', *Management Science*, 52(2): 293–308.

Arora A., Fosfuri, A., and Gambardella, A. (2001). *Markets for Technology: The Economics of Innovation and Corporate Strategy*. Cambridge: MIT Press.

Arora, A., and Fosfuri, A. (2003). 'Licensing in the Presence of Competing Technologies'. *Journal of Economic Behavior and Organization*, 52: 277–95.

Arora A., and Gambardella, A. (1994). 'The Changing Technology of Technical Change: General and Abstract Knowledge and the Division of Innovative Labour', *Research Policy*, 23: 523–32.

Arora A., and Gambardella, A. (1998). 'Evolution of Industry Structure in the Chemical Industry', In Arora, A., Landau, R., and Rosenberg, N. (eds) *Chemicals and Long-Term Economic Growth*. New York: John Wiley & Sons, 379–414.

Arora A., and Gambardella, A. (2010). 'Ideas for Rent: An Overview of Markets for Technology', *Industrial and Corporate Change*, 19: 775–803.

Arora A., and Merges, R. (2004). 'Specialized Supply Firms, Property Rights and Firm Boundaries', *Industrial and Corporate Change*, 13(3): 451.

Arrow, J. K. (1962). 'Economics Welfare and the Allocation of Resources for Invention'. In Nelson, R. (ed.), *The Rate and Direction of Inventive Activity*. Princeton: Princeton University Press, 164–81.

Athreye, S., and Cantwell J. (2007). 'Creating Competition?: Globalisation and the Emergence of New Technology Producers', *Research Policy*, 36(2): 209–26.

Atuahene-Gima, K. (1993). 'Determinants of Inward Technology Licensing Intentions: An Empirical Analysis of Australian Engineering Firms', *Journal of Product Innovation Management*, 10: 230–40.

Atuahene-Gima, K., and Patterson, P. (1993), 'Managerial Perceptions of Technology Licensing as an Alternative to Internal R&D in New Product Development: An Empirical Investigation', *R&D Management* 23(4): 327–36.

Barney, J. B. (1986). 'Strategic Factor Markets: Expectations, Luck, and Business Strategy', *Management Science*, 32(10): 1231–41.

Beard, T. R., and Kaserman, D. L. (2002), 'Patent Thickets, Cross Licensing, and Antitrust', *The Antitrust Bulletin*, 47(2/3): 345–68.

Bekkers, R., Duysters, G., and Verspagen, B. (2002) 'Intellectual Property Rights, Strategic Technology Agreements and Market Structure: The Case of GSM', *Research Policy*, 31(7): 1141–61.

Bessen, J., and Meurer, M. J. (2008). *Patent Failure: How Judges, Bureaucrats, and Lawyers Put Innovators at Risk*. Princeton: Princeton University Press.

Bresnahan, T., and Gambardella, A. (1998). 'The Division of Inventive Labor and the Extent of the Market', in Helpman, E. (ed.) *General-Purpose Technologies and Economic Growth*. Cambridge MA: MIT Press, 253–81.

Bresnahan, T. and Trajtenberg, M. (1995). 'General Purpose Technologies: Engines of Growth', *Journal of Econometrics*, 65: 83–108.

Caves R., Crookel, H., and Killing, J. P. (1983). 'The Imperfect Market for Technology Licensing', *Oxford Bulletin of Economics and Statistics*, 45(3): 249–67.

Chesbrough, H. (2003). *Open Innovation*. Cambridge, MA: Harvard University Press.

Cohen, W. M., Goto, A., Nagata, A., Nelson, R. R., and Walsh, J. P. (2002). 'R&D Spillovers, Patents and the Incentives to Innovate in Japan and the United States', *Research Policy* 31: 1349–67.

Cohen, W. M., and Levinthal, D. A. (1990). 'Absorptive Capacity: A New Perspective of Learning and Innovation', *Administrative Science Quarterly* 35: 128–52.

Cohen W. M., Nelson R. R., and Walsh, J. P. (2000). 'Protecting their Intellectual Assets: Appropriability Conditions and Why U.S. Manufacturing Firms Patent (or Not)', NBER Working Paper no. 7522.

Corts, K. S. (2000). 'Focused Firms and the Incentive to Innovate', *Journal of Economics & Management Strategy*, 9(3): 339–62.

Davis, L. (2008). 'Licensing Strategies of the New "Intellectual Property Vendors"', *California Management Review*, 50(2): 6–30.

Dierickx, I., and Cool, K. (1989). 'Asset Stock Accumulation and Sustainability of Competitive Advantage', *Management Science*, 35(12): 1504–11.

Dunning, J. H. (1981). *International Production and the Multinational Enterprise*. London: Allen and Unwin.

Dushnitsky, G., and Klueter, T. (2011). 'Is There an eBay for Ideas? Insights from Online Knowledge Marketplaces', *European Management Review*, 8: 17–32.

Farrell, J., and Gallini, N. T. (1988). 'Second Sourcing as a Commitment: Monopoly Incentive to Attract Competition', *Quarterly Journal of Economics*, 103: 673–94.

Fershtman, C., and Kamien, M. I. (1992). 'Cross Licensing of Complementary Technologies', *International Journal of Industrial Organization*, 10: 329–48.

Fosfuri, A. (2006). 'The Licensing Dilemma: Understanding the Determinants of the Rate of Technology licensing', *Strategic Management Journal*, 27(12): 1141–58.

Furman, J. (2003). 'Location and Organizing Strategy? Exploring the Influence of Location on Organization of Pharmaceutical Research', *Advances in Strategic Management*, 20: 49–88.

Gallini, N. T. (1984). 'Deterrence Through Market Sharing: A Strategic Incentive for Licensing', *American Economic Review*, 74: 931–41.

Gambardella, A., and Giarratana, M. (2012). 'General Technologies, Product-market Fragmentation and the Market for Technology', *Research Policy*, 42: 315–25.

Gambardella, A., Giuri, P., and Luzzi, A. (2007). 'The Market for Patents in Europe', *Research Policy*, 36: 1163–83.

Gambardella, A., Harhoff, D., and Verspagen, B. (2008). 'The Value of European Patent', *European Management Review*, 5(2): 69–84.

Gambardella, A., and McGahan, A. (2010). 'Business-model Innovation, General Purpose Technologies, Specialization and Industry Change', *Long Range Planning*, 43: 262–71.

Gambardella, A., and Torrisi, S. (2010). 'Heterogeneity of Technology Licensing Patterns across Europe', *Working Paper. GlobInn Project, EC FP 7 Cooperation Work Programme*, January.

Gans, J. S., and Stern, S. (2003). 'The Product Market and the Market for "Ideas": Commercialization Strategies for Technology Entrepreneurs', *Research Policy*, 32: 333–50.

Giuri, P., Mariani, M., Brusoni, S., Crespi, G., Francoz, D., Gambardella, A., Garcia-Fontes, W., Geuna, A., Gonzales, R., Harhoff, D., Hoisl, K., Lebas, C., Luzzi, A., Magazzini, L., Nesta, L., Nomaler, O., Palomeras, N., Patel, P., Romanelli, M., and Verspagen, B. (2007). 'Inventors and Invention Processes in Europe. Results from the PatVal-EU Survey', *Research Policy*, 38: 1107–27.

Giuri, P., and Torrisi, S. (2011a). 'Cross-licensing, Cumulative Inventions and Strategic Patenting', Paper presented at the *DRUID Society Conference 2011*, Copenhagen Business School, 15–17 June 2011.

Giuri, P., and Torrisi, S. (2011b). 'The Economic Use of Patents", in InnoS&T report of the EC project *Innovative S&T Indicators Combining Patent Data and Surveys: Empirical Models and Policy Analyses*. Grant agreement no.: 217299, Del. 7.3.

Grindley, P. C., and Teece, D. J. (1997). 'Managing Intellectual Capital: Licensing and Cross-licensing in Semiconductors and Electronics', *California Management Review*, 39(2): 8–41.

Hall, B. H., Thoma, G., and Torrisi, S. (2009). 'Financial Patenting in Europe', *European Management Review*, 6: 45–63.

Hall, B. H., and Ziedonis, R. H. (2001). 'The Determinants of Patenting in the U. S. Semiconductor Industry, 1980–1994', *Rand Journal of Economics*, 32: 101–28.

Harhoff, D., and Reitzig, M. (2004). 'Determinants of Opposition against EPO Patent Grants: The Case of Biotechnology and Pharmaceuticals', *International Journal of Industrial Organization*, 22: 443–80.

Harhoff, D., Narin, F., Scherer, F. M., and Vopel, K. (1999). 'Citation Frequency and the Value of Patented Inventions', *Review of Economics and Statistics*, 81: 511–5.

Heller, M., and R. Eisenberg (1998). 'Can Patents Deter Innovation? The Anticommons in Biomedical Research', *Science*, 280: 698–701.

Hill C. W., Hwang, L. P., and Kim, W. C. (1990). 'An Eclectic Theory of the Choice of International Entry Mode', *Strategic Management Journal*, 11: 117–28.

Hofstede, G. (1991). *Cultures and Organizations: Software of the Mind*. Berkshire, UK: McGraw-Hill.

InnoS&T (2011). Final Report of the FP7 Project 'Innovative S&T Indicators Combining Patent Data and Surveys: Empirical Models and Policy Analysis', Available at <http://bcmmnty-qp. unibocconi.it/QuickPlace/innovativest/Main.nsf/h_C93B07E6012A16EBC125775800682 F36/6D9A810AEBB96DDFC1257989002E1F30/?OpenDocument>

Jung T., and Walsh, J. (2010). 'Organizational Paths of Commercializing Patented Inventions', Paper presented at Imperial College London Business School, 16–18 June 2010.

Khazam J., and Mowery, D. (1994). 'The Commercialization of RISC: Strategies for the Creation of Dominant Designs', *Research Policy*, 23(1): 89–102.

Killing, J. P. (1978). 'Diversification Through Licensing', *R&D Management*, 8(3): 159–63.

Kline S., and Rosenberg, N. (1986). 'An Overview of Innovation', in Landau, R. and Rosenberg, N. (eds), *The Positive Sum Strategy*. Washington, DC: National Academy Press, 275–305.

Kogut, B., and Zander, U. (1992). 'Knowledge of the Firm, Combinative Capabilities and the Replication of Technology', *Organization Science*, 3: 383–97.

Lamoreaux, N. R., and Sokoloff, K. L. (2001). 'Market Trade in Patents and the Rise of a Class of Specialized Inventors in the 19th-century United States', *The American Economic Review*, 91(2): 39–44.

Laursen, K., Leone, I., and Torrisi, S. (2010). 'Technological Exploration Through Licensing: New Insights from the Licensee's Point of View', *Industrial and Corporate Change*, 19(3): 871–97.

Lerner, J., and Tirole, J. (2004). 'Efficient Patent Pools', *American Economic Review*, 94(3): 691–711.

Lowe, J., and Taylor, P. (1998). 'R&D and Technology Purchase Through License Agreements: Complementary Strategies and Complementary Assets', *R&D Management*, 28(4): 263–78.

Mendi, P. (2007). 'Trade in Disembodied Technology and Total Factor Productivity in OECD Countries', *Research Policy*, 36: 121–33.

Mowery, D. (1983). 'The Relationship between Intrafirm and Contractual Forms of Industrial Research in American Manufacturing, 1900–1940', *Explorations in Economic History*, 20: 351–73.

Nelson, R. R. (1959). 'The Simple Economics of Basic Scientific Research', *Journal of Political Economy*, 67(2): 297–306.

OECD (2006). *Technology Indicators, Technology Balance of Payment: Payments/Receipts*. Paris: OECD.

Plaines, I. L., Rivette, K. G., and Kline, D. (2000). 'Discovering New Value of Intellectual Property', *Harvard Business Review*, January–February: 54–66.

Razgaitis, S. (2004). 'US/Canadian Licensing in 2003: Survey Results', *Journal of the Licensing Executive Society*, 34: 139–51.

Remsberg, C., and Higdon, H. (1994). *Ideas for Rents: The UOP Story*. Des Plains, IL: Universal Oil Corp.

Rivette, K. G., and Kline, D. (2000). 'Discovering New Value in Intellectual Property', *Harvard Business Review*, 78(1): 54–66.

Robbins, C. A. (2006). Measuring Payments for the Supply and Use of Intellectual Property', Bureau of Economic Analysis U.S. Department of Commerce, Washington, DC.

Rockett, K. E. (1990). 'Choosing the Competition and Patent Licensing', *RAND Journal of Economics*, 21: 161–71.

Roth, A. E. (2008). 'What Have We Learned from Market Design?', Hahn Lecture *Economic Journal*, 118: 285–310.

Sakakibara, M. (2011). 'An Empirical Analysis of Pricing in Patent Licensing Contracts', *Industrial and Corporate Change*, 19: 927–45.

Scotchmer, S. (1996). 'Protecting Early Innovators: Should Second Generation Products be Patentable?', *Rand Journal of Economics*, 27: 322–31.

Scotchmer, S. (2004). *Innovation and Incentives*. Cambridge, MA: MIT Press.

Serrano, C. (2010) 'The Dynamics of the Transfer and Renewal of Patents', *Rand Journal of Economics*, 41(4): 686–708.

Sheehan, J., Martinez, C., and Guellec, D. (2004). *Understanding Business Patenting and Licensing: Results of a Survey*', Paris: OECD.

Shepard, A. (1987). 'Licensing to Enhance Demand for New Technologies', *Rand Journal of Economics*, 18: 360–68.

Siebert, R., and von Graevenitz, G. (2008). 'Does Licensing Resolve Hold Up in the Patent Thicket?' Discussion Papers in Business Administration, 2104, University of Munich, Munich School of Management.

Silverman, B. S. (1999) 'Technological Resources and the Direction of Corporate Diversification: Toward an Integration of the Resource-based View and Transaction Cost Economics', *Management Science*, 45(8): 1109–24.

Somaya, D., Kim, Y., and Vonortas, N. S. (2011). 'Exclusivity in Licensing Alliances: Using Hostages to Support Technology Commercialization', *Strategic Management Journal*, 32: 159–86.

Stigler, G. (1951). 'The Division of Labor is Limited by the Extent of the Market', *Journal of Political Economy*, 59: 185–93.

Teece, D. J. (1986). 'Profiting from Technological Innovation', *Research Policy*, 15(6): 285–305.

Teece, D. J. (1998). 'Capturing Value from Knowledge Assets: The New Economy, Markets for Know-how, and Intangible Assets', *California Management Review*, 3: 55–79.

Torrisi, S. (1998). *Industrial Organization and Innovation: An International Study of the Software Industry*. Cheltenham: Edward Elgar.

von Hippel, E. (1990). 'Task Partitioning: An Innovation Process Variable', *Research Policy*, 19: 407–18.

Winter, S. (1987). 'Knowledge and Competence as Strategic Assets', in Teece, D. J. (ed.), *The Competitive Challenge: Strategies for Industrial Innovation and Renewal*. Cambridge, MA: Ballinger, 159–184.

WIPO (2011). 'World Intellectual Property Indicators', World Intellectual Property Organization Economics and Statistics Division. Available at<http://www.wipo.int/ipstats> (accessed 21 July 2013).

Ziedonis, R. H. (2004). 'Don't Fence Me In: Fragmented Markets for Technology and the Patent Acquisition Strategies of Firms', *Management Science*, 50: 804–20.

Zuniga, M. P., and Guellec, D. (2009). 'Survey on Patent Licensing: Initial Results from Europe and Japan', STI Working Paper 2009/5. Paris: OECD.

....

CHAPTER 13

CAPITAL MARKETS, INNOVATION SYSTEMS, AND THE FINANCING OF INNOVATION

....

ALAN HUGHES[1]

INTRODUCTION

....

INNOVATION requires the coordination of multiple intra- and inter-organizational relationships including those concerned with access to financial resources internal and external to the business. These relationships drive both technological and organizational innovation as businesses finance the acquisition, re-combination and coordination of the resources required for successful innovation. This chapter focuses on the relationship between business sector innovation and access to financial capital. It does so from an innovation systems perspective. This entails locating business financing within the wider set of resource and coordination patterns that characterize national, regional or sectoral systems of business innovation.

These relationships extend beyond the private sector. They include connections with the substantial stocks and flows of intellectual human and financial capital arising from the activities of the public sector. This includes in particular, access to publicly funded research and to human capital outputs from the higher and secondary educational sectors. There may also be direct or indirect financial support from the state for innovation related expenditures in the private sector.

To the extent that such complementary public investments and support vary across place and sector we may expect different systems of financial relationships to exist. The provision of finance may also be affected by cultural and legal differences in contractual and non-contractual relationships and patterns of corporate governance. It might be expected, therefore, to find differentiated and clustered national configurations of innovative activity and financing. These will be reinforced at regional or local levels if knowledge flows themselves are relatively sticky and dependent upon interpersonal interactions. These possibilities have been explored extensively in the national systems and varieties of capitalism approaches to innovation and finance.

There is at the same time a literature which has focused on more generic aspects of the distinctive financing relationships associated with innovation. This is associated with a view that competition between systems based on different patterns of finance for innovation will lead to a convergence of institutional configurations between nations. Whilst recognizing that innovation requires multiple inputs it typically focuses in its empirical aspects upon the funding of R&D as a key innovation input.

The chapter begins with an outline of the generic market failure approach to the analysis of finance and innovation and relevant empirical findings. It then focuses on international differences and an assessment of the insights arising from empirical research based inter alia on the varieties of capitalism and innovation systems approaches. A final section concludes.

Financing Innovation: Generic Market Failure Issues

The argument that financing innovation activity poses particular constraints (and consequently characteristic difficulties in accessing finance and variations in capital structure across innovative and non-innovative firms) is usually linked to a number of specific characteristics of the innovation process. These arguments usually note at the outset that a very wide range of expenditures is required to support innovation. These range from discovery and invention through to final product or process innovation introduction. The argument is, however, often made that R&D is the most significant component of these and then proceeds to focus on the particular characteristics of R&D expenditure (see, for example, Hall, 2010) and a specific set of characteristics of R&D which pose particular financing issues.

The first of these is the domination of R&D costs by the staff costs associated with highly skilled and qualified labour input. R&D effort is thus seen as essentially concerned with intangible assets linked to human capital. From the point of view of the firm employing this human capital, the value is embedded in employees and will be lost if they leave the firm or are laid off. Even though some sectors where tacit knowledge may become rapidly obsolescent have increased in importance (e.g. the ICT sector) there is evidence that the overall variance of R&D investment growth rates is between a quarter and one-fifth of that for other investment in the US (Hall, 2005, 2010). Firms thus seek to smooth R&D spend and require a higher rate of return to cover the costs of avoiding adjustment. Secondly, the returns from R&D expenditures are highly skewed and have distributions with very wide or undefined variances (e.g. Scherer, 1999). This may be the case in network industries in particular where significant first mover advantages exist and also in those sectors more prone to radical innovation. Finally, to the extent that the tacit knowledge embodied in the intangible human capital created by R&D is specific to the firm in which it was created, bankruptcy will not be readily associated with the ability to salvage value from the business. The implication is that lenders to high R&D -intensive businesses will find it difficult to secure contracts in relation to the value of the underlying assets. This will limit access to the use of debt or loans. R&D-intensive firms may thus be less highly leveraged than other firms.

These particular characteristics of innovation expenditures (viewed through the lens of R&D investment) exacerbate more general characteristics of financial markets which make them prone to market failures. These more general effects are usually summarized under the headings of asymmetric information and moral hazard.[2] The uncertainty posited in relation to R&D investment can be coupled with the argument that potential innovators may have more information and be better informed about the likelihood of success than potential investors. Attempts to price for risk lead bad projects to drive out good projects. In the case of innovation, firms that have new and challenging ideas are likely to be particularly unwilling to share that information in the market place. Public knowledge about the level of R&D expenditure *per se* may help, but will not reduce the uncertainty in relation to outcomes.

An additional argument is that the separation of ownership and management may lead the latter (who wish to preserve their jobs) to be relatively risk averse in their investment strategies. This will therefore bias their use of internal cash flows. They will favour investments which offer greater likelihood of job security. To the extent that long-term R&D investments imply higher risks of failure this will conflict with their need for security and will be avoided.[3]

The financing of innovation has also been approached from a management strategy rather than a market failure perspective. Work in this vein is well summarized in O'Brien (2003), who extends the notion of the importance of investment smoothing in relation to R&D to integrate behavioural models of business strategy. These are linked to the notion of financial slack. He argues that firms which are relatively R&D intensive and are thus assumed to be basing their competitive strategy on innovation will maintain high levels of financial slack and have relatively low leverage ratios (see also Vincente-Lorente (2001).

The specific characteristics of R&D investment will thus lead to a relatively heavy reliance on internal funds to support smoothed R&D investment activity compared to equity and debt. Limited access to financial markets will therefore lead R&D-intensive firms to be more constrained (given internal cash flow) than other firms.

There is a large literature that develops empirical estimates of the nature, extent, and form of financial constraints facing R&D-intensive firms. This literature may be broadly divided into two types. The first estimates investment financing equations. Tests are then carried out to see if the impact of cash flow variations on investment differs across firms classified in terms of their R&D intensity. The second broad type addresses the question of whether it is possible to identify differences in the financing characteristics of R&D-intensive firms compared to other types of business. This has focused on whether there are differences in capital structure between R&D-intensive and other firms in terms of equity, or debt and in the sources of finance used to fund growth.

A particular sub-set of the literature has focused on the financing of early stage knowledge intensive or innovative start-up firms. Here attention has focused, in particular, on the role of venture capital. This form of funding is hypothesized to be a particular response to the need for investors to reduce information asymmetries and uncertainty about outcomes. There are distinctive features of venture capital finance in these contexts. It combines extensive pre-investment analysis with the input of management expertise to permit careful monitoring of performance. New management and particular CEOs can be appointed when appropriate. Specialized funding instruments (e.g. convertible debt) permit the adoption of a real options approach to financing investment. Successive tranches of financing are conditional on progress in meeting specified performance objectives and may be combined with options to convert loans to equity.

Each of these literatures has been extensively reviewed (see, for example, Hall and Lerner, 2010; Hall, 2010). Hall and Lerner (2010) conclude that there is some evidence that smaller firms are more likely to face information asymmetries and uncertainties than large established firms. In the case of start-ups in R&D-intensive industries this may be a particularly acute problem. They point to the existence of the venture capital industry as an indication of the nature and extent of the problem. They also highlight, however, that the role of venture capital as an effective solution to financing problems in this area may be limited for example in terms of its capacity to focus on a few individual sectors. Its health may also depend critically on exit routes through buoyant equity markets either focused particularly on this type of firm or in terms of overall main stock market performance. This type of funding may therefore be particularly sensitive to overall trends in sentiment and stock market movements. It has also often focused on later stage than earlier stage investments (see e.g. Lerner, 2009).

One of the features of this literature, in particular as it has expanded to cover different countries, is the extent to which the relationships between financing patterns and innovative activity vary across countries rather than across firms within a country. Market failure problems may thus be conditional on the wider innovation system characteristics within which markets are embedded.

National Financial Systems and Innovation Finance

One well-known stream of literature to have focused on issues of governance and coordination in the relationship between financing and innovation is the varieties of capitalism approach (e.g. Hall and Soskice, 2001a). This categorizes national political economies on the basis of the way in which firms resolve coordination problems. These problems arise in the spheres of industrial relations, vocational training and education, corporate governance, customer and supplier inter-firm relations and, finally, internal employee coordination. Financial aspects of innovation emerge most closely in the analysis of corporate governance. This is seen as having a critical impact on the nature of finance sought; the way in which investors and the suppliers of finance interact; and the way the latter seek to monitor and assure returns on their investments (see, for example, Hall and Soskice, 2001b).

A core distinction in the varieties of capitalism literature is between the ideal types of 'liberal market economies' and 'coordinated market economies'. In the former coordination activities are primarily by a combination of competitive markets and inter-firm hierarchies. Market relationships are arms' length and set in a competitive and formal contracting framework. In coordinated market economies, non-market relationships are more important as coordinating devices. This implies much more inter-organizational relational activities, and less complete contracting. Monitoring is based not upon market signals, but on the exchange of insider information of various kinds. In liberal market economics, equilibrium outcomes in firm behaviour are moderated by adjustments to market prices. Strategic interaction amongst firms and coordinated outcomes are seen to be the key determinants of movements towards stable outcomes in coordinated market economies. In coordinated systems, particular sets of organizations and institutions (rules of conduct, norms of behaviour) are focused on reducing the uncertainty associated with the behaviour of others so that mutual credible commitments can be made. The institutional rules of behaviour include substantial exchange of information, behavioural monitoring, and sanctions for defectors from corporate behaviour. This implies strong networks across employers and labour organizations. In relation to financing this means, in particular, the development of patterns of cross-firm shareholdings and close relationships between banks and the businesses they fund.

Proponents of the varieties of capitalism hypothesis contend that there will be systematic differences in corporate strategy, including innovation behaviour, between varieties of capitalism. These are based on differences in the overall institutional framework within which those firms operate. A further point which emerges from this approach is that there are important complementarities between institutions in different parts of the economy. In a financial system in a liberal market based economy, the responsiveness of financing to short-term movements in profitability will not work well with a labour market in which firms seek to maintain long-term employment contracts. The latter would prejudice the ability of a firm to make short-term flexible reallocations or reductions of its labour inputs. In assessing the extent to which different forms of finance and different types of financial coordination are effective in inducing differences in innovation performance, it is essential, therefore to consider simultaneously the nature of coordination in labour markets. Empirically this leads to the view that economies should cluster into broad groups. Those in which the employment and financing spheres are relatively highly dominated by market transactions on the one hand and those where direct coordinated activities dominate on the other.

In relation to the financial system (and the closely related way in which corporate governance institutions work), it is argued that access to long-term 'patient capital' is complementary to labour market coordination based on the long-term retention of a skilled workforce and to investment in generating long-term returns. Information considered private, or insider information in a liberal market based system must be available in a coordinated market system to those whose investments in the business are expected to lead to long-term gains. The result is highly networked activities within the corporation and between firms. It is also argued that this implies less scope for unilateral decision-making by top management in organizations in coordinated market economies than in liberal market economies.

Finally, there is the argument that coordinated market economies will be better suited to supporting incremental innovation. In this case continuous, but small improvements are made to what are relatively stable slowly changing sets of products and processes. In liberal market economies, on the other hand, the capacity for rapid top executive policy change and flexibility in the re-allocation of human and other capital means that they should be better at supporting radical innovation in sectors where there are rapid and discontinuous changes in technology (see, for example, Hall and Soskice, 2001a).[4]

Lazonick has, however, argued that financial commitment of a patient kind which is coordinated with other elements in the corporation is essential for the support of a productive innovation process in a stock market-based system. In his view an appropriate framework for analysing the function of the stock market must be broken down into the analysis of five sub-functions, namely the creation, control, and combination of assets, patterns of compensation and the role of cash and the implication of these for high technology industries in particular. In a series of contributions he has argued that the way these functions operate may vary significantly both over time, in a particular national system and within corporations in different sectors. His analysis in particular points to the view that the US stock market in recent decades has been over-focused on cash and compensation in the pursuit of managers' self-interests. This has been at the expense of the development of a framework of financing and governance capable of supporting long-term investment in high-risk innovative environments (see, for example, Lazonick, 2007, 2009).

This distinction between liberal market economies and coordinated market economies is readily linked conceptually to analyses of financial markets. Here a distinction between bank-based and stock market-based systems is typically drawn (Rajan and Zingales, 1995; Allen and Gale, 1999). In this literature arms'-length relationships in the stock market-based systems are contrasted with coordinated, long-term relationship banking in the bank-based systems. In the bank-based systems a significant role is played by banks as key intermediaries in channelling household savings to the business sector. They are also seen as playing a significant role in equity markets as holders of large blocks of stock in industrial companies. This distinction in turn relates to another approach to the analysis of corporate stock holding. This approach emphasizes the distinction between 'outsider' and 'insider' patterns of corporate control and govern-ance. In 'outsider', stock market-based systems dispersed shareholder influence is exer-cised through relative price signals. Impersonal buying and selling of shares in response to good or poor performance alters prices and cost of capital. In extreme cases of bad performance, takeovers are an ultimate sanction for failing firm management. In con-trast, in coordinated systems 'insider' block holdings of shares are common. Influence is exercised directly rather than by indirect price signals and transfers of ownership on an open market. In the insider/outsider dichotomy, the block holding insiders can include financial *and* non-financial businesses. In addition, non-shareholding stake holders, such as the labour force and labour unions, may be included in corporate influence on decision-making through their involvement in particular models of corporate govern-ance, including, for example, in the German case two tier boards. This state may also play a coordinating role in shareholdings as part of wider patterns of industrial or eco-nomic development strategy (see, for example, Zysman, 1983).

Although these insider/outsider classifications have strong complementarities with the liberal market and coordinated market models, they emphasize different compo-nents of the system. These components may move in different directions at least in principle. It is, thus, possible for bank intermediation to decrease in importance and for bank shareholdings to decline too whilst other insider block holding relationships could increase or remain the same and vice versa (see, for example, Deeg, 2009). Equally, the way in which the institutions in countries placed within these broad typologies may operate their financial systems, may also be affected by the way in which the overall legal systems within which they operate have developed.

A body of literature based on the quantitative analysis of variations in legal systems across countries has also developed. This has, in particular, examined the link between the 'efficiency' of the legal framework within which the financial governance and insol-vency systems of countries and their overall economic performance operate (La Porta et al., 1998, 2008). In its original form this approach too has aspects which echo the coor-dinated market and liberal market typologies of the varieties of capitalism approach. Here, however, the contrast is made between English-law origin economies (e.g. UK, Commonwealth, USA) and Civil-Law economies (typically East Asia and most of mainland Europe). It is argued that the English-law origin economies have developed greater contract and property rights protection than the Civil-law origin states. The for-mer, as a result, have a 'comparative competitive advantage' in the development of their financial markets. They may be expected to be better attuned than civil-law systems to deliver financial flows on the scale and in the form required for the efficient allocation of resources between alternative uses and for the overall innovative performance and rate of growth of companies and the economy as a whole. Recent contributions using richer metrics have cast doubt on this. They emphasize institutional complementarity so that the 'efficiency' of financial markets must be seen in the context of legal regulation else-where (e.g. labour markets) (Ahlering and Deakin, 2007; Acharya et al., 2010a, 2010b).

Deakin and Mina (2012) provide a useful review of the upshot for finance and innova-tion which is summarized in Table 13.1.

Table 13.1 Complementarities between corporate governance and modes of innovation

	Shareholder protection	Creditor protection	Worker protection	Mode of innovation
Liberal market systems	High (legal support for hostile takeover bids, share buy-backs shareholder activism)	Medium or weak (debtor in possession laws, laws favouring corporate rescue over liquidation)	Weak (minimal legal support for employment protection, no co-determination)	• Strong venture capital market • 'Schumpeterian' creative destruction regime • Higher-risk investment • High incidence of radical innovation • Efficient labour market matching
Coordinated market systems	Weak (minimal legal support for market for corporate control, limited minority shareholder rights)	Medium or strong (legal recognition of priority for secured creditor's rights)	Strong (effective legal support for employment protection and co-determination)	• Limited use of venture capital • Slower creative destruction dynamics • Investment risk more spread • Incremental technology development • Continuous employee learning

Source: Deakin and Mina, 2012.

In summary, taken as a whole, the literature suggest that the analysis of the financing of innovation requires a holistic approach. In particular, it requires consideration of the institutional complementarity between labour markets and financial markets and an assessment, not only of patterns of financial intermediation in the economy, but also the relationship between patterns of shareholding and the overall nature and sources of financial flows available to firms. Moreover, given the extent to which large corporations make extensive use of capital markets, it also leads to an important distinction between the financing patterns of firms both at different stages of development at different scales of innovation activity and in the context of specific innovation systems.

INNOVATION SYSTEMS AND FINANCE OF INNOVATION

Dosi (1990) argues that innovative environments are characterized by learning and selection as twin core system elements. Differences in the way that firms are organized and perform can be directly related to different underlying patterns and types of learning and selection. This means that the influence of different patterns of financial structures and institutions on innovation is via the way in which they influence firm learning and selection patterns across firms and technologies. Dosi builds on the work of Zysman (1983) and Hirschman (1970) to contrast the systems. The first consists of economies based on "institutionalized ownership, control relationships and the exercise of voice" as ways of influencing the direction and scale of the supply of finance and the monitoring of performance with an emphasis on bank finance and loan contracts. The second system is based on use of "impersonal ownership control relationships" relying on entry and exit mechanisms of selection through trading on financial markets in systems typically reliant on stock and bond market financing.

This leads to the argument that coordinated or credit-based systems will place more importance on learning as opposed to selection. In addition, reliance on credit-based and coordinated-based systems of learning may lead to exclusion effects, so that particular competences become the full focus of attention and a trade-off between the width of the information and the depth of processes and competences arises. The emphasis will be on relatively well-established technological trajectories whereas in market-based systems there is a greater likelihood for the exploration of new technological trajectories.

Neither Dosi nor the varieties of capitalism literature argue that one system or the other will necessarily produce higher rates of innovation taken as a whole nor that one will progress faster in relation to the conversion of inventive ideas into innovative outcomes and productivity growth. The particular way in which any credit- or stock market-based system operates depends on its own internal institutional characteristics and the thickness and effectiveness of market and non-market connections between agents in the system.

Guerrieri and Tylecote (1997) and Tylecote (2007) extend the argument to sectoral systems of innovation (see Chapter 10 by Malerba and Adams). They list the financial system amongst their external requirements for technological competitive advantage and innovation performance. This sits alongside technically trained manpower and estimates of the quality of the science base. They make a distinction between investments which are visible to external sources of finance, for example in capital equipment and buildings, followed by increasingly less visible expenditures ranging from research and development through training, marketing, design and servicing. Industries with heavy visible investment requirements in their innovation systems will be more appropriate for outsider dominated systems. Shareholders mostly operating on the basis of public visible information, with diversified portfolios and relatively limited commitments to individual firms will be the primary source of financing for innovation.

On the other hand, increasing emphasis on lower visibility investments will be more appropriate for inside dominated systems linked to coordinated market economies. Here managers and large shareholders have stable relationships with each other and with banks. This then allows a cross classification of countries in terms of the relative emphasis of their systems on firm specific perceptiveness as opposed to industry-specific expertise. The former is characteristic of coordinated or credit-based systems and the latter is characteristic of stock market-oriented systems.

They apply their analysis to Germany, Japan, France, Italy, and Switzerland alongside Sweden, the USA, and the UK. Their qualitative assessment of the data leads them to find broad patterns of sectoral specializations which are consistent with their national system characteristics with the exception of the relatively small economies of Sweden and Switzerland. In those cases they argue that outcomes depend on specialized decision-making by relatively few organizations. Guerreri and Tylecote also note that the reliance on stock exchange- or outsider-dominated systems, which provide relatively low levels of support for low visibility activities, may offer substantial advantages in relation to the identification and exploitation of novelty. Here insider systems may be weakest unless venture capital market is developed which can provide both inside expertise and finance alongside appropriate governance structures in newly emerging technology areas.

R&D Expenditure, Finance for R&D, Share Ownership and the Structure of Capital markets: Empirical Evidence[5]

In this section a brief overview is provided of the main empirical dimensions along which countries vary in their R&D expenditures, the broad pattern of financing of those expenditures and the nature of their financial systems. This is an essential backdrop against which empirical analyses of the link between finance and innovation can be assessed. This section draws upon data for countries referenced most frequently in the literature, including the US, Scandinavian countries, UK, Japan, Korea, Germany, and France. (The following sections draw extensively on OECD data discussed more fully in Hughes and Mina, 2012.)

The United States and Japan dominate gross expenditure on research and development (GERD). The most R&D-intensive economies (GERD/GDP) are, however, Finland, Korea, Sweden, and Japan. These are also the economies that are at the top of the list in business expenditure as a percentage of GDP (BERD). Business expenditure on R&D is the primary component of overall R&D on expenditure in all these economies. The way that it is funded is clearly therefore a central feature of concern for the innovative performance of the economies concerned.

Expenditure on R&D in the higher education sector and the government sector is also substantial in each economy and in some cases, such as Germany, is equivalent to around half of the overall business expenditure effort and in the case of France is even higher in relative terms. The same is true for the UK and Denmark. The overall innovative impact of BERD irrespective of how it is financed may also depend on these typically publicly funded expenditures.

Investments in human capital, brand equity, and especially fixed capital equipment outstrip expenditure on R&D by a substantial margin in these countries. Insofar as the innovation process as a whole is critically dependent on complementary investments in addition to R&D per se, then a wider perspective than the financing of business R&D alone is important in understanding the relationship between finance and innovation. It is necessary to consider the way in which innovation as a whole is financed and the inter-relationship between R&D and other innovation system components.

Countries differ significantly in the share of their GDP which is accounted for by the service sector. They also vary significantly in the extent to which their economic activity in manufacturing may be classified as high, medium or low technology. The first of these factors may affect the overall R&D intensity of an economy and is one of the reasons why economies, such as the US and the UK, which have relatively large service sectors, may have a downward bias in their ratio of R&D to GDP.

OECD data is broadly consistent with the varieties of capitalism approach in that the US and the UK have R&D relatively concentrated in high technology sectors whilst Germany is relatively concentrated in the medium to high technology sectors. Similarly, Japan, as an example of a coordinated market economy, has a somewhat higher share of its manufacturing R&D in the medium high technology sectors than in the high technology sectors, although the differences are much smaller than in the case of Germany. Korea, however, is closer to the UK and France than it is to either Germany or Japan. Finland is the most high technology-intensive economy amongst these nations.

There are major and significant variations across countries in terms of the way in which business expenditure on R&D is funded and in the way in which gross expenditure on R&D as a whole is financed. In most countries the financing of GERD as a whole is relatively dominated by domestic funding sources. In Japan and Korea, hardly any overseas funding supports domestic GERD. On the other hand, the UK is an extreme example of a country dependent upon external financing to fund its domestic R&D. Sweden also has a relatively high overseas funding influence. Government funding also varies substantially across countries. It is striking that the percentage of BERD that is financed by government is highest in the US given the liberal free market credentials typically attributed to that country. The role of the public sector as a source of venture capital in the US has been typically underplayed in interpretations of that country's innovation performance. The role of major departments of state, in particular through programmes such as the Small Business Innovation Research Programme (SBIR), have played a major role in the direct development of early stage technologies and helped to de-risk investments by later stage private sector venture capitalists (see, for example, Lerner, 1998; Connell, 2006; and Hughes, 2008). There are therefore significant differences across countries in their innovative inputs and in the extent to which the public sector funds and influences private sector R&D.

Table 13.2 reveals a number of differences in the relative importance of stock markets, bank leverage and venture capital.[6]

Differences between the UK and the US on the one hand and Germany and Japan on the other are again apparent. Thus, the two 'Anglo-Saxon' economies have substantially

Table 13.2 Financing R&D: stock markets, bond markets, leverage, and venture capital

Country	Stock Market Capitalization/GDP		Stock Market Total Value Traded/GDP		Private bond market capitalization/GDP		Listed Companies Median Leverage Ratios		Venture Capital% GDP			
	Average (2001–2010)	Rank	Average (2001–2010)	Rank	Average (2001–2010)	Rank	Average (1991–2006)	Rank	Average (2000–2003)	Rank	2008	Rank
Finland	106.5	3	129.5	4	23.8	9	0.3	4	0.20	5	0.23	2
Korea	68.4	7	143.2	3	57.7	3	0.5	1	0.27	2	0.07	8
Sweden	102.4	4	125.5	5	45.5	4	0.2	7	0.24	3	0.21	3
Japan	77.8	6	88.8	6	41.8	6	0.3	3	0.03	10	0.01	10
Denmark	62.3	8	50.3	10	140.9	1	NA	NA	0.13	6	0.30	1
Germany	46.4	10	65.5	8	39.6	7	0.2	7	0.10	9	0.05	9
US	124.4	2	259.3	1	107.5	2	0.2	9	0.38	1	0.12	6
France	81.2	5	81.8	7	44.0	5	0.3	5	0.11	8	0.09	7
UK	127.5	1	175.7	2	16.0	10	0.2	8	0.22	4	0.21	3
Norway	52.9	9	60.1	9	25.2	8	0.4	2	0.12	7	0.13	5

Source: Calculated from World Bank Global Financial Development Database (GFDD), The World Bank (Cols 1–3), Fan et al., 2010 (Col 4) and OECD Science and Technology Indicators (Col 5).

higher stock market capitalizations and stock market turnover relative to GDP compared to Japan and to Germany in particular. Germany and Japan are also much less reliant on private bond market activity than the US. In this case, however, the UK differs significantly from the United States. It has one of the lowest ratios of bond market capitalization to GDP of the sample of countries as a whole. Levels of leverage in the UK and the US are relatively low, especially compared to Japan and the Nordic countries. There is not, however, a particularly important difference between these two countries and Germany.

Finally, the table shows venture capital as a percentage of GDP. The largest market for venture capital in absolute terms is to be found in the US, followed some distance behind by the UK. This form of finance is exceptionally sensitive to the state of the stock market. The final two columns therefore show VC funding as a percentage of GDP pre- and post-the global financial crisis. Prior to the crash the US was indeed the most VC-intensive country consistent with its stock market orientation. It lost the position, however, after the crash. The Nordic economies were strikingly able to maintain their VC intensity as did the UK (though this reflects a collapse in both the numerator and the denominator). Japan and Germany have low VC intensity consistent with their bank-dominated financial systems.

Overall, these broad indicators suggest that there are significant, but complex variations across countries which do not always correspond to simple two way ideal type divisions. This suggests that analyses of the financing of innovation need to be rooted in detailed contextual approaches of a country's overall innovation and economic system.

In an influential series of studies comparing international financing patterns in the 1970s and 80s it was argued that firms overwhelmingly rely on internal financial sources in a wide range of countries including the UK, the US, Germany, and Japan and that clear cut bank versus stockmarket classification systems were not supported by the evidence (see, for example, Mayer, 1988; Edwards and Fischer, 1994).

In a study focusing on balance sheet structures using recent harmonized national accounts data and covering later years Byrne and Davis (2002) show, however, that significant differences in financial market structures persist between, for example, the UK and Germany. These differences are broadly consistent with the latter falling into the coordinated insider system and the UK into the more dispersed shareholder liberal market economy system. The UK household sector held 52 per cent of its assets in the form of equities compared to only 27 per cent in the case of Germany. Bank deposits accounted for 45 per cent of such holdings in Germany and only 25 per cent of assets of the household sector in the UK. Similarly, in the case of Germany loans amounted to 42.8 per cent of the company sectors' liabilities in 2000 compared to 22.5 per cent in the case of the UK. By contrast, 70 per cent of the UK company sectors' liabilities took the form of equity compared to only 55 per cent in Germany in 2000. Whilst this difference is relatively clear cut, the differences between the UK, France, and Italy in terms of these patterns is much less clear with the UK much more like the latter two than Germany (see also Deeg, 2009 and Schmidt et al., 2002).

Governance systems and interpretations of insider/outsider models also emphasize the dispersion of shareholdings. Financial institutional holdings, for example, are much more important in the UK, the US (especially the largest firms), Norway, and Sweden than is the case in Denmark, Finland, or Germany and, especially, in Japan and South Korea. A particularly significant feature of mainland European systems compared to the UK and the US has been the relative persistence in several countries of blocks of large non-financial cross shareholdings with concomitant implications for coordinated strategies to the financing of innovation (See e.g. Gordon and Roe, 2004; Gugler et al., 2004).

Recent trends in the globalization of shareownership may destabilize coordinated national block holding systems. In general, the internationalization of stock markets in the past two decades has been reflected in an increase in foreign shareholdings, particularly in the case of the UK, Germany, and France. There have been notable changes in the role of the public sector and foreign sectors and overseas short-term focused hedge fund activism in France in particular (see, for example, the discussion in O'Sullivan, 2003; Goyer, 2011; and Culpepper, 2005).

One of the most striking developments in the internationalization of equity investment has been the growth of sovereign wealth funds. The increasing significance of sovereign wealth funds, in particular those based on the Middle East economies and in China, has led to major changes in equity holding patterns in Europe and the USA. In the case of the USA, this has led to pressure to protect the liberal market-based US economy from what is interpreted as a mercantilist intrusion by state-owned or strategically focused sovereign wealth fund investment activity (see, for example, Pistor, 2009, and Gilson and Milhaupt, 2008).

These developments reinforce the need to take a dynamic view of changes of financial systems. They have been used to reinforce the view that globalization of capital flows and intensified international trade competition will lead to a convergence in financial structures for innovation over time. Structural differences clearly remain. Moreover, although broad distinctions remain especially in relation to bank funding and patterns of block holding of shares it is apparent that there are multiple varieties of capitalism within and between the two broad ideal types. These differences suggest that differences in financing innovation and constraints therefore may be expected to occur across countries and the firms located within them. Businesses fund R&D and innovation in distinctive national contexts. These patterns may, however, change as the underlying structures of financial and other markets evolve. They are likely to be particularly affected by the differential impact of the global financial crisis on different national systems and which have yet to be fully worked out.

Do National Financial Systems matter for the Financing of Innovation?

Despite the wide interest in the subject there have been relatively few comparative international attempts to relate measures of innovation performance to financial structures at a national level. Jaumotte and Pain (2005) explore the impact of patterns of financial development on R&D and patenting activity in a large sample of OECD economies. They include measures of internal financing available to companies, a measure of stock market capitalization plus credit provided by domestic financial institutions as a measure of total financial development as well as the proportion of the latter accounted for by stock market capitalization alone. They conclude that it is difficult to disentangle the separate influences of these three factors. All are significantly positively related to R&D intensity at the national level. They find a significantly negative coefficient on the interaction between profit share and aggregate financial development suggesting that when profits are high, there is less requirement to access stock market and other sources. They do not assess venture capital impact because of a lack of data. Furman et al. (2002) include access to venture capital as a potential determinant of the innovative capacity of a large sample of OECD proxied by patenting activity. They find that venture capital has no impact. Finally, Fu and Yang (2009) find a significantly *negative* effect of venture capital on the efficiency with which R&D and other inputs are translated into patenting activity in a cross-section of twenty-one OECD countries in the period 1990 to 2002.

In summary, it appears that although financing innovation poses major challenges market and that coordinated based systems address them in different ways this is not associated with significant differences in innovation performance across countries.

A number of studies have explicitly focused on micro-analyses of cross country financing effects at the corporate level. They cover both VC activity and IPOs as well as corporate and industry R&D activity.

Black and Gilson (1998) emphasize the interdependence between the scale of venture capital markets and the possibility of exit through initial public offerings on stock markets. They conclude that the US has become a world leader in biotechnology and computer-related high technology in which venture capital market funding is most predominant in the US. They also note, in keeping with the variety of capitalism literature, that labour market institutions that hinder the breaking of labour contracts and 'flexible hiring and firing' also inhibit venture capital activities in, for example, Germany and Japan. In contrast, Manigart et al. (2002) find that venture capital firms require higher not lower rates of return in the UK and the US compared to France, Belgium, and the Netherlands. Moreover, Mayer et al. (2005) analyse qualitative data for 500 venture capital funds in the UK, Israel, and Germany in 2000 and for Japan in 1999. Their results are not consistent with simple market- versus bank-based analyses. They differ in emphases from those of Black and Gilson (1998) and Allen and Gale (1999) in showing for example that the UK and Germany are more alike than the UK and Japan. Kim and Weisbach (2007) examine the motivation for initial public offerings and seasoned equity offerings from 38 countries in the period 1990 to 2003. They find that that legal origin impacts on the relationship between equity offerings and subsequent R&D expenditures.

Studies of corporate and industrial R&D are summarized in Table 13.3.

Taken together the studies reviewed do not suggest a superior performance in funding R&D and innovation for liberal market economies or stock market-based systems of financing. If anything the reverse is true with systems based on more coordinated bank

Table 13.3 Cross-country studies of corporate effects on R&D

Authors	Focus of study	Sample	Illustrative findings
Bhagat and Welch (1995)	Determinants of corporate R&D.	1,484 large companies in US, Canada, UK, Europe and Japan, 1985–90	Few differences between firms in the influence of debt, stock returns, cash flow or tax liabilities, but significant differences in Japanese companies where high leverage is positively linked to R&D.
Bah and Dumontier (2001)	Corporate policy choices of firms that spend a high proportion of their net revenue on R&D.	900 R&D intensive and non R&D intensive firms from the US, the UK, Japan, and Europe, 1996	European, UK and US similar in reliance on short-term debt financing which is higher in the case of R&D-intensive than non-R&D intensive firms. This is not the case in Japan, which has longer-term debt.
Bond, Harhoff, and Van Reenen (2003)	Relationship between cash flow, investment in fixed capital and R&D.	900 German and UK manufacturing firms, 1985 and 1994	Cash flow is positively related to investment in R&D-intensive firms in the UK, but not in Germany
Carlin and Mayer (2003)	Comparative growth R&D and investment characteristics of sectors in different financial systems	27 industries in 14 OECD countries, 1970–95	Concentrated, rather than dispersed, ownership is associated with faster growth of equity and with skill dependent industries and with higher R&D shares in industries dependent on equity. Equity dependent industries have lower R&D shares in countries with highly concentrated banking systems
Hall, Mairesse, Brandstetter, and Crepon (1999)	Relationship between cash flow and investment and R&D	204 US firms, 156 French firms and 221 Japanese firms, 1978–89	Cash flow has high predictive power in relation to R&D expenditures in the US but not in France and Japan.
Honoré, Munari, and van Pottelsberghe de la Potterie (2011)	Relationship between governance ratings and R&D intensity	279 R&D active European firms, 2003–7	Governance practices responding to short-term financial market expectations are detrimental to long-term R&D investments.
Belloc (2013)	Shareholder protection and R&D investment	48 countries, 1993–2006	Strong shareholder protection will weaken rather than encourage R&D investments.

(Continued)

Table 13.3 (*Continued*)

Authors	Focus of study	Sample	Illustrative findings
Munari et al. (2010)	Owner identity and governance and R&D	1,000 publicly quoted companies in France, Germany, Italy, Norway, Sweden, and the UK, mid-1990s	Widely held businesses tend to have higher R&D activity than more tightly held, and in particular family held, businesses. This positive impact is much weaker in the UK than in other European countries due to absence in the widely held group in the UK of large block shareholders to act as a buffer against short-term performance pressures.
Lee and O'Neill (2003)	Ownership structures and R&D investment	US and Japanese publicly traded companies in seven industries, 1995	Japanese firms are more characterized by norms of stewardship in relation to the management of their businesses and have more patient long-term attitudes to risky R&D.
Miozzo and Dewick (2002)	Relationship between corporate governance and innovation	Interviews with major contractors in the construction industry in Denmark, Sweden, Germany, France, and the UK	National differences in share ownership and pressures to pay dividends affects investments in new technologies adversely esp. in the UK

based, insider finance, and control producing superior performance at company level. More interesting is perhaps the variety that exists *within* the ideal types rather than differences between them.

Convergence and the Finance of Innovation: Concluding Thoughts

The idea that convergence in financial and governance systems across capitalist economies was inevitable has been widely canvassed (see, for example, Baumol, 2002, and Hansmann and Kraakman, 2004). The evidence reviewed suggests that this convergence and the triumph of a particular system of stock market financial relationships and governance is exaggerated. Significant differences remain between financial systems. Whilst differences persist they do not lead to simple characterizations in terms of ideal types of varieties of capitalism, but they do suggest significant differences between nation states. The global financial crisis and its origins in the US system have led to renewed interest in systems more focused on long-term commitment to the real economy.

Differences between national systems therefore remain significant in relation to policy debates about the future structure of industrial economies. Economies, such as the UK and the US, which are seeking to rebalance their economies away from the services sector, face major challenges in the financing of long-term R&D. The evidence reviewed here suggests that businesses with access to more coordinated patient capital structures are more productive in investment in R&D and innovation. At the same time the global financial crisis suggests has that the predicted hegemony of the ideal type liberal stock market based system was premature.

Even if market and/or socio-political forces for convergence to a liberal stock market system re-emerge, the evidence suggests that there will be significant obstacles to be overcome in imposing a one-size-fits-all market-driven solution. Aggressive attempts to impose shareholder activism through, for example, hedge fund activity in the case of Japan led to a reassertion of the benefits of firm-centric governance structures. This has re-emphasized in Japan the importance of the firm and its long-term performance rather than the short-term financial needs of particular groups of equity holder (see, for example, Buchanan et al., 2012) and the role of bank finance in restructuring Japanese firms (see e.g Ahmadjian, 2007; Arikawa and Miyajima, 2007; Jackson and Miyajima, 2007). More generally, analyses which focus on issues on complementarity between institutional forms in different components of the economic system have emphasized that change in one dimension will be ineffective unless combined with, or are congruent with, changes in other sectors. Thus, for example, the introduction of shareholder norms of behaviour associated with dispersed stock market systems may be ill-suited to development inside systems emphasizing more coordinated forms of labour market process. They will sit uneasily alongside governance structures which embed stakeholder representation and participation. The process by which new or changing norms of behaviour associated with shareholder maximization may infiltrate previously coordinated or insider systems will also be diverse. They will depend on the role played by groups with varying elements of power, both in the corporate governance system and in the political system more generally (see, for example, the discussions in Gordon and Roe, 2004; Amable, 2009; and Dore, 2000; Aoki, 2010). To the extent that these differences persist and influence the financing of R&D and innovation, we may expect differences in innovation behaviour across firms and their national contexts to also persist.

NOTES

1. The author acknowledges financial support from ESRC, TSB, NESTA and BIS through the UK Innovation Research Centre (UK~IRC) and excellent research assistance from Alberto García Mogollón and Robert Hughes.

2. In addition tax arrangements introduced by governments may alter the choice between types of finance for innovation and alter the cost of finance, for example, between external and internal (cash flow) sources or as in the case of R&D tax credits the relative cost of types of investment expenditure.

3. In relation to asymmetric information, the argument that those internal to the firm are better informed than those external to the firm may be reversed in the case of new businesses. Here it may be argued that start-up entrepreneurs may be over-confident and suffer from hubris in relation to the likelihood of success of their activities. In this case the problem for the external market is to exercise effective monitoring and control.

4. See for example Dore, 2000; and for a less pessimistic view for champions of the coordinated market economies Berger and Dore, 1996.

5. International comparisons of financial and governance systems are fraught with empirical and conceptual difficulties. Divergent results can occur both because conceptual categories differ or are very loosely defined.

6. The exhibit focuses on the first decade of the current century. It is therefore affected by the financial 2008/09 crash. A separate calculation for the period 1991–2000, however, revealed almost identical rankings so that the characterization based on the first decade of this century is a relatively stable one.

References

Acharya, V., Baghai-Wadji, R., and Subramanian, K. (2010a). 'Labor Laws and Innovation', *NBER Working Paper* No. 16484.

Acharya, V., Baghai-Wadji, R., and Subramanian, K. (2010b). 'Wrongful Discharge Laws and Innovation', Working Paper, NYU-Stern Business School.

Ahlering, B., and Deakin, S. (2007). 'Labour Regulation, Corporate Governance and Legal Origin: A Case of Institutional Complementarity?', *Law & Society Review*, 41: 865–98.

Ahmadjian, C. (2007). 'Foreign Investors and Corporate Governance in Japan', in Aoki, M., Jackson, G., and Miyajima, H. (eds), *Corporate Governance in Japan: Institutional Change and Organizational Diversity*. New York: Oxford University Press, 125–150.

Akerlof, G. A. (1970). 'The Market for 'Lemons': Quality, Uncertainty, and the Market Mechanism', *Quarterly Journal of Economics*, 84(3): 488–500.

Allen, F., and Gale, D. (1999). *Comparing Financial Systems*. Cambridge, MA: MIT Press.

Amable, B. (2009). *The Diversity of Modern Capitalism*. Oxford: Oxford University Press.

Aoki, M., Jackson, G., and Miyajima, H. (eds) (2007). *Corporate Governance in Japan: International Change and Organizational Diversity*. Oxford: Oxford University Press

Aoki, M. (2010). *Corporations in Evolving Diversity*. Oxford: Oxford University Press.

Arikawa, Y., and Miyajima, H. (2007). 'Relationship Banking in Post-bubble Japan: Co-existence of Oft- and Hard-budget Constraints', in Aoki, M., Jackson, G., and Miyajima, H. (eds) *Corporate Governance in Japan: Institutional Change and Organizational Diversity*, New York: Oxford University Press, 51–78.

Armour, J., Deakin, S., Mollica, V., and Siems, M. (2009). 'Law and Financial Development: What we are Learning from Time-series Evidence', *BYU Law Review*: 1435–1500.

Bah, R., and Dumontier, P. (2001). 'R&D Intensity and Corporate Financial Policy: Some International Evidence', *Journal of Business Finance & Accounting*, 28(5/6): 671–92.

Baumol, W. J. (2002). *The Free-market Innovation Machine: Analysing the Growth Miracle of Capitalism*. Princeton, NJ: Princeton University Press.

Belloc, F. (2013). 'Law, Finance and Innovation: The Dark Side of Shareholder Protection', *Cambridge Journal of Economics* (forthcoming): 1–26.

Berger, S., and Dore, R. (eds) (1996). *National Diversity and Global Capitalism*. Ithaca, NY, and London: Cornell University Press.

Bhagat, S., and Welch, I. (1995). 'Corporate Research & Development Investments: International Comparisons', *Journal of Accounting and Economics*, 19: 443–70.

Black, B. S., and Gilson, R.J. (1998). 'Venture Capital and the Structure of Capital Markets: Banks versus Stock Markets', *Journal of Financial Economics*, 47: 243–77.

Bond, S., Harhoff, D., and Van Reenen, J. (2003). *Investment, R&D and Financial Constraints in Britain and Germany*. Centre for Economic Performance, London School of Economics and Political Science.

Buchanan, J., Chai, D. and Deakin, S. (2012). *Hedge Fund Activism in Japan: The Limits of Shareholder Primacy*. Cambridge: Cambridge University Press.

Byrne, J. P., and Davis, E. P. (2002). 'A Comparison of Balance Sheet Structures in Major EU Countries', *National Institute Economic Review* No. 180: 83–95.

Carlin, W. (2009). 'Ownership, Corporate Governance, Specialisation and Performance: Interpreting Recent Evidence for OECD Countries', in Touffut, J.-P., *Does Company Ownership Matter?* Chelmsford: Edward Elgar.

Carlin, W., and Mayer, C. (2003). 'Finance, Investment, and Growth', *Journal of Financial Economics*, 69: 191–226.

Connell, D. (2006). *'Secrets' of the World's Largest Seed Capital Fund: How the United States Government Uses its Small Business Innovation Research (SBIR) Programme and Procurement Budgets to Support Small Technology Firms*. Cambridge: Centre for Business Research, University of Cambridge.

Culpepper, P. D. (2005). 'Institutional Change in Contemporary Capitalism: Coordinated Financial Systems since 1990', *World Politics*, 57: 173–199.

Deakin, S., and Mina, A. (2012). 'Institutions and Innovation: Is Corporate Governance the Missing Link?', in Pittard, M., Monotti, A. and Duns, J. (eds) *Business Innovation: A Legal Balancing Act—Perspectives from Intellectual Property, Labour and Employment, Competition and Corporate Laws*. Cheltenham: Edward Elgar, 456–82.

Deeg, R. (2009). 'The Rise of Internal Capitalist Diversity? Changing Patterns of Finance and Corporate Governance in Europe', *Economy and Society*, 38(4): 552–79.

Dore, R. (2000). *Stock Market Capitalism: Welfare Capitalism: Japan and Germany versus the Anglo-Saxons*. Oxford: Oxford University Press.

Dosi, G (1990). 'Finance, Innovation and Industrial Change', *Journal of Economic Behavior and Organization*, 13: 299–319.

Edwards, J., and Fischer, K. (1994). *Banks, Finance and Investment in Germany*. Cambridge: Cambridge University Press.

Fan, J. P. H., Titman, S., and Twite, G. (2010). 'An International Comparison of Capital Structure and Debt Maturity Choices', *NBER Working Paper* No. 16445.

Fu, X., and Yang, Q.G. (2009). 'Exploring the Cross-country Gap in Patenting: A Stochastic Frontier Approach', *Research Policy*, 38: 1203–13.

Furman, J. L., Porter, M. E., and Stern, S. (2002). 'The Determinants of National Innovative Capacity', *Research Policy*, 31: 899–933.

Gilson, R. J., and Milhaupt, C. J. (2008). 'Sovereign Wealth Funds and Corporate Governance: A Minimalist Response to the New Mercantilism', *Stanford Law Review*, 60(5): 1345–70.

Gordon, J. N., and Roe, M. J. (eds) (2004). *Convergence and Persistence in Corporate Governance*. Cambridge: Cambridge University Press.

Goyer, M. (2010). 'Corporate Governance', in Morgan, G., Campbell, J. L., Crouch, C., Pedersen, O. K. and Whitley, R. (eds) *The Oxford Handbook of Comparative Institutional Analysis*. Oxford: Oxford University Press.

Goyer, M. (2011). *Contingent Capital: Short-term Investors and the Evolution of Corporate Governance in France and Germany*. Oxford: Oxford University Press.

Guerrieri, P., and Tylecote, A. (1997). 'Inter-industry Differences in Technical Change and National Patterns of Technological Accumulation', in Edquist, C. (ed.) *Systems of Innovation: Technologies, Institutions and Organisations*. London and New York: Routledge, 107–29.

Gugler, K., Mueller, D. C., and Yurtoglu, B. B. (2004). 'Corporate Governance and Globalization', *Oxford Review of Economic Policy*, 20(1): 129–56.

Hall, B. H. (2005). 'Measuring the Returns to R&D: The Depreciation Problem', *Annales d'Economie et de Statistique*, 79/80: 341–381.

Hall, B. H. (2010). 'The Financing of Innovative Firms', *Review of Economics and Institutions*, 1(1): Article 4.

Hall, B. H., and Lerner, J. (2010). 'The Financing of R&D and Innovation', in Hall, B. H. and Rosenberg, N. (eds) *Handbook of the Economics of Innovation*. Elsevier-North Holland.

Hall, B. H., Mairesse, J., Branstetter, L., and Crepon, B. (1999). 'Does Cash Flow Cause Investment and R&D: An Exploration using Panel Data for French, Japanese, and United States Firms in the Scientific Sector', in Audretsch, D. and Thurik, A. R. (eds) *Innovation, Industry Evolution and Employment*. Cambridge: Cambridge University Press, 129–56.

Hall, P. A., and Soskice, D. (2001b). 'An Introduction to Varieties of Capitalism', in Hall, P. A. and Soskice (eds) *Varieties of Capitalism: The Institutional Foundations of Comparative Advantage*. Oxford: Oxford University Press, 1–68.

Hall, P. A., and Soskice, D. (eds) (2001a). *Varieties of Capitalism: The Institutional Foundations of Comparative Advantage*. Oxford: Oxford University Press.

Hansmann, H., and Kraakman, R. (2004). 'The End of History for Corporate Law', in Gordon, J. N. and Roe, M. J. (eds) *Convergence and Persistence in Corporate Governance*. Cambridge: Cambridge University Press, 33–68.

Hirschman, A. (1970). *Exit, Voice and Loyalty*. Cambridge, MA: Harvard University Press.

Honoré, F, Munari, F., and van Pottelsberghe de la Potterie, B. (2011). 'Corporate Governance Practices and Companies' R&D Orientation: Evidence from European Countries', *Bruegel Working Paper* 2011/01, January.

Hughes, A., and Martin, B. R. (2012). *Enhancing Impact: The Value of Public Sector R&D*. Cambridge and London: UK~IRC and CIHE. Available at < http://www.cbr.cam.ac.uk/pdf/Impact%20Report%20-%20webversion.pdf >

Hughes, A. (2008). 'Innovation Policy as Cargo Cult: Myth and Reality in Knowledge-led Productivity Growth', in Bessant, J. and Venables, T. (eds), *Creating Wealth from Knowledge: Meeting the Innovation Challenge*, Cheltenham: Edward Elgar, 80–104.

Hughes, A., and Mina, A. (2012). *The UK R&D Landscape*. UK~IRC and CIHE Cambridge and London March Available at <http://www.cbr.cam.ac.uk/pdf/RDlandscapeReport.pdf>

Jackson, G., and Miyajima, H. (2007). 'Introduction: The Diversity and Change of Corporate Governance in Japan', in Aoki, M., Jackson, G., and Miyajima, H. (eds) *Corporate Governance in Japan: International Change and Organizational Diversity*. Oxford: Oxford University Press, 1–47.

Jaumotte, F., and Pain, N. (2005). 'From Ideas to Development: The Determinants of R&D and Patenting', *OECD Economics Department Working Papers*, No. 457, Paris: OECD.

Kim, W., and Weisbach, M. S. (2007). 'Motivations for Public Equity Offers: An International Perspective', *Journal of Financial Economics*, 87: 281–307.

La Porta, R., Lopez-de-Silanes, F., and Shleifer, A. (2008). 'The Economic Consequences of Legal Origins', *Journal of Economic Literature*, 46: 285–332.

La Porta, R., Lopez-de-Silanes, F., Shleifer, A., and Vishny, R. (1998). 'Law and Finance', *Journal of Political Economy*, 106: 1113–55.

Lazonick, W. (2007). 'The US Stock Market and the Governance of Innovative Enterprise', *Industrial and Corporate Change*, 16(6): 983–1035.

Lazonick, W. (2009). *Sustainable Prosperity in the New Economy? Business Organisation and High-tech Employment in the United States.* Kalamazoo, MI: Upjohn Institute Press.

Lee, P. M., and O'Neill, H. M. (2003). 'Ownership Structures and R&D Investments of US and Japanese Firms: Agency and Stewardship Perspectives', *Academy of Management Journal*, 46(2): 212–25.

Lerner, J. (1998). *The Government as Venture Capitalist: The Long-Run Impact of the SBIR Program.* Cambridge, MA: NBER.

Lerner, J. (2009). *Boulevard of Broken Dreams: Why Public Efforts to Boost Entrepreneurship and Venture Capital have Failed—and What to Do about it.* Princeton, NJ: Princeton University Press.

Manigart, S., De Waele, K., Wright, M., Robbie, K., Desbrières, P., Sapienza, H. J., and Beekman, A. (2002). 'Determinants of Required Return in Venture Capital Investments: A Five-Country Study', *Journal of Business Venturing*, 17(4): 291–312.

Mayer, C. P. (1988). 'New Issues in Corporate Finance', *European Economic Review* 32: 1167–1183.

Mayer, C., Schoors, K., and Yafeh, Y. (2005). 'Sources of Funds and Investment Activities of Venture Capital Funds: Evidence from Germany, Israel, Japan, the UK and the United States', *Journal of Corporate Finance*, 11(3): 586–608.

Miozzo, M., and Dewick, P. (2002). 'Building Competitive Advantage: Innovation and Corporate Governance in European Construction', *Research Policy*, 31: 989–1008.

Morgan, G. (2010). 'Money and Markets', in Morgan, G., Campbell, J. L., Crouch, C., Pedersen, O.K., and Whitley, R. (eds) *The Oxford Handbook of Comparative Institutional Analysis.* Oxford: Oxford University Press.

Munari, F., Oriani, R., and Sobrero, M. (2010). 'The Effects of Owner Identity and External Governance Systems on R&D Investments: A Study of Western European Firms', *Research Policy*, 39: 1093–1104.

O'Brien, J. P. (2003). 'The Capital Structure of Implications of Pursuing a Strategy of Innovation', *Strategic Management Journal*, 24: 415–31.

O'Sullivan, M. (2003). 'Recent Developments in the Financing Role of the Stock Market for French Corporations', *Working Paper* No. 03-36, European Integration, Financial Systems and Corporate Performance (EIFC).

Pistor, K. (2009). 'Sovereign Wealth Funds, Banks and Governments in the Global Crisis: Towards a New Governance of Global Finance?', *European Business Organisation Law Review*, 10(3): 333–52.

Rajan, R. G., and Zingales, L. (1995). 'What do we Know About Capital Structure? Some Evidence from International Data', *The Journal of Finance*, 50(5): 1421–60.

Rajan, R. G., and Zingales, L. (2001). 'Financial Systems, Industrial Structure, and Growth', *Oxford Review of Economic Policy*, 17: 467–82.

Scherer, F. M. (1999). *New Perspectives on Economic Growth and Technological Innovation*. Washington, D.C.: The Brookings Institution.

Schmidt, R. H., Hackethal, A. and Tyrell, M. (2002). 'The Convergence of Financial Systems in Europe: Main Findings of the DFG Project', *Schmalenbach Business Review*, May: 7–53.

Tylecote, A. (2007). 'The Role of Finance and Corporate Governance in National Systems of Innovation', *Organisation Studies*, 28: 1461–81.

Vincente-Lorente, J. D. (2001). 'Specificity and Opacity as Resource-based Determinants of Capital Structure: Evidence for Spanish Manufacturing Firms', *Strategic Management Journal* 22 (2): 157–177.

Wehinger, G. (2012). 'Bank Deleveraging, the Move from Bank to Market-based Financing, and SME Financing', *Financial Market Trends* 2012(1): 1–15.

Zysman, J. (1983). *Governments, Markets and Growth*. Ithaca, NY: Cornell University Press.

CHAPTER 14

....

CONSUMPTION OF INNOVATION

....

RITSUKO OZAKI AND
MARK DODGSON

INTRODUCTION

....

INNOVATIONS diffuse when numerous individual decisions are made to adopt them, and appreciating why consumers make their choices is therefore essential to understanding how innovations become widely used and successful. Diffusion results from aggregate adoption behaviours, which in turn depend on consumption decisions. To understand diffusion—and thereby the level of success of innovation—we have to understand consumption. This has broad implications for the management of innovation. Consumers are increasingly involved, and demanding to be engaged, in the process of creating innovation (von Hippel, 2005), and lifestyle choices on issues such as sustainability and well-being are becoming more important. Concerns for a sustainable lifestyle, for example, influence the choice of a hybrid car (Heffner et al., 2007) and protectiveness towards future generations associated with using renewable energy can affect the adoption of energy-efficiency and 'green' technologies, such as solar water heating and compact fluorescent lamps (Caird and Roy, 2008). It is therefore crucial for innovation management to recognize the nature and significance of the act of consumption.

This chapter begins by illustrating the interrelationships between innovation and consumption by using an example of an entrepreneur in the Industrial Revolution whose products leveraged and contributed to the broad social, economic, and cultural changes of the period during which the phenomenon of mass consumption emerged. It then offers a literature review that argues that studies of the adoption of innovation, a field of study that draws on economics and the applied sciences of design, can be valuably supplemented by research into the consumption of innovation, a field of study that draws on sociology, anthropology, and social psychology. We then offer two contemporary cases—the Toyota Prius and green tariff electricity—to illustrate the multiplicity of factors affecting, and the constraints on, the consumption of innovations. We then draw the lessons of better understanding consumption for the management of innovation.

CONSUMPTION AND INNOVATION—LESSONS FROM THE PAST

We begin our examination of the importance of consumption for innovation management by using an example of a renowned innovator and the way his innovations contributed and responded to changing patterns of consumption. The period is the Industrial Revolution, when consumption and lifestyle patterns changed dramatically as industrial wages were paid and new industries and businesses created novel sources of wealth. The population of England doubled in the eighteenth century and the new manufacturing towns that emerged during this period of the Industrial Revolution brought significant expansion in purchasing power.

As a result the nation witnessed a 'consumer explosion' (McCracken, 1990). There were '…new developments in the frequency with which goods were bought, the influences brought to bear on the consumer, the numbers of people engaged as active consumers, and the tastes, preferences, social projects, and cultural co-ordinates according to which consumption took place' (McCracken, 1990: 16). Berg (2005) refers to the reconfiguration of consumption in the eighteenth century from needs to desires. She argues, for example, how an increased desire for porcelain amongst the middle ranks and for fine earthenware amongst the labouring poor, small artisans, and tradespeople reflected the growing taste for luxury. Uglow (2002: xvii) writes about how during this period the country was 'rethinking the whole relationship of "luxury" to culture'. These changed tastes resulted from exposure to luxury goods derived from increasing international trade, and a growing appreciation of the 'sociabilities of commerce and shopping' beyond merchants and young ladies (Berg, 2004, 2005: 36). Agnew (1993: 25) has people at the time describing the pattern of consumption as manic and addictive.

The size and sophistication of the consumer market developed throughout the eighteenth century. Stylish table accessories, for example, were in huge demand in the burgeoning industrial cities and increasingly wealthy colonies. Tea drinking, and more fashionable coffee and hot chocolate, was becoming a national characteristic (McKendrick, 1960). Hundreds of coffee shops opened in London during the eighteenth century.

McCracken (1990: 17) locates this consumption in the 'viciously hierarchical nature of eighteenth century England (where) goods had suddenly become tokens in the status game'. Fashion had its role to play. Koehn (2001: 25) refers to the strong thirst for novelty at the time: 'In furniture, pottery, fabrics, and millinery, consumers insisted on new fashions'. Robinson (1986: 108) argues 'fashion helped both strengthen the hierarchical system of status and at the same time to persuade members of the mid-dling classes and even of the lower orders that they could at least to some extent imitate their betters'.

The consumption of luxury stimulated a 'significant source of innovation in technolo-gies, products, marketing strategies, and commercial and financial institutions' (Berg, 2004: 92). In the case of pottery, for example:

> Fine earthenware was developed for tableware, and here new qualities of taste and aesthetics, manners, and eating cultures could be combined with technology and industrial development. The result was the huge opportunities offered by a new commodity...
>
> (Berg, 2005: 130).

Into this context, and very much contributing to it, was the great entrepreneur and successful industrialist, Josiah Wedgwood (Dodgson, 2011). His many product, pro-cess, and organizational innovations were informed by his reading of cultural changes, and complemented by success at insinuating his goods into the upper classes and his mastery of the 'trickle down' into the lower classes. This strategy has been argued to be part of a radical change in the definition of status and the use of goods to express status (McCracken, 1990). Wedgwood aimed at what he called the 'Middling Class'—the new and aspiring market of consumers 'who wanted to enjoy wares with a flavour of metro-politan styles but who could not aspire to buy China' (Young, 1995: 10).

Wedgwood assiduously sought patronage from politicians and aristocracy: what he called his 'lines, channels and connections' (McKendrick, 1960: 418). His clients included King George III and Queen Charlotte, and Catherine the Great of Russia. 'A thousand parcels, containing £20,000 worth of pottery, were dispatched to the minor nobility of Europe in an attempt to imitate the strategy of starting at the top of the social pyramid and proceeding downwards' (Tames, 2001: 22). His products excited the increasing middle class as they differentiated them from the coarser earthenware used by the lower classes and displayed some features of the fine porcelain used by the upper class. When his relentless pursuit of the acclaim of the aristocracy was near completion, he began on the minor nobility.

The consumption of Wedgwood's goods by the aristocracy was immensely valuable. 'They praised his ware, they advertised it, they bought it, and they took their friends to buy it.... In the small, interconnected, gossip-ridden world of the English aristocracy in the eighteenth century, such introductions were vital, for even a very few sales could have an important effect' (McKendrick 1960: 414–15). His 'appeals are to price, quality and fashion, to self-interest and self-esteem' (Robinson, 1986: 105).

For Wedgwood, therefore, his many innovations in the pottery he produced and the way in which they were manufactured and marketed, occurred within the context of massive changes in patterns of consumption. Their success depended on using his deep appreciation of the nature of the social and cultural changes occurring at the time and using them for commercial advantage.

The Literature on Innovation Adoption and Consumption—The Case for Their Synthesis

In the classical economics literature, consumption is seen as 'exchange value' (purchase and re-sale prices) and 'use value' (utility, or satisfaction of needs and wants) (du Guy et al., 1997). In this perspective of consumption there is no space for human agency (du Guy et al., 1997). However, consumption is not a mere appendage of economic production, but an important social issue (Featherstone, 1991). Alan Warde (2005: 137) defines consumption as 'a process whereby agents engage in appropriation and appreciation...over which the agent has some degree of discretion'. This definition presents a view that the consumer is not passive, but active, and that consumption cannot be reduced to economic value. Consumption has symbolic significance.

Current empirical consumer marketing studies mainly look at quantitative differences between consumer groups through examining behavioural constructs (e.g. novelty-seeking, risk-taking), time and money spent on a particular activity, and demographic and geographical attributes. The results of focusing on such quantitative differences are non-contextual and 'static' pictures of consumers and do not really explain 'how' consumers form their opinions about certain products and services (Holt, 1997; Ozaki and Dodgson, 2010). The depth to which adoption is embedded in social practices, and its cultural dimension, need to be understood. Although qualitative and interpretive marketing research does exist (e.g. Alvesson, 1994; Belk, 1995; Thompson, 1997), it does not focus on innovation adoption. The objective of innovation management is for innovations to be consumed and eventually diffused; their providers therefore need to uncover 'why' people prefer and want certain things. 'Who wants what' is not enough. Thus, looking at consumption through the qualitative lens complements quantitative consumer adoption research and adds to the insights provided by the marketing approaches describe by Prabhu in Chapter 3.

Rindova and Petkova (2007) identify a gap between the 'intended value' of an innovation, expected by its producers and reflective of the ambitions of their designers and engineers, and 'perceived value', expected by consumers. As long as this gap exists, it restricts innovation diffusion and to create a bridge it is necessary to understand how consumers perceive a particular innovation and what motivates them to adopt it.

Effective innovators convince consumers that a specific object possesses not only useful functions, but also a certain cultural meaning that consumers can identify with and quality that they value. The role of designers, therefore, is to inform consumers of the qualities intended of the object. For that, innovators need to understand how consumers see their innovative product or service and what meaning they attach to it.

An example is the Sony robot, AIBO (Rindova and Petkova, 2007), introduced to the Japanese market in 1999. When Sony launched it on the American market, reaction was lukewarm. Sony changed its outer form from a dog to a human shape in order to shift consumer opinions, hoping that this new shape would induce a perception that this product was a companion, not a toy. This created new perceived value and the AIBO started to sell (Rindova and Petkova, 2007). Another illustration is the Sony Walkman (du Guy et al., 1997). The Sony Walkman was originally created for urban youths listening to their music. It had two headphone jack sockets for listening with friends simultaneously, because solitarily listening to music in public places was seen to be impolite. However, many more people bought a Walkman, ranging in age from 18 to 60, particularly those engaging in outdoor activities such as jogging and bike riding; and people listened to the music individually. The way people used the Walkman was more personal than shared. The Sony Walkman II had a new design with only one headphone jack socket. The image of outdoor activities was also incorporated into the product's advertising. The Sony Walkman II was a great success because the ways it was consumed was well understood by designers, who acted as intermediaries between production and consumption (du Guy et al., 1997).

Understanding consumption therefore needs to draw on quantitative and qualitative analysis, the economic and functional intent of their producers, and their perceived value by consumers. This need to refer broadly for explanations why innovations diffuse is further explored in the following review of literature on adoption and consumption.

Adoption

The classic study of innovation diffusion remains Everett Rogers' *Diffusion of Innovations* (2003), first published in 1962, who states on the first page of the book: 'getting a new idea adopted, even though it has obvious advantages, is often very difficult. Many innovations require a lengthy period, often of many years, from the time they become available to the time they are widely adopted'. Decisions on whether or not to take up an innovation are not instantaneous, but a process that occurs over time, consisting of a series of different actions. As a result, Rogers argues that diffusion research should focus more on the consumer of innovations and that the degree to which people adopt new ideas (i.e. overt behavioural change, or action, rather than cognitive change, or intent) should become the main dependent variable in diffusion research.

Rogers identifies five sequential stages in innovation adoption. An individual: (1) gains knowledge of an innovation (the knowledge stage), (2) forms an attitude towards it (the persuasion stage), (3) decides to adopt or reject it (the decision stage), (4) implements it (the implementation stage) and (5) confirms the decision (the confirmation stage). A range of prior conditions bring consumers into the process in the first place, including their previous experiences, existing needs and problems, norms of their social systems (e.g. their social groups), and general 'innovativeness'.

Innovation adoption is therefore strongly bounded by the social context in which it occurs. For Rogers, the innovation-decision is a social and psychological process as much as an economic one. Indeed, this is widely understood in the innovation adoption literature. One of the best-known case studies of the diffusion of innovation, for example, Morison's (2004, first published in 1966) study of gunfire at sea shows that despite obvious intrinsic benefits, much depends upon the social context in which innovations are introduced and attitudes towards their source.

Others argue that social influences, such as network effects (Bikhchandani et al., 1992), herd behaviour (Banerjee, 1992) and social interaction and learning (Bandura, 1986), play a significant role in accelerating adoption. According to Rogers (2003), once 10 to 20 per cent of the population adopt an innovation, there is relatively rapid adoption by the remaining population, forming an S-shaped curve.[1] This resonates with other similar models in innovation studies, such as Foster (1986), and Abernathy and Utterback (1978), but Rogers places particular emphasis on the importance of the social network of the potential adopter, and the influence of opinion leaders and peer groups.

It is in Rogers' persuasion stage where a general attitude towards and perception of the innovation develops. An individual becomes more psychologically involved with it and actively seeks, interprets, and assesses the credibility of information about the innovation. The most important factors at this stage are perceptions of the innovation's characteristics or attributes. He argues that most of the variance in adoption rate is explained by five perceived attributes: relative advantage (e.g. economy and status), compatibility (e.g. values, norms, and practices), complexity (difficulty in understanding and use), trialability (the degree to which an innovation can be experimented with) and observability (the degree to which effects of adoption are visible). Rogers regards the first two attributes as the most important.

It is the perception of these attributes that affects individuals' decision whether or not to adopt an innovation. Ostlund (1974) showed in his study of innovation attributes that perceptual variables (consumer perceptions of products) are better predictors of adoption than adopters' personal characteristics and demographics. In addition, given that innovations can involve an element of uncertainty, Ostlund, and other researchers, have added perceived risk to Roger's five innovation attributes as an expected probability of economic or social loss resulting from innovation (Labay and Kinnear, 1981; Ostlund, 1974). Lunsford and Burnett's (1992) study of barriers to innovation adoption for the elderly, for example, identifies that it is perceived relative advantage, product usage (complexity), compatibility with values and risk together that influence their adoption decisions. This shows that an analysis of innovation attributes provides more depth to understanding than that provided by demographic and psychographic analysis alone.

Another widely used approach to innovation adoption, the Technology Acceptance Model (TAM), focuses on the utility and usability aspects of innovations to explain how consumers choose to adopt a particular technology. TAM argues the most important factors influencing decisions on if and how to use technology are: perceived usefulness and perceived ease of use (Davis et al., 1989; Bagozzi et al., 1992). These factors are defined as the degree to which a person believes that using a particular technology would 'enhance his or her performance' and 'be free from effort' (Davis, 1989: 320). TAM's emphasis on utility and usability corresponds to the relative advantage (usefulness) and complexity (ease of use) attributes of Rogers' framework. Venkatesh and

Davis (2000) later developed the Technology Acceptance Model 2 (TAM2), which incorporates social influences, such as subjective norms. According to TAM2, an individual's innovation adoption can be predicted by (a) their belief about the consequence of adopting a new technology and (b) how they think other people would think of them if they adopt.

TAM is an adaptation of intention models from social psychology that study the processes by which consumers' beliefs form attitudes towards certain behaviour ('intention to behave') and then lead to the performance of the behaviour (Davis et al., 1989), specifically of Ajzen and Fishbein's (1980) Theory of Reasoned Action (TRA). TRA assumes that human behaviour is *rationally* selected by practitioners and that decisions are made *intentionally* based on a particular goal.

This focus on the influence of social environments, usefulness, and usability is in line with Rogers' framework, and particularly the effect of social networks, compatibility with norms, and relative advantage and complexity of the innovation. TAM and TAM2 present a rational relationship between consumers' perception of an innovation and their adoption decision, highlighting the process in which evaluation of information about an innovation forms attitudes towards the innovation and leads to an adoption decision.

However, these rational and cognitive approaches provide rather limited perspectives for understanding consumer adoption behaviour, compared to Rogers' framework. This is because the adoption and diffusion of innovation is a social process (Rogers, 2003) and, beyond cognitive assessment and rational choice, there are non-rational influences and cultural issues that impinge on consumers' adoption behaviour (see Faiers et al., 2007). Indeed, potential adopters react to innovation in many different ways, and a consumer's decision to adopt is informed by a wide range of personal and social factors. In his theory of interpersonal behaviour, Triandis (1977), like TAM and TRA, considers both the effect of attitudes and social norms as the antecedents to intentions to behave, and he also includes the influence of 'affect', such as unconscious, intrinsic responses to a particular behaviour and the role of habits, as mediators of actual behaviour. Similarly, Fitzmaurice (2005) argues that people's purchasing behaviour can be hedonistic, self-expressive, and identity-congruent, and that these elements should be incorporated into TRA. The explanation of why some people choose an iPod over their technically equivalent and cheaper competitors, for example, lies mainly in affective aspects of the product.

To understand human conduct holistically, more dynamic, contextual, and emotive pictures of behaviour need to be considered in understanding consumer adoption behaviour. We need to go beyond the remit of the rational and cognitive approach when exploring what an innovation means to the consumer in their everyday contexts and how this motivates adoption. Of these approaches to innovation adoption, we find Rogers' (2003) framework is more comprehensive and suitable for understanding where consumers' evaluations of innovations come from and how motivations to adopt innovations are formed. And yet, although Rogers' theory does touch upon issues such as values, practices, and status, these are, nevertheless, better explained in the sociological, anthropological, and social-psychological approaches in consumption studies. To complement Rogers' diffusion theory, we now turn our attention to the consumption of innovation literature.

Consumption

Contemporary society is a consumer society, defined by *Blackwell's Dictionary of Twenty-Century Social Thought* (1993) as a society organized around the consumption, rather than the production, of goods and services. That is not to underestimate the social significance of production, but simply to highlight that members of consumer society treat high levels of consumption as symbolic of social success and personal happiness and hence choose consuming as their overriding life goal (Campbell, 1995: 100).

There are different views on the symbolic meaning of consumption. One view of symbolic meaning sees it as a reflection and expression of existing social orders (e.g. class and wealth) (Bourdieu, 1984). Another suggests contemporary consumers have 'choice'. Consumerism in this sense represents the idea that our identity is not defined by our past or inheritance. Consumption is an identity-building exercise (Lash and Urry, 1994), so consumers can go beyond where they come from, and their gender, age, and ethnicity. It is 'achieved' identity, not 'ascribed' identity (Dittmar, 1992). Thus, consumption brings not only economic 'exchange value' and 'utility value', but also 'sign value' that represents symbolic meanings. Consumption distinguishes and communicates values, identities, and memberships (Slater, 1997). Hirschman (1982) shows that symbolism can be a source for the generation of innovations and that a certain symbolic innovation (e.g. styles of clothing) can be adopted when consumers find the innovation is compatible with their self-identity and image. An innovation generated primarily through symbolic changes communicates a different social meaning than it did previously.

Dittmar (1992) claims consumption is about 'reflexivity': we, contemporary consumers, choose, construct, display, and maintain who we are and who we like to be seen as. In the age of the consumer society, identities are negotiated through consumption (Slater, 1997): we define ourselves by what we consume. We may want to keep up with Joneses, or we may want to remain different from the Joneses. It is up to us.

Consumed objects communicate meanings attached both inwardly and outwardly. Inward communication carries personal meanings that do not have to be conveyed to others. Outward communication, on the other hand, communicates with others. Such outward meanings include 'conspicuous consumption' to show off one's wealth and status (Veblen, 1899) as a display of status symbols (e.g. a Rolex watch) and advocate a particular belief or a social movement (e.g. environmentally sustainable household products). Timmor and Katz-Navon's (2008) study of how people adopt new products shows that their need for assimilation and differentiation depends on the degree of the need for being distinct from others (or similar to a social group) and on the perceived group size. The reason why one acquires an iPod, for example, when s/he already has an alternative device can be differentiation or membership. Outward meanings can also reproduce and represent social relationships, social bonds, and moral obligations (e.g.

gift exchange; Mauss, 1990): people relate to each other through the goods they acquired or were given, and thus goods are considered to constitute social processes (Miller, 1987). Consumption is therefore where we try to achieve our goals and desired images, such as wealth, healthy eating, environmentalism, and sound relationships.

Seen this way, it is clear that meaning is both inscribed and attached. In the case of healthy eating or environmentalism, products are already inscribed with particular meaning and consumers subscribe to it. However, there are cases where meaning is attached by consumers during the process of products being consumed. Blackberry's message ('Sent from my BlackBerry®'), for example, was considered originally to show a status of the owner as the device was uncommon and was used only by business people. But now, many more people have Blackberrys and these kinds of messages are being replaced by notes such as 'excuse any typos' which can be interpreted as saying that the owner is concerned about you but is so busy that they cannot respond properly. As Baudrillard (1988) puts it, meaning does not necessarily reside in an object, but in how the object is used. So, by understanding how an object is consumed and what meaning is attached to it, a firm can bring the consumer goods and a representation of cultural meaning together, the cultural meaning that reflects what consumers value, as in the case of the Sony Walkman. Designing products with meaning is important for innovation managers; and for that, understanding consumers and consumption is vital.

Among the concepts consumption studies offer that are useful for innovation management is the 'trickle-down effect' (Simmel, 1904). This is downward diffusion created by a subordinate social group that hunts upper-class status makers. As we saw earlier, Wedgwood's strategy to seek to insinuate his goods into the upper-class lifestyles and thereafter trickle-down to lower classes is a good illustration (McCracken, 1988). A contemporary interpretation of the trickle-down effect shows that there is an upward and sideways movement, as well as a downward movement. Here, people use consumption for differentiation, as well as imitation, expressing not only status and power, but also other identity elements. This perspective helps us understand social contexts of innovation diffusion (e.g. symbolic meaning, purpose, nature of difference, etc.) (Slater, 1997).

Similarly, the 'Diderot effect' is a cultural phenomenon that innovative organizations can exploit. It is a force that encourages a person to maintain a cultural consistency in his/her possessions (McCracken, 1988). The story goes as follows. The Editor of the French encyclopaedia, Denis Diderot, receives a scarlet dressing gown from a friend as a gift. Well pleased, Diderot displaces his old, comfortable gown with this new arrival. Wearing the elegant gown, he looks around in his study, which is filled with bric-a-brac, and decides that his desk is not good enough. He then replaces the tapestry on the wall as it looks a little ragged, and this process continues. In the end, he misses his old gown and the harmony that the study and its contents created. He concludes that it was the work of the scarlet gown (McCracken, 1988: 119). The entry of a new object, whose cultural significance is inconsistent with that of the whole of the current possessions, introduces the entirely new set of consumer goods. Consumer goods are linked by 'unity'. Apple's 'i' series is a good example. An individual adopts a new thing, such as an iPod, which encourages them to maintain a 'cultural consistency' in their complement of goods, such as a MacBook, an iPhone, and an iPad.

Perspectives in consumption studies not only apply to innovation in products, but also to innovation in services. Consuming services, such as going to the theatre or taking a holiday cruise, has meaning. Undertaking these activities can reflect the actor's personal identity for self-expression ('I support the arts'), social identity to seek reassurance from peers concerning the actor's identity ('I am part of this community'), rituals that respect social organization ('I don't like this playwright, but I'd better go as it shows I care about our group's support for new works'), and pleasure-seeking as a form of imaginative hedonism ('it was heavenly') (e.g. Campbell, 1995; Holt, 1995). Thus, consumers seek and attach meaning to services, and service innovators can benefit from understanding that meaning.

In summary, the purchase of new goods and services is considered to represent both personal and social meanings, because aspiring consumers 'adopt a learning mode towards consumption and the cultivation of a lifestyle' (Featherstone, 1991: 19). As Lash and Urry (1994: 57) put it, 'inasmuch as consumption has taken on heightened significance in contemporary identity-building, choice here should not be understood in a simply utilitarian sense'. The consumption of a hybrid car, for example, is not only about reducing petrol usage, but also is about self-expression of being part of a green community (Kahn, 2007; Ozaki and Sevastyanova, 2011). Understanding meaning attached to consumed objects and activities will help innovators to increase their customers' perceived value. This is shown in the following case on the multiplicity of contributors to decisions to consume. This is followed by a case illustrating the difficulties involved in engaging consumers in innovative services.

The Case of the Toyota Prius

Examining the reasons why consumers buy hybrid cars, such as the Toyota Prius, reveals a complexity of factors. We report here on a study examining the reasons why consumers bought Prius cars (Ozaki and Sevastyanova, 2011). Financial concerns—initial and subsequent running costs—were found to be centrally important, with issues of fuel economy and reduced road taxes being especially valued by consumers. Affectional factors, such as size, comfort, quietness, and ease of use, add to the practical, rational, and utilitarian dimensions of consumption decisions. The reputation of the company for reliability, and consumers' past experiences of driving Toyota cars or a hybrid car (e.g. through a test drive—trialability in Rogers' sense—or previously driven cars) are influential. As Rogers (2003) puts it, knowledge about an innovation can provide the motivation to learn more about and ultimately to adopt it.

Also rated highly by purchasers are the car's perceived environmental benefits (e.g. 'driving a hybrid car will reduce carbon emissions') and compatibility with environmental values/beliefs (e.g. 'driving a hybrid car means doing the right thing'). These also reflect personal and social expressions through consumption. Expressing personal identity and a stylish, fashionable self-image are highly significant, with hybrid car ownership reflecting them through green values. People's identities are reflected in their consumption (Lash and Urry, 1994), which also constructs their desired image, in this case, being 'different' and 'trendy'. Personal interest in technology is also highly relevant. Some people are intrinsically attracted to technology and have a positive attitude towards technical novelty, such as a combination of electric and petrol engines in hybrid cars. Current Prius owners are, therefore, early adopters according to Rogers' (2003) categorization because they are able to deal with remote ideas, such as the environment, and are also favourable towards science and technology and open to new ideas.

The expressive aspect of consumption assists compliance with social norms. People keen to comply with the norms of their groups need to perceive an innovation as consistent with these norms and its adoption as adherence to them. Compliance with social norms, expressed in such statements as 'socially desirable behaviour', 'being considerate to others', 'sharing common values', and 'being socially responsible', were found to be important. This points to the significant role of social norms and pressure in the adoption of sustainable behaviour.

Innovation management benefits from the systematic examination of the multidimensionality in consumers' hybrid car purchase motivations as it highlights the range of important elements in adoption decision-making and points to ways to increase adoption rates. Ownership of a hybrid car is a signal of financially motivated consumption, but purchasers' preferences also emphasize practical, expressive, and experiential aspects and social pressures.

The Case of Green Electricity Tariffs

Green electricity, generated from renewable sources such as wind, solar, and biomass, is an environmental innovation that has not, to date, been widely adopted by consumers. Signing up to a green electricity tariff can help domestic consumers reduce their carbon emissions, but less than 1 per cent of UK households have done so (Graham, 2007). Green electricity requires little or no behavioural change for householders to integrate it into their everyday practices. This 'easy-to-adopt' service innovation might be expected to demonstrate a smooth translation of consumer values into the adoption of innovation. But consumer behaviours in energy use are not as 'green' as might be expected (Ozaki, 2011). Many explanations have been suggested as to why environmentally friendly products diffuse slowly into markets (e.g. Fraj-Andrés and Martínez-Salinas, 2007; Rehfeld et al., 2007). Green alternatives might be considered as being too expensive, not offering the same functionality as existing products, or they might require consumers to deal with unacceptable inconveniences.

We report here on a study of the adoption of green tariffs amongst a sample of staff at Imperial College London (Ozaki, 2011). The sample had a strong bias towards green consumers, and many respondents were actively engaged in environmentally friendly activities, such as recycling, and had memberships of, and made donations to, green movements. Despite the group's 'green' bias, there was great hesitation amongst them about adopting a green electricity tariff, and even those with high adoption intentions were indecisive. Positive attitudes towards pro-environmental behaviours do not necessarily translate into the performance of the behaviours. People are capable of being contradictory or hypocritical.

Switching to a green tariff can be seen as an inconvenience. It requires not only time to fill in a form, but also to contact the supplier, change payment settings, and other actions. Most people are busy in their daily lives, and this is not an attractive proposition. Costs are also a problem. Even a slight increase in cost is unappealing when energy prices are rapidly rising and affecting every household. Thus, the cost and (in)convenience of signing up significantly affect the adoption of green tariffs. This problem is compounded when consumers do not have sufficient and accurate information and are uncertain about the quality of green electricity (e.g. 'is it really generated from renewable sources?' and 'is it reliable?'). The nature of the contract and costs can also cause some anxiety, which in turn leads to rejection. Perceived relative advantage (Rogers, 2003) and risk (Ostlund, 1974) clearly play an important role in this case.

LESSONS FROM THE CASES

The cases provide a number of lessons for innovation management by revealing the factors that encourage and constrain the consumption of innovative products and services. The case of hybrid vehicles shows the importance of financial benefits and effective communication with the public. Such communication constructs a discourse that purchasing a particular innovation is not out of the ordinary and informs consumers not only about functional, but also aesthetic, practical, and experiential aspects of innovations. The case of green electricity poses the challenge of how to fill the gap between intentions and actual behaviour, and this requires a deep appreciation of why people consume innovations. What pushes people from 'intention to adopt' to 'actual adoption' is a combination of: a sense of control over costs and associated inconveniences; perceived personal benefits compatible with people's values and identity; strong social influences and normative beliefs; and good information that helps mitigate perceived risk/uncertainty. For innovation managers to understand the drivers of and barriers to the uptake of innovations, it is helpful to combine adoption and consumption perspectives so as to fully appreciate the following range of influential factors.

Costs and Financial Benefits

The studies identify the value of financial incentives for consumers to overcome their resistance to the cost and inconvenience of switching products and services, even when consumers are knowledgeable and receptive to the innovations. In the example of green tariffs, innovators have to recognize the scale of the deleterious consequences of the cost premium. Energy bills are already expensive and even environmentally aware people are not keen to pay extra. The switch-over procedure was also a barrier, especially when switching from one supplier to another, and minimizing such costs affects consumption. In the case of hybrid cars, we saw the benefit of accentuating reduced congestion and parking charges, and the availability of any subsidies.

Personal Benefits

The cases show the value of the perceived benefits from adoption having personal relevance. In the case of environmental technology, for example, emotional appeals encourage potential adopters to take action. Concerns for one's children's future can trigger emotional reactions and thus change perceptions and attitudes. Environmental issues may be seen as abstract and not immediate, and the reaction of many people to them may be less acute than their responses to personal benefits and social norms, which offer more concrete indications of what is accepted and expected. Compatibility relates with personal identity (e.g. 'I am green and act pro-environmentally') and with social norms to social identity (e.g. 'I am part of the group that is concerned about the environment').

Social Influences

The cases show green values and awareness on their own do not seem to convince people to adopt a green tariff, or drive a hybrid car, so strong messages from producers, suppliers, and policy-makers that our behaviours can make a difference is needed. On the social level, a guarantee of social benefits, such as a promise that electricity suppliers will make a donation to an environmental charity when customers adopt a certain innovative technology or service, may incentivize potential consumers. Consumer education fosters public recognition of the positive consequences of adopting innovations and creates shared societal norms among consumers.

Information Provision

Potential adopters need accurate information to evaluate and make a decision about the value, risks, and uncertainties of innovation. We saw how green electricity information is fragmented and inaccurate. It is still at an early stage of the diffusion process and information about it has to be provided with clarity and consistency to consumers. Consumers become confused about innovation when there is not enough information or it contains differing and inaccurate messages. Precise information, for example, about how the electricity is generated, and how the premium prices consumers pay are used, allows potential adopters to compare suppliers, choose one that suits them, and encourage them to sign up. More user-friendly websites from innovators and reports from consumer organizations could especially help consumers with high adoption intentions to decide. In the case of environmental technology, stakeholders can learn from the example of eco-labelling. Eco-labelling was originally developed by NGOs and the European Union now legislates for its use. Labels are not only a message about a product or a service, but also validate claims about sustainability standards verified by a formally recognized and accredited independent third party (de Boer, 2003). Eco-labelling encourages companies that want to differentiate themselves based on their sustainable product attributes and helps consumers identify more environmentally friendly products/services and suppliers (Gunne and Anders, 2007; Sammer and Wüstenhagen, 2006). Standardized information would help consumers with high adoption intentions to consider green electricity and build their trust in green electricity suppliers.

The Prius case identifies the range of information that is useful for consumers. It encompasses aesthetic, experiential, and practical values associated with the technology, as well as the role of trial/past experience in purchasing decisions, which highlights the importance of information on what these innovations offer. This knowledge would help people overcome fears or doubts about their technical performance and practical aspects, and create demand. This suggests that there should be more communication regarding innovation and its potential 'value added'. More affective and practical information would increase consumers' positive perceptions of new technologies.

Conclusions

Consumers' decisions on whether or not to adopt an innovation are affected by much more than instrumental evaluations of utility and technical qualities. Consumers make decisions to adopt innovations for a variety of reasons that can be socially influenced or personal. Today's consumption decisions are becoming ever more complex (see Gabriel and Lang, 1995; Kotler and Caslione, 2009), making innovation management increasingly challenging.

Understanding how innovations are consumed is therefore vital for innovation management. Marketing research can usefully distinguish differences between groups in their personal and demographic characteristics in relation to the adoption of innovations. However, successful adoption and diffusion depends on the fit between consumer contexts and motivations and innovations, and there is a need for innovation management research to study the way these underlying factors affect innovation adoption decisions. The perspectives of innovation consumption studies offer a broader contextual and emotive picture of consumers that includes not only demographic and personality traits affecting customer requirements, which is the focus of marketing research, but also the dynamic contexts where consumers form their opinions and their underlying values govern their actions. The ways motivations are formed, and the meaning ascribed to consumption, need to be incorporated into our understanding of innovation adoption. By combining two traditions—innovation adoption and consumption studies—the existing understanding by managers of the demand side in innovation is broadened. Innovation diffusion is the poor relative in innovation studies, with substantially greater focus on the creation of innovation, rather than its patterns of use (Ozaki and Dodgson, 2010; Ozaki, Shaw, and Dodgson, 2012). Given the increasing interest in 'market facing' innovation, it becomes essential to move beyond a superficial understanding of what people consume to a deep appreciation of why they consume.

As a result, there remain many interesting innovation management research questions to explore in the relationships between the adoption and consumption of innovation. Two will be proposed here. First, there would be value in a greater understanding of the priority of motivations in innovation adoption and consumption. The consumption literature shows us, for example, how norms and the influence of social networks (e.g. pressure from peer groups or opinion leaders) can play a big part in the decision to adopt an innovation. The questions are whether, how, and when these social dimensions assume greater significance compared to factors such as cost and utility. The key to understanding the process of innovation adoption involves exploring more completely the combinations of and relationships between emotive and instrumental motivations.

Second, in a similar vein, Rogers' theory argues relative advantage and observability confers social status. Consumption studies provide deep insights into the status-conferring nature of innovation, such as the way the expression of self plays an important part in the process by which meaning is attached to objects and consumption activities. Some innovation can help a person believe that they achieve a higher status in society. By exploring the process of gaining such meaning from both adoption and consumption perspectives, the way consumers come to adopt an innovation will be better contextualized and this will help theoretical understanding and practically improve innovators' capacity to market and position their products and services better.

Consideration of consumption perspectives also needs to be extended to business-to-business transactions. The normative implications for the management of innovation are clear. The most important decision made during the innovation process is that made by the consumer. Markets are created, profits produced, and innovative firms survive and grow only when individuals and organizations decide to adopt innovations. Firms that wish to improve their innovation performance have to address the 'supply-side' inputs to their innovation processes, such as market and technological knowledge, product development, and R&D investments. But it is also essential for them to understand the 'demand-side' consumption of innovation and how adopters influence the innovation process. The identification of determinants of consumer adoption behaviours allows firms to measure and forecast the economic effects of innovations, which then helps them to improve the positioning of their innovations. Understanding the distinctive characteristics and motivations of consumers helps to explain why one product is chosen over another one that has the same price, function, and utility. This requires study of the *meaning* that is attached to the product and the *context* in which the adoption decision was made. Few of today's organizations can prosper without understanding the motivations and actions of consumers towards their innovative products and services.

NOTE

1. Unlike Rogers' (2003) argument of smooth and continuous increase, Moore (1998) discusses the difficulty in 'crossing the chasm', the transition from early adopters' adoption to early majority's adoption, which would push up the S curve.

REFERENCES

Abernathy, W., and Utterback, J. (1978). 'Patterns of Industrial Innovation', *Technology Review*, 80(7): 40–7.

Agnew, J.-C. (1993). 'Coming Up for Air: Consumer Culture in Historical Perspective', in J. Brewer and R. Porter (eds) *Consumption and the World of Goods*. New York: Routledge, 19–39.

Ajzen, I., and Fishbein, M. (1980). *Understanding Attitudes and Predicting Social Behaviour*. Englewood Cliffs, NJ: Prentice-Hall.

Alvesson, M. (1994). 'Critical Theory and Consumer Marketing', *Scandinavian Journal of Management*, 10(3): 291–313.

Bagozzi, R. P., Davis, F. D., and Warshaw, P. R. (1992). 'Developing and Test of a Theory of Technological Learning and Usage', *Human Relations*, 45: 660–86.

Bandura, A. (1986). *Social Foundations of Thoughts and Action: A Social Cognitive Theory*, Englewood Cliffs, NJ.: Prentice-Hall.

Banerjee, A. V. (1992). 'A Simple Model of Herd Behaviour', *Quarterly Journal of Economics*, 107(3): 797–817.

Baudrillard, J. (1988). *Selected Writings*. Stanford: Stanford University Press.

Belk, R. W. (1995). 'Studies in the New Consumer Behaviour', in D. Miller (ed.) *Acknowledging Consumption: A Review of New Studies*. London: Routledge, 53–94.

Berg, M. (2004). 'In Pursuit of Luxury: Global History and British Consumer Goods in the Eighteenth Century', *Past and Present*, 182: 85–142.

Berg, M. (2005). *Luxury and Pleasure in Eighteenth-Century Britain*. Oxford: Oxford University Press.

Bikhchandani, S., Hirshleifer, D., and Welch, L. (1992). 'A Theory of Fads, Fashion, Custom, and Cultural Change as Information Cascades', *Journal of Political Economy*, 100(5): 992–1026.

Bourdieu. P. (1984 [1974]), *Distinction: A Social Critique of the Judgement of Taste*. Cambridge, MA: Harvard University Press.

Caird, S., and Roy, R. (2008). 'User-centred Improvements to Energy Efficiency Products and Renewable Energy Systems: Research on Household Adoption and Use', *International Journal of Innovation Management*, 12(3): 327–55.

Campbell, C. (1995). 'The Sociology of Consumption', in D. Millar (ed.), *Acknowledging Consumption: A Review of New Studies*. London: Routledge, 96–126.

Davis, F. D., Bagozzi, R. P., and Warshaw, P. R. (1989). 'User Acceptance of Computer Technology: A Comparison of Two Theoretical Models', *Management Science*, 35(8): 982–1003.

de Boer, J. (2003). 'Sustainability Labelling Schemes: The Logic of their Claims and their Functions for Stakeholders', *Business Strategy and the Environment*, 12(4): 254–64.

Dittmar, H. (1992). *The Social Psychology of Material Possessions: To Have is To Be*. Harvester Wheatsheaf St Martin's Press.

Dodgson, M. (2011). 'Exploring New Combinations in Innovation and Entrepreneurship: Social Networks, Schumpeter, and the Case of Josiah Wedgwood (1730–1795)', *Industrial and Corporate Change*, 20(4): 1119–51.

du Guy, P., Hall, S., Janes, L., Mackay, H., and Negus, K. (1997). *Doing Cultural Studies: The Story of the Sony Walkman*. Thousand Oaks: Sage.

Faiers, A., Cook, M., and Neame, C. (2007). 'Towards a Contemporary Approach for Understanding Consumer Behaviour in the Context of Domestic Energy Use', *Energy Policy*, 35(8): 4381–90.

Featherstone, M. (1991). *Consumer Culture and Postmodernism*. London: Sage.

Fitzmaurice, J. (2005). 'Incorporating Consumers' Motivations into the Theory of Reasoned Action', *Psychology and Marketing*, 22: 911–29.

Foster, R. N. (1986). *Innovation: The Attacker's Advantage*. London: Macmillan.

Fraj-Andrés, E., and Martínez-Salinas, E. (2007). 'Impact of Environmental Knowledge on Ecological Consumer Behaviour: An Empirical Analysis', *Journal of International Consumer Marketing*, 19(3): 73–102.

Gabriel, Y., and Lang, T. (1995). *The Unmanageable Consumer*. London: Sage.

Graham, V. (2007). *Reality of Rhetoric? Green Tariffs for Domestic Consumers*. London: National Consumer Council.

Gunne, G., and Anders, B. (2007). 'The Impact of Environmental Information on Professional Purchasers' Choice of Products', *Business Strategy and the Environment*, 16(6), 421–29.

Heffner, R., Kurani, K., and Turrentine, T. (2007). 'Symbolism in California's Early Market for Hybrid Electric Vehicles', *Transportation Research Part D: Transport and Environment*, 12:396–413.

Hirschman, E. C. (1982). 'Symbolism and Technology as Sources for the Generation of Innovations', *Advances in Consumer Research*, 9: 537–41.

Holt, D. B. (1995). 'How Consumers Consume: A Typology of Consumption Practices', *Journal of Consumer Research*, 22: 1–16.

Holt, D. B. (1997). 'Poststructuralist Lifestyle Analysis: Conceptualizing the Social Patterning of Consumption in Postmodernity', *Journal of Consumer Research*, 23: 326–50.

Kahn, M. E. (2007). 'Do Greens Drive Hummers or Hybrids? Environmental Ideology as a Determinant of Consumer Choice and the Aggregate Ecological Footprint', *Journal of Environmental Economics and Management*, 54: 129–45.

Koehn, N. (2001). *Brand New: How Entrepreneurs Earned Consumers' Trust from Wedgwood to Dell*. Cambridge, MA: Harvard Business Press.

Kotler, P., and Caslione, J. A. (2009). *Chaotics: The Business of Managing and Marketing in the Age of Turbulence*. New York: Amacon.

Labay, D. G., and Kinnear, T. C. (1981). 'Exploring the Consumer Decision Process in the Adoption of Solar Energy Systems', *Journal of Consumer Research*, 8: 271–8.

Lash, S., and Urry, J. (1994). *Economies of Signs and Space*. London: Sage.

Lunsford, D. A., and Burnett, M. S. (1992). 'Marketing Product Innovations to the Elderly: Understanding the Barriers to Adoption', *Journal of Consumer Marketing*, 9(4): 53–63.

McCracken, G. (1990). *Culture and Consumption: A New Approach to the Symbolic Character of Consumer Goods and Activities*. Bloomington: Indiana University Press.

McKendrick, N. (1960). 'Josiah Wedgwood: An Eighteenth-century Entrepreneur in Salesmanship and Marketing Techniques', *The Economic History Review*, 12(3): 408–33.

Mauss, M. (1990), *The Gift*. New York: Norton.

Miller, D. (1987). *Material Culture and Mass Consumption*. New York: Basil Blackwell.

Moore, G. A. (1998). *Crossing the Chasm: Marketing and Selling Technology Products to Mainstream Customers,* 2nd edn. Bloomington, MN: Capston.

Morison, E. E. (2004 [1966]). 'Gunfire at Sea: A Case Study of Innovation', in R. A. Burgelman, C. M. Christensen, and S. C. Wheelwright (eds), *Strategic Management of Technology and Innovation*, 4th edn. Boston: McGraw-Hill, 431–40.

Ostlund, L. E. (1974). 'Perceived Innovation Attributes as Predictors of Innovativeness', *Journal of Consumer Research*, 1: 23–9.

Outhwaite, W. (ed.) (1993). *The Blackwell Dictionary of Twentieth-century Social Thought*. Oxford: Wiley-Blackwell.

Ozaki, R. (2011). 'Adopting Sustainable Innovation: What Makes Consumers Sight up to Green Electricity?' *Business Strategy and the Environment*, 20(1): 1–17.

Ozaki, R., and Dodgson, M. (2010). 'Adopting and Consuming Innovations', *Prometheus: Critical Studies in Innovation*, 28(4): 311–26.

Ozaki, R., and Sevastyanova, K. (2011). 'Going Hybrid: An Analysis of Consumer Purchase Motivations', *Energy Policy*, 39: 2217–27.

Ozaki, R., Shaw, I., and Dodgson, M. (2013). 'The Co-production of "Sustainability": Negotiated Practices and the Prius', *Science, Technology and Human Values*, 38(4): 518–41.

Rehfeld, K. M., Rennings, K., and Ziegler, A. (2007). 'Integrated Product Policy and Environmental Product Innovations: An Empirical Analysis', *Ecological Economics*, 61(1): 91–100.

Rindova, V. P., and Petkova, A. P. (2007). 'When is a New Thing a Good Thing? Technological Change, Product Form Design, and Perceptions of Value for Product Innovations', *Organization Science*, 18(2): 217–32.

Robinson, E. (1986). 'Matthew Boulton and Josiah Wedgwood: Apostles of Fashion', *Business History*, 28(3): 98–114.

Rogers, E. M. (2003). *Diffusion of Innovations*, 5th edn. New York: The Free Press.

Sammer, K., and Wüstenhagen, R. (2006). 'The Influence of Eco-labelling on Consumer Behaviour: Results of a Discrete Choice Analysis for Washing Machines', *Business Strategy and the Environment*, 15(3): 185–99.

Simmel, G. (1904). 'Fashion', *International Quarterly*, 10: 130–50.

Slater, D. (1997). *Consumer Culture and Modernity*. Cambridge: Polity Press.

Tames, R. (2001). *Josiah Wedgwood*. Princes Risborough: Shire Publications.

Thompson, C. J. (1997). 'Interpreting Consumers: A Hermeneutical Framework for Deriving Marketing Insights from the Texts of Consumers' Consumption Stories', *Journal of Marketing Research*, 34(4): 438–55.

Timmor, Y., and Katz-Navon, T. (2008). 'Being the Same and Different: A Model Explaining New Product Adoption', *Journal of Consumer Behaviour*, 7: 249–62.

Triandis, H. (1977). *Interpersonal Behaviour*. Monterey, CA: Brooks & Cole.

Uglow, J. (2002). *The Lunar Men: Five Friends Whose Curiosity Changed the World*. New York: Farrar, Straus and Giroux.

Veblen, T. (1994 [1899]). *The Theory of the Leisure Class: An Economic Study in the Evolution of Institutions*. New York: Penguin.

Venkatesh, V., and Davis, F. D. (2000). 'A Theoretical Extension of the Technology Acceptance Model: Four Longitudinal Field Studies', *Management Science*, 46(2): 186–204.

von Hippel, E. (2005). *Democratizing Innovation*. Cambridge, MA: MIT Press.

Warde, A. (2005). 'Consumption and Theories of Practice', *Journal of Consumer Culture*, 5(2):131–53.

Young, H. (1995). 'Introduction', *The Genius of Wedgwood*. London: Victoria and Albert Museum.

CHAPTER 15

SUSTAINABLE INNOVATION MANAGEMENT

FRANS BERKHOUT

INTRODUCTION

INNOVATION is classically focused on improving value to consumers and reducing the private costs of production, so providing greater profits for the owners of business. But increasingly business firms are broadening the way value and costs are defined to take account of the environmental and social dimensions of the production and use of goods and services. All goods and services have environmental and social costs, and associated risks, because they consume natural resources and environmental services during production and consumption. The central problem is that these externalities, spillovers, and risks represent 'public bads' which private producers and consumers share with others. In an unregulated state there are few incentives to address them. This also applies to the management of innovation. Few of the normal incentives driving innovation exist to address environmental and social externalities. This also means that firms have not traditionally developed strategies, structures, resources, and capabilities to handle environmental and sustainability concerns. Classically, governments have served as the carriers of the public interest, and have regulated firms' behaviour to secure public goods such as clean air and safe water, or the protection of species, forests, or the climate system. Many of these regulations have been technology-based (including subsidies, standards, and bans) and they have therefore had an impact on the innovation and adoption of new technologies by firms. Compliance with regulations has implied a degree of innovative effort by firms.

This traditional view of technology and the environment is no longer appropriate for a number of reasons:

- Firms have become increasingly active in managing the environmental and social impacts of goods and services and this has translated into their technology and innovation strategies, moving 'beyond compliance';

- Consumers have become more aware and demanding about the environmental and social performance of goods and services, seeking pro-environmental and pro-social choices and behaviours;
- Governments are no longer the only guarantor of the public will, with a wide range of organized social interests, including business itself, influencing that development of public discourses, technical standards, and the direction of technical change more generally.

This means that business has become not only more aware of the social and environmental impacts of its activities, directly or indirectly, but also much more active—internally and externally—in seeking to manage them. For many businesses, sustainability-driven innovation in products, services, and business models forms a central part of the response to this more complex social and political context (cf. Kiron et al., 2013). In addressing such challenges, firms are confronted with the question of the extent to which an incremental or radical innovative response is needed. This is because many of the underlying technologies and systems in key sectors such as energy, transport, food and agriculture, and the built environment are now viewed as fundamentally unsustainable because they depend on non-renewable or scarce resources, or because they are at the heart of major global environmental problems, such as biodiversity loss or climate change. There is, therefore, a societal demand for radical alternatives.

Key questions for innovation management within business firms are:

1. What are the known externalities or risks associated with goods, services, and operations? What are the uncertainties and controversies, and what are potential new risks that could emerge in future?
2. Can these externalities and risks be managed through incremental innovation, or is a more radical reconfiguration of the value proposition called for? What does this mean for technology and innovation strategy, and for the business model of the firm?
3. How should structures, resources, capabilities, and processes within the firm be changed to embed environmental and social performance into innovation management? What does it mean for relationships up and down supply and value chains, and for relationships with customers and other stakeholders?
4. Given the new governance context of sustainable innovation, how should firms engage with stakeholders (internally and externally) in innovation management, and in seeking to influence the context for innovation (discourses, policies, behaviours)?
5. How is sustainability becoming integrated across the identity, activities, and performance of a business? What are the advantages of a leadership position relative to a follower strategy?

The premise of this chapter is that firms have a 'social licence' to operate, determined by legal rules and standards for the performance of their processes and products (Gunningham et al., 2004). These rules set a baseline of process, transparency, and performance that business needs to achieve. Through time, as knowledge develops about the environmental and social impacts of business activities and as social norms change, there has been a progressive accumulation and strengthening of such rules. New rules and standards often require an innovative response from firms. But firms are also increasingly active in seeking to shape the regulatory and governance context in which they operate. Business plays an important role in setting new rules by responding to public debate by developing innovative sustainable solutions and by participating actively in the public sphere through marketing, brand management, communications, and corporate social responsibility activities. Environmental and social rule-setting has become an important source of competitive advantage in many sectors globally.

Firms therefore go 'beyond compliance' for a wide range of reasons: in response to market demand; in search of cost advantage; as part of a differentiation strategy; as part of an upgrading strategy; to build intangible brand value around green reputation; to reduce regulatory and other risks; and so on. The overall corporate strategy, including the business model, will determine how firms pursue beyond compliance strategies. A range of environmental strategies are available to the firm, ranging from good housekeeping to setting sustainability goals central to all business activities. To implement a chosen strategy, firms need to build capabilities, including innovation, and marketing and communications capabilities, as well as to manage relationships with regulators, customers, suppliers, and opinion formers in society, including civil society organizations. The problem of innovation management is therefore deeply related to the way in which a firm manages its external relationships. And the analysis of innovation management needs to be seen as being embedded in a network of relationships related to (global) value chains, regulatory and social governance, and wider socio-technical regimes.

The following section outlines the relationship between technological innovation and the environment, arguing that this is complex and reciprocal. It stresses the transformative potential of system innovation as a way of meeting the challenge of sustainability transitions in the socio-technical regimes underpinning some of the major global industries, including resources, energy, food, transport, and construction industries, that are now expected over coming decades. The following section briefly discusses theories of system innovation which emphasize the co-evolution of technology with institutions, infrastructures, and consumer preferences and practices, seeing these as major challenges for innovation management in business firms. Drawing on this context, the chapter then discusses the literature on sustainable innovative capabilities, supply and value chains, stakeholder relationships, and business model innovation. The link between corporate strategy and sustainable innovation strategy is discussed, followed by a concluding section.

Innovation and the Environment

Technical change can be characterized as having three main influences on the environmental performance of business: sensing and providing information about the state of the environment (information and knowledge effects); improving the efficiency with which resources are harvested or extracted and transformed into outputs (efficiency effects); and providing the means for transforming resource-use and environmental impacts of business (transformational effects). The relevance for corporate environmental and innovation management of each is reviewed here.

INFORMATION AND KNOWLEDGE EFFECTS

Sensors, monitors, and other instruments collect information about the distribution of resources (for instance, seismic surveys for oil and gas), the environmental impacts of productive activities (for instance, emissions concentrations at production sites), and the state of the environment (for instance, a census of fish stocks) at all levels from the planetary and global to the microscopic. There has been a vast increase in environmental monitoring efforts by business, governments, and international organizations over the past decades, covering all aspects of environmental dynamics and quality. Environmental monitoring is fundamental because much that is important about the environmental impacts of business activities is intangible, not immediately observable, or open to experience. Specialized science-based analytical capabilities are frequently needed to make sense of environmental data and to place the risks they reveal into some social and economic context. These are competencies which few firms will to develop themselves. To a large degree, business knowledge of its own environmental and social impacts will depend on scientific instruments and knowledge.

Tangible environmental impacts of business activities (such as smoke from a power plant, or the landscape impacts of wind turbines) are an important aspect of public awareness and were the focus of early regulatory effort. But many of the most challenging environmental impacts of business are now largely intangible or only revealed through damage to human health or ecosystem functioning at a remove in space and time. Dioxins, for example, are cancer-causing chlorinated compounds that were found in the emissions of many industrial processes, including from the chemicals and pulp and paper industries. Their detection requires sampling and laboratory analysis, while the assessment of health impacts relies on a large body of toxicological and epidemiological evidence. Not only are the externalities often not directly tangible, their effects may also be far-reaching and accessible only through scientific analysis. Establishing a link between chlorinated compounds used as refrigerants (CFCs) and the destruction of stratospheric ozone (the 'ozone hole'), for instance, required a combination of long-run observational data and Nobel Prize-winning atmospheric chemistry to be detected. Much of modern understanding and social perception of environmental problems is therefore mediated through scientific instruments and analysis. What science is not able to do is to define whether an environmental or health problem that has been detected and analysed is important, and needs to be addressed. For that a social and political process is needed in which both science and business play an important role.

Three important considerations follow for corporate strategy. First, firms have had to develop new skills and capabilities to understand, measure, and manage the often complex environmental impacts for which they are (often only partly) responsible. Second, firms have become actively involved, frequently as a result of new rules on public disclosure, on reporting about their environmental and social performance to regulators and the general public (Haddock-Fraser and Fraser, 2008). Such disclosures have, in turn, become a new spur for innovation and investment, principally as a result of the reputational effects that negative (or positive) environmental performance information can have on customers, investors, regulators, and neighbours (Bebbington et al., 2008). Third, firms have become actively engaged in the public and scientific discourse around harms and risks associated with their processes and products. To the extent that there often continue to be uncertainties about the scale or importance of impacts—with climate change perhaps the paradigmatic example—firms often play a role, either directly or via proxies, in the social and political processes that 'frame' environmental problems as important and urgent, and which may lead to social pressure for controls.

Efficiency Effects

Technology plays a fundamental role in enabling the transformation of resource inputs into outputs. Innovation, by improving the efficiency with which inputs are converted to outputs, is also an important driver of environmental performance improvement of business. Such eco-efficiency improvements may be seen as a natural consequence of general efficiency upgrading by business through incremental innovation, investment in new capital, improved control of processes and supply chains, and the provision of additional services to customers. How to improve the rate at which such incremental efficiency gains are achieved has been a key academic debate for over twenty years, sparked by a small commentary piece by Michael Porter (Porter, 1991). Counter to prevailing economic wisdom that profit-maximizing firms would ignore profitable opportunities to become more efficient, Porter argued that firms could gain competitive advantage if they were pressed by 'smart' regulation to improve the resource efficiency of their processes and reduced the environmental footprint of their products. Underlying this claim was the finding in a series of case studies that firms were wasteful of resources.

Empirical research has indeed shown that even in sectors with high energy and resources costs, like pulp and paper, where the competitiveness of firms depends on the efficient use of resources, there are wide differences in the resource-efficiency of companies (Tyteca et al., 2002). Porter argued that strict but flexible regulations would generate new innovation leading to better environmental performance, and, in many cases, improved business performance. Porter and van der Linde (1995) argued that there were five reasons for this. Regulation:

- signals to companies that there are resource inefficiencies and the potential for innovation
- stimulates information gathering that can lead to benefits by raising corporate awareness
- reduces the uncertainty that investments in environmental protection will generate economic value
- creates pressure that motivates learning and innovation, and
- levels the playing field as businesses transition to higher environmental performance.

Even now, the 'Porter Hypothesis' remains a touchstone for debates about business, innovation, environmental performance, and the role of public policy. It encapsulates neatly the central concern about the role of innovation and how this interacts with regulation, which itself is dynamic. The key question is whether a firm should invest scarce innovative resources in achieving environmental performance improvements that may be costly and whose benefits may not be fully appropriable. Firms under-invest in sustainable technologies due to what Rennings (2000) calls the 'double externality' problem. Besides the familiar problem faced by firms of capturing knowledge spillovers from their investments in innovation, sustainable technologies create additional environmental spillovers that may be even more difficult to capture by firms. Positive environmental benefits, such as cleaner air or a safer climate, are shared with the rest of (global) society. Investments in innovation by businesses create social value which is largely captured by society in general. Value slips away, dampening firms' incentives to invest in sustainable technologies (Lepak et al., 2007).

Under such circumstances, policy and regulation can signal broader societal expectations about environmental performance improvements, while also providing a basis for the appropriation of market rents by business. Central to Porter's argument is the notion of 'innovation offsets'; the idea that the additional costs imposed by regulation are often (though not always) compensated for by cost reductions due to the elimination of waste or the creation of consumer value (for instance, through a willingness to pay a premium for products and services with a higher environmental quality). These benefits may not be immediately available, but over the longer term, as regulations tighten, or as relative prices change, first-movers will benefit from having developed new processes and products that meet emerging requirements ahead of their competitors. The challenge lies in good investment opportunities being missed because they are viewed by the manager as too risky, too costly, or outside his or her habits and routines.

The many empirical studies of the effect of tougher environmental regulation on innovation, productivity, and business performance have provided mixed evidence (cf. Jaffe and Palmer, 1997; Popp, 2006). In one of the more recent studies, Lanoie et al (2011) surveyed 4,200 manufacturing plants in seven OECD countries, finding that environmental regulations lead to greater R&D investment by firms. Regulation therefore spurs innovation, with benefits for business performance. But they also found that the added costs of complying with regulation had a net negative effect on business profitability, so that the positive effect on innovation did not outweigh the negative cost effect of the regulation itself, at least over the short term.

As Ambec et al. (2011) argue, this kind of analysis tends not to take fully into account the dynamic dimensions of the Porter Hypothesis. It takes time for the imposition of a regulation to feed through to innovative responses in firms and on to impacts on the costs and competitiveness of firms. Lanoie et al. (2008) used lagged data on productivity and investments in environmental technologies for manufacturing plants in Quebec. They found that environmental regulations had a significant positive effect on productivity growth rates, especially in sectors highly exposed to external competition. The 'win-win' of tougher environmental regulations and business competitiveness may therefore emerge only over the longer run.

A distinctive feature of both the Porter Hypothesis and the debate it fostered is that it has been entirely concerned with private costs to business. No account is taken of the wider environmental or health benefits that may follow from tougher regulations. If these social benefits were included, the potential costs to business competitiveness would probably look quite different. In developing their environmental innovation strategy, such wider social benefits would also be considered in a social cost–benefit analysis.

Transformational Effects

Beyond improving the resource efficiency and reducing the environmental impacts of business activities, firms may also look further to become a part of the solution to environmental problems. At base, all business activities draw on natural and energy resources, which will return to environmental 'sinks'—water, land, the biosphere, and the atmosphere—in the form of wastes and metabolites. To a large extent, modern economies remain 'linear' in their handling of resources and are powered by fossil fuels. Fossil energy represents more than 85 per cent of primary energy use in the world today (BP, 2012). The radical alternative to fossil-based linearity of resource use is solar-powered cyclicity of resource use. Under this ideal, first developed by visionaries like Amory Lovins in the mid-1970s (Lovins, 1976), economies would be powered directly and indirectly mainly by solar power, as sunlight, wind, and chemically stored renewable solar energy in the form of biomass. In this spectrum, the role of nuclear power as a renewable energy remains controversial. Material resources in technical-economic systems would be conserved, upgraded, and recycled continuously in a decarbonized economy (Hawken et al., 1999). A closed-loop or 'cradle-to-cradle' economy (McDonough and Braungart, 1998) would emerge in which waste was treated as a resource.

While such a vision remains far from being realized, it has increasingly become the 'meta-narrative' by which sustainability initiatives in business are judged. The ultimate objectives of decarbonization and the closing of resource loops have also been central to much environmental policy over past decades. According to the best analysis, the UN Framework Convention on Climate Change (1992), the landmark piece of international law under which over 190 countries have now agreed to act to prevent 'dangerous anthropogenic interference with the climate', implies that global emissions of greenhouse gases need to peak and decline to zero by the end of the twenty-first century (Rogelj et al., 2011). Although not formalized in international law, a long-term global target for complete decarbonization of energy has been accepted by many governments.

To give a sense of the scale of the challenge, it is worth taking the European electric power generation system as an example. This is the sector in which it is commonly assumed most carbon emissions reductions in Europe will be realized (Deetman et al., 2012). Halving emissions by 2050—the current EU goal—would require the construction by then of about 3,000 TWh of renewable power generation capacity (Jacobsson and Karltorp, 2011). This is ten times the size of the current power generation capacity of the United Kingdom.

The technological and innovation challenge of moving from a fossil-fuel driven and linear economy to a solar-powered and closed-loop economy is radical, universal, and urgent. Decarbonization, for instance, represents a transformation of the global energy production industry (oil, gas, and coal), as well as a transformation in energy-using sectors, transport and mobility, the built environment, and food and agriculture, among others. For such transitions to be realized, whole industrial sectors would need to be overturned and a completely new configuration of technologies, businesses, sectors, and consumer preferences and practices would need to emerge in coming decades. Much innovation research and policy on energy and materials is therefore now focused on large-scale, transformative changes in economies, unfolding over long periods of time and leading to radically new socio-technical regimes (Geels, 2002; Smith et al., 2005). For businesses across many sectors technological and business transitions represent complex new dilemmas. While it suggests enormous potentials for returns on investment in R&D and innovation in the long term, the prospect of technological, business, and market discontinuities also generates large strategic risks and uncertainties. Technological and market path-creation is not something firms can do by themselves.

SYSTEM INNOVATION AND TECHNOLOGICAL REGIMES

Seeing sustainable innovation as being ultimately concerned with the wholesale transformation of technological regimes has major consequences for how we see innovation management. Rather than being concerned with incremental innovation of processes, products, and services, or even with the substitution of specific components of existing technological systems, the goal becomes a reconfiguration of systems of provision that are fundamental to the structure and performance of entire economies. Innovation management therefore also becomes oriented to system change. This concern with radical system innovation is not a new concern for innovation scholars. When Henderson and Clark's (1990) system-level architectures need to change—what they call architectural innovation—firms must develop new sets of knowledge, capabilities, and networks. But beyond this, a wider set of changes will also take place in supply and value chains, in infrastructures, in institutional rules determining how a technology is produced and used, in preferences and practices of customers, and so on. Recent research on system innovation has sought to understand a complex set of co-evolutionary changes in which firms are one actor among many others. Seen from this perspective, innovation is an inherently uncertain, collaborative, and open-ended process taking place over a long period, in which besides firm competitiveness, a set of social or environmental expectations and goals are central. Reconciling these often conflicting private and social objectives is a field of tension within which sustainable innovation management is now situated.

The nature, rate, and direction of change in technological regimes differs from change in specific and known technologies, business models, and value chains. Regimes are composed of stable assemblages of technical artefacts, organized in co-evolving market, institutional, social, and knowledge systems. Because of the interrelated and interlocking nature of technological regimes, change tends to be slow and path-dependent. For a regime change to be carried through it must be recognized as necessary, feasible, and advantageous by a broader range of actors and institutions than would normally be the case for a discrete technological change.

Major progress has been made in the analysis of innovation in technological systems. Rip and Kemp (1998) and Geels (2002), in developing a multilevel perspective on system innovation, argued that technological innovation and change could not be seen separately from social, economic, and institutional change. Technologies co-evolve with societal and private preferences, behaviours, and norms, each shaping the other in 'socio-technical systems'. Sustainable innovation therefore needs to be seen as being socially embedded, involving engagement, collaboration, and communication with a wide range of societal actors and in societal arenas. Innovation implies shaping what are often uncertain and radical perspectives on the future. Major questions arise, such as: How does a predominantly coal-based electricity utility respond to the societal objective to eliminate carbon from energy systems? What should be the response of a major car-maker to the challenge of moving to a non-gasoline-based propulsion system for automobiles? What does a global food business do as wild fisheries become depleted and animal proteins come to be seen as unsustainable, forcing the search for new sources of protein?

For analysts such as Kemp, Schot, and Hoogma (1998) it is the transformation of socio-technical regimes (systems transitions) that lies at the heart of achieving more sustainable production and consumption systems. In simple terms, existing regimes constituted by incumbent firms, conventional business models, and patterns of skills, capabilities, and innovation, come under pressure due to inherent and irresolvable problems and social pressures to address these. Examples include the management of horses and manure in large Western cities by the end of the nineteenth century which created pressure to find cleaner forms of transport (Geels, 2004), or contemporary social objections to inhumane treatment of animals bred for meat production. It may be argued that all socio-technical regimes face a number of inherent drawbacks or weaknesses (congestion and accidents on road systems, for instance), but that during periods of stable development, these are not threatening to the stable, path-dependent development of the system. In some cases however, these pressures generate tensions in an existing regime that force the directed search for alternatives, while also leading to increased effort to protect conventional practices through defensive adjustments.

More or less radical alternatives that respond to the incumbent regime's weaknesses emerge in technological and market niches, conceived as networks of actors outside the mainstream, including innovative firms, practitioners (often associated with a values-based movement or community), civil society organizations, and public authorities who may sponsor or facilitate experimentation (Kemp and Rip, 2001). These niches may come to disrupt incumbent regimes to a greater or lesser extent. Smith (2006) investigated the emergence over three decades of the organic food movement in the UK from a peripheral niche promoted by counter-cultural groups to emerge in the late 1990s as a mainstream part of the value offer of major food retailers. He argues that the organic movement has had a major influence on the adoption of a range of new methods in food production, transparency across the supply chain, a growth in local sourcing, and food labelling affecting part of the food retail market. Substantial innovative effort has been made by firms through global value chains, but without these changes altering fundamentally the agriculture and food regime.

Niche actors are seen as performing three key functions leading to the emergence of more stable configurations of technologies, practices, and rules that meet customer preferences that can begin to compete as realistic alternatives to the conventional regime:

- *the articulation of expectations and visions* to orient collaborative learning processes, attract attention, and legitimize protection;
- *the building of social networks* providing a new constituency, facilitating interaction and leveraging resources; and
- *learning* across a multitude of dimensions: about technical and design specifications; market preferences and costs/prices; industry and production networks; societal and environmental effects; and so on.

Actors participating in technological or market niches face a number of strategic dilemmas. Schot and Geels (2008) define these as related to: whether to have flexible expectations or directed goals in the face of uncertainty; whether to encourage technological variety to encourage learning or restrict it because it generates new uncertainty; whether to work with incumbents with resources and broad capabilities or with new entrants untied to vested interests; how long to maintain protection of the niche; and whether to wait for 'windows of opportunity' to emerge in existing regimes, or seek to force the issue by creating a sustainable alternative and bring pressure from the outside. How firms, as participants in innovative sustainable niches, handle each of these dilemmas will depend on their innovative capabilities, their current and envisaged business models, and their capacity to manage collaboration across supply chains and with societal stakeholders.

Environmental Innovation Capabilities, Business Models, and Linkages

Environmental Innovation Capabilities

Whether environmental innovation requires specific and new capabilities has been widely debated in the literature. Wagner and Llerana (2011) find in a survey of European and North American firms that existing innovative capabilities are fundamental to sustainable innovation. Likewise, Chakrabarty and Wang (2012) argue that strong innovation capabilities associated with intensive R&D and strong market orientation linked to high levels of internationalization are fundamental to developing and maintaining sustainability practices in large firms. Nidumolu et al. (2009) and Padgett and Galan (2009) also find that R&D provides a strong basis for pro-sustainability improvements in products and processes. This is also true for small and medium-sized firms. Hofmann et al. (2012), in an analysis of 239 US manufacturing small and medium-sized enterprises, find a strong positive association between the adoption of high technology (such as Computer Aided Design and statistical process control) and environmental practices (such as an environmental management system).

In developing a general argument about the development of dynamic capabilities (Teece et al., 1997, and see Chapter 1 by Dodgson, Gann and Phillips) for managing innovation for sustainability, Seebode et al. (2012) define four broad approaches, coupled with specific configurations of strategy and capabilities. The approach taken will depend on whether the innovation challenge is incremental or radical, and whether existing frames or new frames are called for (see Figure 15.1). In the left column established routines for search, selection, and implementation will suffice. But where new frames are called for, more open-ended and collaborative innovation management is needed, emphasizing search in unfamiliar technological fields, selection under conditions of high uncertainty, the need to mobilize new skills, structures, and languages internally, and an innovation strategy that involves road-mapping over the longer term and potentially a new corporate paradigm. They illustrate this argument by describing Phillips' evolving innovation strategy which is seen as having moved progressively towards reframing and co-evolution.

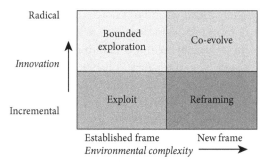

FIGURE 15.1 Map of innovation space (Seebode, Jeanrenaud, and Bessant, 2012)

While embedded in and drawing on existing innovative capabilities, environmental innovation also requires the development of new capabilities by firms which are related to environmental management, new forms of collaboration with suppliers and customers, and stakeholder engagement. The underlying motivations for adopting environmental practices and pursuing sustainable innovation are also different, feeding a more outward-facing orientation to the management of innovation. The most common business environmental management practices identified in studies include: employee involvement programmes, internal audits, supplier audits, environmental reporting, environmental management systems (EMS), ISO 14000 (the International Standardization Organization standard for environmental management), total quality management, just-in-time manufacturing, life-cycle analysis, total cost accounting, pollution prevention plans, and designated environmental management staff (Florida and Davison, 2001; Simpson and Samson, 2010). The growth of EMS systems, starting in the 1990s, followed the widespread adoption of 'quality management' standards across global industries, and was motivated by a similar conviction that standardized management control and independent certification would bring environmental performance improvement, matched by cost savings and therefore competitiveness gains for firms. Berkhout (1997) found that environmental assessment capabilities, such as life-cycle assessment, have become conventional in many sectors ranging from process-based commodity producers (paper, steel), complex products (automobiles, electronics), and simple consumer products (health and home care). A common theme in this literature is the great variation that exists between the motivations, practices, and environmental performance of business organizations adopting environmental management practices.

Building environmental management or corporate responsibility capabilities, regardless of the motivation, presents firms with a number of management challenges. First is the degree to which capabilities should be developed in-house or externally. While many firms have established separate environmental or corporate responsibility competencies over the last two decades, collaboration with other organizations and stakeholders is frequently inevitable because of the specialist knowledge and capabilities required in, for instance, conducting environmental assessments or developing novel sustainable products and processes. These sources of external knowledge and competences require new skills in managing relationships and in knowledge integration (Foster and Green, 2000). Petruzzelli et al. (2011), in a study of 'green' and other patenting behaviour by global companies in four industrial sectors, found that green innovations are characterized by higher levels of inter- and intra-firm collaborations compared to other innovations developed by the same firms. They also found that green innovations represented a higher level of complexity and novelty than other innovations.

Another reason why external collaboration is important is that environmental management and reporting is typically outward-facing, towards the supply chain, customers, regulators, and broader civil society. The environmental quality of a process or product is not typically evident to the customer or the regulator, but needs to be underpinned by systematic evidence of the production process, the supply chain, or the fate of a product. Such evidence needs to be transparent and reliable. Third-party verification plays an important role in countering charges of 'greenwashing'—the attempt by firms to gain market or reputational advantages by making unfounded claims about the environmental quality of products or processes (Laufer, 2003; Ramus and Montiel, 2005). Managing this interface with external suppliers of environmental knowledge and with external stakeholders in the construction of legitimacy and reputation around environmental claims is a key new capability which many firms (and other organizations) have developed.

A second consideration is the degree to which environmental management is integrated across corporate functions, including R&D, quality management, supply chain management, and marketing (Shrivastiva and Hart, 1995; S. L. Hart, 1995). An environmental management function may provide the basis for developing more specialized environmental skills, but as environmental issues grow in strategic importance, there are likely to be growing benefits from integration across business functions. Hall and Wagner (2012) find a positive association of the integration of strategic issues with environmental management with both product and process innovation. They also find that this feeds through to better relative environmental performance. Alignment of a firm's strategic goals with sustainability depends on top management support, but needs to be coupled with decentralization of decisions to managers across the organization. Wagner and Llerena (2011) stress the importance of 'bottom up' eco-efficiency initiatives championed by key individuals in the companies they surveyed. They also argue that for leading companies, environmental criteria come to be introduced earlier in innovation management—in pre-development and stage-gate processes—as sustainability is recognized as a strategic issue within businesses..

MANAGING SUPPLY AND VALUE CHAINS

In recognition of the outward-facing and collaborative setting for sustainable innovation, one of the key developments over the past two decades has been a growing focus on sustainable supply and value chain management (Lamming and Hampson, 1996). Carter and Rogers (2008: 368) define sustainable supply chain management as 'the strategic, transparent integration and achievement of an organization's social, environmental and economic goals in the systemic coordination of key inter-organizational business processes for improving the long-term economic performance of the individual and its supply chain'. In the context of increasingly global markets, Seuring and Müller (2008) argue that firms may be seeking to manage risks (such as new public or private environmental performance standards) and environmental and economic performance (since environmental performance of products needs to be considered across the full life cycle, and waste increases costs), or seeking to exploit opportunities for competitive advantage through innovation of sustainable products. Ageron et al. (2012), in a survey of supply chain managers in French companies, show that regulatory requirements, customer demand, competitive pressure, and external stakeholders provide the key motivations for sustainable supply chain management, with top management support an important enabling factor. A key innovative capability underpinning sustainable innovation is therefore a capacity to create, maintain, and improve collaboration across supply and value chains, for global leading firms and for their suppliers.

Keeping abreast of multiple, complex, and dynamic supply and value chains—to a large extent outside the direct management scope of firms—and leveraging sustainable innovation strategies through them has become a critical concern for business and innovation management. It has also become a major focus for the development of new innovative capabilities within leading firms and across supply and value chains. Focal companies, whether these are retailers or final goods suppliers, play a pivotal role in fostering environmental improvements in their suppliers. In order to achieve these changes in suppliers, such companies enact knowledge transfers and technical cooperation in innovation, adding to reciprocal dependencies and causing shifts from 'arm's length' towards more complex governance structures (Simpson et al., 2007; Marchi et al., 2013). Such inter-firm linkages may already exist for conventional reasons of quality, speed, dependability, and cost, but they are deepened by the need to manage environmental and social quality as well. Studies of global textiles and apparel supply chains, for instance, show that suppliers of organic cotton had to make substantial investments to improve production facilities and processes in order to meet environmental standards (Chouinard and Brown, 1997; Goldbach et al., 2004; Seuring and Müller, 2008). Substantial new knowledge flows were required to become what King and Lenox (2002) characterize as 'lean and green' suppliers, as well as cooperation between a wider range of companies along the supply chain than would conventionally be necessary (De Bakker and Nijhof, 2002).

The global value chain (GVC) framework has been developed as a way of analysing the global division of labour between independent actors in developed and developing countries. It emphasizes the role of global leading firms in shaping value chains, determining their governance and the organization of innovative activities (Gereffi, 2005). GVC is concerned with analysing international inter-firm linkages from the perspective of the de-integration of production from vertically integrated firms into globally dispersed value chains requiring coordination. The approach is also useful to understanding environmental upgrading across global value chains (Marchi et al., 2013). By requiring suppliers across the world to comply with environmental standards, and by transferring technology and specialized knowledge related to the sustainability of products and processes, leading firms can play a decisive role in shaping their suppliers' environmental performance. Indeed, it may be argued that while environmental performance in rich countries is driven primarily by regulation, stakeholders, and competitive pressures, in emerging and developing economies value chain pressures exerted by lead firms may be the most significant (Jeppeson and Hansen, 2004).

Managing Stakeholders

Hall and Vredenburg (2003) argue that sustainable development innovation is more complex than conventional, market-driven innovation, because companies have to consider a wider range of stakeholders and their often contradictory demands. Broadly, two related areas of scholarship have sought to interpret business relationships with these 'secondary' stakeholders: stakeholder theory and the literature on corporate social responsibility (CSR). Freeman (1984), in proposing a 'stakeholder model' of the firm, argued that managers were concerned not only with shareholders, but also with the views and interests of employees, suppliers, and customers and disparate members of societal organizations, including local community organizations, consumer advocates, environmentalists, and so on. Firms engage with stakeholders in order to foresee and manage societal risks to their business, but also to profit from positive relationships with stakeholders (Donaldson et al., 1995; Jones and Wicks, 1999). Good stakeholder relationships generate competitive advantage through reputational benefits, regulatory relief through building trust, and as a source of new ideas underpinning innovation.

Establishing a link between stakeholder engagement by firms and innovativeness is complex. Ayuso et al. (2011) frame sustainable innovation as an organizational capability which combines the ability of a firm to establish strong and interactive relationships with stakeholders with the capacity to manage the acquired stakeholder knowledge and transform it into socially and environmentally sound innovations. In a large econometric study using data for 983 firms in the Dow Jones Industrial Index across different sectors, they find that knowledge sourced from employees and external stakeholders did affect innovativeness. But this was mediated by the nature of stakeholder engagement and knowledge management routines within the firm. Firms are only able to leverage strong stakeholder relationships into greater innovativeness if they have capabilities to capture and absorb the information and trust they receive as a result.

The debate about corporate social responsibility (CSR) has a long history and turns on the question of whether firms (or the managers who run them on behalf of shareholders) have responsibilities to society beyond making profits. Some, like Milton Friedman (1970), argue that the practical impact of 'social responsibility' must always be to divert management effort away from normal competition, and to impose new and unnecessary costs on business. Yet, concern about CSR is seen in the widely available pronouncements by business leaders and the proliferation of initiatives, many sponsored by governments, that have emerged in recent years. O. Hart (1995) was one of the first to argue that CSR represented a resource or capability leading to sustained competitive advantage in firms. A link with R&D and innovation was made by McWilliams and Segal (2000), who argued that CSR principles applied in R&D and innovation management lead to new or improved products and processes. They paid particular attention to the role of CSR in differentiation strategies by firms through innovation, for instance by developing products based on recyclates. Several studies have found a positive relationship between R&D intensity and CSR practices (Bansal, 2005; Husted and Allen, 2007). Lopez et al. (2008, cited in Gallego-Alvarez et al., 2011) find that company R&D is positively affected by adoption of CSR-oriented goals, and that companies that adopt CSR practices tend to have higher R&D expenditures. Not all studies are equally sanguine about the CSR-innovation link. Gallego-Alverez et al. (2011) compared R&D intensity in companies listed within the Dow Jones Sustainability Index (DJSI) with those not listed. They found a short-term negative relationship between CSR practices and innovation, although this effect varied widely across business sectors. The link between CSR and innovation therefore appears to emerge over the longer term and is sectorally differentiated.

It is not enough to see CSR as being directly linked to innovative performance alone. CSR is also a dynamic capability generating a range of intangible advantages enabling firms to be innovative and successful. By emphasizing broader social values and norms, besides only economic value (what Porter and Kramer (2011) call 'shared value') CSR also mediates across the internal–external boundary of the firm. CSR makes a number of contributions to firm performance. First, it is a way in which firms have sought to build a consistent picture of their corporate values as an ideological system that socializes employees to strategic objectives. It may also substitute for hierarchical approaches to management control by helping to bind employees to these corporate goals. CSR is therefore frequently internally directed. Second, structural and market changes have profoundly influenced the competitive environment in which many firms operate (Baron, 2001; Maxfield, 2008). Greater competition, especially in commoditized markets, has forced businesses to seek new ways of differentiating their products and services (and their brand) with the final consumer. CSR is one way of supporting the creation and reproduction of a corporate reputation and brand around sustainability as a social value. This type of signalling to the customer is also oriented at shareholders. As shareholding has become more distributed, both the intensity and variety of public signalling by corporations has increased. Shareholder value is still primarily defined by the growth in share prices, but a number of other means of signalling to shareholders have been developed by business. These may be seen as a way of demonstrating management quality and the prospects for future share performance, even during periods of weak share performance. The greater vulnerability of firms (and their managers) in more competitive markets has produced a need for new forms of communication about management performance. Evidence of CSR has been one sort of response to this need.

Third, corporate social responsibility may be seen as a response to the changing context of social regulation within which firms operate (Matten and Moon, 2008). The 'statist' model of regulation in which governments—international, regional, national or local—impose legally enforceable standards on firms is slowly being complemented by a model of social regulation that is more interactive and distributed. New information-rich voluntary and market-based regulatory measures are being developed to operate beside classical systems of 'command and control' regulation. While national environmental policy styles remain highly specific, the capacity of governments to secure the public interest in the environmental field has been reshaped. Increasing voluntarism, 'partnership' between business and government, and a more influential role for non-governmental organizations are all signs of this process of 'ecological modernization'.

This new context of social regulation poses challenges for business, which has sought to develop new capabilities and roles in response. Although firms in many industries have secured greater economic freedoms as a result of the market liberalization and deregulation, in many cases this has been matched by a new set of pressures to demonstrate conformance with social norms and expectations. Corporate social responsibility can therefore be seen as a way of 'filling the space' that has been opened in the reshaping of environmental and social governance of firms (Kolk, 2005). Paradoxically, many firms now operate in a more difficult and insecure social environment, with a wider range of constituencies to relate to and more diverse expectations to meet. Formal regulation, while often inflexible and procedurally onerous, presents firms (especially large ones) with clear objectives and a simple set of external relationships to manage. In a more fluid and crowded regulatory context these certainties are removed, and firms seek to re-establish order by becoming more active in setting objectives and managing relationships.

In short, corporate social responsibility can be argued as being shaped by a bundle of needs in many firms: new needs for internal cultural cohesion and management control; new needs to 'signal' about management quality to customers and shareholders; and new needs to engage actively in the new context of social regulation of firms. A fully-fledged corporate sustainability strategy, including the potential to realize sustainable innovation and a radical reframing of value chains and business models, needs to be based on strong CSR and social values capabilities.

SUSTAINABLE BUSINESS MODEL INNOVATION

Given the potentially radical implications of sustainable innovation in many business sectors, especially those heavily dependent on non-renewable, harmful, or scarce resources, a new debate has emerged about the need for business model innovation as firms come to see sustainability as a core strategic goal (Nidumolu et al., 2009, and see chapter 21 by Massa and Tucci). Research suggests that there is growing awareness in global firms of the link between sustainability-driven business model innovation and long-term profitability. In a survey of 2,600 executives, Kiron et al. (2013) find that 48 per cent of firms had changed business models for sustainability reasons, and that of these 75 per cent saw a positive link between business model innovation and profitability. Moreover, the deeper the changes to the business model (especially in target segments and value chain process change), the greater the likelihood of an association with greater profitability. They also show a 'stakeholder effect', with those firms increasing their collaboration with stakeholders showing a greater association with profitability. These are indicative results, but they are part of a growing focus in the business literature on the need for and benefits of strategic reorientation in business as a response to sustainability challenges. Such reorientation is more effective when placed in the context of stakeholder engagement and CSR.

DEFINING A SUSTAINABLE INNOVATION STRATEGY

It is clear that there is a deep connection between corporate and innovation strategy in many business sectors, and this is also the case for sustainable innovation strategy. A large literature exists on the way in which, across different firms and through time, businesses have adjusted to the pressures and opportunities presented by the need to become less polluting, more eco-efficient, and to develop greener products and services. In general, these may be seen as 'stages models' in which firms become increasingly more aware of the environmental and social impacts of the activities, build capabilities to manage them, and begin integrating sustainability into their business models, and innovation and corporate strategies. Through this integration and embedding of sustainability into the core strategic goals of the firm, sustainability becomes more closely aligned with the competitive strategy of the business. The whole literature is not reviewed here, but two more recent and influential contributions are highlighted.

Nidulomu et al. (2009), using case study examples from firms in a range of sectors, describe how corporate sustainability strategies evolve through time from a starting position in which compliance with existing regulations and standards is the basis for engagement with environment and sustainability, though a series of five stages to a strategic outlook which views the logic of a business through a sustainability lens (see Table 15.1). At each stage new dimensions of business model innovation are called for. They detail for each stage a series of specific innovation opportunities that are in turn linked to the upgrading of innovative capabilities in firms.

A second type of approach to sustainability strategy is illustrated by Orsato (2006). This is less concerned with the dynamics of strategy over time, seeking instead to couple the underlying competitive strategy of firms with their environmental strategy. He argues that a variety of environmental strategies are available to firms, driven by the specific competitive and innovation strategy they pursue. Based on Porter's (1980) competitive advantage models, which identify low costs and differentiation as two generic types of competitive advantage, Orsato (2006) proposes four types of competitive environmental strategy, determined by the firm's competitive focus (whether on organizational or on product/service innovation) and its competitive advantage

Table 15.1 Stages in corporate sustainability strategies and innovation opportunities (Nidulomu et al., 2009)

STAGE 1 Viewing Compliance as Opportunity	STAGE 2 Making Value Chains Sustainable	STAGE 3 Designing Sustainable Products and Services	STAGE 4 Developing New Business Models	STAGE 5 Creating Next Practice Platforms
Central challenge	**Central challenge**	**Central challenge**	**Central challenge**	**Central challenge**
To ensure that compliance with norms becomes an opportunity for innovation.	To increase efficiencies throughout the value chain.	To develop sustainable offerings or redesign existing ones to become eco-friendly.	To find novel ways of delivering and capturing, which will change the basis of competition.	To question through the sustainability lens the dominant logic behind business today.
Competencies needed	**Competencies needed**	**Competencies needed**	**Competencies needed**	**Competencies needed**
>> The ability to anticipate and shape regulations.	>> Expertise in techniques such as carbon management life-cycle assessment.	>> The skills to know which products or services are most unfriendly to the environment.	>> The capacity to understand what consumers want and to figure out different ways to meet those demands.	>> Knowledge of how renewable and non-renewable resources affect business ecosystems and industries.
>> The skill to work with other companies, including rivals, to implement creative solutions.	>> The ability to redesign operations to use less energy and water, produce fewer emissions, and generate less waste.	>> The ability to generate real public support for sustainable offerings and not to be considered as 'greenwashing'.	>> The ability to understand how partners can enhance the value of offerings.	>> The expertise to synthesize business models, technologies, and regulations in different industries.
	>> The capacity to ensure that suppliers and retailers make their operations eco-friendly.	>> The management know how to scale both supplies of green materials and the manufacture of products.		

(Continued)

Table 15.1 (*Continued*)

STAGE 1 Viewing Compliance as Opportunity	STAGE 2 Making Value Chains Sustainable	STAGE 3 Designing Sustainable Products and Services	STAGE 4 Developing New Business Models	STAGE 5 Creating Next Practice Platforms
Innovation opportunity	Innovation opportunities	Innovation opportunities	Innovation opportunities	Innovation opportunities
>> Using compliance to induce the company and its partners to experiment with sustainable technologies, materials, and processes.	>> Developing sustainable sources of raw materials and components.	>> Applying techniques such as biomimicry in product development.	>> Developing new delivery technologies that change value-chain relationships in significant ways.	>> Building business platforms that will enable customers and suppliers to manage energy in radically different ways.
	>> Increase the use of clean energy sources such as wind and solar power.	>> Developing compact and eco-friendly packaging.	>> Creating monetization models that reveals to services rather than products.	>> Developing products that won't need water in categories traditionally associated with it, such as cleaning products.
	>> Finding innovative uses for returned products.		>> Devising business models that combine digital and physical infrastructures.	>> Designing technologies that will allow industries to use the energy produced as a by-product

(cost or differentiation). If the focus is on organizational innovation, firms will pursue eco-efficiency and 'beyond compliance' leadership strategies (Marchi et al., 2013). In an eco-efficiency strategy, companies compete by offering lower costs, so improving competitiveness while reducing impacts on the environment. When competitive advantage is sought by the company through differentiation, greater advantage will be achieved through a strategy of demonstrating beyond compliance performance, enabling firms to capture final markets through eco-labelling or green marketing. If the firm's focus is on products and services innovation (rather than organizational innovation), then a strategy of eco-branding and environmental cost leadership will be more appropriate.

Under this strategy a firm seeks to differentiate itself from its competitors by offering clearly distinct eco-friendly products and services, differentiated from unsustainable or polluting alternatives. A premium price generates opportunities for green market rents. With an environmental cost leadership strategy the firm competes by offering both novel green products or services, while also achieving lower prices than unsustainable alternatives. This break-through strategy will typically involve a radical rethinking of the product or service. Orsato (2009) also proposes a strategic approach called the 'blue ocean strategy', where the focus is on resetting the competitive landscape through an innovative strategic vision for the firm based on sustainability similar to the position of Nidulomu et al. (2009).

Conclusions

All business activities have environmental and social externalities that may need to be managed. For much of the post-war period the assumption amongst scholars, business, and regulators was that government—seeking to secure public goods such as a safe and clean environment on behalf of society—would set standards for business and ensure compliance. Regulations evolved and grew more stringent, but always through an inter-action with business and taking assumed negative competitiveness impacts into account. A stable process of incremental environmental upgrading of industry took place, lead-ing over time to significant reductions in many polluting emissions from processes and products. With emerging global scarcity of critical resources, including energy, water, phosphorus and land, and in the context of global environmental problems such as cli-mate change and ecosystem and biodiversity loss, there is broad agreement that radi-cal, rather than incremental, change is needed in production and consumption systems. Technological innovation will play a key role in these systems innovations, and inno-vation management in business is central to achieving such radical changes over the longer term. Understanding sustainable innovation is therefore an urgent and impor-tant research task. There is also agreement that the role of government in governing the environmental quality of growth has changed. Other social actors, including citizens, civil society organizations, and consumers, have come to play a significant role in shap-ing innovation towards sustainability. Increasingly firms, large and small, have begun to embed sustainability goals in their technology, innovation and corporate strategies. This is no longer only in search of short-term competitive gain through eco-efficiency, but also as a way of positioning the firm on a sustainable strategic path.

In summary, this chapter develops a number of arguments:

1. Sustainable system innovation needs to be seen as a co-evolutionary process involving not just innovative firms, but a broader context of institutions, infrastructures, consumer practices, and so on, emerging over periods of decades.

2. Innovation in this context needs to be conceived of as open, collaborative, dynamic, and uncertain, stretching far beyond the boundaries of individual firms. Competition needs to be matched by cooperation, and firms need to actively engage in complex new global governance arrangements linked to system transformations. Managing multiple relationships with the external business, government, and societal environment, and the intangible assets connected with these relationships, therefore becomes a core part of corporate and innovation strategies.

3. System innovation is concerned with creating pathways to new socio-technical configurations, as well as decisively abandoning old path dependencies and established capabilities and routines.

4. Sustainable innovation in business is based on many conventional dynamic capabilities, but they also require the development of new capabilities linked to environmental management, intensified collaboration across supply and value chains, improved management of relationships with a range of stakeholders beyond suppliers, customers, and shareholders, and the management of intangibles linked to reputation, risk, and brand value.

5. More sustainable innovation is linked to corporate sustainability strategy, which itself has become increasingly embedded in core corporate strategies. Many businesses will move through a series of stages in the integration and alignment of environmental and sustainability strategies with corporate strategies. There is a variety of sustainability strategies that may be followed across different businesses and industries.

6. Establishing the short-term competitiveness or environmental performance benefits of sustainability strategies—the promise of 'win-wins'—has proven difficult, with a diverse range of results. Yet, a strategic commitment to sustainability helps to ensure the long-term sustainability and profitability of the business as a whole—to make it 'future resilient'—by evolving business models and innovations in response to social and environmental needs, and emerging resources and environmental constraints.

Sustainability has moved from being a fringe management issue to being a core feature of corporate and innovation strategies. Technological innovation will play a fundamental role in securing sustainable growth pathways. The challenge for far-sighted businesses will be to catch the wave of creative destruction which will shape these futures.

REFERENCES

Ageron, B., Gunasekaran, A., and Spalanzani, A. (2012). 'Sustainable Supply Management: An Empirical Study', *International Journal of Production Economics*, 140(1), 168–82. doi:10.1016/j.ijpe.2011.04.007

Ambec, S., Cohen, M. A., and Elgie, S. (2011). *The Porter Hypothesis at 20: Can Environmental Regulation Enhance Innovation and Competitiveness?* Working Paper DP11-01, Resources for the Future: Washington, DC.

Ayuso, S., Rodríguez, M. Á., García-Castro, R., and Ariño, M. Á. (2011). 'Does Stakeholder Engagement Promote Sustainable Innovation Orientation?', *Industrial Management & Data Systems*, 111(9), 1399–417. doi:10.1108/02635571111182764

Bansal, P. (2005). 'Evolving Sustainably: A Longitudinal Study of Corporate Sustainable Development', *Strategic Management Journal*, 26(3): 197–218. doi:10.1002/smj.441

Baron, D. P. (2001). 'Private Politics, Corporate Social Responsibility and Integrated Strategy', *Journal of Economics & Management Strategy*, 10(1): 7–45.

Bebbington, J., Larrinaga, C., and Moneva, J. M. (2008). 'Corporate Social Responsibility and Reputation Risk Management', *Accounting, Auditing and Accountability Journal*, 21(3): 337–61.

Berkhout, F. (1997). 'The Adoption of Life-Cycle Approaches by Industry: Patterns and Impacts', *Resources, Conservation and Recycling*, 20: 71–94.

BP Statistical Review of World Energy June 2012 (bp.com/statisticalreview).

Carter, C. R., and Rogers, D. S. (2008). 'A Framework of Sustainable Supply Chain Management: Moving toward New Theory', *International Journal of Physical Distribution & Logistics Management*, 38(5): 360–87. doi:10.1108/09600030810882816

Chakrabarty, S., and Wang, L. (2012). 'The Long-Term Sustenance of Sustainability Practices in MNCs: A Dynamic Capabilities Perspective of the Role of R&D and Internationalization', *Journal of Business Ethics*, 110(2): 205–17. doi:10.1007/s10551-012-1422-3

Chouinard, Y., and Brown, M. S. (1997). 'Going Organic: Converting Patagonia's Cotton Product Line', *Journal of Industrial Ecology*, 1(1): 117–29. doi:10.1162/jiec.1997.1.1.117

De Bakker, F., and Nijhof, A. (2002). 'Responsible Chain Management: A Capability Assessment Framework', *Business Strategy and the Environment*, 11(1), 63–75. doi:10.1002/bse.319.

Deetman, S., Hof, A. F., Pfluger, B., Van Vuuren, D. P., Girod, B., and Van Ruijven, B. J. (2012). 'Deep Greenhouse Gas Emission Reductions in Europe: Exploring Different Options', *Energy Policy*, 55: 152–64.

Donaldson, T., and Preston, L. E. (1995). 'The Stakeholder Theory of the Corporation: Concepts, Evidence, Implications', *Academy of Management Review*, 20(1): 65–91.

Florida, R., and Davidson, D. (2001). 'Gaining from Green Management: Environmental Management Systems inside and outside the Factory', *California Management Review*, 43(3): 64–84.

Foster, C., and Green, K. (2000). 'Greening the Innovation Process', *Business Strategy and the Environment*, 9: 287–303.

Freeman, R. E. (1984). *Strategic Management: A Stakeholder Approach*. Boston: Pitman.

Friedman, M. (1970). 'The Social Responsibility of Business is to Increase its Profits', *New York Times*. Available at < http://www.umich.edu/~thecore/doc/Friedman.pdf > (accessed 8 July 2013).

Gallego-Álvarez, I., Prado-Lorenzo, J. M., and García-Sánchez, I.-M. (2011). 'Corporate Social Responsibility and Innovation: A Resource-Based Theory', *Management Decision*, 49(10): 1709–27. doi:10.1108/00251741111183843

Geels, F. W. (2002). 'Technological Transitions as Evolutionary Reconfiguration Processes: A Multi-Level Perspective and a Case-Study', *Research Policy*, 31: 1257–74.

Geels, F. W. (2004). 'From Sectoral Systems of Innovation to Socio-Technical Systems: Insights about Dynamics and Change from Sociology and Institutional Theory', *Research Policy*, 33(6–7): 897–920.

Gereffi, G. (2005). 'The Global Economy: Organization, Governance, and Development', in N. J. Smelser and R. Swedberg (eds), *Handbook of Economic Sociology*. Princeton: Princeton University Press, 160–82.

Goldbach, M., Seuring, S., and Back, S. (2004). 'Co-ordinating Sustainable Cotton Chains for the Mass Market', *Greener Management International*, 43: 65–79.

Gunningham, N., Kagan, R., and Thornton, D. (2004). 'Social License and Environmental Protection: Why Businesses go beyond Compliance', *Law & Social Inquiry*, 29(2): 307–41. doi:10.1111/j.1747-4469.2004.tb00338.x

Haddock-Fraser, J., and Fraser, I. (2008). 'Assessing Corporate Environmental Reporting Motivations: Differences between "Close-to-Market" and "Business-to-Business" Companies', *Corporate Social Responsibility and Environmental Management*, 15: 140–55.

Hall, J., and Vredenburg, H. (2003). 'The Challenges of Innovating for Sustainable Development', *MIT Sloan Management Review*, 45(1): 61–8.

Hall, J., and Wagner, M. (2012). 'Integrating Sustainability into Firms' Processes: Performance Effects and the Moderating Role of Business Models and Innovation', *Business & Society*, 21(August 2011): 183–96. doi:10.1002/bse

Hart, O. (1995). 'Corporate Governance: Some Theory and Implications', *Economic Journal*, 105(430): 678–89.

Hart, S. L. (1995). 'A Natural Resource Based View of the Firm', *Academy of Management Review*, 20(4): 986–1014.

Hawken, P., Lovins, A., and Lovins, H. (1999). *Natural Capitalism: The Next Industrial Revolution*. London: Earthscan.

Henderson, R. M., and Clark, K. B. (1990). 'Architectural Innovation: The Reconfiguration of Existing Product Technologies and the Failure of Established Firms', *Administrative Science Quarterly*, 35(1): 9–30.

Hofmann, K. H., Theyel, G., and Wood, C. H. (2012). 'Identifying Firm Capabilities as Drivers of Environmental Management and Sustainability Practices—Evidence from Small and Medium-Sized Manufacturers', *Business Strategy and the Environment*, 21(January): 530–45. doi:10.1002/bse

Husted, B. W., and Allen, D. B. (2007). 'Strategic Corporate Social Responsibility and Value Creation among Large Firms', *Long Range Planning*, 40(6): 594–610. doi:10.1016/j.lrp.2007.07.001

Jacobsson, S., and Karltorp, K. (2011). 'Formation of Competences to Realise the Potential of Offshore Wind Power in the European Union', draft Available at: < http://cvi.se/uploads/pdf/Jacobsson%20and%20Karltorp%202011.pdf > (accessed 8 July 2013).

Jaffe, A. B., and Palmer, K. (1997). 'Environmental Regulation and Innovation: A Panel Study', *Review of Economics and Statistics*, 79(4): 610–19.

Jeppeson, S., and Hansen, M. (2004). 'Environmental Upgrading of Third World Enterprises through Linkages', *Business Strategy and the Environment*, 13(4): 261–74.

Jones, T. M., and Wicks, A. C. (1999). 'Convergent Stakeholder Theory', *Academy of Management Review*, 24(2): 206–21.

Kemp, R., and Rip, A. (2001). 'Constructing Transition Paths through the Management of Niches', in R. Garud and P. Karnoe (eds), *Path Dependence and Creation*. Mahwah, NJ: Lawrence Erlbaum, 269–99.

Kemp, R., Schot, J., and Hoogma, R. (1998). 'Regime Shifts to Sustainability through Processes of Niche Formation: The Approach of Strategic Niche Management', *Technology Analysis & Strategic Management*, 10(2): 175–98.

King, A., and Lenox, M. (2002). 'Exploring the Locus of Profitable Pollution Reduction', *Management Science*, 48(2): 289–99. doi:10.1287/mnsc.48.2.289.258

Kiron, D., Kruschwitz, N., and Reeves, M. (2013). 'The Benefits of Sustainability-Driven Innovation', *MIT Sloan Review*, 54(2): 69–73.

Kolk, A. (2005). 'Corporate Social Responsibility in the Coffee Sector: The Dynamics of MNC Responses and Code Development', *European Management Journal*, 23(2): 228–36.

Lamming, R., and Hampson, J. (1996). 'The Environment as a Supply Chain Management Issue', *British Journal of Management*, 7(March): S45–S62.

Lanoie, P., Laurent-Lucchetti, J., Johnstone, N., and Ambec, S. (2011). 'Environmental Policy, Innovation and Performance: New Insights on the Porter Hypothesis', *Journal of Economics & Management Strategy*, 20(3): 803–42. doi:10.1111/j.1530-9134.2011.00301.x

Lanoie, P., Patry, M., and Lejeunesse, R. (2008). 'Environmental Regulation and Productivity: Testing the Porter Hypothesis', *Journal of Productivity Analysis*, 30: 121–8.

Laufer, W. S. (2003). 'Social Accountability and Greenwashing', *Journal of Business Ethics*, 43(3): 253–61.

Lepak, D. P., Smith, K. G., Taylor, M. S., and Smith, K. E. N. G. (2007). 'Value Creation and Value Capture: A Multilevel Perspective', *Academy of Management Perspectives*, 32(1): 180–94.

Lopez, M. V., Perez, M. C., and Rodrıguez, L. (2008). 'Strategy, Corporate Social Responsibility and R&D Expenditure: Empirical Evidence of European Convergence', Paper presented at the 31st Annual Congress of the European Accounting Association, Rotterdam.

Lovins, A. (1976). 'Energy Strategies: The Road Not Taken?', *Foreign Affairs*, 65: 65–95.

McDonough, W., and Braungart, M. (1998). 'The Next Industrial Revolution', *Atlantic Monthly*, (4): 82–92.

McWilliams, A., and Segal, D. (2000). 'Corporate Social Responsibility and Financial Performance: Correlation or Misspecification', *Strategic Management Journal*, 21(5): 603–9.

Marchi, V. De, Maria, E. Di, and Micelli, S. (2013). 'Environmental Strategies, Upgrading and Competitive Advantage in Global Value Chains', *Business Strategy and the Environment*, 22(May 2012): 62–72. doi:10.1002/bse

Matten, D., and Moon, J. (2008). ' "Implicit" and "Explicit" CSR: A Conceptual Framework for a Comparative Understanding of Corporate Social Responsibility', *Academy of Management Review*, 33(2): 404–24.

Maxfield, S. (2008). 'Reconciling Corporate Citizenship and Competitive Strategy: Insights from Economic Theory', *Journal of Business Ethics*, 80: 367–77.

Nidumolu, R., Prahalad, C. K., and Rangaswami, M. R. (2009). 'Why Sustainability is now the Key Driver of Innovation', *Harvard Business Review*, September: 2–9.

Orsato, R. J. (2006). 'Competitive Environmental Strategies: When does it Pay to be Green?', *California Management Review*, 48(2): 127–44.

Padgett, R. C., and Galan, J. I. (2009). 'The Effect of R&D Intensity on Corporate Social Responsibility', *Journal of Business Ethics*, 93(3): 407–18. doi:10.1007/s10551-009-0230-x

Petruzzelli, A. M., Deangelico, R. M., Rotolo, D., and Albino, V. (2011). 'Organisational Factors and Technological Features in the Development of Green Innovations: Evidence from Patent Analysis', *Innovation: Management, Policy & Practice*, 13: 291–310.

Popp, D. (2006). 'International Innovation and Diffusion of Air Pollution Control Technologies: The Effects of NOx and SO2 Regulation in the US, Japan, and Germany', *Journal of Environmental Economics and Management*, 51: 46–71.

Porter, M. E. (1991). 'America's Green Strategy', *Scientific American*, 264: 168.

Porter, M. E., and Kramer, M. R. (2011). 'Creating Shared Value', *Harvard Business Review*, January–February: 3–17.

Porter, M. J., and Linde, C. van der (1995). 'Toward a New Conception of the Environment-Competitiveness Relationship', *Journal of Economic Perspectives*, 9(4): 97–118.

Ramus, C. A., and Montiel, I. (2005). 'When are Corporate Environmental Policies a Form of Greenwashing?', *Business & Society*, 44(4): 377–414.

Rennings, C. (2000). 'Redefining Innovation—Eco-Innovation Research and the Contribution from Ecological Economics', *Ecological Economics*, 32: 319–32.

Rip, A., and Kemp, R. (1998). 'Technological Change', in S. Rayner and E. L. Malone (eds), *Human Choice and Climate Change*, ii: *Resources and Technology*. Columbus, Oh: Battelle Press, 327–99.

Rogelj, J., Hare, W., Lowe, J., Vuuren, D. P. van, Riahi, K., Matthews, B., Hanaoka, T., Jiang, K., and Meinshausen, M. (2011). 'Emissions Pathways Consistent with a 2°C global Temperature Limit', *Nature Climate Change*, 1: 413–18.

Rotmans, J., Kemp, R., and Asselt, M. van (2001). 'More Evolution than Revolution', *Foresight*, 3(1): 15–31.

Schot, J., and Geels, F. W. (2008). 'Strategic Niche Management and Sustainable Innovation Journeys: Theory, Findings, Research Agenda, and Policy', *Technology Analysis & Strategic Management*, 20(5): 537–54. doi:10.1080/09537320802292651

Seebode, D., Jeanrenaud, S., and Bessant, J. (2012). 'Managing Innovation for Sustainability', *R&D Management*, 43(2): 195–206.

Seuring, S., and Müller, M. (2008). 'From a Literature Review to a Conceptual Framework for Sustainable Supply Chain Management', *Journal of Cleaner Production*, 16(15): 1699–710. doi:10.1016/j.jclepro.2008.04.020

Shrivastiva, P., and Hart, S. (1995). 'Creating Sustainable Corporations', *Business Strategy and the Environment*, 4: 154–65.

Simpson, D., Power, D., and Samson, D. (2007). 'Greening the Automotive Supply Chain: A Relationship Perspective', *International Journal of Operations & Production Management*, 27(1): 28–48.

Simpson, D., and Samson, D. (2010). 'Environmental Strategy and Low Waste Operations: Exploring Complementarities', *Business Strategy and the Environment*, 118: 104–18. doi:10.1002/bse

Smith, A. (2006). 'Green Niches in Sustainable Development: The Case of Organic Food in the United Kingdom', *Environment and Planning*, 24: 439–59. doi:10.1068/c0514j

Smith, A., Stirling, A., and Berkhout, F. (2005). 'The Governance of Sustainable Socio-Technical Transitions: A Quasi-Evolutionary Model', *Research Policy*, 34: 1491–510.

Teece, D. J., Pisano, G., and Shuen, A. (1997). 'Dynamic Capabilities and Strategic Management', *Strategic Management Journal*, 18(7): 509–33.

Tyteca, D., Carlens, J., Berkhout, F., Hertin, J., Wehrmeyer, W., and Wanger, M. (2002). 'Corporate Environmental Performance: Evidence from the MEPI Project', *Business Strategy and the Environment*, 11: 1–13.

Wagner, M., and Llerena, P. (2011). 'Eco-innovation through Integration, Regulation and Cooperation: Comparative Insights from Case Studies in Three Manufacturing Sectors', *Industry and Innovation*, 18(8): 747–64.

....

CHAPTER 16

MANAGING SOCIAL INNOVATION

....

THOMAS B. LAWRENCE,
GRAHAM DOVER, AND
BRYAN GALLAGHER

INTEREST in managing social innovation from the private, public, and not-for-profit sectors has been growing at a dramatic rate. There has also been a recent spate of writing on social innovation, much of it from just outside the academy—published by foundations and think tanks—that has provided a useful set of ideas and insights (Mulgan, 2006 a; Phills, Deiglmeier, and Miller, 2008). At the same time, a survey of the social innovation literature found 'little serious research, no widely shared concepts, thorough histories, comparative research or quantitative analysis', which limits the impact potential of social innovation research (Mulgan, Tucker, Rushanara Ali, and Sanders, 2007a: 7). Although scholarly commitment to understanding how to manage social innovation has not kept pace with the practical commitments of other sectors, the research that has been done provides a fascinating account of an emerging arena (Nilsson, 2003; Mulgan, 2006b; Phills et al., 2008). In this chapter, we make two contributions to the scholarly discussion of managing social innovation. First, we review an important set of themes in the social innovation literature, summarizing in each case key points of progress and pointing to areas of weakness. Second, we present a conceptualization of social innovation that addresses key weaknesses in the existing literature and provides a platform for the future.

RESEARCH ON MANAGING SOCIAL INNOVATION

Writing on social innovation spans a wide array of topics and disciplines. Relevant scholarship is found under the topic headings of social entrepreneurship, social enterprise, social economy, social capital, social finance, and corporate social responsibility.

This work occurs in management and organization studies (Drucker, 1987; Westley, 1991; Kanter, 1999), and also in urban studies (Moulaert, Martinelli, Gonzalez, and Swyngedouw, 2007), technology management (Dawson, Daniel, and Farmer, 2010), and sociology (Gerometta, Haussermann, and Longo, 2005). Unsurprisingly, the eclectic interest in social innovation has led to a proliferation of definitions which tend to emphasize very different characteristics (Goldenberg, Kamoji, Orton, and Williamson, 2009; Nilsson, 2003). Thus, rather than begin with a definition, we review the existing literature organized around four themes that characterize a broad understanding of social innovation (Phills et al., 2008): (1) a focus on social problems; (2) an interest in finding novel solutions to those social problems; (3) the absence of a particular organizing model; and (4) the benefits of solutions distributed beyond the innovators.

Social Problems as the Starting Point

Perhaps the most widely shared focus in the literature on social innovation is a concern for solving social problems. In a review of the social innovation literature, Phills et al. (2008: 3) argue that this is a coalescing point for those with different views: 'there tends to be greater consensus within societies about what constitutes a social need or problem and what kinds of social objectives are valuable (for example, justice, fairness, environmental preservation, improved health, arts and culture, and better education)'. When we move beyond abstractions, such as justice or health, however, the literature on managing social innovation reveals significant challenges in identifying social problems.

The variation in social problems addressed in the social innovation literature is illustrated by the range of dimensions employed to describe them. One critical dimension is scale: some scholars argue that social innovation is concerned with global problems that affect everyone (Cooperrider and Pasmore, 1991) such as climate change, while others focus on locally situated problems such as deprivation within specific neighborhoods (Nussbaumer and Moulaert, 2004; Drewe 2008). Social problems also vary dramatically in terms of the abstractness with which they are specified, ranging from 'the economic crisis' (Mulgan, 2009) to obesity and addictions in specific populations (Mulgan, 2006b). Similarly, while some writing on social innovation frames problems as urgent and immediate (Mulgan et al., 2006), others describe social problems associated with social innovation as longer-term, multigenerational issues (Alexander, 2008).

Despite the variety in kinds of social problems examined, the social innovation literature tends to present them in objective terms, such that their incontrovertible existence makes them clearly identifiable targets for novel solutions. There appears little critical awareness that the identification and description of problems might itself be contentious. Mulgan et al. (2007a: 9), for example, identify 'addiction to alcohol, drugs and gambling' under the label 'behavioural problems of affluence', as if what constitutes addiction and its causes are generally accepted and understood. The literature generally ignores conflict around social problems and frequently cites examples, such as poverty, where it seems that there is a consensus on the need to act. But even examples such as poverty are contentious, let alone issues such as sexual and reproductive rights or the unequal distribution of global financial resources. Relatively unexplored are questions such as how these problems have emerged over other issues, how these problems are defined and understood, and how focusing on these problems in this way might privilege some and penalize others (Loseke, 2003). Social problems set the scene for action but then play passive roles—often found in the opening paragraph of a case study that provides the setting and stage for the innovator and their solution. It is as though the social innovator emerges from and operates in a politics-free space, where social problems exist as independent entities, do not change as they are examined or discussed, and are understood independently of the solutions proposed to address them.

For the scholarly understanding of managing social innovation to progress, the role of social problems must be fundamentally reconsidered. Interest in the effectiveness of innovations to address social problems needs to be balanced with an understanding of the identification of social problems as a process of social construction (Loseke, 2003; Gergen, 2009; Abdelnour and Branzei, 2010). What constitutes a problem, its boundaries, its effects, and its importance are all socially negotiated phenomena that reflect the norms, values, and beliefs of those involved, as well as the power relations and distributions of resources among them (Clegg, Courpasson, and Phillips, 2006; Gergen, 2009). This includes negotiations around such issues as the scale of issues that count as social problems, whether they must be global or societal in nature, or whether the social problems of concern to social innovation scholars also include those affecting small communities or neighbourhoods. It also includes negotiations focused on the relationship between social problems and individual failings. Some of the most contentious debates of our time are rooted in more basic debates around the roles of choice and social systems in contributing to individual suffering, and consequently whether and how that suffering constitutes a 'social problem'. One need only consider the global variation in approaches to drug use and addiction to see the centrality of how social problems are defined in shaping the efforts of social innovators: the US-based 'war on drugs' has at its core a conceptualization of drug use as an individual choice that is morally wrong and reflects weakness of character on the part of those involved; in contrast, 'harm reduction' approaches in Europe, Canada, and Australia are based on an understanding of drug use and addiction as rooted in social structures and systems that manifest as a health issue at the individual level, and hence require interventions that minimize the health, social, and economic harms to communities and individuals (Single, 1995; Inciardi and Harrison, 1999).

That social problems are socially constructed does not mean that they are 'not real' or somehow unimportant. The 'realness' of social constructions, including social problems, is embedded in the communities that jointly construct those problems and who act in response to those constructions (Gergen, 2009). Thus, social construction as a lens on social problems is not a basis for inaction, but rather a way of understanding how and why we act at times, and do not act at others. If anything, a social constructionist lens heightens the responsibility of individuals, communities, and societies to not only respond to the social problems they face, but to recognize their own roles in identifying only some issues and situations as social problems. For scholars of social innovation, a constructionist lens is also valuable in understanding and studying social problems, as it shifts the question of what counts as a social problem from being a boundary condition of social innovation scholarship to an integral element of managing social innovation, and consequently a necessary part of the processes and practices examined.

A Focus on Novel Solutions

In October 2009, the Lien Centre for Social Innovation at the University of Singapore announced the winners of its $1 million (Singapore dollars) social innovation competition. The winning entries included rats able to sniff out landmines, a social enterprise that recycles unwanted clothes for use by the rural poor in India, an online platform to fund scholarships for Cambodian and Vietnamese children, and interlocking bricks to reduce the costs of assembling basic housing (A world of winning ideas to lift up Asia 2009). These initiatives illustrate the second major theme in research on social innovation—interest in novel solutions to social problems (Mulgan, Tucker, et al., 2007a; Phills et al., 2008; Murray, Caulier-Grice, and Mulgan, 2010). As illustrated by the winners of the University of Singapore competition, novel solutions associated with social innovation can take many forms, including products, process, and technologies, as well as broader responses to social problems such as 'a principle, an idea, a piece of legislation, a social movement, an intervention, or some combination' (Phills et al., 2008: 36). Often important to social innovation is a new form of organizing, such as Charter schools (Phills et al., 2008) and farm agents (Drucker, 1987). Despite the diversity of these examples, discussions of novel solutions in the literature tend to emphasize three characteristics.

The first characteristic emphasized in the literature is the degree to which novel solutions can be shared or diffused across a variety of settings (Cooperrider and Pasmore, 1991; Mulgan, Tucker, et al., 2007a; Pearson, 2006; Leadbeater, 2008). This can be achieved, argues Christensen et al. (2006), when novel solutions to social problems are just 'good enough'—easy to replicate, more convenient, and less expensive than rival services attracting the under-served, and a simpler proposition to those over-served by the market. The Aravind Eye Care System in India, for example, has managed to streamline eye surgery and reduce the cost of manufacturing lenses from $200 to $3 so that it can treat 300,000 of the world's poorest people and still make a profit (Rangan and Thulasiraj, 2007).

A second characteristic of novel solutions emphasized in the social innovation literature is their relationship to technology (e.g. Huddart, 2010). In some cases a new technology itself is enough to qualify as a novel solution, such as low-power laptops to educate children in areas without electricity (Buchele and Owusu-Aning, 2007). More commonly, novel solutions embed technologies in new environments, mesh with them, or create the conditions for new technologies to emerge (Kinder, 2010).

One set of technologies—the Internet—is often cast as playing a central role in novel solutions (Leadbeater, 2008; Morino, 2009). The Internet and its associated network technologies have been used as a resource to generate a wide range of social innovations, including new social networks such as Tyze, an online support system connecting those with disabilities to family and friends (Huddart, 2010), and The School of Everything, an online service that puts people in touch with those in their area who can help them 'learn whatever, whenever and wherever…[f]rom Biology to Beekeeping, History to Hula hooping' (School of Everything, 2013). Initiatives connected to this technology are often described as being particularly collaborative and 'open', in contrast to some new technologies that are focused on narrow objectives, produced by 'experts', and protected by private intellectual property rights (Leadbeater, 2008; Murray et al., 2010).

A third characteristic the literature emphasizes in its descriptions of novel solutions is the participation of users in development and implementation processes (Rodin, 2010). Users are argued to have first-hand knowledge of a social problem and legitimacy among their peers (Svensson and Bengtsson, 2010), which makes them well suited to generate and diffuse new solutions. Technology can facilitate participation, as illustrated by Ohmynews, which uses web-based technology to involve citizen journalists in South Korea, or ReachOut!, an Australian web-based peer-to-peer approach to tackling depression among young people. More generally, a design-focused approach to social innovation (Burns, Cottam, Vanstone, and Winhall, 2006; Brown and Wyatt, 2010) works to implement user-centered processes (Manzini, 2009) that start from 'the presumption that people are competent interpreters of their own lives and competent solvers of their own problems' (Mulgan, Tucker, et al., 2007a: 22).

In describing novel solutions, the scholarly and practice-oriented literatures have tended to over-simply the relationships among potential solutions and their connections to the contexts in which they emerge. New practices, processes, and products can complement or contradict each other and interact significantly with existing solutions to the same and other social problems. Discussions of novel solutions in the social innovation literature tend to gloss over disagreements as to their effectiveness, as well as the values and beliefs that underpin them (Bacon, Faizullah, Mulgan, and Woodcraft, 2008): Charter Schools are, for example, cited as a defining example of social innovation (Phills et al., 2008), but have critics who view them as novel but inappropriate (Renzulli and Roscigno, 2008). The complex interactions among novel solutions can also be a source of strength, underpinning greater effectiveness in dealing with social problems than would be possible with isolated solutions. In the 1980s, Band Aid and Live Aid raised more funds for famine relief in Ethiopia than thought possible and raised worldwide consciousness of these issues, based significantly on their blending of 'music, as one symbolic language, with television imagery' at an opportunistic time for the music industry (Westley, 1991: 1020).

Furthering our understanding of the role of novel solutions in managing social innovation depends on a conception of these solutions as dynamically embedded in networks of innovation that interact in complex ways with their contexts and with the histories of the problems they address and the solutions that have preceded them. All innovation occurs in an ecology of problems, solutions, and actors working to connect them, as well as a history of successful and failed attempts to innovate in response to social problems (Hilgartner and Bosk, 1988; Wallace, 1999). Novel solutions may also play roles in fostering social problems within or outside of the domains in which they were created (Froud, Johal, Montgomerie, and Williams, 2010), and even their success may create resistance to change in the future that entrenches social problems (Westley, Zimmerman, and Patton, 2006). Mumford and Moertl (2003: 265) argue that social innovation should be seen as a series of solutions that interact with each other over time to create 'chains of innovation'. The individual solution may not appear complete, coherent, or comprehensive, but these 'timely, more limited solutions…address key issues while laying an organizational foundation for more long-term efforts' (Mumford, 2002: 258). Thus, studies of the effectiveness or scalability of individual solutions are of limited value unless they incorporate the complex ecologies and histories of those solutions. From this perspective, managing social innovation is not the creating of individual solutions to social problems, but a continuing, reflexive, responsive set of practices that revolve around the identification and interpretation of the ecologies and histories of social problems and novel solutions.

Organizing Models

A third important theme in the literature on social innovation concerns how it can and should be organized, including which sectors are likely to contribute, the role of individuals, how groups and networks are involved, and the impact of context on the social innovation process. In response to the question of what sectors are key, dominant responses tend towards pluralistic answers that suggest social innovation can emerge from any sector and is especially associated with cross-sector inter-organizational relationships (Mulgan, Tucker, et al., 2000a; Bacon et al., 2008). The case for multi-sector and cross-sector organizing of social innovation is rooted in a number of different arguments. The first is a practical one, suggesting that no single organization or sector has the resources, money, and expertise to fix social problems, such as climate change, with impacts that extend beyond any arbitrary boundary (Osborn, 2009; Goldsmith, Georges, and Burke, 2010). A second strand to this argument highlights the creativity that can emerge through cross-sector partnerships, illustrated by innovations such as fair trade, urban farming, and restorative justice (Murray et al., 2010). Although largely convincing, a commitment to an eclectic approach to organizing can mask the different objectives and logics that characterize different sectors, as well as the potential power imbalances and structurally motivated resistance (Antadze and Westley, 2010; Goldsmith, Georges, Burke, and Bloomberg, 2010). Partnerships between organizations from different sectors are often extremely difficult to sustain (Berger, Cunningham, and Drumwright, 2004), with much depending on each partner's motivation and ability to 'recalibrate their roles as the relationship unfolds' (Le Ber and Branzei, 2010: 167).

The role of individuals in managing social innovation has been highlighted in the growing literatures on social and civic entrepreneurship (e.g. Bornstein, 2007; Light, 2008; Goldsmith, Georges, and Burke, 2010), as well as the social innovation literature (Westley and Antadze, 2010). A review of historical cases, such as the initiatives of Benjamin Franklin (Mumford, 2002), identify the importance of individuals who were able to develop groundbreaking ideas out of their everyday experiences and a willingness to experiment. Two overlapping skills stand out. Social innovators appear able not only to diagnose causes of social problems, but also to consider the 'downstream consequences' of any proposed solution. This diagnostic ability may come from having a unique combination of outsider and insider knowledge (Marcy and Mumford, 2007). Second, successful social innovators seem distinctively able to garner elite support and financial resources: Franklin, for instance, was able to enroll supportive elites who provided him with ideas and finances (Mumford, 2002).

Along with individuals, the literature points to the importance of networks and collaborations in managing social innovation, with many social innovations only possible through the combined efforts of many (Murray et al., 2010). A key issue is the involvement of individuals and groups most disadvantaged by existing arrangements (Westley, 2008). Although understanding of this challenge is only emerging, several themes appear important. First is an emphasis on organizing social innovation in ways that privilege the 'lived experience' of groups previously overlooked or under-represented. This means finding ways to collaborate and co-create solutions with the people intended to benefit from social innovation (Burns et al., 2006), as well as encouraging experts to closely observe the experience of such groups as 'smallholder farmers, school children and community health workers as they improvise their way through their daily lives' (Brown and Wyatt, 2010: 33). Central to this approach is the idea that solutions can be found in a community's 'assets' (Kretzmann and McKnight, 1993). One method rooted in asset-based social innovation is the search for 'positive deviants' who seem somehow to overcome difficulties that defeat others (Brown and Wyatt, 2010; Rodin, 2010). Engaging highly diverse sets of interests also requires organizing mechanisms that facilitate connections among otherwise separate communities, which depend on involving individuals and organizations who operate as connectors or intermediaries (Goldsmith, Georges, Burke, et al., 2010; Kinder, 2010), creating safe spaces to share and experiment with ideas, and extending networks that can support and diffuse innovations (Bacon et al., 2008; Murray et al., 2010).

Although the literature on social innovation tends to emphasize the agency of individuals, organizations, and networks, there is also a recognition of the important role of context in shaping social innovation processes and outcomes. Westley et al. (2006: 20) argue that a 'successful social innovator is, intentionally or not, part of the dynamics of transformation rather than the heroic figure leading the change'. The impact of context can be managed in part through efforts to create geographical and political contexts that are amenable to and supportive of social innovation. Research on the differential success of cities and regions in generating social innovation has shown that it emerges from a complex interaction of outside pressures and resources

with local creativity and leadership skills (Bacon et al., 2008). The possibility of fostering social innovation has led policy-makers and civic leaders to design and construct
specific places to encourage social innovation creativity and connections, such as the
'Social Innovation Park'[1] in northern Spain, 'The Hub', a network of twelve city office
locations for 'people who believe they can change things…change minds, change
lives, and ultimately change the world a little',[2] and the 'Centre for Social Innovation'
in Toronto, that seeks to build a community of innovation. Places that act as 'demonstration sites' to show the feasibility of a proposed innovation can also attract expertise and provide training for those who will disseminate the solution (Mumford and
Moertl, 2003).

One important issue largely missing in discussions of organizing social innovation is the challenge of managing the political dynamics that inevitably accompany
social change efforts. While the literature recognizes 'tensions' in the process (Phills
et al., 2008: 41) and the struggle against vested interests (Pol and Ville, 2009), there is a
dearth of explicit attention to managing these dynamics and how competing interests
might affect organizing. Resistance to change is rooted in people's investments—material (time and money), cognitive (assumptions and values), and relational (social capital and networks)—in existing arrangements (Mulgan, Tucker, et al., 2007a), and thus
lessened when maintaining the status quo is seen as less efficient, or when individuals
are exposed to different sets of interests, ideas, and relationships that make them question current practices and ways of thinking. How such shifts occur, however, is less
clear. Managing social innovation effectively may require being much more upfront
about the winners and losers of existing arrangements before and after the implementation of a social innovation, as well as adopting a more nuanced approach to resistance. The idea of resistance is currently associated with 'the old order' that pioneers
need to sidestep (Murray et al., 2010: 13), but this overlooks its embedded nature
and the contributions of existing arrangements to the organizing of novel solutions.
Change is often initiated by individuals with 'atypical experiences' and 'marginalized
backgrounds' that allow them to see problems that others overlook (Mumford and
Moertl, 2003: 262), and to deal with and exploit paradoxes in the system (Pina e Cunha
and Capose e Cunha, 2003).

Looking across writing that addresses how social innovation is organized—by individuals, organizations, and networks—there is a recognition of the complexity of managing social innovation, the need for participation by highly diverse sets of actors, and
the significant impact of context on how social innovation occurs and what it produces.
To further understanding of the interaction of managing social innovation, how it is
organized, and where it occurs, insights from agency-focused writing (that emphasizes
the work of individuals and groups) need to be integrated with context-focused analysis
(that highlights the role of social and material structures), as well as incorporate a more
explicitly political lens. Adopting a view of organizing social innovation as a process of
structuration (Giddens, 1984) would help accomplish such integration. The concept of
structuration connects agency and structure as a duality, such that structure is 'both a
product of and a constraint on human action' (Barley and Tolbert, 1997: 97). Actors are

not only situated within structures, but also use their knowledge of structures to engage in purposeful action that may be intended to reshape those same structures (Lawrence, 1999). Thus, a structuration lens on the organizing of social innovation highlights the interplay of social innovators' strategies and tactics with the social, organizational, and technological structures that motivate, constrain, and enable them.

Benefits Distributed Beyond the Innovators

The fourth important theme in the literature on social innovation is concerned with its effects, and particularly the distribution of benefits emerging from an innovation. Phills et al. (2008: 39) argue that a novel solution to a social problem qualifies as a social innovation if it is 'more effective, efficient, sustainable, or just than existing solutions and for which the value created accrues primarily to society as a whole rather than private individuals'. This argument, however, under-appreciates the complexity of such evaluations. First, there are significant difficulties in identifying how a single novel solution affects a complex social problem where the processes are not easily reduced to simple cause–effect relationships (Westley and Antadze, 2009). Second, the distribution of benefits is not always easy to untangle: solutions may have multiple private and public benefits (Aiken, 2010), and many innovations, such as in health and education, are largely experienced on an individual basis (Auerswald, 2009), and so focusing on general societal benefits can overlook the costs experienced by a minority.

The argument for prioritizing public benefits responds specifically to market-based solutions, which raises the question of how to make sense of the role of the market and business in social innovation. The issue is not whether or not businesses can innovate in ways that alleviate social problems: in the area of unemployment and job creation, for example, it has been argued that the 'contribution of Wal-Mart stores and McDonald's Corporation dwarf those of a dozen Grameen banks' (Auerswald, 2009: 55). The emerging fields of social finance and impact investing represent attempts to meet social and financial objectives, and private investors, accustomed to weighing up returns, are said to bring important 'rigour' to the process (The Economist, 2010: 56). JP Morgan's $165 million Urban Renaissance Property Fund, for example, provides affordable housing in some of America's poorest communities and also earns a market rate return of 15 per cent for its investors (Rodin, 2010). Market-based approaches to social innovation are especially challenging when social problems are the result of market failures or where economic benefits are both hard to measure and potentially peripheral to an innovator's mission.

Perhaps the most fundamental problem with assessing the distribution of benefits of social innovation is the politics and ethics of such a process. As Westley and her colleagues (2006) argue, social innovation is an inevitably political process that involves the redistribution of power. Consequently, any calculus of benefits and their distribution will be a political one, with its outcome significantly dependent on who is the arbiter and what is the context. Moreover, the processes and impacts of social innovation are never 'ethically neutral' (Pol and Ville, 2009: 822), and so even discounting the politics of social innovation, assessing the distribution of benefits can never be done from an objective position. These dynamics make somewhat hollow any sweeping statements regarding the distribution of benefits to public versus private interests. This is a key issue for studying managing social innovation, as it forces scholars to wrestle with the relationships among the wide variety of private-, voluntary-, and public-sector responses to social problems, as well as hybrid organizations and inter-organizational collaborations. To restrict the concept of social innovation to those initiatives that explicitly declare their benefits are for the public good is to adopt a naïve, uncritical stance on the role of rhetoric and politics in managing social innovation.

A Theoretical Framework for Managing Social Innovation

Having reviewed the social innovation literature, we now make an argument for a specific conception of what it is to manage social innovation, and from that conception we suggest directions for future scholarship in the area. The literature on managing social innovation reveals an understanding of social innovation that includes: (1) social problems as socially constructed; (2) novel solutions as socially and historically embedded; (3) the organization of social innovation as a process of structuration; and (4) the distribution of benefits as an inherently political and ethically contestable issue. Bringing these together suggests that managing social innovation needs to be understood as a socially embedded process in which definitions of social problems, their solutions, and the consequences of those solutions are negotiated and re-negotiated on a continuing basis. Although this description provides a helpful foundation, it is insufficient to provide theoretical traction in the study of managing social innovation. Thus, one more element that is central to both the social innovation literature and the practice of managing social innovation is introduced: the transformation of social systems.

Transformation and Social Innovation

The key to understanding the practical appeal and scholarly importance of social innovation is its connection to social transformation. Social innovation scholars have argued that what distinguishes social innovation from other types of innovation is the intention of the innovator to transform social arrangements (Mumford, 2000; Westley et al., 2006; Cahill, 2010; Goldenberg, 2010; Goldsmith, Georges, and Burke, 2010). In contrast to social entrepreneurship that may focus on individual actions, and social enterprise that addresses organizational issues, managing social innovation focuses on effecting system change (Westley and Antadze, 2009). The cumulative effect of social innovations leads to an increase in a society's innovative capacity (Pearson, 2006), with the focus not on 'Band-Aid solutions' that address immediate symptoms of social problems, but on solutions that tackle their underlying causes (Westley, 2008). The issue of what counts as transformation is, like other facts of social innovation, contested. In the evaluation of transformation, one measure is the innovation's *impact*—its demonstrable effect on the social problem (Pearson, 2006; Westley and Antadze, 2010). A second, much admired dimension is *scale*—the number of beneficiaries of the social innovation (Dees, Anderson, and Wei-Skillern, 2002; Moore and Westley, 2009). From this perspective, transformation through social innovation only occurs when it is adopted nationally or globally (Bacon et al., 2008). Other criteria for the evaluation of transformation include the creation of 'new social relationships', such as greater participation of a marginalized group in society (Nussbaumer and Moulaert, 2004) or a substantial disruptive effect on social arrangements (Westley and Antadze, 2009).

The issues highlighted in the review—social innovation as a socially and historically embedded process of social construction—lead to a focus on transformation that emphasizes less how many individuals were affected and more on how they were affected. Figure 16.1 summarizes the theoretical framework proposed for the study of managing social innovation. At the heart of this framework is the relationship between how social problems are understood in a community, and novel solutions proposed or implemented to address those problems. In this framework, social problems are socially constructed by interested sets of actors including those afflicted, those working to address them, those in the community they affect directly and indirectly, and a potential, diverse network of actors who influence the social construction process although they operate at a considerable distance from the issues. This is not meant to suggest a pluralistic or democratic process: the social construction of social problems can be a process with intense, highly conflictual politics as actors work to maintain or disrupt entrenched conceptions of social problems. In trying to achieve a social transformation, *Mothers Against Drunk Driving*, for example, had to tackle the 'fact of life' view that 'men drank, [and] drank to excess' that was 'deeply embedded in cultural perceptions and expectations' (Westley et al., 2006: 195). From this perspective, social innovations require 'changing how societies think' (Mulgan, Tucker, et al., 2007a: 22) about social problems in ways that may require 'a significant, creative and sustainable shift' (Nilsson, 2003: 3). Success occurs when the new ideas become a 'changed common sense' made possible by 'a series of reinterpretations by practitioners, beneficiaries, funders and the wider public' (Mulgan, Rushnara Ali, Halkett, and Sanders, 2007: 22–23). One social innovator describes this kind of transformation as occurring when ideas get into the 'water supply' (Etmanski, 2011).

Novel solutions are, within the framework proposed, understood as socially and historically embedded sets of practices, routines, and technologies. This conception of novel solutions pushes towards focusing on the complexities associated with their implementation as well as their potentially diverse and unexpected consequences. To make a difference necessitates work that seeks to substantially alter social arrangements;

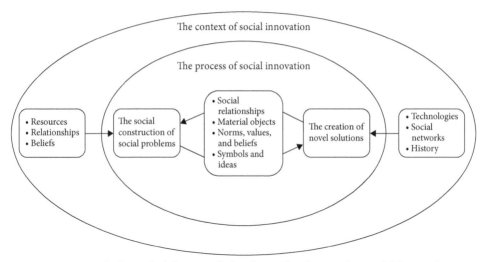

FIGURE 16.1   A theoretical framework for the study of managing social innovation

social innovation is seen as 'innovation in social relations, as well as in meeting human needs' (MacCallum, Moulaert, Hillier, and Vicari, 2009: 2). Implementing novel solutions can involve changing the 'rules of the game' at the local level (Moulaert et al., 2007; Goldsmith, Georges, Burke, et al., 2010), where selective targeting of established patterns of behaviour and thought can have transformational impacts without a high 'volume of adoption' (Westley, 2008: 5). The embeddedness of novel solutions brings into focus the degree to which they challenge, reinforce, or complicate social systems.

The approach proposed has some significant advantages for the study of managing social innovation. First, it emphasizes how the social problem is understood, rather than starting and ending with the novel solution. The advantage of focusing on a social problem is that it not only offers a way to explore the impact of a novel solution, but also helps to avoid solution myopia, an exclusive focus on the characteristics of the novel solution that detaches it from the environment it seeks to change. Second, it is grounded in a particular social context, a community. This sets boundaries around a social problem and its solutions but also provides a means to explore transformation as a process and not just an outcome. A community focus provides a way to explore how a social problem is experienced by a particular group of people and how novel solutions might interact with these ways of thinking and behaving towards a social problem in ways that can lead to significant change.

A third advantage of this framework is that it highlights the complex relationship between novel solutions to social problems and existing ways of thinking about those social problems. The literature on managing social innovation tends to present this relationship in one-directional terms, where the introduction of a novel solution in itself leads to a change in existing ways of thinking about social problems. The introduction of micro-credit, for example, is often presented as an innovation that single-handedly changed attitudes towards some of the world's poorest peoples. Seeing the relationship in these terms can lead those in search of societal change to focus on replicating solutions, as more of the same solution is equated with more change. Another common frame in the literature is the recognition of how novel solutions are influenced and shaped by existing ways of thinking, but this is, for the most part, presented in rather negative terms (e.g. Murray et al., 2010), equating it with the self-interested responses of individuals and organizations defending the status quo that privileges them. A much less explored dynamic occurs when innovators are embedded in the communities that they seek to change (Mumford, 2002). While existing ways of thinking about a social problem may have constraining effects, they may also play an enabling role in creating the conditions from which new solutions emerge and are developed.

Thus, the relationship between novel solutions and existing ways of thinking about social problems is a recursive one. Consequently, managing social innovation involves attending to how social problems are constructed in a community, how those constructions affect potential solutions, and in turn how novel solutions might affect the ways that problems are understood, both directly and indirectly, through the complex of ideas, beliefs, values, relationships, technologies, and places that connect them. Transformation from this perspective is a multifaceted process that might involve a social problem and a novel solution, but is more likely to involve a nested set of processes, practices, and actors, such that transformation occurs through the reconfiguration of social, discursive, and material connections. Moreover, the framework suggests that transformation through social innovation is likely to occur as an iterative process, with both intended and unintended consequences, rather than as some kind of hyper-successful implementation of an intended strategy. Novel solutions that significantly shift the way a community understands a complex and contested social problem are likely to provoke responses, both supportive and resistant. Transformational novel solutions are also likely to interact with communities' experiences of past and existing novel solutions that are interwoven into existing ways of thinking about social problems.

Studying the managing of social innovation and social transformation from a perspective that highlights the dynamic, recursive relationship between novel solutions and existing ways of thinking about social problems requires methods and theories that can cope with the complexity of the phenomenon. Some writers on social innovation have argued for approaches that involve learning by engaging in and with practice (Mulgan, Tucker, et al., 2007a; Dover and Lawrence, 2010). Others have drawn on complexity theory with its interest in exploring the dynamics of complex social systems (Westley, 2008), and highlighted the importance of exploring an institutional perspective: the established meanings, rules, and practices underpinning social arrangements (Westley et al., 2006; Heiskala, 2007). Westley (2008) argues that these 'taken for granted institutions are often the source of intractable problems. Real innovation without change in these institutions is therefore unlikely'. Of importance are institutional entrepreneurs: those with the skills to recognize local institutional dynamics and seize 'windows of opportunity' (Westley, 2008). How individuals and organizations are able to do this work and how it might shape the transformative impact of the social innovation remains under-explored.

CONCLUSION

The study of managing social innovation represents an exciting, emerging field in which important insights have been generated, but where there is far more to explore and discover. Our review of the literature on managing social innovation identifies four prominent themes: social problems as the starting point, a focus on novel solutions, a variety of potential organizing models, and benefits distributed beyond the innovators. From the review, a set of issues is identified to which scholarly research on managing social innovation should attend. First, an understanding of the social construction of social problems. Such a shift heightens the responsibility of individuals and groups to recognize their roles in identifying only some issues and situations as social problems, and shifts the question of what counts as a social problem from being a boundary condition of social innovation scholarship to an integral element of managing social innovation. Second, the study of novel solutions should focus on their social and historical embeddedness, rather than imagining them as springing from green fields. This means that managing social innovation needs to be understood as a set of practices that involve identifying and interpreting the ecologies and histories of social problems and novel solutions. Third, in examining the organizing of social innovation, there is value in scholars adopting a structuration (Giddens, 1984) perspective that conceptualizes organization as a process of managing the recursive duality of agency and structure. Finally, a scholarly examination of the benefits of managing social innovation must include an explicit concern for its politics and ethics, with a careful consideration of how benefits are distributed, the multi-sector negotiations that decide that distribution, and the interests that underpin what is considered a benefit and what is not. Drawing on these four proposed shifts, we articulate a theoretical framework for the study of managing social innovation. Central to our framework is the idea of transformation, understood as a recursive process linking the creation of novel solutions to the social construction of social problems. This framework could provide a foundation for research on managing social innovation that would attend to its complexity but also provide guidance for practice.

NOTES

1. For more info on the Social Innovation Park–<http://socialinnovator.info/connecting-people-ideas-and-resources/innovation-intermediaries/hubs/social-innovation-parks> (accessed 24 July 2013).
2. The Hub. (n.d.). Invitation. Available at <http://the-hub.net/invitation.html>

References

Abdelnour, S., and Branzei, O. (2010). 'Fuel-Efficient Stoves for Darfur: The Social Construction of Subsistence Marketplaces in Post-Conflict Settings', *Journal of Business Research*, 63(6): 617–29.

Aiken, M. (2010). 'The Impact of Multi-Purpose Community Organisations: Towards a Conceptual Framework for Research' [online]. *Assessing the Impact of Multi-Purpose Community Organisations on Communities Using Action Research: Insights from a Collaboration between Practitioners and Researchers.* Presented at the NCVO/VSSN Researching the Voluntary Sector Conference, Leeds, UK, 6 September, p. 12.

Alexander, B. K. (2008). *The Globalisation of Addiction: A Study in Poverty of the Spirit.* Oxford: Oxford University Press.

Antadze, N., and Westley, F. (2010). 'Funding Social Innovation: How Do We Know What to Grow?', *The Philanthropist*, 23(3): 343–56.

Auerswald, P. (2009). 'Creating Social Value'. *Stanford Social Innovation Review*, 7(2): 51–5.

'A World of Winning Ideas to Lift up Asia' (2009, October). *The Straits Times*, A1, D1–D9 Singapore.

Bacon, N., Faizullah, N., Mulgan, G., and Woodcraft, S. (2008). *Transformers: How Local Areas Innovate to Address Changing Social Needs* [online]. (Research Report). London, UK: National Endowment for Science, Technology and the Arts, pp. 1–135. Available at <http://www.nesta.org.uk/library/documents/Report%20-%20Local%20Social%20Web.pdf> (accessed 24 July 2013).

Barley, S. R., and Tolbert, P. S. (1997). 'Institutionalization and Structuration: Studying the Links between Action and Institution', *Organization Studies*, 18(1): 93–117.

Berger, I. E., Cunningham, P. H., and Drumwright, M. E. (2004). 'Social Alliances: Company/Nonprofit Collaboration', *California Management Review*, 47(1): 58–90.

Bornstein, D. (2007). *How to Change the World: Social Entrepreneurs and the Power of New Ideas.* Updated. Oxford: Oxford University Press.

Brown, T. and Wyatt, J. (2010). 'Design Thinking and Social Innovation'. *Stanford Social Innovation Review*, 8(1): 31–5.

Buchele, S. F., and Owusu-Aning, R. (2007). 'The One Laptop per Child (OLPC) Project and its Applicability to Ghana'. *Proceedings of the 2007 International Conference on Adaptive Science and Technology* Accra, Ghana: 113–18.

Burns, C., Cottam, H., Vanstone, C., and Winhall, J. (2006). *Transformation Design* [online]. (No. 02). London, UK: Design Council: 1–33. Available at <http://www.designcouncil.info/mt/RED/transformationdesign/TransformationDesignFinalDraft.pdf> (accessed 24 July 2013).

Cahill, G. (2010). 'Primer on Social Innovation: A Compendium of Definitions Developed by Organizations Around the World' [online]. *The Philanthropist*, 23(3). Available at <http://www.thephilanthropist.ca/index.php/phil/article/viewArticle/846> (accessed 24 July 2013).

Christensen, C. M., Baumann, H., Ruggles, R., and Sadtler, T. M. (2006). 'Disruptive Innovation for Social Change', *Harvard Business Review*, 84(12): 94–101.

Clegg, S. R., Courpasson, D., and Phillips, N. (2006). *Power and Organizations.* London, UK: Sage.

Cooperrider, D. L., and Pasmore, W. A. (1991). 'Global Social Change: A New Agenda for Social Science?', *Human Relations*, 44(10): 1037–55.

Dawson, P., Daniel, L., and Farmer, J. D. (2010). 'Introduction to a Special Issue on Social Innovation', *International Journal of Technology Management*, 51(1): 1–8.

Dees, J. G., Anderson, B. B., and Wei-Skillern, J. (2002). *Pathways to Social Impact: Strategies for Scaling out Successful Social Innovations* [online]. CASE Working Paper Series No. 3. Center for the Advancement of Social Entrepreneurship: 1–15 Available at <http://www.caseatduke.org/documents/workingpaper3.pdf> (accessed 24 July 2013).

Dover, G., and Lawrence, T. B. (2010). 'A Gap Year for Institutional Theory: Integrating the Study of Institutional Work and Participatory Action Research', *Journal of Management Inquiry*, 19(4): 305–16.

Drewe, P. (2008). 'The Challenge of Social Innovation in Urban Revitalization', in P. Drewe, J.-L. Klein, and E. Hulsbergen (eds), *The Challenge of Social Innovation in Urban Revitalization*. Amsterdam: Techne Press, 9–16.

Drucker, P. F. (1987). 'Social Innovation: Management's New Dimension', *Long Range Planning*, 20(6): 29–34.

Etmanski, A. (2011). *Social Innovation and Finance: What is it and Why is it Going On in Canada?* Presented at the Social Innovation and Finance Tour, Vancouver, BC, 19 January.

Froud, J., Johal, S., Montgomerie, J., and Williams, K. (2010). 'Escaping the Tyranny of Earned Income? The Failure of Finance as Social Innovation', *New Political Economy*, 15(1): 147–64.

Gergen, K. J. (2009). *An Invitation to Social Construction*, 2nd edition. London, UK: Sage.

Gerometta, J., Haussermann, H., and Longo, G. (2005). 'Social Innovation and Civil Society in Urban Governance: Strategies for an Inclusive City', *Urban Studies*, 42(11): 2007–21.

Giddens, A. (1984). *The Constitution of Society: Outline of the Theory of Structuration*. Berkeley, CA: University of California Press.

Goldenberg, M. (2010). 'Reflections on Social Innovation' [online]. *The Philanthropist*, 23(3). Available at <http://journals.sfu.ca/philanthropist/index.php/phil/article/viewArticle/844> (accessed 24 July 2013).

Goldenberg, M., Kamoji, W., Orton, L., and Williamson, M. (2009). *Social Innovation in Canada: An Update* (Research Report). Ottawa: Canadian Policy Research Networks: 68.

Goldsmith, S., Georges, G., and Burke, T. G. (2010). *The Power of Social Innovation: How Civic Entrepreneurs Ignite Community Networks for Good*, 1st edn. San Francisco: Jossey-Bass.

Goldsmith, S., Georges, G., Burke, T. G., and Bloomberg, M. R. (2010). *The Power of Social Innovation: How Civic Entrepreneurs Ignite Community Networks for Good*. San Francisco: John Wiley and Sons.

Hämäläinen, T. J., and Heiskala, R., (eds) (2007). *Social Innovations, Institutional Change, and Economic Performance: Making Sense of Structural Adjustment Processes in Industrial Sectors, Regions, and Societies*. Cheltenham, UK: Edward Elgar Publishing.

Hilgartner, S., and Bosk, C. L. (1988). 'The Rise and Fall of Social Problems: A Public Arenas Model', *American Journal of Sociology*, 94(1): 53–78.

Huddart, S. (2010). 'Patterns, Principles, and Practices in Social Innovation', *The Philanthropist*, 23(3): 221–34.

Inciardi, J. A., and Harrison, L. D. (1999). *Harm Reduction: National and International Perspectives*. Thousand Oaks, CA: SAGE.

Kanter, R. M. (1999). 'From Spare Change to Real Change: The Social Sector as Beta Site for Business Innovation', *Harvard Business Review*, 77(3): 122–32.

Kinder, T. (2010). 'Social Innovation in Services: Technologically Assisted New Care Models for People with Dementia and Their Usability', *International Journal of Technology Management*, 51(1): 103–20.

Kretzmann, J. P. and McKnight, J. L. (1993). Building Communities From the Inside Out: A Path toward Finding and Mobilizing a Community's Assets [online]. ACA Publications. Available at <http://www.units.muohio.edu/servicelearning/sites/edu.servicelearning/files/BuildingCommunitiesInsideOut.pdf>

Lawrence, T. B. (1999). 'Institutional Strategy'. *Journal of Management*, 25(2): 161–87.

Le Ber, M. J., and Branzei, O. (2010). '(Re)Forming Strategic Cross-Sector Partnerships', *Business & Society*, 49(1): 140–72.

Leadbeater, C. (2008). *Understanding Social Innovation* [online]. Melbourne, Australia: Center for Socail Impact:1–5. Available at <http://www.charlesleadbeater.net/cms/xstandard/Understanding%20Social%20Innovation.pdf> (accessed 24 July 2013).

Light, P. C. (2008). *The Search for Social Entrepreneurship*. Washington, DC: Brookings Institution Press.

Loseke, D. R. (2003). *Thinking about Social Problems: An Introduction to Constructionist Perspectives*, 2nd edition. New Brunswick, NJ: Aldine Transaction.

MacCallum, D., Moulaert, F., Hillier, J., and Vicari, S. (eds) (2009). *Social Innovation and Territorial Development*. Farnham, UK: Ashgate.

Manzini, E. (2009, December 19). 'Designing for Social Innovation: An Interview with Ezio Manzini' [online]. Available at <http://johnnyholland.org/2009/12/designing-for-social-innovation-an-interview-with-ezio-manzini/> (accessed 24 July 2013).

Marcy, R. T., and Mumford, M. D. (2007). 'Social Innovation: Enhancing Creative Performance through Causal Analysis', *Creativity Research Journal*, 19(2–3): 123–40.

Moore, M.-L., and Westley, F. (2009). *Surmountable Chasms: The Role of Cross-Scale Interactions in Social Innovation* [online]. Waterloo, ON: Social Innovation Generation, University of Waterloo. Available at <http://www.ipg.vt.edu/resilience/docs/MooreML&WestleyF_CrossScaleInteractions_v1.docx> (accessed 24 July 2013).

Morino, M. (2009, April 16). 'The Innovation Imperative' [online], *Social Innovations Stanford Social Innovation Review*. Available at <http://www.ssireview.org/blog/entry/the_innovation_imperative/> (accessed 24 July 2013).

Moulaert, F., Martinelli, F., Gonzalez, S., and Swyngedouw, E. (2007). 'Introduction: Social Innovation and Governance in European Cities: Urban Development between Path Dependency and Radical Innovation', *European Urban and Regional Studies*, 14: 195–209.

Mulgan, G. (2006a). 'The Process of Social Innovation', *Innovations*, 1(2), 145–62.

Mulgan, G. (2009). Post-Crash, Investing in a Better World: Video on TED.com [online]. Available at <http://www.ted.com/talks/geoff_mulgan_post_crash_investing_in_a_better_world_1.html> (accessed 3 January 2012).

Mulgan, G., Ali, R., Halkett, R., and Sanders, B. (2007). *In and Out of Sync: The Challenge of Growing Social Innovations*. London, UK: NESTA.

Mulgan, G., Tucker, S., Ali, R., and Sanders, B. (2007a). *Social Innovation: What it is, Why it Matters and How it Can be Accelerated* [online] (Working Paper). London, UK: The Young Foundation, pp. 1–52. Available at <http://youngfoundation.org/publications/social-innovation-what-it-is-why-it-matters-how-it-can-be-accelerated/>

Mulgan, G., Wilkie, N., Tucker, S., Ali, R., Davis, F., and Liptrot, T. (2006). *Social Silicon Valleys: A Manifesto for Social Innovation: What It Is, Why It Matters and How It Can be Accelerated*. London: The Young Foundation.

Mumford, M. D. (2002). 'Social innovation: Ten cases from Benjamin Franklin', *Creativity Research Journal*, 14(2): 253–66.

Mumford, M. D., and Moertl, P. (2003). 'Cases of Social Innovation: Lessons from Two Innovations in the 20th Century', *Creativity Research Journal*, 15(2–3): 261–6.

Murray, R., Caulier-Grice, J., and Mulgan, G. (2010). *The Open Book of Social Innovation*. London, UK: Young Foundation.

Nilsson, W. O. (2003). *Social Innovation: An Exploration of the Literature*. Montreal, QC: McGall-Dupont Social Innovation Initiative.

Nussbaumer, J., and Moulaert, F. (2004). 'Integrated Area Development and Social Innovation in European Cities', *City*, 8(2): 249–57.

Osborn, T. (2009). 'Toward Commonality: Thoughts on Diversity in a New Era of Change', *National Civic Review*, 98(3): 25–9.

Pearson, K. (2006). *Accelerating Our Impact: Philanthropy, Innovation and Social Change* [online]. Montreal, QC: The J.W. McConnell Family Foundation:1–32. Retrieved from <http://angoa.org.nz/docs/Accelerating_our_Impact.Pearson.pdf>

Phills, J. A. J., Deiglmeier, K., and Miller, D. T. (2008). 'Rediscovering Social Innovation', *Stanford Social Innovation Review*, (Fall), 34–43.

Pina e Cunha, M., and Capose e Cunha, R. (2003). 'The Interplay of Planned and Emergent Change in Cuba', *International Business Review*, 12(4): 445–59.

Pol, E., and Ville, S. (2009). 'Social Innovation: Buzz Word or Enduring erm?', *Journal of Socio-Economics*, 38(6): 878–85.

Rangan, V. K., and Thulasiraj, R. D. (2007). 'Making Sight Affordable (Innovations Case Narrative: The Aravind Eye Care System)', *Innovations: Technology, Governance, Globalization*, 2(4): 35–49.

Renzulli, L., and Roscigno, V. (2008). 'Charter schools and the public good', in J. H. Ballantine and J. Z. Spade (eds), *Schools and Society: A Sociological Approach to Education*, 3rd edn. London, UK: Pine Forge Press, 363–9.

Rodin, J. (2010). *Social Innovation: What it is and What it Means for Philanthropy* [online]. Denver, CO. Available at <http://www.rockefellerfoundation.org/uploads/files/e0ebe32d-7 95a-444f-ae03-bdb09e0b3acc-social.pdf> (accessed 24 July 2013).

School of Everything (2013). *About us* [online]. London, UK:1–18. Available at <http://school-ofeverything.com/about>

Single, E. (1995). 'Defining Harm Reduction', *Drug and Alcohol Review*, 14(3): 287–290.

Svensson, P. and Bengtsson, L. (2010). 'Users' Influence in Social-Service Innovations: Two Swedish Case Studies', *Journal of Social Entrepreneurship*, 1(2): 190–212.

The Economist (2010, August 12). 'Social Innovation: Let's Hear Those Ideas' [online]. Available at <http://www.economist.com/node/16789766> (accessed 24 July 2013).

Wallace, J. M. (1999). 'The Social Ecology of Addiction: Race, Risk, and Resilience', *Pediatrics*, 103(Supplement 2): 1122–7.

Westley, F. (1991). 'Bob Geldof and Live Aid: The Affective Side of Global Social Innovation', *Human Relations*, 44(10): 1011–36.

Westley, F. (2008). *The Social Innovation Dynamic* [online], 1st edition. Waterloo, ON: Social Innovation Generation. Available at <http://sig.uwaterloo.ca/sites/default/files/documents/TheSocialInnovationDynamic_001_0.pdf> (accessed 24 July 2013).

Westley, F., and Antadze, N. (2009). 'Making a Difference: Strategies for Scaling Social Innovation for Greater Impact', *The Innovation Journal: The Public Sector Innovation Journal*, 15(2): Article 2.

Westle, F., and Antadze, N. (2010). *From Total Innovation to System Change: The Case of the Registered Disability Savings Plan, Canada* [online]. Waterloo, ON: Social Innovation Generation, University of Waterloo: 1–14. Available at <http://sig.uwaterloo.ca/sites/default/files/documents/Westley,%20Antadze%20-%20RDSP%20Case%20Study_VMarch 1502010.pdf> (accessed 24 July 2013).

Westley, F., Zimmerman, B., and Patton, M. (2006). *Getting to Maybe: How the World is Changed.* Mississauga, ON: Random House Canada.

CHAPTER 17

INNOVATION MANAGEMENT IN JAPAN

TAKAHIRO FUJIMOTO

INTRODUCTION

THIS chapter describes and analyses basic characteristics of product and process innovations and their management in post-war Japan. Instead of examining Japan's national innovation system as a whole, including the government and universities (Nelson, 1993), the chapter focuses only on industrial innovations by manufacturing firms because one of the distinctive characteristics of Japan's system is that private enterprises occupy a larger share of its innovation activities compared with other advanced countries.

Overall, Japan in the late twentieth and the early twenty-first century has been known as a country that produces numerous industrial innovations. Its R&D expenditures as a proportion of GDP of over 3 per cent since the mid-1990s have been constantly higher than that of the USA, Germany, UK, and other advanced nations (OECD 2012) Japanese firms and individuals acquired around 285,000 patents worldwide between 1995 and 2010, which was followed by their US counterparts with 186,000 patents (World Intellectual Property Organization statistics database). Japan's technology export/import ratios (460 per cent in 2010), as well as technology trade surpluses ($22 billion in 2010; second to the USA: $33 billion in 2009) have been among the highest in major advanced nations. Toyota became the world's largest R&D spender in 2012 (EU Commission, 2013). All these indicators suggest that, even after the 1990s when the Japanese economy was stagnant, innovation activities have been continuing at an internationally competitive level. It also means, however, that Japan's industrial innovations were insufficient to contribute to its overall economic growth during the same period. Their contribution to the profitability and growth of the Japanese firms also declined (Sakakibara and Tsujimoto, 2003). Thus, in other words, industrial innovations in Japan have been frequent over recent decades, but they have not been so effective in boosting the national economy as a whole.

THE NATURE OF EFFECTIVE INNOVATIONS IN JAPAN

Japan's 'first in the world' innovations include the Toyoda automatic loom, monosodium glutamate (Ajinomoto), quartz watches, compact disc (CD) Karaoke systems, and parallel-hybrid vehicles, amongst others, Simply listing such cases does not answer the question about Japan's distinctive characteristics in innovation, which are influenced by many factors ranging from chance events to the innovator's personality.

It should be noted here that certain Japanese manufacturing industries, such as the automobiles, machine tools, robots, and functional chemicals, have been relatively competitive internationally even as the country's overall economy suffered from low growth in the post-Cold-War era. Innovations have been reasonably frequent in these industries. Thus, to the extent that such innovations contributed to the relative growth of Japan's competitive sectors, it can be said that industrial innovations were *effective* in these particular parts of the country.

This chapter will therefore focus on the nature of Japanese firms' innovations that contributed to industrial competitive performance, or *effective innovations*. Special attention will be paid to two main aspects of innovation-related characteristics: what kind of *products* Japanese firms innovate effectively, and what kind of *capabilities* enable them to innovate such products effectively. By applying an evolutionary framework of *design-based comparative advantage* (Fujimoto, 2007, 2012a), which predicts that dynamic fit between organizations' capabilities and products' architectures bring about their competitive performance, this chapter will identify the basic characteristics of effective innovations and their management in Japan (Figure 17.1).

Our analysis based on a capability-architecture framework, suggests that firms in post-war Japan tend to make active and effective innovations in products with relatively *integral architectures* (Ulrich, 1995) by mobilizing their rich endowment of *coordinative capabilities* (Fujimoto, 1999).

In this, special attention will be paid to Japan's automobile industry, mainly because (i) innovations are relatively numerous in this sector, (ii) the industry still enjoys international competitive advantages, (iii) architectures of the Japanese cars tended to be relatively integral, and (iv) the industry is known for building strong coordinative capabilities including the Toyota Production System (Monden, 1983; Ohno, 1988; Womack et al., 1990). As such, product and technology developments in this particular industry seem to be a good example of effective innovation in Japan.

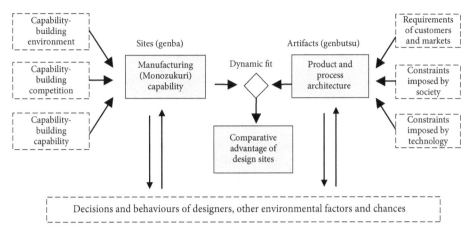

FIGURE 17.1 Design-based comparative advantage

The Capability-Architecture-Performance Framework

Before examining this proposition, it is useful to examine certain key concepts and logic for explaining why a country's firms tend to innovate certain types of products by using certain types of capabilities. The idea is that innovation and manufacturing are both design-based, creating and utilizing new designs that contribute to additional economic value, while *manufacturing* in a broad sense implies effective flows of value-carrying design information to the market.

Design-as-information, as axiomatic design theory suggests (Suh, 1990), consists of a bundle of interconnections between an artefact's functional elements (e.g. its performance goals) and structural elements (e.g. components' design parameters). As a product is nothing but a tradable artefact, a product innovation can be interpreted as a new combination of its functional-structural design elements and their interconnections.

Design-as-activity is a series of efforts for coordinating an artefact's functional and structural design elements. The result of effective design-as-activity is a product innovation, or a new combination of design-as-information.

Today's product (or process) innovations are in many cases repetitive creations of a series of new products (or processes) by competing organizations. In other words, effective innovation is executed mostly by a certain system of interrelated organizational routines, or *organizational capability* (Nelson and Winter, 1982; Clark and Fujimoto, 1991). As an organization is a system of coordinated activities (Barnard, 1938), the key dimensions of its capability include the degrees and types of coordination.

On the other hand, we can define an artefact's *architecture* as the formal aspect of its design-as-information, or the abstract pattern of interconnections between its functional and structural elements (Ulrich, 1995). Whereas product technologies, or concrete causal knowledge between the artefact's structures and functions, are specific to different industries, the concept of product-process architectures, or abstract patterns of functional-structural interconnections, can be applied to analyses across industries.

It follows from the above argument that a country's patterns of industrial innovations may be characterized from at least two points of view: (i) the *capabilities for innovations* in its firms, and (ii) the *architectures* of the new products that are innovated by that country's firms.

Applying this capability-architecture framework to the case of Japanese industries in a dynamic way shows that post-war Japanese industries tended to possess rich endowment of coordinative capabilities (e.g. teamwork of multi-skilled engineers/workers) mostly for certain historical reasons, and that Japan's *coordination-rich* industries tended to enjoy design-based comparative advantage in *coordination-intensive* products. In other words, Japan's industrial innovations tended to be more active in industries with relatively *integral* (i.e. coordination-intensive) architectures, including automobiles and functional chemicals, rather than those with *modular* (i.e. coordination-saving) architectures, such as digital products and package software.

This chapter proposes an evolutionary framework of design-based comparative advantage, or that of dynamic fit between capabilities and architectures, for explaining certain characteristics of innovations in post-war Japan (Fujimoto, 1999, 2007, 2012b, Shiozawa and Fujimoto, 2010). A summary of this approach is as follows:

- It complies with the conventional framework of comparative advantage theories, which emphasizes country-industry fit and relative productivity advantage across countries. It adopts design-based concepts of comparative advantage by infusing the broad view of manufacturing ('monozukuri' in Japanese) into existing trade theories. While maintaining the comparative advantage framework, it focuses on the fit between the manufacturing organization's capability and the product's architecture.
- Both capabilities and architectures are treated as endogenous and dynamic. It assumes that a certain evolutionary process results in uneven distribution (i.e. endowment) of a given organizational capability across countries and firms and that capabilities are affected by the availability and scarcity of resources. In this sense, history matters.
- It assumes that organizational capabilities are more difficult to move across borders than money, capital, goods, and services even in the era of globalization, and tend to become country-specific. The country's capability-building environments (e.g. resource scarcity), intensity of competition in industry's capability-building, and firms' capability-building capability (i.e. evolutionary capability) all affect the prevalent nature of capabilities of its manufacturing sites ('genba' in Japanese).
- It also assumes that the product's macro (overall) architecture is selected *ex post* by markets and the society more broadly, whereas its micro architectures are generated *ex ante* by engineers. When the product faces demanding functional requirements and/or strict constraints (e.g. safety and environmental

regulations), its macro architecture tends to become integral, other things being equal. By contrast, when the requirements and constraints are less strict, it tends to become more modular. Thus, a product's architecture is not given and it evolves though the micro–macro loops between design selections made by engineers and markets.

- It tries to explain why certain products are imported or exported within the framework of the overall trend of the twenty-first century towards intra-industry trade of differentiated products.

Why are Innovations in Japan Still Important?

A Balanced View Between Integral and Modular Architectures

This analysis presents additional insights on industrial innovation and competition in the twenty-first century in general. While there have been arguments that praised the power of coordinative capability for integral products (e.g. lean production: Womack et al., 1990; Clark and Fujimoto, 1991) or the power of combinatorial capability for modular products (e.g. Baldwin and Clark, 2000) around the turn of the century, the world in this century will be more complicated.

In the world of digital artefacts, for instance, including electronics hardware, package software, information services, and financial products, which are basically driven by such weightless things as electrons and logic, the power of modularity in creating almost infinite functional variety for the customers has been enormous (Baldwin and Clark, 2000).

At the same time, in the world of weighty artefacts with high functionalities, such as the automobile, power stations, and sophisticated industrial equipment, products tend to become even more complex and integral because global resource constraints, environmental protection and safety issues, and other technical-physical-environmental constraints imposed upon them have become stricter since the beginning of the twenty-first century.

Likewise, the growth of large emerging economies (e.g. China, India, and Brazil) at the beginning of this century brought about explosive expansion of 'volume zone' segments with a large number of entry customers, who are relatively price-oriented and thus prefer simpler and modular products. The sudden emergence of huge lower-end segments adversely affected many of the performance-oriented innovators in Japan, as their over-engineered products rapidly lost cost competitiveness vis-à-vis their rivals with reverse-engineered or reverse-innovated products (Christensen, 1997).

On the other hand, as customers accumulate experiences with the product in question, they may become more performance-oriented in both advanced and emerging economies. Environmental and safety regulations may also become gradually tougher even in emerging countries. It is therefore difficult to predict the long-term trend of average simplicity/complexity of a given product as mixed trends of both simplification and complication have been observed, depending upon the nature of the products, segments, and countries in question.

For the above reason, this chapter returns to conventional trade theory of comparative advantage, incorporates certain insights from design theory (architecture thinking), and makes it dynamic by introducing the evolutionary framework of capability-building (Aoki, 1994, 2001; Langlois and Robertson, 1995; Ulrich, 1995; Fujimoto, 1999, 2007, 2012b; Baldwin and Clark, 2000). Effective innovations in Japan can be explained by this comparative and contingent framework that deals with both strength and weakness of integral and modular architectures, as well as coordinative and specific (or combinatorial) capabilities.

In this context, the rest of the chapter will explore: (i) how and why coordinative capabilities were accumulated in post-war Japan; (ii) in what conditions a given product or process becomes integral or modular; and (iii) whether we can apply the capability-architecture hypothesis to the automobile and other cases.

EVOLUTION OF COORDINATIVE CAPABILITIES

Capability-Building as an Historical Phenomenon

With the above framework and questions in mind this chapter examines the capability, architecture, and performance aspects of industrial innovations in Japan. First, the dynamics of capabilities are examined.

One of the major characteristics of the Japanese manufacturing organization ('genba') has been its rich endowment of *coordinative capability*, or *teamwork of multi-skilled employees*. Although the group-orientation of the Japanese people has often been discussed as a cultural phenomenon (Cole, 1971; Hazama, 1971; Dore, 1973; Hofstede, 1980), this chapter argues that coordinative capability as a source of the Japanese industrial competitiveness has basically been a result of certain historical imperatives (Fujimoto, 1999). With a notable exception of pre-war Japan's cotton spinning and weaving industry, which enjoyed international competitive advantage not only through low wages but also by productivity improvements (Kuwahara, 2004; Shimokawa and Fujimoto, 2009; Koike, 2012), it was after the Second World War when Japan's industrial competitiveness and coordinative manufacturing capabilities (e.g. Toyota Production System) were recognized internationally. The historical effects on Japanese manufacturing (or tradable goods) industries since the latter half of the twentieth century therefore warrants attention.

Manufacturing Capability for Creating 'Good Flows of Good Design'

An organization's *capability* is defined as a system of organizational routines that generate sustainable performance advantage over its rivals (Nelson and Winter, 1982; Grant, 2005). *Manufacturing* in a broad sense ('monozukuri') involves creating flows of good design information (i.e. value-adding information) to customers, encompassing product development, production, purchasing, and sales (Clark and Fujimoto, 1991; Fujimoto, 1999). It follows that *coordinative capability in manufacturing* is a system of development–production–purchasing–sales routines that are mutually adjusted for creating competitive flows of value-carrying design information through the collaboration of workers, engineers, and managers.

Japan's manufacturing sites and industries in post-war Japan are known for establishing various coordinative capabilities, including the Toyota Production System (TPS), Total Quality Control (TQC), Total Productive Maintenance (TPM), Team-Oriented Product Development, Collaborative Supplier Systems, and so on (Monden, 1983; Takeuchi and Nonaka, 1986; Womack et al. 1990; Clark and Fujimoto, 1991; Nishiguchi, 1993; Dyer, 1994; Cusumano and Nobeoka, 1998; Fujimoto, 1999; Liker, 2004), and they contributed to Japan's competitive advantages in relatively coordination-intensive industries (i.e. products with integral architecture) such as low-emission and fuel-efficient cars, analogue television sets, high-precision machine tools, functional chemicals, and so on. The emergence of such capabilities themselves may be regarded as a type of process innovation.

Advantages and Improvements in Productive Performance

Generally speaking, *competitiveness* (competitive performance) refers to something's ability to be selected by someone in the situation of free selection. Three layers of competitive performances can be distinguished: *profit performance* is a firm's ability to be selected by investors in the capital market (e.g. return on investment, stock price); *market performance* is its product's ability to be selected by customers in the product market (e.g. market share, price competitiveness); *productive performance* is its manufacturing site's ability (e.g. productivity, lead times, defect rates).

Among the three, it was the productive performance of Japanese factories and development projects ('genba') for coordination-intensive (or complex integral) products that generated high levels of international competitive advantages from the 1980s and thereafter. Womack et al. (1990), for example, found that the Japanese car assembly factories' productivity (assembly person-hours per vehicle) was on average 1.5 times higher that their North American counterparts in the late 1980s. Although assembly productivity improved by roughly 1.5 times on both sides in about twenty years, Japan's productivity advantage still remained in the 2010s (Holweg and Pil, 2004, Oshika and Fujimoto, 2011).

In the area of product development in the car industry, Clark and Fujimoto (1990) found that Japan's average development productivity (engineering person-hours per project) was about two times higher than the North Americans in the late 1980s. Unlike the aforementioned case of assembly productivity, development productivity in both Japan and Western countries became somewhat lower in the next decades due partly to the increasing product complexity that cancelled out firms' capability building, but a significant scale of productivity gap remained in the late 2000s (Higashi and Heller, 2012). The same studies also identified significant lead-time advantages of Japanese projects.

Similar patterns of development performance advantages were found in coordination-intensive products such as super computers and large-scale enterprise software (Cusumano, 1991; Iansiti 1997).

In summary, Japan's manufacturing industries and factories tended to enjoy relatively high levels of productive performance in coordination-intensive (integral) products due partly to relatively rich endowment of coordinative capability in its manufacturing sites, even in the middle of long-term recession and deflation after 1990.

'Economy of Scarcity' and the Origin of Coordinative Capabilities

The next question is where this rich endowment of coordinative capabilities came from. Explanations may include such candidates as Japan's group-oriented culture, tight coordination processes for rice making in the pre-modern agricultural community, the family-oriented business disciplines of merchants since the seventeenth century, as well as the operation of the pre-war cotton spinning-weaving industry as a prototype of the Toyota-style manufacturing system. This chapter emphasizes historical imperatives after the war, particularly its high-growth era between the 1950s and the early 1970s.

It should be noted here that Japan's high-growth era (mostly 1950s and 1960s) preceded the period when Japan's industrial performance in coordination-intensive sectors became recognized worldwide (1970s and 1980s). The former was a period of chronic labour shortage; unemployment rates were around 1 per cent in most of the high-growth era. This was partly because Japan relied neither on the massive inflow of immigrants found in the USA in its high-growth period, nor on large-scale migration of labour force from agricultural regions in China in the 1990s and 2000s.

When rapid economic growth is accompanied by large-scale inflows of labour into industry, firms find it economically rational to make quick use of the inflow of the people by dividing production processes into narrower tasks, simplifying individual work content, shortening training periods, standardizing operational rules, and thereby making the production systems coordination-saving and specialization-oriented. The

Ford System and Taylor System were both typical examples of such coordination-saving efforts that supported America's rapid growth era (Nelson, 1980; Hounshell, 1984). China's rapid growth in the 1990s, led by the southern regions, such as Canton province, also relied on coordination-saving systems that utilized a single-skilled workforce arriving from the inland provinces for relatively short periods.

When rapid growth is achieved without massive immigration, such as in post-war Japan, the chronic shortage of workers makes labour switching cost higher, which motivates firms to select longer-term approaches to employment systems that nurture their coordinative capabilities. Labour shortages also make narrow specialization less effective. In order for the limited number of existing employees to cope with growing output, they need to become multi-skilled with task-overlapping just like a modern soccer team (Drucker, 1998, 2004), where attacking players also contribute to defence and vice versa. Long-term employment also provides workers and engineers with opportunities to become multi-skilled to the extent that it is accompanied by skill-based wage and promotion systems.

Labour shortages may also force individual firms to rely more on subcontractors. The switching cost of these suppliers, however, will also become high, so this will facilitate longer-term transactions between them. And long-term employment and transactions, in turn, promote accumulation of coordinative capabilities within and between manufacturing firms and suppliers.

To summarize, when an economy grows rapidly without massive inflow of labour from abroad or rural areas, the resulting shortage of labour inputs will tend to (i) *constrain intra-firm division of labour* inside firms (i.e. multi-skilled employees), (ii) *promote inter-firm division of labour* between firms (i.e. supplier systems), and (iii) *promote coordinative capabilities* within and between firms (i.e. team work and relational transactions). Japan's manufacturing innovation activities can be characterized by 'teamwork of multi-skilled employees' and 'relational transactions with suppliers'.

This kind of historical imperative, of input scarcity during the high-growth era, may have become a source of Japan's industrial competitiveness and innovativeness, particularly in coordination-intensive sectors such as automobiles (Fujimoto, 1999). Such a phenomenon of emergent capability building, driven by input shortage, may be called the *economy of scarcity*. The scarcity of labour inputs in the high-growth era played a role of 'initial push' which resulted in the emergence of many manufacturing sites with high levels of coordinative capabilities in post-war Japan across regions and industries.

Once established, domestic manufacturing sites try to survive intensifying competition and tougher environments (e.g. appreciation of the currency, lower economic growth, emergence of low wage countries) by continuing further capability building, which is what has happened in Japan since the 1970s. In this way, the coordinative capability of Japan's manufacturing sites became a sustainable source of post-war Japan's productive performance such as higher productivity and quality, as well as shorter lead-times, of both operations and innovations.

CAPABILITY BUILDING AND COMPARATIVE
ADVANTAGE IN DESIGN

Productive/Market/Profit Performance in the
Cold-War Period

This, however, does not necessarily mean that Japanese firms and industries during the same period enjoyed constantly higher market and profit performance.

International market performance, or trade structure, is affected not only by physical productivity (productive performance) but also by other factors such as exchange rates and international wage differences. As the basic 200-year-old formula of David Ricardo puts it: $p_{iJ} = a_{iJ} w_J e_J (1+r_J)$, where p_{iJ} is production price of goods i in country J in dollars (key currency), a_{iJ} is labour input coefficient (inverse of physical labor productivity) of that industry, w_J is average hourly wage rate in domestic currency, and e_J is exchange rate of country J's currency versus key currency, and r_J as average profit rate of firms in country J (Ricardo assumes this as equal among all firms).

Assume for simplicity's sake that productivity of the same product category in the same country is identical among surviving firms, international differences in quality and delivery are already adjusted, tariffs and transportation costs can be ignored, and competitive advantages of product i in country J is determined by the Ricardian natural price p_{iJ} vis-à-vis that of the competing country A, or p_{iJ}/p_{iA}. Using this simple formula, we can see how the long-term trend of relative productive performance (a_{iJ}/a_{iA}), market performance (p_{iJ}/p_{iA}), and profit performance (r_J/r_A) are influenced by such 'handicap' factors as relative wages (w_J/w_A) and exchange rates (e_J/e_A). So trade is a handicapped competition in the above sense.

In the 1950s and 1960s, both exchange rates ($1/360 yen in the 1960s) and wage rates (roughly 1/6 in 1965 dollars vis-à-vis the USA) were low, but so were Japan's productivity levels of trading goods. As a result, Japan had a chronic trade deficit until the 1960s. In the 1970s and 1980s, both exchange rates ($1/240 yen in 1985) and wage rates (roughly 1/2 in 1975 dollars) started to increase. By the 1990s, Japan's average international wage rate fully caught up with that of major Western countries ($[w_J e_J]/[w_A e_A] \fallingdotseq 1$). And yet Japanese manufacturing sectors, and coordination-intensive goods such as cars and analogue TV sets in particular, started to enjoy international price advantages, which resulted in Japan's significant trade surpluses and trade frictions with Western nations in the 1980s. Profit rates of the leading Japanese firms vis-à-vis their Western rivals (r_J/r_A) were, however, modest, at best.

This historical evidence infers that the capability-building and productivity increase of Japanese manufacturing in the 1970s and 1980s was significant enough ($a_{iJ}/a_{iA}\downarrow$) to overcome the large increase in both relative wages and exchange rates ($[w_J\,e_J\,]/[w_A\,e_A]\uparrow$) and gained price-competitiveness ($p_{iJ}/p_{iA}\downarrow$) without sacrificing profit rates (r_J/r_A) Also, judging from the fact that Japan's trade surplus concentrated in automobiles and complex and analogue machines and appliances, it can be inferred that Ricardian comparative advantages in design cost tended to exist in coordination-intensive goods (I) rather than coordination-saving ones (S), that is, $a_{IJ}/a_{IA} < a_{SJ}/a_{SA}$ (Fujimoto and Shiozawa, 2011).

Productive/Market/Profit Performance in the Post-Cold-War Period

In the 1990s, however, two major events, unfavourable to the Japanese trading goods sectors, occurred back to back. One was the end of Cold War and the entry of Chinese industries into the global market, which meant the abrupt emergence of a huge and neighbouring low-cost country, whose relative wage rates ($[w_J\,e_J\,]/[w_A\,e_A]$) were less than one-twentieth of Japan's throughout the 1990s. The other was revolutionary innovation in the Internet and other digital communication technologies since the mid-1990s, which caused rapid substitution of analogue (relatively coordination-intensive) devices and equipment by digital (relatively coordination-saving) ones. Furthermore, Japan's exchange rates continued to increase (about $1/80 yen in 1995 and 2012).

Because of the above changes in the wage handicaps and architecture of Japanese products, the competitive gaps in market performance between coordination-intensive (I) goods (e.g. cars, machine tools, functional chemicals) and coordination-saving (S) ones (e.g. bicycles, PCs, memory semiconductors) became increasingly evident. Although physical productivities continued to increase ($a_{iJ}/a_{iA}\downarrow$) in both I and S sectors, many Japanese manufacturing sites in S industries have closed down since the 1990s. A domestic assembly factory of Japan's major consumer electronics company, producing video cameras, for example, increased productivity by over five times between 2000 and 2010 ($a_{iJ}/a_{iA}\downarrow$), while lowering the wage rate ($w_J/w_A\downarrow$) by shifting its main workforce to low-wage temporary workers, but the company eventually decided to close down this factory in 2012. As TV sets became a digital product, Japan's major TV manufacturers (Panasonic, SONY, and Sharp) suffered huge deficits in 2012. Despite the wage decrease and productivity increase, Japan's coordination-saving products found it difficult to survive in the twenty-first century.

Many manufacturing firms in coordination-intensive industries coped with the adversarial environment by reducing production capacity and profits, yet many of their domestic factories survived while keeping most of their regular workforce. Japan's domestic automobile production continued to be around 10 million units throughout most of the post-Cold-War period (1990s–2010s), about half of which was exported. In the period of the US financial bubble economy, Japan's major automobile firms enjoyed relatively high profit rates, although average profit ratios of Japanese manufacturing firms continued to deteriorate in the same period.

Further empirical research on the topic of design-based comparative advantages is warranted, but some exploratory statistical analyses indicate that architectural integrality (i.e. coordination-intensity) is positively correlated with international market performance (i.e. export ratios) (Fujimoto, 2007), and that the coexistence of factories' coordinative capability and products' coordination intensity is positively correlated with productive performance in product development (Tsuru and Morishima, 2012). It is also known that Japanese productivity advantages in coordination-rich products in both production and product development (a_{iJ}/a_{iA}) continued to be large between 1980 and 2010 (Womack et al., 1990; Clark and Fujimoto, 1992; Holweg and Pil, 2004; Higashi and Heller 2012).

From all this evidence, the changes in Japanese industrial and company performance can be summarized as follows:

- Relative productive performance of better Japanese manufacturing sites continued to increase rapidly from 1950 to 1990 ($a_{iJ}/a_{iA}\downarrow$), which more than compensated for the increasing burden of the wage-currency handicaps ($[w_J\,e_J\,]/[w_A\,e_A]\uparrow$) in creating competitive advantages in many of Japan's manufacturing sectors.
- As a result, relative market performance (export competitiveness) of Japanese tradable goods continued to increase in the 1970s and 1980s ($p_{iJ}/p_{iA}\downarrow$), causing a trade surplus and frictions despite the heavier handicaps. Since the wage rates of major advanced countries were similar in the 1980s, productivity advantages of leading Japanese firms, based on their coordinative innovations and capability building, directly contributed to their market performance in the world market.
- The end of the Cold War and the beginning of the digital era totally changed the nature of global competition for leading Japanese firms. The magnitude of the wage difference between Japan and the emerging countries (e.g. China) during the 1990s was so large ($[w_J\,e_J\,]/[w_A\,e_A] \gg 1$) that Japan's manufacturing sites for coordination-saving products (S) quickly lost their relative market performance ($p_{SJ}/p_{SA} > 1$), despite their continued capability building and productivity increase ($a_{SJ}/a_{SA}\downarrow$). The profit rate of major Japanese firms in such sectors sharply declined as well ($r_{SJ}\downarrow$).
- The coordination intensive sectors (I), however, kept their quality-adjusted price competitiveness ($p_{IJ}/p_{IA} < 1$), and thereby continued to export their products, as well as expanding overseas production. Their average wage rates, core worker ratios, and profit rates tended to be higher than the firms and factories in coordination-saving sectors ($w_{IJ} > w_{SA}$; $r_{IJ} > r_{SJ}\downarrow$). As a result, despite Japan's stagnant economic growth since the 1990s, it continued to expand its exports and produce trade surpluses for most of the period between 1990 and 2010.

Evolution of Integral Architectures

Requirements, Constraints, and Macro-Architecture

This chapter has so far argued that Japan's innovations and operations may be characterized by the organizational capability of Japanese manufacturing firms and factories. It is important to look at the market or social side as well. Architecture of a certain product is not a given but is endogenous, affected not only by physical laws, technological limitations, and the like but also by the intensity of customers' functional requirements, social constraints, and government regulations,

When these requirements and constraints are more demanding, the artefact's macro-architecture, or its overall patterns of function-structure interconnections, tends towards the integral type, whether it is a product or process (Ulrich, 1995; Fujimoto, 2007). When they are not so demanding it tends to become more modular, partly because most engineers are constantly trying to simplify (i.e. modularize) the artefact's micro-architecture of its detailed function-structure links wherever possible. This again recognizes that, in the evolutionary framework of artefacts and architectures, variation of an artefact's micro-architecture is made *ex ante* by engineers, whereas selection of its macro-architecture is made *ex post* by markets and societies (Fujimoto, 2012b).

In this context, additional factors may make Japanese products and processes more integral than many other countries. First, Japan is a country that does not have abundant natural resources and spaces. High energy costs, population density of large cities, space limitations, and fear of natural disasters all tended to raise the level of safety-energy-environmental constraints and regulations imposed by the Japanese government and society. In response to serious pollution and safety problems in the high-growth era, for example, the Japanese government started to adopt strict environmental regulations, which stimulated many energy-saving and emission-reducing innovations by Japanese car makers, including emission-controlling devices and catalysts, lean-burn engines, hybrid vehicles, and so on.

Second, the general (presumably cultural) tendency of the average Japanese customers to be strict about the details of product appearance, compactness, and functional preciseness might have forced the Japanese manufacturers to develop integral products, which optimize and customize their design parameters to such demanding domestic markets. The Japanese enthusiasts in various fields ('otaku' in Japanese) also raised functional requirements very high. Japanese firms, for example, lost competitive advantages in many digital products, but still enjoy a large global market share in high-end digital cameras. It is often said that the subtle but strict requirements of amateur photographers, train-watchers, bird-watchers, and other enthusiasts support a significant portion of its domestic demand.

Third, Japanese engineers themselves tend to prefer design optimization and to self-impose very high goals not only in functions but also in compactness and lightness.

Japan, for example, is no longer a major producer of personal computers but, as of 2012, the lightest full-function PC in the world is designed and produced in Japan.

The problem is that these extremely functional, precise, light, and compact products by the Japanese firms are not the best sellers in the global markets. Many of Japan's high-function products, which might satisfy a small portion of Japanese enthusiasts and perfectionists, have tended to suffer from over-specification, over-design, and over-quality, and thereby lost cost competitiveness (Christensen, 1997), particularly in emerging markets with price-sensitive entry customers.

It may be true that these price-oriented customers might eventually become function-oriented as they accumulate user experiences, but the passive strategy of waiting in the high-end segment creates huge opportunity costs and allows lower-end competitors to enter into high-end segments in future. Many of Japan's leading manufacturing firms have to choose the best balance between the defence of function-oriented segments, their traditional strongholds, and entry into price-oriented ones of the emerging and growing countries.

To summarize, the high levels of requirements and constraints imposed by Japanese society, customers, and engineers themselves all stimulated innovations in products and processes with relatively integral architectures, and this became a source of Japanese industrial competitiveness. To the extent that Japanese requirements are excessive from the global market's point of view, however, manufacturers have to do at least two things. First, they could build up additional product development capability that suits simpler and modular products either inside or outside Japan. Second, they could strengthen global marketing and branding strategies to educate entry users in emerging markets, raise their functional sensitivity, and thereby facilitate upward movement of these 'volume' segment users into 'function-oriented' segments.

Effective Innovation in Japan in Integral Products

Returning to the basic concepts of design theory (Suh, 1990; Ulrich, 1995), integral architecture, as an ideal type, refers to the case where interconnections between the artefact's functional design elements (e.g. its performance goals) and structural design elements (e.g. its component's shapes and materials) are complex with many-to-many connections; whereas modular architecture, in its pure form, refers to the case where they are simple with one-to-one correspondence.

Design activities in this sense are in essence coordination of an artefact's functional and structural elements prior to its production. It follows that products and processes with more integral architectures are more coordination-intensive, whereas those with more modular architecture are more coordination-saving. Adopting axiomatic design theory's analogy of design as solving a functional-structural simultaneous equation (Suh, 1990), it can be suggested that by a simple economic model and simulations, a design organization with a higher coordinative capability will get Ricardian comparative advantage in design cost in relatively integral (coordination-intensive) products rather than modular ones (Okuma and Fujimoto, 2006; Fujimoto, 2012b).

When manufacturing firms or project teams in the same country share the same types of organizational routines which speed up the collaborative problem-solving process mentioned above, this set of routines can be called a *coordinative capability for product innovations*. Such coordinative routines may include compact and coherent project teams of multi-skilled engineers (Clark and Fujimoto, 1991; Yasumoto and Fujimoto, 2005; Koike, 2012), supplier involvement for co-design and joint problem solving based on trust relationship (Asanuma, 1989; Helper, 1990; Clark and Fujmoto, 1991; Cusumano and Takeishi, 1991; Sako, 1992; Nishiguchi, 1993; Dyer, 1994; Takeishi, 2002, 2003), overlapping or front-loading problem solving inside or between the projects (Clark and Fujimoto, 1991; Nonaka and Takeuchi, 1995; Cusumano and Nobeoka, 1998; Thomke and Fujimoto, 2000), and heavyweight product managers for concept-driven coordination (Fujimoto, 1989; Clark and Fujimoto, 1991).

When the firms in question gradually create coordinative routines over time (Fujimoto, 1999), this capability-building process can be considered as a series of effective process innovations. And these process innovations for coordinative capability building, in turn, will bring about design-based comparative advantage in relatively coordination-intensive (i.e. integral) products, such as the high-performance or low-emission automobiles, high-precision machine tools and components, functional chemicals, high-tension steel, and so on.

In this way, effective product innovations for integral product improvements and effective process innovations for coordinative capability-building seem to have been reinforcing each other in important competitive sectors of post-war Japan.

This is the case in the Japanese auto industry. First, the Japanese auto industry in the early twenty-first century still maintains certain levels of advantages in productivity, costs, and lead times in product development as well as production productivity (Fujimoto and Nobeoka, 2006; Oshika and Fujimoto, 2011; Higashi and Heller, 2012). As a result, despite unfavourable exchange rates and international wage differences, vehicles designed by Japanese firms still retain significant market shares in the global market (about 30 per cent, or 24 million units in 2012).

Second, product innovations continue to flourish in this industry, compared with international competitors, as well as other Japanese industries such as digital products. The auto industry, for example, occupied 54 per cent of Japan's total technology exports (1.8 trillion yen) in 2004, whereas the share of the electric/electronics machine industry was 17 per cent. The technology export/import ratio of the Japanese auto industry was 137.8 in 2004 (1.6 in electric machinery, 1.5 in electronics parts, 1.0 in information-communication machinery). The auto industry is also Japan's second-largest R&D-spending industry, after the electric-electronics industry.

Third, the architecture of automobiles in Japan continues to be relatively integral. The percentage of product-specific parts in the average 'all-new' model, for example, was about 80 per cent (Higashi and Heller, 2012), which was higher than that in the 1990s (about 60 per cent). The complexity and integrality of the high-function vehicles seems to be increasing further in the early part of this century (Fujimoto, 2012a).

Fourth, the Japanese auto industry and firms have still continued to build their coordinative capability in recent years, and most of the literature on Japan's coordinative capabilities and routines for product development is based on researchers' observations and data collections in the automobile industry (Asanuma, 1989; Clark and Fujimoto, 1991; Cusumano and Nobeoka, 1998; Thomke and Fujimoto, 2000).

These findings infer that innovations have been both frequent and effective in the Japanese auto industry of the late twentieth and early twenty-first century. These innovations have been effective in at least two areas—*process innovations for building coordinative capability* (e.g. Toyota Production System), and *product innovations for developing integral and complex artefacts*—an observation that is consistent with our capability-architecture-performance view of effective industrial innovations.

CONCLUSION AND FUTURE AGENDA

This chapter explored the basic characteristics of innovations and their management in post-war Japan. After pointing out the relative performance of Japanese innovations at the national level in recent years, it focused on particular types of innovations that have effectively contributed to the international competitiveness of certain Japanese industries. Based on the framework of design-based comparative advantage, the analysis hypothesized that effective innovations in post-war Japan have tended to be concentrated in products with coordination-intensive (integral) architectures, as well as processes with coordinative capabilities, due partly to choices of engineers/designers, selection by markets/societies, and certain historical imperatives.

Statistical evidence supports the above framework and hypothesis. By regression analyses of first-hand data of selected Japanese firms at the project level, Oshika and Fujimoto (1999) indicated that there are positive and statistically significant correlations between architectural integrality and export ratios of those Japanese products (after controlling for labour intensity) in both assembly and process industries (see Figure 17.2). Tsuru and Nakajima (2012) analysed similar hypotheses in Japanese, Korean, and Chinese firms and found the statistically significant results that profit performances tend to be higher when personnel policies, which could be re-interpreted as the level of their coordinative capabilities, fit their products' architectural integrality/modularity.

There are alternative ways of measuring architectural integrality/modularity. Some are more precise theoretically but difficult to measure; others are easier to measure but oversimplified from the theoretical point of view (Yasumoto and Fujimoto, 2005; Fujimoto, 2007). An academically acceptable measurement systems needs to be

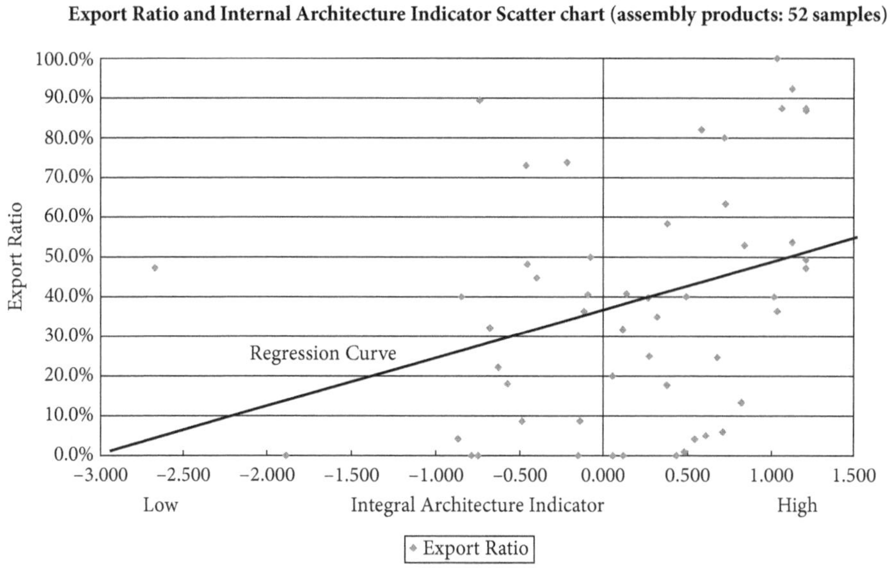

Export Ratio and Internal Architecture Indicator Scatter chart (assembly products: 52 samples)

FIGURE 17.2 Architectural integrality and export ratio of Japanese products

developed and further accumulated statistical evidence is needed regarding the relationships between architectures, capabilities, and performances.

After the First World War, Alfred Marshall, in his book *Industry and Trade* (Marshall, 1919), analysed the nature of industrial leaderships of England (i.e. the established industrial nation) as well as Germany and the United States (i.e. emerging nations of those days), and emphasized the Ricardian trade benefit between those countries with different industrial leadership. A century later, analysis of the Marshallian industrial leaderships of advanced industrial nations (Japan, USA, EU, etc.) and emerging ones (Korea, China, India, etc.) is needed once again.

To the extent that the global economy of the early twenty-first century witnesses the expansion of free intra-industrial trades with product differentiation, the framework of design-based comparative advantage might increase its importance in explaining various trade phenomena between countries of different types of industrial leadership (Fujimoto, 2012b). In this context, Japan's industrial innovations, although they were probably overemphasized in the 1980s and unduly de-emphasized since then, will continue to bring about certain types of industrial leadership particularly in the sectors of coordination-intensive goods.

REFERENCES

Aoki, M. (1994). 'Contingent Governance of Teams: Analysis of Institutional Complementarity', *International Economic Review*, 35: 657–76.

Aoki, M. (2001). *Towards a Comparative Institutional Analysis*. Cambridge, Mass.: MIT Press.

Asanuma, B. (1989). 'Manufacturer–Supplier Relationships in Japan and the Concept of Relation Specific Skill', *Journal of the Japanese and International Economies*, 3: 1–30.

Asanuma, B. (1997). *Nihon no kigyo soshiki: Kakushinteki tekio no mekanizumu* (in Japanese). Tokyo: Toyo Keizai Shinposha.

Baldwin, C. Y., and Clark, K. B. (2000). *Design Rules: The Power of Modularity*. Cambridge, Mass.: MIT Press.

Barnard, C. I. (1938). *The Functions of the Executive*. Cambridge, Mass.: Harvard University Press,

Christensen, C. M. (1997). *The Innovator's Dilemma: When New Technologies Cause Great Firms to Fail*. Boston: Harvard Business School Press.

Clark, K. B., and Fujimoto, T. (1990). 'The Power of Product Integrity', *Harvard Business Review*, 68(6): 107–18.

Clark, K. B., and Fujimoto, T. (1991). *Product Development Performance*. Boston: Harvard Business School Press.

Clark, K. B., and Fujimoto, T. (1992). 'Product Development and Competitiveness', *Journal of the Japanese and International Economies*, 6: 101–43.

Cole, R. E. (1971). *Japanese Blue Collar: The Changing Tradition*. Berkeley and Los Angeles: University of California Press.

Cusumano, M. A. (1991). *Japan's Software Factories: A Challenge to U.S. Management*. New York: Oxford University Press.

Cusumano, M., and Nobeoka, K. (1998). *Thinking Beyond Lean: How Multi-Project Management is Transforming Product Development at Toyota and Other Companies*. New York: Free Press.

Cusumano, T., and Takeishi, A. (1991). 'Supplier Relations and Management: A Survey of Japanese, Japanese-Transplant, and U.S. Auto Plants', *Strategic Management Review*, 12(8): 563–88.

Dore, R. (1973). *British Factory, Japanese Factory: The Origins of National Diversity in Industrial Relations*. Berkeley and Los Angeles: University of California Press.

Drucker, P. F. (1998). *Peter Drucker on the Profession of Management*. Boston: Harvard Business School Press.

Drucker, P. F. (2004). *Jissensuru keieisha* (in Japanese) [Advice for entrepreneurs], trans. A. Ueda. Tokyo: Daiyamondosha.

Dyer, J. H. (1994). 'Dedicated Assets: Japan's Manufacturing Edge', *Harvard Business Review*, 72(6): 174–8.

EU Commission (2013). 'The 2012 EU Industrial R&D Scoreboard'.

Fujimoto, T. (1989). 'Organizations for Effective Product Development: The Case of the Global Automobile Industry' (unpublished DBA dissertation, Harvard University Graduate School of Business Administration, Boston).

Fujimoto, T. (1999). *The Evolution of a Manufacturing System at Toyota*. New York: Oxford University Press.

Fujimoto, T. (2007). 'Architecture-Based Comparative Advantage: A Design Information View of Manufacturing', *Evolutionary and Institutional Economics Review*, 4(1), 55–112.

Fujimoto, T. (2012a). 'Manufacturing Capability and the Architecture of Green Vehicles', in G. Calabrese (ed.), *The Greening of the Automotive Industry*. Basingstoke: Palgrave Macmillan.

Fujimoto, T. (2012b). 'An Economic Analysis of Architecture and Coordination: Applying Ricardian Comparative Advantage to Design Costs and Locations', *Evolutionary and Institutional Economics Review*, 9(1): 51–124.

Fujimoto, T., and Nobeoka, K. (2006). Kyosoryokubunseki niokeru keizokuno chikara [Competitive strength-analysis and the power of duration] (in Japanese), *Soshiki Kagaku* [Organizational Science], 39(4): 43–55.

Fujimoto, T., and Oshika, T. (2006). 'Empirical Analysis of the Hypothesis of Architecture-Based Competitive Advantage and International Trade Theory', MMRC Discussion Paper Series No. 71, Manufacturing Management Research Center, the University of Tokyo, <http://merc.e.u-tokyo.ac.jp/mmrc/dp/pdf/MMRC71_2006.pdf>.

Fujimoto, T., and Shiozawa, Y. (2011). 'Inter and Intra Company Competition in the Age of Global Competition: A Micro and Macro Interpretation of Ricardian Trade Theory', *Evolutionary and Institutional Economics Review*, 8(1–2): 1–37, 193–231.

Grant, R. M. (2005). *Contemporary Strategy Analysis*, 5th edn. Oxford: Blackwell.

Hazama, H. (1971). *Nihonteki keiei* [Japanese management] (in Japanese). Tokyo: Nihon Keizai Shinbunsha.

Helper, S. R. (1990). 'Competitive Supplier Relations in the U.S. and Japanese Auto Industries: An Exit/Voice Approach', *Business and Economics History*, Second Series, 19: 153–62.

Higashi, H., and Heller, D. A. (2012). 'Thirty Years of Benchmarking Product Development Performance: A Research Note', MMRC Discussion Paper Series No. 395, Manufacturing Management Research Center, University of Tokyo.

Hofstede, G. H. (1980). *Culture's Consequences: International Differences in Work-Related Values*. Beverly Hills: Sage.

Holweg, M., and Pil, F. K. (2004). *The Second Century: Reconnecting Customer and Value Chain through Build-To-Order*. Cambridge, Mass.: MIT Press.

Hounshell, D. A. (1984). *From the American System to Mass Production: 1800–1932*. Baltimore: Johns Hopkins University Press.

Iansiti, M. (1997). *Technology Integration: Making Critical Choice in a Dynamic World*. New York: Free Press.

Koike, K. (2012). *Kohinshitsu Nihon no kigen*. Tokyo: Nihon Keizaishinbun Shuppannsha.

Kuwahara, T. (2004). 'Zaikabono soshikinoryoku: Ryotaisenkino Naigaiwatakaisha [Organizational capabilities of zaikabo (Japanese owned cotton spinning mills in China): Naigaiwata & Co. in the inter-war years]', *Ryukoku Daigaku Keieigaku Ronshu*, 44(1): 45–65.

Langlois, R. N., and Robertson, P. L. (1995). *Firms, Markets and Economics Change: A Dynamic Theory of Business Institutions*. London: Routledge,.

Liker, J. K. (2004). *The Toyota Way*. New York: McGraw Hill.

Marshall, A. (1919). *Industry and Trade*. London: Macmillan.

Monden, Y. (1983). *Toyota Production System*. Norcross, Ga.: Industrial Engineering and Management Press, Institute of Industrial Engineers.

Nelson, D. (1980). *Frederick W. Taylor and the Rise of Scientific Management*. Madison: University of Wisconsin Press.

Nelson, R. R. (ed.) (1993). *National Innovation Systems: A Comparative Analysis*. New York: Oxford University Press.

Nelson, R. R., and Winter, S. G. (1982). *An Evolutionary Theory of Economic Change*. Cambridge, Mass.: Belknap Press of Harvard University Press.

Nishiguchi, T. (1993). 'Competing Systems of Auto Components Development', Presented at the Annual Sponsors' Briefing Meeting, International Motor Vehicle Program (MIT), June.

Nonaka, I., and Takeuchi, H. (1995). *The Knowledge-Creating Company: How Japanese Companies Create the Dynamics of Innovation*. New York: Oxford University Press.

OECD (2012). *Main Science and Technology Indicators*.

Ohno, T. (1988). *Toyota Production System: Beyond Large-Scale Production*. Cambridge, Mass.: Productivity Press.

Okuma, S., and Fujimoto, T. (2006). 'Sekkei purosesu to akitekucha no kyoso yui' (in Japanese), MMRC Discussion Paper Series No. 70, Manufacturing Management Research Center, the University of Tokyo, <http://merc.e.u-tokyo.ac.jp/mmrc/dp/pdf/MMRC70_2006.pdf>.

Oshika, T., and Fujimoto, T. (2011). 'Comparative Analysis on Productivity in Asian Automaker Plants: IMVP Round 4 Studies (2006)', MMRC Discussion Paper Series No. 351, Manufacturing Management Research Center, the University of Tokyo, <http://merc.e.u-tokyo.ac.jp/mmrc/dp/pdf/MMRC351_2011.pdf>.

Sakakibara, K., and Tsujimoto, M. (2003). 'Why did R&D Productivity of Japanese Firms Decline?' ESRI Discussion Paper Series No. 47, Economic and Social Research Institute, Cabinet Office.

Sako, M. (1992). *Prices, Quality and Trust: Inter-Firm Relations in Britain and Japan*. Cambridge: Cambridge University Press.

Schumpeter, J. A. (1912/1934). *Theorie der wirtschaftlichen Entwicklung*. (1912, in German) Leipzig: Duncker & Hamblot. *The Theory of Economic Development*. (1934) Cambridge, Mass.: Harvard University Press.

Shimokawa, K., and Fujimoto, T. (eds) (2009). *The Birth of Lean*. Cambridge, Mass.: Lean Enterprise Institute.

Shiozawa, Y., and Fujimoto, T. (2010). Sekaikyosojidainiokeru kigyokan-kigyonaikyoso (in Japanese), *Keizaigaku Ronshu*, 76(3): 22–63.

Suh, N. P. (1990). *The Principles of Design*. New York: Oxford University Press.

Takeishi, A. (2002). 'Knowledge Partitioning in the Interfirm Division of Labor: The Case of Automotive Product Development', *Organization Science*, 13(3): 321–38.

Takeishi, A. (2003). *Bungyo to Kyoso* (in Japanese). Tokyo: Yuhikaku.

Takeuchi, H., and Nonaka, I. (1986). 'The New New Product Development Game', *Harvard Business Review*, 64(1): 137–46.

Thomke, S., and Fujimoto, T. (2000). 'The Effect of "Front-Loading" Problem Solving on Product Development Performance', *Journal of Product Innovation Management*, 17: 128–42.

Tsuru, T., and Morishima, M. (eds) (2012). *Sekai no kojo kara sekai no kaihatsu kyoten he: Seihin kaihatsu to jinzai management no nichi chu kan hikaku* [From Global Factories to Global R&D Centers: An International Comparison of Product Development and Human Resource Management in Japan, Korea, and China; in Japanese]. Tokyo: Toyo Keizai Shinposha.

Tsuru, T., and Nakajima, K. (2012). 'Product Architecture and Human Resource Management: Comparing Japanese, Chinese, and Korean Firms Based on a Questionnaire Survey', Discussion Paper Series A No. 563, Institute of Economic Research, Hitotsubashi University.

Ulrich, K. T. (1995) 'The Role of Product Architecture in the Manufacturing Firm', *Research Policy*, 24(3): 419–40.

Womack, J., Jones, D. T., and Roos, D. (1990). *The Machine that Changed the World*. New York: Rawson Associates.

World Intellectual Property Organization (2011). 'WIPO Statistics Database', December.

Yasumoto, M., and Fujimoto, T. (2005). 'Does Cross-Functional Integration Lead to Adaptive Capabilities? Lessons from 188 Japanese Product Development Projects', *International Journal of Technology Management*, 30(3–4): 265–98.

CHAPTER 18

INNOVATION MANAGEMENT IN CHINA

MARINA YUE ZHANG

INTRODUCTION

JOSEPH Needham, the English biochemist who became the world's leading expert on Chinese scientific history, spent almost his entire life investigating and documenting China's scientific achievements (Winchester, 2008). The ancient inventions of paper, gunpowder, movable printing templates, and the compass are only a few of the best known. The fact that for almost two thousand years, up to the sixteenth century, China generated scientific and technological developments that far exceeded any other society in the world is less known. However, since the European Renaissance and into modern times, Western countries have taken the leading role in scientific discovery and inventions. Needham's famous Grand Question asks why China was left behind in science and technology in modern history. This question has engendered much discussion and debate, with explanations ranging across the differences between China and the West in history, civilization, culture, ideology, society, and politics.

Today, China's re-emergence as a major power in the world economy is one of the most significant developments in modern history. The economic reforms and 'open door' policy adopted by the country more than three decades ago have prepared the ground for China's economic development and have yielded outstanding results in a number of areas. China has made great progress in its innovation capability. China has been recognized, for example, as a source of R&D for multinational enterprises. However, the challenge of integrating the innovation capability from China and other emerging markets into the global innovation network has been a widely recognized issue in innovation management (von Zedtwitz et al., 2004). On the other hand, China's impressive economic performance has come with a price. Its economic development has followed a pattern of specialization characterized by intensive utilization of low-skilled and low-cost labour and aggressive exploitation of natural resources.

Chinese enterprises may maintain their competitive advantages in manufacturing in a relatively stable technical environment; however, they are often stuck in a technology catch-up mode whenever the technological trajectory moves to a new paradigm because of a lack of absorptive capacity and learning ability (Hu and Mathews, 2008).

China's innovation management is still in its infancy. Policy intervention still plays a critical role in the development of the country's innovation capability. It is, in fact, a complex process which is driven by the interplay of multiple factors (economic, social, political, cultural, and technological, among others) at different levels (country, industry, and enterprise). Such complexity is embedded in the country's unique institutional setting, which is considered opaque, if not mysterious, to outsiders (Peng, 2000; Peng et al., 2009). Following Needham's Grand Question and in the context of China's rise as a significant player in the world economy, a few central questions remain unanswered: How have China's economic reforms affected its innovation capabilities? Will China be able to restore its position as a world leader in science and technology and transfer that capacity into innovation? And, if so, how?

To provide some answers to these questions, this chapter seeks to unfold the modern development of innovation in China from three different levels: country, industry, and firm. The country-level analysis focuses on China's innovation policy framework and the evolution of the country's National Innovation System (NIS); the industry-level analysis uses examples of specific industries to analyse the achievements made and challenges faced by China's industry policy-makers; and the firm-level analysis investigates innovation management practices in Chinese enterprises, including state-owned enterprises, entrepreneurial enterprises, and multinational enterprises operating in China. This chapter explores the emerging significance of innovation management in China's broad social, political, economic, and cultural context, demystifying how historical practice has come into being and shedding light on the future for managers who are interested in leveraging China's innovation to their advantage.

INNOVATION AT THE COUNTRY LEVEL

China's innovation activity almost stood still between the eighteenth and mid-twentieth centuries amid a succession of internal turmoil and external invasions. After the foundation of the People's Republic of China in 1949, China adopted a Soviet Union-styled innovation system, in which science and technology (S&T) activities at research institutes and universities were undertaken to serve the nation's needs (often political in nature), and production at state-owned enterprises was designed to meet quotas decided by a central-planning regime that was largely out of touch with market realities. In this system, the research institutes under the governance of the Chinese Academy of Sciences, and a small group of universities under the administrative control of the Ministry of Education, were engaged in basic research in science and technology. A large number of specialized universities and research institutes, which were under the administrative umbrella of specific industrial ministries, provided S&T support to, and solved practical problems for, their corresponding industries (Gao et al., 2011). The links between research and industry were controlled by the government. In this regime, neither knowledge producers (research institutes and universities) nor knowledge users (industry and enterprises) had incentives to improve their innovation capability. Meaningful interplay between research and industry was rare. However, during this period, China achieved several mission-critical S&T breakthroughs, including the launch of a Chinese-made missile, the development of an atom bomb and of satellites, which demonstrated China's scientific power, especially its military might, to the world.

This model was disrupted during the Cultural Revolution between 1966 and 1976 when the whole country was engaged in intense political struggles. Order was restored to universities and research institutes only in 1978 when Deng Xiaoping, the pioneer reformer of China, proclaimed that 'science and technology is the primary productivity'. Since then, China has introduced a series of innovation policies and started to build its NIS with the aim of rebuilding its innovation capability at the national level.

An Overview of China's Innovation Capability

In the past three decades, China has made impressive progress in building an innovation policy framework at the national level. As the result of such an effort, the R&D/GDP ratio in China reached 1.49 per cent in 2007, which was more than double that of a decade before. Given the rapid growth of the country's GDP during that period, China's R&D expenditures increased at an average annual rate of about 21 per cent between 1998 and 2007, and reached RMB 371 billion yuan (about USD 48.8 billion) in 2007,[1] making China the fourth largest country in R&D spending, after the USA, Japan, and Germany (OECD, 2010).

The higher education and public research sector has been another area of great growth. After the consolidation of the higher education sector, China now has 1,552 universities and colleges, of which 678 are considered research universities. There are 49 university science parks in the country which host more than 4,000 technology-based start-ups employing more than 70,000 people. More than 570,000 people work in China's 4,169 public research institutes. At the national level, China has 139 Key State Laboratories, 87 of which reside in well-known research universities and 52 in research institutes (Gao et al., 2011). Those Key State Laboratories undertake mission-critical basic research projects for the country. In terms of absolute numbers of R&D personnel, China has been ranked among the top three in the world since 1998 and in 2004 China overtook Japan to become the top country on this measure. Not only the sheer number, but also the age structure and educational level (e.g. more than 50 per cent of China's R&D staff are below the age of 35 and 45 per cent have postgraduate degrees), make China stand out as one of the world's leading countries measured by R&D personnel input (Stening and Zhang, 2011). However, some indicators of output from these substantial increases are rather disappointing. For instance, scientific publications by Chinese researchers, especially measured per capita, are among the lowest in the world.[2]

Overall, the largest proportion of innovation in the past three decades or so has focused on building the country's innovation system 'hardware'. While substantial funds have been invested in upgrading innovation-supporting infrastructures, including buildings, facilities, and equipment, little attention has been given to developing innovation 'software' such as patents, know-how, absorptive capacity, and learning capability. Most innovation in China concentrates on 'D'—development—rather than 'R'—research. Regional disparity is another drawback to innovation at the national level. Such regional disparities reflect China's notorious imbalance in regional economic development. The innovation capability and efficiency in the economically developed coastal provinces and municipalities (e.g. Zhejiang, Guangdong, Beijing, and Shanghai, among others) are much higher than in the economically backward inland western and central provinces (The Developmental Report on China's NIS, 2008).

Innovation Policy Framework

China has traditionally followed a top-down policy-making approach, including in innovation, in which the process is broken down into the following elements: identifying key issues in a certain policy area; defining the policy rationale; designing policy measures; implementing the policy; and evaluating its effect. This approach was monopolized by government, with little involvement from industry and society at large. However, recent decades have seen the development of bottom-up innovative activities, and innovation policy-making has shifted to a combination of top-down and bottom-up approaches.

The 1980s witnessed several major changes in S&T policy. As the economy moved away from central planning, the government reduced its funds for public research institutes and introduced a more market-driven funding mechanism. Under the new mechanism, personnel with innovative ideas at public research institutes and universities were granted financial rewards. Commercialization of their innovations was officially made possible through the establishment of 'technology markets'.[3] Technology markets refer to service platforms on which technology transfers and technology commercialization are driven by economic logic rather than policy rules. Technology markets are an important mechanism linking technology supply and technology demand in a market system. The legitimization of technology markets is considered the most critical institutional reform in enhancing China's technology-based entrepreneurship. To promote technology-based entrepreneurship, China launched its 'Spark Program' and 'Torch Program' as 'cradles' to nurture and develop new technologies and innovation capability. Towards the end of the 1980s, the country started to promote a more supportive environment for encouraging innovative enterprises. Science parks and university incubators started to emerge in the country to host technology-based entrepreneurial firms. This created a group of high-tech entrepreneurial enterprises.

Founder Group of Peking University, Unigroup of Tsinghua University, and Lenovo Group of the Chinese Academy of Sciences were early pioneers and became globally recognized enterprises.

The 1990s witnessed large-scale mergers and consolidations of research institutes and universities. Apart from a small group of elite universities, many universities went through a process of decentralization in which their administrative reporting line changed from central industrial ministries to local governments. This allowed local universities and research institutes to become a critical source of innovation at local levels. As a result, the interplay between research institutes/universities and local enterprises became closer and more market-driven. Meanwhile, a special technological innovation fund for technology-based small- and medium-sized enterprises (SMEs) was established to promote bottom-up innovation. At the 15th National Congress of Communist Party in 1997, Jiang Zemin, the then President of China, announced a new S&T policy of 'revitalizing the nation through science and education' ('ke jiao xing guo'). In August 1999, a National Innovation Congress was held in Beijing which identified new and high-tech sectors, such as information technology, biotechnology, space technology, energy technology, and new materials, as priority sectors by which to enhance China's 'technological innovation, development of high-tech industries and industrialization'. The country also started to promote the notion of 'empowering the nation through talent' ('ren cai cing guo'), which triggered the first wave of 'reverse brain drain'—talented persons, born in China but educated in the West, returning to the motherland.

The beginning of the twenty-first century witnessed mushrooming growth of technology-based entrepreneurship, with 'new economy' companies such as sina. com, Tencent Group, and Alibaba Group emerging, among others. In 2000, China's R&D expenditure reached 1 per cent of GDP, a tipping point for the country's S&T development (Gao and Jefferson 2007). China's fifteenth five-year plan between 2001 and 2006 identified technological innovation as one of the pillar policies by which to achieve 'more balanced and sustainable growth in the new century'. Other policy initiatives included introducing multiple sources of funding for S&T-related activities, special channels to attract S&T talent (for example, continuing to attract overseas Chinese returning home), and better conditions to support high-tech entrepreneurship. Foreign funds started to pour into China's innovation system. Foreign direct investment (FDI) in the manufacturing sector and venture capital in high-tech industries provided not only much-needed financial resources, but also advanced technologies, lessons in corporate governance and management practices. A practice widely used during this period was a so-called 'market for technology' policy—using technology transfer as a condition for foreign firms to enter into China's lucrative markets through Sino-foreign joint ventures or technology licensing agreements. While foreign technology and investment helped Chinese manufacturers build their reputation as the world workshop of production, Chinese enterprises had to pay large sums as royalties for importing advanced technologies.

The year 2006 marked a significant shift in the innovation policy framework. Before that, China largely relied on the supply of foreign technology for its industrial innovations. To offset this limitation, the 2006 National Science and Innovation Conference officially announced the adoption of a 'Medium- to Long-Term Strategic Plan for the Development of Science and Technology'. This document set the tone for innovation policy for the decades to come. Central to this policy was the government's determination to shift the country's developmental pattern to a more sustainable one focusing on building indigenous innovation capability ('zi zhu chuang xin'). President Hu Jintao noted the need for the country to develop the 'capacity for independent innovation', as 'the core to our nation's developmental strategy'. The overarching goal of the policy is to make China an 'innovation-oriented' society by the year 2020 and, over the longer term, to become one of the world's leading 'innovation economies'.

At the extremes, there are two schools of thought in relation to China's innovation policy. One stream believes that in the context of globalization, China does not need to develop its own intellectual property and can rely on imported technologies (except those used in critical areas such as national defence), arguing that maximizing China's competitive advantage in low-cost production can sustain the country's growth for a long time to come. This school of thought, however, fails to account for how as a result China's environment will be further damaged, social problems intensified, and the long-term sustainability for growth severely challenged. The other school of thought believes that China should develop its indigenous innovation, excluding any foreign participation in the value chain of innovation, from scientific discovery, to technological innovation and applications. This school of thought is similar to 'new techno-nationalism'—a policy adopted by Germany after the Second World War (Ostry and Nelson, 1995). Though many Chinese scholars argue that China's indigenous innovation policy is very different from 'new techno-nationalism', the implementation of such a policy can sometimes diverge from its original objectives and runs the danger of leading China down a road of isolation in its innovation. Indeed, with the rise of globalization, more pragmatic policy-makers recognize that it becomes harder for any innovation to be undertaken in isolation, within national boundaries. These pragmatists furthermore realize that indigenous innovation is not just about original innovation; re-innovation based on absorption and integration of imported ideas is perhaps a more important means by which Chinese enterprises can enhance their innovation capability.

China's National Innovation System (NIS)

The concept of an NIS has become an important policy tool shaping the context for innovation management in China. In recent years, this concept has come to be recognized in China to include a system with a purposeful combination of market and non-market mechanisms aiming to optimize the production, development, and use of new knowledge for sustainable growth, through institutionalized processes in the public and private sector (OECD, 2008). Following this logic, a nation's innovation capability is determined not only by the institutional setting and policy framework, but also by the *interplay* of multiple players in the system, including businesses.

The evolution of the Chinese NIS can be traced back to the mid-1980s when the country started the reform of its S&T system. The changes brought about by the reforms released the creativity and productivity of individual scientists. They became the primary actors engaged in technology transfer and innovation commercialization aimed at solving practical problems in industries at local levels. These bottom-up innovative activities called for further institutional reforms that eventually formalized many such practices. As a result, S&T-based industrial parks, university science parks, and technology business incubators were started, and spin-offs from public research institutes emerged. In the 1990s, the government started to play a more active role in policy-making to promote innovation using an NIS approach. An embryonic innovation system started to form and this system was developed through the combined efforts of the public and private sectors. A closer link between universities/research institutes and enterprises was established and, for the first time in the country's history, enterprises were recognized as key players in the country's innovation system. Towards the second half of the 1990s, foreign-funded R&D laboratories and technology-based entrepreneurial ventures became the driving force for innovation, while state-owned enterprises went through restructuring and became significant consumers of imported technologies. Between 2000 and 2006, a more systemic innovation regime started to form at regional levels. From 2006 onward, China has entered a new stage of building an NIS in which enterprises are key actors driving innovation, and an integrated innovation value chain is being formed linking industry, research institutes, and universities.

China's NIS is still in its infancy. The linkages between actors and among subsystems (regional and industry) remain weak (Gu and Lundvall, 2006; Dodgson and Xue, 2009). Synergies within the system are limited and spillover effects rare. The idea of public–private partnership is not yet well understood by many actors in the system. There is a lack of a long-term vision for most R&D activities. Most innovative activities focus on developing near-market technologies that can be commercialized quickly, rather than on nurturing pre-competitive, basic scientific discovery, which may not bear any short-term commercial benefits but can contribute to knowledge accumulation of benefit to the economy and society.

Numerous policy challenges remain. An NIS is a complex and multifaceted massive social network. China's NIS needs to be expanded beyond the fences of S&T parks, university science parks, and incubators and develop regional- and industry-based innovative clusters and sub-networks in the country. To make the system more sustainable, three key transformations are required: (1) the key innovative actors of the NIS need to be changed from research institutes/universities to enterprises; (2) the focus of innovation needs to be moved from technology acquisition to technology creation/innovation; and (3) the driving force for innovation needs to be shifted from a technology-push model to one that better integrates market demands.

INNOVATION AT THE INDUSTRY LEVEL

China does not have a long history of industrialization. Throughout this history policy intervention has been a critical feature of industrial innovation, reflecting the top-down mentality inherited from China's central-planning system. China's industries, especially those in infrastructure such as telecommunications, public transport, or national defence, have been shaped to represent the outcome of carefully planned economic or political aspirations in the country. For a long time, innovation outputs were measured by indicators of technology transfer at the industry-level using aggregated data. As a result, technology acquisition was the focus of industry policy, with little attention paid to building absorptive capacity of the industry. For example, China's automobile industry, which was built largely on the 'market for technology' policy, did not develop significant innovation capability of its own. While Sino-foreign joint ventures in this industry produce a large volume of automobiles for China's booming market, the core technologies are mainly retained by a handful of the world's automobile giants (The Developmental Report on China's NIS, 2008).

Industry Standards

Today, most of China's significant industry actors are recently reformed, reorganized, and reconstituted state-owned enterprises (SOEs). Over the past twenty years, these SOEs have become the economic pillars of the country's various industries, as well as fulfilling significant social and political functions in the country. In the short history of China's industrialization, with the support of the government many state-owned enterprises have undergone rapidly accelerated growth. In the first decade of their development, most SOEs obtained critical technologies from the West by helping foreign firms gain access to China's market. In the second decade, most of those enterprises privatized part of their assets on major stock exchanges abroad and domestically. Privatization, and the urging of government, required those Chinese SOEs to build more sustainable innovation capability and to adopt modern management practices and corporate governance mechanisms. The side effect of the rapid industrialization during the past three decades, however, is over-capacity in production, over-exploitation of the natural environment, and a severe shortage of innovative talent.

The standardization of network standards in infrastructure systems, such as in the telecommunications and energy industries, is an important institutional instrument used in government intervention, especially for those strategically critical industries in China. Industry policy has actively encouraged Chinese enterprises to participate in and contribute to the formulation of international technology standards. The logic is simple: market size, dynamism, and its rapidly evolving technological capabilities give China a good chance to enact and nurture its industry standards. For an industry standard to achieve any commercial value, however, it has to rely on building a whole industry value chain, rather than on only a few isolated technological break-throughs (see Chapter 28 by Leiponen). In other words, the development of comple-mentary infrastructure, applications, and institutional settings is vital for the success of an industry standard. China has striven, for example, to promote its home-grown third-generation (3G) mobile communications technology—TD-SCDMA—as an industry standard. In its effort to do so, the government appointed China Mobile Communications Group—the nation's largest mobile operator—to commercialize this standard after helping construct the value chain by creating the time and space nec-essary to nurture indigenous technology. The value chain for TD-SCDMA, however, which relies on the collaboration of a large array of complementary assets controlled largely by incumbent technology suppliers, was far more complex and took far longer than initially anticipated. This policy delayed the country for almost a decade in the commercialization of 3G, and is yet to achieve its goal of promoting China's core tech-nology globally (Zhang and Stening, 2010).

Emerging Industries

China's current industry policy has identified twenty state-driven emerging indus-tries as vital to the country's long-term sustainability. They include clean energy, next-generation Internet, high-end generic microchips, biotechnology, new-concept vehicles, high-end manufacturing technologies, and new materials, among others. Emerging industries have the characteristics of high uncertainty in markets, technolo-gies, and competition, providing a fluid institutional setting with few clear-cut rules of the game, and abundant opportunities and risks. China's industry policies aim to help Chinese enterprises take a pre-emptive position in industry standards setting.

The solar photovoltaic (PV) industry in China, for example, which will be so impor-tant for the nation to address its pressing environmental problems, has proven a success in building an industry standard for an emerging industry. In 2010, China accounted for about 50 per cent of the world solar PV cell and solar panel production, yet consumed only 2.2 per cent of the world's solar energy. The relative success of China's solar PV industry can be attributed to its low-cost advantage in manufacturing, a relatively short and new value chain, and less complex ownership of the core IPRs in the value chain. Furthermore, the innovation cycle of the solar PV industry is longer, in comparison with those of the 3G communications industry, which allows Chinese enterprises to main-tain their competitive advantage in technology for much longer. Local governments have also played a critical role in building large regional clusters centred on several star enterprises on the industry value chain. For example, a cluster centred on SunTech (NYSE: STP), Canadian Solar Inc (NASDAQ: CSIQ), and Trina Solar (NASDAQ: TSL) emerged in Jiangsu province specializing in solar panel production; a cluster centred on ReneSola (NYSE: SOL) in Zhejiang province and LDK Solar (NYSE: LDK) in Jiangxi Province specializing in silicon purification and solar system installation; and a cluster centred on JA Solar (NASDAQ: JASO) and Yingli Green Energy (NYSE: YGE) in Hebei province specializing in solar cell production.

In recent decades, technology convergence—the phenomenon of increasingly common technologies, user interfaces, applications, and distribution channels—has become an important force behind the emergence of new industries. This is especially true in the digital age where technological fields rarely exist in isolation; they form larger or higher-level networked systems where products/services are interdependent. Cross-industry convergence can lead to changes in the nature of technological innovation, product development, value-creating or value-adding processes, or consumer behaviour. Such convergence, driven by architectural innovations, requires changes in the existing institutional configuration or creation of new ones to accommodate different needs of different industries (Zhang, 2011). This is a more complex endeavour. To benefit from opportunities arising in this area, more flexible industry policy is needed to facilitate the creation of new market structures, new norms for stakeholders, and new rules for cooperation/competition across industry boundaries. To benefit from industry convergence, Chinese industry needs to develop an integrative innovation capability.

INNOVATION AT THE ENTERPRISE LEVEL

Enterprises are key actors in a nation's innovation system in most industrialized countries. In China, although the importance of enterprises in the NIS has been recognized and is increasing, enterprises are not yet playing a significant and independent role in the NIS. There are three major types of enterprise engaging in innovation: multinational enterprises operating in China, technology-based entrepreneurial firms (publicly, privately, or collectively owned), and state-owned enterprises. These three types of enterprise follow different rules of the game in corporate governance and management practices, as well as innovation strategies. On average, Chinese enterprises spend less than 1 per cent of their sales revenues on innovation compared to about 3 per cent of the world average (OECD, 2008). Overall, very few Chinese enterprises have yet developed their core technological base, a world brand, and indigenous innovation capability.

Multinational Enterprises (MNEs)

R&D activities funded by FDI (largely in MNEs) now account for 25–30 per cent of the total R&D of enterprises in China. China has sought to attract FDI not primarily for foreign capital per se but rather as a means of acquiring advanced technologies and modern management practices (von Zedtwitz et al., 2007).

There is a divergence of opinion in China regarding MNEs' role in helping Chinese enterprises upgrade their innovation capability. One stream of opinion regards MNEs as being helpful in building China's industry, by introducing not only advanced technologies, but also modern management practices and a wide range of other soft skills. The spillover effects created through labour mobility are considered important in having helped China build its technological base in some industries, such as software development. Another stream of opinion, however, believes that core technologies mostly remain controlled by foreign partners in their joint ventures with Chinese enterprises or by foreign headquarters. According to this view, technology transfer from foreign partners to Chinese counterparts is limited and spillover effects are rare. The second school of thought has gained prominence in recent years. Some leading academic researchers and policy-makers, for example, have pointed out that MNEs 'crowd out' local enterprises for highly skilled manpower and also, through their control of core technology and industry standards, charge unduly high licence fees for their technologies (von Zedtwitz et al., 2004).

FDI in China may have benefited those MNEs by enhancing their competitiveness in the host country. Furthermore, the R&D output resulting from the R&D investment of MNEs in China has been recognized as a critical means of enhancing their competitive advantage globally. In recent years, the investment by MNEs in China has changed from getting a ticket for market entry to gaining access to the large talent pool in R&D. MNEs are essentially changing their innovation strategy in China from exploitation to exploration (March, 1991) to take advantage of their investment in local talent. The real challenge for MNEs today, however, is to integrate the locally developed 'component competence' into their global R&D systems, because such an integration largely relies on 'architectural competence'[4] which is often missing in those local R&D centres (Qi et al., forthcoming).

Motorola, for example, despite its global difficulties, has built a successful R&D centre for its mobile phones in China. This centre developed a strong local innovation capability, initially serving global products sold in China, then moving on to provide independent product design and development to both local and global markets, and finally integrating its local innovation capability into its global system. In order to achieve a balance between local adaptability and global integration, this centre built its innovation capability in both component competencies and architectural competencies. The spillover effects have been significant. The centre is regarded as a 'training school' for R&D talent in the mobile phone industry in China. In fact, the rapid development of the local mobile phone industry (including a large number of 'copycats') can be attributed to such spillover effects of talent mobility (Qi et al., forthcoming).

Technology-Based Entrepreneurial Firms

In the thirty years since China embarked upon its economic reforms, government policy has been instrumental in the emergence and development of a multitude of collective and private enterprises, which are the home for many indigenous innovations. In more recent years, a wave of innovative Chinese enterprises, such as Lenovo in personal computers and Huawei in the telecommunications industry, have emerged and some have developed innovation capabilities, a global brand, and have expanded their operations abroad. Some have tapped into foreign pools of critical technologies through mergers and acquisitions and the establishment of overseas R&D units. The most innovative enterprises in China are technology-based, capital-intensive enterprises.

Huawei, for example, is one of China's most innovative enterprises. It is a privately owned technology-based company. Founded by Ren Zhengfei, a military retiree, in Shenzhen in the mid-1980s, the company was created to provide PBX (private branch exchange) switching systems to telecom operators in China. Today, the company is the largest telecom equipment supplier (delivering a large variety of products and services) in the world, connecting one third of the world's mobile phones. It generates 70 per cent of its sales revenues from the global market and nearly 20 per cent of its 110,000 employees are non-Chinese.

For a long time, Huawei's innovation was based on a 'catching up' model. In the 1980s, it merely provided low-cost (periphery) switching products to telecom operators and did not have much innovation of its own. It developed its initial competitive advantage by integrating different components (or technologies) to achieve low-cost solutions for its customers, sometimes using reverse engineering techniques. In the 1990s, it moved into the mobile communications sphere, but its innovation was still based on creative imitation. Since 2000, it has greatly strengthened its innovation capability, moving from imitation to innovation. Its innovation capability lies in its ability to respond to its customers and the changing environment faster than its competitors, as well as its strong adaptability in business models. Huawei was the first Chinese enterprise to dedicate not less than 10 per cent of its sales revenues and more than 45 per cent of its manpower to R&D. In 2008, with 1,737 PCT (Patent Cooperation Treaty) patents filed, Huawei became the world leader in enterprise patents, surpassing Panasonic of Japan (WIPO, 2009).

One technique Huawei uses in its innovation management is to employ a 'blue army'—mock attackers—to critically examine the company's strategy and develop counter-proposals. This mechanism helps the company to identify potential risks and provide solutions ahead of its competition. The 'blue army' enjoys the authority to report directly to the company's senior management team of seven executives. Such techniques are valuable and add to the company's innovation skills, but the challenge Huawei continues to face is how to bring real technological breakthroughs to the market and become a real industry leader in the global mobile communications industry.

There are many technology-based entrepreneurial firms such as Huawei that are driving industry upgrades and evolution in China. As technology followers, they do not yet have the capability to lead disruptive innovations. The pervasive use of ICT, and increasing modularization in R&D and manufacturing, have caused fragmentation in industry value chains. This fragmentation trend has offered technology-based entrepreneurial firms unprecedented opportunities to drive innovation in business models. These firms are agile in responding to diverse, complex, and dynamic customer demands faster than their competitors. They have the ability to locate and integrate dispersed resources from all corners in business (globally and domestically) to deliver effective, though not necessarily the most efficient, solutions to customers. The innovation in business models introduces novel transactional logics that can bridge disparities in supply and demand. These firms are not alone. Once a business model has proven a success, it will attract a large group of imitators. Together, they can cause disruptions in an industry by making certain components in the value chain redundant and therefore uprooting incumbent firms whose very survival is built upon those components. This effectiveness-centred approach in business model innovations, which is different from efficient-centred approaches commonly used in the West, is the driving force for industry disruptions in China.

In recent years, venture capital (VC) funded enterprises, especially in the Internet, IT, and other high-tech industries, are injecting new blood to China's innovation capability. Though China has abundant liquidity in its financial system, it lacks both the expertise and the necessary legal and regulatory conditions for a healthy VC system to work. Domestic VC firms which have been set up by the government or by SOEs, at national or provincial levels, are run by government officials who do not always have adequate technical, commercial, or managerial skills in venture investment projects. Foreign VC-funded technology start-ups have introduced advanced technologies and business models as well as modern corporate governance and management practices into China. This group of firms has created a model of integrating imported technologies and local market insights to serve Chinese users. This model is gradually diffusing into China's economy and supports the general trend for bottom-up innovation, centred on collaboration between the research community and industry, focused on nurturing entrepreneurial start-ups, and relying on capital market funds.

State-Owned Enterprises (SOEs)

All of China's core industries are occupied almost exclusively by SOEs, but few of them can be regarded as innovative. Among 28,000 SOEs, only 25 per cent have their own R&D centres (The Developmental Report on China's NIS, 2008). However, there are some exceptions whose experiences in innovation are notable.

China Mobile Communications Group (CMCG) runs the largest mobile networks in the world, with nearly 800 million mobile users. CMCG's early success could be attributed to favourable policies from the government, the take-off of customer demand for mobile communications, and its innovation capabilities in marketing, customer services, network operations/optimization, services, and business models. It successfully transformed itself from a 'communications expert' to an 'integrated information expert' during the second-generation (2G) technology era. Since the beginning of 2009, it has offered 3G services based on the TD-SCDMA standard, but as witnessed, in many ways (technological maturity, commercial experience, availability of terminals, network efficiency, and so on) the TD-SCDMA standard is widely considered to lag its competitors, WCDMA and cdma2000, whose licences were granted to CMCG's much weaker competitors, China Unicom and China Telecom, respectively. This policy intervention was

368 MARINA YUE ZHANG

aimed at creating a more balanced market structure in the industry by breaking down CMCG's monopoly. Despite being handicapped by the TD-SCDMA standard, CMCG strengthened its innovation capability by leveraging its mammoth user base and marketing muscle to its advantage. The company did not lose many of its high-end, core subscribers,[5] and became one of the earliest mobile operators in the world to actively lead the industry consortium to move on to 4G networks.

DISCUSSION

To respond to Needham's Grand Question, in his own later work he himself claims (1969: 190), 'the answer to all such questions lies, I now believe, primarily in the social, intellectual, and economic structures of the different civilizations'. Indeed, differences between Chinese and Westerners in appreciation of scientific discovery and technological innovation should be situated in a broader cultural context, a manifestation of Needham's forms of civilization. Culture has different facets and is revealed in the broader society, including the business system, the institutional structure, and even the managerial behaviour and cognitions of individuals. National culture, therefore, has a strong influence on entrepreneurial and innovative activities in a society. We will now examine the implications of Chinese innovation management within its cultural context at three levels: values, institutions, and cognitions of its people.

Values

In countries where Confucianism has a powerful presence, its influence has often been used to explain why there is a lack of original science and technology (Redding, 1995). According to this argument, since Confucianism advocates such cultural values as respect for elders, respect for knowledge and learning, the pursuit of harmony and order, risk reversion, a strict social hierarchy with large power distance, and collectivism, societies based on it may develop in a mentality of respect for authority and the status quo, low tolerance of mistakes, failures, and diversity, following feelings rather than enquiring into truth, respect based on relationships rather than merit, and so on. This, it is said, creates a barrier for innovative ideas, especially those arising from a younger person lacking status.

This view has been challenged, however, by other researchers who point to China's historical achievements in scientific discovery under the influence of Confucianism and the scientific accomplishments of other East Asian countries, such as Japan, Korea, and Taiwan, which share similar Confucian cultural heritages but continue to make great progress in science and technology in modern times. That evidence suggests that Confucianist values do not directly constrain innovation. After the Second World War, Japan, Korea, and Taiwan went through a similar developmental path to China's today, from simple imitation to creative imitation to innovation (Kim, 1997). It was their governments' policies and their institutional frameworks and business systems that created differences in how to manage innovation. For instance, the business system in Taiwan is based mainly on family businesses. These small and nimble organizations are able to change quickly and innovatively. By contrast, in Japan and Korea, competitiveness in innovation is based on conglomerate-based industrial structures that rely heavily on supportive government policies.

In summary, it is not the culture per se in China that directly affects its innovation practice. Rather, core cultural values coexist with the country's institutional framework, market mechanism, business system, industrial structure, education, and cognition of individuals, all of which impact upon the country's innovation capability.

Institutional Setting

The institutional setting is one of the most important manifestations of a nation's culture. At the macro-economic level, the lack of a suitable institutional framework that facilitates the creation and diffusion of new ideas can be blamed for China's lack of competitiveness in innovation. Weak enforcement of the legal protection for the holders of intellectual property rights (IPRs), the lack of an efficient market mechanism to minimize the transaction costs of knowledge sharing and technology transfer, an educational system which encourages knowledge digestion rather than creation, and a lack of transparency and justice in the reward system for innovations are often-noted deficiencies in China's institutional setting, impeding the country's innovation capability (Zhang and Stening, 2010). In Chinese organizations, R&D and innovation activities are largely controlled by star personnel, rather than relying on an innovation system through which everybody is allowed to make a contribution as in most Western enterprises. The analogy of cooking Chinese food in a Chinese restaurant, which relies on some special and often secret recipes held by a few star chefs, versus running a Western-style fast food restaurant, relying on a system and standard recipes, illustrates the differences between innovation management in Chinese and Western organizations. As we know from examples of Western companies such as IBM (Gerstner, 2002), innovation does not just come from the vision of top leaders; it is embedded in a firm's DNA—included in the core values shared by each employee and scattered through the whole process of its business, rather than centred on one or a few star employees. For Chinese enterprises, one lesson is clear: the need to invest in building the DNA of innovation and to leverage the collective wisdom of each employee, business partner, and even customers, in the innovation value chain.

In Western society, the market mechanism often facilitates the commercialization of innovations based on 'creative destruction'. In China, however, the institutional order makes transaction costs so high that any individual or organization will probably not survive the institutional changes or the challenges to the status quo induced by such innovations before they have a chance to be commercialized. In the Chinese institutional setting, knowledge is conveyed through human interactions, rather than through formal processes and systems, which makes risk-taking innovative activities rare, and transaction costs for knowledge sharing high. In traditional Chinese organizations, especially SOEs, the reward system punishes those who initiate changes that bear risks of mistakes and failure, and rewards those who maintain the status quo yet generate mediocre outcomes. Overall, a lack of trust among individuals in Chinese society makes knowledge sharing and dissemination more difficult than in Western societies (Gu and Lundvall, 2006).

The Chinese educational system emphasizes learning by memorizing and then internalizing existing knowledge rather than enquiry and creative destruction in thinking. This, too, is often cited as a reason for China's lagging behind the West in scientific discovery and technological innovation in modern times (Stening and Zhang, 2011). The bias in Chinese education towards theoretical learning means that many university graduates have poor creative and practical skills, which are critical for innovation.

The success of Silicon Valley is built upon a formula of linking technology and capital. China has successfully copied the model of building all kinds of science parks and technology incubators, and they are, in fact, producing technological innovations. On the other hand, there is abundant private capital in places like Zhejiang and Shaanxi. The capital cannot find the right investment destination so it has been used to speculate in the real estate markets in and out of China. If there were a market-based financial system, which includes promoting efficiency and transparency of the securities market, locally developed technological innovations would have a better chance to be commercialized in China.

Cognitive Patterns

The cognitive pattern of the people in a nation is perhaps the deepest manifestation of the deep-down culture of a nation, which is the result of its history, social structure, language, and political regime. Cognition is an enduring feature of a nation's culture that has a long-lasting and profound impact on the nation's development of science, technology, and innovation. Chinese holistic thinking treats complex problems as being embedded deeply in their context. This type of cognition is in stark contrast with the classical Newtonian scientific approach that values logical rigour and causal relationships built by experiment, tending, instead, to emphasize the 'relatedness' of variables in scientific enquiry. One consequence of such a cognitive pattern is Chinese 'customary thinking', which discourages posing challenges to authority, supports the status quo and views representing the 'middle way' (collective yet often compromised views), and encourages the tolerance of high levels of ambiguity rather than enquiring after the truth and pursuing precision. Respecting power rather than merit, pursuing harmony, and achieving collective mentalities in organizations and in the society limit enquiring minds, diminishing the innovation capability of organizations. It is difficult for people dominated by this kind of thinking to take the initiative or to make changes, and to engage in a collaborative yet constructively competitive teamwork, both of which are essential for successful innovative activities. Another disadvantage of this cognitive pattern is that it can lead to oversimplified, formulae-based, fixed modes of thinking, which are inadequate in solving dynamic and complex problems.

The development of a society whose culture fosters the sharing and use of knowledge is a complex endeavour, but a critical element in innovation. The continuing significance of cognitions that are deeply embedded in cultural roots can be illustrated by the fact that many overseas returnees, after living abroad for a considerable amount of time, readily readjust their attitudes and behaviour to conform with Chinese culture once they return to China.

CONCLUSIONS

China has made substantial progress in its innovation management capabilities, reflecting major changes occurring at the national, industry, and firm levels. In this process, government policy frameworks have played a significant role in shaping industry structure, market competition, investment and even technology standards. After very rapid development during the past thirty or so years, it is an opportune time for government policy-makers to reflect on the effectiveness of the country's innovation policies, and for business decision-makers to adjust their innovation strategies, with the objective of helping China achieve its goal of building the country into an innovative society by the middle of this century.

One apparent dilemma China is facing is that, at the industry level, the government is pushing for indigenous innovation; however, at the firm level, many enterprises believe their competitive advantage (at least in the short term) still largely comes from imported technologies. This dilemma is created because in mixed economies, such as China's, industries and enterprises often have different motivations for innovation. For industries, the strategic importance for the country, the positions in global value chains, long-term sustainability, and social responsibility are priorities; however, for enterprises, innovation, ultimately, should be driven by the efficiency of input/output ratios, the bottom line being to enhance competition and profitability. This is especially true in China where industries are controlled by the government while most enterprises (even state-owned ones) are, at least partially, public or private companies. It is not an easy task to synchronize the different interests of different stakeholders.

In recent years, the fragmentation of industry value chains has enabled firms with innovative business models to be integrated into the value chain more easily than ever before. Many of such business model innovations nowadays come from interactions between firms and their customers and other stakeholders on the value chain (Chesbrough, 2003). Consumer insights are becoming a critical source for innovation in the future, and China has the world's largest interactive networks, upon which users not only share ideas about which products are more desirable, but also *how* to make improvements in those less desirable. China has, for example, over one billion mobile phone users and 550 million Internet users. Companies that can build a mechanism to benefit from the 'collective intelligence' of China's consumers have the potential to create and capitalize on the value in their massive interactions with users, contributing to their global innovation value chains.

In the context of globalization, one critical challenge any country faces in innovation is to integrate its innovation system into international networks. Over the past three decades, China has focused on building domestic infrastructures for innovation; however, little attention has been paid to developing a global perspective in its R&D and innovation policies. Chinese cultural values, institutional setting, and the cognition of its people are still very much in favour of a zero-sum game, 'I win-you lose', mentality. The danger of this is that it can lead to a situation in which China faces hostility from all quarters of the world. The antitrust lawsuits against China's solar PV cell manufactures, the blockade of China's telecommunications equipment, and the scuttling of merger and acquisition attempts by Chinese enterprises for foreign business entities are all examples of the predicament China and its enterprises are facing in the world. It is simple to blame foreigners for treating Chinese enterprises with bias; however, it may be a good time for the Chinese government and Chinese enterprises to reconsider their approach when entering the global arena. They should consider adopting a more global perspective that would benefit all of humankind by scientific discovery and technological innovation, de-emphasizing the national origins of such achievement. While building its indigenous innovation capability for the long run, China should utilize its unique advantages in innovation, such as its integrative capability, mass production capacity, business model innovation coming from serving a large user base, and huge talent pool in science and engineering, to push the frontiers of science and technology with their applications in innovations.

Notes

1. This calculation is based on data published by the National Bureau of Statistics and the Ministry of Science and Technology of the People's Republic of China in various years.
2. Though China ranked number three, after the USA and the UK, in international publications measured by the number of papers included in the *Science Citation Index* (SCI) and *Social Science Citation Index* (SSCI), according to a bibliometric database by Scientific Business of Thompson Reuters (2007), the average output per researcher is very low.
3. The earliest technology market appeared in Shenyang, Liaoning province, in the early 1980s, when a type of 'technology service company' was established with the aim to channel technological innovations from research institutes to enterprises.
4. Henderson and Cockburn (1994) propose two categories of competences in R&D: 'component competence' refers to local abilities and knowledge that are fundamental to routine problem-solving tasks, while 'architectural competence' refers to abilities to integrate component competences effectively and to develop fresh innovation.

5. The reason why CMCG was able to prevent the loss of its high-end core subscribers was a protective policy which prevents users retaining their mobile phone numbers when shifting mobile networks. Many high-end core subscribers are pioneer users of mobile communications and their mobile numbers are part of their identity.

References

Chesbrough, H. (2003). *Open Innovation: The New Imperative for Creating and Profiting from Technology*. Cambridge, MA: Harvard Business School Press.

The Developmental Report on China's NIS (2008). *The Developmental Report on China's National Innovation System* (in Chinese). Beijing: Strategic Research Group of China's NIS Construction.

Dodgson, M., and Xue, L. (2009). 'Innovation in China.' *Innovation: Management, Policy and Practice*, 11: 2–5.

Gao, J., and Jefferson, G. (2007). 'Science and Technology Take-Off in China? Sources of Rising R&D Intensity', *Asia Pacific Business Review*, 13: 357–71.

Gao, J., Liu, X., and Zhang, M.Y. (2011). 'China's NIS: The Interplay Between S&T Policy Framework and Technology Entrepreneurship', in S. Mian (ed.), *Science and Technology Based Regional Entrepreneurship: Global Experience in Policy & Program Development*. Cheltenham: Edward Elgar.

Gerstner, L. (2002). *Who Says Elephants Can't Dance? Inside IBM's Historic Turnaround*. New York City: HarperCollins Publishers.

Gu, S., and Lundvall, B. A. (2006). 'Policy Learning as a Key Process in the Transformation of the Chinese Innovation System', in B. A. Lundvall, P. Intarakumnerd, and J. Vang (eds), *Asia's Innovation Systems in Transition*. Cheltenham: Edward Elgar, 293–312.

Henderson, R., and Cockburn, I. (1994). 'Measuring Competence? Exploring Firm Effects in Pharmaceutical Research', *Strategic Management Journal*, 15(S1): 63–84.

Hu, M. C., and Mathews, J. A. (2008). 'China's National Innovative Capacity', *Research Policy*, 37: 1465–79.

Kim, L. (1997). *Imitation to Innovation: The Dynamics of Korea's Technological Learning*. Boston: Harvard Business School Press.

March, J. G. (1991). 'Exploration and Exploitation in Organization Learning', *Organization Science*, 2: 71–87.

Needham, J. (1969). *The Grand Titration: Science and Society in East and West*. London: George Allen and Unwin.

OECD (2008). *OECD Review of Innovation Policy: China*. Paris: Organization for Economic Co-operation and Development (OECD).

OECD (2010). *OECD Factbook 2010: Economics, Environment and Social Statistics*. Paris: Organization for Economic Co-operation and Development (OECD).

Ostry, S., and Nelson, R. (1995). *Techno-Nationalism and Techno-Globalism*. Washington, DC: The Brookings Institution.

Peng, M. W. (2000). *Business Strategies in Transition Economies*. Thousand Oaks, Calif.: Sage Publications.

Peng, M. W., Sun, S. L., Pinkham, B., and Chen, H., (2009). 'The Institution-Based View as a Third Leg for a Strategy Tripod', *Academy of Management Perspectives*, August: 63–81.

Qi, M., Wang, Y., Zhu, H., and Zhang, M. Y. (forthcoming). 'The Evolution of R&D Capability in Multinational Corporations (MNCs) in Emerging Markets: Evidence from China', *International Journal of Technology Management*.

Redding, G. (1995). 'Overseas Chinese Networks: Understanding the Enigma', *Long Range Planning*, 28: 61–9.

Stening, B. W., and Zhang, M. Y. (2011). 'Challenges Confronting Higher Education in China', in Z. Szalai (ed.), *Higher Education in Hungary and the World: Tendencies and Potentialities*. Budapest: Mathias Corvinus Collegium, 56–69.

von Zedtwitz, M., Gassmann, O., and Boutellier, R. (2004). 'Organizing Global R&D: Challenges and Dilemmas', *Journal of International Management*, 10: 21–49.

von Zedtwitz, M., Ikeda, T., Gong, L., and Hamalainen, S. (2007). 'Managing Foreign R&D in China', *Research Technology Management*, 50: 19–27.

Winchester, S. (2008). *The Man Who Loved China: The Fantastic Story of the Eccentric Scientist Who Unlocked the Mysteries of the Middle Kingdom*. New York: HarperLuxe.

WIPO (2009). *Global Economic Slowdown Impacts 2008 International Patent Filings*. World Intellectual Property Organization (WIPO). http://www.wipo.int/pressroom/en/articles/2009/article_0002.html. Geneva, January 27, 2009.

Zhang, M. Y. (2011). 'The Intersection of Institutional Entrepreneurship and Industry Convergence: The Evolution of Mobile Payments in Korea', *Academy of Management Conference*, August. San Antonio, Tex.

Zhang, M. Y., and Stening, B. W. (2010). *China 2.0: The Transformation of an Emerging Superpower . . . and the New Opportunities*. Singapore: Wiley & Sons.

CHAPTER 19

···

TECHNOLOGY AND INNOVATION

···

MARK DODGSON AND DAVID M. GANN

INTRODUCTION

THIS chapter addresses the consequences of technology for the management of innovation. In contrast to the more common studies of how innovation produces new technologies, it explores how technology affects the processes of innovation. We examine the effect of technology on the creation of new ideas and the search for and design of solutions to problems, and on the development and implementation of key aspects of innovation strategies.

There is always a danger when discussing the importance of technology of succumbing to technological determinism. This is a view that holds, first, that technology emerges independently of the society, economy, and political systems in which it develops and, second, that it by itself causes social and economic change. The first of these assumptions is easily dismissed and has been discredited by extensive research in Science and Technology Studies into the broad ranging factors affecting the development of particular technologies (e.g. Hackett et al., 2007). The second, also much-criticized perspective still holds currency by being seen as 'common sense' and according with people's everyday experiences (Wyatt, 2007). The Internet, mobile communication devices, and social networking technologies, for example, have changed our lives profoundly. The dangers of this view is it fails to account for the choices that can be made and the multiplicity of factors that contribute to shaping how technologies are configured and eventually succeed by being widely used (Tuomi, 2002). Within organizations it can lead to the perilously simplistic opinion that changes required by managers—for example, in reforming work practices, productivity improvements or managing knowledge—can be brought about simply by investments in technology.

While cautious about overstating the impacts of technology, and accepting that its effects are inseparably linked to its social, economic, political, and cultural contexts, we do ascribe it considerable potential in its transformative powers. It is not coincidental that technology is often used to define the historical period we live in: the 'iron', 'steam', 'mass production', or 'digital' age. Throughout history, particular technologies in the form of instruments and tools—such as the telescope, the microscope, and the scalpel—have played a crucial role in helping humanity understand ourselves and our places in the world (Sennett, 2008).

There are numerous examples of the close connections between technological instruments and the progress of ideas (Crump, 2001; Galison, 2003). Faraday was an expert maker of experimental machinery (Hamilton, 2003). In his role as a Swiss patent officer, Einstein was deeply engaged in examining patent applications for electric clocks and distributed electrical time signals, thinking practically alongside his development of the theory of relativity. The development and use of X-ray crystallography by Franklin and Wilkins at King's College London, played a major role in the discovery of the structure of DNA by Crick and Watson. In his history of the development of the laser, Nobel Prize-winner Robert Townes relates how the theory and practice of lasers developed hand in hand (Townes, 1999). He describes his experience of using wartime radar equipment to conduct basic research at Bell Labs, and how other laboratories at the time benefited from cheap war service scientific instruments. New instruments continue to create new methods and different ways of seeing the world. The electron-tunnelling microscope, for example, allows us to 'see' and manipulate atoms in nanotechnology (Jones, 2004). As Galison (2003: 325) says: 'Clocks, maps, telegraphs, steam engines, computers all raise questions that refuse a sterile either/or dichotomy of things versus thoughts. In each instance, problems of physics, philosophy and technology cross'. Knowledge and technological invention are inextricably linked.

Contemporary technologies extend, of course, beyond remarkable improvements being seen in instrumentation used for sensing and measuring, and include automation used to transfer, transform, and control material, and computerization used for generating, processing, and storing data. These technologies provide essential tools for research and experimentation, means for prototyping and testing, and mechanisms for producing, delivering, and maintaining goods and services. Technology in all these forms is a centrally important contributor to productivity in every part of economies, in agriculture, manufacturing, and services, and has underpinned the massive shifts the world has experienced from agrarian to industrial to post-industrial societies (Bell, 1976). Digitalization[1] and the Internet, for example, have dramatically changed economies, creating opportunities for increased experimentation, communication, and collaboration. They create value for organizations by enhancing knowledge and stimulating learning, thereby improving their ability to produce new products, processes, and services, and increasing their productivity.

The introduction of digital technologies into manufacturing industry has been described by *The Economist* as presaging a third industrial revolution as we move from mass-scale to individualized production. This, it is argued, will reverse the flow of man-ufacturing jobs to nations with lower labour costs.

> things (will be) made economically in much smaller numbers, more flexibly and with a much lower input of labour, thanks to new materials, completely new processes such as 3D printing, easy-to-use robots, and new collaborative manufacturing services online. And that in turn could bring some of the jobs back to rich countries...
>
> (*The Economist Special Report: Manufacturing and Innovation*, 21 April 2012).

Innovation scholars have long recognized the consequences of technology for industrial dynamics. These include the classic insights of Schumpeter (1934) on capitalism being an evolutionary process of continual innovation and 'creative destruction', and the contribution of sequences of techno-economic paradigms, or historical surges of technological changes, with profound impacts on economic growth (Freeman and Perez, 1988). Our concern here is with the effect of technology at the organizational level and the factors affecting the nature and focus of innovation within organizations. We begin by discussing how managers have choices over the way technology is used by reference to the question of skills, and then analyse the role of technology in problem solving, a core element of innovation management.

TECHNOLOGY, WORK, AND PROBLEM SOLVING

The use of technology for innovative and productive ends depends on how it is configured along with the way people work. Craftspeople intimately apply tools in the conduct of their craft; industrial workers monitor, react to, and control machines to produce their various outputs; knowledge workers develop and rely on computer power, data creation and storage, networking and processing capacities to inform their decisions. The ways technologies are used affects the ability of organizations to innovate, learn, and create new knowledge, and technological change provides an opportunity to redefine who does what and how things should be done around them. Amongst the enduring themes in research into the consequences for innovation of introducing new technologies are the extent to which they reduce skills by removing discretion and creativity from labour, on the one hand, and liberate work from tedious and repetitive tasks on the other.

Debate on the consequence of technology for work and skills, and its use in increasing, decreasing, or sharing managerial control and discretion, has a long history. Marx was unequivocal in his conviction that technological changes removed skills and controls from workers (Marx, 1981: 545). This view of the use of technology as a form of capitalist control was re-stated with considerable effect at the time by Braverman (1974). He argued the use of technology was a dominant method of modern management to remove craft-based work with high levels of expertise and discretion, and hence expense, by 'de-skilling'. History, however, reveals a more complex picture. In his *Story of the Engineers* (Jefferys, 1946), tells, for example, of the millwrights during the Industrial Revolution. Prior to the arrival of large-scale manufacturing in the 1820s, millwrights were responsible for virtually all tasks associated with building and maintaining factory machinery. The development of steam power applied to industries such as textiles had a massive effect on their skills. But their demise corresponded with the growth of new, specialized skills required by larger machines, greater speed, intricacy, and accuracy (Jeffreys, 1946: 16).

> The chief characteristic of this new economy was the emphasis on planning against improvisation. Fast production with high-speed steel on a turret lathe was of no value unless plans had been made to utilise the time saved. Up-to-date milling and drilling machines were liabilities unless the cutters, jigs and gauges had been planned and made before the job went into production and steps taken to ensure the continuities of production and elimination of hold-ups. New methods would end in confusion unless all parts of the factory worked at a smooth tempo with a steady cooperative discipline.
>
> (Jefferys, 1946: 124).

The impact of technological changes for work and organization are, therefore, profound and likely to both destroy old skills and create new ones, and the requirements of new technologies place new demands on management, in technical design, planning, and organization. Throughout these processes of change, managers retain choices on how technologies are configured and used. In one of the earliest empirical studies of the influence of technology on management, Woodward (1965) found that work organization and managerial styles and authority were affected by technological complexity in different productions systems based on small batch and unit, large batch and mass, and process production. There were, however, choices to be made. Research in the 1970s (e.g. Touraine, 1962; Kerr et al., 1973) argued that the consequences of technology for work and skills was contingent on a number of factors, such as the nature of labour markets, and the number of exceptional cases in the system (Perrow, 1970).

A classic study of the choices that can be made around the development of technology was Noble's (1986) research into the history of numerical control machine tools. In essence, he found this technology had the capacity to retain skills amongst the operators of machines, or remove them to separate planning offices. Due to the influence of the US Department of Defense on the design of the nascent technology, the latter course was chosen. As this technology progressed and became computerized, research showed the advantages for responsiveness and productivity of increasing the operators' computer skills drawing on their tacit knowledge of machines and materials (Dodgson, 1984).

The lessons of these studies are that the ways technologies are created and used in the workforce are contingent on a number of factors, only one of which is the functional requirement of the technology. Managers wishing to use technologies to innovative ends have choices to make about how they are configured and used. As new technologies make old skills redundant, they create demand for new skills, and, when the onus on managers is to improve innovativeness, there is advantage in using and enhancing the knowledge and skills of people working with the technology. This is especially important where work requires creativity and collaboration, as we shall see in the following section on problem solving in engineering services.

Engineering services are involved in designing, producing, and maintaining artefacts in almost every facet of society and the economy. The way that goods and services are designed is being profoundly affected by technology. Traditional, laborious hand-calculations, drafting, physical testing, and model building approaches to many design tasks have, since the 1960s, been replaced by computer-based modelling and simulation. Over recent years this technology has significantly expanded its capabilities, aided by more powerful processors, greater visualization capacities, and more sophisticated software allowing new functions across multiple and more widely diffused technological platforms, and better data on physical properties. These technologies are increasingly being used to support innovation by leading organizations (Schrage, 2000; Tuomi, 2002; Thomke, 2003; Brynjolfsson and Saunders, 2010).

To understand the impact of these technologies on problem solving, we need to appreciate how engineers design. Engineering knowledge builds on scientific principles, but relies heavily on 'rules of thumb', 'informed guesses', 'routines', and 'norms' of problem solving that are built up through engineering education and experience with real world problems. Constant (2000) refers to development of engineering knowledge as the development of *recursive practice*, slow and steady accretion of knowledge about how things work and how they fit together. Engineers often experiment and improvise, reflecting on knowledge learnt in a range of previous experiences (Schön, 1991). The combination of theory and practice, reflection and application, builds engineering judgement over time (Petroski, 1985; Vincenti, 1990; Simon, 1996; Perlow, 1999).

To solve the problems that confront them, engineers often rely on social and informal cultures, networks and patterns of communication (Allen, 1977). Project work carried out in diverse, multidisciplinary teams that span a number of different organizations benefit from sharing experiences, telling stories, sketching, and playing with artefacts from related engineering work in order to develop solutions to problems (Hargadon and Sutton, 1997; Kelley, 2001). Added-value in design and development processes comes from creative and schematic work where designers use 'conversations' and 'visual cues', interacting with one another and their clients and suppliers (Schrage, 2000; Whyte, 2002). This schematic work involves seeing the interfaces between technological components and obtaining engagement and commitment from different specialists and stakeholders.

The technologies of modelling and prototyping have always played a central role in engineering problem solving. For engineers, models provide a mechanism for learning about artefacts before and after they have been built (Petroski, 1985). Models allow engineers to examine different options and weigh the choices of structural elements, materials and components against one another (Thomke, 2003). Physical prototypes are expensive and time consuming to create and often unreliable, and digital tools are cheap and easily available (Schrage, 2000; Thomke, 2003). Digital models assist in abstracting physical phenomena, allowing engineers to experiment, simulate, and play with different options whilst engaging with a range of interested parties.

These models can also help manage innovative, complex, multidisciplinary, and collaborative projects by providing the means by which various forms of integration can occur. If as Simon (1996) suggests, engineers solve complex problems by breaking them down into smaller components, the technologies can assist in both analysis at the decomposed level and in the re-integration of those components. It helps ensure technical compatibility in systems and their interfaces, and provides cost-effective ways of manipulating and playing with potential solutions to problems.

The social and technical integrating potential of digital technologies remains novel and its consequences for the skills of engineers are still being enacted. It is possible to envisage a de-skilling trajectory, similar to that predicted by the application of computers to industrial technology (Braverman, 1974). Whilst some firms may embark on that road, the result is likely to be similar to those firms that unsuccessfully tried to use industrial technology to de-skill in circumstances where innovation is required. Productivity depends upon the merger of the new technological possibilities with the appreciation and understanding of the old theories and principles, and 'ways of doing things': the established rules of thumb and norms. The technologies of modelling and simulation are being combined with more traditional forms of engineering design and problem solving in novel ways. The continued importance of craft knowledge limits attempts to routinize some aspects of design, and deep craft expertise can provide innovative solutions when aided by new digital tools. The famously innovative architect, Frank Gehry, has commented upon the combination of new technology and design processes that have allowed him to produce his innovative, malleable, plastic form of building design saying that:

> This technology provides a way for me to get closer to the craft. In the past, there were many layers between my rough sketch and the final building, and the feeling of the design could get lost before it reached the craftsman. It feels like I've been speaking a foreign language and now, all of a sudden, the craftsman understands me. In this case, the computer is not dehumanising, it's an interpreter.
>
> (F. Gehry in S. Pacey, *RIBA Journal*, November 2003, 78–9)

Digital technologies have the potential to provide the basis for a more open, transparent approach to engineering design and decision-making. It could also allow faster results with greater concurrency of engineering work on different parts of a problem simultaneously, allowing near real-time decision-making. Its capacity to assist in interdisciplinary decision-making by allowing people from different backgrounds to understand potential solutions opens a whole range of possibilities for a more informed and democratic working praxis.

TECHNOLOGY AND INNOVATION IN SERVICES

Much of our knowledge of innovation management, and much of the discussion above, is based on our understanding of manufacturing industries. While the boundaries between manufacturing and services are blurred, services raise different challenges for innovation management (see Chapter 30 by Tether). Some of the major differences in innovation between the industrial and services economy are contrasted in Table 19.1. The focus in the industrial economy is on understanding and manipulating natural and engineered physical objects and systems. In the services economy, innovators place greater emphasis on understanding and adapting organizational systems comprising information, people, and processes. The key objectives of industrial innovation are to develop new products, increase product quality and functionality, and improve production. The overriding objective of innovation in services, in contrast, is to provide a positive customer experience that is often highly personalized. Personalization of services represents a major shift in focus in decision-making from producers to consumers, and advantages innovators capable of utilizing technologies that integrate consumer demands and requirements into decisions about innovations.

As Tether argues in Chapter 30, services are not produced in the laboratories and factories of the industrial R&D era where they can be physically prototyped, fully tested and optimized. They are often produced at the point at which they are consumed and so the act of consumption rather than invention is the focal point for innovation. Innovative offerings are therefore developed using a 'market-facing' approach, often connected to those people and organizations whose requirements and demands are articulated and expressed during the process of consuming innovation. This is done through prototypes and shared experiments with users in real time, including with employees, partners, clients, and the public at large, increasing the need for collaborative skills.

Table 19.1 The changing nature of innovation

	Industrial economy	Services economy
Focus	Natural and engineered physical objects and systems	Organizational systems: people, information and processes
Design objectives	Production oriented, product excellence, competitive costs	Consumption oriented, market-facing, positive customer experiences
Organization and culture	Siloed within disciplines, narrow, sequential, deep, Proprietary	Multidisciplinary, holistic, concurrent, broad, collaborative, open

Source: Dodgson, Gann, and Wladawsky-Berger, 2010

The implication is that innovations with the potential to create social and economic value from services will not lie exclusively within the technological functions in companies, but will emerge through their effective collaboration with their users (see Chapter 5 by Franke). The consequences for innovation management are that interdisciplinary and collaborative skills are therefore often of critical importance to the success of outcomes. Teams are needed for innovation that brings different skills to bear on common system design, development, and management. This requires expertise in understanding individual, group, and organizational behaviour. Engineers need to engage effectively with people and organizations with which they have had little contact in the past, requiring the ability to span across traditional boundaries.

TECHNOLOGY AND STRATEGY

Technology provides a core source of strategic competitive advantage in organizations in almost every element of economies. This extends beyond firms that are obviously technology-based in, for example, information and communications technology and pharmaceuticals, into sectors as diverse as food, construction, and transportation. Technology also provides important competitive advantages in services, such as health, retailing, banking, and insurance. The manner in which technology is strategically selected, accumulated, and protected is examined throughout this volume. Here we use the examples of three companies: Arup, IBM, and P&G, to illustrate the way technology can support corporate and innovation strategies that involve high levels of internal and external connectivity to assist business diversification.

Arup is a global design and engineering services company, recognized for its technical expertise. It has contributed to some of the world's greatest modern buildings, including the Sydney Opera House and the Paris Pompidou Centre. Through its work, it has developed knowledge about many different areas including structural, mechanical, electrical and electronic systems, earthquakes, fire, acoustics, and environmental engineering. Arup works on several thousand projects simultaneously, for a wide range of clients. The firm started in structural engineering and as it expanded, it diversified into different areas encompassing over fifty specialist groups. Technology is used to support the strategically critical activities of sharing knowledge in and between its projects and providing the basis for creating new capabilities and businesses. Its extensive knowledge management system, for example, which includes websites on corporate information, projects, people, interest groups and networks, and insights from staff, is used widely and drawn upon to deal with planned and unexpected events (Criscuolo et al., 2007). The company's capacity to develop new technologies and services is seen in its contribution to the creation of fire engineering—a new discipline that involves widespread and deep collaborations with architects, builders, regulators, and firefighters—and its manifestation as a new business (Dodgson, Gann, and Salter, 2007).

IBM is an information technology and services company. It is a leader in innovation, with a multi-billion R&D budget, and is consistently ranked as the world's most prolific patentee. Its resurrection following near bankruptcy in the early 1990s is described in Gerstner (2002), and key to its revival was an improved capacity to respond to changes in markets and technologies by more effectively using its internal technological capabilities and external connections. It aimed to break away from its past introspection and 'not-invented-here' syndrome by breaking down its dependence on large, semi-autarkic R&D laboratories and becoming more 'market-facing'.

A number of aspects of IBM's strategy affect its use of technology. It has rapidly progressed into the business of IT services, which now account for the majority of its sales, and this move has depended upon its technological leadership. IBM's leadership is characterized by a 'research and engineering mindset', which maintains core strength in capabilities for searching for science and technology and the capacity in turbulent times to absorb new ideas from outside (Gerstner, 2002). It has an Academy of Technology, with 300 elected Fellows—representatives of a diverse range of science and engineering—who are highly esteemed within and outside the company. Their objective is to advise IBM on technical trends, directions and issues, and raise awareness about emerging trends throughout the organization by building and sharing knowledge. As an example of this Academy's use of new technology, it has held two of its annual meetings in a virtual world. To continually learn about new ideas so as to avoid the problems experienced in the early 1990s, IBM makes extensive use of networks and alliances (Dittrich et al., 2007), combined with a global perspective (Palmisano, 2006). To assist this connectivity, IBM makes widespread use of social networking and collaborative technologies for service innovation, as seen in its intranet, ibm.com, and has experimented extensively with virtual worlds (Dodgson, Gann, and Phillips, 2013). The company runs a web portal, ThinkPlace, where staff can submit, share, and be rewarded for new ideas. It has also conducted a number of 'Jams' where it opens and manages Internet portals to encourage and prioritize ideas from staff and external stakeholders. Its Innovation Jam attracted 40,000 suggestions from 150,000 staff.

Procter and Gamble (P&G) is one of the world's largest and most successful consumer businesses. It has a substantial R&D organization and patent portfolio. Throughout the late 1990s, it experienced lower than expected sales growth and attributed this to shortcomings in its ability to produce new products—defined in the company as having billion dollar plus sales—to satisfy consumers' changing needs. No new major product had been developed for over two decades. P&G recognized that to meet sales growth targets its innovation rate would need to increase significantly. P&G's management also realized that the cost of investments in R&D, technology, and innovation were increasing faster than sales growth, and that this was unsustainable. The recognition that the vast majority of solutions to P&G's problems lay outside of P&G was a critical first step in the development of Connect and Develop, its open innovation strategy (see Chapter 22 by Alexy and Dahlander).

The Connect and Develop project is facilitated by the technologies that assist the creation, transfer, and utilization of knowledge across organizational boundaries (Dodgson et al., 2006). P&G employees distributed around the world can communicate through an internal website called InnovationNet. InnovationNet is automated and Artificial Intelligence is used to support data-mining that acts in a similar way to Amazon.com by taking account of users' interests when sending back information on material the user may be interested in, and connecting people with the same interests. One of the purposes of these Internet-based systems, also used by Arup and IBM, is to facilitate communications within and between 'communities of practice' (Brown and Duguid, 2000). P&G has numerous communities of practice, such as bleach, polymers, analytical chemistry, flexible automation and robotics, technology entrepreneurs, fast cycle development, and organic chemistry. Its website also facilitates extranet communication with external business partners and serves as a link to external databases (Sakkab, 2002). It is estimated that the number of innovations sourced externally has increased as a result of its open innovation strategy from 10 to 50 per cent.

In an interview with P&G's CEO (Chui and Fleming, 2011), the range of influence of digital technologies on the company and its competitive advantage becomes clear. Digitalization affects the control of operations, transport, and logistics, relationships with retailers, and interactions with its customers using advanced analytical tools on consumer feedback. It also, as we shall see below, changes the way the company innovates by the use of simulating, modelling, and virtual prototyping.

These examples illustrate some ways technology provides opportunities for corporate strategic development by facilitating business diversification and improving internal and external connectivity and engagement. We now turn to the impact of technology on the organizational processes that support the development of innovations.

TECHNOLOGIES AND THE INNOVATION PROCESS

Technologies not only support strategies for increasing openness and connectivity. Dodgson, Gann, and Salter (2005) refer to a range of digital technologies that are being used in modern innovation processes and which they describe as 'innovation technology' (IvT). IvT includes visualization, virtualization, simulation, and rapid prototyping technologies. Dodgson, Gann, and Salter argue these technologies are becoming ubiquitous, with extensive use in a wide range of sectors, from pharmaceuticals to the mining and construction industries and across many different types and sizes of organization. They are being used to intensify the innovation process, and increase the speed of innovation: the 'automation' of drug discovery at GlaxoSmithKline (GSK), for example, is estimated to reduce the time it takes to get a new drug to market by up to two years, and its use in car design is argued to reduce the time it takes from four to three years (Gann and Dodgson, 2007). We now turn to a range of technologies that can be used by innovation managers to improve innovation processes. In all discussion about technology, it is necessary to recognize its capacity for rapid and unpredictable changes. The technology discussed in this chapter is presently being applied, but there is no presumption of its future relevance and continued use in the face of new and potentially unforeseen new advances.

Digital Infrastructure

Advances in the capacities of technology to support innovation build upon the enormous increase and improvement in computing processing power and storage, connecting infrastructure of the Internet and broadband, and mediating software that enables connections to be made between different systems. These technologies are in themselves major innovations, and they provide for new business opportunities through improved communications and business-to-business and business-to-consumer transactions. Our concern here lies with the possibilities this infrastructure provides for intensifying innovation processes.

Management decisions are informed by the quality of evidence available to them and the insights they proffer. A key element of the digital age is the availability of masses of data, the so-called 'data deluge'—created by scientific experimentation, pervasive sensors (connecting what is called the 'internet of things', devices connected to devices), and mobile phones that relay text and images—and the capacities of technologies to help with the search for, analysis of, and representation of that data. They include eScience, or Grid technologies, which have their basis in high volume data transmission and computation. The technology includes software that allows shared diagnosis and analysis of data by teams working in different locations and in different parts of the research and development process. They involve 'middleware' that allows visualization of merged data sets to improve the shared understanding and operation of virtual research organizations. An example of the way eScience supports the search and experimentation process can be seen in the pharmaceutical sector where firms such as GSK deploy powerful computing systems to analyse patterns in cell chemistry in the search for new 'leads' which are then taken to the next stage of testing and development. In the aerospace sector, firms such as Rolls Royce use Grid technologies to connect research teams in different locations, including in their University Technology Centres: for example, a project called DAME (distributed aircraft maintenance environment) uses high bandwidth computer networks to connect engineers developing real-time monitoring of jet engines. Research conducted at Rolls Royce's collaborative University Technology Centres is accessed in this way, bringing wider and deeper perspectives and knowledge into its decision-making.

The data deluge is not only used for research and technology-driven innovation processes. The banking and insurance company, HBOS, uses data-mining techniques to analyse large volumes of customer data, identifying market segmentation for novel, highly targeted service offerings. Egg Banking has deployed sophisticated simulation tools to model customer life cycle profitability for new pricing and loyalty approaches. Deutsche Bank makes extensive use of risk modelling to identify major risks and the likely impact of alternative risk reduction strategies. Many of these companies use the data to attempt large-scale process optimization and how to improve large-scale fore-casting. This requires new analytical skills for decision-making in inherently complex and unpredictable social systems, such as urban environments, healthcare, and financial systems. The new sources of data provide opportunities for real-time analysis of con-sumer demand for services, and in the case of P&G's CEO claims that 'with digital tech-nology, it's now possible to have a one-on-one relationship with every consumer in the world', (Chui and Fleming, 2011).

The availability of enormous amounts of digitalized data provides the opportunity to connect the planning of different kinds of system and integrate physical and digital systems. This is seen in the case of managing innovation in city infrastructure systems. The engineering tools and models used in the design, construction, and operation of a city's infrastructure can now utilize a suite of technologies and processes to assist this integration. The future successful operation of the increasing number of cities in the world, with their rapidly expanding populations, depends on physical infrastructure (e.g. buildings, bridges, and roads) becoming increasingly interlinked with and oper-ated using new digital technologies. Traditionally, city systems have been developed and operated independently, but they are interconnected. The transportation system, for example, is intimately linked to the energy-provision system. Increased use of electric vehicles reduces petrol consumption but increases the demand for electricity. Improved telecommunications infrastructure increases opportunities for working at home and for telemedicine, which affect transportation and health systems. The provi-sion of potable water depends on high levels of energy provision for processing and transfer. Large amounts of water are needed to cool power stations. The ability to inter-link such systems offers opportunities for innovation in improving existing services and creating completely new types of services. The integration of physical-digital sys-tems provides an example of the ways that cities of the future will make use of advanced technologies, instruments, and sensors to integrate within and between their various component systems (Dodgson and Gann, 2011).

Modelling and Simulation

The modelling and simulation described earlier uses a variety of technologies, such as Computer Aided Design (CAD), and a number of techniques such as Computational Fluid Dynamics (CFD) and Finite Element Analysis (FEA). CAD systems are routinely used in the design of complex new products, such as aircraft, and the relatively simple design of toothbrushes and shoes. Modelling and simu-lation is also conducted using combinations of readily available computer software packages, using spreadsheets and visualizations such as heatmaps. Simulations are conducted on models that represent key parameters and relationships in a system and provide an abstraction with some explanatory powers relating to cause and effect. Simulation technology simplifies reality, facilitating analysis of how parts of a system work together.

These tools are increasingly used to conduct rapid analysis of multiple designs, improving the opportunity to assess the possible performance of different solutions. They allow experiments to be conducted on system performance, allowing design teams to ask 'what if?' questions about aspects of a virtual artefact. They can also be used for the design of logistics networks. P&G, for example, uses optimization and simulation techniques for the design of supply networks. These technologies allow the selection of the most effective solution from amongst thousands of options simultaneously to determine the most efficient supply chain structure.

Visualization

Technologies developed in the film and games industries have improved capacities to visualize products, services, and processes in virtual reality. Using a virtual environment allows experimentation that would often be physically impossible or prohibitively expensive to undertake in reality (Thomke, 2003). Virtual reality is being used in the fashion and clothing industries to improve designs and customize production. P&G uses virtual reality shopping malls to test consumer reaction to its products compared to competitors. These tools provide information, for example, from factors such as eye movement analysis which are not easily determined from traditional focus group and telephone product evaluation. It uses a visualization suite to test representations of package designs with consumers, and through a 3D computer aided design (CAD) system that can simulate and model prototypes, linked to rapid prototyping technologies.

'Mixed-reality' simulation studios combine real and virtual objects to create virtual prototypes in 'immersive' studios with a high degree of detail. These allow firms and their customers to experience products and services before they are produced in reality. The studios combine technologies from the computer games world with advanced engineering and design tools, including the use of holographic imaging and touch-sensitive virtual models. Firms such as GE and Boeing use immersive studios to explore design options, allowing them to cut costs and time taken in traditional prototyping activities.

Augmented Reality (AR) is another example of digital/physical integration by combining physical and virtual models in 3D with interactivity in real time. It overlays (or augments) virtual models (objects and information) over real world physical models. AR involves taking images of the real physical world with a camera which is then overlaid with 3D CAD models using mobile computers.

As an example of its use, AR has the potential to revolutionize innovation in the built environment through the way in which information about buildings is provided. With the help of AR technology, the information about the surrounding real world of the user becomes interactive and digitally usable. By tracking the user's position and orientation, complicated spatial information can be directly registered to the physical model where it applies. Extensive amounts of data on a building or piece of infrastructure can be easily accessed. Hidden infrastructure, for example, such as optical fibre, sewer, and gas lines can be shown beneath a road surface. This is especially valuable for the construction, improvement, or maintenance of infrastructure in busy city environments. The construction company, Laing O'Rourke is using augmented reality systems for architectural applications. The system overlays a 3D digital model of a portion of a building's structural system over a user's view of the physical view in which they are standing. The augmented reality systems and detailed 3D models allow Laing O'Rourke engineers and designers to see existing mechanical appliances and electrical wiring through ceilings, walls, roofs, and floors (Gann et al., 2011).

When Arup's fire engineering team discovered through their modelling and simulations that the best way of evacuating people in tall buildings in extreme events is using the lift (provided smoke could be removed from lift shafts), it had to confront years of popular understanding that 'in case of fire, do not use the lift'. Promoting a highly innovative solution to sceptical fire authorities, fire fighters, building residents and insurers, architects and builders, required a considerable degree of cross disciplinary and professional comprehension. By producing a visual tool with which these various groups could interact and engage, this understanding was much more easily reached than it would have been from discussions centred around technical drawings and data sets (Dodgson et al., 2007).

Rapid Prototyping

In a series of articles, *The Economist* has argued that rapid prototyping—sometimes called 3D printing—will transform manufacturing industry. Rapid prototyping technologies, which assemble components additionally, layer by layer, rather than conventional cutting techniques, are used extensively. P&G, for example, uses virtual and rapid prototyping systems around a philosophy of 'Exploring Digitally: Confirming Physically'. Once a digital model is produced, virtual product testing begins, with teams of people from around the world examining the virtual model and commenting on what they do and do not like. These technologies are interconnected with manufacturing technology. P&G uses rapid prototyping machines that were originally developed for Formula 1 motorsport, where it is essential to be able to create high precision parts quickly. The group works with local rapid prototyping tool developers and shares experiences with them about how best to develop prototypes. It uses virtual and rapid prototyping to test whether new packaging containers can be opened by disabled or elderly people, and how appealing they are likely to be in the particular market sector they address prior to scaling up to mass manufacture. In the case of its development of a new package for a men's fragrance, it tested the appeal of various bottle shapes with groups of women (the major purchasers of men's fragrances), who had accurate representations of the options available to them without P&G having to go to the expense of producing glass prototypes. Some engineering industries use rapid prototyping machines to produce 3D prints of parts from materials that can be used in the final artefact, leading to a new form of bespoke 'manufacture'.

Conclusions

Technology profoundly affects the way work is undertaken and innovation occurs, but the directions of its influence are not determined purely by technical factors. Managers have choices about how technologies are configured and used. There was a continuing thread of concern throughout the industrial era about the capacity of technology to de-skill and remove craft skills from workers, and this has been extended into service industries and the digital age (Sennett, 2008). Yet, there are distinct advantages in retaining discretion and craft in the configuration of jobs. Making jobs interesting and rewarding has the obvious benefits of improving staff satisfaction and retention, but it can also improve innovation and productivity more directly. By removing the repetitious and boring it can allow more time for the creative and fun. Just as in the industrial age, choices about technological configuration around digital technologies have implications for work, skill, and craft. As one observer of the digital age puts it:

> Embracing digital making as a craft enables space for experimentation and collaborative production as well as an in-depth understanding of how things work, and how to build and repair them
>
> (Carpenter, 2011: 51).

When considering the case of innovation technologies—the digital modelling, simulation, and visualization tools used to support the innovation process—appreciation of the importance of analytical skills and judgement based on craft knowledge and experience remains critical.

There are some obvious dangers with such technologies that innovation managers need to appreciate. Simulations rely on data in which small errors or misjudgements can be magnified. There may be problems with the interpretation and validation of results. Slick use of virtual reality visualizations can present results in a convincing manner, masking problems in the ways in which models are built and potential errors in the data underpinning them, resulting in misunderstanding and misinterpretation.

The skills needed to capture the benefits and overcome the limitations of virtual and rapid prototyping and the use of social and collaborative technology platforms require the continual development of deep knowledge and craft. The value of use of these tools depends upon the quality of questions being addressed, understanding the reliability of data put in, and clear understanding of the assumptions and simplifications built into the models and results they portray. The value of these technologies lie in their use by organizations keen to realize and supplement the skills and creativity of their employees, encouraging better communications across organizational and technological boundaries, and enhancing their commitment to and engagement with their work.

The technologies should not be used to de-skill, but to allow the expansion and integration of different skills so necessary for creating solutions to complex and complicated problems.

Technology connects with innovation through innovation strategies. Continuous deep technological excellence allows companies such as IBM and Arup to build their core technological strengths and diversify their businesses. The use of technologies valuably assists integration and engagement within and outside of an organization. Innovation strategies merge internal and external strengths. Technology assists a wider range of participants to engage with innovation—customers, suppliers, members of innovation ecosystems—shaping and representing potential solutions to problems. This may result in better understanding of the wider external factors that affect innovation and design, and by engaging the range of potential contributors, determine the eventual success of the solution. Highly complex problems in which it has been traditionally difficult for interested parties with limited expertise to engage may be assisted through the use of modelling and simulation techniques. By operating in virtual design environments, and with the appropriate skills and awareness of the limitations of the technology, and with the engagement of interested parties, designers can have greater confidence in the appropriateness, accuracy, and efficacy of their innovations.

The types of models and prototypes we have discussed are, in some cases, encouraging a new form of collective experimentation and enquiry about different ideas, visions, and stories. Customers and end-users can actively engage in the design of new products and services, shaping outcomes in 'market-facing' innovation processes using the design tools described in Chapter 5 by Franke. Connecting to the Internet is easy, done by tapping, typing, speaking, or taking a photograph, and it can then offer real-time experimentation and feedback on services being offered. IvT is also providing new opportunities for communities of innovators to evolve, creating flatter structures, subverting the role of 'experts' through discussions in online communities and other collaborative design and social networking spaces (Ball, 2004; Surowiecki, 2004). We can anticipate that this will maintain the momentum of changes that are breaking down barriers between organizations, disciplines, professionals, and the public/private realm, thereby creating an entirely new environment for innovation and opportunities for innovation managers.

This involves the transformation of the product-centric perspectives of much contemporary innovation towards the design of organizational systems driven by human and market objectives. The design of complex systems delivering services, such as financial or digital media businesses, or city utilities, benefits from skills in new experimental approaches using the opportunities to simulate and model provided by a range of visualization and virtual reality technologies.

The impact of technology is rarely unambiguous, and more concurrency of people working on multiple projects and increased speed of delivery of information and feedback may lead to less time for reflection and thinking about problems before they arise. Such continuing challenges highlight how, despite the contributions technology can make in intensifying innovation processes, they remain risky. Understanding the limitations of the use of technology by its users is essential for its successful application. It has to be recognized that technology is but a tool whose development and use is a result of human choices, reflecting personal differences, and the preferences and political realities within organizations.

Note

1. The conversion into binary code of diverse forms of information, such as text, sound, image, or voice, therefore massively increasing their capacities to be processed and merged.

References

Allen, T. (1977). *Managing the Flow of Technology: Technology Transfer and the Dissemination of Technological Information Within the R&D Organization.* Cambridge, MA.: MIT Press.

Ball, P. (2004). *Critical Mass: How One Thing Leads to Another.* New York: Farrar, Strauss and Giroux.

Bell, D. (1976). *The Coming of Post-industrial Society.* New York: Basic Books.

Braverman, H. (1974). *Labor and Monopoly Capital: The Degradation of Work in the Twentieth Century.* New York: Monthly Review Press, 465.

Brown, J. S., and Duguid, P. (2000). *The Social Life of Information.* Boston, MA: Harvard Business School Press.

Brynjolfsson, E., and Saunders, A. (2010). *Wired for Innovation: How Information Technology is Reshaping the Economy.* Boston, MA: MIT Press.

Carpenter, E. (2011). 'Social Making', in Charny, D. 'The Power of Making', Exhibition Catalogue, London, Victoria and Albert Museum.

Chui, M., and Fleming, T. (2011). 'Inside P&G's Digital Revolution', *McKinsey Quarterly.* November.

Constant, E. (2000). 'Recursive Practice and the Evolution of Technological Knowledge', in Z. Ziman (ed.), *Technological Innovation as an Evolutionary Process.* Cambridge: Cambridge University Press, 219–30.

Criscuolo, P., Salter, A., and Sheehan, T. (2007). 'Making Knowledge Visible: Using Expert Yellow Pages to Map Capabilities in Professional Services Firms', *Research Policy*, 36(10): 1603–19.

Crump, T. (2001). *A Brief History of Science: As Seen Through the Development of Scientific Instruments.* London: Constable.

Dittrich, K., Duysters, G., and de Man, A.-P. (2007). 'Strategic Repositioning by Means of Alliance Networks: The Case of IBM', *Research Policy*, 36(10): 1496–511.

Dodgson, M. (1984). 'New Technology, Employment and Small Engineering Firms', *International Small Business Journal*, 3(2): 118–19.

Dodgson, M., and Gann, D. (2011). 'Technological Innovation and Complex Systems in Cities', *Journal of Urban Technology*, 18(3): 99–111.

Dodgson, M., Gann, D., and Salter, A. (2005). *Think, Play, Do: Technology, Innovation and Organization.* Oxford: Oxford University Press.

Dodgson, M., Gann, D., and Salter, A. (2006), 'The Role of Technology in the Shift Towards Open Innovation: The Case of Procter and Gamble', *R&D Management*, 36 (3), 333–46.

Dodgson, M., Gann, D., and Salter, A. (2007). '"In Case of Fire, Please Use the Elevator": Simulation Technology and Organization in Fire Engineering', *Organization Science*, 18(5): 849–64.

Dodgson, M., Gann, D., and Wladawsky-Berger, I. (2010). 'Engineers and Services Innovation', *Ingenia—Journal of the Royal Society of Engineers*, 44: 33–5.

Dodgson, M., and Gann, D. (2011). 'Technological Innovation and Complex Systems in Cities', *Journal of Urban Technology*, 18(3): 99–111.

Dodgson, M., Gann, D., and Phillips, N. (2013). 'Organizational Learning and the Technology of Foolishness: The Case of Virtual Worlds in IBM', *Organization Science*, 24(5): 1358–76.

Freeman, C., and Perez, C. (1988). 'Structural Crises of Adjustment: Business Cycles and Investment Behaviour', in Dosi G., Freeman, C., Nelson, R., Silverberg, G. and Soete, L. (eds), *Technical Change and Economic Theory*. London: Pinter Publishers, 38–66.

Galison, P. (2003). *Einstein's Clocks, Poincaré's Maps: Empires of Time*. London: Sceptre: Hodder and Stoughton.

Gann, D., and Dodgson, M. (2007). *Innovation Technology: How New Technologies are Changing the Way we Innovate*, London: National Endowment for Science, Technology and the Arts.

Gann, D., Dodgson, M., and Bhardwaj, D. (2011). 'Physical–digital Integration in City Infrastructure', *IBM Journal of Research and Development*, 55(1&2): 1–10.

Gerstner, L. (2002). *Who Says Elephants Can't Dance: Inside IBM's Historic Turnaround*. New York: Harper Business.

Hackett, E., Amsterdamska, O., Lynch, M., and Wajcman, J. (2007). *The Handbook of Science and Technology Studies*, 3rd edn. Cambridge, MA: MIT Press.

Hamilton, J. (2003). *Faraday: The Life*. London: Harper Collins.

Hargadon, A., and Sutton, R. (1997). 'Technology Brokering and Innovation in a Product Development firm', *Administrative Science Quarterly*, 42(4): 716–49.

Jefferys, J. (1946/1970). *The Story of the Engineers, 1800–1945*. New York: Johnson Reprint Corp.

Jones, R. (2004). *Soft Machines: Nanotechnology and Life*. Oxford: Oxford University Press.

Kelley, T. (2001). *The Art of Innovation: Lessons in Creativity from IDEO, America's Leading Design Firm*. New York: Doubleday.

Kerr, C., Dunlop, J., Harbison, F., Myers, C. (1973), *Industrialism and Industrial Man* (London: Penguin).

Marx, Karl (1981). *Capital, Volume 1*. Harmondsworth: Pelican, 1136.

Noble, D. F. (1986). *Forces of Production: A Social History of Industrial Automation*. New York: Oxford University Press.

Palmisano, S. (2006). 'The Globally Integrated Enterprise', *Foreign Affairs*, 85(3): 127–36.

Perlow, L. A. (1999). 'The Time Famine: Toward a Sociology of Work Time', *Administrative Science Quarterly*, 44(1): 57–81.

Perrow, C. (1970). *Organizational Analysis: A Sociologial View*. London: Tavistock Publications.

Petroski, H. (1985). *To Engineer is Human: The Role of Failure in Successful Design*. New York: University of Columbia Press.

Sakkab, N. (2002). 'Connect and Develop Complements Research & Develop at P&G', *Research-Technology Management*, 45(2): 38–45.

Schön, D. A. (1991). *The Reflective Practitioner*. Aldershot: Ashgate Publishing.

Schrage, Michael (2000). *Serious Play: How the World's Best Companies Simulate to Innovate*. Boston, MA: Harvard Business School Press.

Schumpeter, J. A. (1934). *The Theory of Economic Development: An Inquiry Into Profits, Capital, Credit, Interest and the Business Cycle*. Cambridge, MA: Harvard University Press.

Sennett, R. (2008). *The Craftsman*. New Haven, CT: Yale University Press.

Simon, H. A. (1996). *The Sciences of the Artificial*, 2nd edn. Boston, MA: MIT Press.

Surowiecki, J. (2004). *The Wisdom of Crowds*. London: Abacus.

Thomke, S. (2003). *Experimentation Matters*. Boston: Harvard Business School Press.

Touraine, A. (1962). 'A Historical Theory in the Evolution of Industrial Skills', in C. Walker (ed.), *Modern Technology and the Organisation*. London: McGraw Hill.

Townes, C. (1999). *How the Laser Happened: Adventures of a Scientist*. New York: Oxford University Press.

Tuomi, I. (2002). *Networks of Innovation: Change and Meaning in the Age of the Internet*. New York: Oxford University Press.

Vincenti, W. (1990). *What Engineers Know, and How They Know It*. Baltimore: John Hopkins Press.

Whyte, J. (2002). *Virtual Reality and the Built Environment*. Oxford: Architectural Press.

Woodward, J. (1965). *Industrial Organization: Theory and Practice*. London: Oxford University Press.

Wyatt, S. (2007).'Technological Determinism is Dead: Long Live Technological Determinism', in E. J. Hackett, O. Amsterdamska, M. Lynch and J. Wajcman (eds), *The Handbook of Science and Technology Studies*, 3rd edn. Cambridge, MA: MIT Press.

PART IV

STRATEGY, MANAGEMENT, AND ORGANIZATION

CHAPTER 20

....

INNOVATION, STRATEGY, AND HYPERCOMPETITION

....

RITA GUNTHER MCGRATH AND JERRY KIM

STRATEGY scholars have viewed innovation as distant from their primary concerns. They have tended to focus on issues such as where and how organizations should compete; where positions of advantage come from; and how industries evolve. Strategy models generally assume some period of stability within which analytics can inform managerial choices. The ultimate goal of strategy has long been to understand how a firm can create a 'sustained' competitive advantage, which has been defined as 'enduring firm differences in above-normal returns' (Oliver, 1997: 697). When advantages disappear quickly, however, the premise of relative stability vanishes. To understand performance, scholars in strategic management will need to venture into the uncertain, volatile, and rapidly changing environment characterizing innovation (McGrath, 2013). In this review, we propose that with the advent of what some have called 'hypercompetition' in more portions of our economy, those in the field of strategic management would be well advised to move issues of innovation to a more prominent place on their research agenda.

STRATEGIC MANAGEMENT AS A FIELD
....

Before delving into why strategic management needs to increasingly take heed of innovation, it is instructive to examine how strategy developed as a field. It has been a remarkable success, if success is defined in terms of the growth in the number of scholars who identify themselves with the field, and with membership in such organizations as the Academy of Management and the Strategic Management Society. 'Business policy', as it was once called, originally targeted as its distinctive domain the job of the general manager. As Hambrick and Chen point out, at the core of strategic management was the idea that 'the work of top executives, or general managers, was qualitatively different from the work of managers at lower hierarchical levels' (Hambrick and Chen, 2008: 40). Early work focused on critical decisions that general managers made. Rather than competing with or attempting to invalidate more established fields such as organizational behaviour, marketing, or economics, strategy scholars drew extensively from them, with the idea that strategic management's distinctive contribution was to integrate the theories and findings from more established fields to address the unique perspective of the general manager.

Thus, strategy scholars borrowed from marketing (Hatten et al., 1978) to look at questions of industry structure and market demand. They borrowed from organization theory to understand issues such as legitimacy, social acceptability, and how firms might best navigate institutional concerns (Miles and Cameron, 1982). They borrowed from economics to adapt tools of economic analysis to questions of managerial decision-making (Caves and Porter, 1977). This did, however, give the field something of an identity crisis. If it could not claim uniqueness with respect to methods, underlying theoretical assumptions, or even boundary conditions, what, then, made it unique, other than its focus on general managers? Nag, Hambrick, and Chen (2007) attempted to answer this question and after an exhaustive investigation of both published literature and the perspective of scholars in the field came up with the following description:

> The field of strategic management deals with the major intended and emergent initiatives taken by general managers on behalf of owners, involving utilization of resources to enhance the performance of firms in their external environments.

> (Nag et al. 2007)

Even with this definition in hand, Nag et al. raised a concern that process, the *management* part of strategic management, did not at the time of their study appear to be of much interest to the strategy community; at least in the sense that their respondents embraced strategy processes as within the legitimate domain of the field. Similarly, the activities of people with respect to strategy and the societal impacts on and consequences for strategy were not defined as central to the field. The field by the time of their study had thus left behind the core qualitative and human concerns that animated its earlier stages (Barnard, 1938; Ansoff, 1965; Andrews, 1971). As we hope to demonstrate, this is a significant gap because the systematic management of innovation has a great deal to do with process and with the social infrastructure within and between organizations.

THEORETICAL AND EMPIRICAL TRADITIONS IN THE FIELD OF STRATEGY

Two broad theoretical traditions run through much of the strategy literature. The first, adopted from industrial organization economics, views firm performance largely as a function of the structure of the underlying markets in which firms compete, and different firms' positions within those markets. It emphasizes industry-level attributes such as concentration ratios, decisions to invest in advertising, choices about the setting prices, and so on. The second, which seems to have eventually crystallized in a perspective variously called the 'resource-based' view of the firm or the 'dynamic capabilities' view of strategy, focuses on how firms build capabilities over time, and emphasizes factors such as capabilities, learning, leadership, and the cumulative effects of decisions over time.

Theory Borrowed from Economics

Michael Porter was a successful champion for the industrial economics view in strategy (Porter, 1980). His work was the most highly cited of all in the 1980–85 period (Hambrick, 1990). The brilliant thing about industrial economics (or I/O) from an academic point of view is that it offered the prospect of a wealth of quantifiable variables that might be linked to performance, and a set of underlying assumptions about how those variables might be related.

The theory built on the structure-conduct-performance paradigm, which assumed that the performance of an industry was determined by the conduct of the firms within it, which will yield different results depending on the structure of underlying markets (Mason, 1949; Bain, 1959). A critically important assumption was that, as Porter himself observed, the structure of an industry reflected 'relatively stable' economic and technical dimensions, meaning that these could be analysed over a reasonably long period of time. A further simplifying assumption in this paradigm was, as he put it, 'because structure determined conduct (strategy), which in turn determined performance, we could ignore conduct and look directly at industry structure in trying to explain performance. Conduct merely reflected the environment' (Porter, 1981: 611).

Ironically, one of the intellectual fathers of the industrial economics perspective had an entirely different view, which has been for the most part ignored in economics-grounded work in strategy. Edward Mason, in 1939, asserted that '... firms are not, regardless of what economic theory may suppose, undifferentiated profit maximizing agencies which react to given market situations in ways which are independent of their organizations' (Mason, 1939: 66). Schmalensee, in an address at the International Industrial Organization Conference in 2012 remarks that 'this view has long been out of fashion in economics, in part because economic theory has not had much to say about exactly how organization should matter' (Schmalensee, 2012: 160).

Nonetheless, a vast amount of strategy work built upon this theoretical tradition. Initially, researchers studied features of industries such as concentration, product differentiation, and barriers to entry, finding that entry barriers could be created by economies of scale, product differentiation advantages, and cost advantages (Bain, 1956). Later research examined game-theoretic problems, such as what choices firms should make under conditions of information asymmetry (Saloner, 1991). Other influential studies examined first-mover advantages and switching costs (Lieberman and Montgomery, 1988; Farrell and Klemperer, 2007). Transactions cost theory sought to explain why firms organized as they did, whether within a firm or by trading on the open market (Williamson, 1975). Agency theory tried to explain how managerial incentives affected decision-making and outcomes (Jensen and Meckling, 1976; Campbell et al., 2012). Work on competition employed game-theoretic and other concepts to develop theories about when firms would respond to a competitive move and when not (Chen and MacMillan, 1992; Gimeno and Woo, 1996; Ferrier et al., 1999; Haleblian et al., 2012).

Perhaps the most famous framework from this tradition was Porter's synthesis of the core idea that one can understand a firm's position best by analysing five competitive forces—power over buyers and suppliers, entry barriers, lack of substitutes, and the general state of rivalry (Porter, 1980). The key point with respect to the economic tradition in strategy is the use of structural variables at the industry level, to examine strategic decisions made by firms.

Theory Borrowed from a Large Number of Traditions: The Resource-Based View

Critics of the I/O perspective objected to its relatively static approach. Instead, proponents of the so-called resource-based view argued that as firms evolve, they acquire capabilities and resources which are path-dependent, relatively enduring, and which create the heterogeneity which underlies sustainable competitive advantage (Wernerfelt, 1984; Barney, 1991; Teece et al., 1997; Hoffman, 2000). The resource-based view traces its intellectual tradition to writers such as Edith Penrose, who advocated thinking of firms in terms of the 'services' their capabilities provide (Penrose, 1959), although the perspective also borrows a number of ideas from other fields, such as the behavioural theory of the firm, learning theory, and increasingly theories involving cognition and individual mental processing (Levitt and March, 1988; Hodgkinson and Healey, 2011).

While seldom tackling exactly the same questions as researchers from the I/O perspective, scholars in the resource-based view focused far more on what was going on within firms than in their environments (McGrath, 1993; Galunic and Rodan, 1998; Ahuja and Katila, 2004). Although not always included explicitly under the resource-based umbrella, research streams in the areas of organizational learning (Cohen and Levinthal, 1990); organizational evolution and adaptation (Levinthal, 1997); the management of knowledge (Helfat and Raubitschek, 2000); the path-dependent accumulation of assets (Dierickx and Cool, 1989); and organizational structure (Robins and Wiersema, 1995) are examples of strategy research in which scholars looked at internal structures and processes as being important to performance. There has also been much criticism of the resource-based view, with observers complaining that it is fundamentally tautological and that its methods lack rigour (Priem and Butler, 2001; Edward and Hermann Achidi, 2006).

The Advent of Empiricism

Theories, of course, do not emerge in a vacuum. To understand the way in which theories in strategic management developed, it is useful to consider the context in which the (largely US-based) field emerged. American industry in the post-Second World War period enjoyed unprecedented growth and a lack of competition as companies from other countries struggled to regain their footing after the devastating conflict. Many potential competitors were behind the 'iron curtain' of Soviet influence. Companies from countries such as China and India were still either not participating in world trade or were labouring under regulations that made it nearly impossible to do so. In other words, American companies enjoyed a period of relatively stable competitive exchange (Whitman, 1999). Assumptions born in those times would be severely shaken as the pace of competition increased.

During this era, what Walter Kiechel called the 'Lords' of strategy in both consulting and academia began to stake out their intellectual domain (Kiechel, 2010). Kiechel argues that strategy itself was a new idea which emerged in this period, and that concurrently the idea appeared that one could use quantitative analysis to come up with ways to enhance firm performance. This was exciting: rather than making decisions on the basis of long experience or learning through observation of others' situations, there were analytical techniques that could be applied, yielding hopefully superior and replicable solutions.

Thus, we had the development of the Boston Consulting Group's work on the learning curve which eventually led to the product-portfolio matrix (Henderson, 1980), a model that attained an astonishing degree of ubiquity among the leaders of America's corporations (Lubatkin and Lane, 1996). McKinsey, BCG, and Bain explored analytical frameworks to help companies make pricing and production decisions and to go through their operations with what some called the 'new Taylorism' in search of efficiencies. All of this reinforced the idea that firms could gain competitive advantages through analysis.

Kiechel argues that the consultants gave voice to the concept of strategy in business first, in the 1960s, with the academics following behind. Business policy was rebranded 'strategic management' when a group of ninety-three academics, consultants, and businesspeople came together in 1977 under the leadership of Dan Schendel and Chuck Hofer in Pittsburgh (Schendel and Hofer, 1979; Hambrick and Chen, 2008). Strategy scholars sought to adopt the style of research that more established fields used. Hambrick found that the bulk of the most influential articles published between 1980 and 1985 had adopted the trappings of hard science, including explicit hypothesis tests, multivariate analysis, and significant samples (Hambrick, 1990).

The prayers of many had been answered, therefore, with the development of one of the first large-scale empirical databases designed to measure the outcomes of specific strategic choices. The Profit Impact of Market Strategies database, or PIMS, was launched by Sidney Schoeffler, a senior executive at General Electric in the 1960s. With the analytical fervour of the age, GE's leaders were eager to see if they could analyse why some of their business units were more profitable than others. The company funded a comprehensive survey in which it asked business unit managers to provide data on various attributes of their operations, the strategic choices that they made, and the results that ensued.

The variables might seem obvious today—relative product quality, local market share, costs, investment intensity, market growth rate, and so on. It was groundbreaking at the time because it was a large-sample, empirical database that captured such variables and their effects on performance. PIMS was eventually expanded beyond GE and became a project of the Marketing Science Institute in the 1970s, who administered an annual survey of 2,600 business units from 200 companies. Some of these data were made available to academic researchers, opening a wealth of possibilities for empirical work on strategy issues (Buzzell et al., 1975; Anderson and Paine, 1978; MacMillan and Day, 1987). Not only PIMS, but other large-scale digital databases such as Compustat, FTC data, and data coded by SIC codes, were becoming available to scholars anxious to find statistical regularities in the relationship between management decisions and subsequent outcomes, what Richard Bettis has, with concern, called 'the search for asterisks' (Bettis, 2012).

Historical Assumptions in the Strategy Literature

Both the economics and the resource-based view of strategy share certain assumptions. The first is that industries have clear boundaries, are relatively stable, and are slow to change. This is more prominent in the I/O perspective. It is present as well in the RBV, especially as scholars attempt to compare firms to find the impact of possessing specific resources (Collis, 1991; Carr, 1993). The second follows, which is that the most significant competitor for any firm is other firms within its industry—firms that offer close substitutes. A third is that resources (with a few exceptions) are properties of organizations and are linked to them.

These assumptions increasingly fail to reflect reality in contexts in which a firm's business model, or competitive advantage, or viability, are challenged more quickly than they had been before. A host of factors from globalization to the digital revolution are eliminating entry barriers, empowering new competitors, levelling the technological playing field, and otherwise creating what some have called 'hypercompetitive' environments for firms. Under fast-moving competitive conditions, as Ian MacMillan laid out years ago, seizing the strategic initiative—or innovating—becomes critical to long-term organizational survival (MacMillan, 1982).

THE RISE OF HYPERCOMPETITION AND THE COMPETITIVE 'WAVE'

In the 1990s, D'Aveni and colleagues (D'Aveni and Gunther, 1994) spearheaded a challenge to the concept of sustainable advantage which sparked a stream of research that questions the dominant sustainability paradigm and suggests that there are many environments in which hypercompetition, rather than competitive equilibrium, is the norm (Craig, 1996; Gimeno and Woo, 1996; Ilinitch et al., 1996). The assumption made by these scholars is that all positions of advantage are temporary, doomed to be swept away by Schumpeter's famous 'waves of creative destruction' (Schumpeter, 1942). Advantages are characterized as temporary or transient, and innovation (rather than positioning and resource accumulation) drive performance outcomes (D'Aveni and Gunther, 1994; Christensen and Bower, 1996).

Evidence that Competition is Eroding Advantages More Quickly

Some have questioned whether hypercompetition is confined to fast-moving industries such as consumer electronics (McNamara et al., 2003). We examined the top 100 firms in the *Fortune* 500 (by revenue) as of 1955. Most were in the energy business, producing steel, making food, and in gritty industries like manufacturing batteries. Of the top 100 in the 1955 list, there was only one company (CBS) that didn't make or handle physical things. It's a very different story when you have a look at the 2012 list. Only 39 of the top 100 in the 2012 list were involved in the making of things (this is also a function of *Fortune* changing its criteria for inclusion to allow service firms to appear on the list in 1994).

What replaced the manufacturers were retailers (Wal-Mart, Costco), banks, and other financial services companies (Citigroup, TIAA-CREF, Wells Fargo). Many of the manufacturers on the list today are high-technology companies (Cisco, Google, Apple). The companies on the 2012 list, in other words, represent sectors with shorter product life cycle offerings such as consumer electronics, retailing, and technology services. What this suggests is that more economic activity, at least among very large firms, involves either services or fast-moving consumer products. These in turn are more vulnerable to rapid competitive attack as they often have fewer entry barriers and suffer substitution risk.

Other researchers have found similar patterns. Richard Foster, a former McKinsey director and author of a book called *Creative Destruction* (Foster and Kaplan, 2001) on this theme, recently found that the lifespan of companies in the Standard and Poors 500 (an index chosen to represent the total economy) has been steadily shrinking for decades (Foster, 2012). In 1960, a typical S&P company had been around for sixty-plus years. By 2010, the typical S&P company was 16 or 17 years old. Further, the index is extremely dynamic. According to Foster's research, a new firm is added and an existing one removed roughly every two weeks! Thus, firms such as Express Scripts, Juniper Networks, and Google enter the index while firms such as Sears, The New York Times, and Radio Shack exit it. Churn is not limited to the United States. We also examined the composition of London's FTSE 100, and found similar volatility—on average, 15 per cent of that index churns each year. In a year of major change (2000) fully 30 per cent of the places on the index changed.

The share of US national income generated in 'effectively competitive' markets rose from 56 to 77 per cent overall between 1958 and 1980 (Shepherd, 1982). In many industries, further acceleration is becoming the norm. Just as an example, an automotive design company observed that a typical automotive design cycle five years ago (approximately in 2008) was approximately 60 months, or five years. Today, it is 24–36 months or two to three years (Altera Corporation, 2012). The evidence is pretty clear that in many sectors of the economy, competition has become more aggressive than it was at one stage.

We know far less about how a firm can perform well even as its resources and product-market positions regularly collapse in value than we do about how firms can exploit durable positions of advantage. The very notion of high performance from exploiting purely temporary advantages seems to fly in the face of both the economics-oriented schools of thought and of the resource-based view. With the advent of Clayton Christensen's ideas of disruptive innovation, the need to integrate strategy and innovation theories seems clear (Christensen, 1997).

The Competitive 'Wave'

MacMillan (1988) described transient advantages in terms of 'waves' which would emerge, be available for exploitation, and then disappear again. Figure 20.1 provides an illustration, with sales or some other attractive outcome variable on the vertical dimension and time on the horizontal. This profile comprises four distinct temporal stages. The first is the launch period from T_0 to T_1, when the firm is investing funds and resources to put a new competitive advantage in place. The second is the period of ramp up from T_1 to T_2, when the offering is presented to the market, and as the result of creating value for a customer segment, the firm increases its sales from zero along line AB to sell d units before sales level off. Then, the firm can enjoy an exploitation period from T_2 to T_3, when sales level off and maintain a steady level until competitive attack or other change leads to erosion of sales. Finally, there is the period of erosion from T_3 to T_4, when competitors attack or technology changes or the basis of competition shifts. Sales decline from d units along line CD until the firm elects to abandon the eroded advantage.

At this point in a hypercompetitive environment, to survive, a firm would need to launch, ramp up, and begin to exploit an additional advantage, depicted by the dotted line in Figure 20.1, and the dynamic cycle is repeated. Note that in order to do this, the follow-on wave must have been launched before the grace period from T_G to T_3, during which time the follow-on advantage must be ramped up in order to sustain profit flows, now from the follow-on cycle. The dilemma for many organizations is that extracting resources from a successful business to re-purpose them to an uncertain, innovative new business is, politically and in power terms, very, very difficult for most organizations. Relatively few have figured out how to do this on an ongoing basis (McGrath, 2012).

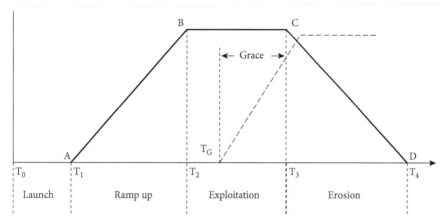

FIGURE 20.1 The 'wave': temporal profile of transient advantage

Systematic Innovation as a Strategic Process

Now, we come to the crux of our argument about the need to better integrate theories of strategic management and theories of innovation. For the most part, strategic management theory focuses on the 'exploitation' part of the wave, in which an advantage is well established. Strategy has thus to do with manoeuvring, rivalry, figuring out where your resources will stand you in good stead, achieving attractive cost positions relative to rivals, and the like. Innovation, from this vantage point, is an activity that can happen episodically. When advantages last for a long time, constant innovation isn't really necessary.

In a transient advantage context, however, the exploitation period is relatively short. There are fewer and fewer conditions in which the boundary conditions of industrial organization theory match the realities of the strategic environment. Scholars working with the resource-based view have done a better job of attempting to introduce dynamism into their models, some even incorporating a life cycle perspective and attempts to distinguish among 'dynamic' and 'operational' capabilities (Helfat and Raubitschek, 2000; Helfat and Peteraf, 2003; Helfat and Winter, 2011). The specific insight about what is required to make innovation a systematic process has not so far figured very prominently.

Consider just a few challenges to extant theories that transient advantage presents. Take the argument that industries are the fundamental building blocks of structural analysis. Today, the most significant competitors may well be launching their initiatives from other industries. Recently for instance, a reporter for the *Wall Street Journal* found that family spending on cell phones was absorbing resources that previously might have gone to dining out, automotive expenses, travel, and apparel (Troianovski, 2012).

If Innovation Were More Central to Strategy Research

Innovation has become far more central to the practice of strategic management; but our theories of innovation strategy have not kept up. It is in many cases still seen as a special situation for strategy.

In an extreme view of the transient advantage economy, organizations themselves would be temporary. Consider movie-making, political campaigns, or running the Olympics. Organizations designed to do these things are focused on specific projects—capabilities are brought to bear as needed and dissolved when the need has passed. Capabilities are not built up in organizations that contain them, rather they are created by individuals and networks with loose organizational affiliations. Access to capabilities, rather than the painstaking development of capabilities that are tied semi-permanently to a firm, can become the norm. A firm today can use computing power from Amazon, programmers from oDesk, desk space from the Regus Group, and talented specialists from guru.com. This has important implications for what we measure, who we focus our attention on, and what we believe to be the core task of managers

WHAT WOULD WE MEASURE?

One advantage of the I/O tradition in strategy has been an extremely rich set of variables that are measurable, typically having to do with industry structure or competitive exchange. The dilemma with these measures is that they often reify the concept of 'industry' and that they are so broad that individual firm behaviour—actual strategic choices—gets averaged away in the analysis. If we are to take innovation-fuelled hypercompetitive behaviour into account, our measures need to change.

We Would Use a Different Measure of Performance: Entrepreneurial Rents, not Comparative Returns to Others in an Industry

The goal of transient advantage strategies is not that different from that of strategies which seek sustainability. It remains the generation of 'rents'—commonly defined as profits in excess of those made on average by competitors (Alchian, 1991). In the case of transient advantages, however, it may not be practical to use this concept of rents, since it is hard to know what an 'industry normal' return would be in a temporary situation. Indeed, the very definition of 'sustainable' competitive advantage implies that there is a known comparison group whose performance one could benchmark against.

Rather, we propose that a better metric for the achievement of transient advantage may be to capture what Rumelt called 'entrepreneurial rents', namely the returns to a firm of seeing and exploiting opportunities that others missed (Rumelt, 1987). Rumelt distinguishes among the classic concept of rents, representing a firm's ability to take advantage of a scarce resource it possesses (such as proprietary access to raw materials), and entrepreneurial rents, which result from the discovery of new combinations of resources that are to some extent proprietary to their discoverers. He defines entrepreneurial rent as follows:

> ...the difference between a venture's ex post value (or payment stream) and the ex ante cost (or value) of the resources combined to form the venture. If we posit expectational equilibrium (ex ante cost equals expected ex post value) then expected entrepreneurial rents are zero. The basic thrust of this definition is to identify those elements of profit that are the result of ex ante uncertainty. (1987: 143)

What is useful about this definition is that it lends itself to an industry-agnostic perspective on how to measure the returns to strategic innovation. Theoretically, one could develop such a measure by gathering data on expectations for what an innovation might be able to accomplish and comparing this with what actually happens. Although this approach does suffer from the drawback that it would by definition be an *ex post* measure of the presence of entrepreneurial rents, it would at least be distinguished from measures that artificially impose an industry lens on sets of transactions. One might even be able to take this further and measure the ability of a firm to capture a portion of the available discretionary resource relative to competing offers. Thus, our example of telecommunications payments crowding out restaurants would be accommodated by such a measure.

We Would Assess Firm Performance in Terms of Their Ability to Make Innovation a Key Systematic Capability

In many firms designed to exploit long-lived advantages, innovation is an episodic activity (Burgelman and Valikangas, 2005). Champions emerge for a new idea, it gathers momentum, the structures are put in place to support it (such as a corporate venturing group), and then all too often the organization turns its attention elsewhere and the new ventures are dismantled. The cycle begins again when another champion decides to begin to advocate for an idea, and the process repeats itself over and over again. While to some extent this may be a function of organizations first exploring new opportunities then focusing on those it discovers (Burgelman and Grove, 2007); all too often such a positive pattern is not the case. Instead, in companies such as Sony, Nokia, Xerox, and Kodak, the on-again, off-again pattern of innovation activity demoralizes the innovators and wastes precious resources.

In a transient advantage economy, the ability to create a pipeline of innovations on a systematic basis will be key. Similarly, the ability to abandon and exit exhausted opportunities will be crucial, as resources are extracted from those businesses and re-purposed to fund innovation.

Be Less Wedded to Stock-Market Based Metrics for Performance

Innovation is resource-intensive, unpredictable, uncertain, and does not arrive on a schedule. Restructuring is often painful, and costs resources in the near term while organizations make the necessary changes to reconfigure their assets to new competitive realities. Only during the 'exploitation' phase of an organization's strategy do the conventional stock market metrics of predictable performance, steady returns, and accurate earnings forecasts make any sense at all. And yet, in an enormous number of strategy studies, measures of performance, such as total shareholder return, earnings per share, stock price, and change in market capitalization, are used as the dependent variable. Indeed, there are some in strategy who have argued that only total returns to shareholders matter when thinking about corporate performance (Friedman, 1970).

The problem is that our public markets don't work very well for managing the complete wave. The financial approach that best suits the innovation process can be thought of in terms of option value—in the early stages, small investments that convey the right but not the obligation to make larger investments down the road (McGrath, 1997). The idea is to place small bets, most of which may not work out, with the idea of finding where the opportunities are via hypothesis testing and experimentation. This is antithetical to the desire for predictability reflected in the public markets. The financial entities most comfortable with these kinds of investments are venture capital firms or innovation consortia.

When a competitive advantage, or set of advantages, are waning, the restructuring that is necessary is extraordinarily difficult to navigate in the context of being a publicly traded firm. The analysts and market researchers are brutally unkind to companies trying to reconfigure their assets to reflect new competitive realities. Consider, for instance, the criticism that was heaped on Verizon's Ivan Seidenberg when he exited businesses such as physical phone books and land-line networks to concentrate on FIOS-style services and wireless expansion (Brown and Latour, 2004). Only some years after his so-called 'gutsy bet' did the financial results justify the bold portfolio changes (Rosenbush et al., 2003; Ward, 2009). More typically, firms facing the need to restructure seek to take shelter from the relentless glare of the public markets, find a private equity firm to cover their financing needs, and give the company some breathing room. Such a pattern reflected the restructuring of companies such as Burger King, Alliance Boots, and Toys R Us. As of this writing, the iconic computer manufacturer Dell, which once could do no wrong, looks to be seeking the shelter of a private existence once again.

This division of labour in the financial markets is not reflected in the performance variables that strategists have typically worked with. Part of this is a data availability problem—it's a lot easier to get information about publicly traded firms than it is about venture investments or private equity deals. Nonetheless, continuing to act as though stock market measures for shareholders are the best measure of performance is a bit like the apocryphal story of the gent looking for his lost keys under a lamp, not because he lost them there, but because that's where the light is shining. The vision that value-oriented shareholders should be rewarded by management teams seeking to do the best long term for the company's owners simply no longer reflects reality. Indeed, the *New York Times* recently reported that 'The average holding period of a stock was eight years in 1960; today, it's four months' (Einsinger, 2012). A four-month relationship with a company is hardly conducive to the longer-term view managing successive waves of advantage demands. Further, an excessive focus on shareholder metrics has, in the eyes of some observers, promulgated extreme short-termism (Stout, 2012). What are needed are metrics that assess performance over the complete 'wave' of competitive advantage.

WHO WOULD WE STUDY?

Unlike strategy scholars who have tended to focus on the comparative performance of individual firms, innovation scholars have long recognized that breakthroughs rarely happen in isolation. Organizations such as firms, corporate and government research labs, or universities play a critical role in innovation, and much scholarly work has been done on the organizational processes and structure that promote innovation (as referenced in earlier chapters of this Handbook). But with the advent of hypercompetition, the network of actors *outside* the boundaries of the organization has become critical in understanding innovation and strategic performance.

We Would See that New Capabilities are Created in a Network; Firms No Longer the Central Focus

Searching for, and integrating knowledge from, the external environment has always been a key strategic activity for innovating firms (March, 1991; Kogut and Zander, 1992). What has changed in recent years is the increasing shift of the main locus of innovation from inside the firm to the network of actors that comprise the firm's environment. Academic theories and corporate programmes such as open innovation paradigms (Chesbrough, 2003), Procter & Gamble's Connect and Develop model for innovation (Huston and Sakkab, 2006), and the 'wide lens' approach to ecosystems (Adner, 2012) all reflect this radical change in our understanding of the locus of innovation.

Two trends in particular have contributed to the locus of innovation moving outside the boundary of the firm: First, the complexity of knowledge required for innovation, and the speed with which it must be implemented, have far surpassed the internal capabilities of even the most innovative firms. As knowledge in basic science continues to deepen over time, it becomes excessively costly for even the most well-resourced firms to maintain expertise in all relevant fields. But more importantly, as recent studies of innovation and creativity have shown, the most radical innovations are more likely to emerge when distant domains of knowledge are bridged together (Hargadon and Sutton, 1997; Fleming, 2001; Burt, 2004). Network structures provide the flexibility and speed required to succeed, and competitive markets will thus lead to the adoption of innovation strategies with an inter-organizational focus.

The second trend that pushes the locus of innovation to the space between organizations is the frequency with which technology-driven markets exhibit network externalities (Katz and Shapiro, 1985). When the adoption of other customers increases the value of the product independent of the attributes, as seen in contexts such as telephones, personal computing, and technology standards, cultivating a vibrant ecosystem of complementary products becomes an essential strategic tool for initiating and maintaining the virtuous cycle of growth. The rise of Microsoft's Windows as the dominant operating system for PCs is a prime example of how partnership with third-party developers (i.e. complements) is the key driver of innovation, despite (arguably) lacklustre innovation within the boundaries of the firm. Focusing on Microsoft in isolation would have severely underestimated the value creation that occurred in the network of firms around the WINTEL standard.

In studying the network, it is also important to remember that we should not be confined to for-profit firms. In many cases, actors within the broader community of organizations are critical for success. For example, in the field of biotechnology, a 'diverse cast of organizations including universities, public research organizations' contributed to the evolution of the field (Powell et al., 2005) with non-market actors such as the University of California San Diego and Stanford University serving as 'anchor tenants' supporting nascent innovating firms. Besides research organizations (oftentimes uncompensated), developers of complementary products are important to consider, especially in markets where switching costs are high, and consumer lock-in is likely (for a review, see Farrell and Klemperer, 2007).

We Would Recognize that the Roles of Competitor and Collaborator are Fuzzy and Often Overlap

In traditional strategy research, the key actors that comprise a firm's competitive environment have clear and well-defined roles. Suppliers provide inputs; buyers purchase the outputs; and competitors harm the performance of the firm. However, in a competitive environment focused on speed and agility, the roles of key market participants become more dynamic and multifaceted.

The role of suppliers in innovation performance has always been a well-recognized fact from the earliest innovation studies (Rothwell et al., 1974), but as competition intensifies and technological change accelerates, suppliers become more than just makers of parts specified to clients' specifications. They are increasingly becoming joint problem solvers involved in design and prototyping (Dyer and Singh, 1998). Downsizing and outsourcing of manufacturing have made innovation partnerships with suppliers an even greater competitive advantage for firms (for example, many of the innovations in Apple Inc's iPhone and Macbook computers have resulted from breakthroughs at supplier firms such as Corning.)

End-users of products are also key actors in the external network of innovation, leading to unexpected and innovative products (von Hippel, 1986, 1988). User inventions have the potential to enrich the innovation ecosystem, but they can also provide valuable feedback and ideas for the corporation. For example, a recent study of the medical device industry (Chatterji and Fabrizio, 2012) found that corporate inventions that integrate user knowledge are more broadly diffused and have a greater impact. Finally, competitors are increasingly viewed as potential alliance partners in facilitating and extending innovations (Mitchell et al., 2002). Cross-licensing (Grindley and Teece, 1997) and patent pools (Shapiro, 2001) are a common strategy for firms in high-technology industries. Such coordination facilitates the formation of technology standards, which, in turn, can lead to faster rates of innovation.

WHAT WOULD WE CONSIDER THE CORE TASKS OF GENERAL MANAGERS?

Orchestrating and Architecting Rather than Grand Strategizing

The image of the strategist from the 1960s and 1970s was that of an analytical professional. Armed with projections and statistics, these individuals were able to build multi-year strategic plans which would favourably position their organizations relative to rivals. In a transient advantage environment, this no longer characterizes the role of the general manager. Instead, what we believe we will increasingly observe is the general manager as orchestrator.

A recent study of ten companies who had successfully grown net income by 5 per cent or more for a ten-year period (out of a sample of 4,793 publicly traded firms with market capitalizations of $1 billion or more as of the end of 2009) found evidence of this orchestrational role (McGrath, 2012). The firms combined stable systems which provided continuity with dynamic actions that created resilience in the face of change. Of particular note is that during the entire ten-year study period, while the firms experienced tremendous change, none of them engaged in large-scale downsizing or restructuring. Instead, what they appeared to do was continually adjust their resources and structures as advantages emerged, were exploited, and waned. There was no need for 'change management' as these firms were continuously changing and renewing their capabilities. Innovation took its place in these companies alongside other management disciplines and was managed systematically.

We Would See that Managing Relationships in the Network is a Key Management Skill

In a hypercompetitive environment, the capacity to mobilize the right relationships and alliances quickly becomes critical to a firm's being able to generate the next 'wave' of advantage. But how should firms balance the composition of actors within the network to create a firm-specific competitive advantage? Partnerships are likely to emerge where internal knowledge is weak, but markets are likely to fail (Pisano, 1989). The implication is that networks should be comprised of actors that best compensate for skills lacking in the organization, and that more partnerships will generally translate into more knowledge and potential innovation. However, recent work on organizational learning has challenged this 'alliance as complementarity' view, and suggests a broader composition of partnerships, even if the direct economic benefits are not obvious. Experience partnering with a diverse range of organizations has been shown to lead to even more relationships, enhancing the firm's absorptive capacity and its ability to adapt in new environments. As Powell, Koput, and Smith-Doerr (1996) note, collaboration is emergent, and participation is an 'admission ticket' to more diverse types of collaboration in the future.

Viewing the network as the key unit of analysis also calls attention to the importance of managing external perceptions in a hypercompetitive and uncertain environment. Sociologists have long argued that technologies are 'socially constructed' in that attributes besides the quality of the solution—most prominently, the identity of the inventor—play a major role in the evolution of technologies (Pinch and Bijker, 1987). Connections with prominent and highly reputed organizations lead to more favourable perceptions of the firm's innovation by key actors within the firm's network (Podolny and Stuart, 1995; Stuart and Podolny, 1996), and this in turn leads to even more high-status partners willing to join the network of relationships (i.e. ecosystem) created by the firm. In this view, more partners are not necessarily a strategic advantage and, in fact, can detract from the appeal of a particular innovation if they signal low quality to other parts of the network.

Taken together, the outward shift in the locus of innovation requires managers to dedicate considerable attention to building rich relationships with a diverse set of actors, many of whom (such as competitors or complementors) would not have been considered strategic partners decades ago. But beyond securing partnerships with direct resource or knowledge implications, innovation strategies must take into account the indirect reputational and legitimizing effects that a potential partnership may bring.

Conclusion

We have suggested in this chapter that the advent of hypercompetition challenges the traditional separation between the conception of strategic behaviour of firms in the strategy literature and the processes required to create an ongoing stream of new advantages while simultaneously abandoning exhausted advantages.

Shifting from a primary emphasis on sustainable advantage and an industry level of analysis to temporary advantages and firm- or even unit-level of analysis has implications for the conduct of strategic management research. First, many established research approaches aren't going to shed much light on the phenomenon of transient advantage because they operate at too aggregate a level. Researchers often study industries and firms over long periods of time, using data captured by financial or analyst-driven databases. Transient advantages, in contrast, suggest the need for fine-grained looks at particular firm categories and the competitive moves and counter-moves in each. The idea of a competitive 'arena' is likely to become increasingly important as a central unit of analysis for strategy, where an arena is a particular concatenation of a market segment, an offer and a location (McGrath et al., 1998).

Even as we extol innovation as centrally important to creating a pipeline of advantages, it is worth remembering that we should resist the temptation to regard the consequences of innovation as only positive. Indeed, Rogers many years ago pointed out that scholars seldom pay attention to the negative or unintended consequences of innovations (Rogers, 1983). Recent research suggests that it is entirely possible for innovations with negative effects to be introduced and adopted, much to the regret of firms making those decisions (Greve, 2011). Few consumers would argue that innovations such as CDO's, airline baggage fees, automated phone chains, and the abysmal level of service in general represent positive changes from the perspective of the customer experience. Hypercompetition is not an entirely positive development, even for consumers who theoretically benefit from the increased incentives it creates for companies to compete for their business.

Transient advantage brings the 'resource allocation' school of strategy very much to the forefront once more (Bower, 1970; Christensen and Bower, 1996; Noda and Bower, 1996; Gilbert and Bower, 2002). To return to a traditional concern of strategy—the job of the general manager—increasingly, the task will involve orchestrating competitive arenas that may themselves be at different developmental stages—some just beginning to launch, others already exhausted. Strategy thus has an opportunity of sorts to return to its traditional roots and provide insights to general managers tasked with making these decisions quickly in high-stakes environments.

References

Adner, R. (2012). *The Wide Lens: A New Strategy for Innovation*. New York: Portfolio/Penguin.

Ahuja, G., and Katila, R. (2004). 'Where do Resources Come from? The Role of Idiosyncratic Situations', *Strategic Management Journal*, 25(8/9): 887–907.

Alchian, A. A. (1991). *Rent: The World of Economics*, ed. J. Eatwell, M. Milgate, and P. Newman. New York: W. W. Norton, 591–7.

Altera Corporation (2012). 'Automotive Industry Trends', Available at <http://www.altera.com/corporate/about_us/abt-index.htmlhttp://www.altera.com/end-markets/auto/industry/aut-industry.html#fast>.

Anderson, C. R., and Paine, F. T. (1978). 'PIMS: A Reexamination', *The Academy of Management Review*, 3: 602–612.

Andrews, K. R. (1971). *The Concept of Corporate Strategy*. Homewood, IL: Dow Jones-Irwin.

Ansoff, H. I. (1965). *Corporate Strategy*. New York: McGraw-Hill.

Bain, J. S. (1956). *Barriers to New Competition*. Cambridge, MA: Harvard University Press.

Bain, J. S. (1959). *Industrial Organization*. New York: Wiley.

Barnard, C. (1938). *The Functions of the Executive*. Cambridge, MA: Harvard University Press.

Barney, J. B. (1991). 'Firm Resources and Sustained Competitive Advantage', *Journal of Management*, 17(1): 99–120.

Bettis, R. A. (2012). 'The Search for Asterisks: Compromised Statistical Tests and Flawed Theories', *Strategic Management Journal*, 33(1): 108–13.

Bower, J. L. (1970). *Managing the Resource Allocation Process: A Study of Corporate Planning and Investment*. Boston: Division of Research Graduate School of Business Administration Harvard University.

Brown, K., and Latour, A. (2004). 'Heavy Toll: Phone Industry Faces Upheaval As Ways of Calling Change Fast; Cable, Internet, Wireless Hurt the Value of Old Networks, Threaten a Business Model; Echoes of Railroads' Ordeal', *Wall Street Journal*, 25 Aug.: A1.

Burgelman, R. A., and Grove, A. S. (2007). 'Let Chaos Reign, Then Rein in Chaos—Repeatedly: Managing Strategic Dynamics for Corporate Longevity', *Strategic Management Journal*, 28(10): 965–79.

Burgelman, R., and Valikangas, L. (2005). 'Managing Internal Corporate Venturing Cycles', *Sloan Management Review*, 46(4): 26–34.

Burt, R. S. (2004). 'Structural Holes and Good Ideas', *American Journal of Sociology*, 110(2): 349–99.

Buzzell, R. D., Gale, B. T., and Sultan, R. G. M. (1975). 'Market Share: A Key to Profitability', *Harvard Business Review*, 53(1): 97–106.

Campbell, J. T., Campbell, T. C., Sirmon, D. G., Bierman, L., and Tuggle, C. S. (2012). 'Shareholder Influence over Director Nomination via Proxy Access: Implications for Agency Conflict and Stakeholder Value', *Strategic Management Journal*, 33(12): 1431–51.

Carr, C. (1993). 'Global, National and Resource-Based Strategies: An Examination of Strategic Choice and Performance in the Vehicle Components Industry', *Strategic Management Journal*, 14: 551–68.

Caves, R. E., and Porter, M. E. (1977). 'From Entry Barriers to Mobility Barriers: Conjectural Decisions and Contrived Deterrance to New Competition', *Quarterly Journal of Economics*, 91: 241–62.

Chatterji, A. K., and Fabrizio, K. (2012). 'How do Product Users Influence Corporate Invention?', *Organization Science*, 23(4): 971–87.

Chen, M.-J., and MacMillan, I. C. (1992). 'Nonresponse and Delayed Response to Competitive Moves: The Roles of Competitor Dependence and Action Irreversibility', *Academy of Management Journal*, 35: 359–70.

Chesbrough, H. W. (2003). *Open Innovation: The New Imperative for Creating and Profiting from Technology*. Boston: Harvard Business Press.

Christensen, C., and Bower, J. (1996). 'Customer Power, Strategic Investment, and the Failure of Leading Firms', *Strategic Management Journal*, 17(3): 197–219.

Christensen, C. M. (1997). *The Innovator's Dilemma: When New Technologies Cause Great Firms to Fail*. Boston: Harvard Business School Press.

Cohen, W. M., and Levinthal, D. A. (1990). 'Absorptive Capacity: A New Perspective on Learning and Innovation', *Administrative Science Quarterly*, 35: 128–52.

Collis, D. J. (1991). 'A Resource-Based Analysis of Global Competition: The Case of the Bearings Industry', *Strategic Management Journal*, 12: 49–68.

Craig, T. (1996). 'The Japanese Beer Wars: Initiating and Responding to Hypercompetition in New Product Development', *Organization Science*, 7(3): 302–21.

D'Aveni, R. A., and Gunther, R. E. (1994). *Hypercompetition: Managing the Dynamics of Strategic Maneuvering*. New York: The Free Press.

Dierickx, I., and Cool, K. (1989). 'Asset Stock Accumulation and Sustainability of Competitive Advantage', *Management Science*, 35(12): 1504–13.

Dyer, J. H., and Singh, H. (1998). 'The Relational View: Cooperative Strategy and Sources of Interorganizational Competitive Advantage', *Academy of Management Review*, 23(4): 660–79.

Edward, L., and Hermann Achidi, N. (2006). 'What to Do with the Resource-Based View: A Few Suggestions for what Ails the RBV that Supporters and Opponents Might Accept', *Journal of Management Inquiry*, 15(2): 135.

Einsinger, J. (2012). 'Challenging the Long-Held Belief in "Shareholder Value"', *The New York Times*.

Farrell, J., and Klemperer, P. (2007). 'Coordination and Lock-in: Competition with Switching Costs and Network Effects', *Handbook of Industrial Organization*, 3: 1967–2072.

Ferrier, W. J., Smith, K. G., and Grimm, C. M. (1999). 'The Role of Competitive Action in Market Share Erosion and Industry Dethronement: A Study of Industry Leaders and Challengers', *Academy of Management Journal*, 42(4): 372.

Fleming, L. (2001). 'Recombinant Uncertainty in Technological Search', *Management Science*, 47(1): 117–32.

Foster, R. (2012). 'Creative Destruction Whips through Corporate America', *Innosight Executive Briefing*, February, 10(1): 1–6.

Foster, R., and Kaplan, S. (2001). *Creative Destruction: Why Companies that are Built to Last Underperform the Market—and How to Successfully Transform Them*. New York: Doubleday.

Friedman, M. (1970). 'A Friedman Doctrine: The Social Responsibility of Business is to Increase its Profits', *The New York Times*.

Galunic, D. C., and Rodan, S. (1998). 'Resource Recombinations in the Firm: Knowledge Structures and the Potential for Schumpeterian Innovation', *Strategic Management Journal*, 19(12): 1193.

Gilbert, C., and Bower, J. L. (2002). 'Disruptive Change: When Trying Harder is Part of the Problem', *Harvard Business Review*, 80(5): 94.

Gimeno, J., and Woo, C. Y. (1996). 'Hypercompetition in a Multimarket Environment: The Role of Strategic Similarity and Multimarket Contact in Competitive De-escalation', *Organization Science*, 7(3): 322–41.

Greve, H. R. (2011). 'Fast and Expensive: The Diffusion of a Disappointing Innovation', *Strategic Management Journal*, 32(9): 949–68.

Grindley, P. C., and Teece, D. J. (1997). 'Managing Intellectual Capital: Licensing and Cross-licensing in Semiconductors and Electronics', *California Management Review*, 39(2): 8–41.

Haleblian, J., McNamara, G., Kolev, K., and Dykes, B. J. (2012). 'Exploring Firm Characteristics that Differentiate Leaders from Followers in Industry Merger Waves: A Competitive Dynamics Perspective', *Strategic Management Journal*, 33(9): 1037–52.

Hambrick, D. C. (1990). 'The Adolescence of Strategic Management, 1980–1985: Critical Perceptions and Reality', in J. W. Fredrickson, *Perspectives on Strategic Management*. New York: Harper & Row, 237–53.

Hambrick, D. C., and Chen, M.-J. (2008). 'New Academic Fields as Admittance-Seeking Social Movements: The Case of Strategic Management', *Academy of Management. The Academy of Management Review*, 33(1): 32–54.

Hargadon, A., and Sutton, R. I. (1997). 'Technology Brokering and Innovation in a Product Development Firm', *Administrative Science Quarterly*, 42(4): 716–49.

Hatten, K. J., Schendel, D. E., and Cooper, A. C. (1978). 'Strategic Model of the U.S. Brewing Industry: 1952–1971', *Academy of Management Journal*, 21: 592–610.

Helfat, C. E., and Peteraf, M. A. (2003). 'The Dynamic Resource-Based View: Capability Lifecycles', *Strategic Management Journal*, 24(10): 997.

Helfat, C., and Raubitschek, R. S. (2000). 'Product Sequencing: Co-evolution of Knowledge, Capabilities and Products', *Strategic Management Journal*, 21(10–11): 961–80.

Helfat, C. E., and Winter, S. G. (2011). 'Untangling Dynamic and Operational Capabilities: Strategy for the (N)Ever-Changing World', *Strategic Management Journal*, 32(11): 1243–50.

Henderson, B. D. (1980). 'The Experience Curve Revisited', Boston: The Boston Consulting Group, Perspective No. 229.

Hodgkinson, G. P., and Healey, M. P. (2011). 'Psychological Foundations of Dynamic Capabilities: Reflexion and Reflection in Strategic Management', *Strategic Management Journal*, 32(13): 1500–16.

Hoffman, N. P. (2000). 'An Examination of the "Sustainable Competitive Advantage" Concept: Past, Present, and Future', *Academy of Marketing Science Review*: 4. Available at <http://thoughtleaderpedia.com/Marketing-Library/Sustainable%20Competitive%20Advantage/SustainableCompetitveAdvantage_hoffman04-2000.pdf>.

Huston, L., and Sakkab, N. (2006). 'Connect and Develop', *Harvard Business Review*, 84(3): 58–66.

Ilinitch, A. Y., D'Aveni, R. A., and Lewin, A. Y. (1996). 'New Organizational Forms and Strategies for Managing in Hypercompetitive Environments', *Organization Science*, 7(3): 211.

Jensen, M., and Meckling, W. (1976). 'The Theory of the Firm: Managerial Behavior, Agency Costs and Ownership Structure', *Journal of Financial Economics*, 11: 5–50.

Katz, M. L., and Shapiro, C. (1985). 'Network Externalities, Competition, and Compatibility', *American Economic Review*, 75(3): 424–40.

Kiechel, W. (2010). *The Lords of Strategy: The Secret Intellectual History of the New Corporate World*. Boston: Harvard Business Press.

Kogut, B., and Zander, U. (1992). 'Knowledge of the Firm, Combinative Capabilities and the Replication of Technology', *Organization Science*, 3: 383–97.

Levinthal, D. (1997). 'Adaptation on Rugged Landscapes', *Management Science*, 43(7): 934–50.

Levitt, B., and March, J. G. (1988). 'Organizational Learning', *Annual Review of Sociology*, 14: 319–40.

Lieberman, M. B., and Montgomery, C. B. (1988). 'First Mover Advantages', *Strategic Management Journal*, 9: 41–58.

Lubatkin, M. H., and Lane, P. J. (1996). 'Psst... The Merger Mavens still have it Wrong!', *Academy of Management Executive*, 10(1): 21–39.

McGrath, R. G. (1993). 'The Development of New Competence in Established Organizations: An Empirical Investigation (Ph.D. thesis, University of Pennsylvania).

McGrath, R. G. (1997). 'A Real Options Logic for Initiating Technology Positioning Investments', *Academy of Management Review*, 22(4): 974–96.

McGrath, R. G. (2012). 'How the Growth Outliers do it', *Harvard Business Review*, 90(1): 110–16.

McGrath, R. G. (2013). *The End of Competitive Advantage: How to Keep your Strategy Moving as Fast as your Business*. Boston: Harvard Business Review Press.

McGrath, R. G., Chen, M.-J,. and MacMillan, I. C. (1998). 'Multi-market Maneuvering in Uncertain Spheres of Influence: Resource Diversion Strategies', *Academy of Management Review*, 23(4): 724–40.

MacMillan, I. C. (1982). 'Seizing Competitive Initiative', *Journal of Business Strategy*, 2(4): 43–57.

MacMillan, I. C. (1988). 'Controlling Competitive Dynamics by Taking Strategic Initiative', *Academy of Management Executive*, 2(2): 111–18.

MacMillan, I. C., and Day, D. (1987). 'Corporate Ventures into Industrial Markets: Dynamics of Aggressive Entry', *Journal of Business Venturing*, 2: 19–39.

McNamara, G., Vaaler, P. M., and Devers, C. (2003). 'Same as it Ever Was: The Search for Evidence of Increasing Hypercompetition', *Strategic Management Journal*, 24(3): 261.

March, J. G. (1991). 'Exploration and Exploitation in Organizational Learning', *Organization Science*, 2(1): 71–87.

Mason, E. S. (1939). 'Price and Production Policies of Large-Scale Enterprise', *American Economic Review Supplement*, 29(29): 1.

Mason, E. (1949). 'The Current State of the Monopoly Problem in the U.S.', *Harvard Law Review*, 62: 1265–85.

Miles, R. H., and Cameron, K. S. (1982). *Coffin Nails and Corporate Strategies*. Englewood Cliffs, NJ: Prentice-Hall.

Mitchell, W., Dussauge, P., and Garrette, B. (2002). 'Alliances with Competitors: How to Combine and Protect Key Resources?', *Creativity and Innovation Management*, 11(3): 203–23.

Nag, R., Hambrick, D. C., and Chen, M.-J. (2007). 'What is Strategic Management, Really? Inductive Derivation of a Consensus Definition of the Field', *Strategic Management Journal*, 28(9): 935–55.

Noda, T., and Bower, J. L. (1996). 'Strategy Making as an Iterated Process of Resource Allocation', *Strategic Management Journal*, 17(Special Issue: Evolutionary Perspectives on Strategy): 159–92.

Oliver, C. (1997). 'Sustainable Competitive Advantage: Combining Institutional and Resource-Based Views', *Strategic Management Journal*, 18(9): 697–713.

Penrose, E. (1959). *The Theory of the Growth of the Firm*. Oxford: Oxford University Press.

Pinch, T. J., and Bijker, W. E. (1987). 'The Social Construction of Facts and Artifacts', in W. E. Bijker, T. P. Hughes, and T. J. Pinch (eds), *The Social Construction of Technological Systems*. Cambridge, MA: MIT Press, 17-50.

Pisano, G. P. (1989). 'Using Equity Participation to Support Exchange: Evidence from the Biotechnology Industry', *Journal of Law, Economics, & Organization*, 5(1): 109–26.

Podolny, J. M., and Stuart, T. E. (1995). 'A Role-Based Ecology of Technological Change', *American Journal of Sociology*, 100(5): 1224–60.

Porter, M. (1980). *Competitive Strategy: Techniques for Analyzing Industries and Competitors*. New York: The Free Press.

Porter, M. E. (1981). 'The Contributions of Industrial Organization to Strategic Management', *Academy of Management Review*, 6(4): 609–20.

Powell, W. W., Koput, K. W., and Smith-Doerr, L. (1996). 'Interorganizational Collaboration and the Locus of Innovation: Networks of Learning in Biotechnology', *Administrative Science Quarterly*, 41(1): 116–45.

Powell, W. W., White, D. R., Koput, K. W., and Owen-Smith, J. (2005). 'Network Dynamics and Field Evolution: The Growth of Interorganizational Collaboration in the Life Sciences', *American Journal of Sociology*, 110(4): 1132–205.

Priem, R. L., and Butler, J. E. (2001). 'Tautology in the Resource-Based View and the Implications of Externally Determined Resource Value: Further Comments', *Academy of Management: The Academy of Management Review*, 26(1): 57.

Robins, J. A., and Wiersema, M. (1995). 'A Resource Based Approach to the Multibusiness Firm', *Strategic Management Journal*, 16: 277–99.

Rogers, E. M. (1983). *The Diffusion of Innovations*, 3rd edn. New York: Free Press.

Rosenbush, S., Lowry, T., Crockett, R. O., and Kunii, I. M. (2003). 'VERIZON'S GUTSY BET. (cover story)', *Business Week* (3844): 52–62.

Rothwell, R., et al. (1974). 'SAPPHO Updated-Project SAPPHO Phase II', *Research Policy*, 3(3): 258–91.

Rumelt, R. P. (1987). Theory, Strategy and Entrepreneurship, in Teece, D. J. (ed.) *The Competitive Challenge: Strategies for Industrial Innovation and Renewal*. New York: Harper & Row, 137–58.

Saloner, G. (1991). 'Modeling, Game Theory and Strategic Management', *Strategic Management Journal*, 12: 119–36.

Schendel, D. E., and Hofer, C. W. (1979). *Strategic Management*. New York: Little Brown.

Schmalensee, R. (2012). ' "On a Level with Dentists?" Reflections on the Evolution of Industrial Organization', *Review of Industrial Organization*, 41(3): 157–79.

Schumpeter, J. (1942). *Capitalism, Socialism, and Democracy*. New York: Harper Perennial.

Shapiro, C. (2001). 'Navigating the Patent Thicket: Cross Licenses, Patent Pools, and Standard Setting', *Innovation Policy and the Economy*, 1: 119–50.

Shepherd, W. G. (1982). 'Causes of Increased Competition in the U.S. Economy, 1939–1980', *Review of Economic Statistics*, 64: 613.

Stout, L. (2012). *The Shareholder Value Myth: How Putting Shareholders First Harms Investors, Corporations, and the Public*. San Francisco: Berrett-Koehler Publishers.

Stuart, T. E., and Podolny, J. M. (1996). 'Local Search and the Evolution of Technological Capabilities', *Strategic Management Journal*, 17: 21–38.

Teece, D. J., Pisano, G., and Shuen, A. (1997). 'Dynamic Capabilities and Strategic Management', *Strategic Management Journal*, 18(7): 509–33.

Troianovski, A. (2012). 'Cellphones are Eating the Family Budget', *Wall Street Journal*. New York: Dow Jones Incorporated.

von Hippel, E. (1986). 'Lead Users: A Source of Novel Product Concepts', *Management Science*, 32(7): 791–805.

von Hippel, E. (1988). *The Sources of Innovation*. Cambridge, MA: MIT Press.

Ward, S. (2009). 'You Can Really Hear Them Now', *Barron's*, 89(5): 24.

Wernerfelt, B. (1984). 'A Resource-Based View of the Firm', *Strategic Management Journal*, 5: 171–80.

Whitman, M. N. (1999). *New World, New Rules: The Changing Role of the American Corporation*. Boston: Harvard Business School Press.

Williamson, O. (1975). *Markets and Hierarchies: Analysis and Antitrust implications*. New York: Free Press.

CHAPTER 21

....

BUSINESS MODEL INNOVATION

....

LORENZO MASSA AND
CHRISTOPHER L. TUCCI

INTRODUCTION

In the past fifteen years, the business model (BM) has become an increasingly important unit of analysis in innovation studies. Within this field, a consensus is emerging that the role of the BM in fostering innovation is twofold. First, by allowing managers and entrepreneurs to connect innovative products and technologies to a realized output in a market, the BM represents an important *vehicle* for innovation. Second, the BM may be also a *source of innovation in and of itself*. It represents a new dimension of innovation, distinct, albeit complementary, to traditional dimensions of innovation, such as product, process or organizational.

This chapter serves to introduce the notion of *Business Model Innovation* (BMI) and has four main objectives, to: (1) clarify the origins and notion of the BM; (2) organize the literature on BMI around emerging literature streams; (3) offer an overview of the various tools that have been proposed in supporting managers and entrepreneurs in dealing with BMI; and (4) offer a discussion of the principal managerial challenges related to managing BMs and BMI.

As a starting point, we seek to introduce and clarify the notion of the BM. We review the received literature, highlight the origins of the BM and clarify the nature of the construct. Next, we introduce the notion of BMI and define it as the activity of designing—that is, creating, implementing and validating—a new BM and suggest that the process of BMI differs if an existing BM is already in place vis-à-vis when it is not. Accordingly, for analytical purposes, we distinguish between *BM design* in newly formed organizations, and *BM reconfiguration* in incumbent firms. We note that these literatures tend to focus on the *antecedents* and *mechanisms* at the background of BMI. We suggest that a new literature is emerging focusing on the *consequences* of BMI, and pointing to the role of BMI in unlocking the private sector potential to contribute to solving environmental and social issues. We offer a review. Finally, we analyse various tools and perspectives that seek to make sense of the BM and support BMI and business modelling (the set of activities supporting BM representation, sense-making and strategic planning for BMI). We highlight the complementary nature of different perspectives and organize them in a conceptual meta-framework. In doing so, we also provide evidence of the current use of BM-related ideas in practice. We conclude with a discussion of some of most salient managerial challenges related to managing BMs and BMI.

What is a Business Model? Definition and Emergence of the Concept

In the past several years, interest in the concept of BMs has virtually exploded, attracting the attention of managers and academics alike. Zott and colleagues (Zott, Amit, and Massa, 2011) searched for the use of the term *business model* in general management articles and noted a dramatic increase in the incidence of the term in the fifteen-year period between 1995 and 2010, in parallel with the popularization and broad diffusion of the Internet.

Teece (2010) notes that BMs have been an integral part of economic behaviour since pre-classical times. Indeed, firms have always operated according to a *business model* but, until the mid 1990s, firms traditionally operated following similar logics, typical of the industrial firm, in which a product/service—typically produced by the firm (and its suppliers)—is delivered to a customer from which revenues are collected. Even if instances of firms and organizations adopting innovative BMs have been recognized in business history (cf. Osterwalder and Pigneur, 2010), it is only recently that the scale and speed at which innovative BMs are transforming industries and, indirectly, civil society, has attracted the attention of scholars and practitioners. Thus, BMs seek to make sense of these novel forms of 'doing business'. According to Magretta (2002), the BM is a story that answers Peter Drucker's age old questions: (1) who is the customer, (2) what does the customer value, (3) how do we make money in this business, (4) and what is the economic logic that explains how we can deliver value to customers at an appropriate cost? The emergence of novel logics employed by firms in doing business as they go to market has increasingly popularized the notion of BM.

Several scholars agree that the Internet, together with related advances in information and communication technologies (ICTs), acted as catalysis for BM experimentation and innovation (e.g, Timmers, 1998; Amit and Zott, 2001; Afuah and Tucci, 2001), opening up new opportunities for organizing business activities. Entire industrial sectors have evolved along radically new trajectories of innovation and offered new logics of value creation not seen in recent business history.

Casadesus-Masanell and Ricart (2010) have observed that two other phenomena have been accompanied by considerable innovation in the way firms 'do business'. These are: (1) the advent of post-industrial technologies (Perkmann and Spicer, 2010), and (2) efforts by the corporate sector to enter new markets in developing or underdeveloped countries and reach customers at the Bottom of the Pyramid[1] (BoP) (Prahalad and Hart, 2002; Prahalad, 2005). A third one, related to BoP, is represented by 'sustainability'. Let us discuss each of these in turn.

Scientific and technological advances in so-called post-industrial technologies (e.g., software or biotechnology) have been accompanied by a surge of organizational architectures and governance structures which radically differ from those observed in traditional manufacturing organizations (e.g., Bonaccorsi, Giannangeli, and Rossi, 2006). Firms, for example, have emerged that host and maintain IT applications across the Internet, offering software as a service (rather than a product: Susarla, Barua, and Whinston, 2009). The Open Source software movement has been accompanied by the emergence of new governance structures (Bonaccorsi et al., 2006) and novel forms of *collaborative entrepreneurship* (Miles, Miles, and Snow, 2006). Similarly, the biotechnology sector has been the locus of considerable BMI (e.g., Pisano, 2006), with firms emerging that focus on specific tasks and relative services along the product development value chain (Konde, 2009). Innovative BMs are observed in other sectors as well. Firms such as ARM (semiconductors), Dolby (sound systems), CDT (electronic displays) or Plastic Logic (plastic materials) all have specialized in the management of intellectual property and operate in the 'market for ideas' (see Chapter 12 by Gambardella, Giuri, and Torrisi) by licensing the rights of their innovative technologies and solutions rather than commercializing the products themselves. Whether post-industrial technologies can be properly considered an antecedent of BM innovation vis-à-vis other explanations such as the intellectual property revolution (Pisano, 2006), the disintegration of the value chain observed in many industries, or the institutionalization of Open Innovation as a way to organize innovation activities outside the traditional boundaries of the firm (see Chapter 22 by Alexy and Dahlander), remains an unexplored research question. Certainly, however, post-industrial technologies have been accompanied by the emergence of novel ways to conduct business.

Opportunities to address economic needs at the BoP in emerging markets (Ricart, Enright, Ghemawat, Hart, and Khanna, 2004) have also pointed researchers and practitioners towards the systematic study of BMs. The core argument in the BoP literature is that the vast, untapped market of the world's poor represents a large opportunity for companies to both serve customers and make a profit. However, business opportunities at the BoP challenge conventional ways of doing business. Due to the fundamentally different social, economic, and cultural environments that characterize emerging markets, companies are urged to rethink every step in their supply chain and develop novel BMs (Prahalad and Hart, 2002). In addition, existing models may have limited applicability and need to be adapted (Seelos and Mair, 2007). Chesbrough, Ahern, Finn, and Guerraz (2006), studying product deployment in the developing world, highlight that while the 'right' product design is a necessary condition for penetrating low income markets, those companies that ultimately succeed in generating commercially sustainable operations are those that put in place the right BM. These BMs play a crucial role in creating key elements, such as distribution channels, supplies and sales channels necessary for the successful execution of business transactions. Thus, enterprises that aim at reaching the BoP constitute an important source of BM innovation (Prahalad, 2005).

While the study of BMs has traditionally focused on business activities, the emergence of new organizational architectures designed for purposes other than economic profits, such as solving social problems and sustainability issues, has started to attract the attention of scholars studying BMs (Seelos and Mair, 2007; Yunus, Moingeon and Lehmann-Ortega, 2010). For example, Nobel laureate Mohamed Yunus has been pioneering the concept of microfinance and designed a novel organization, the Grameen Bank, whose main purpose is the eradication of poverty (cf. Yunus et al., 2010), a critical issue in the discussion on sustainability (cf. WCED, 1987). Scholars increasingly employ the term BM in referring to the way such organizations operate and in capturing instances of value creation whose nature is not necessarily economic (see Chapter 16 by Lawrence et al.).

To conclude this section, the BM is an elusive concept allowing for considerable *interpretative flexibility* (Bijker, Hughes, and Pinch, 1987). Zott et al. (2011) have recently reviewed the most recent literature on the topic and noted that various conceptualizations of the BM exist that often serve the scope of the particular phenomenon of interest to the researcher. There are, however, some emerging common themes that act as a common denominator among the various conceptualizations of the BM that have been provided. In particular scholars seem to recognize—explicitly or implicitly—that the BM is a 'system level concept, centered on activities and focusing on value' (2011: 1037). It emphasizes a systemic and holistic understanding of how an organization orchestrates its system of activities for value creation. In addition, they noted that the phenomenon of value creation as depicted by the BM typically occurs in a value network (cf. Normann and Ramirez, 1993; Parolini, 1999), which can include suppliers, partners, distribution channels, and coalitions that extend the company's resources. Therefore they suggest that the BM also introduces a new unit of analysis in addition to the product, firm, industry, or network levels. Such a new unit of analysis is nested between the firm and its network of exchange partners.

These considerations suggest that, at first glance, the BM may be conceptualized as depicting the rationale of how an organization (a firm or other type of organization) creates, delivers, and captures value (economic, social, or other forms of value) in relationship with a network of exchange partners (Afuah and Tucci, 2001; Osterwalder, Pigneur, and Tucci, 2005; Zott et al., 2011). This broad definition is elastic with respect to the nature of the value created, and serves the scope of introducing the topic of BMI.

BUSINESS MODELS AND INNOVATION

The literature at the intersection of the BM concept and the domain of innovation has advanced two complementary roles for the BM in fostering innovation. First, BMs allow innovative companies to commercialize new ideas and technologies. Second, firms can also view the BM as a source of innovation in and of itself, and as a source of competitive advantage.

The first view is mainly rooted in the literature on technology management and entrepreneurship. It is recognized that innovative technologies or ideas per se have no economic value. It is through the design of appropriate BMs that managers and entrepreneurs may be able to unlock the output from investments in R&D and connect it to a market. By allowing the commercialization of novel technologies and ideas, the BM becomes a vehicle *for* innovation. Xerox invented the first photocopy machine, but the technology was too expensive and could not be sold. Managers at Xerox solved the problem by leasing the machine, inventing a new BM for doing so. In this view, the BM is a manipulable device that mediates between technology and economic value creation (Chesbrough and Rosenbloom, 2002).

The second view is that the BM represents a new dimension of innovation itself, which spans the traditional modes of process, product, and organizational innovation. This new dimension of innovation may be source of superior performance, even in mature industries (Zott and Amit, 2007). Dell in the computer industry, Southwest in the airline industry, or Apple with iPod and iTunes in the music industry, just to mention a few known cases, secured impressive growth rates and outperformed the competition by establishing innovative BMs.

This suggests that firms can compete through their BMs (Casadesus-Masanell and Ricart, 2007). Novel BMs may be a source of disruption (Christensen, 1997), changing the logic of entire industries and replacing the old way of doing things to become the standard for the next generation of entrepreneurs to beat (Magretta, 2002). According to Chesbrough, BMI may have more important strategic implications than other forms of innovation, in that 'a better BM will beat a better idea or technology' (2007: 12).

Building on the literature at the nexus between BMs and innovation, we propose that BMI may refer to (1) the design of novel BMs for newly formed organizations, or (2) the reconfiguration of existing BMs. We refer to the first phenomena by employing the term *business model design (BMD)*, which refers to the entrepreneurial activity of creating, implementing and validating a BM for a newly formed organization. We use the term *business model reconfiguration (BMR)* to capture the phenomenon by which managers reconfigure organizational resources (and acquire new ones) to change an existing BM. Thus, the process of reconfiguration requires shifting, with different degrees of radicalism, from an existing model to a new one. We contend that both phenomena are change phenomena and could lead to BMI. Xerox's design of a leasing model to market the Xerox 914 in the late 1950s or Gillette's design of the 'Razor and Razor Blade' model can

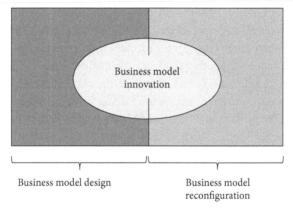

FIGURE 21.1 Business model innovation as a subset of business model design and reconfiguration

be considered innovative designs. Indeed, they led to the emergence of new BMs not seen before. Not all design or reconfiguration efforts will necessarily lead to BMI, however. To be a source of BMI, the output of design or reconfiguration activities should be characterized by some degree of novelty or uniqueness. In other words, while in principle BMI *may* result as the product of design and/or reconfiguration of new and existing BMs, respectively, it constitutes a subset of the larger set comprising the whole product of BM design and reconfiguration activities (see Figure 21.1).

While sharing the potential for the same outcome (namely BMI), reconfiguration and design are two distinct activities that imply important differences. For example, because reconfiguration assumes the existence of a BM, it involves facing challenges that are idiosyncratic to existing organizations, such as organizational inertia, management processes, modes of organizational learning, modes of change, and path dependent constraints in general, which may not be an issue in new firms. On the other hand, newly formed organizations may face other issues such as considerable technological uncertainty, lack of legitimacy, lack of resources and, in general, liability of newness, which do influence the design and validation of new BMs (cf. Aldrich and Auster, 1986; Bruderl and Schussler, 1990). Because of these differences and for analytical purposes, we treat the above phenomena separately in the next two sections.[2]

BUSINESS MODEL DESIGN

As mentioned above, BMD refers to the very first instance of a new BM and is usually associated with entrepreneurial activity. The process of BMD could be considered a process of entrepreneurial venture creation involving the design of the content, structure, and governance of the transactions that a firm performs in cooperation with a network of exchange partners so as to create and capture value (Amit and Zott, 2001). It involves

traditional entrepreneurial activities such as internally and externally stimulated opportunity recognition, organization creation, linking with market (Bhave, 1994), and also
the design of boundary spanning organizational arrangements (Zott and Amit, 2007).
According to Zott and Amit (2007), the latter is a critical feature of BMD in that 'a BM
elucidates how an organization is linked to external stakeholders, and how it engages
in economic exchanges with them to create value for all exchange partners' (2007: 181).
Thus, BMD is concerned with traditional entrepreneurial choices (product/market mix,
organizational design, control systems, etc.) as well as the design of a boundary spanning activity system, so as to link an offering (technology or service) to a realized output
market. In a nutshell, BMD includes designs that take place across as well as within firms
(McGrath, 2010).

Uncertainty associated with the viability of new BMs may be considerable.
Uncertainty arises not only because of entrepreneurs' inability to predict customers'
response to their offering, future market conditions and dynamics, but also because of
the computational and dynamic complexity associated with BM planning and design.
Computational complexity arises because of the large number of logically possible
combinations between BM components (Afuah and Tucci, 2001), activities (Zott and
Amit, 2010) and/or choices (Casadesus-Masanell and Ricart, 2010). Dynamic complexity arises because of the non-linear interdependencies—including delays and feedback
loops—between BM components, activities, and/or choices. Both computational as well
as dynamic complexity increase uncertainty surrounding BMD. Even if it were possible
to detect future trends and changes, uncertainty would not be entirely eliminated, only
reduced.

Uncertainty affects modes of BMD and associated entrepreneurial tasks. According
to McGrath (2010), unlike traditional concepts in entrepreneurship, such as business
planning and business plan design, 'strategies that aim to discover and exploit new models must engage in significant experimentation and learning' (2010: 247). BMs cannot
be fully planned *ex ante*. Rather, they take shape through a discovery-driven process;
this process places a significant premium on experimentation and prototyping (Sosna,
Trevinyo-Rodríguez, and Velamuri, 2010; McGrath, 2010). Hayashi (2009) noted that
many companies have had original BMs that did not work. This does not necessarily
imply failure if companies are able to shift to *Plan B*. Hayashi proposes that in order
to shift to plan B and 'find' the right business model, managers and entrepreneurs
should engage in experimentation and challenge their initial assumptions. Investigating
numerous 'what if' questions may be a useful strategy. The discovery-driven nature of
BMs also affects the effectiveness of different design and planning approaches. Financial
tools that make sense in an experimental world (e.g. real-options reasoning) may be
more appropriate than more deterministic ones (e.g. projected economic value added
and net present value) in supporting BMD (McGrath, 2010).

While many new BMs may fail before a viable model is 'found', these (new BMs) may
be a source of abnormal returns. As Ireland, Hitt, Camp, and Sexton (2001) note, entrepreneurs are often interested in finding fundamentally new ways of doing business
and work on new models that have the potential to disrupt an industry's competitive

rules. According to McGrath (2010), BM disruption may occur following Christensen's model of disruptive innovation. At the beginning, these new models are more like experiments than proven business ideas and may not attract the scrutiny of incumbent firms. Newly formed ventures employing novel BMs often operate in market niches, serve customers that incumbents do not serve, and at price points they would consider unattractive, and rely on novel resources that are not necessarily under the control of incumbents. The latter may ignore the threats coming from innovative BMs.[3] And entrants could progressively experiment with their businesses and find disruptive channels.

BUSINESS MODEL RECONFIGURATION

The increasing consensus that BMI is key to firm performance (e.g. Ireland et al., 2001; Chesbrough, 2007; Johnson, Christensen, and Kagermann, 2008) has brought scholars working on the BM to focus on issues related to BM renewal and innovation in incumbent firms.

Considerations of issues related to BMI in incumbent firms were already present in Chesbrough and Rosenbloom's study (2002) of the Xerox Corporation and its research center at Palo Alto (PARC). According to the authors, the BM as an *heuristic logic* might act as a mental map, which mediates the way business ideas are perceived by filtering information as valuable or not. This filtering process within a successful established firm is likely to preclude the identification of models that differ substantially from the firm's current BM. In its cognitive dimension, the BM concept is similar to Prahalad and Bettis' (1986) notion of a *dominant logic*. The dominant logic represents prevailing wisdom about how the world works and how the firm competes in the world. It can act as a filter of information, preventing managers from seeing opportunities and removing certain possibilities from serious consideration, when they fall outside of the prevailing logic and driving firms into a *dominant logic trap* (Chesbrough, 2003).

Bouchikhi and Kimberly (2003) have referred to a similar phenomenon as the *identity trap*. In their view, an organization's identity can become a trap when it so constrains strategic options that the organization cannot cope effectively with a changing environment. Attempts to change that are in conflict with this core identity are often doomed to failure. Chesbrough (2010) suggests two types of barriers to BMI in existing firms. The first type of barrier is structural. Barriers exist in terms of conflicts with existing assets and BMs (i.e. inertia emerges because of the complexity required for the reconfiguration of assets and operational processes). The second type of barrier is cognitive. It is manifested by the inability of managers who have been operating within the confines of a certain BM to understand the value potential in technologies and ideas that do not fit with the current BM.

Three tools are suggested that could help to overcome these barriers (Chesbrough, 2010). The first consists of constructing maps of BMs to clarify the processes underlying them which then become a source of experiments to consider alternative combinations of the processes. The second involves conferring authority for experimentation within the organizational hierarchy. The third is experimentation itself. Experimentation is conceptualized as a process of discovery aimed at gaining cumulative learning from (perhaps) a series of failures before discovering a viable alternative to the BM. Sosna et al. (2010) have analysed the case of a Spanish family-owned dietary products business facing a threat to their BM of obsolescence from unforeseen external changes. The company was able to successfully reconfigure its BM thanks to experimentation, evaluation and adaptation—a trial and error learning approach—involving all echelons of the firm.

Giesen, Berman, Bell, and Blitz (2007) have proposed that BMI in incumbent firms can be classified into three groups: (1) industry model innovation, which consists of innovating the industry value chain by moving into new industries, redefining existing industries, or creating entirely new ones; (2) revenue model innovation, which represents innovation in the way revenues are generated, for example through reconfiguration of the product-service value mix or new pricing models; and (3) enterprise model innovation, changing the role a firm plays in the value chain, which can involve changes in the extended enterprise and networks with employees, suppliers, customers, and others, including capability/asset configurations. They analyse each type of innovation with respect to firm performance, and report two key findings: (1) each type of BMI can generate success, and (2) innovation in enterprise models focusing on external collaboration and partnerships is particularly effective in older companies relative to younger ones. Zott and Amit (2010), who view the BM as a system of boundary spanning interdependent activities, have built on their decade-long research program on the BM and recently proposed that managers can fundamentally innovate a BM in three ways: by (1) adding new activities, (2) linking activities in novel ways, or (3) changing which parties perform an activity (Amit and Zott, 2012). In other words, from a managerial standpoint, BMI consists of innovating the content (i.e. the nature of the activities), the structure (i.e. linkages and sequencing of activities) or the governance (the control/responsibility over an activity) of the activity system between a firm and its network (Zott and Amit, 2010).

To become BM innovators, companies need to create processes for making innovation and improvements (Mitchell and Coles, 2003). Doz and Kosonen (2010) have proposed a leadership agenda for accelerating BM renewal. To overcome the rigidity that accompanies established BMs, companies should be made more agile, which can be achieved by developing three meta-capabilities: strategic sensitivity, leadership unity, and resource flexibility. Doz and Kosonen point to the importance of the top management team to achieve collective commitment for taking the risks necessary to venture into new BMs and abandon old ones. Santos, Spector, and Van der Heyden (2009) have proposed a theory of BMI within incumbent firms in which they emphasize the importance of the behavioural aspects involved through mutual engagement and organizational justice. BMI, they argue, should not only consider the structural aspects of the formal organization (typically activity sets), but should also focus on informal organizational dynamics.

More recently, Bock, Opsahl, and George (2010) have linked the research on the BM with the notion of strategic flexibility (Shimizu and Hitt, 2004) and proposed that firms engage in BMI to gain strategic flexibility by enhancing capabilities to respond to environmental complexity while decreasing formal design complexity.

The fragmented and young literature on BMR in incumbent firms implicitly offers a snapshot of the theoretical richness and challenges associated with studying the phenomenon as well as with carrying out the managerial tasks associated with the process of reconfiguration. Johnson et al. (2008), perhaps not surprisingly, have noted that during the past decade of the major innovations within existing corporations 'only a precious few have been business model related' (2008: 52). BMR may well represent an extension of what Henderson and Clark (1990) initially conceived as 'architectural' innovation, that is, complex innovations that require a systemic reconfiguration of existing organizational and technological capabilities. Indeed, BMR is a complex art. As Teece (2007) notes, it requires 'creativity, insight and a good deal of customer-competitor and supplier information and intelligence' (2007: 1330). Additional complexity is added in incumbent firms by the existing repertoire of capabilities that constrain managers' ability to innovate the BM, either blinding it 'from seeing novel opportunities to innovate or acting upon those opportunities when they see them' (Pisano, 2006: 1126).

BUSINESS MODELS AND SUSTAINABILITY

While studies of the BM have traditionally emphasized its importance for firms' success, a new literature is emerging that studies the role of BMI in promoting sustainability (cf. WCED, 1987), analysing BMI from the point of view of its consequences in terms of either corporate social and environmental impact or as a strategic implication for sustainability (i.e. a way to align firms' search for profits with innovations that would ultimately benefit society and help solve sustainability issues—see Chapter 15 by Berkhout).

Firms could create value for sustainability in two ways: (1) by adopting more sustainable practices and processes that would reduce (or prevent the occurrence of) 'end-of-pipe' negative impacts (for example, reducing energy, water consumption and material intensity or social problems such as work place stress); or (2) by engineering and marketing new technologies that would help solve sustainability problems (for example, renewable energies, electrical vehicles [EVs], or green materials). In other words, value for sustainability may exist in a firm's practice(s) or in a firm's product(s), or both.

While there are different strategic alternatives along the product-practice mix to develop solutions for sustainability issues and improve corporate sustainability performances, market externalities of various forms could prevent profit-seeking firms from fully embracing sustainability and dilute the effectiveness of their initiatives. Activities that have an adverse environmental impact (such as pollution) or a negative social impact (such as exploitation of labour in marginalized and disadvantaged groups) are not fully internalized in the costs of enterprises' products/services (Cairncross, 1993).

Accordingly, firms wanting to improve their environmental and social performance face a structural constraint (i.e. the risk associated with the implementation of sustainability when the market does not reward sustainability initiatives). Similarly, in the absence of appropriate government incentives or market regulation, green technologies may be more expensive than traditional ones—for decades an impediment to the market diffusion of certain technologies related to renewable energies.

Another problem is related to a different type of network externality in complex technological systems. Many technologies provide no value for customers unless necessary complements are also available and this problem applies to traditional and sustainable technologies alike. EVs are of no value if there is no necessary complementary technology (e.g. batteries) or infrastructure such as battery charging stations. The same two- (or three-) sided market argument can be reversed. For example, developing an infrastructure for EVs makes no economic sense if there is no available technology for producing reliable and cost-efficient EVs. The market diffusion of greener technologies may be hampered because firms may not control the full technological architecture necessary to realize the value of a technology.

Some authors have suggested that by innovating their BM, firms could overcome these barriers and make profits while benefiting the environment. For example, service-based BMs (selling a service rather than a product) could contribute to aligning firms' search for profitability with innovating for sustainability. For example, when the carpet company Interface shifted from selling carpets to a 'floor covering service' (Lovins, Lovins, and Hawken, 1999), it started to research, design, and manufacture more recyclable carpets. Under the new BM, when the carpets become worn, Interface replaces them and re-introduces the old ones back into the supply chain; if carpets are highly recyclable, the firm profits from this operation while benefiting the environment. The carpets themselves are eventually designed as tiles so that only consumed parts need to be replaced. The recyclable, modular carpets significantly reduce material and energy consumption, allowing Interface to deliver a better service that costs far less to create and capture the value arising from the new operations (Lovins et al., 1999). Offering services rather than products, and working innovative pricing strategies and novel revenue streams, would also help firms marketing more expensive green technologies and spread their market adoption. As Wüstenhagen and Boehnke (2006) have noted, 'Given the capital intensity of sustainable energy technologies . . . reducing the upfront cost for consumers is one of the key concerns in marketing innovation in this sector' (2006: 256). BMs based on leasing or contracting, or a mix of products/services may represent a solution to the problem (Wüstenhagen and Boehnke, 2006). Finally, new BMs could also help overcome issues of strategic complementarities and solve coordination issues. Better Place, the global provider of EV networks and services, worked to accelerate the transition to sustainable transportation by facilitating the market diffusion of EVs. The company's BM was not based on EV manufacturing; rather, it was based on alliances with EV manufacturers, utility companies, governments, battery manufacturers, and others to produce a market-based transportation infrastructure that would support EV diffusion. By positioning itself upstream in the value system, and by orchestrating the network with a unique BM, the company attempted to solve two-sided market issues in sustainable transportation. However, Better Place's bankruptcy in May 2013 demonstrates some of the complexity associated with developing new BMs for sustainability.

Business Model-related Ideas: The Theory and Practice of BMI

The BM is a systemic and conceptually rich construct, involving multiple components, several actors (boundary spanning) and complex interdependencies and dynamics. Because of that, the managerial cognitive effort required to visualize and explore possibilities for BMI as well as the effort for orchestrating (implementing and managing) the architecture of innovative BMs may be considerable.

Awareness of the complexities associated with BM cognition—description of existing BMs or design of new ones—coupled with the increasing relevance of BMs and business modelling for practice (cf. Zott et al., 2011), have led academics and practitioners to propose several avenues and tactics in support of BMI. Different tools such as perspectives, frameworks, and ontologies have been proposed that employ a mix of informal textual, verbal, and ad hoc graphical representations. These tools ascribe, with varying degrees, to three core functions at the nexus between the theory[4] and practice of BMI. First, they offer a *'reference language'* that fosters dialogue, promotes common understanding, and supports collective sense-making (cf. Amit and Zott, 2012). Second, by offering scaled-down simplified representations of BMs, they allow for graphical representations that simplify cognition and offer the possibility of virtually experimenting with BMI (for example, by supporting the formulation and elaboration of important 'what if' questions and the evaluation of strategic alternatives: Osterwalder and Pigneur, 2010). Third, they offer representations—both graphical as well as verbal—that allow managers and entrepreneurs to articulate and instantiate the value of their venture and to support the engagement of external audiences so as to gain legitimacy, activate resources, and foster action. We note that different tools and perspectives tend to emphasize certain functions while overlooking others. For example, the strength of certain perspectives resides in their simplicity and parsimony. As such, these perspectives are particularly effective in supporting collective sense-making around a BM. Other perspectives are more articulated; their development may be slightly more arduous but allow for a better appreciation of the dynamics occurring between the various components of a BM (cf. Casadesus-Masanell and Ricart, 2007, 2010).

More broadly, we note and illustrate in Figure 21.2 that tools supporting BMI could be structured into several levels of decomposition with varying depth and complexity depending on the degree to which they abstract from the reality they aim to describe.[5]

At the highest level of abstraction is a view of the BM as a *narrative* (Perkmann and Spicer, 2010). According to Magretta (2002), the BM is a story, a verbal description of how an enterprise works. It should be noted that BM narratives not only entail a descriptive function, but also a normative one. According to Brown (2000), narratives represent an important way in which people seek to infuse ambiguous situations with meaning and persuade sceptical audiences that their account of reality is believable. Perkmann and Spicer (2010) have suggested that because of their forward-looking character, BM narratives play an important role in inducing expectations among interested constituents about how a business's future might play out. Narratives of the BM can be constructed by managers and entrepreneurs and used not only to simplify cognition, but also as a communicative device that could allow achieving various goals, such as persuading external audiences, creating a sense of legitimacy around the venture (for example, by drawing analogies between a venture's BM and the BM of a successful firm) or guiding social action (for example, by focusing attention on what to consider in decision-making and instructing how to operate).

The recognition of patterns in the structure of BMs has led to the introduction of typologies and *BM archetypes*. An archetype can be understood as an ideal example of a type, in this case a BM. A well-known example is the *Freemium* BM, adopted by firms such as Acrobat: its core logic lies in delivering a basic version of the product for free and charging for a premium version. Gillette popularized what today is known as the *Razor and Razor Blade* BM, which rests on 'selling cheap razors to make customers buy its rather expensive blades' (Zott and Amit, 2010: 218). This model is now popular in other industries where products such as printers (and cartridges) or game consoles (and software games) are brought to market relying on a similar logic. Archetypes are often presented with an identifying label (a 'title' that identifies the BM type) followed by a short description of the core essence of the BM. Archetypes perform several functions, including offering descriptions of 'role models', that is, models to be followed and imitated (Baden-Fuller and Morgan, 2010).

While narratives and archetypes may serve several important purposes, they tend to be difficult to manipulate and manoeuvre (e.g. it is difficult to evaluate the likely consequences of changes in one part of the BM on the entire system on the basis of a narrative or an archetype). Higher descriptive accuracy, and perhaps a more rigorous approach to structuring and organizing plans for BMI, are offered by *graphical frameworks* of the BM, which are conceptualization and formalization of the BM obtained by enumerating, clarifying and representing its essential components (see Figure 21.2). A popular example among managers and practitioners is represented by the Business Model Canvas[6] (Osterwalder and Pigneur, 2002). The Business Model Canvas offers a scaled-down representation of the generic BM that is obtained by enumerating and visualizing what the authors consider to be the nine critical components of a BM. Similarly, Johnson and colleagues (Johnson, Christensen, and Kagermann, 2008; Johnson, 2010) have proposed a simple framework comprising four interdependent elements; customer value proposition, profit formula, key resources and key processes. By focusing on these elements the framework offers a synthetic 'representation

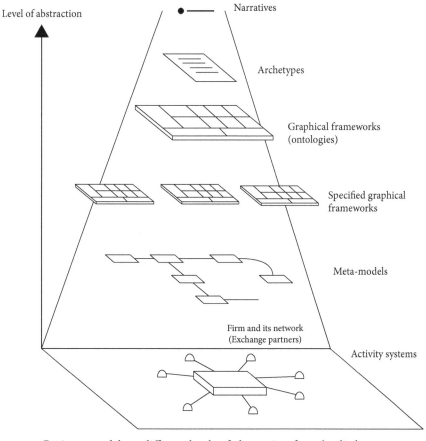

Level of abstraction

Narratives

Archetypes

Graphical frameworks
(ontologies)

Specified graphical
frameworks

Meta-models

Firm and its network
(Exchange partners)

Activity systems

FIGURE 21.2 Business models at different levels of abstraction from 'reality'

of how a business creates and delivers value, for both the customer and the company'
(Johnson, 2010: 22).

We contend that the power of frameworks and archetypes, and perhaps the explanation of their popularity among practitioners, stands in their simplicity and parsimony,
which, however, come at the expense of descriptive depth. In particular, frameworks
and archetypes have shortcomings in their inability to offer a full account of the dynamic
aspects associated with a particular BM. *Meta-models*[7] of the BM may help to overcome
this limitation. Casadesus-Masanell and Ricart (2010) have built on system dynamics
(Sterman, 2000) and offered a way to conceptualize and represent BMs based on choices
and consequences, and on an evaluation of the degree to which consequences are flexible vs. rigid (an important aspect to consider in dealing with BM reconfiguration).
Causal loops (both damping and self-reinforcing) support understanding of how the
architecture of choices drives the overall behaviour of a BM and leads to a configuration
of consequences. This perspective allows for a more fine-grained description of existing BMs supporting the use of 'theories' to describe and understand the link between
choices and likely consequences.

Gordijn and Akkermans (2001) have proposed a conceptual modelling approach that they call the 'e3-value ontology', designed to help define how economic value is created and exchanged within a network of actors. This modelling technique takes a value viewpoint, unlike other traditional modelling tools that take either a business process viewpoint (typical of operations management) or a system architecture viewpoint (typical of the information systems literature). The proposed meta-model borrows concepts from the business literature such as actors, value exchanges, value activities, and value objects, and uses these notions to model networked constellations of enterprises and end-consumers who create, distribute, and consume things of economic value.

In a similar vein, Zott and Amit (2010) have proposed an activity system perspective for supporting the design of new BMs. This perspective relies on an understanding of the BM as a system of interdependent activities (rather than choices and consequences) centered on a focal firm and including those conducted by the focal firm, its partners, vendors or customers, and so on. As such, it allows describing and conceptualizing BMs with considerable depth and accuracy. According to the authors, 'an activity in a focal firm's BM can be viewed as the engagement of human, physical and/or capital resources of any party to the BM (the focal firm, end customers, vendors, etc.) to serve a specific purpose toward the fulfillment of the overall objective' (2010: 217). To better understand the BM as a set of interdependent activities, Zott and Amit differentiate between design elements (i.e. content, structure, and governance) and design themes (efficiency, novelty, complementarities, and lock-in). Design elements comprise the selection of activities (content), the sequencing between them (structure) and choices concerning *who* performs them (governance) within the network. Taken together, design elements comprise the *infrastructural logic* of a BM's architecture. In addition, managers could structure the activity system around different *design themes*. For example, 'efficiency-centred' design (with efficiency being a design theme) refers to how firms use their activity system design to aim at achieving overall greater efficiency through reducing transaction costs. Other design themes are 'novelty' (innovation in the content, structure, or governance of the activity system), 'lock-in' (BM whose central feature is the ability to keep third parties attracted as a BM participant) or 'complementarities' (bundling activities within a system so as to produce more value than running activities separately).

MANAGING BUSINESS MODELS

Challenges associated with managing BMI go beyond the complexities related to managerial cognition and sense-making. While BMI has the potential for transformative growth and exponential returns for the innovator, it is a highly risky move that may involve changing the entire architectural configuration of a business. Accordingly, a critical challenge for managers is understanding when new BMs are needed (Johnson, 2010). Once opportunities have been identified whose exploitation requires the development of new BMs, managers in incumbent firms may be confronted with problems related to simultaneously managing multiple BMs (Markides and Charitou, 2004). Firms entering the BOP, or addressing new needs or new customer segments, are challenged by the potential conflicts between dual BMs (cf. Markides and Charitou, 2004). In this section we describe some of the key findings and insights from research in this important area of organization studies.

BMI can support companies in exploiting new opportunities (seizing 'white space') in three different ways (Johnson, 2010): (1) by supporting the development of new value propositions that would address an unsatisfied 'job-to-be-done' for existing customers; (2) by tackling new customer segments that have traditionally been overlooked by existing value propositions; or (3) by entering entirely new industries or a 'new terrain'. These instances present different managerial challenges and opportunities related to BMI.

First, the extent to which the development of new value propositions for an existing customer base requires BMI is a function of the shifts in the basis of competition (cf. Moore, 2004). At different stages in market development companies compete and innovate on different dimensions as illustrated in Figure 21.3.

At early stages, customers' unsatisfied needs mostly relate to product features and functions. Companies compete accordingly on functionality and focus on product innovation. When functionality-related needs are mostly fulfilled, the basis of competition shifts as customers require higher quality and reliability. In these cases, innovation is mostly process-oriented. When quality and reliability have improved sufficiently, customer value is provided by convenience, customization, and finally, when the market starts becoming commoditized, by lower costs. According to Johnson, managers should focus on BMI at these stages in market evolution, in that innovative BMs may allow developing new customer value propositions in response to commoditization in a way that product and process innovation would not. Innovative BMs may allow developing entirely new value propositions tailored to the customers' individual needs, or may be able to lower costs significantly.[8]

Second, innovative BMs may unlock opportunities to serve entirely new customer segments. These instances correspond to a process of democratization (cf. Osterwalder and Pigneur, 2010) in that they allow extending products and services to potential customers who are non-consumers, for instance because existing offerings are too expensive (with respect to potential customers' wealth), complicated (with respect to potential customers' skills) or because potential customers lack access (both geographical distances, lack of information and/or time) to them. Attempts to reach customers at the BOP, as previously discussed, fall into this category.

While serving existing customers in innovative ways or reaching new potential customers may require developing new BMs in response to identifiable and somehow predictable market-driven circumstances, a third category of opportunities is offered by less predictable *tectonic* industry changes, resulting, for example, from technological discontinuities or dramatic shifts in government policy and regulation. BMI can support companies creating new business platforms uniquely suited to the radically altered terrain, such as the innovative BM developed by Better Place, which attempted

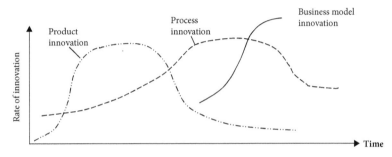

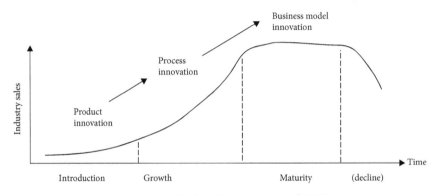

Introduction	Growth	Maturity	Market development stages
Small Sales
High Cost
Low Quality	Emergence of a Dominant Design
Sales start increasing
Market penetration	Commoditization
Market Saturation	Characteristics
Compete on functionality/ performance	Compete on quality/services	Compete on convenience
Compete on prices	Competition
Product	Process	Business Model	Loduc of innovation

FIGURE 21.3 Market development and BMI

(unsuccessfully, in hindsight) to exploit opportunities arising from a complex plethora of forces increasingly supporting demand for sustainable greener transportation.

Kaplan (2012) proposes that BMI should in fact be treated with equal importance to product innovation and provides a practical guide for incumbents, starting with fifteen principles for BMI divided into three major categories: connect, inspire, and transform. *Connect* concerns the 'team sport' nature of BMI and how to nurture it, for example enabling chance meetings between innovators outside of normal 'silos', putting into place structures enabling flexible collaborative networks from across the company, and emphasizing collaborative design thinking. *Inspire* refers to injecting a sense of meaning into developing new ideas, encouraging systems-level thinking, challenging current assumptions, and experimenting rapidly. *Transform* is about encouraging large-scale rather than incremental changes, constantly trying new things, and building urgency to innovate. Kaplan also tackles the important problem of how to conduct R&D for BMI, returning to the theme of experimentation and proposing a 'BMI factory' that is a 'connected adjacency' with the current one (rather than trying to destroy the current one), championed by top management, explicitly desired, staffed with innovators taking on diverse roles (such as idea generators, ethnographers, and BM designers) and maintained as a separate activity from product innovation that supports the current BM.

A critical managerial challenge related to the management of BMs is represented by the conflicts arising from multiple BMs (Markides and Oyon, 2010). Seizing new opportunities by developing new BMs may involve, for existing firms, operating (or considering) two BMs at the same time (Markides and Oyon, 2010). For example, to tap potential customers in India, Hindustan Unilever (the Indian subsidiary of Unilever) operates with a BM that is different from the parent's BM. ING Group started the highly successful ING Direct to tap into not only Internet financial services users, but the different ways those users utilize such services. Singapore Airlines has launched SilkAir to appeal to customers in the low-cost segment of the market in addition to its traditional operations.

According to Markides and colleagues, there are serious tradeoffs involved in competing with dual BMs, as a new BM risks cannibalizing existing sales and customer bases, destroying or undermining the existing distributor network, compromising the quality of services offered to customers, or simply defocusing the organization by trying to do everything for everybody (Markides and Charitou, 2004). To manage these tradeoffs, strategy experts have traditionally proposed keeping the two BMs separated in two different units (cf. Christensen, 1997). Instead, Markides and Charitou (2004) propose a contingency approach, according to which the quest for the best strategy is understood as fundamentally depending on (1) the degree to which the two BMs are in conflict, and (2) the degree to which the two markets related to the BMs are perceived to be strategically similar. Reducing these two dimensions to dichotomous situations (serious vs. minor conflicts and high vs. low strategic relatedness) leads to four logically possible situations with four different strategies. The latter includes both pure strategies (i.e. separate vs. integrate) as well as hybrid ones (e.g. start with integrated BMs while preparing the conditions for future 'divorce' or start separate preparing conditions for future 'marriage'). Hybrid strategies require the organization to become more ambidextrous. Operational tactics for managing dual BMs include conferring operational and financial autonomy to separate units, allowing units to develop their own culture and budgetary system and to have their own CEOs, while, at the same time, encouraging cooperation by means of a common incentive and reward system and by transferring the CEO from inside the organization rather than appointing one from outside.

CONCLUSION

In this chapter, we have reviewed the small but rapidly growing literature on BMs and BMI. In the course of most industrial sectors and humanitarian undertakings, there will come a time when the traditional way of creating, delivering, and capturing value is no longer valid, efficient, useful, or profitable. In such moments (or perhaps just before!), organizations that embrace BMI will embrace the possibility to reshape industries and possibly change the world. As this exciting field is expanding every day with increasing scholarly and managerial interest, we hope this chapter helps establish a better and more uniform understanding of BMI, and helps bridge the gap between theory and practice.

NOTES

1. In economics and business management the Bottom of the Pyramid (or 'Base of the Pyramid' or simply 'BoP') is the term used to refer to the largest but poorest socio-economic group. The expression is used in particular by people developing new models of doing business that deliberately target that demographic, often using new technology.
2. As previously noted, the process of reconfiguration also comprises creating, implementing, and validating a BM. In this sense the set comprising reconfiguration activities could be considered a superset of design activities.
3. However, note the caveats to this aspect of the theory developed in, among others, King and Tucci (2002).
4. The term 'theory' as related to business model and BMI is used here quite deliberately as resembling van Aken's notion of Mode 2 knowledge production as the product of a Design Science Research approach (van Aken, 2005) or as comprising an articulated body of knowledge in the form of what Simon (1969) understood as *criteria for the design* of man-made social artifacts (in this case organizations).
5. Common across these tools is an (often implicit) understanding of the business model as a *model* (Baden-Fuller and Morgan, 2010), i.e., a simplified representation of a reality that exists at the level of the firm and its network of exchange partners.
6. Initially known as the 'Business Model Ontology', the framework developed by Osterwalder and Pigneur has become increasingly popular with managers under the label 'Business Model Canvas'.
7. We borrow the term *meta-model* from the literature on systems engineering. In systems engineering, meta-modelling is generally understood as the analysis, construction, and development of the frames, rules, constraints, models, and theories applicable and useful for modelling a pre-defined class of problems.
8. Johnson (2010) provides several real examples of companies competing through their BM. Zipcar offers car sharing services and competes with traditional car rental companies on convenience. IKEA mixed some degree of convenience and customization with radically lower costs for home furniture.

REFERENCES

Afuah, A., and Tucci, C. L. (2001). *Internet Business Models and Strategies: Text and Cases.* New York: McGraw-Hill.

Aldrich, H. E., and Auster, E. (1986). 'Even Dwarfs Started Small: Liabilities of Size and Age and their Strategic Implications', in *Research in Organizational Behavior*, vol. 8, B. M. Staw and L. L. Cummings (eds). Greenwich, CT: JAI Press, 165–98.

Amit, R., and Zott, C. (2001). 'Value Creation in e-business', *Strategic Management Journal*, 22: 493–520.

Amit, R., and Zott, C. (2012). 'Creating Value Through Business Model Innovation', *MIT Sloan Management Review*, 53(3): 41–9.

Baden-Fuller, C., and Morgan, (2010). 'M. S. Business Models as Models', *Long Range Planning*, 43: 156–171.

Bhave, M. P. (1994). 'A Process Model of Entrepreneurial Venture Creation', *Journal of Business Venturing*, 9: 223–42.

Bijker, W. E., Hughes, T. P., and Pinch, T. J. (1987), *'The Social Construction of Technological Systems: New Directions in the Sociology and History of Technology'*, Cambridge MA: MIT Press.

Bock, A., Opsahl, T., and George, G. (2010). 'Business Model Innovations and Strategic Flexibility: A Study of the Effects of Informal and Formal Organization', Working paper Imperial College SSRN 1533742.

Bonaccorsi, A., Giannangeli, S., and Rossi, C. (2006). 'Entry Strategies under Competing Standards: Hybrid Business Models in the Open Source Software Industry', *Management Science*, 52(7): 1085–98.

Bouchikhi, H., and Kimberly, J. R. (2003). 'Escaping the Identity Trap', *MITSloan Management Review*, 44: 20–26.

Brown, A. D. (2000). 'Making Sense of Inquiry Sensemaking', *Journal of Management Studies*, 37(1): 45–75.

Bruderl, J., and Schussler, R. (1990). 'Organizational Mortality: The Liabilities of Newness and Adolescence', *Administrative Science Quarterly*, 35: 530–47.

Cairncross, F. (1993). *Costing the Earth*. Boston, MA: Harvard Business School Press.

Casadesus-Masanell, R., and Ricart, J. E. (2007). 'Competing through Business Models', Working Paper 713, IESE Business School, Barcelona.

Casadesus-Masanell, R., and Ricart, J. E. (2010). 'From Strategy to Business Models and to Tactics', *Long Range Planning*, 43: 195–215.

Chesbrough, H. W. (2003). *Open Innovation: The New Imperative for Creating and Profiting from Technology*. Boston, MA: Harvard Business School Press.

Chesbrough, H. W. (2007). 'Business Model Innovation: It's Not Just About Technology Anymore', *Strategy and Leadership*, 35: 12–17.

Chesbrough, H. W. (2010). 'Business Model Innovation: Opportunities and Barriers', *Long Range Planning*, 43: 354–63.

Chesbrough, H., Ahern, S., Finn, M., and Guerraz, S. (2006). 'Business Models for Technology in the Developing World: The Role of Non-governmental Organizations', *California Management Review*, 48: 48–61.

Chesbrough, H. W., and Rosenbloom, R. S. (2002). 'The Role of the Business Model in Capturing Value from Innovation: Evidence from Xerox Corporation's Technology Spinoff Companies', *Industrial and Corporate Change*, 11: 533–4.

Christensen, C. M. (1997). *The Innovator's Dilemma*. Boston: Harvard Business School Press.

Doz, Y., and Kosenen, M. (2010). 'Embedding Strategic Agility: A Leadership Agenda for Accelerating Business Model Renewal', *Long Range Planning*, 43: 370–82.

Giesen, E., Berman, S. J., Bell, R., and Blitz, A. (2007). 'Three Ways to Successfully Innovate your Business Model', *Strategy and Leadership*, 35: 27–33.

Gordijn, J., and Akkermans, H. (2001). 'Designing and Evaluating e-business Models', *Intelligent E-Business*, July/August: 11–17.

Hayashi, A. M. (2009). 'Do You Have a Plan "B"?' *MIT Sloan Management Review,* 51: 10–11.

Henderson, R. M., and Clark, K. B., (1990). 'Architectural Innovation: The Reconfiguration of Existing Product Technologies and the Failure of Established Firms', *Administrative Science Quarterly,* 35: 9–30.

Ireland, R. D., Hitt, M.A., Camp, M., and Sexton, D.L. (2001). 'Integrating Entrepreneurship and Strategic Management Actions to Create Firm Wealth', *Academy of Management Executive,* 15: 49–63.

Johnson, M. W. (2010). *Seizing the White Space: Business Model Innovation for Growth and Renewal.* Boston: Harvard Business Press.

Johnson, M. W., Christensen, C. C., and Kagermann, H. (2008). 'Reinventing Your Business Model', *Harvard Business Review,* 86: 50–59.

Kaplan, S. (2012). *The Business Model Innovation Factory: How to Stay Relevant When The World is Changing.* Wiley: New York.

King, A., and Tucci, C. L. (2002). 'Incumbent Entry into New Market Niches: The Role of Experience and Managerial Choice in the Creation of Dynamic Capabilities', *Management Science,* 48(2): 171–86.

Konde, V. (2009). 'Biotechnology Business Models: An Indian Perspective', *Journal of Commercial Biotechnology,* 15: 215–26.

Lovins, A. B., Lovins, H. L., and Hawken, P. (1999). 'A Roadmap for Natural Capitalism', *Harvard Business Review,* May-June.

Magretta, J. (2002). 'Why Business Models Matter', *Harvard Business Review,* May.

Markides, C., and Charitou, C. D. (2004). 'Competing with Dual Business Models: A Contingency Approach', *Academy of Management Executive,* 18(3): 22–36.

Markides, C. and Oyon, D. (2010). 'What to Do Against Disruptive Business Models (When and how to Play Two Games at Once).' *MIT Sloan Management Review,* 51(4): 25–32.

McGrath, R. G. (2010). 'Business Models: A Discovery Driven Approach', *Long Range Planning,* 43: 247–61.

Mitchell, D., and Coles, C. (2003). 'The Ultimate Competitive Advantage of Continuing Business Model Innovation', *Journal of Business Strategy,* 24: 15–21.

Miles, R. E., Miles, G., and Snow, C. C. (2006). 'Collaborative Entrepreneurship: A Business Model for Continuous Innovation', *Organizational Dynamics,* 35: 1–11.

Moore, G. A. (2004). 'Darwin and the Demon: Innovating Within established enterprises', *Harvard Business Review,* 82: 86–92.

Normann, R., and Ramirez, R. (1993). 'From Value Chain to Value Constellation: Designing Interactive Strategy', *Harvard Business Review,* July-August: 65–77.

Osterwalder, A., Pigneur, Y., and Tucci, C. L. (2005). 'Clarifying Business Models: Origins, Present and Future of the Concept', *Communications of the Association for Information Science (CAIS),* 16: 1–25.

Osterwalder, A., and Pigneur, Y. (2010). *Business Model Generation.* Hoboken, NJ: John Wiley and Sons.

Parolini, C. (1999). *The Value Net: A Tool for Competitive Strategy.* Chichester, UK: John Wiley and Sons Ltd.

Perkmann, M., and Spicer, A. (2010). 'What are Business Models? Developing a Theory of performative Representation', in M. Lounsbury (eds), *Technology and Organization: Essays in Honour of Joan Woodward (Research in the Sociology of Organizations,* 29: 265–75), Emerald Group Publishing Limited.

Pisano, G. (2006). 'Profiting from Innovation and the Intellectual Property Revolution', *Research Policy*, 35: 1122–30.

Prahalad, C. K. (2005). *The Fortune at the Bottom of the Pyramid: Eradicating Poverty Through Profits*. Philadelphia: Wharton School Publishing.

Prahalad, C. K., and Bettis, R.A. (1986). 'The Dominant Logic: A New Linkage Between Diversity and Performance', *Strategic Management Journal*, 7: 485–511.

Prahalad, C. K., and Hart, S. (2002). 'The Fortune at the Bottom of the Pyramid', *Strategy & Business*, 26: 2–14.

Ricart, J. E., Enright, M. J., Ghemawat, P., Hart, S. L., and Khanna, T. (2004). 'New Frontiers in International Strategy', *Journal of International Business Studies*, 35(3): 175–200.

Santos, J., Spector, B., and Van der Heyden, L. (2009). *Toward a Theory of Business Model Innovation Within Incumbent Firms*. Working paper no. 2009/16/EFE/ST/TOM, INSEAD.

Seelos, C., and Mair, J. (2007). 'Profitable Business Models and Market Creation in the Context of Deep Poverty: A Strategic View', *Academy of Management Perspectives*, 21: 49–63.

Shimizu, K., and Hitt, M. (2004). 'Strategic Flexibility: Organizational Preparedness to Reverse Ineffective Strategic Decisions', *Academy of Management Executive*, 18: 44–59.

Simon, H. 1969. *The Sciences of the Artificial*. Cambridge, CA: MIT press.

Sosna, M., Trevinyo-Rodríguez, R. N., and Velamuri, S.R. (2010). 'Business Models Innovation Through Trial-and-error Learning: The Naturhouse Case'. *Long Range Planning*, 43: 383–407.

Sterman, J. D. (2000). *Business Dynamics: Systems Thinking and Modeling for a Complex World*. New York: Irwin Professional McGraw–Hill.

Susarla, A., Barua, A., and Whinston, A. B. (2009). 'A Transaction Cost Perspective of the "Software as a Service" Business Model', *Journal of Management Information Systems*, 2: 205–40.

Teece, D. J. (2007). 'Explicating Dynamic Capabilities: The Nature and Microfoundations of (Sustainable) Enterprise Performance', *Strategic Management Journal*, 28: 1319–50.

Teece, D. J. (2010). 'Business Models, Business Strategy and Innovation', *Long Range Planning*, 43: 172–94.

Timmers, P. (1998). 'Business Models for Electronic Markets', *Electronic Markets*, 8(2): 3–8.

van Aken, J. E. (2005). 'Management Research as a Design Science: Articulating the Research Products of Mode 2 Knowledge Production in Management', *British Journal of Management*, 16: 19–36.

WCED. (1987). *Our Common Future*. Oxford: Oxford University Press

Wüstenhagen, R., and Boehnke, J. (2006). 'Business models for sustainable energy', in Andersen, M., and Tukker, A. (eds), *Perspectives on Radical Changes to Sustainable Consumption and Production* (SCP) (Proceedings). Roskilde and Delft: RISO and TNO, 253–9.

Yunus, M., Moingeon, B., and Lehmann-Ortega, L. (2010). 'Building Social Business Models: Lessons from the Grameen Experience', *Long Range Planning*, 43: 308–25.

Zott, C., and Amit, R. (2007). 'Business Model Design and the Performance of Entrepreneurial Firms', *Organization Science*, 18: 181–99.

Zott, C., and Amit, R. (2010). 'Designing your Future Business Model: An Activity System Perspective', *Long Range Planning*, 43: 216–26.

Zott, C., Amit, R., and Massa, L. (2011). 'The Business Model: Recent Developments and Future Research', *Journal of Management*, 37(4): 1019–42.

CHAPTER 22

....

MANAGING OPEN
INNOVATION

....

OLIVER ALEXY AND
LINUS DAHLANDER

INTRODUCTION

....

IN recent years, few terms have gained as much attention in innovation management as *open innovation*. The concept was coined by Henry Chesbrough, who defines it as 'the use of purposive inflows and outflows of knowledge to accelerate internal innovation, and expand the markets for external use of innovation, respectively. Open innovation assumes that firms can and should use external ideas as well as internal ideas, and internal and external paths to market, as they look to advance their technology', (2006a: 1). Although it has certainly made standalone contributions, the concept is also tightly related to previous concepts of innovation, such as user innovation (von Hippel, 1988; Bogers and West, 2012) or cumulative innovation (Scotchmer, 1991; Murray and O'Mahony, 2007). Accordingly, for this chapter, we use a slightly different definition of open innovation: all flows of knowledge across the boundary of the firm, independent of the form or direction, that are deliberate and that aim to create and capture value for the firm. These knowledge flows can be both inbound (arriving at the company) and outbound (leaving the company). Finally, they can involve both monetary exchange and more informal relationships (Dahlander and Gann, 2010).

At the core of the open innovation concept is a simple yet gripping idea: knowledge flows that transcend the walls of a company can allow firms to become more efficient and effective, especially in their research and development efforts. This has several potential implications for the management of innovation. First, it contests the assumption that the firm should be the nexus of all innovation it eventually introduces to the market (see Chapter 2 by Salter and Alexy). Second, opening up the innovation process has to be anchored concomitantly in an overall innovation strategy that combines internal and external sources of innovation. Finally, implementing open innovation brings tremendous challenges to people involved in R&D. It is not trivial to combine and integrate existing expertise *within* a company, and this challenge is further exacerbated when it also comes to converting *external* knowledge to new products and services.

Whether or not open innovation is a new concept has been debated—after all, firms have long used collaborative behaviour to improve the outcomes of their R&D efforts (see Chapter 23 by Dodgson). Both Allen (1983) and Nuvolari (2004), for example, have shown that industries shared research findings and documents more than a hundred years ago. Hargadon's (2003) work on the Edison laboratory shows an intricate web of relations across organizations. Freeman (1974) made a similar remark, suggesting that R&D labs are not 'castles on the hills'.

Even though the idea has a long range of precursors, two recent trends have drastically increased the ease of collaborating across firm boundaries, rendering the topic of open innovation potentially more important: globalization and the emergence of the Internet. Both trends affect the ease of collaboration by reducing transaction costs, increase the potential number of partners and the extent of the market, and lower barriers to communication. Accordingly, in the last few decades, firms have been exposed to more options for collaboration across their boundaries.

Importance of the Issue

Since its introduction in 2003, the term *open innovation* has left quite a mark on both the business and the academic world. Ignoring academic volumes published on the topic (e.g. Chesbrough, Vanhaverbeke, and West, 2006), Figure 22.1 plots the number of papers published annually in journals listed by Thomson's ISI Web of Science, illustrating that the number of papers has increased rapidly. In 2010, close to 100 papers were published on the topic.

This literature abounds with success stories of open innovation, but we have now moved beyond anecdotal evidence. Recent research shows that a firm's engagement in open innovation positively affects its financial performance and market value (Stam, 2009; Waguespack and Fleming, 2009). Also, it features prominently the need to include open innovation activities strategically into larger business models—processes and practices the firm employs to create and capture value—an issue central in Chesbrough's original conception (2006b) and extensions (Chesbrough and Appleyard, 2007), and built upon by several subsequent authors (see Chapter 21 by Massa and Tucci). Nonetheless, as we will argue, there has been little scholarly work on the conditions under which open innovation is a practical strategy, and how organizations can implement those practices. There are clearly costs of both collaboration and coordination when a firm adopts open innovation (Grant, 1996). Coordination costs stem from

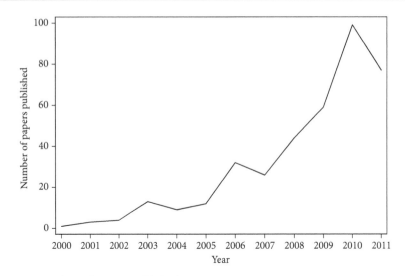

FIGURE 22.1 Number of papers published on open innovation

Note: The number of papers declines in 2011 as an artifact of the construction of the database, which was completed in October 2011.

differences between partners when innovation processes transcend organizational boundaries, and costs of competition arise in protecting ideas from actors that can act opportunistically. It is therefore important to consider both the advantages and the disadvantages of implementing open innovation, as the tradeoffs reveal when this is the best strategy to pursue.

The concept of open innovation has also made a striking impact on the practitioner world. For example, as a result of P&G's renowned 'Connect + Develop' initiative, P&G nowadays sources more than 50 per cent of innovations from outside its corporate boundaries. GlaxoSmithKline's 'Centre for Excellence in External Drug Discovery' manages a pipeline of options on drugs in development by external parties that rivals its in-house pipeline in size. Yet GSK has more than 10,000 employees in R&D; CEEDD has only about 20 (Alexy, Criscuolo, and Salter, 2009).

RESEARCH THEMES

Even though the number of papers using the open innovation concept has increased, it has been difficult to compare their findings because their interpretation of 'open' varies. Dahlander and Gann (2010) extended the work of Gassmann and Enkel (2006), devising a simple categorization based on the flow of knowledge and type of exchange, leading up to the two-by-two matrix illustrated in Table 22.1. This table outlines two forms of inbound innovation—Acquiring and Sourcing—as well as two forms of outbound— Selling and Revealing.

Table 22.1 Overview of types of openness

	Inbound innovation	Outbound innovation
Pecuniary	*Acquiring* Acquiring inventions and input to the innovative process through informal and formal relationships	*Selling* Out-licensing or selling products in the market place
Non-pecuniary	*Sourcing* Sourcing external ideas and knowledge from suppliers, customers, competitors, consultants, universities, public research organizations, etc.	*Revealing* Revealing internal resources to the external environment

Source: Dahlander and Gann (2010).

Acquiring

Acquiring relates to buying inputs to the innovation process in the marketplace. Work in this vein explores how openness can be understood by investigating how licensing and acquiring knowledge from outside constituents affect companies. The challenge is viewed as one of combining the acquisitions of knowledge with internal expertise to search for and evaluate potential inputs. In accordance with transaction cost economics (Williamson, 1975; Arora et al., 2001), firms will find that, sometimes, it might just be cheaper to license or buy an existing and available outside technology than to develop it from scratch.

Sourcing

Sourcing stems from how firms scan and use the external environment as input to the innovation process. Scholars who use this lens focus on how firms explore their environment as a complement to their internal knowledge production, which might in fact be limiting them in their quest for new ideas to use in innovation. First, a lack of assets, such as skilled employees, cash, or machines and equipment, may constrain the solution space from which the firm can choose (Chesbrough, 2003). Second, because of the evolutionary nature of their knowledge base (Cohen and Levinthal, 1990; Kogut and Zander, 1992), firms might look in vain in some areas for solutions which could be readily available in others (Jeppesen and Lakhani, 2010), and their inclusion might be highly beneficial to the firm (Rosenkopf and Nerkar, 2001). The story of the longitude prize provides a fitting example. A clockmaker beat astronomers, including Sir Isaac Newton, to the solution to the measurement of longitude at sea (Sobel, 1995). Similarly, von Hippel's work (1988, 2005) shows that for a wide array of innovations, their functional source is not the focal firm (manufacturer) but a downstream (user) or upstream (supplier) actor. In particular, he notes the importance of lead users, pointing out how their foreshadowing general market trends and developing of prototypes can assist corporations in designing new and highly successful innovative products (see Chapter 5 by Franke).

Laursen and Salter (2006) show that there are limits to the benefits from sourcing knowledge externally. They find that openness with respect to making use of external knowledge sources has a U-shaped effect on innovative performance. Their result highlights the fact that openness may be advantageous to the firm but also raises the issue of the appropriate degree of openness, by showing that both too little and too much sourcing from outside may be detrimental to the firm.

Selling

This type of openness refers to how firms commercialize their inventions and technologies through selling or licensing the resources they have developed to other companies. Underlying this research is the idea that some companies have 'Rembrandts in the attic' that are unused (Rivette and Kline, 2000). By selling or licensing those through agreements, companies can increase their own profits while allowing technologies to be commercialized by someone more suitable. Some scholars argue that the potential of selling technologies has not been fully exploited: Gambardella, Giuri, and Luzzi (2007) even suggest that the market for technology could be close to 70 per cent larger if some obstacles could be overcome (see Chapter 12 by Gambardella, Giuri, and Torrisi).

Revealing

Revealing proprietary knowledge will be beneficial if the firm can appropriate value from its diffusion (von Hippel, 1988). In general, knowledge ought to be revealed selectively— that is, only when it is advantageous to the firm (Henkel, 2004, 2006). Firms reveal knowledge strategically and at the same time appropriate the rents of other parts of their business privately to gain advantages over their competitors (Chesbrough, 2006b).

Particularly for information technologies (Varian and Shapiro, 1999), sharing underlying technological specifications is central to the generation of standards. The existence of a standard might lead to an increase in the size of the market, thereby enlarging the pie from which the firm can appropriate its piece (von Hippel, 1988; Varian and Shapiro, 1999; and see Chapter 28 by Leiponen).

If the company defines (parts of) a standard—even if it is an open one—it is highly likely that it includes parts that are beneficial to only that one company, either because only this firm knows of them or because they are already optimally realized in the firm's existing products (Henkel, 2004). Spencer (2003) finds positive effects on the innovative performance of firms actively choosing to reveal some of their intellectual property (IP). She argues in particular that a firm will do so in order to attract other organizations to join its technological trajectory, which may help in both standard-setting and the development of industry structure. Gawer and Cusumano (2002) observe similar behaviour in Intel's platform development strategy, which makes use of openness to establish new markets and increase adoption of the PC platform in certain areas.

If one strong standard or a dominant one already exists, releasing related know-how can make it part of the standard or at least increase compatibility of the standard with existing intra-firm knowledge (Harhoff, Henkel, and von Hippel, 2003; Henkel, 2004). Such increased compatibility will create network effects that encourage diffusion and adoption not only of the released know-how, but also of related innovations and second-generation innovation built on it (e.g. Shepard, 1987; Farrell and Gallini, 1988). Information may also be leaked strategically to commoditize layers of the product architecture in which the focal firm is weak and to shift competition to layers where the firm has a competitive advantage (Raymond, 2001; West, 2003).

Coupled Models: Combining Different Types of Openness

The preceding categorization distinguishes between types of openness, although they are often combined in practice. But only a few studies have analysed how different types of openness are interrelated (see e.g. Acha, 2007; van de Vrande et al., 2009), possibly because of confusion about what openness entails. With a better understanding of different types of openness, it is possible to discuss and analyse how they are interrelated in specific initiatives that organizations adopt.

For instance, voluntarily revealing and sourcing ideas from the environment often go hand-in-hand. Companies co-create new knowledge through forging and sustaining collaboration with external partners. von Hippel (1988) argued that firms often trade knowledge to achieve mutual benefits if the rent a firm can expect from sharing is higher than when it keeps the knowledge proprietary, especially when the knowledge holds little competitive advantage (Allen, 1983). This concept is formalized in Henkel's (2005) model. It shows that when two firms voluntarily reveal developments of complementary technologies, both firms' profits and product quality can increase.

Furthermore, exchange of knowledge across organizational boundaries enables organizations to benefit from the improvements made by other parties. Such inputs help organizations to tailor existing and future products to the market, leading to faster product development and higher success rates (von Hippel, 1988; Dalle and Jullien, 2003). In addition, if customers are allowed to make changes to products, they will be more likely to commit fully to the product and to impactful improvements (von Hippel, 2001; Jeppesen and Frederiksen, 2006). In this way, the company receives knowledge that is often difficult to find, transfer, or acquire otherwise, so-called 'sticky information' (von Hippel 1994). Also, people outside a company sometimes even do mundane tasks such as user support or documentation (Lakhani and von Hippel, 2003). And finally, even in situations where the relevant knowledge is vital to the firm's competitive advantage, know-how trading can be beneficial. For example, two competitors, each of which needs the other's technical know-how, might decide to offer a joint bid for a government contract because they would be unable to complete the respective technologies independently (von Hippel, 1988).

Companies often combine selling and acquiring as they seek to streamline their own R&D processes. Unused technologies are sold to external partners that are better suited for commercializing the technology, providing additional sources of revenue. On the flip side, companies also acquire outside technologies they seek as inputs in the innovation process. In all these cases the underlying rationale is to find the company most suitable to commercialize a technology.

CONTINGENCIES OF OPEN INNOVATION: WHEN IS IT BENEFICIAL?

Managers pose the question about open innovation: Does it pay off? Researchers ask a related question: Under what conditions is open innovation worth pursuing? The answers to those questions require an understanding of contingent factors that affect whether and when openness is beneficial for the organization. Some of these contingencies are internal to the firm, others external, a distinction we use below.

Internal Contingencies

Cognitive Bias—Not Invented Here

To find the best seed for innovation is not easy because people who work in organizations are subject to cognitive bias. People simply like what they are. Employees inside an organization often reject innovations from the outside because they challenge established wisdom inside the company: the 'not invented here syndrome' (Katz & Allen, 1982).

The not invented here syndrome is often an outcome of repeated external collaboration with unsatisfying outcomes (Alexy, Criscuolo, and Salter, 2012). In particular, firms at the forefront in a technology often struggle to identify external parties who can add value to their innovative efforts and consequently develop a negative attitude towards external collaboration. In other cases, 'not invented here' is caused by a fear that working with external partners will expose individual weakness, and that associated organizational change will disrupt individuals' work routines and even put their jobs at risk. In support of this argument, Alexy, Henkel, and Wallin (2013) find that less-skilled employees exhibit higher degrees of skepticism towards open innovation. Their reluctance implies that further educating people inside the company not only increases the human capital, but also employees' ability to recognize and work with important knowledge from the outside.

Skillsets

Alexy, Henkel, and Wallin (2013) show that engagement in open innovation trickles down to the micro-level of how people organize inside companies. Some tasks within the R&D processes undergo significant changes in both the skills required to execute them and the degree of control individuals have to share with outside actors. These changes are strongest for people whose job roles are of 'executing' character, such as developers in software, R&D engineers, and project managers. Consequently, companies intending to engage in open innovation need to address potential concomitant employee concerns. They may choose, for example, to adapt existing reward systems to reflect the different nature of project planning and execution necessary to benefit from knowledge that stems from outside the organization. Other organizations find it easier to adapt their HR policies and hire external people with the necessary skills or train their internal staff to develop them.

Attention

Laursen and Salter (2006) depict open innovation as a search process. Companies can use various sources; each of the search channels provides different sources of information but requires different norms of exchange. Thus, while each source brings novel information, accessing it comes at a cost. Companies that use more search channels are more innovative, but some organizations spread the attention of the top managers and R&D experts too thinly and thus perform worse (Laursen and Salter, 2006).

Many companies seek to tap existing outside knowledge by running so-called unsolicited idea submission processes, but struggle because of attention-related issues. Specifically, although such processes often succeed in attracting external ideas, existing company selection processes are overstrained by the increase, preventing any improvement in overall innovation performance (Alexy et al., 2012). Simply put, while it seems as if open innovation answers the problems of limited reach and paucity of ideas available inside the firm, it may only replace them with an equally complex problem of selecting the right ideas.

A potential solution may be to involve specialist intermediaries to decrease transaction costs. Several authors describe how companies such as InnoCentive and NineSigma support businesses by presenting their internal problems to a broad range of knowledgeable outside parties and facilitating the transfer of only a selected number of highly probable suggested solutions (Jeppesen and Lakhani, 2010). In doing so, they may also assist in solving the issue of contamination described below.

Compatibility

One of the main sources of coordination costs stems from a lack of compatibility between partners which can affect both content (matching solutions to problems) as well as structure (different languages and norms of exchange).

Regarding inbound open innovation, the higher the content compatibility of external knowledge—the better external knowledge fits current or anticipated future needs—the higher its potential value to the firm. Thus, the general availability of an external partner providing such matching knowledge (universities doing research in the area, for example) is a mandatory precursor for inbound innovation to be valuable. Second, the greater the structural compatibility—that is, the more closely aligned are the parties' internal and external structures and the language used to comprehend and describe certain problems and solutions—the less effort the firm will need to invest in making external knowledge usable and the higher will be the potential net benefit resulting from engaging in open innovation.

For outbound open innovation, content and structural compatibility are important in identifying firms who would be suitable recipients (i.e. buyers) of the firm's internal knowledge. The better internal knowledge fits an external party's needs and the smaller the cost to them to make the knowledge usable for themselves, the more willing they should be to adopt it, and the higher the price they should be willing to pay to acquire or license it from the focal firm. Notably, if the focal firm realizes that no external party has a need for its knowledge, or that costs of adoption are prohibitive, internal measures can be taken to increase the attractiveness of the to-be-opened knowledge to outside parties. Many software firms, for example, provide additional documentation or toolkits together with the actual software they open up in order to show potentially multifaceted applications of their software to external parties and smooth the adoption process.

Boundary-Spanning and Internal Legitimacy

Much of the early work on boundary-spanning (Allen, 1977; Tushman and Scanlan, 1981; Ancona and Caldwell, 1992) analysed how people within R&D labs forged connections externally. They found that 'communication stars' who were connected with people both inside and outside the organization were better able to innovate. They gained fresh insights from engaging with outsiders, while retaining credibility inside the organization, to bring ideas to fruition.

Cultivating broad networks within as well as outside an organization is not easy and has the potential for people to spend more time on forging connections than getting work done. Time is a scarce resource: seeking more external engagement can have unintended consequences if it distances employees from their own company. Firms need to ensure that knowledge and partnerships identified by their outside scouts are in fact being harnessed by the internal organization, and then to build the necessary absorptive capacity (Cohen and Levinthal, 1990).

Complementary Assets

Complementary assets are needed to commercialize an innovation, but they do not include the assets directly tied to the innovation. They include access to distribution, service, and manufacturing facilities and often explain who profits from innovation

(Teece, 1986). In open innovation settings, complementary assets are particularly important because their ownership radically mitigates concerns about loss of intellectual property (Dahlander and Wallin, 2006; Henkel, 2006). Regarding outbound innovation, if a firm holds complementary assets, releasing internal knowledge related to them might ultimately be profitable.

Sometimes sharing internal knowledge fuels diffusion in such a way that the value of possessing complementary assets increases more than the loss the firm incurs from leaking to competitors. On the other hand, if a firm has weak complementary assets it may be forced to enter relationships with partners enjoying stronger positions, evidenced by the central importance of alliances in the biotechnology industry (between small biotechnology and big pharmaceutical firms). On the inbound side, firms holding strong complementary assets may undertake systematic identification of external parties that need them to increase the utilization and profitability of existing investments.

External Factors

Appropriability Regimes and the Problem of Contamination

Teece (1986) describes the appropriability regime as 'environmental factors, excluding firm and market structure.... The most important dimensions of such a regime are the nature of the technology and the efficacy of legal mechanisms of protection' (Teece, 1986: 287). Legal mechanisms in particular may be crucial in the establishment of formal relationships. The more strongly enforced the legal mechanisms that define ownership over intellectual property along clearly demarcated boundaries, the easier it will be for two parties to contract over the exchange of innovation. Many firms even refuse to speak to externals who have not filed for intellectual property protection, because their ideas may 'contaminate' internal R&D (Chesbrough, 2006b; West, 2006; Alexy, Criscuolo, and Salter, 2012).

The problem here is that if no documentation or boundary exists to delineate what exactly the external has to offer, it will be difficult in the case where collaboration does *not* proceed for the firm to prove in the future that the external knowledge did *not* influence internal R&D. And in those situations where firms decide in favour of collaboration, IP rights are important to facilitating negotiations by mitigating the paradox of disclosure—the problem that once valuable knowledge is disclosed to facilitate collaboration, its market price drops to zero—and to enable transfer of ownership by a simple trade of the IP right for money.

Issues such as these clearly show that markets for technologies depend on the existence of tradable assets demarcated and protected by IP rights. Nevertheless, some fields reject private ownership of knowledge, creating an issue of moral repugnance (Roth, 2008): 'open science' (Gans and Stern, 2010) and 'open source software'

(O'Mahony, 2003) are spaces where some actors reject the establishment of tradable IP rights, making financially based transactions difficult to achieve.

Whereas strong IP regimes are typically positively related with pecuniary forms of exchange, weak IP regimes can facilitate more informal means of interaction. von Hippel and von Krogh (2003) submit that an inability to protect knowledge developed in-house because of weak appropriability regimes may spur companies to identify strategies in which the selective disclosure of such knowledge will be to their advantage.

Cumulative Nature of Knowledge

Most early work on open innovation focused on high-technology industries with a significant manufacturing component, and on the pharmaceutical and biotechnology industry. This selection, as well as the general importance of collaboration in these industries, reflects that the areas of technology they rest on build on a science base that is cumulative. Invention is usually built on the current knowledge base, allowing inventors to 'stand on the shoulders of giants' when they make progress.

These processes of knowledge production naturally lend themselves to joint inventive activity because knowledge flowing between actors with different but related knowledge bases will lead to an increased likelihood of scientific advancement. The innovation process is viewed as a box of Lego, where the child with the greatest diversity of pieces can recombine them into the most novel outcomes.

More recent work has begun to shed light on open innovation in project-based industries (Davies, Gann, and Douglas, 2009) as well as the services sector (Chesbrough, 2011), highlighting that principles of open innovation should be applicable to more business sectors. When the knowledge frontier is less cumulative, it becomes more difficult to divide tasks, resulting in significant coordination problems.

Technological maturity (Abernathy and Utterback, 1978) also plays an important role in open innovation (Christensen, Olesen, and Kjær, 2005). The earlier in its life cycle a technology is, the more likely that rallying a crowd behind one technology can give one company a lead in establishing a dominant design—but sometimes an existing dominant design is displaced through strategic engagement in open innovation (Varian and Shapiro, 1999). In turbulent technology environments, companies will often struggle to identify partners on their journey. On the reverse side, when stormy seas have settled and technologies become well understood, recognizing suitable collaborators and establishing mutually beneficial partners will be relatively easy.

Type of Technological Advance

An extensive literature points to the fact that many, perhaps most, radical innovations have been made not by established companies (Schumpeter, 1942), but by individual inventors, start-ups, or university researchers, or as part of government programmes. The Internet is a popular example.

Existing companies, on the other hand, very often specialize in improving the technology they already have in-house—to use March's (1991) terminology, they exploit rather than explore—and few companies have the skill to do both, especially at the same time. Here, open innovation may be a useful pathway to allow companies to continuously renew themselves through new technologies (Dodgson, Gann, and Salter, 2006). Practices such as the lead user method (von Hippel, 1988) involve external actors in the design of novel and imaginative products and services. GSK's CEEDD milestone-based approach to collaboration is a further design model to harness the environment as a source of radical innovation (Alexy, Criscuolo, and Salter, 2009). And Intel has developed 'Intel Research' as an open-innovation-based sensor network to discover nascent trends around its Silicon platform (MacCormack and Herman, 2004).

If designed appropriately, open innovation activities can also be used to sustain incremental innovation. The continuous and joint development of open source software platforms such as the Apache web server or the Linux operating systems, for example, allows companies such as IBM to save hundreds of millions of dollars each year.

Number of Partners Available

Implicit in much reasoning on open innovation is that potential partners are just 'out there' ready and waiting to collaborate with other companies. Any company that has sought to collaborate with other partners knows this is a cumbersome task. The focal organization may not be an attractive collaboration partner, leading companies to invest in internal R&D to gain a 'ticket of admission' to a collaboration network (Nelson, 1991).

One further problem is that companies often do not know who holds 'matching' knowledge (i.e. compatible in its content and structure), such as solutions to the problems they are facing. Here, intermediaries have emerged to bring together two parties to their mutual benefit. In contrast, companies that have a rough understanding of where to look will need to ask themselves what it is they are actually looking for when they are seeking external knowledge. Is it one specific partner with certain assets or attributes? Or is it rather the 'crowd', from which companies want to identify average opinions, general needs, or market trends? Depending on these choices, extant research has made clear that different patterns of interaction and IP protection will be mandated to achieve increased innovation performance (Alexy, Criscuolo, and Salter, 2009).

DISCUSSION

Open innovation promises to leverage internal R&D and allow organizations to benefit from engagement with external partners. The question is why some companies are more successful than others in implementing open innovation?

While the early literature on open innovation argued that being open was generally beneficial, more recent work has moved on to study under what specific conditions openness can contribute to improving firms' innovative performance. This literature has looked at contingencies both internal and external to firms. In turn, these contingencies will also determine their decision whether or not to engage in related practices. We have built on related work on the topic and specified some of these contingencies, and it is our hope that future work will refine them further.

Future Research on the Phenomenon

A logical next step is to improve our understanding of how firms active in this new, open environment can be managed—specifically, which 'standard' management practices (known from closed innovation) apply 'as is' and which do not work? Which can be adapted? Which new ones have to be developed?

Incentives function differently for people outside an organization. These people are often motivated to take part, but for reasons different from members of the focal organization, suggesting that one cannot implement the same practices as used within organizations. The role and effect of monetary rewards differ widely across varying types of arrangements. People solving problems for firms via the intermediary InnoCentive expect payment, for example, and money may be seen as a prime motivator (Jeppesen and Lakhani, 2010). For problem-solvers in a similar intermediary, TopCoder, however, status within the peer group is a far more important motivator (Boudreau, Lacetera, and Lakhani, 2011). In open source communities, financial rewards may even have negative effects under certain conditions (Alexy and Leitner, 2011). In short, providing incentives to people outside the corporate boundaries is much more complex than to internal actors. It requires a lot of careful planning and deliberation—the same should apply to other 'closed innovation' management practices.

The competitive dynamics of being open is poorly understood. Although existing work has identified the crucial importance of this question (Christensen, Olesen, and Kjær, 2005), the specific effect of being open today on a company's future competitive position needs further work. Although some research describes how companies seek to benefit in the present from open and hybrid business models (Chesbrough, 2006b; Bonaccorsi, Giannangeli, and Rossi, 2006), it is unclear how this translates to competitive advantage in the future. Competitive dynamics also surface between outside actors that participate in the focal firm's open innovation initiatives. Boudreau and Lakhani (2009), for example, show that outcomes are substantially affected by whether open innovation activities are designed to facilitate collaboration or competition between contributors.

Careful deliberation is important when one considers being more open. Although some firms gain from opening up, it remains unclear that others can and will reciprocate such moves. Opening up innovation processes promises to involve a broad range of external people, speeding up innovation. But coordinating between external people with different interests who are also beyond hierarchical reach is a challenge for many organizations. To balance and reach agreement among a wide array of partners requires attention and a deep understanding of each others' needs. These challenges should not be trivialized, especially because alliances between companies, as well as interpersonal relationships between individuals, often wither and die (Burt, 2002; and see Chapter 23 by Dodgson).

More research is needed on the competencies and practices companies need to implement in-house to benefit from being open. Questions arise about suitable models for allowing individuals and departments to collaborate externally, where the famous slack models propagated by firms such as 3M or Google have caught the public's attention, but have not been scrutinized for their efficacy and efficiency. It is unclear whether these success stories translate into benefits for more 'average' companies. The expectations of job and skill profiles are also changing in an open environment—how much should individuals collaborate? What positions should they hold in their companies? How can these people bring externally acquired knowledge into the firm? And finally, there is the crucial issue of aggregation, linking to work on absorptive capacity (Cohen and Levinthal, 1990): How does the external and internal engagement of all actors in the organization actually combine to lead to beneficial outcomes for the focal organization?

Improving Management Theory

The number of papers on the topic has increased at a fast rate, but it does not constitute a field on its own. It is therefore important to discuss how it links to other streams of literature and find points of common interest (see Dahlander and Gann, 2010; Bogers and West, 2012). We elaborated on contingencies that explain when certain open innovation strategies are beneficial to a focal firm. A comprehensive explanation of when and to what extent firms share valuable resources with others will enhance explanatory power.

Sustained and substantial engagement in open innovation may be somewhat counterintuitive as firms increase the number of external actors they depend on for innovative success. Some recent research has, however, argued that the strategic use of open innovation allows changing the nature and potential impact of such external constraints (Alexy and Reitzig, 2013). Other work describes solutions to the problem of competing institutional logics when open and closed innovation clash (Perkmann, 2009). We encourage more research on questions such as how open innovation is shaped by institutional pressures, as well as how the strategic use of open innovation allows companies to shape institutions around them.

Open innovation may be seen as a challenge to established classical theories of management. The resource-based view emphasizes that resources have to be rare, inimitable and non-substitutable to create competitive advantage (Barney, 1991; Grant, 1991). An implication of open innovation is that many of the most essential inputs to the innovation process are neither owned nor controlled by the firm. Yet some firms are able to reap rewards from using novel ways of extending and leveraging their resource base. More work is needed for scholars of open innovation to connect the idea to these theories of management. Relatedly, transaction cost economies explains when activities will be organized through market transactions or within the organization (Williamson, 1975). But openness is more than extending the markets for technology and also includes informal interactions with external participants (Dahlander and Gann, 2010). Continuing personal interactions are often required to transfer complex knowledge in order to develop innovations. The literature on openness has been criticized for not paying enough attention to its implications for other theories, and there are many opportunities to do this.

Managerial Implications

Open innovation places fundamental new requirements on the management of innovation. Firms that want to engage successfully in open innovation have to develop and hone new capabilities or adapt and extend existing ones. First and foremost, managers need to foster a culture of openness that accepts that the most important ideas for the company often originate from the outside. Incentive structures and career paths have to be adapted to encourage individual employees to look for external commercialization opportunities for ideas that do not match the firm's trajectory. For instance, in some companies employees receive bonuses for the number of patents they produce without considering their business relevance. As a result, many inventions are put on the shelf. This can potentially be overcome by rewarding individuals for the revenues their inventions generate, even from licensing to other organizations. Second, engagement in open innovation presents companies with a plethora of options for collaboration. It is then critical for the company to winnow down the harvest of opportunities by carefully selecting the most promising ones. Having access to a broad range of external sources of innovation is only beneficial to the extent that companies have the ability to focus on a few. Third, much research suggests that building and cultivating both internal and external networks is important to developing innovations. Internal ties increase the awareness of the organization's needs and build legitimacy among colleagues, whereas external ones are more likely to tap into novel sources of knowledge (Dahlander et al., 2011). If organizations map the internal and external networks of their employees (Criscuolo et al., 2007), they can be more systematic about how they compose teams consisting of some individuals with strong internal and others with strong external networks. On a final note, it is clear that open innovation can disrupt the routines of managers and regular R&D employees. To ingrain the importance of open innovation strategies in employees' mindsets will require high upfront as well as continuous smaller investments by the company.

CONCLUSION

The coexistence of internal and external idea generation and selection and internal and external commercialization of knowledge holds great promise for firms and other contributors to innovation, whether they be individual inventors, universities, research centres, or governments. As we have argued, this has implications for the management of innovation, including our understanding of where innovation comes from, how it is brought to practice, as well as managed on a daily basis. The questions of when and how to engage in open innovation are interwoven with other aspects of the management of innovation discussed elsewhere in this Handbook.

With ever-increasing interconnectedness between and among these different actors, we would expect that open innovation will begin to permeate the activities of even more firms. We would expect the greatest increase in 'new' users of open innovation practices to come from small firms, from non-high-tech settings, and probably also from geographic regions on which current academic work does not necessarily focus: the developing and underdeveloped world. The last point in particular will raise new and interesting questions with regard to the institutional environments and legal frameworks prevalent in these countries, and how those might affect open innovation activities in turn. In short, we remain convinced that there will remain ample opportunity to study interesting and relevant questions about open innovation to gain valuable insights for theory and practice about this important topic over the years to come.

REFERENCES

Abernathy, W. J., and Utterback, J. M. (1978). 'Patterns of Industrial Innovation', *Technology Review*, 80: 41–7.

Acha, V. (2007). 'Open by Design: The Role of Design in Open Innovation', Working Paper. Imperial College Business School.

Alexy, O., Criscuolo, P., and Salter, A. (2009). 'Does IP Strategy Have to Cripple Open Innovation?' *Sloan Management Review*, 51(1): 71–7.

Alexy, O., Criscuolo, P., and Salter, A. (2012). 'No Soliciting: Managing Unsolicited Ideas for R&D', *California Management Review*, 53(3): 116–39.

Alexy, O., Henkel, J., and Wallin, M. (2013). 'From Closed to Open: Job Role Changes, Individual Predispositions, and the Adoption of Commercial Open Source Software Development', *Research Policy*, forthcoming.

Alexy, O., and Leitner, M. (2011). 'A Fistful of Dollars: Are Financial Rewards a Suitable Management Practice for Distributed Models of Innovation?' *European Management Review*, 8(3): 165–85.

Alexy, O., and Reitzig, M. (2013). 'Private-collective Innovation, Competition, and Firms', Counterintuitive Appropriation Strategies.' *Research Policy*, 42(4): 895–913.

Allen, T. (1977). *Managing the Flow of Technology: Technology Transfer and the Dissemination of Technological Information Within the R and D Organisation.* Cambridge, MA: MIT Press.

Allen, R. C. (1983). 'Collective Invention', *Journal of Economic Behaviour and Organization*, 4: 1–24.

Ancona, D. G., and Caldwell D. F. (1992). 'Bridging the Boundary: External Activity and Performance in Organizational Teams', *Administrative Science Quarterly*, 37(4): 634–665.

Arora, A., Fosfuri, A., and Gambardella, A. (2001). *Markets for Technology*. Cambridge, MA: MIT Press.

Barney, J. (1991). 'Firm Resources and Sustained Competitive Advantage', *Journal of Management*, 17(1): 99–120.

Bogers, M., and West, J. (2012). 'Managing Distributed Innovation: Strategic Utilization of Open and User Innovation', *Creativity and Innovation Management*, 21(1): 61–75.

Bonaccorsi, A., Giannangeli, S., and Rossi, C. (2006). 'Entry Strategies under Competing Standards: Hybrid Business Models in the Open Source Software Industry', *Management Science*, 52(7): 1085–98.

Boudreau, K. J., Lacetera, N., and Lakhani, K. R. (2011). 'Incentives and Problem Uncertainty in Innovation Contests: An Empirical Analysis.' *Management Science*, 57(5): 843–63.

Boudreau, K., and Lakhani, K. (2009). 'How to Manage Outside Innovation'. *Sloan Management Review*, 50(4): 69–75.

Burt, R. S. (2002). 'The Social Capital of Structural Holes', in M. F. Guillén, R. Collins, P. England, and M. l. Meyer (eds), *The New Economic Sociology: Developments in an Emerging Eeld*. New York, NY: Russell Sage Foundation.

Chesbrough, H. (2003). *Open Innovation: The New Imperative for Creating and Profiting from Technology*. Boston, MA: Harvard Business School Press.

Chesbrough, H. (2006a). 'Open Innovation: A New Paradigm for Understanding', in H. Chesbrough, W. Vanhaverbeke, J. West (eds.), *Open Innovation: Researching a New Paradigm*. Oxford: Oxford University Press, 1–12.

Chesbrough, H., (2006b). *Open Business Models: How to Thrive in the New Innovation Landscape*. Boston, MA: Harvard Business School Press.

Chesbrough, H. (2011). *Open Services Innovation: Rethinking your Business to Grow and Compete in a New Era*. San Francisco, CA: John Wiley & Sons.

Chesbrough, H. W., and Appleyard, M. M. (2007). 'Open Innovation and Strategy', *California Management Review*, 50(1): 57–74.

Chesbrough, H., Vanhaverbeke, W., and West, J. (eds) (2006). *Open Innovation: Researching a New Paradigm*. Oxford: Oxford University Press.

Christensen, J. F., Olesen, M. H., and Kjær, J. S. (2005). 'The Industrial Dynamics of Open Innovation: Evidence from the Transformation of Consumer Electronics', *Research Policy*, 34(10): 1533–49.

Cohen, W. M., and Levinthal, D. A. (1990). 'Absorptive Capacity: A New Perspective on Learning and Innovation', *Administrative Science Quarterly*, 35(1): 128–52.

Criscuolo, P., Salter, A., and Sheehan, T. (2007). 'Making Knowledge Visible: Using Expert Yellow Pages to Map Capabilities in Professional Services Firms,' *Research Policy*, 36(10): 1603–19.

Dahlander, L., and Gann, D. M. (2010). 'How Open is Innovation?' *Research Policy*, 39(6): 699–709.

Dahlander, L., and Wallin, M. W. (2006). 'A Man on the Inside: Unlocking Communities as Complementary Assets', *Research Policy*, 35(8): 1243–59.

Dalle, J.-M., and Jullien, N. (2003). '"Libre" Software: Turning Fads into Institutions?' *Research Policy*, 32(1): 1–11.

Davies, A., Gann, D., and Douglas, T. (2009). 'Innovation in Megaprojects: Systems Integration at London Heathrow Terminal 5', *California Management Review*, 51(2) 101–25.

Dodgson, M., Gann, D., and Salter, A. (2006). 'The Role of Technology in the Shift Towards Open Innovation: The Case of Procter and Gamble', *R & D Management*, 36(3): 333–46.

Farrell, J., and Gallini, N. T. (1988). 'Second-sourcing as a Commitment: Monopoly Incentives to Attract Competition', *Quarterly Journal of Economics*, 103(4): 673–94.

Freeman, C. (1974). *The Economics of Industrial Innovation*. London: Pinter.

Gambardella, A., Giuri, P., and Luzzi, A. (2007). 'The Market for Patents in Europe', *Research Policy*, 36(8): 1163–83.

Gans, J. S., and Stern, S. (2010). 'Is There a Market for Ideas?' *Industrial and Corporate Change*, 19(3): 805–37.

Gassmann, O., and Enkel, E. (2006). 'Towards a Theory of Open Innovation: Three Core Process archetypes', *R & D Management Conference*.

Gawer, A., and Cusumano, M. A. (2002). *Platform Leadership: How Intel, Microsoft, and Cisco Drive Industry Innovation*. Cambridge, MA: Harvard Business School Press.

Grant, R. M. (1991). 'The Resource-based Theory of Competitive Advantage: Implications for Strategy Formulation', *California Management Review*, 33(3): 114–35.

Grant, R. M. (1996). 'Towards a Knowledge Based Theory of the Firm', *Strategic Management Journal*, 17(Winter Special Issue): 109–22.

Hargadon, A. B. (2003). *How Breakthroughs Happen: The Surprising Truth about How Companies Innovate*. Cambridge, MA: Harvard Business School Press.

Harhoff, D., Henkel, J., and von Hippel, E. (2003). 'Profiting from Voluntary Information Spillovers: How Users Benefit by Freely Revealing their Innovations', *Research Policy*, 32(10): 1753–69.

Henkel, J. (2004). 'Open Source Software from Commercial Firms: Tools, Complements, and Collective Invention', *ZfB-Ergänzungsheft*, 74(4).

Henkel, J. (2006). 'Selective Revealing in Open Innovation Processes: The Case of Embedded Linux', *Research Policy*, 35(7): 953–69.

Jeppesen, L. B., and Frederiksen, L. (2006). 'Why Do Users Contribute to Firm-hosted User Communities? The Case of Computer-controlled Music Instruments', *Organization Science*, 17(1): 45–63.

Jeppesen, L. B., and Lakhani, K. R. (2010). 'Marginality and Problem-solving Effectiveness in Broadcast Search', *Organization Science*, 21(5): 1016–33.

Katz, R., and Allen, T. J. (1982). 'Investigating the Not Invented Here (NIH) Syndrome: A look at the Performance, Tenure, and Communication Patterns of 50 R&D Project Groups', *R&D Management*, 12(1): 7–19.

Kogut, B., and Zander, U. (1992). 'Knowledge of the Firm, Combinative Capabilities, and the Replication of Technology', *Organization Science*, 3(3): 383–97.

Lakhani, K., and von Hippel, E. (2003). 'How Open Source Software Works: 'Free' User-to-user Assistance', *Research Policy*, 32(7): 923–43.

Laursen, K., and Salter, A. J. (2006). 'Open for Innovation: The Role of Openness in Explaining Innovation Performance Among UK Manufacturing Firms', *Strategic Management Journal*, 27(2): 131–50.

MacCormack, A. D., and Herman, K. (2004). 'Intel Research: Exploring the Future', *Harvard Business School Case*, 9: 605–051.

March, J. G. (1991). 'Exploration and Exploitation in Organizational Learning', *Organization Science*, 2(1): 71–87.

Murray, F., and O'Mahony, S. (2007). 'Exploring the Foundations of Cumulative Innovation: Implications for Organization Science', *Organization Science*, 18(6): 1006–21.

Nelson, R. R. (1991). Why Do Firms Differ, and How Does It Matter? *Strategic Management Journal*, 12: 61–74.

Nuvolari, A. (2004). 'Collective Invention During the British Industrial Revolution: The Case of the Cornish Pumping Engine', *Cambridge Journal of Economics*, 28(3): 347–63.

O'Mahony, S. (2003). 'Guarding the Commons: How Community Managed Software Projects Protect their Work', *Research Policy*, 32(7): 1179–98.

Perkmann, M. (2009). 'Trading Off Revealing and Appropriating in Drug Discovery: The Role of Trusted Intermediaries', *Academy of Management Proceedings*: 1–6.

Raymond, E. S. (2001). 'The Cathedral and the Bazaar', in E. S. Raymond, *The Cathedral and the Bazaar: Musings on Linux and Open Source by an Accidental Revolutionary* (2nd ed.). Sebastopol, CA: O'Reilly, 19–63.

Rivette, K., and Kline, D. (2000). *Rembrandts in the Attic: Unlocking the Hidden Value of Patents*. Boston, MA: Harvard Business School Press.

Rosenkopf, L., and Nerkar, A. (2001). 'Beyond Local Search: Boundary Spanning, Exploration, and Impact in the Optical Disk Industry', *Strategic Management Journal*, 22(4): 287–306.

Rosenkopf, L., and P. Almeida. (2003). 'Overcoming Local Search through Alliances and Mobility', *Management Science*, 49(6): 751–66.

Roth, A. E. (2008). 'What Have We Learned From Market Design?' *Economic Journal*, 118(527): 285–310.

Schumpeter, J. A. (1942). *Capitalism, Socialism, and Democracy*. New York, NY: Harper and Brothers.

Scotchmer, S. (1991). 'Standing on the Shoulders of Giants: Cumulative Research and the Patent Law', *Journal of Economic Perspectives*, 24(3): 29–41.

Shepard, A. (1987). 'Licensing to Enhance Demand for New Technologies', *RAND Journal of Economics*, 18(3): 360–68.

Sobel, D. (1995). *Longitude: The True Story of a Lone Genius who Solved the Greatest Scientific Problem of his Time*. New York, NY: Walker.

Spencer, J. W. (2003). 'Firms' Knowledge-sharing Strategies in the Global Innovation System: Empirical Evidence from the Flat Panel Display Industry', *Strategic Management Journal*, 24(3): 217–33.

Stam, W. (2009). 'When does Community Participation Enhance the Performance of Open Source Software Companies?' *Research Policy*, 38(8): 1288–99.

Teece, D. J. (1986). 'Profiting from Technological Innovation: Implications for Integration Collaboration, Licensing and Public Policy', *Research Policy*, 15(6): 285–305.

Tushman, M. L., and Scanlan, T. J. (1981). 'Boundary Spanning Individuals: Their Role in Information Transfer and their Antecedents', *Academy of Management Journal*, 24(2): 289–305.

Van de Vrande, V., de Jong, J. P. J., Vanhaverbeke, W., and de Rochemont, M., (2009). 'Open Innovation in SME's: Trends, Motives and Management Challenges', *Technovation* 29(6/7): 423–37.

Varian, H. R., and Shapiro, C. (1999). *Information Rules: A Strategic Guide to the Network Economy*. Boston, MA: Harvard Business School Press.

von Hippel, E. (1986). 'Lead Users: A Source of Novel Product Concepts', *Management Science*, 32(7): 791–805.

von Hippel, E. (1988). *The Sources of Innovation*. New York, NY: Oxford University Press.

von Hippel, E. (1994). '"Sticky Information" and the Locus of Problem Solving: Implications for Innovation', *Management Science*, 40(4): 429–39.

von Hippel, E. (2001). 'User Toolkits for Innovation', *Journal of Product Innovation Management*, 18(4): 247–57.

von Hippel, E. (2005). *Democratizing Innovation*. Cambridge, MA: MIT Press.

von Hippel, E., von Krogh, G. (2003). 'Open Source Software and the 'Private-collective' Innovation Model: Issues for Organization Science', *Organization Science*, 14(2): 209–23.

Waguespack, D. M, and Fleming L. (2009). 'Scanning the Commons? Evidence on the Benefits of Startups Participating in Open Standards Development'. *Management Science*, 55: 210–23.

West, J. (2003). 'How Open is Open Enough? Melding Proprietary and Open Source Platform Strategies'. *Research Policy*, 32(7): 1259–85.

West, J. (2006). 'Does appropriability enable or retard open innovation?' in H. Chesbrough, W. Vanhaverbeke, and J. West (eds.), *Open Innovation: Researching a New Paradigm*. Oxford: Oxford University Press, pp. 109–133.

Williamson, O. E. (1975). *Markets and hierarchies: Analysis and antitrust implications*. New York, NY: Free Press.

CHAPTER 23

COLLABORATION AND INNOVATION MANAGEMENT

MARK DODGSON

INTRODUCTION

COLLABORATION is a *sine qua non* of innovation management because innovation invariably involves many and diverse contributors. Whereas a large, research and development (R&D)-based corporation of the past might have been autarkic in its innovations, very few organizations today, if any, can innovate without collaborating in some form or another. Even companies with the business and technological prowess of IBM rely heavily on collaboration (Dittrich et al., 2007). This chapter describes what collaboration is and the forms it takes. It explains why collaboration is such an important issue for innovation and its management, and what encourages it to occur. It then explores the ways it can be effectively managed to improve innovation performance. Collaboration can be very challenging for organizations—the process itself can be unstable and troublesome, and disputes can arise over the ownership of its outcomes—so the ways it is managed critically affects its success.

Collaboration is defined as the shared commitment of resources to the mutually agreed aims of a number of partners. Although collaboration does not occur unless individuals and teams work effectively together, our level of analysis will be the organization. Collaboration is used to develop new markets, gain access to production and distribution networks, and most commonly to address issues related to research, technology, and innovation. Mowery and Grodal (2005: 59) explain the predominant focus on the latter issues as a result of the increasing importance of the creation and commercialization of knowledge for competitiveness and economic growth. Geographically they can have a local, regional, national, and international focus. Collaboration can take many forms: strategic alliances and joint ventures, R&D consortia, university–industry and government–industry partnerships. Strategic alliances are partnerships between organizations that, unlike joint ventures, do not involve shared equity in third parties.

R&D consortia are larger groups of companies that fund and share common research projects. Collaboration can take place in 'vertical' arrangements within supply or value chains, such as when a car manufacturer partners with a supplier of components, and in 'horizontal' structures involving parties engaged in similar activities, such as when one chemical company works with another. It is only collaboration in our definition if all contributors commit resources to it and mutually determine its objectives. They are different from broadly defined 'networks', (see Chapter 6 by Kastelle and Steen) which often exist for purposes of information search.[1] Transactions within single organizations or conducted by means of contracts to purchase goods and services are not collaborations. It differs, therefore, as a form of organizing from hierarchies or markets (Williamson, 1975), while providing some of the controls of the former and incentives of the latter.

Collaboration is not a new phenomenon. An attempt was made to form a consortium of competing companies to address a research problem of mutual concern in the Staffordshire potteries in England in 1775 (Dodgson, 2011). In the 1920s and 1930s, an appreciation of the value of collaboration led to the creation of Research Associations to address common technological problems in numbers of UK industries. Funded by levies from industry, the Research Associations built expertise in specific sectors, such as ceramics and textiles, and in industrial processes such as welding. Having gone through various changes in structure and funding, a number of these organizations are still operating today. In the 1980s, driven by government policies, large collaborative programmes in information and communications technology were instigated in Europe, Japan, and the USA. More contemporarily, some databases of numbers of collaborations formally identified in the business and trade press show increasing collaborative activity (Hagedoorn, 2002). These databases are not comprehensive (Schilling, 2009), many collaborations are not publicized, and some are secretive, so these databases may under-estimate actual levels. Schilling's (2009) review of some of these databases shows some inconsistency in temporal trends, but they all highlight the high proportion of them in information technology, transportation, and chemicals. Studies of the emergence of new technologies, such as biotechnology and new materials (Rothaermel and Deeds, 2004; Faems et al., 2010), and of the dynamics of entire industries, regions and national economies, show the importance of collaboration. The phenomenal success of Taiwanese industry, for example, has depended on high levels of collaboration (Mathews, 2002; Dodgson et al., 2008).

Why Organizations Collaborate
to Innovate

A number of academic disciplines have addressed the question of why organizations collaborate. Economists, for example, theorize why organizations collaborate according to the issue of transaction costs in different organizational structures (Williamson, 1975), incentives using game theory (Binenbaum, 2008), and evolutionary and complex systems approaches to accessing knowledge in uncertain conditions (Foster and Metcalfe, 2004; see also Dodgson, 2007). Organization theorists consider the social capital of networks, and address the role of collaboration in filling 'structural holes' in networks (Walker et al., 1997; Ahuja, 2000; Burt, 2005). Sociologists consider the legitimation and status that collaboration confers and issues of power within them (Podolny and Page, 1998). Political economists address the differences in varieties of capitalism between, for example, analyses of 'stock market' and 'welfare' capitalism (Dore, 2000) and 'liberal market' and 'coordinated market' economies (Hall and Soskice, 2001), and argue their different characteristics can affect the closeness of inter-firm collaboration. Strategy academics bring their perspectives on the value provided in collaboration through access to resources and organizational learning (Eisenhardt and Schoonhoven, 1996). Mowery and Grodal (2005: 67) contend that organizational sociology and business management show that 'networks and innovation constitute a virtuous cycle. External linkages facilitate innovation, and at the same time innovative outputs attract further collaborative ties'.

These different viewpoints reflect widely divergent concerns. Economists tend to focus on the cost considerations of collaboration, and especially on the impact of any public subsidy for them (e.g. Cohen, 1994). Sociologists and political scientists are concerned with the group and national dynamics underlying the formation and conduct of collaboration. Management strategists view collaboration from the standpoint of accessing strategic resources for competitive advantage. Our approach here is to take a corporate and innovation perspective, and we consider the role of collaboration in contributing to organizations' abilities to attain complementarities, encourage learning and develop their capabilities, and deal with uncertainty and complexity.

Complementarities

As referred to throughout this book, according to Schumpeter, innovation involves new combinations. In modern-day economies, organizations cannot develop the range of expertise needed—in research, technology, and marketing—to realize the opportunities for new combinations by themselves. When the resources and capabilities necessary to develop and benefit from innovation are distributed amongst different organizations, these new combinations can be forged through collaboration. In Teece's (1986) formulation, collaboration allows innovators access to the complementary assets necessary to appropriate returns to their investment. Access to these assets overcomes many of the bottlenecks that organizations face in getting their ideas successfully applied in markets. This may involve access, for example, to manufacturing and distribution networks or knowledge of regulations.

Collaboration is also a source of synergies between complementary partners in platforms (see Chapter 32 by Gawer and Cusumano) and ecosystems (see Chapter 11 by

Autio and Llewellyn). When we use a mobile telephone, for example, we rely on the different types of expertise of infrastructure providers of communications towers and digital switches, handset manufacturers, telecommunication carrier and service firms, and content providers. The integration of such systems requires collaboration in a number of forms, including in the creation of technical standards. Another example is seen in the increased amount of information technology in automobiles. Nowadays, cars are sophisticated communications devices and carmakers and technology firms are collaborating to make them safer and better connected with the Internet and information and communications systems. Microsoft, for example, collaborates with Ford and Nissan in areas such as in-car information and entertainment and the use of cloud computing,

Other forms of synergies can arise from different kinds of organization working together. Small firms, for example, can be flexible and responsive to new market and technological opportunities in ways large firms cannot, and one of the benefits of collaboration is that it combines the entrepreneurial behavioural advantages of the former with the structures and resources of the latter in a form of 'dynamic complementarity' (Rothwell and Dodgson, 1991).

The choice to collaborate to attain these complementarities, rather than by alternatives such as market contracts or merger and acquisitions, is influenced by the nature of the knowledge being transferred. Knowledge that is not codified, that is, written down in documents and drawings, and is tacit within individuals or engrained in the particular cultures of organizations, is difficult to acquire, and its transfer is better facilitated by the deep and long-term engaged communications that can be encouraged by collaboration (Mowery and Grodal, 2005). Collaboration may also provide a cheaper option than its alternatives. Developing capabilities in-house, or using mergers and acquisitions (see Chapter 29 by Ahuja and Novelli), which have a poor record of success, may be less attractive than a partnership with clearly defined budgetary limits and proscribed timelines for performance. When ideas and knowledge are at their early stages of development, furthermore, they are difficult to value and price and so collaboration may be a more effective means of exchange than market mechanisms.

Learning and Capabilities

Over time, well-established routines cause organizations to become introspective (March, 1991, 2006), restricting their ability to innovate. Organizations learn to do the things they already do better, learn to do new things, and learn about the need to learn (Argyris and Schön, 1978). Procedures and routines for learning about what organizations already do are more easily established, but they can inhibit other forms of learning (Morgan, 1986; March, 2006). Organizations furthermore cling to outmoded identities that thwart 'higher level' or radical learning (Brown and Starkey, 2000), and introspection results from processes of institutional isomorphism (Di Maggio and Powell, 1983) that make organizations that work together over time more similar. March (1991) argues, based on his dichotomy of exploration and exploitation in organizational learning, that the latter often restricts the former. Focusing on what is already known produces returns that are positive, proximate, and predictable; focusing on the novel produces returns that are uncertain, distant, and often negative. As a result, this produces a tendency to substitute exploitation of known alternatives for the exploration of unknown ones (March, 1991). In Leonard-Barton's (1992) terms, core capabilities become core rigidities (see also Chapter 7 by Leonard and Barton). For these reasons, organizations fail to learn in ways that allow them to innovate in response to the changes and disruption that surround them. Here collaboration plays a role as a stimulus to learning about the practices, technologies, and strategies in other organizations (Child and Faulkner, 1998). It presents an opportunity to question the status quo and examine and assess the relevance and applicability of different approaches and practices in partners.

Uncertainty and Complexity

The uncertainties that confront innovation managers are much greater when innovations are radical or disruptive and when, as a result of rapidly changing technologies and volatile markets, their development processes are unpredictable. That is, when managers are uncertain about their journey and destination. Knight (1921) classically distinguishes between risk and uncertainty. Whereas risk is a known chance (or measurable uncertainty), uncertainty is an unmeasurable or truly unknown outcome. In more radical and disruptive innovations it may not be possible to quantify risks because the issues being dealt with are highly unique and likely to be influenced by a range of findings and events that can produce a variety of unpredictable outcomes. All types of organizations face the challenge of preparing for an unknown future, distinguishing between the risks of a possible range of outcomes occurring and uncertainties that cannot be predicted. Collaboration can help reduce uncertainties by making sense of rapid changes and by building shared expectations and approaches to innovation challenges. The development of collaboratively agreed technical standards, for example, reduces the risk that an organization's products or services do not comply with emerging new platform structures. When new markets are being formed, there are benefits in not going it alone, as Sony discovered with its Betamax video recorder that lost out in the market to the technically inferior VHS system because of Matsushita's better complementary partnerships (Rosenbloom and Cusumano, 1988). Podolny and Page (1998: 66) point out that organizations can alleviate sources of external constraint or uncertainty by strengthening their relationship with the particular sources of dependence. In this sense, collaboration is a strategy of sharing control in order to retain it.

The sheer complexity of many innovations requires inputs from a diversity of disciplines, professions, and capabilities that one organization, no matter how large, cannot possess. The construction of modern day commercial aircraft, for example, incorporates highly specialized expertise in aerodynamics, propulsion, avionics, and materials. Aircraft manufacturers coordinate inputs from providers that often have deeper expertise in these areas than they do themselves, and collaboration helps to integrate innovations in each area.

During the 1980s the phenomenon of the dedicated biotechnology firm emerged: usually small firms basing their business around new advances in molecular biology (Dodgson, 1991). Many large pharmaceutical firms at the time were uncertain about the potential of the technology and wary of the disruption it would cause to their traditional drug discovery methods. As a result, many large firms chose to collaborate with the biotechnology firms as a way of keeping a watching brief. In this way the learning that occurs in collaboration provides options for organizations (Powell et al., 1996). It is not as irreversible as acquisitions, and can build knowledge internally to allow better future decisions on whether and how to innovate.

BUSINESS AND UNIVERSITY COLLABORATION

Compared to companies' staff, customers, and suppliers, universities are relatively low in the hierarchy of sources of innovation for firms. They can, however, be sources of more radical and disruptive innovations (Belderbos et al., 2004). The challenge of collaboration between universities and business lies fundamentally in the way the former are concerned with research problems and the latter with business problems. There are enhanced opportunities to collaborate where the two coincide, although differences often remain over time horizons and, potentially, over protection and ownership of intellectual property. Reference is also often made to the cultural differences between businesses and universities, in, for example, preferences for theory compared to practice. Many leading businesses, however, recognize the value of these differences and their combination, and choose to collaborate with universities precisely because their staff have different approaches and means of addressing problems. Universities increasingly see collaboration as part of their 'engagement' roles, along with teaching and research (see Chapter 4 by McKelvey), although there are many voices concerned with the need for universities to retain their distinctive identities and not become too 'business-like' (e.g. Bok, 2003).

Universities undertake extensive research and consulting projects for businesses, but for these to be considered formal collaborations in our definition there would need to be shared resource commitments and agendas. These are sometimes encouraged and formalized by government policies, such as in the case of the Cooperative Research Centres (CRCs) in Australia, which involve partnerships between researchers and end-users of research, including businesses. The Minister for Science and Technology claimed at the launch of the CRC programme in 1991 that they: 'will help Australia to achieve closer linkages between science and the market'. Nearly 200 CRCs have operated during the programme's existence, and while there is conflicting evidence of their value, and some have succeeded far better than others, there is a general consensus that they have generated beneficial research and economic outcomes (O'Kane et al., 2008).

These benefits have, however, tended to be geared towards large, research-intensive organizations.

One of the challenges of business–university collaboration is the management of their interfaces. Many universities have industrial liaison and/or technology transfer offices. The performance of these organizations has been very mixed. There are notable successes, such as MIT and Stanford, but few around the world have generated more income than they cost to run. Those that operate primarily on the basis of protecting then trading university-generated intellectual property often limit the amount of collaboration occurring, acting as a barrier and disincentive to business. While in some limited circumstances this strategy can be useful—for example in giving a biotechnology spin out a patent position from which to attract investment—more beneficial is an approach that facilitates broad communications, dialogue, and exchange. This requires a concern for customer relationship management. Many large multinational companies have executives responsible for liaising with particular universities. There is value in universities appointing senior academics with similar responsibilities. These kinds of positions help build deep, multilevel, and long-term strategic relationships.

GOVERNMENT POLICIES FOR COLLABORATION

Governments, as seen in the case of the Australia's CRCs, can play an important role in encouraging collaboration. They support collaboration for a variety of reasons, including concern for diminishing international competitiveness, increasing the scale of resources devoted to innovation in key sectors and technologies, and reducing duplication in scarce investments. Policy-makers over recent decades have increasingly referred to the concept of national innovation systems. Governments are the only institutions that can take a whole national innovation systems approach, and collaboration is an important element in ensuring cohesiveness in their various components (Lundvall, 2007).

Outside of wartime, one of the earliest government schemes for collaboration, begun in 1982, was the Fifth Generation Computer Systems project, an initiative of Japan's Ministry of International Trade and Industry to create a 'fifth generation computer'. In reaction to this project the British government instigated the Alvey Programme, a sponsored collaborative research programme in information technology that ran from 1983 to 1987. At the same time, the European Commission initiated the European Strategic Program on Research in Information Technology (ESPRIT). ESPRIT was a series of collaborative information technology research and development projects. Five ESPRIT programmes ran from 1983 to 1998. An example of an American equivalent was Sematech, formed in 1987 as a result of the concern for growing Japanese competitiveness in the industry, when the US government brought together fourteen US-based semiconductor companies to solve common manufacturing problems and share risks from research investments.

The results from such government programmes are mixed. There are conflicting views on their economic contribution, and actual research outcomes have been argued to vary in quality and significance, but a common theme in their evaluation has been how it improves the propensity of otherwise competing firms to collaborate, and increases their skills to do so. As Zahra and George (2002) argue, firms with greater diversity in their partnerships develop skills to improve their ability to absorb valuable information from outside sources.

In one of the most comprehensive economic studies of the value of R&D consortia, Branstetter and Sakakibara (2002) found that participation in Japanese government-sponsored consortia had positive effects on research performance and these were more marked in those focusing on basic research. In the case of Europe's Eureka programme—an internationally collaborative initiative, begun in 1985 to promote closer cooperation between firms and research centres in 'market-led' innovation—participation has been shown to increase the return on assets compared to non-participating firms (Bayona-Saez and Garcia-Marco, 2010).

Sematech has been extensively studied. Browning et al.'s (1995) analysis shows that it overcame early disorder and ambiguity of purpose and developed a 'moral community' where individuals made contributions to the industry without regard to immediate and specific feedback. They argue that Sematech provided a 'sort of neutral ground on which "blood enemies" can cooperate within certain agreed-upon boundaries' (1995: 145). Carayannis and Alexander (2004) argue that Sematech shows how difficult it is to create and sustain R&D consortia, but argue its evolution helped the industry regain competitiveness. Grindley et al. (1994) argued that Sematech increased the competitiveness of the already strong members, and their success depended on their having other strengths such as marketing. Irwin and Klenow (1996) found that members of Sematech cut their own R&D spending as a result of joining the consortia, evidence of what they call the sharing objective of reducing duplication. Link et al. (1996) conclude that government funding was helpful because Sematech advanced its members and semiconductor technologies in ways which could not have been justified on economic grounds outside of a collaborative research arrangement. Perhaps the most important indication of its contribution is that government-supporting funds were removed in 1996, but the consortium continues to operate.

Government support for such collaborations can be controversial. US President Ronald Reagan had to introduce new legislation to allow research consortia to be formed, as President Woodrow Wilson's antitrust laws of 1914 had previously banned them. Any such collaborations pose a challenge for governments about what to do with excluded firms, and how they should treat foreign-owned companies. To some extent, funding collaboration prevents governments from being accused of attempting to 'pick winners' in technologies and firms: something most have been notoriously bad at. However, in some cases, the amount of collaboration actually instigated by schemes introduced under the rubric of increasing collaboration has been minimal. It can hide a form of corporate welfare.

COLLABORATION AND TECHNICAL STANDARDS

Proprietary standards can confer considerable competitive advantage. While a few widely accepted standards are *de facto*, that is, established by individual companies and diffused because of their market power, the vast majority are *de jure*, that is, developed in standards bodies. Establishing *de jure* standards inevitably involves collaboration. Zhang and Dodgson (2007) show the complexity of these collaborations in the case of mobile payments. Mobile payments essentially involve using mobile telephones to pay bills, for example with utilities, restaurants and shops, transportation tickets, and parking. The potential of mobile payments has excited the attention of numerous interests in banking and finance and mobile hardware and software producers, and the competitiveness between them all has led to a lack of consensus on technology and security standards, which in turn has curtailed the development of the industry.

There are numerous industry consortia designed to promote mobile payment standards, and these reflect widely differing concerns. They reveal broad differences in choices in technology and whether the consortia are led by financial institutions; mobile operators or mobile equipment manufacturers; Americans, Europeans, or Asians; and a variety of engagement mechanisms with small start-up companies. The existence of these different consortia, and the preparedness in some cases to belong to a number of them, reflects the uncertainties that exist around the development of a new industry. Decisions about who to collaborate with in 'standards wars' are an important element of strategic positioning in emerging industries.

INSTABILITIES AND TENSIONS

Collaborations experience tensions and unplanned disruptions. Some studies report failure rates of 30–50 per cent on the 'bumpy road' of partnerships (Lokshin et al., 2011). Some terminations—such as in the case of a two-year alliance between Suzuki and Volkswagen—can lead to mutual and highly public recriminations. Three out of fourteen members of Sematech exited after their first five-year commitment because their objectives were not realized. Das and Teng (2000) argue that strategic alliances are sites where three conflicting forces develop: between cooperation versus competition, rigidity versus flexibility, and short-term versus long-term orientation. They argue the tensions of cooperation are greater between than within organizations (although some organizations have been known to report that it is easier to collaborate externally than internally), and in contrast to single organizations, collaborations find it more challenging to maintain a balance between flexibility and rigidity and their short- and long-term needs.

Echoing the findings of a classic study of innovating firms (Rothwell et al., 1974), Lokshin et al. (2011) show the value of consistency of innovation strategy and external orientation in overcoming the malfunctions of alliances. They found that:

> firms that persistently rely on a product focus innovation strategy through developing new products, new markets and improving product quality face better chances in dealing with a "bumpy road." A strategy based on the exploration of these new opportunities, where firms persistently engage in joint innovation projects, rather than jointly exploiting cost efficiencies, lowers the probability of inter-organizational malfunctioning.
>
> (Lokshin et al., 2011: 305)

They also found that avoiding instabilities improved innovation performance, and working with a diverse portfolio of partners developed the experience required to avoid malfunctions.

THE STRENGTHS AND WEAKNESSES OF STRONG AND WEAK TIES

Key questions for innovation management are the relative virtues of a few deep collaborations compared to a large number of shallow ones, and the balance to be maintained by building cohesive, long-term partnerships whilst being receptive to new ones. Much thinking in this area results from a sociological study of job seekers by Granovetter (1973; and see discussion in Chapter 6 by Kastelle and Steen) who, in a classic paper, argues that strong ties in networks are likely to communicate information of limited value whereas weak ties convey potentially useful novel information. Many weak ties for individuals who span multiple social worlds are argued to lead to a person having access to a great diversity of information. The question of what this means for individuals has been extensively reconsidered for organizations. Radical innovations especially are often based on new combinations of diverse knowledge domains, and are argued to benefit from diverse networks of weak ties (Elfring and Hulsink, 2007: 1853). However, weak ties have a number of disadvantages, for instance they are less useful for mobilizing resources and conveying tacit knowledge. Whereas strong ties build trust that overcomes these disadvantages, they can suffer from inertia and lock-in (Maurer and Ebers, 2006).

There is also some uncertainty about the contribution of structural holes in collaborative networks for innovation. A structural hole (Burt, 2004) describes the situation where an actor not only spans social worlds, but spans otherwise poorly connected worlds. The value of the diversity of information resulting from this structural position is argued to put these actors—or brokers—at a distinct advantage. Brokers within structural holes are believed to lead to improved innovation performance (Burt, 2005), but in some situations, however, networks rich in structural holes are shown to damage innovation performance (Ahuja, 2000). In such contexts, the social capital found in dense networks is more advantageous for innovation.

There is an emerging view that both strong and weak ties are beneficial to firms, but under different conditions, for different purposes and at different times in a firm's development (Rowley et al., 2000; Hite and Hesterly, 2001; Elfring and Hulsink, 2007). Thus, for example, embeddedness—where social relationships influence economic actions (Uzzi, 1999)—is argued to be more common in the earlier stages of the development of the firm because it helps overcome challenges of resource access and limited awareness of available resources and opportunities (Hite and Hesterly, 2001: 279). In the case of new product development, Hansen (1999) shows weak ties assist the search for innovation opportunities and strong ties assist the transfer of complex knowledge. The debate on the relative merits of strong and weak ties for innovation and entrepreneurship, however, is continuing (see, e.g. Reagans and Zuckerman, 2008), with some arguing that evidence remains inconclusive (Kastelle and Steen, 2010b). It may be the case, however, that as suggested by some research (Rosenkopf, Metiu, & George, 2001), that informal collaborations between firms tend to lead to more formal collaborations over time. Weak ties are a conduit and filter to future strong ties.

THE ROLE OF THE BROKER

Sometimes organizations that might benefit from collaboration cannot find partners themselves and brokers facilitate their connection. These brokers may include consultants and advisors and a range of public and private sector organizations (see discussion in Chapter 9 by Hargadon). These are not mechanisms designed to connect organizations for particular projects—such as the internet-based Innocentive or Yet2.com—but to facilitate long-term strategic engagement. Two examples will be provided.

Taiwan's Industrial Technology Research Institute (ITRI) has been crucial for that country's industrial development. Established in 1973, it employs over 6,000 people with the broad objective of increasing Taiwan's international competitiveness in technology industries. ITRI developed a successful strategy focused on using innovation networks for technology diffusion. An international market opportunity is identified by ITRI, which then assembles a collaboration with a combination of several large multinational firms and a number of small Taiwanese firms. Access to the Asian market for the multinational is traded for access to technology. The local part of the alliance is structured to maximize the flow of knowledge to constituent firms, which cooperate to build products based on this technology but also compete to bring these products to market. Over time, two or three dominant firms grow out of this network, and they will have the technical and learning skills needed to compete successfully in international markets (Dodgson et al., 2008).

AMIRA International, a not-for-profit private sector company was established in 1959 as an independent association of mineral companies, to facilitate the technical advancement of its members in the mineral, coal, petroleum, and associated industries. Based in Melbourne, with offices in Perth, Cape Town, and Johannesburg and affiliations with similar bodies in North America and Europe, it was created to develop, broker, and facilitate collaborative research projects, with a focus on assisting the uptake of leading edge science and technology by the members and on strengthening their business and development. By taking a partnership approach to R&D, AMIRA International manages and enhances the competitive position of its members by providing access to leading edge technology they would not otherwise be exposed to. AMIRA operates by developing and managing jointly funded research projects on a fee-for-service basis. AMIRA estimates that companies sponsoring the research projects enjoy financial leverage of 10 to 20 times the funds contributed, simply by sharing the costs and benefits of the research (Dodgson and Steen, 2008).

The Management of Collaboration

Partner Selection

The most important issue in collaboration is partner selection. Amongst the many challenges of collaboration, organizations sometimes partner with others with which they normally compete. So, for example, Hitachi, Toshiba, and Panasonic collaborate to make LCD panels; BMW and Toyota share low emission vehicle battery technology; and Apple and Motorola compete over smartphones and collaborate over digital music, so it is important to determine the demarcation between areas where relationships are collaborative and where they are competitive. Selecting the most appropriate partners depends on the objectives of the collaboration. When the objectives are *economizing*, by sharing costs and building scale, then the sort of collaborator will be different to when they are *strategizing*, or sharing resources to, for example, assist diversification or improve organizational learning. In the former case, economizing usually requires compatible and sympathetic resources, practices and technologies to ensure swift and seamless connections. Although in all cases there are advantages in having complementary cultures, when learning is an objective there are occasions when the benefits are derived from collaborating with different kinds of organization. An example might be the opportunities for mutual learning between a large, bureaucratic pharmaceutical company and a small, flexible start-up biotechnology firm. Diversity is a stimulus to learning. One of the challenges of managing collaboration lies with the collaboration paradox that you learn more from organizations that are more difficult to work with (Dodgson, 1993).

Although obviously limited as an analogy, there is some resonance in considering partner selection as being akin to selecting a life partner. Early flushes of attraction need to evolve into deep compatibilities and complementary growth paths for both partners, with frequent re-affirmation of the contribution of the bonds. Should the partnership not deliver on expectations, a failure to amicably negotiate separation terms over the fruits of the relationship results only in painful arguments, with the major beneficiaries being the legal profession.

When selecting a partner it is not only important to assess their immediate attractions—in resources and capabilities—and their record as a collaborator—which can be an important asset in its own right—but also their culture and traditions. As Hibbert and Huxham (2010: 547) put it, the possibility of imagining a common future may well rest on the discovery of common ancient traditions, albeit perhaps re-appropriated and re-interpreted. One of the reasons that Philips in the Netherlands and Matsushita in Japan collaborated successfully for forty years—and helped develop, for example, the VHS recorder—lay with the similar guiding philosophies of their founders.

Collaborations are instigated by various means. They can emerge in response to environmental change or be engineered for specific purposes (Doz et al., 2000). They might be created through a systemic search for a partner or an opportunistic meeting between two CEOs at a conference. They could emerge from engineering teams getting to know one another through common interests and then building shared agendas. They might be mediated by consultants or advisors, or by brokering organizations, such as ITRI and AMIRA.

No matter their progenitors, in each case, for a variety of reasons, the capacity to build mutual trust and empathy is important. Trust is needed in partners' expertise, professionalism, and discretion, for example in not leaking valuable information to competitors outside of the collaboration. When collaborations extend over many years, such as Rover and Honda's fifteen-year joint development of new technologies, at any one stage one partner may be up on the deal, and trust is needed that things will balance out over time. Collaborations often begin on the basis of empathy between individuals. People, however, move to other jobs and organizations and to prevent the breakdown of collaboration when this occurs, inter-personal trust needs to evolve into inter-organizational trust. This requires a community of interest, organizational cultures receptive to external ideas, and widespread and continually supplemented knowledge among employees of the status and purpose of the collaboration (Dodgson, 1993). Building trust, or 'relational capital', encourages learning from partners and protects against opportunistic behaviour (Kale et al., 2000).

Organizations get selected as partners when they possess distinctive capabilities and resources, and this includes the capacity to develop these over time. So the capacity to evolve and grow can be just as attractive as what currently exists: partners are chosen because of their future potential.

Structuring and Organizing the Collaboration

Collaborations work best when there is mutual respect amongst peers in the various partners. Researchers and technologists generally prefer to work with counterparts with similar levels of knowledge and expertise, and holding equivalents in esteem is important throughout all aspects of the partnership, for example with business development, marketing, and legal departments.

The nature of the reporting relationship of the collaboration into the home institutions has to be clarified. A joint venture between a British telecommunications company, BT, and a US chemical company, Du Pont, faced difficulties as it reported into a technical function in the former and a marketing function in the latter (Dodgson, 1993). It was being judged by different criteria. The question of who is responsible for managing the collaboration is similarly germane. In the latter case, a rotation of chief executives seconded from the partners caused difficulties as they were perceived to represent their home employer, and these were not overcome until an independent manager was recruited. Collaborations often need highly talented managers who may be required, for example, to encourage performance in partner organizations without the means to use the incentives and sanctions available in their own. Such talented managers may be in short supply and those there are may already enjoy considerable status, running large parts of a company with clear career progression. The opportunity to divert to smaller and comparatively riskier ventures may not be appealing, in which case persuasive incentives are required.

The review procedures for collaboration need, on the one hand, to clearly identify criteria for assessment and performance and, on the other, be capable of adapting to changing circumstances and opportunities. Collaborators have to assess what their objectives are, that is, a particular level of technological development, the creation of a market of a certain size, or the transfer of better practices. There may need to be flexibility in these objectives as unforeseen obstacles or opportunities might emerge. Sematech's objective evolved, for example, from building next generation manufacturing technology to generic technology and equipment industry infrastructure. Clear determination of criteria and time-scales for termination are required. A successful technological outcome may be celebrated by one partner, and decried by another if it fails to meets its market objectives. Termination is not always a sign of failure, but can reveal how original objectives have been met.

Assessing the returns from collaboration are as notoriously difficult as those from investments in R&D. Distinctions can be made between quantified and tangible and qualitative and intangible returns. In their evaluation of Sematech, Link et al., (1996) found the costs of participating much easier to identify than the benefits. Only in a few cases did partners endeavour to quantify their benefits. They found some companies seemed to rely mostly on 'resonance' benefits—defined as industry-wide focusing effects—to justify their continued participation. These authors suggest that the intangible benefits weighed more heavily in participants' decisions to continue to participate in the consortia.

Ownership of intellectual property that results from collaboration is a potential source of conflict between partners that can be mitigated by clear early determinations. Agreeing mutual licensing rights, or geographical separation of target markets, for example, can prevent disputes.

ARCHITECT THE INNOVATION ECOSYSTEM

All the actors that can influence the ways an organization innovates can be described in combination as its ecosystem (see Chapter 11 by Autio and Thomas). These include customers and suppliers; manufacturers and suppliers of financial and legal services; governmental, university, and commercial providers of R&D; means of transportation and logistics; and government regulations. Relationships within these ecosystems can involve contracts, spot trading, and collaborations. Ecosystems provide an opportunity set for collaborations within them. The value of the concept of innovation ecosystems lies in the ways, like their natural counterparts, they emphasize the multiple and interrelated factors affecting their continual evolution, and their vulnerability to dramatic and disruptive events. Businesses identify the value of ecosystems: in an email to all staff, for example, Nokia's CEO explained the problem confronting the company at that time was a failure to build its ecosystem.

The case of smartphones provides insights into the ways ecosystems can provide the basis for future competitiveness. Smartphones are comprised of hardware (the handset), operating system, applications, and carrier services. Various kinds of ecosystem have evolved. In the case of Apple's iPhone, for example, the company controls the development of the handset and operating system, curates applications, and deals with specific carriers, such as AT&T in the USA. In contrast to this relatively closed system, Google's Android operating system is very open. It is given away free. Handset manufacturers such as Motorola, HTC, and Samsung, and carriers such as Verizon, work closely with Google, but the intention is for Android to be used with hundreds of device manufacturers and numerous service providers. There is no attempt to control the development and introduction of applications, although they can be removed if complaints arise. Another smartphone ecosystem is the joint development and marketing venture between Nokia and Microsoft. Nokia will eventually replace Symbian, its operating system, with Microsoft's Windows 7. Along with Microsoft's Office, Bing, and Xbox Live, this will be combined with Nokia's handsets, applications store, and its Maps service. The companies are collaborating over the development of a bill payment system. These smartphone ecosystems are themselves part of larger ecosystems. Apple's smartphones, for example, depend on their relationships with its other products and with iTunes.

It is impossible to judge which of these different models for smartphones, if any, will be most successful, but it is clear that future competitiveness in the industry will result from battles of ecosystems, rather than between individual products and services. Cohesiveness and clarity of objectives amongst collaborative organizations will be crucial for the success of these ecosystems, as will the capacity to engage more broadly, for example with providers of payment systems, telecommunications infrastructure, and semiconductors. The ability to architect the ecosystem—which itself will be built around a base of collaborations—will be the source of future competitiveness.

Conclusions

As argued by Salter and Alexy in Chapter 2, collaboration is a stylized fact of innovation. The role that collaborations play—that is, the close partnerships between a number of organizations—within emerging and evolving innovation ecosystems is a field of research ripe for future exploration. Similarly, the consequences for collaboration of open innovation strategies (see Chapter 22 by Alexy and Dahlander), platforms (see Chapter 32 by Gawer and Cusumano), and new digital technologies for connectivity (see Chapter 19 by Dodgson and Gann) all warrant future research. What effect on collaboration is there, for example, if much of an organization's intellectual property is freely available? How is trust and empathy built in inter-personal networks based on the use of social media? Do visualization technologies allow more diverse collaborations to be managed? While all these changes pose new questions for innovation management, there are some features of collaboration that, while continually challenging, remain constant.

The benefits of collaboration are achieved by careful partner selection, the building of long-term trust-based relationships, effective structures and managerial processes, and clear, but flexible, assessment and evaluation criteria. Collaboration is a key feature of open innovation processes involving many and varied contributors (see Chapter 22 by Alexy and Dahlander). The capacity to attract collaborators, through possession of valued and distinctive resources and capabilities, and then work effectively with them, despite the difficulties this entails, is central to the ability of organizations to create, capture, and deliver value, and hence their continuing competitiveness. There remains an unresolved problem for innovation managers attempting to quantify the economic returns from collaboration, and their justification will invariably rest with strategic considerations of developing resources and capabilities, and increasing future options. Participation in Sematech reduced partner companies' intra-mural R&D expenditures. Nokia used substantial reductions in its R&D expenditures to justify its joint venture with Microsoft. There remain many interesting research questions on the strategic value of reducing these investments and on the wisdom of seeing internal and collaborative R&D as alternatives. It is likely in future that the best strategic opportunities lie with those organizations most adept at balancing their internal and external innovation efforts and their capacities to succeed both as competitor and collaborator.

Note

1. Where the term 'network' is used elsewhere in this chapter, it will be used in the sense of our definition of collaboration.

References

Ahuja, G. (2000). 'Collaboration Networks, Structural Holes, and Innovation: A longitudinal study', *Administrative Science Quarterly*, 45(3): 425–55.

Argyris, C., and Schön, D. (1978). *Organizational Learning: Theory, Method and Practice*. London: Addison Wesley, 305.

Bayona-Saez, C. and Garcia-Marco, T. (2010). 'Assessing the Effectiveness of the Eureka Program', *Research Policy*, 39(10): 1375–86.

Belderbos, R., Carree, M., and Lokshin, B. (2004). 'Cooperative R&D and Firm Performance', *Research Policy*, 33(10): 1477–92.

Binenbaum, E. (2008). 'Incentive Issues in R&D Consortia: Insights from Applied Game Theory, *Contemporary Economic Policy*, 26(4): 636–50.

Bok, D. (2003). *Universities in the Marketplace*. Princeton, NJ: Princeton University Press.

Branstetter, L., and Sakakibara, M. (2002). 'When do Research Consortia Work Well and Why? Evidence from Japanese panel data', *American Economic Review*, 92(1): 143–59.

Brown, A., and Starkey, K. (2000). 'Organizational Identity and Learning: A Psychodynamic Perspective', *Academy of Management Review*, 25(1): 102–20.

Browning, L., Beyer, J., and Shetler, J. (1995). 'Building Cooperation in a Competitive Industry: SEMATECH and the Semiconductor Industry', *Academy of Management Journal*, 38(1): 113–51.

Burt, R. (2004). 'Structural Holes and Good Ideas', *American Journal of Sociology*, 110(2): 349–99.

Burt, R. S. (2005). *Brokerage & Closure: An Introduction to Social Capital*. Oxford: Oxford University Press.

Carayannis, E., and Alexander, J. (2004). 'Strategy, Structure, and Performance Issues in Precompetitive R&D Consortia: Insights and Lessons Learned from SEMATECH', *IEEE Transactions on Engineering Management*, 51(2): 226–32.

Child, J., and Faulkner, D. (1998). *Strategies of Co-operation: Managing Alliances, Networks and Joint Ventures*. Oxford: Oxford University Press, 371.

Cohen, L. (1994). 'When can Government Subsidize Research Joint Ventures? Politics, Economics, and Limits to Technology Policy', *American Economic Review*, 84(2): 159–63.

Das, T., and Teng, B.-S. (2000). 'Instabilities of Strategic Alliances: An Internal Tensions Perspective', *Organization Science*, 11(1): 77–101.

DiMaggio, P., and Powell, W. (1983). 'The Iron Cage Revisited: Institutional Isomorphism and Collective Rationality in Organizational Fields', *American Sociological Review*, 48: 147–60.

Dittrich, K., Duysters, G., and de Man, A.-P. (2007), 'Strategic Repositioning by Means of Alliance Networks: The Case of IBM', *Research Policy*, 36(10): 1496–511.

Dodgson, M. (1991). *The Management of Technological Learning: Lessons from a Biotechnology Company* (de Gruyter Studies in organization 29). Berlin: Walter de Gruyter.

Dodgson, M. (1993). *Technological Collaboration in Industry: Strategy Policy and Internationalization in Innovation*. London: Routledge.

Dodgson, M. (2007). 'Technological Collaboration', in H. Hanusch and A. Pyka (eds), *Elgar Companion to Neo-Schumpeterian Economics*. Cheltenham: Edward Elgar, 193–200.

Dodgson, M. (2011). 'Exploring new Combinations in Innovation and Entrepreneurship: ZSocial Networks, Schumpeter, and the Case of Josiah Wedgwood (1730–1795)', *Industrial and Corporate Change*, 20(4): 1119–51.

Dodgson, M., and Steen, J. (2008). 'New Innovation Models and Australia's Old Economy', in J. Bessant and T. Venables (eds), *Creating Wealth from Knowledge: Meeting the Innovation Challenge*. Cheltenham: Edward Elgar.

Dodgson, M., Mathews, J., Kastelle, T., and Hu, M.-C. (2008). 'The Evolving Nature of Taiwan's National Innovation System: The Case of Biotechnology Innovation Networks', *Research Policy*, 37(3): 430–45.

Dore, R. (2000). *Stock Market Capitalism: Welfare Capitalism: Japan and Germany versus the Anglo-Saxons*. Oxford: Oxford University Press, 264.

Doz, Y., Olk, P., and Ring, P. (2000). 'Formation Processes of R&D Consortia: Which Path to Take? Where does it Lead?', *Strategic Management Journal*, 21(3): 239–66.

Eisenhardt, K., and Schoonhoven, C. (1996). 'Resource-based View of Strategic Alliance Formation: Strategic and Social Effects in Entrepreneurial Firms', *Organization Science*, 7: 136–50.

Elfring, T., and Hulsink, W. (2007). 'Networking by Entrepreneurs: Patterns of Tie-Formation in Emerging Organizations', *Organization Studies*, 28(12): 1849–72.

Faems, D., Janssens, M., and Van Looy, B. (2010). 'Managing the Cooperation–Competition Dilemma in R&D Alliances: A Multiple Case Study in the Advanced Materials Industry', *Creativity and Innovation Management*, 19(1): 3–22.

Foster, J., and Metcalfe, J., (2004). *Evolution and Economic Complexity*. Cheltenham, Edward Elgar.

Granovetter, M. (1973). 'The Strength of Weak Ties', *American Journal of Sociology*, 78: 1360–90.

Grindley, P., Mowery, D., and Silverman, B. (1994). 'SEMATECH and Collaborative Research: Lessons in the Design of High-technology Consortia', *Journal of Policy Analysis and Management*, 13(4): 723–58.

Hagedoorn, J. (2002). 'Inter-firm R&D Partnerships: An Overview of Major Trends and Patterns since 1960', *Research Policy*, 31(4): 477–92.

Hall, P. A., and Soskice, D. (eds) (2001). *Varieties of Capitalism: The Institutional Foundations of Comparative Advantage*. Oxford: Oxford University Press, 540.

Hansen, M. T. (1999). 'The Search–Transfer Problem: The Role of Weak ties in Sharing Knowledge across Organizational Subunits', *Administrative Science Quarterly*, 44(1): 82–111.

Hibbert, P., and Huxham, C. (2010). 'The Past in Play: Tradition in the Structures of Collaboration', *Organization Studies*, 31(5): 525–54.

Hite, J. M., and Hesterly, W. S. (2001). 'The Evolution of Firm Networks: From Emergence to Early Growth of the Firm', *Strategic Management Journal*, 22(3): 275–86.

Irwin, D., and Klenow, P. (1996). 'High-tech R&D Subsidies: Estimating the Effects of Sematech', *Journal of International Economics*, 40(3–4): 323–44.

Kale, P., Singh, H., and Perlmutter, H. (2000). 'Learning and Protection of Proprietary Assets in Strategic Alliances: Building Relational Capital'. *Strategic Management Journal*, 21(3): 217–37.

Kastelle, T., and Steen, J. (2010). 'Network Analysis Application in Innovation Studies', *Innovation: Management, Policy and Practice*, 12(1), Special Edition: 2–117.

Knight, F. (1921). *Risk, Uncertainty and Profit*. Boston, MA: Houghton Mifflin Co.

Leonard-Barton, D. (1992). 'Core Capabilities and Core Rigidities: A Paradox in Managing New Product Development', *Strategic Management Journal*, 13 (Special Issue: Strategy Process: Managing Corporate Self-Renewal): 111–25.

Link, A., Teece, D., and Finan, W. (1996): 'Estimating the Benefits from Collaboration: The case of SEMATECH', *Review of Industrial Organization*, 11: 737–51.

Lokshin, B., Hagedoorn, J., and Letterie, W. (2011). 'The Bumpy Road of Technology Partnerships: Understanding Causes and Consequences of Partnership Mal-functioning', *Research Policy*, 40(2): 297–308.

Lundvall, B.-A. (2007). 'National Innovation Systems: Analytical Concept and Development Tool', *Industry and Innovation*, 14(1): 95–119.

March, J. (1991). 'Exploration and Exploitation in Organizational Learning', *Organization Science*, 2(1): 71–87.

March, J. (2006). 'Rationality, Foolishness, and Adaptive Intelligence', *Strategic Management Journal*, 27(3): 201–14.

Mathews, J. (2002). 'The Origins and Dynamics of Taiwan's R&D consortia', *Research Policy*, 31 (2): 633–51.

Maurer, I., and Ebers, M. (2006). 'Dynamics of Social Capital and their Performance Implications: Lessons from Biotechnology Start-ups', *Administrative Science Quarterly*, 51 (2): 262–92.

Morgan, G. (1986). *Images of Organization*. New York: Sage.

Mowery, D., and Grodal, S. (2005). 'Networks of Innovators', in J. Fagerberg, D. Mowery, and R. Nelson (eds), *The Oxford Handbook of Innovation*. Oxford: Oxford University Press.

O'Kane, M., and Australia. Dept. of Innovation, Industry Science and Research (2008). *Collaborating to a Purpose [electronic resource]: Review of the Cooperative Research Centres Program* (PANDORA electronic collection; Canberra, A.C.T.: Dept. of Innovation, Industry, Science and Research).

Podolny, J., and Page, K. (1998). 'Network Forms of Organization', *Annual Review of Sociology*, 24(1): 57–76.

Powell, W., Koput, K., and Smith-Doerr, L. (1996). 'Interorganizational Collaboration and the Locus of Innovation: Networks of Learning in Biotechnology', *Administrative Science Quarterly*, 41(1): 116–45.

Reagans, R., and Zuckerman, E. (2008), 'Why Knowledge does not Equal Power: The Network Redundancy Trade-off', *Industrial and Corporate Change*, 17(5): 903–44.

Rosenbloom, R., and Cusumano, M. (1988). 'Technological Pioneering and Competitive Advantage: The Birth of the VCR Industry', in M. Tushman and W. Moore (eds), *Readings in the Management of Innovation*. USA: Harper Business, 3–22.

Rosenkopf, L., Metiu, A., and George, V., (2001). 'From the Bottom Up? Technical Committee Activity and Alliance Formation', *Administrative Science Quarterly*, 46(4): 748–72.

Rothaermel, F., and Deeds, D. L. (2004). 'Exploration and Exploitation in Biotechnology: A System of New Product Development', *Strategic Management Journal*, 25(3): 201–221.

Rothwell, R., and Dodgson, M. (1991). 'External Linkages and Innovation in Small and Medium-Sized Firms', *R&D Management*, 21(2): 125–37.

Rothwell, R., et al. (1974). 'SAPPHO updated: Project SAPPHO, Phase II', *Research Policy*, 3: 258–91.

Rowley, T., Behrens, D., and Krackhardt, D. (2000). 'Redundant Governance Structures: An Analysis of Structural and Relational Embeddedness in the Steel and Aemiconductor Industries', *Strategic Management Journal*, 21(3): 369–86.

Schilling, M. A. (2009). 'Understanding the Alliance Data', *Strategic Management Journal*, 30(3): 233–60.

Teece, D. (1986). 'Profiting from Technological Innovation: Implications for Integration, collaboration, licensing and Public Policy', *Research Policy*, 15: 285–305.

Uzzi, B. (1999). 'Embeddedness in the Making of Financial Capital: How Social Relations and Networks Benefit Firms Seeking Financing', *American Sociological Review*, 64(4): 481–505.

Walker, G., Kogut, B., and Shan, W. (1997). 'Social Capital, Structural Holes, and the Formation of an Industry Network', *Organization Science*, 8(2): 109–25.

Williamson, O. (1975). *Markets and Hierarchies: Analysis and Anti-trust Implications*. New York: Free Press.

Zahra, S. A., and George, G. (2002). 'Absorptive Capacity: A Review, Reconceptualization, and Extension', *Academy of Management Review*, 27(2): 185–203.

Zhang, M., and Dodgson, M. (2007). *High-Tech Entrepreneurship in Asia: Innovation, Industry and Institutional Dynamics in Mobile Payments*. Cheltenham: Edward Elgar.

CHAPTER 24

ORGANIZING INNOVATION

NELSON PHILLIPS

INTRODUCTION

ECONOMICS was the source of much early insight into innovation. Work by authors such as Schumpeter (1950), Williamson (1975), and Nelson (1993) provided important theoretical understandings of the role of innovation in industry evolution, and of the importance of innovation to economic growth and the well-being of societies. This stream of work remains an influential perspective on innovation, with economists working in this area continuing to provide a stream of important insights into a broad range of innovation topics.

At the same time, within the economics literature, the actual process of innovation 'has been treated more or less as a "black box"' (Fagerberg, 2005: 3). Despite its extensive contributions, economics has therefore provided little in the way of insight into what makes individual efforts at innovation more likely to be successful. As a result, a second stream of literature—innovation studies—has developed, focusing directly on the black box of innovation. This literature grew initially out of science and technology policy and its concern with the factors that drive innovation; but it rapidly became highly interdisciplinary reflecting the complex nature of innovation processes (Fagerberg, 2005). From rather humble beginnings, the innovation studies literature is now very large and offers a comprehensive understanding of innovation and the innovation process.

Most recently, innovation management as a specialized area of innovation studies has developed. Scholars working in this area have adopted a highly practical approach to innovation and have focused on the strategies and practices decision-makers can use to increase the organizational benefits of innovation. This area of scholarly investigation now provides extensive prescriptions for managers to increase the likelihood of innovation success[1] reflecting the variety of forms of innovation that exist and the complexity of the contexts within which it takes place.

At the same time, even this literature has been criticized for devoting 'little attention to the internal dynamics and social processes within organizations' that frame and shape innovation processes (Lam, 2005: 122). The Organization and Management Theory (OMT) literature, on the other hand, provides deep insight into organizations and organizing. It is therefore an important potential source of additional insight into innovation management at the level of the individual organization. This perspective is both highly complementary to existing work that has been developed in innovation management, and critical to deepening our understanding of how to best understand and manage innovation.

In particular, the OMT literature provides valuable insights into the complex functioning of the organizational context in which innovation activities take place. As Dougherty and Hardy (1996: 1121) found in their study of innovation in large organizations, when firms are 'not organized to facilitate innovation: occasionally innovation did occur, but it occurred in spite of the system, not because of it' (Dougherty and Hardy, 1996: 1121). Even the most flawlessly executed innovation process has little chance in a highly unfavourable organizational environment. Much research has shown that the organizational context can make the difference between innovation success and failure, and that organizational characteristics like team composition (e.g. Hoegl and Gemuenden, 2001), network connections (e.g. Powell et al., 1996), leadership (e.g. Quinn, 1985), and culture (e.g. Kanter, 1983) require close attention when managing innovation.

At the same time, insights into innovation in the OMT literature are scattered across a broad literature and remain largely unconnected. In addition, there are a number of themes in the literature that are important for understanding innovation, but where the ramifications for understanding innovation have not been clearly drawn. To help make these insights more accessible, this chapter draws on a range of sources in OMT to identify and discuss some of the key organizational factors that influence the effectiveness of innovation processes. The discussion is intended to focus attention on some of the important dimensions of organization upon which innovation depends, even though the extent of the link may still require further investigation, and to highlight the interdependences among these different aspects of organization. It is the combination of these dimensions of organization that affects the effectiveness of attempts at innovation, and efforts to manage innovation need to include a consideration of the interrelationships among the different dimensions to maximize innovation performance and outcomes.

More specifically, three dimensions of organization have been identified in the literature as being of particular importance to innovation: culture, leadership, and the role of teams. Each of these characteristics will be discussed in turn, followed by an explanation of what existing research has found about their role in shaping the effectiveness of innovation, and areas for further research will be identified. In addition, three areas of research will be discussed that have particular potential for contributing to the understanding of innovation but where little has been done to connect to innovation management despite its clear relevance: institutional theory, theories of practice adoption, and organizational identity. The chapter concludes with a few comments about the potential of innovation research to contribute to OMT.

Current Research in OMT with Implications for Innovation Management

Innovation is an important theme in a number of areas of OMT. In addition, innovation management scholars have drawn on a number of streams of literature in OMT to good effect. In this section, three areas where connections between OMT and innovation have been made will be summarized and the question of where research in these areas might profitably focus in the future will be discussed. Definitions of the three areas and exemplary citations are provided in Table 24.1.

Table 24.1 Summary of OMT theories with existing connections to innovation

Dimension of Organizing	Key Insights	Exemplary Citations
Culture	Both organizational and national cultures are important and need to be considered when managing innovation. Culture shapes both the motivation to innovate and the impact of prescriptions for improving innovation. An 'innovation culture' in organizations has been linked to success in innovation.	Kaasa and Vadi (2010) Naranjo-Valencia et al. (2011) Van Everdingen and Waarts (2003)
Leadership	Leaders directly affect the innovation process. Leadership behaviour may encourage or discourage risk taking and innovation. Transformational leadership is most closely associated with innovation in the existing literature. At the same time, there are many other areas of leadership research that hold potential and should be explored further.	Jung et al. (2003) Oke et al. (2009)
Teams	Innovation is largely carried out by teams. Therefore, the large literature on team effectiveness is an important potential source of insight into innovation management. Work to date has shown that team cognitive styles and the nature of team interaction are both strong predictors of innovation effectiveness.	Miron-Spektor et al. (2011) Hoegl and Gemuenden (2001)

Culture and Innovation

Culture is a core concept in social science. But, like many of the core concepts, it lacks an agreed definition. At the same time, some generally agreed elements of culture can be identified. In one classic and often cited definition, Geertz (1973: 5) defined culture quite poetically: 'Believing with Max Weber, that man is an animal suspended in webs of significance that he himself has spun, I take culture to be those webs, and the analysis of it to be therefore not an experimental science in search of a law but an interpretive one in search of meaning.' While not all culture scholars would agree with this definition in its entirety, it usefully highlights what is central to most discussions of culture: culture is a symbolic system constructed by people as they interact and communicate and this system of meaning has important consequences for the members of the groups who produce it. It also highlights the fact that the study of culture is about understanding the meaningful world of a group of actors.

In the OMT literature, studies of culture occur in two quite distinct literatures. On the one hand, a group of scholars has spent considerable effort understanding the role of national culture in organizational processes. This area is often referred to as cross-cultural management and scholars have begun to connect ideas from this area of work to innovation. On the other hand, another group of scholars has focused on culture at the organizational level. This stream of literature looks at how distinctive cultures develop in organizations and the ramifications of organizational culture for organizational functioning. This stream of research has also been connected explicitly to the management of innovation.

National Culture and Innovation. Researchers interested in cross-cultural management have focused primarily on understanding how national cultures, defined as 'the collective mental programming' of the people of a particular nation (Hofstede, 1980: 43), vary and the impact this has on processes of organizing and on the applicability of management practices developed in one national context in another national context. To understand this complex question, scholars have focused on developing a set of dimensions along which national cultures can be compared. Armed with this understanding of cultural difference, researchers work to measure some of the variations in national culture.

While a number of typologies have been proposed, the most influential and commonly used framework for comparing national cultures is the one proposed by Hofstede (1980, 1989, 1991). Hofstede worked with IBM in the early 1970s to survey more than 100,000 people in forty countries about their beliefs and values. Based on the results of this survey, Hofstede proposed five dimensions for understanding differences in national cultures. The first dimension, power distance, measures the degree to which the members of a particular cultural group accept the fact that power is distributed unequally. If power distance is high, then members accept that some members of society are in charge and others are consigned to being followers. The second dimension, uncertainty avoidance, refers to the degree to which members of a society feel that uncertainty and ambiguity are acceptable or whether they must be managed. Cultures that are high on uncertainty avoidance find change difficult and prefer to follow familiar rules. Third, individualism-collectivism refers to the degree to which a culture focuses on the group versus the individual. Cultures that are high on individualism believe that individuals are responsible for themselves, and are characterized by a more loosely knit social network. Fourth, masculinity refers to the extent to which the dominant values of a society include a focus on the acquisition of money, assertiveness, and a lack of concern for quality of life and other people. Finally, long-term orientation is a measure of the degree to which people focus on the future versus focusing on the here and now.

Hofstede's dimensions have been used in many different streams of research, including exploring themes related to innovation management. Kaasa and Vadi (2010), for example, investigated the relationship between the ability to initiate innovation and national culture using his work. They report their study shows 'significant support for the argument that the capability of a country or region to initiate innovation is related to its culture' (Kaasa and Vadi, 2010: 594). They go on to provide more nuanced explanations of the role of each of the four dimensions of culture identified by Hofstede in the ability to initiate innovation. Similarly, Van Everdingen and Waarts (2003) used Hofstede's dimensions to explore the link between national culture and willingness to adopt new innovations. They find strong support for the influence of culture on the willingness to adopt innovations. In another study of the impact of national culture on innovation, Steensma et al. (2000) investigate how three dimensions of culture—tolerance for uncertainty, competitiveness, and individualism—impact the formation of technology alliances. They find that 'national culture directly and indirectly affects the formation of technology alliances' (Steensma et al., 2000: 966).

Based on the existing literature, work on national culture has two distinct lessons for innovation management. On the one hand, there is evidence that the degree of innovativeness, or at least the likelihood of engaging in innovation or adopting innovative ideas and practices, varies from one national culture to another. The characteristics of some cultures make change more difficult and innovation more of a challenge. The role of national culture in innovation therefore requires careful consideration when thinking about how to manage it and theories of innovation management need to be highly sensitive to questions of national culture.

On the other hand, what also is clear is that the impact of innovation management practices and techniques will vary from context to context. At a general level, the core lesson of the cross-cultural management literature is that management techniques are not universal. In fact, the results from any management approach are highly conditioned by the cultural context. As Trompenaars and Hampden-Turner (2012) argue, '[r]ather than there being "one best way of organizing", there are several ways, some much more culturally appropriate than others'. Innovation management must therefore include a sensitivity to differences in national culture and care needs to be taken when approaches from one cultural context are adopted in another. There is no one best way of organizing innovation and a careful concern for cultural difference is necessary in its management.

Organizational Culture and Innovation. Much like conceptualizations of national culture, early conceptualizations of organizational culture described it as a relatively stable set of taken-for-granted elements that shape members' thoughts and actions in a coherent and predictable way, and provide the structural stability fundamental for the everyday functioning of an organization (Schein, 1992). Of particular relevance to innovation management, research in this tradition emphasizes how, due to their socialization into the organization's culture, members of organizations with 'strong' cultures will find it difficult to accept changes that clash with the existing culture (Gagliardi, 1986; Ogbonna and Wilkinson, 2003).

In what is perhaps the most influential work in the area, Schein (1992) distinguishes three interconnected layers that express culture in organizations. The deepest level is composed of *basic underlying assumptions*, those unconscious, taken-for-granted beliefs and perceptions that are strongly held and rarely questioned by organizational members (Schein, 1992: 21–2). These assumptions entail both cognitive and emotional understandings of organizational members and serve as the foundation of the other two layers.

Espoused values are openly expressed organizational values, ethical rules, and strategies. These are socialized within organizations and often expressed in public documents. Founders and organizational leaders play a fundamental role in the development of this layer of culture, selecting and recounting the organizational narratives that cement underlying beliefs (Smirnich, 1983).

Artefacts lie above espoused values, being the most visible and tangible expressions of a culture. They include organizational processes, organizational structures, language, and jargon, as well as physical arrangements such as architecture and dress codes. Although artefacts per se may appear to be value free to someone approaching the organization for the first time, Schein's studies show that organizational members link visible features of an organization to underlying basic assumptions.

Building on the idea that organizations have distinct cultures, a number of scholars have argued that there are particular types of organizational cultures that predispose organizations to be more innovative. The reason for this is simple: 'Since [organizational culture] influences employee behaviour, it may lead them to accept innovation as a fundamental value of the organization' (Naranjo-Valencia et al., 2011: 56). Bock et al. (2012: 292), for example, find 'that creative culture is positively associated with strategic flexibility across geographies and sectors'. Similarly, De Tienne and Mallette (2012: 7) find that 'cultural forces appear at least as important as other factors in predicting product innovation'. In a very large study comparing thirty-one drivers of innovation, Tellis, Prahbu, and Chandy (2009: 15) find that the 'internal corporate culture is an important driver of radical innovation'. In a related study of 471 Spanish companies, Naranjo-Valencia, Jimenez-Jimenez, and Sanz-Valle (2011: 55) find that '[organizational] culture is a clear determinant of innovation strategy' and that different cultures are appropriate for imitation strategies versus innovation strategies.

More recent work on organizational culture in OMT, building on an increasingly influential perspective in cultural sociology (Swidler, 1986; DiMaggio, 1997), has started questioning the view of culture as a relatively unified system of beliefs constraining behaviour, and has shifted the emphasis to the notion of culture as a repertoire or 'toolkit' (Weber, 2005; Osterman, 2006; Zilber, 2006; Rindova et al., 2010). According to this view, culture is not a set of normative prescriptions, but rather a repertoire of cognitive and symbolic resources that members can flexibly draw upon to develop 'strategies of action'. Culture influences behaviour by providing individuals with meaning structures that shape the courses of action available in a given situation, along with the 'worldviews' that support them (Swidler, 2001: 73–4). Cultural repertoires are described as composed of diverse and heterogeneous resources that individuals use—frequently shifting among multiple and contradictory ideas—to make sense of their experiences, develop courses of action, and account for them (Swidler, 2001). At an organizational level, cultural repertoires have generally been studied by exploring the sets of cognitive categories that members draw upon in order to interpret reality and to formulate strategies of action (Weber, 2005; Rindova et al., 2010).

While this view has gained considerable ground in sociology and in OMT, this view of culture and its ramifications for our understanding of innovation are only beginning to be explored and this is an obvious area where further insight into innovation and its management may be available given this very different conceptualization of culture and its effect on individuals within an organization. One example of research in this area is the work of Leonardi (2011: 348), who uses this conceptualization of culture to explore 'why members of one organizational department are blind to the reasons why others do not share their view of what the functionality of a new technology should be'. His work provides fundamental insight into what is a critical question in innovation management and shows the usefulness and power of this view of organizational culture. He also focuses on the strategies that organizations can use to reduce this problem by 'reintroducing ambiguity into a process that had become relatively concrete, and by reorganizing boundaries in ways that provided structural content in which ambiguity could succeed, they were eventually able to produce a working technology' (Leonardi, 2011: 363).

LEADERS AND INNOVATION

Leadership, defined as 'a social process that takes place in a group context in which the leader influences his or her followers' behaviours so that desired organizational goals are met' (Oke et al., 2009: 65), has been a central preoccupation of OMT scholars for decades. The resulting literature is very large, with thousands of articles and books already published and many more being added each year (see Rumsey, 2013 for a useful overview). The literature spans many traditional academic disciplines with contributions being made in psychology, organizational behaviour, international business, strategy, and other disciplines. The literature also includes a large body of work written for and by practitioners (e.g. Giuliani, 2003), and a very practical literature focused on leadership development and training (e.g. Goleman et al., 2003).

This literature has important implications for innovation. The importance of leaders to organizational success and their central role in defining organizational processes, designing structures, and shaping culture mean they have a critical role in innovation (Denti and Hemlin, 2012). Even more fundamentally, and reflecting back on the definition of leadership above, if a leader can influence 'his or her followers' behaviours so that desired organizational goals are met', then there is no reason why those goals can't be related to innovation. Anecdotal evidence also supports the view of leaders as core to innovation with discussions of innovative leaders from Thomas Edison to Steve Jobs focused on this topic. And this fact has not gone unnoticed by innovation scholars, leading to the observation that 'leading innovation remains one of the most challenging aspects for contemporary leaders' (Oke et al., 2009: 65). Yet, as the same authors point out, at the same time 'there has been very little done to address the link between leadership and innovation'.

One of the areas where some attention has been paid is the connection between innovation and transformational leadership (Bass and Avolio, 1994). From this influential theoretical perspective, there are two types of leadership behaviours that correspond to two types of leaders: transformational leaders and transactional leaders. Transformational leaders are characterized by four components: charismatic influence representing the degree to which leaders are admired, respected, and trusted; inspirational motivation describing the ability of leaders to inject meaning and challenge into followers activities by crafting an attractive vision of the future; intellectual stimulation referring to the degree to which a leader encourages followers to challenge existing assumptions and reframe problems; and individual consideration which involves paying attention to followers' individual needs through coaching and mentoring.

Transactional leaders, in contrast, do not base their leadership activities on inspiring followers and helping them to grow and develop. In general, transactional leadership 'is defined as emphasizing the transaction or exchange of something of value the leader possesses or controls that the employee wants in return for his/her services' (Oke et al., 2009: 66). The two key behaviours of transactional leaders are contingent reward and management by exception. But transactional and transformational leadership should not be thought of as substitutes or opposites. In fact, they are highly complementary and both are required. But at the same time, in many organizations transformational leadership is, according to its advocates, in short supply.

A number of scholars have used this framework to explore the link between leadership and innovation. In an early paper, Jung, Chow, and Wu (2003) use data from thirty-two Taiwanese companies to explore how leadership, and in particular transformational leadership, affects creativity in organizations. They find, 'transformational leadership by the top manager can enhance organizational innovation directly and also indirectly by creating an organizational culture in which employees are encouraged to freely discuss and try out innovative ideas and approaches' (Jung et al., 2003: 539).

Similarly, Oke, Munshi, and Walumbwa (2009) argue that transformational and transactional leadership are each more effective with particular types of innovation (such as, for example, exploration versus exploitation in James March's terms) and that there are organizational variables that affect this link. Drawing on a broad range of literature, they argue that there are clear differences with transformational leadership being more effective when a greater degree of creativity is required.

Finally, Chen, Lin, Lin, and McDonough (2012) investigate the relationship between transformational leadership and innovation at the Strategic Business Unit (SBU) level of analysis based on data from 102 Taiwanese companies. They find that transformational leadership does increase SBU innovation and also find that a stronger innovative culture can act as a substitute in facilitating innovation. This study, once again, shows the complex relationship between leadership and culture.

From this discussion, it is clear that the connection between leadership and innovation is one that has the potential to be very rich and to provide deep insights to innovation management. The importance of leadership to innovation, and the very practical nature of the leadership literature, makes this an obvious fit. These initial studies point to the strength of the relationship and reveal some early findings. Yet, given the vast literature on leadership there is still considerable work yet to do at the intersection of innovation and leadership.

THE ROLE OF TEAMS IN INNOVATION

The literature on teams is rich and varied. While much of the research that has been conducted shares an interest in team effectiveness, a diverse range of variables has been identified that impact effectiveness including 'task, group, and organizational design factors, environmental factors, internal processes, external processes, and group psychosocial traits' (Cohen and Baily, 1997). Given the ubiquitous use of teams in innovation, and the practical bent of the literature on team effectiveness, this literature has obvious potential to increase our understanding of an important aspect of innovation management. To put it simply, if teams involved in innovation can be better managed, then innovation processes should become more effective and the outcomes of innovation improve for the companies carrying them out.

While many of the determinants of team effectiveness are deserving of attention from innovation researchers, just two are considered here as examples of what can be achieved when the literatures on teams and innovation management are brought together. First, there is a growing literature on how members' cognitive styles—that is, 'an individual's stable and preferred cognitive strategy for acquiring, processing, maintaining, and using knowledge for problem solving' (Miron-Spektor et al., 2011: 741)—affect team performance.

Early research in this area has shown that team members with dissimilar cognitive styles differ in their focus on idea implementation versus idea identification (Kirton, 1989). Miron-Spektor, Erez, and Navez (2011) extended this work and compared teams with different proportions of members with creative, conformist, and attentive-to-detail cognitive styles. They found that 'in keeping with prior theory and research, we have shown that creative team members are essential for team radical innovation…conformist members additionally contribute to team innovation… [but] attentive-to-detail members, in contrast, negatively influenced team radical innovation in our study' (Miron-Spektor et al., 2011: 753). In other words, cognitive styles have an important impact on innovation by teams.

Second, there is a more modest yet significant literature on teamwork quality as a predictor of team effectiveness in innovation. In an early article on the topic, Hoegl and Gemuenden (2001) pose two important questions. First, they ask what teamwork actually is and can it be measured. They go on to ask how strong the relationship is between teamwork and project success. They construct a measure of teamwork quality based on six dimensions: communication, coordination, balance of member contributions, mutual support, effort, and cohesion. Using their concept, and based on a study of 575 individuals working in 145 German software teams, they find strong evidence that teamwork quality is associated with team performance. While much more research needs to be done, it is clear that teamwork quality is something that needs careful management to ensure that best outcomes from team-based innovation efforts.

Based on these examples, it is clear that the literature on team effectiveness holds real promise for a deeper understanding of innovation management. In fact, what is surprising is that more of the potential connections between the literature on teams and innovation management have not been explored. There is a real opportunity here for scholars in innovation management to rapidly improve their understanding of innovation effectiveness and to provide practical and implementable guidance for practitioners.

FUTURE RESEARCH DIRECTIONS IN INNOVATION AND OMT

While the above section described some of the ways in which OMT has contributed to our understanding of innovation, there are also a number of areas that are potentially fruitful for our understanding of innovation, but where little has been done in connecting ideas about organizations to questions of how to best manage innovation. In particular, the connection between innovation and institutional theory, the role of practice adoption in innovation, and the connection between innovation and organizational identity all deserve further attention. While the topics discussed in this section

Table 24.2 Summary of OMT theories with potential for improving our understanding of innovation

Dimension of Organizing	Key Insights	Exemplary Citations
Institutional Context	Organizations are embedded in organizational fields characterized by sets of institutions and an institutional logic. This institutional context shapes expectations and actions and is therefore important in understanding innovation processes at the organizational level.	DiMaggio and Powell (1983) Greenwood et al. (2011)
Practice Adoption	Organizations differ greatly in their readiness to adopt new innovation practices. Also, new practices may lack political, technical, or cultural fit making adoption unlikely without adaptation of the practice. The adoption of new innovation practices may therefore require a careful analysis of the 'fit' of the new practices and a strategy for managing misfit to ensure extensive and high fidelity adoption that results in the maximum benefit.	Ansari et al. (2010)
Organizational Identity	Organizational members' understanding of an organization's identity will shape their understanding of the type and rate of innovation that 'we do around here'. Managing innovation may therefore require the effective and timely management of organizational identity.	Albert and Whetten (1985) Dutton and Dukerich (1991)

are necessarily more speculative, they are useful to include as they point to fruitful areas for future research. A summary of the main ideas presented in this section is provided in Table 24.2.

INSTITUTIONS AND INNOVATION

From rather modest beginnings in the 1970s, institutional theory has become the dominant theoretical perspective in OMT.[2] Broadly speaking, institutions are defined as conventions that are self-policing (e.g. Douglas, 1986). Within the tradition of new institutional theory, institutions are defined more specifically as 'historical accretions of past practices and understandings that set conditions on action' through the way in which they 'gradually acquire the moral and ontological status of taken-for-granted facts which, in turn, shape future interactions and negotiations' (Barley and Tolbert, 1997: 99). Institutional theory has a number of important implications for our understanding of innovation and how to manage its challenges, and the focus here is on three concepts from institutional theory that have particular importance for understanding innovation: organizational fields, institutional logics, and institutional distance.

Organizational Fields. Institutions influence behaviour as departures from them 'are counteracted in a regulated fashion, by repetitively activated, socially constructed, controls' (Jepperson, 1991: 145). In other words, deviations from the accepted institutional order are costly in some way, and the more highly institutionalized a particular social pattern is, the more costly deviations become. Institutions involve mechanisms that associate non-conformity with increased costs in several different ways: 'economically (it increases risk), cognitively (it requires more thought), and socially (it reduces legitimacy and the access to resources that accompany legitimacy)' (Phillips et al., 2000: 28). Thus, institutions can be differentiated from other patterned forms of social action that are not subject to these sorts of self-regulating controls.

Central to institutional theory in OMT is the idea of an organizational field, defined as 'those organizations that, in the aggregate, constitute a recognized area of institutional life: key suppliers, resource and product consumers, regulatory agencies and other organizations that produce similar services' (DiMaggio and Powell, 1983: 143) and that 'interact more frequently and fatefully with one another than with actors outside of the field', develop mutual awareness, and see themselves as part of the same community and involved in a common enterprise (Scott, 1995: 207–8). Organizational fields are an important analytical tool as the organizations within a field share sets of institutions including practices, organizational forms, and institutional logics.

The nature of the institutions that characterize a particular organizational field, and the degree of institutionalization of those institutions (that is, how strong the reinforcing mechanisms are), is therefore critical to considerations of innovation. More specifically, to the degree that an innovation is a deviation from the accepted social order, institutional pressures will increase the challenge of innovation and, in some cases, make it impossible unless the institutional context is managed in some way. The social mechanisms that produce the patterned behaviour associated with institutions will increase the difficulty of envisaging and introducing an innovation and make success less likely in direct proportion to their degree of institutionalization.

In a highly institutionalized organizational field, innovation will become difficult as even once someone is able to think beyond the highly institutionalized world in which he or she is embedded, the innovation itself will be seen as illegitimate and inappropriate. This is a form of the infamous 'not invented here' syndrome that is important in understanding why innovation fails and that needs much more investigation and discussion.

Institutional Logics. Research in new institutional theory has increasingly highlighted the importance of institutional logics—'the socially constructed historical patterns of material practices, assumptions, values, beliefs and rules by which individuals produce and reproduce their material subsistence, and provide meaning to their social reality' (Thornton and Ocasio, 1999: 804). Early work on institutional logics focused on the importance of the broad organizing principles that characterize modern societies such as the family, the state, and the market to institutional processes (Friedland and Alford, 1991). More recently, work has focused on how these abstract principles are drawn on and elaborated at the field level and, in particular, how logics change and evolve and the effect this has on the field and its members (e.g. Ocasio, 1997; Thornton, 2002; Lounsbury, 2007).

Institutional logics, as taken-for-granted, resilient social prescriptions, specify the boundaries of a field, its rules of membership, the role identities, and the appropriate organizational forms of its constituent communities (Friedland and Alford, 1991; Thornton, 2004; Greenwood and Suddaby, 2006). They are the 'broader cultural templates that provide organizational actors with means-ends designations, as well as organizing principles' (Pache and Santos, 2010; see also Friedland and Alford, 1991). Thornton and Ocasio (1999) refer to logics as 'the formal and informal rules of action, interaction, and interpretation that guide and constrain decision-makers'. As institutional logics provide social actors with vocabularies of motives and senses of self, they not only direct what social actors want (i.e. shape their interests) and how they are to proceed (i.e. provide guidelines for action), but also who or what they are (i.e. their identity).

In fields where innovation has become institutionalized, therefore, the types and rate of innovation will be shaped by institutionalized expectations. While innovation will occur, it will occur in familiar ways, at familiar rates, and produce familiar results. Deviations from familiar forms of innovation will face social sanctions and increasing resistance. Over time, innovation will become more and more incremental and predictable, consigning the industry or sector to increasingly meaningless and superficial innovation and an ever-increasing probability of an external party from another field introducing a disruptive innovation.

Although the concept of a dominant institutional logic characterizing a field has proven to be an useful tool, an alternative approach has appeared that recognizes that organizational settings are often exposed to different institutional demands at the same time (Kraatz and Block, 2008; Reay and Hinings, 2009; Dunn and Jones, 2010; Greenwood et al., 2010; Pache and Santos 2010; Greenwood et al., 2011). Institutional complexity refers to situations in which a multiplicity of logics, exerting different pressures and influences, are in play in a particular context. Organizations incorporating elements from different institutional logics (Battilana and Dorado, 2010; Pache and Santos, 2010) face the effects of that institutional complexity and 'contend with competing external demands and internal identities' (Jay, 2013). Considering that multiple logics embodied by the organization are independent, not always compatible, and often in conflict (Friedland and Alford, 1991; Pache and Santos, 2010; Greenwood et al., 2011), organizations face heightened challenges in trying to incorporate these antagonistic practices, which may not easily work together (Pache and Santos, 2010; Tracey et al.,

2011). It follows, therefore, that organizations will differ in the responses they give to institutional complexity and in the effectiveness of these responses.

This idea of institutional complexity provides a particularly germane theoretical frame for considering innovation management. While institutions and institutional logics may largely work to limit and shape innovation, institutional complexity provides an opportunity for innovators to see beyond the dominant logic that characterizes their field. Studies of institutional change have identified institutional complexity as one way in which actors can avoid the problem of 'embedded agency',[3] and it seems likely that institutional complexity will provide the same opportunity to other sorts of innovators. If this works and how it works, however, remains to be explored empirically.

Institutional Distance. In international business, institutional distance refers to the degree of difference 'between the regulatory, cognitive and normative institutional environments of the home and the host countries of an MNE (Multi National Enterprise)' (Kostova and Zaheer, 1999: 68). Understanding the 'distance' between business systems on each of these institutional dimensions, and estimating their effect on the ability of a firm to transfer business practices and people between contexts, can help MNEs adjust their strategies to facilitate the transfer of business practices (Kostova, 1999) and to obtain legitimacy in foreign countries (Kostova and Zaheer, 1999). The ease of adjustment will depend on an MNE's familiarity with a country's institutional profile, as this allows the firm to achieve and maintain synergy with both its home country and host country contexts.

This notion can be used more generally to describe differences between organizational fields regarding the regulatory, cognitive, and normative pillars of field-specific institutions (Phillips et al., 2009). Rather than expressing a dichotomous distinction between institutional similarity and disparity, institutional distance allows one to distinguish degrees of difference. The institutional distance between an academic science department and a biotechnology company, for instance, is likely to be smaller than between an academic science department and a small, low-tech firm even though both firms operate in the market sphere. The basic intuition underlying the notion of institutional distance is that increasing distance makes transferring new ideas, practices, and structures more difficult and more costly.

This concept has significant implications for thinking about innovation. Most particularly, this body of work has much to say about the difficulties of transferring innovation practices from one organizational field to another. Much of the innovation management literature looks at best practices in innovative firms and industries and suggests that these can be generalized and transferred to less innovative firms and industries. Yet, research on institutional distance shows that high levels of institutional distance will make the transfer of approaches to innovation from distance fields very difficult. This also means that, in addition to understanding effective innovation management practices, innovation management scholars require a much deeper understanding of the institutional context to appreciate how and when innovation best practice can be successfully transferred.

Practice Adoption and Innovation

A managerial practice can be defined as a bundle of behavioural routines, tools, and concepts to accomplish a certain task (Westphal et al., 1997). In OMT, a body of literature has developed looking at how and when new practices are adopted by organizations. Innovation practices are, clearly, a particular form of managerial practices and this area of research therefore has important ramifications for thinking about the adoption of new innovation practices and particularly why new innovation practices are not adopted or are adopted in only a limited way. In OMT, two bodies of research have investigated the implementation of new practices in organizations: institutional theory and practice theory.

First, research in an institutional theory tradition has shown how, when new practices are adopted by an organization, they are often adapted to increase their fit with the technology, culture, strategy, and politics of the organization by recombining different elements of commonly accepted 'templates' at the time of adoption (e.g. Kraatz and Zajac, 1996; Westphal et al., 1997; Lounsbury, 2001; Kennedy and Fiss, 2009).

Building on this body of research, Ansari and colleagues (2010) theorize practice adaptation as characterized by different degrees of fidelity with the original practice—that is, the extent to which the implementation of the new practice follows a widely acknowledged template rather than tailoring it to the specificities of the organization—and different degrees of extensiveness of implementation across the organization—that is, the extent to which the new practice is adopted across functional, divisional, or geographical organizational units. In particular, their model suggests that technical fit, cultural fit, and political fit are the most important determinants of the outcome of practice adaptation (Ansari et al., 2010).

For innovation management, this again challenges the idea that innovation practices can simply be transferred from one organization to another. Scholars interested in innovation management must carefully consider the insights of the practice adoption literature to understand when a new innovation practice is likely to be adopted in a new context, when it will need to be adapted, and when there is little likelihood of new practices of a particular sort being adopted at all. The characteristics of the adopting firm will play a critical role and all three forms of fit will need to be considered in understanding the probability of the successful adoption of the new innovation practices.

Second, practice-based theories have focused on the micro-level processes that underpin the implementation of new technologies (e.g. Orlikowski, 2000) and related practices (e.g. Feldman, 2000; Rerup and Feldman, 2011). Research in this tradition draws our attention to how new practices are enacted in daily interaction. In fact, research in this tradition suggests that the implementation of new practices is 'situated, dynamic, and emergent' and that any observed configuration is to be considered as 'temporally and contextually provisional' (Orlikowski, 2000).

It also alerts to the fact that when new practices—such as the set of policies and procedures that innovation rests upon—are adopted in a new organizational context, the outcome may be only partly predictable as people will strive to change the practices that seem to 'fall short of ideals' (Feldman, 2000; Rerup and Feldman, 2011). This again highlights the uncertain nature of practice adoption and the need for more research on how and when innovation practices are adopted.

Identity and Innovation

The final area of research in OMT discussed here focuses on organizational identity. An organization's identity is generally understood as those attributes of the organization that members perceive as central, enduring, and distinctive (Albert and Whetten, 1985; Whetten, 2006). An organization's identity provides a framework for members' interpretations of events and therefore influences their subsequent action. Organizational identity helps members of an organization to relate to the broader organizational context within which they act and to make sense of events in relation to their understanding of what the organization is and what it stands for (Fiol, 1991). Dutton, Dukerich, and Harquail (1994) argue that an attractive organizational identity influences the degree to which members identify with the organization, assimilate its core values, and accept its goals as their own.

Gioia and Thomas (1996) argue that decision-makers' perceptions of organizational identity affect the sense-making processes within the organization, as it defines the context for issue interpretation. Their study of strategic change in academia shows how identity and image act as perceptual screens that influence top management teams' information processing and their interpretation of key issues as strategic. Similarly, Dutton and Dukerich's (1991) study of the Port Authority of New York and New Jersey shows how organizational identity limits and directs issue interpretation and action. Their findings suggest that identity constrains the meanings that members give to an issue, constitutes a reference for assessing the importance of an issue, helps to distinguish between aspects of the issue that pose a threat to the organization, and guides the search for solutions that can transform the issue into an opportunity.

The concept of organizational identity, therefore, has important ramifications for thinking about the effects of the organizational context within which innovation takes place. In a similar way to the way culture shapes organizational members' attitude, interests, and motivation towards innovation, identity shapes members' understanding of whether innovation is something central to the organization. If central and enduring understandings of identity include innovation as a defining characteristic then, presumably, innovation will be a common and normal activity within the organization. In fact, if this is so, then a lack of innovation will be seen as being in conflict with the identity and will provoke discussion and action among employees. Conversely, if the identity is one where innovation is not core, then the activity will be seen by organizational members as neither important nor one that they would expect would occur commonly.

Even more profoundly, the identity of the organization will constrain the sorts of innovation that are carried by the organization. As Gawer and Phillips (2013) show in their study of Intel's transition from a traditional supply chain partner to a platform leader, organizational members resist forms of innovation that they do not believe fits with 'who they are'. In this case, Intel managers had to spend considerable amounts of time managing the organization's identity before organizational members accepted the new innovation activities that went along with platform leadership including new forms of innovation. These dynamics need much further research to further our understanding of innovation management.

This points to a further area of particular relevance to our understanding of innovation management. Research on organizational identity suggests that the way organizations are perceived by internal audiences is shaped by a process of 'claim-making' in which influential members and groups try to persuade their audiences—for example, other organizational members—to accept their conceptualizations of the central, enduring, and distinctive features of the organization (Ashforth and Mael, 1996). In most organizations, a set of 'official' identity claims—that is, explicit statements from senior managers of what the organization is and stands for embodied in formal documents, uttered in public speeches, and so on—influence the perceptions of organizational members and external constituents by providing legitimate and consistent narratives through which to make sense of the organization (Ashforth and Mael, 1996; Czarniawska, 1997). Through formal identity claims organizational leaders attempt to influence how organizational members define and interpret the organization. These identity claims provide a coherent guide for how members should behave and how external groups should relate to them (Whetten and Mackey, 2002).

This aspect of identity management has particular ramifications for innovation management. To the degree that managers can manage identity to create the appropriate context for innovation, they can increase the amount of effort and shape the sorts of innovation activities that members of the organization engage in. While much more research is required into the link between the management of identity and the management of innovation, this provides a exciting potential connection for improving the practice of managing innovation.

Conclusions

The goal of this chapter has been to connect discussions of innovation management with the broader literature on OMT. It has balanced areas where significant work has already been completed with more speculative discussion of areas where little work has been done but where there is real potential for advancing understanding of innovation management. The result is a map of sorts for thinking about organizing innovation that will hopefully be helpful for both those interested in innovation management and those interested in OMT.

Tables 24.1 and 24.2 provide a summary of the dimensions of organizing discussed. The complex influence of the organizational context on innovation is, however, more than simply the sum of these dimensions, but includes complex interactions between them. There is an extensive literature, for example, discussing the link between leadership and culture and another discussing culture and structure. How these connections shape innovation require much more investigation and discussion. What is clear, however, is that the connection between OMT and innovation is a fertile area for further investigation with much potential for developing our understanding of innovation and its management.

Research in these areas has the potential to benefit OMT scholarship as well, as the particular context of innovation can often serve as a 'transparent example' (Eisenhardt, 1989) where the nature of innovation reveals important organizational dynamics. Focusing on the role of organizational characteristics and dynamics will reveal much about innovation, but it will also provide an avenue for innovation scholars to contribute to the broader OMT literature.

Many areas of OMT have an interest in innovation that, furthermore, has not yet been connected to the well-developed theories that underpin innovation studies. While institutional theory, for example, was once primarily interested in stability and how and why organizational structures and processes became more and more similar over time, institutional theorists have developed an interest in change and innovation in institutions. The work done in innovation studies has obvious applications in this area and in particular in explaining processes of innovation in institutions where innovation is purposeful and strategic, and perhaps even in providing frameworks for institutional entrepreneurs to use in managing the processes of institutional innovation that they engage in.

In summary, OMT and innovation management have much in common and much to offer each other. The large body of work that OMT scholars have produced holds many insights for innovation management researchers and new ways to think about innovation management that will challenge and improve understandings of this important area. And, in turn, as innovation management and OMT move closer together, the study of innovation also has much to offer OMT.

Notes

1. See Dodgson et al. (2005) for an excellent overview of this literature.
2. For example, more than 40 per cent of the submissions to the Organization and Management Theory Division of the Academy of Management in 2013 were identified by the author(s) as institutional theory.
3. The problem of embedded agency refers to the question of how actors who are embedded in particular institutional context manage to conceive of and create change in that context.

References

Albert, S., and Whetten, D. (1985). 'Organizational Identity', *Research in Organizational Behavior*, 7: 263–95.

Ansari, S. M., Fiss, P., and Zajac, E. (2010). 'Made to Fit: How Practices Vary as they Diffuse', *Academy of Management Review*, 35: 67–92.

Ashforth, B., and Mael, F. (1996). 'Organizational Identity and Strategy as a Context for the Individual', *Advances in Strategic Management*, 13: 19–64.

Backhaus, J. G. (ed.) (2003). *Joseph Alois Shumpeter: Entrepreneurship, Style and Vision*. Dordrecht: Kluwer.

Barley, S. R., and Tolbert, P. S. (1997). 'Institutionalization and Structuration: Studying the Links between Action and Institution', *Organization Studies*, 18: 93–117.

Bass, B., and Avolio, B. (1994). *Improving Organizational Effectiveness Through Transformational Leadership*. New York: Free Press.

Battilana, J., and Dorado, S. (2010). 'Building Sustainable Hybrid Organizations: The Case of Commercial Microfinance Organizations', *Academy of Management Journal*, 53(6): 1419–40.

Bock, A., Opsahl, T., George, G., and Gann, D. (2012). 'The Effects of Culture and Structure on Strategic Flexibility during Business Model Innovation', *Journal of Management Studies*, 49(2): 279–305.

Chen, M., Lin, C., Lin, H., and McDonough, E. (2012). 'Does Transformational Leadership Facilitate Technological Innovation? The Moderating Roles of Innovative Culture and Incentive Compensation', *Asia Pacific Journal of Management*, 29: 239.

Cohen, S., and Baily, D. (1997). 'What Makes Teams Work: Group Effectiveness Research from the Shop Floor to the Executive Suite', *Journal of Management*, 23(3): 239–90.

Czarniawska, B. (1997). *Narrating the Organization*. Chicago: University of Chicago Press.

Denti, L., and Hemlin, S. (2012). 'Leadership and Innovation in Organizations: A Systematic Review of Factors that Mediate or Moderate the Relationship', *International Journal of Innovation Management*, 16(3): 1–20.

De Tienne, D., and Mallette, P. (2012). 'Antecedents and Outcomes of Innovation-Oriented Cultures', *International Journal of Business and Management*, 7(18): 1–11.

DiMaggio, P. J. (1997). 'Culture and Cognition', *Annual Review of Sociology*, 23(1): 263–87.

DiMaggio, P. J., and Powell, W. J. (1983). 'The Iron Cage Revisited: Institutional Isomorphism and Collective Rationality in Organizational Fields', *American Sociological Review*, 48: 147–60.

Dodgson, M., Gann, D., and Salter, A. (2005). *Think, Play, Do: Technology, Innovation, and Organization*. Oxford: Oxford University Press.

Dougherty, D., and Hardy, C. (1996). 'Sustained Product Innovation in Large, Mature Organizations: Overcoming Innovation-to-Organization Problems', *Academy of Management Journal*, 39(5): 1120–53.

Douglas, M. (1986). *How Institutions Think*. Syracuse, NY: Syracuse University Press.

Dunn, M., and Jones, C. (2010). 'Institutional Logics and Institutional Pluralism: The Contestation of Care and Science Logics in Medical Education, 1967–2005', *Administrative Science Quarterly*, 55: 114–49.

Dutton, J., and Dukerich, J. (1991). 'Keeping an Eye on the Mirror: Image and Identity in Organizational Adaptation', *Academy of Management Journal*, 34: 517–54.

Dutton, J., Dukerich, J., and Harquail, C. (1994). 'Organizational Images and Membership Commitment', *Administrative Science Quarterly*, 39: 239–63.

Eisenhardt, K. (1989). 'Building Theories from Case Study Research', *Academy of Management Review*, 14(4): 532–50.

Fagerberg, J. (2005). 'Innovation: A Guide to the Literature', in J. Fagerberg, D. Mowery, and R. Nelson (eds), *The Oxford Handbook of Innovation*. Oxford: Oxford University Press, 1–25.

Feldman, M. (2000). 'Organizational Routines as a Source of Continuous Change', *Organization Science*, 11: 611–29.

Fiol, M. (1991). 'Managing Culture as a Competitive Resource: An Identity-Based View of Sustainable Competitive Advantage', *Journal of Management*, 17: 191–211.

Friedland, R., and Alford, R. R. (1991). 'Bringing Society back in: Symbols, Practices, and Institutional Contradictions', in W. W. Powell and P. J. DiMaggio (eds), *The New Institutionalism in Organizational Analysis*. Chicago: University of Chicago Press, 232–66.

Gagliardi, P. (1986). 'The Creation and Change of Organizational Cultures: A Conceptual Framework', *Organization Studies*, 7(2): 117–34.

Gawer, A., and Phillips, N. (2013). 'Institutional Work as Logics Shift: The Case of Intel's Transformation to Platform Leader', *Organization Studies*. Published online before print 4 July 2013, doi: 10.1177/0170840613492071.

Geertz, C. (1973). *The Interpretation of Cultures*. New York: Basic Books.

Gioia, D., and Thomas, J. (1996). 'Identity, Image and Interpretation: Sensemaking during Strategic Change in Academia', *Administrative Science Quarterly*, 41: 370–403.

Giuliani, R. (2003). *Leadership*. London: Time Warner.

Goleman, D., Boyatzis, R., and McKee, A. (2003). *The New Leaders: Transforming the Art of Leadership into the Science of Results*. London: Time Warner.

Greenwood, R., Díaz, A., Li, S., and Lorente, J. (2010). 'The Multiplicity of Institutional Logics and the Heterogeneity of Organizational Responses', *Organization Science*, 21(2): 521–39.

Greenwood, R., Raynard, M., Kodeih, F., Micelotta, E., and Lounsbury, M. (2011). 'Institutional Complexity and Organizational Responses', *Academy of Management Annals*, 5(1): 317–71.

Greenwood, R., and Suddaby, R. (2006). 'Institutional Entrepreneurship in Mature Fields: The Big Five Accounting Firms', *Academy of Management Journal*, 49(1): 27–48.

Hoegl, M., and Gemuenden, H. G. (2001). 'Teamwork Quality and the Success of Innovative Projects: A Theoretical Concept and Empirical Evidence', *Organization Science*, 12(4): 435–49.

Hofstede, G. (1980). *Culture's Consequences*. Newbury Park, Calif.: Sage.

Hofstede, G. (1989). 'Organising for Cultural Diversity', *European Management Journal*, 7(4): 390–7.

Hofstede, G. (1991). 'Empirical Models of Cultural Differences', in N. Bleichrodt and P. J. D. Drenth (eds), *Contemporary Issues in Cross-Cultural Psychology*. Lisse: Swets & Zeitlinger Publishers, 4–20.

Jay, J. (2013). 'Navigating Paradox as a Mechanism of Change and Innovation in Hybrid Organizations', *Academy of Management Journal*, 56(1): 137–59.

Jepperson, R. L. (1991). 'Institutions, Institutional Effects, and Institutionalism', in W. W. Powell and P. J. DiMaggio (eds), *The New Institutionalism in Organizational Analysis*. Chicago: University of Chicago Press, 143–63.

Jung, D., Chow, C., and Wu, A. (2003). 'The Role of Transformational Leadership in Enhancing Organizational Innovation: Hypotheses and Some Preliminary Findings', *Leadership Quarterly*, 14: 525–44.

Kaasa, A., and Vadi, M. (2010). 'How Does Culture Contribute to Innovation? Evidence from European Countries', *Economics of Innovation and New Technology*, 19(7): 583–604.

Kanter, R. (1983). *The Changemasters*. New York: Simon & Schuster.

Kennedy, M., and Fiss, P. (2009). 'Institutionalization, Framing, and Diffusion: The Logic of TMQ Adoption and Implementation Decisions among U.S. Hospitals', *Academy of Management Journal*, 52: 897–918.

Kirton, M. (1989). *Adaptors and Innovators: Styles of Creativity and Problem Solving*. New York: Routledge.

Kostova, T. (1999). 'Transnational Transfer of Strategic Organizational Practices: A Contextual Perspective', *Academy of Management Review*, 24(2): 308–24.

Kostova, T., and Zaheer, S. (1999). 'Organizational Legitimacy under Conditions of Complexity: The Case of the Multinational Enterprise', *Academy of Management Review*, 24(1): 64–81.

Kraatz, M., and Block, E. S. (2008). 'Organizational Implications of Institutional Pluralism', in R. Greenwood, C. Oliver, K. Sahlin, and R. Suddaby (eds), *Handbook of Organizational Institutionalism*. Los Angeles: Sage, 243–75.

Kraatz, M., and Zajac, E. (1996). 'Exploring the Limits of the New Institutionalism: The Causes and Consequences of Illegitimate Organizational Change', *American Sociological Review*, 61: 812–36.

Lam, A. (2005). 'Organizational Innovation', in J. Fagerberg, D. Mowery, and R. Nelson (eds), *The Oxford Handbook of Innovation*. Oxford: Oxford University Press.

Leonardi, P. (2011). 'Innovation Blindness: Culture, Frames, and Cross-Boundary Problem Construction in the Development of New Technology Concepts', *Organizational Science*, 22(2): 347–69.

Lounsbury, M. (2001). 'Institutional Sources of Practice Variation: Staffing College and University Recycling Programs', *Administrative Science Quarterly*, 46: 29–56.

Lounsbury, M. (2007). 'A Tale of Two Cities: Competing Logics and Practice Variation in the Professionalizing of Mutual Funds', *Academy of Management Journal*, 50: 289–307.

Miron-Spektor, E., Erez, M., and Navez, E. (2011). 'The Effect of Conformist and Attentive-to-Detail Members on Team Innovation: Reconciling the Innovation Paradox', *Academy of Management Journal*, 54(4): 740–60.

Naranjo-Valencia, J., Jimenez-Jimenez, D., and Sanz-Valle, R. (2011). 'Innovation or Imitation? The Role of Organizational Culture', *Management Decision*, 49(1): 55–72.

Nelson, R. (ed.) (1993). *National Innovation Systems: A Comparative Analysis*. New York: Oxford University Press.

Ocasio, W. (1997). 'Towards an Attention-Based View of the Firm', *Strategic Management Journal*, 18: 187–206.

Ogbonna, E., and Wilkinson, B. (2003). 'The False Promise of Organizational Culture Change: A Case Study of Middle Managers in Grocery Retailing', *Journal of Management Studies*, 40: 1151–78.

Oke, A., Munshi, N., and Walumbwa, F. (2009). 'The Influence of Leadership on Innovation Processes and Activities', *Organizational Dynamics*, 38(1): 64–72.

Orlikowski, W. J. (2000). 'Using Technology and Constituting Structures: A Practice Lens for Studying Technology in Organizations', *Organization Science*, 11: 404–28.

Osterman, P. (2006). 'Overcoming Oligarchy: Culture and Agency in Social Movement Organizations', *Administrative Science Quarterly*, 51(4): 622–49.

Pache, A., and Santos, F. (2010). 'When Worlds Collide: The Internal Dynamics of Organizational Responses to Conflicting Institutional Logics', *Academy of Management Review*, 35(3): 455–76.

Phillips, N., Lawrence, T., and Hardy, C. (2000). 'Inter-Organizational Collaboration and the Dynamics of Institutional Fields', *Journal of Management Studies*, 37(1): 23–45.

Phillips, N., Tracey, P., and Karra, N. (2009). 'Rethinking Institutional Distance: Strengthening the Tie between New Institutional Theory and International Management', *Strategic Organization*, 7(3): 339–48.

Powell, W., Koput, K., and Smith-Doerr, L. (1996). 'Interorganizational Collaboration and the Locus of Innovation: Networks of Learning in Biotechnology', *Administrative Science Quarterly*, 41(1): 116–45.

Quinn, J. B. (1985). 'Managing Innovation: Controlled Chaos', *Harvard Business Review*, 63(3): 78–84.

Reay, T., and Hinings, C. R. (2009). 'Managing the Rivalry of Competing Institutional Logics', *Organization Studies*, 30(6): 629–52.

Rerup, C., and Feldman, M. (2011). 'Routines as a Source of Change in Organizational Schemata: The Role of Trial-and-Error Learning', *Academy of Management Journal*, 54: 577–610.

Rindova, V., Dalpiaz, E., and Ravasi, D. (2010). 'A Cultural Quest: A Study of Organizational Use of New Cultural Resources in Strategy Formation', *Organizational Science*, 22: 413–31.

Rumsey, M. (2013). *Oxford Handbook of Leadership*. Oxford: Oxford University Press.

Schein, E. (1992). *Organizational Culture and Leadership*. San Francisco: Jossey-Bass.

Schumpeter, J. (1950). 'The Process of Creative Destruction', in J. Schumpeter (ed.). *Capitalism, Socialism and Democracy*, 3rd edn. London: Allen and Unwin, 81–6.

Scott, W. R. (1995). *Institutions and Organizations: Theory and Research*. Thousand Oaks, Calif.: Sage.

Smirnich, L. (1983). 'Concepts of Culture and Organizational Analysis', *Administrative Science Quarterly*, 28: 339–58.

Steensma, H., Marino, L., Weaver, K., and Dickson, P. (2000). 'The Influence of National Culture on the Formation of Technology Alliances by Entrepreneurial Firms', *Academy of Management Journal*, 43(5): 951–73.

Swidler, A. (1986). 'Culture in Action: Symbols and Strategies', *American Sociological Review*, 51: 273–86.

Swidler, A. (2001). *Talk of Love: How Culture Matters*. Chicago: University of Chicago Press.

Tellis, G., Prabhu, J., and Chandy, R. (2009). 'Radical Innovation across Nations: The Preeminence of Corporate Culture', *Journal of Marketing*, 73: 3–23.

Thornton, P. H. (2002). 'The Rise of the Corporation in a Craft Industry: Conflict and Conformity in Institutional Logics', *Academy of Management Journal*, 45: 81–101.

Thornton, P. (2004). *Markets from Culture: Institutional Logics and Organizational Decisions in Higher Education Publishing*. Stanford, Calif.: Stanford University Press.

Thornton, P. H., and Ocasio, W. (1999). 'Institutional Logics and the Historical Contingency of Power in Organizations: Executive Succession in the Higher Education Publishing Industry, 1958–1990', *American Journal of Sociology*, 105: 801–43.

Tracey, P., Phillips, N., and Jarvis, O. (2011). 'Bridging Institutional Entrepreneurship and the Creation of New Organizational Forms: A Multilevel Model', *Organization Science*, 22: 91–8.

Trompenaars, F., and Hampden-Turner, C. (2012). *Riding the Waves of Culture* London: Nicholas Brealey.

Van Everdingen, Y., and Waarts, E. (2003). 'The Effect of National Culture on the Adoption of Innovation', *Marketing Letters*, 14(3): 217–32.

Weber, K. (2005). 'A Toolkit for Analyzing Corporate Cultural Toolkits', *Poetics*, 33: 227–52.

Westphal, J. D., Gulati, R., and Shortell, S. M. (1997). 'Customization or Conformity? An Institutional and Network Perspective on the Content and Consequences of TQM Adoption', *Administrative Science Quarterly*, 42: 366–94.

Whetten, D. (2006). 'Albert and Whetten Revisited: Strengthening the Concept of Organizational Identity', *Journal of Management Inquiry*, 15: 219–34.

Whetten, D., and Mackey, A. (2002). 'A Social Actor Conception of Organizational Identity and its Implications for the Study of Organizational Reputation', *Business and Society*, 41: 393–414.

Williamson, O. E. (1975). 'Markets and Hierarchies: Analysis and Antitrust Implications: A Study in the Economics of Internal Organization', *University of Illinois at Urbana-Champaign's Academy for Entrepreneurial Leadership Historical Research Reference in Entrepreneurship*. Available at SSRN: <http://ssrn.com/abstract=1496220>.

Zilber, T. (2006). 'The Work of the Symbolic in Institutional Processes: Translations of Rational Myths in Israeli High Tech', *Academy of Management Journal*, 49: 281–303.

CHAPTER 25

....

HUMAN RESOURCE MANAGEMENT PRACTICES AND INNOVATION

....

KELD LAURSEN AND NICOLAI J. FOSS

INTRODUCTION

....

HUMAN capital is a key, and by all accounts increasingly important, part of the resource-base of firms. Human resources have been called the 'key ingredient to organizational success and failure' (Baron and Kreps, 1999), including success and failure in company innovation performance. It is important to understand why and how human capital encourages innovation, and what deployment of human resource management (HRM) practices inside the firm can produce desired levels of innovation performance.

Individual employees, founders, or executives may *directly* give rise to superior innovation performance (Felin and Hesterly, 2007), as in the cases of 'innovative genius' (Glynn, 1996) and 'stars' (Lacetera, Cockburn, and Henderson, 2004) among others. Such human capital is substantially above normal in innovative capacity, whether this is innate (personified, perhaps, by Bill Gates or Steve Jobs) or acquired through training efforts. University researchers who create entrepreneurial start-ups exemplify the direct link between human capital and innovation performance. Superior innovation performance may also be the result of the 'capabilities' stemming from the interactions within a firm's human capital pool (Lepak and Snell, 2002).

The organizational set-up of the firm, notably its human resource management practices, also matter to the contribution of human capital to innovation performance, and it is this effect that we mainly address in this chapter. Thus, management deploys training arrangements, makes decisions on reward structures, sets up teams, allocates decision rights and so on, and these arrangements have implications for the contribution of human capital to innovation.

The influence of these practices may be modelled, both in terms of mediator (human capital mediates the influence from HR practices to innovation performance) and moderator (practices weaken or reinforce the link from human capital to innovation performance) models.[1]

Extant research suggests multiple mechanisms through which such HRM practices influence the relationships between human capital and innovation. Employee communication networks, as partly shaped by organizational structure, may influence innovation (Tsai, 2001). Motivational research demonstrates that the kind of creative behaviours that underlie successful innovation is stimulated by some kinds of rewards but reduced by others (Ryan and Deci, 2000). Managerial styles, the use of feedback, the setting of goals, the use of teams and projects, have all been argued to influence creativity and innovative behaviours.

Organizational practices related to the sourcing, deployment, and upgrading of human capital have been identified in various literatures as influencing innovation performance at the level of firms (Henderson and Cockburn, 1994; Galunic and Rjordan, 1998), networks and industries (Kogut, 2000), and regional or national innovation performance (Almeida and Kogut, 1999; Furman et al., 2002). These practices are important constituent components of 'innovation' or 'dynamic capabilities' (Teece, 2007). A significant part of such practices are those organizational practices that relate to the attraction, selection, training, assessment, and rewarding of employees. They also include organizational practices that may not conventionally be seen as HRM, such as quality circles, extensive delegation of decision rights, management information systems, and formal and informal communication practices in the firm.

In this chapter we survey, organize, and discuss the literature on the role of HRM practices in explaining innovation outcomes. We discuss how individual practices influence innovation, and how the clustering of specific practices matters for innovation outcomes (cf. Ennen and Richter, 2010). Relatedly, we discuss various possible mediators of the HRM/innovation link, such as knowledge sharing, social capital, and network effects. We argue that the causal mechanisms underlying the HRM/innovation links are still ill-understood, calling for further research.

Organizing the Literature

The literature on the relation between HRM practices and innovation performances is vast and not easily identifiable, as relevant papers are not necessarily published in HRM journals and may primarily focus on other issues. There is a choice to be made regarding whether research on, say, the impact of monetary incentives on creativity should be included. We specifically put an emphasis on what is often called 'new' or 'modern' HRM practices (also often called 'High-Performance Work Practices') (Laursen and

Foss, 2003; Teece, 2007; Colombo and Delmastro, 2008) and their relation to innovation performance. We argue that the literature on HRM practices and innovation can be split into five basic sub-literatures (although inevitably there is some overlap). These are shown diagrammatically in Figure 25.1. Link I represents a stream of literature that considers the relationship between HRM practices and firm-level financial performance, using innovation as a theoretical link between these variables. Link II denotes a stream of literature that considers the direct link between HRM practices and innovation, while Link III considers a subsequent literature that in addition to this direct link considers mediating and moderating factors of the HRM–innovation relationship. Link IV comprises a small body of literature that has looked not only at the HRM–innovation relationship, but also at antecedents to HRM practices that led to innovative outcomes. We will discuss these literatures, but first we will identify the most important HRM practices considered in the innovation-related literature.

HRM Practices

The notion of 'modern HRM practices' has become an increasingly common way of referring to high levels of delegation of decisions, extensive lateral and vertical communication channels, high reward systems, often linked to multiple performance indicators, and other practices that either individually or in various bundles are deployed to achieve high levels of organizational performance (Ichniowski et al., 1997; Zenger and Hesterly, 1997; Colombo and Delmastro, 2002; Teece, 2007; Colombo and Delmastro, 2008). In this context, Guthrie (2001: 181) states that: 'The common theme in this literature is an emphasis on utilizing a system of management practices giving employees skills, information, motivation, and latitude and resulting in a workforce that is a source of competitive advantage'.

Following Foss, Laursen and Pedersen (2011), we posit that the HRM practices considered in the literature involve: (a) delegation of responsibility, such as team production; (b) knowledge incentives, such as profit sharing, individual incentives, and incentives for knowledge sharing; (c) internal communication, encouraged for instance by practices related to knowledge sharing or job rotation; (d) employee training, both internal and external; and (e) recruitment and retention, such as internal promotion policies. It can be noted that the first three classes of practices include the practices that are typically included as 'modern' HRM practices in the literature (Teece, 2007), while the latter two classes, in a stylized fashion, can be considered more traditional HRM practices. Table 25.1 provides an overview of our taxonomy and describes the results of a number of representative papers from various parts of the literature.

The early literature was concerned with various 'stand-alone' HRM practices and their effect on organizational performance (e.g. Gerhart and Milkovich, 1990; Terpstra and Rozell, 1993). Most of the empirically based literature since the mid-1990s has focused on the effects of complementary practices, rather than the effect of individual practices

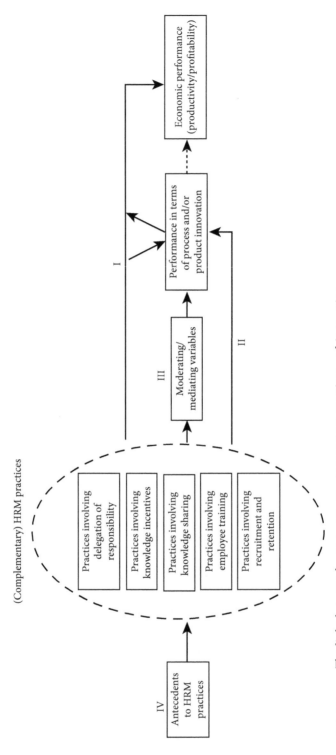

FIGURE 25.1 The links between human resource management practices and innovation

(see the recent overview of the general complementarities literature by Ennen and Richter, 2010). The idea of complementarities in our context implies that the introduction of one HRM practice increases the returns to doing more of other HRM practices related to innovation output (for a general definition of complementarity, see Milgrom and Roberts, 1995: 181). Note that although the notion of 'internal fit' is arguably less precise than the idea of complementarity, this notion is often used in the HRM literature in a similar fashion to that of complementarity (see e.g. Baird and Meshoulam, 1988; Arthur, 1994). Ideas on 'systems' or 'bundles' of HRM practices (see, Subramony, 2009) operate with a similar logic.

The empirical literature on organizational complementarities suggests two approaches: an interaction and a systems approach (cf. Ennen and Richter, 2010). The interaction approach (e.g. Capelli and Neumark, 2001) examines the effect of a few organizational practices, and in contrast, the systems approach (e.g. Ichniowski et al., 1997; Laursen and Foss, 2003) looks at the relative performance outcomes of entire sets of variables. Given the sheer number of individual practices considered in the literature, the systems approach is dominant, even if it only confers an indirect test of complementarity.

Link I: The Role of Innovation

Link I represents a large literature stream that has considered innovation mainly in an indirect fashion. This large body of literature (including for instance, Huselid, 1995; Ichniowski et al., 1997; Ichniowski and Shaw, 1999; Datta et al., 2005) considers HRM practices as explanatory factors (typically complementary) in determining dependent variables such as productivity and profitability. In a typical statement Huselid (1995: 638) notes that the

> ...theoretical literature clearly suggests that the behavior of employees within firms has important implications for organizational performance and that human resource management practices can affect individual employee performance through their influence over employees' skills and motivation through organizational structures that allow employees *to improve how their jobs are performed.*

It should be noted, however, that improving 'how...jobs are performed' may refer to incremental process innovations that are not included in the remit of innovation management as described in this book.

Research within this literature has typically been published in management journals, but some highly influential studies have been published in economics journals (in particular, Ichniowski et al., 1997). As mentioned above this literature has considered the direct effect of (complementary) HRM practices on economic performance, but also moderated relationships between these variables, for example by the type of manufacturing strategy pursued by the respondent's firm (Youndt et al., 1996) or its industry

Table 25.1 Variables in the new HRM practices literature

Authors	Unit of analysis	Dependent variable	Delegation	Internal Communication	Incentives	Employee training	Recruitment and retention	Other HMRP variables
Huselid (1995)	968 publicly listed firms (manufacturing and private services)	Labour productivity, Tobin's q	Labour-management teams, Quality circles	Formal information sharing program, Complaint resolution system	Incentive plans/profit sharing, formal appraisals, merit-based promotion	Hours of training	Formal job analysis, Internal promotion, Employment test prior to recruitment	Attitude survey
Ichniowski et al. (1997)	36 steel finishing lines within 17 firms	Percent uptime	Teamwork (3 items)	Communication (2 items), Job rotation	Line incentives	Skills training (2 items)	High screening recruitment	Employment security
Ichniowski and Shaw (1999)	41 steel finishing lines within 19 firms	Percent uptime, Percent prime yield	Teamwork (3 items)	Labour-management communication (2 items), Job flexibility (2 items)	Incentive pay (2 items)	Training (2 items)	Recruiting (2 items)	Employment security
Mendelson and Pillai (1999)	102 business units from 81 different firms (electronics hardware)	Return on Sales Return on value added, Sales growth	Decentralization (3 items)	Information practices (8 items)	Incentives (3 items)			Focus (3 items), Inter-organizational network (5 items)
Michie and Sheehan (1999)	480 firms (manufacturing and private services)	R&D expenditure, Introduction of advance technological machinery	Teamwork	Flexible job assignment, Communication (4 items)	Profit sharing, Share ownership, Individual pay/Line incentives			Employment flexibility

(Continued)

Study	Sample	Performance	Decision architecture	Knowledge transparency	Decision architecture (incentives)	Training	Activity focus
Mendelson (2000)	60 business units (electronics hardware)	Return on Sales, Return on value added, Sales growth	Decision architecture (6 items, including 3 delegation items)	Knowledge transparency (6 items)	Decision architecture (6 items, including 3 items measuring incentives)		Activity focus (6 items), (External) Information awareness (8 items), Information Age Network (6 items)
Guthrie (2001)	164 firms (manufacturing and private services)	Employee retention rate, Labour productivity	Teams, Employee participatory programs	Information sharing	Skill-based pay, Group-based pay, Performance-based promotion, Employee stock ownership	Training efforts (3 types)	Internal promotion
Capelli and Neumark (2001)	Plants in manufacturing and private services (panel, no. of obs. 433/666)	Labour productivity, Labour costs, Sales less labour costs	Self-managed teams, TQM	Scheduled meetings, Job rotation	Pay for skills and knowledge, Profit sharing		Use of computers, Use of benchmarking vis-à-vis other organizations
Colombo and Delmastro (2002)	438 manufacturing plants (panel data)	Change in the number of managerial layers	Teamwork, number of hierarchical layers	Job rotation	Individual line incentives	Firm pays for training	Type of strategic decision-maker
Laursen and Foss (2003)	1900 firms (manufacturing and private services)	Product innovation	Delegation of responsibility, Inter-disciplinary work groups, Quality circles	Integration of functions, Job rotation, Systems for collection of employee proposals	Pay-for-performance	Firm-internal and firm-external training	

(Continued)

Table 25.1 (*Continued*)

Authors	Unit of analysis	Dependent variable	Delegation	Internal Communication	Incentives	Employee training	Recruitment and retention	Other HMRP variables
Hamilton et al. (2003)	Workers within a single firm (panel data)	Productivity	Team vs. no-team production		Team vs. individual piece rates			
Datta et al. (2005)	132 manufacturing firms	Labour productivity	Self-directed teams	Programs designed to elicit participation and employee input, Complaint resolution system, Provide information to management	Compensation based on group performance. Pay is based on a skill or knowledge-based system, Formal performance feedback	Intensive/ extensive training	Tests administered prior to hiring, Internal promotions, Intensive/ extensive recruiting	
Collins and Smith (2006)	513 high-technology companies	Revenue from new products, Sales growth		Knowledge Exchange and Combination (8 items)	Incentive Policies (3 items)		Selection Policies (4 items)	
Colombo et al. (2007)	109 single plant firms (panel data)	Profitability	Decentralization, Number of plant's hierarchical levels, TQM	Formal team practices, Job rotation	Profit sharing, Individual incentives			
Beugelsdijk (2008)	988 firms (manufacturing and private services)	Incremental, Radical innovation	Job autonomy, Task rotation		Performance-based pay	Training policies, Internal training, External training, Procedures for education of employees	Procedures for recruitment	Procedures for quality maintenance

Study	Sample	Innovation measure						
Chen and Huang (2009)	146 firms (manufacturing and private services)	Administrative innovation (4 items), Technical innovation (3 items)	Participation (3 items)		Appraisal (3 items), compensation (3 items)	Training (6 items)	Staffing (3 items)	
Lopez-Cabrales et al. (2009)	86 firms (manufacturing)	Innovative Activity, Profits	Delegation (two items)	Cross-functional teams, job-rotation	Performance appraisal (4 items), compensation (3 items)	Training activities (two items)	Selection process (4 items), promotion from within	Job-security, socialization program, Tutoring
Zoghi et al. (2010)	3303 firms (panel data) (manufacturing and private services)	Product innovation	Decentralization	Info-sharing	Individual incentive pay, Group Incentive pay, Profit sharing plan			
Foss et al. (2011)	132 firms (manufacturing and private services)	Innovation performance (2 items)	Delegation of Responsibility (two items)	Internal communication (2 items)	Knowledge Incentives (two items)			

Note: Adapted and extended from Foss, Laursen, and Pedersen (2011).

affiliation (Datta et al., 2005). The majority of contributions under this heading adopts a cross-sectional approach, and hypothesizes empirical links between a set of complementary HRM practices and economic performance. There is also research based on panel data within this stream. While initial evidence suggested that these organizational practices (Capelli and Neumark, 2001) had little effect on economic performance such as productivity, more recent panel data evidence has tended to confirm the findings from the studies based on cross-sectional evidence, in that a set of complementary HRM practices have in general been found to have positive influences on economic performance, including productivity and profitability (Van Reenen and Caroli, 2001; Kato and Morishima, 2002; Janod and Saint-Martin, 2004; Colombo et al., 2007). Given this body of literature is only indirectly concerned with innovation management, we will not go in depth into this literature (see Colombo et al., 2013 for an in-depth review of this literature).

Link II: The Direct Link Between HRM and Innovation

Link II refers to literature that has established a direct theoretical and empirical link between HRM practices and innovation outcomes, typically in the form of product or process innovation.

Until the 2000s, the innovation literature was characterized by relatively scant attention being paid to HRM practices and how they influence innovation performance (Laursen and Foss, 2003). The clear exception is some scholars' interest in Japanese organization and how this connects to innovativeness (Aoki and Dore, 1994). Thus, Freeman (1988: 335) explicitly notes how in 'Japanese management, engineers and workers grew accustomed to thinking of the entire production process as a system and of thinking in an integrated way about product design and process design', and he makes systematic reference to quality management, horizontal information flows, and other features of modern HRM practices. The concern with horizontal information flows in Project SAPPHO in the late 1960s demonstrates a long-standing awareness of the relation between HRM practices and innovation performance (Rothwell et al., 1974).

Laursen and Foss (2003) supply a number of theoretical arguments for why HRM practices are favourable to innovative activity. One prominent characteristic of many HRM practices is that they increase decentralization by delegating problem-solving rights to the shop floor. When implemented appropriately, these rights coexist alongside access to relevant knowledge, much of which may be inherently tacit and thus requires decentralization for its efficient use. Increased delegation may better allow for the discovery and utilization of local knowledge within the organization, especially when there are rewards in place that support such discovery (Hayek, 1945; Jensen and Meckling, 1992). The increased use of teams is an important component in the set of modern HRM practices. The use of teams also implies that better use can be made of local knowledge, leading to improvements in processes, and perhaps also to minor product improvements (Laursen and Foss, 2003: 248). Teams have additional benefits, since they are often composed of different human resource inputs. This may imply that teams bring together knowledge that hitherto existed separately, potentially resulting in process improvements when teams are on the shop floor, or 'new combinations' that lead to novel products (Schumpeter, 1912/1934), especially when teams are in product development departments. Increased knowledge diffusion through job rotation, and increased information dissemination facilitated by IT, may also be expected to provide a positive contribution to innovation performance. Training may be a factor leading to a higher rate of process improvements and may also lead to product innovations.

The adoption of a single such practice may sometimes provide a contribution to innovative performance. The increasingly widespread practice of rewarding shop-floor employees for putting forward suggestions for process improvements—such as by giving them a share of the cost savings—is likely to increase incremental innovation activity (Bohnet and Oberholzer-Gee, 2001), regardless of whether or not the firm has employed other organizational practices as well. However, HRM practices should be most conducive to innovation performance when adopted, not singly, but as a system of mutually reinforcing practices. The arguments in favour of this are set out in Laursen and Foss, 2003: 249. The innovation pay-off from giving shop-floor employees more problem-solving rights will likely depend on the level of training of such employees. The converse is also likely to hold: employees may invest more in upgrading their skills if they are also given extensive problem-solving rights, especially if they are provided with intrinsic or extrinsic motivational encouragements. Rotation and job-related training may have complementary impacts on innovative activity. All these practices are likely to complement various incentive-based remuneration schemes—based on individual, team or firm performance—profit sharing arrangements, and promotion schemes. To the extent that implementing HRM practices is associated with extra effort or with the disruption of changing routines, employees will usually demand compensation. From an agency-theory perspective one would expect many HRM practices to work well, in both profits and innovation performance, only if accompanied with new, typically more incentive-based, remuneration schemes.

Arguably the first paper to empirically establish the link between a system of HRM practices and innovative activity was Michie and Sheehan (1999). Using a sample of 480 UK firms drawn from the UK's 1990 Workplace Industrial Relations Survey, the authors investigate the relationships between firms' HRM practices and the level of R&D expenditure. The results suggest that what the authors term 'low road' HRM practices—strict job-description, short-term contracts, and so on—are negatively related to investment in R&D and the adoption of advanced production equipment. In contrast, 'high road' work practices (modern HRM practices) are positively correlated with investment in R&D and modern production equipment.

Laursen and Foss (2003) introduce an innovation–output measure in the HRM literature: the degree of novelty in product innovation. Based on the theoretical arguments presented above regarding complementarities and using data from a Danish survey of 1,900 business firms, the authors conjecture that HRM practices should influence innovation performance positively. Laursen and Foss identify two HRM systems that are conducive to innovation. In the first, seven of a total of nine HRM variables matter nearly equally for the ability to innovate: interdisciplinary workgroups, quality circles, systems for collection of employee proposals, planned job rotation, delegation of responsibility, integration of functions, performance-related pay, and pay-for-performance. The second system is dominated by firm-internal and firm-external training. While only two individual practices are strongly significant in explaining the degree of novelty of product innovation, the two systems are strongly significant in the regressions. The authors interpret these findings as evidence of complementarity.

In a later study in a sample of 240 UK manufacturing firms also using an innovation output measure of product and process innovation as the dependent variable, Michie and Sheehan (2003), find that firms using HRM practices extensively are much more likely to be process and/or product innovators. The so-called 'low road' HRM practices (see above) are found to be negatively associated with process innovation, but appear unrelated to product innovation. On the basis of a data set on approximately 1,400 Swiss firms for the period 1998–2000, Arvanitis (2005) presents findings that are consistent with Michie and Sheehan's findings: a system of HRM practices has a positive effect on firms' probability of introducing process innovation, but not of introducing product innovation. Arvanitis also examines whether there is complementarity between numerical flexibility variables (use of part-time work and temporary work) and HRM practices, and complementarity is found between temporary work and HRM practices in process innovation, but not in product innovation. Jimenez-Jimenez and Sanz-Valle's (2008) study of 173 Spanish firms indicates that product, process, and administrative innovation contribute positively to business performance, and that a comprehensive set of HRM practices enhances innovation. Beugelsdijk (2008) uses a sample of 988 Dutch firms. His results indicate the importance of task autonomy, training, and performance-based pay for generating incremental innovations (share of new-to-the-firm products as a percentage of total sales). For radical innovations (share of new-to-the-industry products as a percentage of total sales), the findings underline the importance of task autonomy and flexible working hours. The use of standby (seasonal/temporary/casual/fixed term) contracts is found to be associated with significantly lower levels of innovativeness. Beugelsdijk also detects significant interaction effects between individual HRM practices, providing further evidence in support of the notion of complementarities between these practices.

Love and Roper (2009), using data on UK and German manufacturing plants, examine the issue of potential complementarities which may arise when cross-functional teams are used in different elements of the innovation process. Using the 'interactions approach', Ennen and Richter (2010), demonstrate that patterns of complementarity are complex; however, they are more marked in the UK than in Germany. The most uniform complementarities are between product design and development and production engineering, with little synergy evident between the more technical phases of the innovation process and the development of marketing strategy. The results point to the value of using cross-functional teams for the more technical elements of the innovation process, but also suggest that the development of marketing strategy should remain the domain of specialists.

While all the above studies report findings based on cross-sectional data, other studies have begun to examine longitudinal variation. This is a difficult task, since research on HRM practices inevitably involves questionnaire-based data that will probably suffer from sample attrition as substantial numbers of firms typically disappear over time. On the other hand, the use of data with a time dimension reduces the concerns one might have regarding endogeneity in cross-sectional studies.

Shipton et al. (2005) provide results based on two waves of a survey in which twenty-seven UK manufacturing firms were present in 1993 and 1995. The dependent variables were measured in 1995 and the independent variables in 1993. Even though the study is small scale and only a limited number of control variables are allowed given the small sample, the authors find that HRM practices, excluding monetary incentives, lead to higher levels of product innovation, but not to higher levels of process innovation. Monetary incentives linked to appraisal appear to yield a negative impact on product innovation, although again it seems that there is no effect on process innovation. These results are only present when the independent variables are lagged: the HRM and incentive variables are insignificant when included in an instantaneous model, indicating that the negative effect is not of a short-term nature.

Zhou et al. (2011) use a merged data set based on four waves of Dutch survey data of 2,044 firms collected between 1993 and 2001, with the dependent variables measured at t and independent variables measured at t-2. Zhou et al. find that functional flexibility (internal labour mobility), training efforts, and highly qualified personnel appear to affect product innovation positively (percentage of sales of products new to the market). Zoghi et al. (2010) use a balanced panel of 3,203 establishments from the Canadian Workplace and Employment Survey. The questions about HRM practices were posed in 1999, 2001, and 2003. The dependent variable is a dummy variable representing whether the given establishment introduced product innovation in the given year. The independent variables include decentralization, information-sharing, and incentive pay (and interactions between them). To mitigate the problem related to time-invariant firm heterogeneity and simultaneity bias, the authors use a fixed-effects model and a model including a lagged dependent variable. The authors find a clear positive link between these factors and product innovation. However, the results suggest that these relationships are not causal (for further discussion of this issue, see the section called 'More time-series evidence'). The results show that the correlation between HRM practices and innovation holds for information-sharing, but is much weaker for decentralized decision-making or incentive pay programmes.

Link III: Moderated and Mediated Relationships Between HRM and Innovation

Link III embodies the literature that has established a mediated or moderated theoretical and empirical link between HRM practices and innovation outcomes.

Laursen (2002) posits that organizational theory suggests that more knowledge-intensive production activities often involve higher degrees of strategic uncertainty for firms and performance ambiguity in relation to individual employees. Therefore he expects that HRM practices perform better within knowledge-intensive industries of the economy than in other industries. Based on a sample of 726 Danish firms with more than fifty employees, the results confirm other findings that HRM practices are more effective in influencing product innovation performance when applied together, compared with situations in which individual practices are applied alone. Furthermore, he found that the application of complementary HRM practices is more effective for firms in 'high' and 'medium' knowledge-intensive industries.

Ritter and Gemünden (2003) examine a model in which 'network competence' mediates the relationship between HRM and a composite encompassing process and product innovation. Network competence is defined as company-specific ability to handle, use, and exploit inter-organizational relationships. Their results, drawn from a sample of 308 German mechanical and electrical engineering companies, reveal that network competence impacts on a firm's product and process innovation success. The organizational antecedents that impact on a company's network competence include intra-organizational communication and the openness of corporate culture.

Lau et al. (2004) outline the role of organizational culture in the link between the HRM system and the development of new products and services. The authors propose that a developmental culture is a missing link in between HRM system and innovation outcomes. It is argued that an HRM system that emphasizes extensive training, performance-based reward, and team development is needed to construct an 'organizational culture' that is conducive to product innovation. Based on a survey of 332 firms in Hong Kong, the empirical results are consistent with the idea that organizational culture acts as a mediator between firms' HRM systems and product innovation outcomes.

Jensen et al. (2007) contrast two modes of innovation. The first, the Science, Technology, and Innovation (STI) mode, is based on the production and use of codified scientific and technical knowledge. The second, the Doing, Using, and Interacting (DUI) mode is akin to a set of HRM practices (except that incentives are not included in the set of HRM practices). Drawing on the results of the 2001 Danish DISKO Survey encompassing 692 firms, analysis shows that firms combining the two modes are more likely to innovate in new products or services than those relying primarily on one mode or the other. In other words, high levels of codified scientific and technical knowledge increase the benefits of HRM practices.

The study by Beugelsdijk (2008) also reports significant interaction effects between HRM practices and firm size, and between HRM practices and R&D intensity, so that the effect of HRM is complementary to other firm-level variables. Based on data from the German Community Innovation Survey, Rammer et al. (2009) find that R&D activities are a main driver of innovation output (number of different types of innovations). However, small and medium-sized firms without in-house R&D can yield a similar innovation success when they apply HRM practices to facilitate innovation processes.

Camelo-Ordaz et al. (2008) examine whether the strategic vision of the top management team and the way employees working in teams are rewarded and assessed affect firms' innovation performance. The study is based on a relatively small sample of ninety-seven Spanish companies from high-tech industries. The results indicate that innovation output requires the existence of compensation practices based on the ideas generated and developed by project teams aligned with top management teams' strategic vision. Using a sample of 188 UK firms, Oke et al. (2012) find that the interaction of innovation strategy execution by top-management and a set of innovation-focused HRM practices are positively related to product innovation performance.

Lopez-Cabrales et al. (2009) examine how two sets of modern HRM practices ('collaborative HRM practices' and 'knowledge-based HRM practices') and employees' knowledge influence the level of innovative activities as they pertain to product innovation. Using a sample of eighty-six Spanish manufacturing firms, the results indicate that HRM practices are not directly associated with innovation unless they take into account employees' knowledge. Specifically, the analyses suggest a mediating role for firm-specific uniqueness of knowledge between collaborative HRM practices and innovative activity. The findings suggest that the so-called knowledge-based HRM practices and innovation output appear not to be linked.

Chen and Huang (2009) examine the role of knowledge management capacity (knowledge acquisition, knowledge sharing, and knowledge use) in the relationship between HRM practices and innovation performance (measured as both technical and administrative innovation). The authors use regression analysis to test the hypotheses in a sample of 146 Taiwanese firms. The empirical findings indicate that HRM practices are positively related to knowledge management capacity which, in turn, has a positive effect on innovation performance. In other words, the results suggest that knowledge management capacity plays a mediating role between HRM practices and innovation performance.

Foss et al. (2011) argue that firms that attempt to leverage user and customer knowledge in the context of innovation must design an appropriate internal organization to support it, and that this can be achieved, in particular, through the use of HRM practices, notably those involving intensive vertical and lateral communication, rewarding employees for sharing and acquiring knowledge, and high levels of delegation of decision rights. Using a data set drawn from a survey of 169 Danish firms among a sample of the largest firms in Denmark, the authors find that the link from customer knowledge to innovation is aided substantially by HRM practices (see also Petroni et al., 2012, for a discussion of the needed changes in R&D organization and personnel management as a consequence of the implementation of the open innovation model).

An important feature of the model proposed by Foss et al. (2011) pertains to the fact that so-called 'knowledge incentives' are part of the organizational variables (positively) mediating the relationship between customer interaction and innovation performance. Somewhat in contrast to this, Fu (2012)—based on a sample of 384 SMEs in the British manufacturing and business services sectors covering the period 1998–2001—finds that while both openness and incentives are positively associated with product innovation efficiency, a substitution effect is found between openness and incentives. Long-term incentives appear to enhance efficiency to a greater extent than short-term incentives, and the substitution effect of openness is stronger in the case of long-term incentives. As measures of long-term incentive schemes, Fu uses the proportion of managers and employees participating in a stock option scheme in the total labour force (alternatively, a dummy variable for firms that have introduced a stock option scheme). Short-term incentive schemes are measured using a dummy variable for firms that have introduced performance-related pay. However, Foss et al. (2011) used incentives related to upgrading one's own skills and to knowledge sharing to measure 'knowledge incentives'. One explanation for these seemingly conflicting results might thus be that internal knowledge sharing is of central importance when it comes to utilizing external knowledge (cf. Cohen and Levinthal, 1990). For this reason it may be advisable to incentivize this type of behaviour when managers want to benefit from external knowledge. On the other hand, incentive schemes that value individuals' personal innovative performance will increase the innovative effort from these individuals, but might discourage the application of external knowledge.

Moving the focus from firm level to the individual level, Binyamin and Carmeli (2010) examine a mediation model that suggests that the relationship between structuring of HRM processes and employee creativity is explained by the intervening variables of perceived uncertainty, stress, and the psychological ability to carry out work-tasks (dubbed 'psychological availability'). Empirically, the study is based on 213 individuals working in knowledge-intensive firms. The results suggest that structuring of HRM processes is negatively associated with perceived uncertainty and stress. Moreover, these perceptions produce a sense of psychological availability, which in turn enhances employee creativity. Arguably, all other things being equal, increased creativity should lead to more innovation at the firm level.

All in all, this more recent part of the literature documents that the relationship between HRM practices and innovation outcomes (in particular related to product innovation) is not only a direct one. The relationship is often found to be conditional on contingent factors and to be fully or partially mediated by other factors related to knowledge creation.

Link IV: Antecedents to HRM Practices

Link IV represents the literature that has established a theoretical and empirical link between HRM practices and their antecedents related to innovation outcomes. The existing literature typically treats HRM variables as being strictly exogenous in explaining innovation outcomes. Accordingly, only a few studies have dealt with this issue. Jackson et al. (1989) examine the driver of the adoption of 'personnel practices' that correspond well to HRM practices. Jackson et al. show that these practices are a function of the principal industry sector, the pursuit of innovation as a competitive strategy, the type of manufacturing technology, and organizational structure.

In a study utilizing data on 1,884 Danish firms examining complementarities between HRM practices, Laursen and Mahnke (2001) confirm that industry affiliation is a key determinant of the adoption of HRM practices. Moreover, they find that innovator strategy, linkages to suppliers and customers, and linkages to knowledge institutions are important determinants of the adoption of (complementary) HRM practices. Laursen and Mahnke (2001) do not, however, consider innovation outcomes. The paper by Foss et al. (2011) already discussed suggests that an open innovation strategy is an important antecedent to the adoption of a set of complementary HRM practices, and that the combination of interacting with customers and HRM practices is a necessary condition for strong innovation performance. Using a sample of 294 Flemish firms, De Winne and Sels (2010) demonstrate that the human capital and the use of a range of external experts are determinants of the adoption of a broad range of HRM practices, and that such a broad range of practices in turn are determinants of innovation output (a composite mixture of items relating to administrative, process, and product innovation).

Research Gaps

The above summary of research involving (modern) HRM practices and innovations reveals considerable activity not only within HRM and innovation research, but also in strategic management and organizational studies. However, several research gaps exist, calling for additional research efforts. In this section we briefly discuss some of these gaps.

More Time-Series Evidence

As noted above, time-series evidence on HRM practices and innovation outcomes is scarce. Zoghi et al. (2010) found that controls for unobserved heterogeneity significantly weakened their results, and moreover, lagged variables did not provide clear evidence that organizational changes pre-date innovation. While these findings are extremely interesting and based on sound econometric method, we still need more investigations taking a longitudinal perspective, not least because the fixed effects estimator and the lagged dependent variable estimator tend to produce rather conservative estimates (see Zoghi et al., 2010: 632–3).

Clustering of Practices

In spite of the prominence in HRM/innovation streams of research on the clustering of practices (cf. Ennen and Richter, 2010), there is still little theorizing that predicts exactly which HRM practices bundle and why, and little empirical work that examines this issue. Empirical work may be somewhat ahead of theory in this area. A good illustration is Laursen and Foss (2003) who find two clusters of HRM practices that are conducive to innovation, but essentially do not theorize why there are differences between them. Empirical work has tended to lump together HRM practices, claiming systems effects. Often empirical research confirms that such systems effects indeed exist, but it may well be that some practices are much more important for the system of practices than others: in other words, that relations of complementarity are stronger between some practices. Clarifying this issue is of obvious practical significance, but extant research has so far had little to say about it.

Specific Practices

Laursen and Foss (2003) found that while systems of HRM practices mattered greatly to innovation performance, the contribution of individual practices was negligible. However, some of the practices they considered were measured rather crudely: for example, rewards are represented with a simple variable representing the share of employees involved in any form of pay for performance (though not piece rates), and their measure for job design only incorporates delegation which is at best an imperfect measure of freedom in the job. Until much more detailed research is conducted, drawing to a much larger extent on the richness of the HRM literature, it is not warranted to conclude that systems of HRM practices matter much more to innovation performance than individual practices.

It should also be noted that single practices may, on conceptual grounds, vary widely with respect to their impact on innovation performance. We have already alluded to potential controversy concerning what kinds of incentives are most likely to drive innovation performance. In addition to the temporal dimension of incentives and the tasks that are incentivized, there is also an issue relating to the levels at which incentives are provided. Are group incentives, for example, more effective than individual incentives on innovation performance? To the extent that groups are capable of mobilizing synergistic advantages with respect to creative problem solving (Paulus, 2000), group-level incentives may make more sense than individual-level incentives.

Moreover, an increasingly prominent argument in motivational psychology asserts that extrinsic motivators, such as monetary incentives, may actually be counterproductive because they tend to drive out the kind of autonomous motivation that is essential for successful problem solving, learning, and creativity (Deci and Ryan, 1985), essential micro-level dimensions of innovative performance. This line of research does not deny that rewards matter, but tends to focus on softer, less controlling rewards (rather than contingent performance rewards). The inclusion of such rewards in future research seems highly promising.

Finer Grained and Richer Causal Stories

While highly attractive because of its emphasis on complementarities between practices, the systems approach that is so influential in research on the HRM/innovation–performance link risks obscuring much of the fine-grained causal texture that links HRM practices and innovation. Thus, individual practices may have an impact that is additional to, and goes beyond, the systems effect.

Consider, for example, job design, one of the most frequently researched practices in the HRM literature. Jobs contain characteristics, such as feedback, the size of the task portfolio, the characteristics of individual task, the ability to carry out a job from the beginning to the end, repetitiveness, and so on, that stimulate different kinds of motivation (Foss et al., 2009). Jobs that imply a greater degree of employee control, autonomy, and non-controlling feedback, for example, are likely to stimulate the autonomous motivation that drives creativity and learning, and, ultimately, innovation performance. Similar arguments may be developed on the basis of other modern HRM practices. Research on teams, for example, has clarified that team problem-solving effectiveness is highly dependent on the clarity with which team members understand the task structure within the team (Kozlowski and Bell, 2003; Ilgen et al., 2005).

There is considerable room for expanding the understanding of how exactly individual HRM practices contribute to innovation performances by unpacking them and examining the contextual variables (i.e. moderators) that influence this contribution. A possible outcome of better understanding of this domain is an improved understanding of systems of HRM practices, because one practice may be a relevant contextual variable influencing the effectiveness of another practice.

What Kind of Innovation?

A final issue concerns the dependent variable in Figure 25.1; that is, performance in terms of product or process innovation. A pertinent question thus is whether there are (modern) HRM practices that are inherently more supportive of one kind of innovation than another. It would seem natural to expect quality circles, for example, to be more conducive to process than to product innovation. Similarly, it could be hypothesized that internal training is also more conducive to process innovation, whereas external training could contribute more to product innovation performance because it gives employees access to larger networks with more diverse knowledge. Other HRM practices may similarly be hypothesized to have a differential impact on innovation performance.

A further way of advancing research is to dehomogenize the basic process and product innovation categories. Thus, the process innovation category includes not only innovations in the basic production process itself, but also innovations in the administrative structure of the firm (Birkinshaw et al., 2008)—including innovations in HRM. While management innovations may mainly be introduced by the higher echelons of the firm, there are HRM practices, notably reward systems, that may positively influence such innovation. Thus, because management innovations are likely to be implemented across the board in the various departments of a firm, and thus affect the financial performance of the entire company, upper echelons are arguably incentivized to implement such innovations by reward instruments that link pay to overall company performance.

In turn, product innovation may be decomposed into innovations of physical products and innovations of services. Service innovation raises distinct HRM challenges. Thus, while the increasing emphasis on user innovation has pointed to the importance of users and customers in the innovation process in general, the importance of heavy customer and user involvement may be particularly important for service innovations, and may therefore particularly empower employees to cooperate with customers and users in the case of these innovations.

Conclusions

The literature on the links between HRM and innovation that we have surveyed in this chapter has expanded considerably over the last one and a half decades. This may partly reflect the fact that both HRM and innovation have been expanding fields in this period. It arguably also reflects trends in the business world that prompt the emerging integration of HRM and innovation research. As firms increasingly adopt open innovation models and engage with external knowledge sources (see Chapter 22 by Alexy and Dahlander), they find that they need to bring new groups of employees into the innovation process. This calls for dedicated training, new performance indicators, new rewards, new ways of communicating with and between employees, and so on. In short, it calls for an active HRM effort. Relatedly, firms may open up the innovation process internally, namely by increasingly sourcing ideas and knowledge from organizational members (Dodgson et al., 2006). Such initiatives are also likely to call for new HRM initiatives.

The link between internal organization and innovation performance has been a frequent theme in innovation research since Schumpeter (1942) and Burns and Stalker (1961). Much of the discussion has involved traditional structural variables, typically drawn from structural contingency theory. The emerging research stream in the intersection of HRM and innovation research represents a new, more fine-grained approach to the understanding of the organizational antecedents of innovation performance. However, as we have shown in this chapter, this is a rather recent undertaking and one that still contains several research gaps.

Notes

1. In general, a moderator is a variable that affects the direction and/or strength of the relation between an independent and a dependent variable. A mediator variable is a variable which

represents a mechanism through which the focal independent variable is able to (indirectly) influence the dependent variable. See Baron and Kenny (1986) for a detailed exposition.

References

Almeida, P., and Kogut, B. (1999). 'Localization of Knowledge and the Mobility of Engineers in Regional Networks', *Management Science*, 45(7): 905–17.

Aoki, M., and Dore, R. (eds). (1994). *The Japanese Firm: The Sources of Competitive Strength*. Oxford: Oxford University Press.

Arthur, J. B. (1994). 'Effects of Human Resource Systems on Manufacturing Performance and Turnover'. *Academy of Management Journal*, 37(3): 670–87.

Arvanitis, S. (2005). 'Modes of Labor Flexibility at Firm Level: Are there any Implications for Performance and Innovation? Evidence for the Swiss Economy', *Industrial and Corporate Change*, 14(6): 993–1016.

Baird, L., and Meshoulam, I. (1988). 'Managing Two Fits of Strategic Human Resource Management', *The Academy of Management Review*, 13(1): 116–28.

Baron, J. N., and Kreps, D. M. (1999). *Strategic Human Resources: Frameworks for General Managers*. New York: John Wiley.

Baron, R. M., and Kenny, D. A. (1986). 'The Moderator-mediator Variable Distinction in Social Psychological Research: Conceptual, Strategic, and Statistical Considerations', *Journal of Personality and Social Psychology*, 51(6): 1173–82.

Beugelsdijk, S. (2008). 'Strategic Human Resource Practices and Product Innovation', *Organization Studies*, 29(6): 821–47.

Binyamin, G., and Carmeli, A. (2010). 'Does Structuring of Human Resource Management Processes Enhance Employee Creativity? The Mediating Role of Psychological Availability', *Human Resource Management*, 49(6): 999–1024.

Birkinshaw, J., Hamel, G., and Mol, M. (2008). 'Management Innovation'. *The Academy of Management Review ARCHIVE*, 33(4): 825–45.

Bohnet, I., and Oberholzer-Gee, F. (2001). '*Pay for Performance: Motivation and Selection Effects*', Cambridge, MA: Harvard Business School.

Burns, T., and Stalker, G. M. (1961). *The Management of Innovation*. London: Tavistock.

Camelo-Ordaz, C., De la Luz Fernandez-Alles, M., and Valle-Cabrera, R. (2008). 'Top Management Team's Vision and Human Resources Management Practices in Innovative Spanish Companies', *International Journal of Human Resource Management*, 19(4): 620–38.

Capelli, P., and Neumark, D. (2001). 'Do "High-performance" Work Practices Improve Establishment-level Outcomes?' *Industrial and Labor Relations Review*, 54(4): 737–75.

Chen, C., and Huang, J. (2009). 'Strategic Human Resource Practices and Innovation Performance: The Mediating Role of Knowledge Management Capacity', *Journal of Business Research*, 62(1): 104–14.

Cohen, W. M., and Levinthal, D. A. (1990). 'Absorptive Capacity: A New Perspective of Learning and Innovation', *Administrative Science Quarterly*, 35(1): 128–52.

Collins, C. J., and Smith, K. G. (2006). 'Knowledge Exchange and Combination: The Role of Human Resource Practices in the Performance of High-technology Firms', *Academy of Management Journal*, 49(3): 544–60.

Colombo, M. G., and Delmastro, M. (2002). 'The Determinants of Organizational Change and Structural Inertia: Technological and Organizational Factors', *Journal of Economics and Management Strategy*, 11(4): 595–635.

Colombo, M. G., and Delmastro, M. (2008). *The Economics of Organizational Design: Theoretical Insights and Empirical Evidence.* Basingstoke, UK: Palgrave Macmillan.

Colombo, M. G., Delmastro, M.. and Rabbiosi, L. (2007). '"High Performance" Work Practices, Decentralization, and Profitability: Evidence from Panel Data', *Industrial and Corporate Change*, 16(6): 1037–67.

Colombo, M. G., Delmastro, M., and Rabbiosi, L. (2013). 'Organizational Design and Firm Performance', in R. T. Christopher, and W. F. Shughart II (eds), *Oxford Handbook of Managerial Economics.* Oxford: Oxford University Press.

Datta, D. K., Guthrie, J. P., and Wright, P. M. (2005). 'Human Resource Management and Labor Productivity: Does Industry Matter?' *Academy of Management Journal*, 48(1): 135–45.

De Winne, S., and Sels, L. (2010). 'Interrelationships between Human Capital, HRM and Innovation in Belgian Start-ups Aiming at an Innovation Strategy', *International Journal of Human Resource Management*, 21(11): 1863–83.

Deci, E. L., and Ryan, R. M. (1985). *Intrinsic Motivation and Self-determination in Human Behavior.* New York: Plenum Press.

Dodgson, M., Gann, D., and Salter, A. (2006). 'The Role of Technology in the Shift Towards Open Innovation: The Case of Procter & Gamble', *R&D Management*, 36(3): 333–46.

Ennen, E., and Richter, A. (2010). 'The Whole is More than the Sum of its Parts: Or Is It? A Review of the Empirical Literature on Complementarities in Organization', *Journal of Management*, 36(1): 207–33.

Felin, T., and Hesterly, W. (2007). 'The Knowledge-based View, Nested Heterogeneity, and New Value Creation: Philosophical Considerations on the Locus of Knowledge', *Academy of Management Review*, 32(1): 195–218.

Foss, N. J., Laursen, K., and Pedersen, T. (2011). 'Linking Customer Interaction and Innovation: The Mediating Role of New Organizational Practices', *Organization Science*, 22(4): 980–99.

Foss, N. J., Minbaeva, D., Reinholt, M., and Pedersen, T. (2009). 'Stimulating Knowledge Sharing among Employees: The Contribution of Job Design', *Human Resource Management*, 48: 871–93.

Freeman, C. (1988). 'Japan: A New National System of Innovation?' in G. Dosi, C. Freeman, R. Nelson, G. Silverberg, and L. L. G. Soete (eds), *Technical Change and Economic Theory.* London: Pinter Publishers, 330–348.

Fu, X. (2012). 'How does Openness Affect the Importance of Incentives for Innovation?' *Research Policy*, 41(3): 512–23.

Furman, J. L., Porter, M. E., and Stern, S. (2002). 'The Determinants of National Innovative Capacity', *Research Policy*, 31(6): 899–933.

Galunic, D. C., and Rjordan, S. (1998). 'Resource Re-combinations in the Firm: Knowledge Structures and the Potential for Schumpeterian Innovation', *Strategic Management Journal*, 19: 1193–1201.

Gerhart, B., and Milkovich, G. T. (1990). 'Organizational Differences in Managerial Compensation and Firm Performance', *Academy of Management Journal*, 33(4): 663–91.

Glynn, M. A. (1996). 'Innovative Genius: A Framework for Relating Individual and Organizational Intelligences to Innovation', *The Academy of Management Review*, 21(4): 1081–111.

Guthrie, J. P. (2001). 'High-involvement Work Practices, Turnover, and Productivity: Evidence from New Zealand', *Academy of Management Journal*, 44(1): 180–90.

Hamilton, B. H., Nickerson, J. A., and Owan, H. (2003). 'Team Incentives and Worker Heterogeneity: An Empirical Analysis of the Impact of Teams on Productivity and Participation', *Journal of Political Economy*, 111(3): 465–97.

Hayek, F. A. (1945). 'The Use of Knowledge in Society'. *American Economic Review*, 35 (September): 519–30.

Henderson, R., and Cockburn, I. (1994). 'Measuring Competence: Exploring Firm Effects in Pharmaceutical Research'. *Strategic Management Journal*, 15(Winter Special Issue): 63–84.

Huselid, M. A. (1995). 'The Impact of Human Resource Management Practices on Turnover, Productivity, and Corporate Financial Performance', *Academy of Management Journal*, 38(3): 635–72.

Ichniowski, C., and Shaw, K. (1999). 'The Effects of Human Resource Management Systems on Economic Performance: An International Comparison of US and Japanese Plants', *Management Science*, 45(5): 704–21.

Ichniowski, C., Shaw, K., and Prennushi, G. (1997). 'The Effects of Human Resource Management Practices on Productivity: A Study of Steel Finishing Lines', *American Economic Review*, 87(3): 291–313.

Ilgen, D. R., Hollenbeck, J. R., Johnson, M., and Jundt, D. (2005). 'Teams in Organizations: From Input-process-output Models to IMOI Models', *Annual Review of Psychology*, 56(1): 517–43.

Jackson, S. E., Schuler, R. S., and Rivero, J. C. (1989). 'Organizational Characteristics as Predictors of Personnel Practices', *Personnel Psychology*, 42: 727–86.

Janod, V., and Saint-Martin, A. (2004). 'Measuring the Impact of Work Reorganization on Firm Performance: Evidence from French Manufacturing, 1995–1999', *Labour Economics*, 11(6): 785–98.

Jensen, M., Johnson, B., Lorenz, E., and Lundvall, B. (2007). 'Forms of Knowledge and Modes of Innovation', *Research Policy*, 36(5): 680–93.

Jensen, M. C., and Meckling, W. H. (1992). 'Specific and General Knowledge and Organizational Sructure', in L. Werin, and H. Wijkander (eds), *Contract Economics*. Oxford: Blackwell, 251–74

Jimenez-Jimenez, D., and Sanz-Valle, R. (2008). 'Could HRM support Organizational Innovation?' *International Journal of Human Resource Management*, 19(7): 1208–21.

Kato, T., and Morishima, M. (2002). 'The Productivity Effects of Participatory Employment Practices: Evidence from New Japanese Panel Data', *Industrial Relations*, 41(4): 487–520.

Kogut, B. (2000). 'The Network as Knowledge: Generative Rules and the Emergence of Structure', *Strategic Management Journal*, 21(3): 405–25.

Kozlowski, S. W. J., and Bell, B. S. (2003). 'Work Groups and Teams in Organizations', in *Handbook of Psychology*. John Wiley & Sons, Inc. 333–75.

Lacetera, N., Cockburn, I. M., and Henderson, R. M. (2004). 'Do Firms Change Capabilities by Hiring New People? A Study of the Adoption of Science-based Drug Discovery', *Advances in Strategic Management*, 21: 133–59.

Lau, C., and Ngo, H. (2004). 'The HR System, Organizational Culture, and Product Innovation', *International Business Review*, 13(6): 685–703.

Laursen, K. (2002). 'The Importance of Sectoral Differences in the Application of Complementary HRM Practices for Innovation Performance', *International Journal of the Economics of Business*, 9(1): 139–56.

Laursen, K., and Foss, N. (2003). 'New Human Resource Management Practices, Complementarities and the Impact on Innovation Performance', *Cambridge Journal of Economics*, 27(2): 243–63.

Laursen, K., and Mahnke, V. (2001). 'Knowledge Strategies, Firm Types, and Complementarity in Human-resource Practices', *Journal of Management and Governance*, 5(1): 1–27.

Lepak, D. P., and Snell, S. A. (2002). 'Examining the Human Resource Architecture: The Relationships among Human Capital, Employment, and Human Resource Configurations', *Journal of Management*, 28(4): 517–43.

Lopez-Cabrales, A., Perez-Luno, A., and Valle Cabrera, R. (2009). 'Knowledge as a Mediator between HRM Practices and Innovative Activity', *Human Resource Management*, 48(4): 485–503.

Love, J., and Roper, S. (2009). 'Organizing Innovation: Complementarities between Cross-functional Teams', *Technovation*, 29(3): 192–203.

Mendelson, H. (2000). 'Organizational Architecture and Success in the Information Technology Industry', *Management Science*, 46(4): 513–29.

Mendelson, H., and Pillai, R. R. (1999). 'Information Age Organizations, Dynamics, and Performance', *Journal of Economic Behavior and Organization*, 38(3): 253–81.

Michie, J., and Sheehan, M. (1999). 'HRM Practices, R&D Expenditure and Innovative Investment: Evidence from the UK's 1990 Workplace Industrial Relations Survey', *Industrial and Corporate Change*, 8(2): 211–34.

Michie, J., and Sheehan, M. (2003). 'Labour Market Deregulation, "Flexibility" and Innovation', *Cambridge Journal of Economics*, 27(1): 123–43.

Milgrom, P., and Roberts, J. (1995). 'Complementarities and Fit: Strategy, Structure, and Organizational Change in Manufacturing', *Journal of Accounting and Economics*, 19(2/3): 179–208.

Oke, A., Walumbwa, F. O., and Myers, A. (2012). 'Innovation Strategy, Human Resource Policy, and Firms' Revenue Growth: The Roles of Environmental Uncertainty and Innovation Performance', *Decision Sciences*, 43(2): 273–302.

Paulus, P. (2000). 'Groups, Teams, and Creativity: The Creative Potential of Idea-generating Groups', *Applied Psychology*, 49(2): 237–62.

Petroni, G., Venturini, K., and Verbano, C. (2012). 'Open Innovation and New Issues in R&D Organization and Personnel Management', *The International Journal of Human Resource Management*, 23(1): 147–73.

Rammer, C., Czarnitzki, D., and Spielkamp, A. (2009). 'Innovation Success of Non-R&D-performers: Substituting Technology by Management in SMEs', *Small Business Economics*, 33(1): 35–58.

Ritter, T., and Gemünden, H. (2003). 'Network Competence: Its Impact on Innovation Success and its Antecedents', *Journal of Business Research*, 56(9): 745–55.

Rothwell, R., Freeman, C., Jervis, P., Robertson, A., and Townsend, J. (1974). 'SAPPHO Updated: Project SAPPHO Phase 2', *Research Policy*, 3(3): 258–91.

Ryan, R. M., and Deci, E. L. (2000). 'Intrinsic and Extrinsic Motivations: Classic Definitions and New Directions', *Contemporary Educational Psychology*, 25(1): 54–67.

Schumpeter, J. A. (1912/1934). *The Theory of Economic Development: An Inquiry into Profits, Capital, Credit, Interest and the Business Cycle* (R. Opie, Trans.). London: Oxford University Press.

Schumpeter, J. A. (1942). *Capitalism, Socialism and Democracy* (paperback edition). London: Routledge.

Shipton, H., Fay, D., West, M., Patterson, M., and Birdi, K. (2005). 'Managing People to Promote Innovation', *Creativity and Innovation Management*, 14(2): 118–28.

Subramony, M. (2009). 'A Meta-analytic Investigation of the Relationship between HRM Bundles and Firm Performance', *Human Resource Management*, 48(5): 745–68.

Teece, D. J. (2007). 'Explicating Dynamic Capabilities: The Nature and Microfoundations of (Sustainable) Enterprise Performance', *Strategic Management Journal*, 28(13): 1319–50.

Terpstra, D. E., and Rozell, E. J. (1993). 'The Relationship of Staffing Practices to Organizational level Measures of Performance', *Personnel Psychology*, 46(1): 27–48.

Tsai, W. P. (2001). 'Knowledge Transfer in Intra-organizational Networks: Effects of Network Position and Absorptive Capacity on Business Unit Innovation and Performance', *Academy of Management Journal*, 44(5): 996–1004.

Van Reenen, J., and Caroli, E. (2001). 'Skill-biased Organizational Change? Evidence from a Panel of British and French Establishments', *Quarterly Journal of Economics*, 116(4): 1449–92.

Youndt, M. A., Snell, S. A., Dean, J. W., Jr., and Lepak, D. P. (1996). 'Human Resource Management, Manufacturing Strategy, and Firm Performance', *The Academy of Management Journal*, 39(4): 836–66.

Zenger, T., and Hesterly, W. S. (1997). 'The Disaggregation of Corporations: Selective Intervention, High-powered Incentives and Molecular Units', *Organization Science*, 8(3): 209–22.

Zhou, H., Dekker, R., and Kleinknecht, A. (2011). 'Flexible Labor and Innovation Performance: Evidence from Longitudinal Firm-level Data', *Industrial and Corporate Change*, 20(3): 941–68.

Zoghi, C., Mohr, R., and Meyer, P. (2010). 'Workplace Organization and Innovation', *Canadian Journal of Economics-Revue Canadienne De Economique*, 43(2): 622–39.

CHAPTER 26

MANAGING R&D AND NEW PRODUCT DEVELOPMENT

MAXIMILIAN VON ZEDTWITZ,
SASCHA FRIESIKE, AND
OLIVER GASSMANN

INTRODUCTION

THE development of new products, processes, and services is the company's response to environmental and market changes (Leonard-Barton, 1992). Firms engage in research and development (R&D) in order to create new insights, technologies, processes, and platforms as the basis for new products. New product development (NPD) includes the conception, generation, analysis, development, prototyping, and testing of new products. R&D and NPD are thus the engine of all innovating firms. The management of R&D and NPD is a cross-functional activity involving input from—and creating output for—marketing, strategy, business development, finance, human resources, sales, legal, IT, and many others. Given the importance of technology for competitiveness in modern times, R&D and NPD have become central functions for most companies, even companies in so-called low-tech industries, such as tourism and services.

The primary purpose of R&D is to develop products that outsell existing market offerings or capture previously unknown market opportunities. There is no simple answer to how firms manage their R&D successfully even though this question has been debated in the literature for decades (Nobelius, 2004). Yet many firms that succeed in the management of R&D command premium price margins, define dominant designs, and often gain significant market share. In this chapter we focus on three central components to a firm's approach to managing R&D and NPD:

1. The Product Development Funnel, or how firms transform an idea or concept into a finished product ready for market launch.

2. R&D Portfolio Management, or how firms strategically identify innovation opportunities and select promising R&D projects.

3. R&D Organization, or how firms structure R&D teams and R&D departments in alignment with business drivers and principal technological needs.

PRODUCT DEVELOPMENT FUNNEL

The concept of the product development funnel is based on the realization that most products are developed following a standard sequence of activities, employing tasks and routines that are also fairly standard across different development projects.

It is called a funnel because, at certain defined intervals and stages, underperforming NPD projects are cancelled or rejected, winnowing the number of projects under development over time. In other words, a wide set of possibilities is transformed into a small set of implementations (Cooper and Edgett, 2007). The funnel represents the entire time of a typical NPD project from initial idea to product launch (Dunphy et al., 1996; O'Sullivan, 2002). The development funnel has its roots in earlier phased NPD models such as NASA's 'Phased Project Planning' in the 1960s (Baker and Sweeney, 1978), which soon led to the formulation of defined gates that projects would need to pass through, after due scrutiny, to advance to the next phase or stage of development, usually involving greater overall resource allocation but also terminating other R&D activities now considered completed based on satisfactory results.

The stage-gate process formulated by Cooper (1985) and his successors is one of the most representative product development funnels and is used as the basis for further illustration here. As shown in Figure 26.1, the development funnel is composed of four distinct phases, or stages:

1. The fuzzy front end/concept development phase, with a primary objective to identify ideas for new products and advance them towards concept readiness;

2. The product definition phase, during which the product concept is matured further and married with market needs;

3. The product development phase, in which the product is designed based on specifications developed in the previous phase; and

4. The testing and launch readiness phase, during which the product is being prototyped, tested, and prepared for production.

At the end of each phase the project is reviewed and must pass specified criteria before being promoted to the next phase. These reviews are called 'gates' and consist of formal project review meetings with key project leaders, product stakeholders, and business decision-makers. Typical review criteria include strategic fit, market and customer

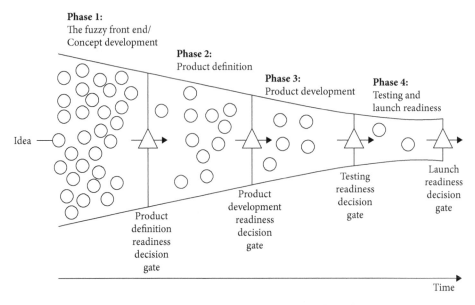

FIGURE 26.1 The product development funnel

input, and technical feasibility, plus any other potential show-stoppers. A project can be approved for the next phase, it can be rejected/recycled, it can be put on hold perhaps to synchronize it with other activities and events, or it can be terminated. Review meetings are not supposed to be problem discussion meetings, although practices across firms diverge. It is not uncommon for project teams to hold mock-up gate meetings, even prep exercises for mock-up gate meetings, redefine specifications to meet gate review criteria, and so on, to improve their chances to pass gate reviews. Gate reviews are also opportunities to train junior team members and expose them to senior management thinking, although this practice distracts from the original purpose of the gate as a project progress review and approval meeting. In some industries firms need to hit tightly defined launch windows such as being able to present a new product at a trade show, and if technical feasibility is difficult to achieve, target specifications are often adjusted downwards.

New methods and technologies, such as agile software development or rapid prototyping, can increase the iteration cycles and speed up the innovation process (see Chapter 19 by Dodgson and Gann). As an example of an 'agile development methodology', a scrum is an iterative and incremental product development method characterized by higher flexibility and speed, partly substituting the rigid gates. Probe-and-learn approaches and less rigid checks are used within small projects or within early phases of the innovation project. Requirements are frozen only during so-called 'sprints', but opened up again for possible revision based on new customer input during the development phase and technical revelations from the R&D work.

Phase 1: The Fuzzy Front End—Developing a Concept

The product concept is developed in phase 1. In the case of a derivative product, an incremental innovation, it is usually quite clear what input has led to the new product concept, and how it was assembled. Derivative product ideas may come from customer feedback, internal product reviews, technology updates, fashion trends, and so on. However, if the new product concept is a departure from the existing base of knowledge, technology, or experience of the firm, that is, if it is a more radical innovation, the creation logic is often less clear and not easily repeatable. In this case we speak of the 'fuzzy front end' of innovation.

Fuzzy front end research has attracted quite a bit of attention as it has been shown that the ultimate success or failure of the new product development process, and indeed the product itself, can often be found in the concept and design of the earliest stages of R&D and NPD, when the core features of a product are crafted (e.g. Stevens and Burly, 2003). However it is still not well understood how the fuzzy front end should be managed, and there is quite a bit of disagreement, both among researchers and practicing managers, about which practices and tools should be used (Khurana and Rosenthal, 1997). Perhaps there are no 'best practices' at the fuzzy front end. Specific contexts can vary greatly as the fuzzy front end combines tacit knowledge, organizational pressure, accidental discoveries and insights, great uncertainty and often complex information (Reinertsen, 1999). Research also shows that how the fuzzy front end is managed has substantial consequences for the project team's vision (Zhang and Doll, 1998).

Koen et al. (2001) outlined four tasks that are generally part of the fuzzy front end:

- *Opportunity identification*—a step during which the product idea comes into being.
- *Opportunity analysis*—when the project idea gets translated into the company specific context. This step includes a first evaluation of its strategic fit, potential customers, and technical details.
- During the *idea genesis*, the company develops the idea further. This step can include the expertise of outsiders (e.g. lead users (von Hippel, 1986 and Chapter 5 by Franke)) or it can be done internally.
- In any case, whether this is a fuzzy front end innovation or a more incremental project idea, the concept phase ends with the *idea selection*. Here, the most promising ideas, those that are best aligned with the company's business and technology goals, are selected, usually also demarcating the formal beginning of the NPD project.

Phase 2: Product Definition—Building the Product Case

The main goal of the product definition phase is to develop the basic aspects of the product's architecture, defining key functional and technical features and integrating initial market specifications. As such, this phase is central in deepening the firm's understanding of the product opportunity and in reducing overall uncertainty. Research shows that an early reduction of market/technology uncertainty and initial planning prior to development (phase 3) have a positive impact on the overall success of NPD projects (Verworn et al., 2008).

The product definition phase translates conceptual goals into explicit operational tasks (Bacon et al., 1994). The form of product definition varies widely, yet most definitions answer similar questions. Among those, the product definition explains the *benefits of the product* in development. This explanation can either be explicit and fact based (for instance, a 30 per cent increase in battery life) as is often the case for incremental innovations, or it can be abstract and even anecdotal (a new electronic device to read newspapers) more common for radical innovations with a fuzzy target market (Seidel, 2007).

The product definition phase also deals with *technical hurdles.* Here the project team tries to anticipate and describe the necessary steps and possible risks in the development of the product. Many projects come to a halt if technical feasibility cannot be proven. Besides the technical analysis, the product definition also includes a *market analysis* during which the competitive landscape of the product in development is assessed. This tends to be easier when the NPD project deals with an already established product that receives an incremental update; the market assessments then focus on quantitative data that compares technical specifications, customer surveys, sales figures, and market share analyses. In NPD projects of more radical nature, the market analysis focuses also on possible substitutes and neighbouring markets. Initial *intellectual property investigations* and patent database checks are also part of this phase, particularly in high-tech industries such as pharmaceuticals and automotives. Depending on their outcome, the project team will weigh options for licensing agreements (see Chapter 28 by Leiponen).

In general, NPD projects in highly dynamic markets require more flexible product definitions that can be altered given changes in the company's environment (Bhattacharya et al., 1998).

Phase 3: Product Development—The Heart of the NPD Process

At the heart of the NPD process is the physical creation and implementation of the product. In this phase the product development team translates all design requirements, based on component and partial prototypes, drawings, mockups, and other inputs, into one integral prototype. Especially in more radical innovation projects, the product concept can still shift during the development phase (Seidel, 2007), in response to new developments in the product's system architecture. Good NPD teams know how to handle such divergences and may even be able to turn them into an advantage for the overall innovation effort. Seidel (2007) shows that about half of all innovation projects still change their product definitions after the completion of the formal definition.

Because of its central position within the overall NPD process, the development phase has been extensively researched and is well covered in the literature on innovation management. In the following, we outline the core aspects of this phase.

First, sufficient access to necessary resources is a key requirement for successful completion of this phase. Of course, resource availability and needs are determined during the previous gate review, but priorities can change during the development phase. Several categories are considered, including access to personnel capable of solving key tasks as required by the NPD project, financial resources, physical infrastructure (as may be the case for extensive experimental work or lab work), and test subjects (e.g. in pharmaceutical development). Access can be granted by product and/ or project champions inside firms (see e.g. Roberts, 1968; Witte, 1973; Chakrabarti, 1974). The influence of these idea champions is based on their hierarchical position (power promoters), their knowledge (technical or functional promoters) or their communication abilities (organizational and process promoters). In most mature stage-gate process implementations, project supervisory committees (PSC), product decision teams (PDT), or product approval committees (PAC) can assume the role of these champions.

Second, cooperation with external partners can be a very valuable building block to complete this phase at greater speed or quality. As such, *collaborations* are of great importance to innovating firms and many new products are co-developed by a consortium of several firms (see Chapter 23 by Dodgson). Users are often integrated into the NPD process (and not just in the development phase) to gain new insights and to create products more fitting to customer needs (see Chapter 5 by Franke). *Open Innovation*, the exchange of knowledge between a firm and outside parties for the purpose of advancing internal innovation, has become a cornerstone of modern NPD (see Chapter 22 by Alexy and Dahlander).

Phase 4: Testing and Launch Readiness—The Final Touches

The fourth and final phase of the product development funnel consists of testing and ensuring market readiness of the product. As the product will eventually be handed over to the business organization responsible for manufacturing, marketing, and sales, a substantial amount of coordination and cross-functional integration is necessary (Annachino, 2007). Throughout the entire NPD process a strong interaction between the development team, sourcing and the marketing department is desirable (Gupta et al., 1986); in reality, maintaining such continuing interaction seems to be difficult. Corporate culture itself often poses a barrier to effective collaboration between marketing and R&D. Individual professional cultures may lead to insulation between disciplines and result in trench wars and failed project hand-overs. Identification with specific functions, such as Marketing or Engineering, impedes effective collaboration between departments. Research shows that a strong orientation towards common goals can help overcome these differences (Mohr et al., 2009). For a more detailed examination of the relationship of new products and marketing please refer to Chapter 3 by Prabhu.

During the final phase of the product development funnel the final product pricing will be decided, as well as distribution channels, market entry date, marketing campaigns, initial production volumes, and so on. Also part of this phase are the tooling (tool development, if the firm intends to manufacture the product itself), supplier selection, production ramp-up, and outbound logistics (Annachino, 2007).

R&D Portfolio Management

Project portfolio management helps firms to make important decisions for a multitude of projects under the constraints of limited resources and time. Using R&D portfolio management, R&D executives decide which R&D projects to invest in, which R&D areas to devote more resources to, and which R&D programmes to abandon. Good R&D portfolio management will aim to achieve a balance between high-risk/high-reward projects and small incremental innovation projects, pay close attention to bottlenecking (when limited resource capacity reduces the capacity of the entire product development process), assure availability of R&D resources to all projects, and generally try to deliver a predictable source of new products to the company.

Key Challenges in R&D Portfolio Management

The underlying purpose of all R&D portfolio management efforts is the same: to provide decision-makers with accurate and timely information on which basis they can judge which project should be undertaken (Cooper et al., 1998). Inherent to all R&D projects is uncertainty about technological feasibility, market acceptance, eventual designs, and possibly even the underlying science. Portfolio management decisions are thus taken under uncertainty, and much effort is carried out to reduce this uncertainty to increase chances of project success and limit costs and waste (Cooper et al., 2004).

Besides these inherent technical challenges to R&D projects, insufficient adoption of good portfolio management practices leads to challenges and shortcomings that often impede a firm's ability to innovate (Cooper, 1988). One typical situation is when the firm commits itself to *too many R&D projects*. Resources for any individual project are too restricted to carry it out successfully. Projects then have to be stopped in mid-development to re-allocate resources to other, now prioritized, projects. Another problem is when *resources are wasted* on projects that do not promise a profitable outcome to the firm. This occurs when the project's fit with the firm's competences or target market is overestimated. Related to this is a *lack of strategic focus*, which leads to high project failure rates or the possible success of projects that are inconsequential to the firm. *Weak selection criteria*, as well as *weak decision criteria*, are at the heart of many of these problems, making it difficult for those in charge to balance risk and reward of the portfolio of projects, and to eliminate politics, irrationalities, nepotism, and other detrimental influences from level-headed decision-making.

R&D Portfolio Techniques

R&D portfolios are a technique to spread the corporate risk that comes with committing to developing new products. Companies that develop several new products concurrently increase their ability to cope with failures in the marketplace. The earnings from one very successful product launch can be used to finance several unsuccessful ones. Seen from a managerial perspective, any NPD project is a unique endeavour with an unpredictable outcome. NPD projects are commonly characterized by a high degree of novelty, complexity, and dynamism. *Novelty* describes the fact that any innovation is inherently new to the developing firm. *Complexity* can be regarded as the efforts and expenses that go into understanding the technological progress, the integration of systems, people, and partners, as well as the corporate understanding of the developed product itself. *Dynamism* reflects the circumstance that innovating firms have an unclear understanding of future market demands and future technological advances. All of these factors lead to the assessment that the development of new products is inherently a high-risk project, and that the ability to manage risk is a crucial building block of successful R&D management (Kwak and LaPlace, 2005; Mu et al., 2009).

Over the last decades scholars and practitioners have developed a multitude of portfolio management techniques. Portfolio techniques range from simple judgements or decision trees to complex quantitative tools with a large number of variables to consider. The form of the R&D portfolio in use varies from case to case. Often, several portfolios are utilized in order to highlight different aspects. Modern visualization tools make it possible to instantly update portfolio representations and see how changes in a single project can affect the entirety of a firm's R&D efforts. Research shows that the selection of individually promising projects does not necessarily lead to the optimal portfolio (Chien, 2002). It is thus necessary to choose wisely among various different portfolio techniques and criteria dimensions to carefully select projects such that overall business objectives are achieved.

One of the more simple representations of an R&D portfolio is the two-dimensional matrix, comparing projects along two principal axes of concern, and often representing by some other means a third integrated dimension, such as the total allocated resource to each project. Figure 26.2 is typical example of an R&D project portfolio with risk plotted against reward and the cost of each project indicated by the diameter of the respective circle.

Other dimensions are frequently used to highlight other relevant aspects such as market attractiveness, expected market growth, market perception (variation of a known product vs. new value proposition), state of technology (known vs. radically new), project complexity, project costs, and so on. The following is a list of some of the more widely used R&D portfolio matrices:

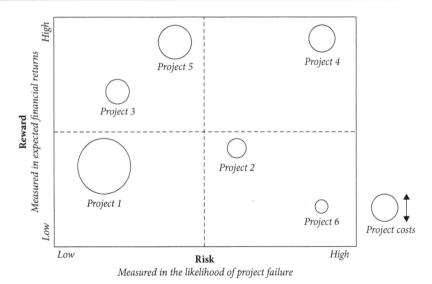

FIGURE 26.2 Bubble chart R&D portfolio

- Market priority vs. technology priority, with each priority dimension determined by a sub-matrix of attractiveness vs. the firm's relative position (also known as the McKinsey matrix).
- Relative technology position vs. significance of the technology, with the technology position usually defined by R&D internal parameters while the technology significance is typically calculated using market criteria; this leads to the popular bet-draw-cash-in-fold portfolio pioneered by the consultants Booz, Allen, and Hamilton.
- Potential for market differentiation vs. market acceptance, with the possibility to track projects as they move through various stages of the technology life cycle and quadrants in the resulting matrix.
- Technology position vs. market/competitive position, with the technology position determined by an analysis of the firm's ability to deploy technology given the target market (also known as the ADL matrix).
- Expected value vs. probability of success, where the expected value is usually given in financial terms and the probability of success is a combination of the probability of both technical and commercial success.
- Market familiarity vs. R&D impact on business, one of the more popular techniques to discuss risks involved in both market and R&D positions of a firm's R&D project portfolio.
- Technical novelty vs. market novelty, which requires an assessment of the firm's presumed competitive position in both technology and market.
- Technological competitive position vs. industry maturity, which not unlike one of the earlier described techniques, maps R&D projects over a life cycle, albeit this time an industrial one.
- Required product changes vs. process changes, leading to a portfolio of breakthrough, platform, and derivative/enhancement R&D projects.

Some portfolio techniques can be used in sequence, making portfolio management more complex but also more powerful, assuming that data input validity can be maintained at a high level of accuracy and relevance. Portfolios are used, directly or indirectly, by product approval committees, usually consisting of senior executives of business divisions and the central R&D office. Depending on the size of the firm and the significance of R&D performed, R&D portfolios may have to be coordinated with business-level NPD portfolios, product roadmaps, and long-term technology roadmaps. In large technology-intensive firms this task is often assumed by an R&D Planning Office. The importance and sophistication of R&D portfolio management has increased significantly over the past 20 years, as described by Roussel et al. (1991) and subsequent literature (Rengarajan and Jagannathan, 1997; Beaujon et al., 2001).

R&D ORGANIZATION

Typically firms institutionalize their R&D activities in a dedicated R&D organization to benefit from the capacity to:

- Nurture and develop R&D specific processes and product knowledge.
- Develop specialized technical career tracks and personnel development opportunities.
- Optimize necessary R&D infrastructure and other capital investments.
- Maintain quality and control over a core source of the firm's competitive advantage.

How the R&D organization is structured internally, and where it is organizationally connected inside the corporation, differs from firm to firm and apart from scientific/technical and market criteria also depends on firm strategy, firm culture, firm organization, and other conditions usually outside the realm of R&D top management (Mintzberg, 1979; Child, 1980).

Input-Driven and Output-Driven R&D Organization

The two fundamental drivers for R&D organization are the source of input or the most efficient output for innovation:

- Input-for-Innovation R&D is organized by *scientific discipline* or by *activities*.
- Output-for-Innovation R&D is organized by *product lines* or by *projects*.

There is no single best way of organizing R&D. Scientific and technical factors, both internal to the firm and external, business needs, principal company structure and culture, and other factors often tip balanced considerations for perfect efficiency or effectiveness. An excellent overview and review of trade-offs of the different forms of R&D organization is provided by Floyd (1997).

Input for Innovation

Firms that compete on the basis of specialization and synergy in science and technology will often organize their R&D around sources of input, that is, scientific discipline or technical areas. Scientists, engineers, designers, and developers are grouped together on the basis of their specialities, mirroring the organization found in universities and research centres, from where this form of organization emerged. For instance, the R&D organization of a company may have a chemistry department, a biology department, a microbiology department, a pharmacology department, and a medicine department.

Organizing by discipline favours specialization and technical expertise over speed and customer orientation. It is best suited for areas in which innovation is the result of R&D in one single field. Other advantages are improved communication and interaction of colleagues with the same technical competence, easy introduction of new scientists and career development for staff, and unproblematic introduction of new knowledge to scientific specialties. This form of organization is, however, quite rigid and inflexible. As it is focused on the scientific discipline rather than innovation, a sense of urgency for meeting deadlines or customer requirements is usually lacking. It requires a great effort of cross-functional coordination to bring together the various disciplines needed in innovation projects, particularly when the focus of attention in innovation shifts away from the science towards the product (Floyd 1997).

Organizing R&D by *type of activity* is another input-based form of R&D coordination. Rather than organizing R&D around sources of scientific knowledge, it focuses on the tasks, activities, and operational procedures of the R&D work to be done. The basic R&D process serves as a template, for example, basic research, applied science, development, design, engineering, prototyping, testing, and innovation marketing. For instance, a pharmaceutical company may have discovery research, pre-clinical R&D, pharmacological clinical R&D, and clinical development. It shares most of the advantages of the discipline-based organization while instilling a greater sense of moving projects 'downstream' along the implied linear innovation path. Interdepartmental coordination and cross-functional innovation are still difficult. The exactness and quality of the science can win over speed and customer involvement.

Output for Innovation

Output-oriented R&D is usually either organized around product lines or around projects. *Product line* R&D mirrors the divisional structure of a corporation. It emphasizes customer-orientation and speed over novelty of R&D. Employees are grouped on the basis of the product lines or business units they are working for. It is a typical organization structure for multi-divisional firms active in various businesses, or corporations active in a single business with different product lines. This organization form comes with a strong orientation towards new products, as employees are organized on the basis of the structure of the customer base the firm wants to serve. There is strong integration with other business activities, greater managerial and organizational flexibility, and continuous attention to time and costs of innovation.

As product lines are managed independently, mismanagement in a single product can often be detected more easily than in R&D organized by discipline. Product line R&D organization, however, is much more liable to duplication of resources, both in terms of administration and the actual R&D discipline, to a lower degree of resource flexibility as resources are tied to specific product lines and businesses and it may be difficult to move them to a different R&D unit (resulting from resistance from business managers), to a lower degree of radical innovation due to strong customer orientation, and to low autonomy of R&D employees due to strong business orientations. Product-based R&D staff are experts on the product they develop, but as products are phased out it may require great effort to integrate such people into a new R&D unit.

R&D can also be organized by project rather than product lines, although few firms use this as their exclusive form of R&D organization and it is most prevalent in research-based organizations. R&D employees are not organized in a stable form by permanent criteria; they are assigned to a single project for its duration or serve as experts on several projects. In between projects, or, time permitting, during projects R&D employees update their competencies or they pursue their own innovative ideas. This organization form attempts to avoid the strong separation between divisions generated by product lines, especially with respect to increasing resource flexibility (Jain and Triandis, 1997; Argyres and Silverman, 2004).

Matrix Organization

There are several trade-offs between the input and the output orientation of R&D organizations. Several factors influence the decision towards one form of structure or the other:

- If the rate of change in technical disciplines is greater than the rate projects can be completed, choose an input-oriented, skill-based organization.
- The higher the level of sophistication and newness of a technology, the greater the need for acquiring knowledge, and the greater the preference for a discipline-based structure.
- If scale economies play a key role, for example when equipment has to be used and its intensive use lowers unit costs, opt for a discipline based structure.
- If the company is highly diversified, or if there are strong interdependencies between technologies in certain product categories, prefer an output-oriented organization.

The *matrix organization* combines input and output orientation of R&D organizations, trying to leverage the advantages of both organizational designs and to avoid their shortcomings (Le Masson et al., 2010). In a matrix organization, all R&D units (staff, tasks, etc.) are associated both with an input and an output organization. This is represented in Figure 26.3.

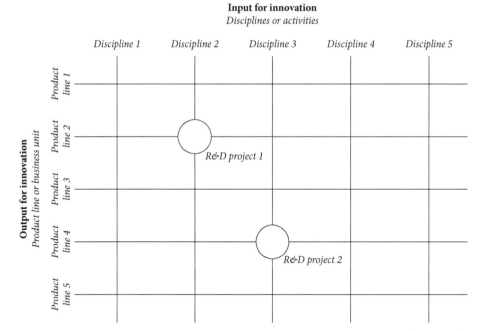

FIGURE 26.3 In a matrix structure each R&D project is part of both an input factor and an output factor

Every R&D employee will have to report to a line manager in charge of a business unit, a profit center, a product, or a product line, but also to a research director/manager in charge of a technology area, a scientific discipline, or a R&D activity. Individual experts can serve on several R&D projects while remaining embedded in their technical discipline and benefiting from specialization advantages. Business units can organize NPD projects, pulling in the best resources available while remaining in charge of the overall innovation process and driving for customer orientation, speed and cross-functional integration. The matrix provides for a clear distinction between managerial and professional responsibilities: R&D managers take care of professional standards, competence development, and the career development of R&D staff, while project managers are responsible for the progress of project work, reaching the overall innovation objective, and responding to customer expectations (Galbraith, 2009).

The main challenges to the matrix are difficulties in the coordination of R&D resources, the lack of clear and accepted separation of responsibilities, and disparate goal alignment. Invariably, line and project priorities will be in conflict with each other. In some organizations this conflict can release a creative tension if an open culture and a clear goal orientation is there. Only mature R&D organizations, usually of sufficient size, with balanced and experienced R&D staff, manage to operate well within a matrix organization. Start-up R&D organizations are usually focused on products and disregard scientific specializations. Small R&D units are unable to accommodate economies of scale required for an organization with two masters in each domain. R&D personnel in countries whose cultures expect strong individual leadership will experience internal conflict over whose authority to follow (the line manager usually wins). Nevertheless, despite its shortcomings and managerial complexity, in either its weak or strong form the matrix is the goal for most mature, multi-business and multi-technology organizations (Le Masson et al., 2010).

R&D Project Organization

R&D Project Team Structures

Innovation activity is more and more carried out in project teams, ranging from individual part-time work to super-sized projects involving thousands of engineers, scientists, supply chain experts, software developers, ad-hoc specialists, designers, marketeers, consultants, and so on (Cooper et al., 2003). Clark and Wheelwright (1992) presented four typical forms of R&D team organizations that we briefly summarize here.

Functional groups, such as analytical chemistry, pharmacology, software development, or industrial design, are organized in an input-oriented fashion around a certain function to be fulfilled, reporting to a functional team manager. The strength of this approach is similar to the input organization: optimizing functional resource allocation, task expertise, depth control, functional accountability, and clarity of career paths. On the downside, this team organization tends to be slow and bureaucratic, susceptible to in-fighting, internal politics, and power struggles, unaware of skills and needs outside its immediate functional realm, and inept at cross-functional innovation. These teams do not own business results—they are owned by the parent R&D organization.

Project teams with a lightweight project manager introduce the first benefits of output-orientation—tapping into available expertise, better coordination across functions, and improved communication. These teams and their project managers are however not empowered to override line management—their projects will be second priority when there are conflicts of time or resources, and they will often carry some blame for their 'distraction' from discipline-based R&D work. This form works best for non-critical NPD projects and projects based on a large number of scientific and technical experts.

The heavyweight project team overrides functional management and gives priority to the project, not the line. Its team members provide commitment and accountability to the project manager and are essentially removed from line responsibility, although still embedded there. Because of the dedication to a single project, this form is harder to staff and more disruptive to the overall R&D work, but sometimes the only sensible team form for large-scale resource-intensive innovation projects. As internal project management is strong, cross-functional coordination is easier, and customer involvement (of both internal and external customers) is greater.

In an autonomous project team, R&D staff are removed from their line organization and placed inside a dedicated project organization under the management of a high-level heavyweight project manager. This team owns the business results of its R&D work, and its primary objective is to deliver the customer requirement on time and quality. Being highly independent for the duration of the project, it is often difficult to reintegrate team members into the original R&D organization. Such R&D projects can be very costly and are only warranted if time and quality are of the essence.

Global NPD Teams

As discussed in Chapter 27 by Håkanson, R&D has become increasingly global, responding to the needs of global customers, localizing products for local markets, following production into low-cost countries and engaging in process R&D for manufacturing, or generally assuming new roles within business units distributed around the globe. This dispersion of R&D has a huge impact on communication and information transfer, one of the critical success factors of good R&D work, and its mediating support facilitators such as team trust, understanding of implicit customer expectations, minimal operating R&D staff size, organizational distance and line of command, and so on. Different national cultures, different language backgrounds, and different referential mental frameworks alone, coupled with the inability to sort out problems face-to-face, are hugely detrimental to efficient R&D operations.

As a consequence, Clark and Wheelwright's (1992) typology has been adapted to geographically dispersed R&D organizations. Gassmann and von Zedtwitz (2003) identified the following four transnational NPD project organizations:

1. *Decentralized self-organization*, in which teams coordinate and perform, often asynchronously, R&D work across different locations, under very weak leadership or sometimes no formal authority whatsoever.
2. *System integration*, in which an R&D coordinator assumes responsibility for project and technical interface management across different R&D contributors and sites, but still has no project authority over any individual R&D team.
3. *Core teams*, in which a central authority acts as an project coordinator and system architect, pulling in experts and R&D sites as needed, usually empowered by a senior executive and operating under the mandate of urgency and immediate business relevance.
4. *Centralized venture teams*, in which distributed R&D resources are centralized to the extent possible, often at extreme costs as time and innovation is critical, and only immobile or inflexible R&D resources are managed internationally on a daily operational basis.

The extent and degree to which R&D projects should be carried out in such a dispersed and decentralized fashion depends on four parameters:

1. The targeted type of innovation: ranging from radical to incremental;
2. The nature of the project between systemic and autonomous;
3. The dominant knowledge mode in the project—tacit or explicit;

4. The possibility to combine resources: accessible, overlapping and reinforcing or mutually exclusive and complementary.

Centralization of NPD projects is necessary for more radical innovations, systemic project work, and where tacit knowledge is prevalent and redundant resources are present. Decentralization of NPD projects is possible for incremental innovation, autonomous project work, and where explicit knowledge is prevalent and complementary resources are used.

Advances in information and communication technologies weaken some of the drawbacks of physical R&D distribution while at the same time permitting further dispersion of R&D to new locations in closer proximity to customers or unique and immobile sources of innovation (see Chapter 27 by Håkanson).

Conclusion

Managing R&D and NPD continues to be a central issue in the field of innovation management. It also remains a source of managerial challenge as well as commercial profit for companies in the market place. In this chapter we focused on three key elements of the management of R&D and NPD: the product development funnel, R&D portfolio management, and R&D organization. Those are the most essential tools that innovation managers and corporate strategists have used for some time now. The future of R&D and NPD management is likely to be shaped by some of the following trends:

Greater *corporate responsibility*: Currently this mainly focuses on manufacturing where both labour conditions and materials in use are becoming important to customers. The social and environmental consequences of how products are made and which materials are used are becoming intimately linked to the management of NPD. Research on this topic remains limited. It is necessary to deepen our understanding of the interrelation between corporate responsibility and NPD.

Improved *flexibility of R&D units*: More and more R&D engineers have never met in person and operate within dispersed virtual teams. This and the dissolving boundaries between internal and external sources of innovation challenge the way R&D and NPD are managed. New management structures are needed that disassemble NPD tasks such that these highly complex and distributed innovation projects can still be carried out efficiently and effectively.

Shift from products to *business models*: The product itself is shifted from being the company's only offering towards being only one element of its offering. Companies are more and more moving towards integrated solutions that include hardware, software, maintenance, and other service offerings (see also Chapter 21 by Massa and Tucci). New products cannot be developed in isolation from consideration of the business models of which they are a part.

Along with these relatively new challenges, there remains the long-term and continuing need for deeper understanding of the ways R&D and NPD are integrated into corporate innovation strategies and are organizationally embedded. We must continually strive to understand better how this can be done successfully.

REFERENCES

Annachino, M. A. (2007). *The Pursuit of New Product Development: The Business Development Process.* Oxford: Butterworth-Heinemann.

Argyres, N. S., and Silverman, B. S. (2004). 'R&D, Organization Structure, and the Development of Corporate Technological Knowledge', *Strategic Management Journal*, 25(8–9): 929–58.

Bacon, G., Beckmann, S., Mowery, D. C., and Wilson, E. (1994). 'Managing Product Definition in High-Technology Industries: A Pilot Study', *California Management Review*, 36(2): 32–56.

Baker, N. R., and Sweeney, D. J. (1978). 'Toward a Conceptual Framework of the Process of Organized Innovation Technological within the Firm', *Research Policy*, 7(1): 150–74.

Beaujon, G. J., Marin S. P., and McDonald, G. C. (2001). 'Balancing and Optimizing a Portfolio of R&D projects', *Naval Research Logistics*, 48(1): 18–40.

Bhattacharya, S., Krishnam, V., and Mahajan, V. (1998). 'Managing New Product Definition in Highly Dynamic Environments', *Management Science*, 44(11): 550–64.

Chakrabarti, A. K. (1974). 'The Role of Champion in Product Innovation', *California Management Review*, 17(2): 58–62.

Chien, C.-F. (2002). 'A Portfolio-Evaluation Framework for Selecting R&D Projects', *R&D Management*, 32(4): 359–68.

Child, J. (1980). *Organisations.* London: Harper & Row.

Clark, K. B., and Wheelwright, S. C. (1992). 'Organizing and Leading Heavyweight Development Teams', *California Management Review*, 34(2): 9–28.

Cooper, R. G. (1985). 'Selecting Winning New Product Projects: Using the NewProd System', *Journal of Product Innovation Management*, 2(1): 34–44.

Cooper, R. G. (1988). 'The New Product Process: A Decision Guide for Management', *Journal of Marketing Management*, 3(3): 238–55.

Cooper, R. G., Edgett, S. J., and Kleinschmidt, E. J. (1998). *Portfolio Management for New Products*, New York: Basic Books.

Cooper, R. G., Edgett, S. J., and Kleinschmidt, E. J. (2003). 'Benchmarking Best NPD Practices— Part I: Culture, Climate, Teams and Senior Management Roles', *Research Technology Management*, 47(1): 31–43.

Cooper, R. G., Edgett, S. J., and Kleinschmidt, E. J. (2004). 'Benchmarking Best NPD Practices— Part II: Strategy, Resource Allocation, and Portfolio Management', *Research Technology Management*, 47(3): 50–59.

Cooper, R. G., and Edgett. S. J. (2007). *Generating Breakthrough New Product Ideas: Feeding the Innovation Funnel.* Canada: Product Development Institute.

Dunphy, S. M., Herbig, P. R., and Howes, M. E. (1996). 'The Innovation Funnel', *Technological Forecasting and Social Change*, 53(3): 279–92.

Floyd, C. (1997). *Managing Technology for Corporate Success.* Aldershot, UK: Gower Publishing.

Galbraith, J. R. (2009). *Designing Matrix Organizations that Actually Work: How IBM, Procter&Gamble, and others Design for Success.* San Francisco: Jossey-Bass.

Gassmann, O., and von Zedtwitz, M. (2003). 'Trends and Determinants of Managing Virtual R&D Teams', *R&D Management*, 33(3): 243–62.

Gupta, A. K., Raj, S. P., and Wilemon, D. (1986). 'A Model for Studying R&D-Marketing Interface in the Product Development Process', *Journal of Marketing*, 50(2): 7–17.

Jain, R. K., and Triandis, H. C. (1997). *Management of Research and Development Organizations: Managing the Unmanageable*. Hoboken: John Wiley & Sons.

Khurana, A., and Rosenthal, S. R. (1997). 'Integrating the Fuzzy Front End of New Product Development', *Sloan Management Review*, 38(2): 103–120.

Koen, P., Ajamian, G., Burkart, R., Clamen, A., Davidson, J., D'Amore, R., Elkins, C., Herald, K., Incorvia, M., Johnson, A., Karol, R., Seibert, R., Slavejkov, A., and Wagner, K. (2001). 'Providing Clarity and a Common Language to the 'Fuzzy Front End', *Research-Technology Management*, 44(2): 46–55.

Kwak, Y. H., and LaPlace, S. (2005). 'Examining Risk Tolerance in Project-Driven Organization', *Technovation*, 25(6): 691–5.

Le Masson, P., Weil, B., and Hatchuel, A. (2010). *Strategic Management of Innovation and Design*. Cambridge: Cambridge University Press.

Leonard-Barton, D. (1992). 'Core Capabilities and Core Rigidities: A Paradox in Managing New Product Development', *Strategic Management Journal*, 13: 111–25.

Mintzberg, H. (1979). *The Structuring of Organizations*. Upper Saddle River, NJ: Prentice-Hall.

Mohr, J., Sengupta, S., and Slater, S. (2009). *Marketing of High-Technology Products and Innovations*, 3rd edn, Upper Saddle River, NJ: Prentice Hall.

Mu, J., Peng, G., and MacLachlan, D. L. (2009). 'Effect of Risk Management Strategy on NPD Performance', *Technovation*, 29(3): 170–80.

Nobelius, D. (2004). 'Towards the Sixth Generation of R&D Management', *International Journal of Project Management*, 22: 369–75.

O'Sullivan, D. (2002). 'Framework for Managing Business Development in the Networked Organization', *Computers in Industry*, 47(1): 77–88.

Reinertsen, D. G. (1999). 'Taking the Fuzziness out of the Fuzzy Front End', *Research Technology Management*, 42(6): 25–31.

Rengarajan, S., and Jagannathan, P. (1997). 'Project Selection by Scoring for a Large R&D Organization in a Developing Country', *R&D Management*, 27(2): 155–64.

Roberts, E. (1968). 'Entrepreneurship and Technology', *Research Management*, 11(4): 249–66.

Roussel, P. A., Saad, K. N., and Erikson, T. J. (1991). *Third Generation R&D: Managing the Link to Corporate Strategy*. Boston: Harvard Business School Press.

Seidel, V. P. (2007). 'Concept Shifting and the Radical Product Development Process', *The Journal of Product Innovation Management*, 24: 522–33.

Stevens, G. A., and Burly, J. (2003). 'Piloting the Rocket of Radical Innovation', *Research Technology Management*, 46(2): 16–25.

Verworn, B., Herstatt, C., and Nagahira, A. (2008). 'The Fuzzy Front End of Japanese New Product Development Projects: Impact on Success and Differences between Incremental and Radical Projects', *R&D Management*, 28(1): 1–19.

von Hippel, E. (1986). 'Lead Users: A Source of Novel Product Concepts', *Management Science*, 32(7): 791–805.

Witte, E. (1973). *Organisation für Innovationsentscheidungen: Das Promotorenmodell*. Göttingen: Schwartz und Co.

Zhang, Q., and Doll, W. J. (1998). 'The Fuzzy Front End and Success of New Product Development: A Causal Model', *European Journal of Innovation Management*, 4(2): 95–112.

....

CHAPTER 27

INTERNATIONALIZATION OF RESEARCH AND DEVELOPMENT

....

LARS HÅKANSON

INTRODUCTION

....

IN multinational corporations (MNCs), international activities provide important stimuli to innovation. Interaction with foreign customers, suppliers, and competitors facilitates timely detection of new market and technological trends. Such opportunities can be enhanced by the establishment of local research and development (R&D) and close technical cooperation with local partners. Performing R&D abroad may also be a means to tap into foreign technical and scientific infrastructures, allowing privileged access to new technological and scientific advances and the possibility to employ engineers with different skills than those available in the home country or at lower costs.

Contrary to common perception, foreign R&D is not a recent phenomenon. While in the past, US and Japanese multinationals performed most of their R&D at home, in the vicinity of the head office and major manufacturing units, many European MNCs were already undertaking substantial development activities abroad in the early twentieth century (Cantwell, 1995). However, most cross-border R&D investments still take place between OECD countries, the vast majority within the triad of the European Union, the United States, and Japan where close to 90 per cent of the world's R&D is performed. More than 80 per cent of these investments originate in only five countries: the United States, Japan, Germany, the United Kingdom, and France; only around 1 per cent comes from emerging countries, including China, Korea, Taiwan, Brazil, and South Africa. However, in recent decades, the volume of R&D undertaken in foreign locations has grown substantially in both absolute and relative terms, is spreading to a larger number of host countries, especially in certain emerging countries, such as China and India (UNCTAD, 2005), and has become common also in MNCs from other parts of the world (von Zedtwitz, 2006; OECD, 2008). Moreover, it appears that international R&D is no longer a phenomenon limited to large and very large multinationals. Casual evidence suggests that small firms in high technology industries increasingly locate some of their development activities to foreign countries, either to maintain better contact with local customers or to engage technical expertise at lower cost than they can at home.

The phenomenon has attained growing significance for government policy-makers in both home and host countries. In host countries, where a high proportion of R&D is performed by foreign affiliates, governments are concerned about the dependency and potential vulnerability of local technical and scientific capabilities; in host countries, whose firms undertake increasing amounts of their R&D in foreign locations, govern-ments fear that such capabilities will be hollowed out and that job opportunities will be lost, in both R&D and associated manufacturing operations. International R&D is increasingly important also for MNC managers, who face new challenges and oppor-tunities in the management and coordination of geographically and organizationally decentralized R&D and in the international exploitation of technical innovations.

GEOGRAPHICAL DECENTRALIZATION OF R&D

Until relatively recently, the establishment of foreign R&D activities was often the result of *local initiatives* on the part of entrepreneurial foreign subsidiary managers. Especially in European MNCs, foreign affiliates traditionally enjoyed a high degree of autonomy. Their managers often had an inducement to allow local engineers to move into design and engineering, rather than exclusively focus on routine technical customer service or support of local production. Such upgrading of local activities was a chance to explore new business opportunities and a means to attract more qualified technical personnel through the provision of more challenging job opportunities. In European MNCs, this tended to be a rather 'spontaneous' process, taking place without much attention or involvement on the part of headquarters managers. Each initiative may, initially at least, have been on a rather limited scale. Over time, however, the cumulative effects of many such initiatives were often quite considerable, gradually shifting the focus of innovation from the home country towards international markets.

A pervasive factor for the geographical decentralization of R&D has been the rise in *international mergers and acquisitions* (M&As) (see Chapter 29 by Ahuja and Novelli). From the 1960s onwards, M&As rather than greenfield establishments have been the major mechanism of MNC growth. Sometimes, of course, M&As are undertaken in order to access new technologies and new technical capabilities, in which case the prospect of extending existing capabilities through the knowledge of acquired units is an important strategic motivation. Often, however, M&As are undertaken for reasons unrelated to R&D—the wish to acquire a brand name, distribution channels, produc-tion capacity or to obtain or strengthen the company's market dominance, for example. In such cases, the knowledge and capabilities of acquired R&D units tend to duplicate those already possessed by the buying firm. Oftentimes, therefore, acquired R&D units are seen to be redundant and ways are sought to close them down without upsetting local constituencies, such as governments, trade unions, and local employees. At other times, acquired R&D units present a welcome opportunity to enlarge existing R&D capacities, in which case they were therefore retained and integrated into the MNC's overall R&D organization.

Over time, the geographical decentralization of R&D to foreign locations has become an increasingly deliberate process with clear strategic aims. From very early on, a significant consideration has been the wish to maintain *proximity to foreign customers* in local environments with different demand structures than those available in the MNC's home location. Efficient innovation typically requires a good understanding of customers and their needs. Most MNCs therefore base their initial growth on products and offerings customized to the needs of their domestic customers, seeking for international expansion countries and markets with similar characteristics to those of their home countries. In the course of internationalization, MNCs nevertheless encounter different types of customers and other demand characteristics than those of the home market, requiring less or greater adaptation of product and service offerings. In undertaking such adaptations, a local R&D department that can engage directly with customers and local production is often a big advantage. The benefits of setting up local R&D are especially promising when local demand conditions can be expected to develop elsewhere as well, placing the MNC in an advantageous position to exploit demand emerging in other countries.

For some MNCs, geographical decentralization of R&D has also been a strategic consequence of a restructuring and *international rationalization of manufacturing*. Reduction of trade barriers and transportation costs, in combination with an increasing homogenization of demand, has in many industries made it possible to exploit scale economies through concentration of production to large, specialized factories serving ever larger geographical markets. In the wake of such rationalization, such specialized manufacturing units were often given group-wide responsibility not only for the production of certain products or components but also for associated R&D.

In the belief that timely access to state-of-the art scientific and technical developments requires physical presence in certain geographical areas, many MNCs, especially in so-called high-tech industries, have established local R&D facilities in foreign 'clusters' of technical excellence in certain industries and technologies (Frost et al., 2002). Encouraged by a large literature in economic geography and industrial economics expounding the benefits of clusters, MNCs have sought to *profit from 'spillovers'* of technical knowledge through various kinds of local establishments in fashionable regions, ranging from fully fledged research laboratories to small 'listening-posts' (Cantwell and Piscitello, 2005). Originally, these kinds of establishments were found primarily in US, European, and Japanese MNCs seeking access to regionally localized knowledge; more lately MNCs from emerging economies are using similar strategies in an effort to technologically 'catch-up' with their competitors (von Zedtwitz, 2006).

The most recent trend contributing to the geographical dispersal of R&D in MNCs has been the establishment of R&D units in countries such as China, India, Brazil, and the former Eastern Bloc countries, in order to employ technical expertise at salary levels below, or sometimes far below, those prevailing in the United States or Western Europe. Although initially established to perform more 'routine' technical tasks, these units over time have moved into more advanced technical development and design, sometimes acquiring the status of 'centres-of-excellence' in their respective fields—a reflection of the political clout they can exercise by virtue of their accumulated technical expertise.

INCREASING NEED FOR COORDINATION

In the past, foreign R&D tended to be on a rather limited scale and most of it was dedicated to the solution to local problems. These could arise, for example, through the need to accommodate the specific requirements of local customers or to attend to technical issues caused by local idiosyncrasies in the availability or quality of components or raw materials. The benefits of information exchange across national borders were therefore limited and in many firms international coordination of R&D was kept at a minimum.

Managing complex, knowledge-intensive activities is costly and difficult. These costs and difficulties are aggravated when associated information exchanges need to take place across geographical distances and time zones and among people from different national and organizational cultures. However, in consequence of rising absolute R&D costs and increasing geographical dispersal of R&D activities, MNCs have been confronted with growing pressures to optimize their global innovation efforts through stronger coordination, collaboration, and more intense information exchange (Santos et al., 2004).

A basic such pressure is the need to *avoid duplication of effort*; few firms are in the happy position that they can—as was once rumoured that IBM did—encourage competition among R&D teams, appointing several groups to pursue the same goal through different means. Most need to ensure that their R&D engineers can profit from lessons learnt in other parts of the organization and that effort and money are not spent 'reinventing the wheel'.

Sometimes, MNCs find it necessary or to their advantage to divide R&D activities relating to the same product or system between several units, often located in different parts of the world. Successful implementation of such division of labour permits the exploitation of diverse localized expertise in a way that may not be possible for firms with geographically more centralized R&D organizations. However, it typically requires not only the efficient *synchronization and harmonization of designs* needed to ensure compatibility of components but also coordination of time schedules and the meeting of development milestones in order to *ensure timely completion and market launches*. The former is an issue not only in cases where technical compatibility is important, but also where marketing considerations dictate that product designs reflect a common corporate image. The latter has become increasingly important as the life cycle of individual design solutions have progressively shortened, especially in high-tech industries characterized by rapid technological change.

Especially difficult coordinative challenges characterize the management of innovations involving the *convergence of technologies*, involving the combination of knowledge elements from different technical disciplines, such as in the case of computer and telecommunications technologies. Similar situations are found in industries producing *complex product systems*, such as automobiles or wind turbines, where product architectures require precise specifications of complex interfaces between individual but mutually interdependent technical components. In addition to the obstacles posed by geographical distances, and differences in national and organizational cultures, knowledge exchanges must then take place between engineers with different areas of expertise and disciplinary backgrounds, overcoming differences in technical vocabularies, theories, and methodological conventions.

DIFFERENT TYPES OF FOREIGN R&D ACTIVITY

The term 'research and development' encompasses a whole range of different types of innovatory activities, each with a different set of characteristics both in terms of focus and in terms of external and internal linkages (Papanastassiou and Pearce, 1999; von Zedtwitz and Gassmann, 2002) (Table 27.1).

Manufacturing activities in foreign subsidiaries often require *technical support* to adapt production processes to local circumstances, such as differences in the cost and quality of components or raw materials, or in the skills and qualifications of the local workforce. Another important task is to support the implementation of new processes associated with the introduction of new product generations. Much such adaptive engineering is often of a relatively routine character, but this is not always the case—sometimes, the degree of novelty of solutions developed to solve local problems is considerable, and sometimes such solutions can find areas of profitable application far from

Table 27.1 Typology of foreign R&D units

Type of unit	Time horizon	Area of application	Dominating functional linkage
Technical support	Short	Plant specific	Production
Adaptive R&D		Market specific	Marketing
Generic R&D		Regional (global)/product specific	Production & marketing
Research	Long	Global/technology/specific	Scientific community

their original context. Unfortunately, many such opportunities go unexploited; since much of this type of work is undertaken within peripheral foreign subsidiaries and without being organized in formal R&D departments they often take place below the radar of headquarters' attention. However, in recognition of their dependent and peripheral status, managers in such units sometimes form informal coalitions across foreign subsidiary borders as a means to obtain more headquarters attention and support.

In many foreign subsidiaries, engineering capabilities are needed not only to support local production but also to provide technical customer service. Performing such service tasks enhances the understanding of customers' needs and problems. This provides impetus to innovatory design efforts, aiming to match product offerings more closely to local customer needs. In the course of such *adaptive R&D*, foreign subsidiaries can evolve into *indigenous technology units* pursuing improvements of existing products and manufacturing processes or new applications of existing technologies. Such initiatives are encouraged by a high degree of local autonomy, but this needs to be balanced by integrative mechanisms and some degree of central coordination in order to ensure that local innovations can be exploited elsewhere in the organization. As in the case of technical support units, there is a risk that local innovations remain undetected or only haphazardly exploited.

Generic innovation geared to the development of new products and production processes sometimes evolves out of adaptive R&D but is more commonly the result of diversification into new product and technology areas through mergers and acquisitions (von Zedtwitz and Gassmann, 2002). In the case of unrelated diversification, firms generally attempt to preserve acquired innovatory capabilities by minimizing the amount of organizational change (Håkanson, 1995). However, when firms acquire complementary technical capabilities with a view to combining them with existing ones, the managerial challenges are typically more substantial. The realization of expected synergies requires that individuals both from the acquiring and from the acquired firm are willing and able to collaborate in the transfer and integration of strategic capabilities. A large literature testifies to the difficulties that need to be overcome in order to create such conditions.

Large MNCs in technology-intensive industries sometimes establish *corporate technology units* outside their countries of origin, dedicated to long-range basic and applied *research* (Edler et al., 2002; von Zedtwitz and Gassmann, 2002). In contrast to other types of foreign R&D units that need to maintain intensive interaction with production and/or marketing related to their areas of specialization, the location of corporate research laboratories is usually the result of careful deliberation. In the expectation that this will benefit the inflow of new knowledge, one key such deliberation is often the geographical proximity to relevant scientific and technical establishments, and to competing and complementary industries, such as can be found in localized geographical 'clusters' in different parts of the world. More tangible and practical concerns are local tax regulations for expatriates, climate, and general living conditions—important when it comes to recruiting foreign specialized expertise. The geographical location in relation to other parts of the corporation may also play a role. It is often considered an advantage, for example, to locate corporate research activities at some distance from operating divisions. This signals a 'neutral' position vis-à-vis other R&D units and may help maintain a long-term perspective through isolation from day-to-day problem solving and 'fire fighting'.

While corporate research units can provide valuable technological input and help sustain a company's long-term innovatory capabilities, there is a danger that their efforts become disassociated from the strategic and commercial needs of operational divisions. The pursuit of new technological capabilities must be balanced with the need to maintain strategic relevance, and long-term objectives with short-term priorities. An efficient mechanism to ensure strong reciprocal ties between corporate research, on the one hand, and operative divisions, on the other, is to require that funding of research activities be wholly or partly funded by the latter. This helps ensure that research portfolios are aligned with commercial realities and do not include areas lacking potential strategic relevance.

COORDINATION AND CONTROL

Over time, MNEs have developed increasingly sophisticated organizational mechanisms to enhance control and coordination of geographically dispersed innovation activities. A key element is usually the design of formal *organizational structures* that define reporting relationships, authority for resource allocation decisions, and accountability for resources spent and results achieved. *Within* any one such organizational structure, coordination can usually be obtained and control exercised without much regard to geographical distances. Coordination *across* such organizational boundaries may be more difficult but can be facilitated by geographical proximity. The attention of an R&D unit co-located with the factory of a foreign subsidiary but reporting to a central R&D function at headquarters will be influenced not only by the latter's formal authority but also—through daily interaction with its managers—by the strategic considerations and tactical needs of the local subsidiary. Conversely, coordination across *both* organizational and international boundaries is generally cumbersome and difficult to achieve.

The design of formal *planning processes, budgeting,* and *project monitoring systems* is a further important mechanism to ensure control and coordination across geographically dispersed R&D efforts. A key element is the personal engagement of officials from different functions and geographical units. Their exposure not only to the plans, results, and activities of other parts of the MNE but also to their concerns and problems helps to ensure that knowledge and capabilities generated in one geographical location can be exploited elsewhere, and that costly duplication of effort can be avoided.

Other common organizational measures include the creation of ad hoc or semi-permanent committees, such as cross-functional *product councils* with members from different geographical locations. Such groups can not only provide valuable market and technical intelligence but also serve as a means to strengthen integration and information exchange between different parts of the organization. In the past, the composition, scheduling, and convening of such group meetings were subject to many restraints in terms of travel and members' opportunity costs; modern computer and telecommunication technologies, such as high-quality video-conferencing, have made such logistic restraints increasingly less important. In their place have appeared challenges associated with the management and effectiveness of cross-functional groups, composed of managers with different backgrounds in terms of national and organizational culture.

Important mechanisms for coordination and control are also various types of *project management systems*, delineating common formats for planning, budgeting, monitoring, and specified 'milestones', at which project progress is reviewed and future activities and outlays are authorized. Making data from such reviews available in digital *project portfolio databases* is an efficient means to internally broadcast information that may be relevant elsewhere in the organization, but where potential recipients are difficult to identify a priori. Although somewhat more costly, similar effects can be obtained by having individual projects manage their own *project websites* on the corporate Intranet, perhaps with selective access to outside partners.

Knowledge Management

Information related to coordination and control is relatively well structured and easy to convey through systematic and impersonal means, such as email, databases, or websites. In contrast, knowledge needed for innovative problem solving is not always well articulated or accessible, and the willingness to seek or share such knowledge is subject to complex organizational and psychological influences. Moreover, it is in the nature of innovatory activities that it is typically difficult to foresee, where, when, and in what context a particular piece of knowledge can be combined with other knowledge in innovative ways. Because of this inherent unpredictability, effective knowledge management in MNEs needs to encompass both IT-supported systems for information exchange and more subtle mechanisms to promote socialization and the creation of joint theoretical understandings and shared codes.

Common IT-supported systems include *document databases, groupware, project web pages, 'yellow pages', intra-organizational online forums*, and the like. Interestingly, the main benefit of such systems is often not the technical information they contain, but as pointers to colleagues around the world who have been involved in relevant projects, solved similar problems, or are active in specific areas of interest. However, the value of this information is contingent on an organizational culture where individual engineers are both willing to seek out the advice of their peers and also to share knowledge with other members of the organization across geographical and functional boundaries. A key element in ensuring such conditions is in the promotion of *socialization* of R&D engineers and other functional managers.

A basic and very effective mechanism to promote such socialization is the *rotation of personnel*. R&D engineers temporarily assigned to R&D laboratories in other countries get to know their foreign colleagues personally, a pre-requisite for effective future communication through impersonal means, such as telephone or email. Such assignments also provide informal knowledge regarding local work practices, individuals with specific expertise, and so on. On returning to their normal places of work, this knowledge enables former expatriates to serve as 'gatekeepers', helping to channel communication among engineers who have no or only little personal knowledge of one another. Traditionally, such foreign placements are designed to last around two to three years, but rotation programmes on such time-scales are difficult to implement. Increasing female employment ratios have made 'dual career' couples more common, significantly adding to other practical family problems of accommodating long-term foreign assignments. Here, short-term foreign assignments over three to four weeks present an inexpensive but highly efficient alternative. Although they cannot substitute for the individual learning experience of longer-term foreign postings, they bring many of the same advantages in promoting intra-organizational communication.

Other common measures to promote knowledge exchanges across national boundaries include the organization of *technical seminars* and *conferences* at different locations. Such events present an opportunity for local R&D managers to advertise and promote themselves vis-à-vis foreign colleagues. This helps to stimulate intra-organizational knowledge exchange and technical cooperation where the reputation and credibility of the cooperating partners are key elements. Equally and sometimes more important are the opportunities for informal interaction that such events permit. In helping to build mutual credibility and trust, such interaction not only serves to facilitate future contact and communication through impersonal means, it occasionally creates unexpected opportunities for innovation. Legendary are the encounters of R&D managers from different parts of the world, discovering over lunch how their respective competences can be combined in new and innovative ways. Stories of the resulting innovations tend to emphasize the chance element of such encounters in the form of unforeseeable outcomes of new combinations of technical expertise. However, the frequency of informal meetings such as the ones occasioned by technical seminars and conferences, and the design of after-hours events, clearly influence the probability that such seemingly serendipitous innovation will occur.

EMERGING AND UNRESOLVED ISSUES

Three-quarters or more of the world's technical and scientific resources are controlled by multinational companies. Their decisions regarding the location and deployment of these resources profoundly affect the technological trajectories that shape current and future productive opportunities and living conditions. The internationalization of R&D and the globalization of innovation have in recent decades become increasingly important features of the environment where these decisions are made and the new technologies underlying innovative products and services are developed.

The systematic collection of relevant empirical data was long hampered by the fact that national statistical offices—with the partial exception of the USA and Sweden—have traditionally limited their interest to entities and activities taking place within their national borders. The most systematic information available on foreign R&D relates, therefore, to the activities of foreign-owned affiliates in individual countries since these are captured in national databases. Only in recent decades have more comprehensive overviews become available through a number of cooperative efforts under the auspices of the OECD (2008, 2010) and UNCTAD (2005). Unfortunately, the cost and bureaucratic difficulties of collecting the required data are considerable and the available surveys are both incomplete as well as few and far between. Moreover, they tend to focus only on the very largest firms; official statistics consistently underestimate the R&D performance of small firms and provide no information on the development activities they undertake in foreign countries.

Given the scarcity of data on international R&D inputs, such as costs and employment in R&D, researchers have turned to the analysis of R&D outputs in the form of patents. Since patent documents include not only the name and address of the person or firm filing the patent but also the name(s) and address(es) of the original inventor(s) as well as citations to other patents, they can be used to trace the role of foreign R&D activity over time (Hall et al., 2001). For this purpose, the historical records of the US Patent Office have provided especially rich opportunities (Zander, 1999; Frost, 2001).

However, there are distinct limits to the type of information that can be obtained through official statistics and other secondary sources. Dedicated surveys directed towards select populations of multinational firms or foreign subsidiaries are needed to throw more light on recent developments. These include, for example, the growing foreign involvement of firms in and from the so called BRIC countries (Brazil, Russia, India, and China) and other emerging economies. Another weakness in our present understanding is the absence of longitudinal studies; with very few exceptions, available data are based on cross-sectional surveys, providing static descriptions of R&D activities undertaken in different places at a particular point in time. Knowledge regarding the evolution (or termination) of R&D in different kinds of units is almost totally absent, leaving unanswered important questions such as the contingencies determining the medium- and long-term effects of foreign acquisitions on host country R&D capabilities.

A further area in need of research is the systematic exploration of organizational and managerial best practice in the management of geographically and organizationally decentralized R&D, including the deployment and effects of new technologies, such as immersive video-conferencing. Here, there is a need both for detailed case studies of individual MNCs and their international R&D networks and for ethnographic case studies of different types of international R&D teams, especially in projects and activities combining knowledge from different disciplines and areas of expertise.

References

Cantwell, J. (1995). 'The Globalisation Of Technology: What Remains of the Product Cycle Model?' *Cambridge Journal of Economics*, 19: 155–74.

Cantwell, J., and Piscitello, L. (2005). 'Recent Location of Foreign-owned Research and Development Activities by Large Multinational Corporations in the European Regions: The Role of Spillovers and Externalities', *Regional Studies*, 39: 1–16.

Edler, J., Meyer-Krahmer, F., and Reger, G. (2002). 'Changes in the Strategic Management of Technology: Results of a Global Benchmarking Study', *R&D Management*, 32: 149–64.

Frost, T. S. (2001). 'The Geographic Sources of Foreign Subsidiaries' Innovations', *Strategic Management Journal*, 22: 101–24.

Frost, T. S., Birkinshaw, J. M., and Ensign, P. C. (2002). 'Centres of Excellence in Multinational Corporations', *Strategic Management Journal*, 23: 997–1018.

Håkanson, L. (1995). 'Learning through Acquisitions: Management and Integration of Foreign R&D Laboratories', *International Studies of Management and Organization*, 25: 121–57.

Hall, B. H., Jaffe, A. B., and Trajtenberg, M. (2001). 'The NBER Patent Citation Data File: Lessons, Insights and Methodological Tools', NBER Working Paper 8498.

Narula, R. (2003). *Globalisation and Technology.* Cambridge: Polity Press.

OECD (2008). *The Internationalization of Business R&D: Evidence, Impacts and Implications.* Paris: OECD.

OECD (2010). *Measuring Globalisation: OECD Economic Globalisation Indicators 2010.* Paris: OECD.

Papanastassiou, M., and Pearce, R. (1999). *Multinationals, Technology and National Competitiveness.* Cheltenham: Edward Elgar.

Santos, J., Doz, Y., and Williamson, P. (2004). 'Is Your Innovation Process Global?', *MIT Sloan Management Review*, 45: 31–7.

UNCTAD (2005). *World Investment Report: Transnational Corporations and the Internationalization of R&D.* New York and Geneva: United Nations.

von Zedtwitz, M. (2006). 'International R&D Strategies of TNCs from Developing Countries: The Case of China', in UNCTAD (ed.), *Globalization of R&D and Developing Countries.* New York and Geneva: United Nations, 117–40.

von Zedtwitz, M., and Gassmann, O. (2002). 'Market versus Technology Drive in R&D Internationalization: Four Different Patterns of Managing Research and Development', *Research Policy*, 31: 569–88.

Zander, I. (1999). 'How Do You Mean "Global"? An Empirical Investigation of Innovation Networks in the Multinational Corporation', *Research Policy*, 28: 195–213.

....

CHAPTER 28

INTELLECTUAL PROPERTY RIGHTS, STANDARDS, AND THE MANAGEMENT OF INNOVATION

....

AIJA LEIPONEN

INTRODUCTION

....

INNOVATION entails the creation of new knowledge for practical application. New products and processes, and new services and business models, all require novel combinations of new or existing knowledge components. The intangible nature of much of this knowledge often enables unintended knowledge flows, or spillovers. Spillovers are 'good' for society, because they ensure that useful knowledge disseminates in the economy and many individuals and organizations receive the benefits from innovation. Innovations tend to build on earlier insights and technologies; we are all standing on the shoulders of, if not giants, then many smart people who went before (Scotchmer, 1991). It is then only fair that the society that created the conditions for innovation gets to benefit from it. However, spillovers are 'bad' because they complicate the provision of incentives for innovation. The more quickly the benefits of innovation dissipate in the economy through learning and imitation by others, the smaller are the returns on investment in creating knowledge, for example in research and development (R&D), for inventors, and the less they are willing to invest. In economic terms, spillovers create a positive externality when other parties benefit from an inventor's investments without paying for it, and as a result, individually optimal investments in R&D are lower than socially optimal ones. An earlier estimate suggested that, in industrialized economies, socially optimal investments might be as high as two to four times the actual investments (Jones and Williams, 1998). This evidence suggests that the economy can develop faster if R&D investments are subsidized (Griffith et al., 2004).

Intellectual property rights and other forms of protection for intellectual assets represent policy tools developed to support the incentives for creativity and innovation. Economically, the most powerful forms of intellectual property rights include patents and copyrights, but trademarks and design registrations (design patents) can also be very valuable in some situations. Additionally, trade secrets are usually not viewed as an intellectual property right (although see Lemley, 2009), but they also legally protect knowledge that is economically valuable, not commonly known, and that is subject to reasonable efforts to protect.

Intellectual property rights were designed to encourage creative production. An early form of patents is said to have existed in ancient Greece; in Britain they were introduced in the fourteenth century and in subsequent centuries in Florence, Italy, and France (Stobbs, 2000: 3–4). In the United States, patents were already being granted in the early colonies, and the first Constitution (1787) included a patent and copyright clause that empowered Congress 'to promote the progress of science and useful arts' through patents and copyrights. Early patents and copyrights were typically ten to twenty years in duration. Trade secrets were initially legally acknowledged through case law in early nineteenth-century Britain and subsequently in the United States and elsewhere (see Lemley, 2009).

The British Statute of Anne (1709/1710), the 'Act for the Encouragement of Learning, by vesting the Copies of Printed Books in the Authors or purchasers of such Copies, during the Times therein mentioned', was the world's first formal copyright law, and it granted book publishers protection for fourteen years. It was also the first instance where regulation of printing did not focus on the crown monopolies or control of critical content, but on the spread and encouragement of education and of 'learned men to compose and write useful books', thereby providing incentives for creation (Rose, 1993).

Thus, several legal instruments have been developed to help innovators capture the returns to their creative efforts. However, these instruments are far from perfect, and in most industries and situations spillovers are alive and well. In fact, most firms report that being faster than competitors to develop and launch new products or services is the most effective way to capture returns from innovative activities (Cohen et al., 2000; Arundel, 2001). This is particularly true of small firms, service firms, and those that cooperate in R&D (Tether and Massini, 2007; Leiponen and Byma, 2009). Thus, many innovators may utilize but do not necessarily substantially benefit from formal legal protections such as patents and copyrights for their knowledge creation.

Although patents and copyrights dominate the policy debates on the implications of intellectual property for innovation, from a managerial perspective, informal strategies toward capturing returns on innovation investments turn out to be the most important, prevalent, and effective ones for firms in most industries. In addition to strategies to increase speed to market, informal protection methods may include attempts to increase the complexity of the product or service design and thus reduce the likelihood of imitation and control of complementary assets in sales, distribution, services, supply relationships, manufacturing, or related products, services, or technologies. Trade secrets are usually also found to be more important methods of appropriation than patents or copyright. Secrecy can be further enhanced by using contractual agreements with external parties or employees. Such non-disclosure and non-compete agreements and informal trade secrets can have strong legal implications, although they are not intellectual property rights as such.

There are, nevertheless, a few very innovative industries where patents or copyrights are extremely important for competitive success, and where the development of technical compatibility standards can further heighten the legal and competitive implications of intellectual property rights. The next section discusses the nature of technological progress and patterns of competitive dynamics in these industries to shed light on their unique strategies with respect to intellectual property rights and technical standards.

The Importance and the Wider Contexts of Intellectual Property Rights and Technical Standards

Patents and copyrights are essential competitive tools in such dynamic and innovative industries as pharmaceuticals, biotechnology, computer hardware and software, wireless communications, and content services. In industries based on chemicals, appropriation of returns to innovation investments tends to rely on patenting to a greater extent than other industries. In complex-technology industries such as many fields of electronics, the reasons for patenting emphasize subsequent licensing negotiations and prevention of lawsuits rather than simply preventing copying of the invention (Cohen et al., 2000). Cohen, Nelson, and Walsh (2000) call this strategic patenting. Because products such as computers and mobile phones consist of hundreds of patented components, firms are forced to negotiate with rivals for rights to use their technologies (Hall and Ziedonis, 2001). To gain beneficial licensing terms, it helps to have a sizeable patent portfolio to cross-license in return.

Information and communication technology (ICT) industries also feature technical standards that accentuate the need for access to patents and cross-licensing. A standardized communication system such as wireless telecommunications may involve thousands of patented inventions that are essential for its functioning. Intellectual property experts at RPX Corporation have estimated that there are 250,000 active smartphone patents granted in the United States.[1] No single network or terminal (handset) vendor can possibly develop all these components and technologies in-house. Moreover, to enable interoperability across vendors, interfaces and communication protocols must be shared and standardized. In many cases, such protocols involve proprietary technologies protected with patents. Consequently, just to enter an industry, a handset innovator (say, Apple in the early 2000s, Research In Motion in the late 1990s, or Nokia in the early 1990s), has to negotiate licenses with incumbents. It has been estimated that the average licensing cost of a 3G ('third-generation') mobile phone is about 13 per cent of its price (Bekkers and Martinelli, 2012). According to these authors, one single patent holder (Qualcomm, which has very fundamental patents related to wireless transmission between terminals and base stations) is thought to receive fully 5 per cent of the handset purchase price through licensing agreements. For incumbents with large patent portfolios of relevant patents, this licensing cost is low because they can cross-license with similarly endowed rivals. In contrast, an entrant with a limited portfolio of relevant patents may actually have to pay the full licensing price. This forms a high barrier, essentially functioning as a significant tax on entry. Emerging innovators thus need to be very strategic about how to enter and what kind of intellectual property rights they can generate to support their competitive strategy. On the other hand, incumbents in such situations can relatively easily utilize their legacy patents to increase costs for competitors or keep them out altogether.

Another highly contested new development in intellectual property is around computer software. Since 1995, the patenting of software and (usually software-based) business methods has been allowed in the United States, whereas in Europe software can only be patented if it is part of another invention. Traditionally, software inventions are governed by copyright, in which case the expression of the code is protected but not its functionality. Patenting in software has been quite controversial, because patents are argued to often be of low quality (cf., Hall et al., 2003). Moreover, software technologies are highly cumulative, the combination and recombination of existing components is essential for efficient software production, and technical standards also feature prominently (Hall and MacGarvie, 2010). Software patents are thus likely to generate many of the strategic complications that are found in complex-product industries. Indeed, Hall and MacGarvie (2010) find that the stock market reaction to the expansion of software patentability was initially mostly negative, and, subsequently, the impact of software patents and patent citation stocks on firms' stock market values has been insignificant for most firms and types of patents. Based on these authors' results, it is difficult to argue that software patents have substantially increased innovation and technical progress in software. Furthermore, software patents have led to a number of legal disputes, including conflicts between patent holders and open-source software projects.

In creative industries, the rise of the Internet has made it incredibly cheap and easy to distribute large amounts of digitized information instantaneously. As a result, digital piracy has emerged and grown into a tidal wave, threatening the profits of creative content industries such as music, books, newspapers, and audio-visual entertainment (TV, movies, etc.). Policy-makers and law-makers in different countries have responded to the calls for reform from industry lobbyists, developing increasingly stringent mechanisms to catch copyright infringers and shut down the services and technologies that aid digital spillovers. It can be argued that in some of these efforts, policies have lost sight of the forest for the trees. It has been difficult, for example, to demonstrate that digital piracy has decreased creativity in music (Waldfogel, 2011), although it is clear that piracy affects the returns to distributors of music (and, more generally, content).

Indeed, Waldfogel (2011) argues that policy-makers should pay attention to maximizing the total welfare of the economy that includes both consumer surplus and producer surplus, rather than focusing on the latter alone.

More generally, the welfare cost of intellectual property rights systems is that they create temporary monopolies. The societal benefits of intellectual property rights arise from cumulative innovation and its dissemination. The goal of intellectual property rights policy should be to minimize the costs and maximize the benefits, rather than propping up old market power structures. It has been argued that the current intellectual property rights system is broken and needs fundamental readjustment (Jaffe and Lerner, 2004). The complex legal issues are further aggravated if interoperability standards are involved, because the values of patents tend to be exceptionally high in those settings. To resolve these tensions, scholars have even proposed systems that directly support creation and dissemination rather than the monopolistic exploitation of new knowledge (Wright, 1983; Gallini and Scotchmer, 2002). As long as intellectual property rights exist in their current form, however, it is essential for innovators to develop skills in navigating these systems, because they are complex, costly, increasingly litigious, and potentially lethal for firms, but may offer great financial rewards for organizations with successful strategies.

Intellectual Property and Innovation Strategy

Considering that intellectual property rights are legal methods, learning to properly use them may require substantial legal expertise. Mowery, Sampat, and Ziedonis (2002) describe how universities began to learn to patent after the Bayh–Dole Act of 1980. The patents of universities that did not patent before Bayh–Dole were initially less important than their peer group that did already patent, but they have caught up over time and closed the gap. Largely through knowledge spillovers from their peers, these universities appear to have developed procedures to recognize which inventions are worth patenting and learned to write more powerful patent documents. Smaller or younger organizations, similarly to universities new to patenting, tend not to have such expertise in-house, and may need to access external service providers.

In addition to not knowing how to write patents and which inventions to patent, a common pitfall is assuming that it is always important to maximize the protection of intellectual property. In reality, protection is costly, and hence most firms will need to be selective about which patents to apply and enforce, and which ones to let expire. In the United States, for example, a patentee must pay additional and escalating fees after three and half, seven and half, and eleven and half years in order to keep a patent in force, whereas in Europe, a patent holder needs to pay annually after two years to maintain their patent. Second, intellectual property rights are control rights that also are associated with incentives. When cooperating with external partners, IP holders are wise to consider when strong IP enforcement is strategically beneficial and when it might be detrimental for continued innovation. For example, engaging in a joint R&D project with an external partner but at the same time strictly appropriating and enforcing any and all IP resulting from or related to such joint work may reduce the partner's incentives for inventive effort in the project (Lerner and Merges, 1998; Leiponen, 2008).

Similarly, firms in network markets such as communication, software, electricity, or transportation networks can often be lenient with their IP licensing terms when they see that they will benefit from others building on or extending the underlying technologies. In this case, the original innovators will benefit mostly from complementary assets related to the broader technical system surrounding the core licensed invention, rather than maximizing intellectual property protection at the level of individual inventions. These 'platform' strategies are extensively discussed in Chapter 32 by Gawer and Cusumano.

Numerous studies suggest that firms in different industries and contexts tend to utilize appropriation methods differently. Appropriation strategies, for example, vary according to the type of innovation (product vs. process); process of innovation (cooperation vs. in-house); firm size; and industrial context (high tech vs. low tech; complex vs. discrete products; services vs. manufacturing). These studies are typically descriptive and assume that a firm would first decide on an innovation strategy, determining, for example, how much to invest in R&D and how to organize the activity; what kinds of innovations to attempt to develop, and in what industrial context to operate (cf. Harabi, 1995; Arundel and Kabla, 1998; Cohen et al., 2000; Arundel, 2001; Tether and Massini, 2007; Leiponen and Byma, 2009). Only after selecting an innovation strategy would the firm choose how to protect the emerging innovations.

One can also imagine, however, that there may be an earlier decision stage that precedes the first stage as described above. Considering the competitive situation, and the available appropriation methods, a firm might choose to organize R&D in a particular way, or decide not to invest in the first place due to low expected returns, or attempt to develop certain types of products or technologies for specific uses or markets in part based on the appropriation considerations. This viewpoint is embedded in Teece's (1986) classic complementary assets framework, where the optimal strategy follows from an assessment of imitability and complementary assets conditions. Gans and Stern (2003) extend this framework to startup innovators, and some empirical evidence is provided by Gans, Hsu, and Stern (2002), suggesting that firms may choose their market strategies based on appropriability conditions.

A related idea is found in the literature on the markets for technology (Arora, Fosfuri, and Gambardella, 2001; and Gambardella, Giuri and Torrisi in Chapter 12). In this stream of research it is argued that enhanced appropriability of returns to R&D investments, for example due to pro-patent policies in the United States since the mid-1980s, improves the conditions for technology trading. When appropriability is weak, firms are reluctant to trade in patents, because their market value is unlikely to reflect the full value of the technology due to market inefficiencies. Under these circumstances, firms may prefer to embed their valuable knowledge assets in organizational processes and tacit knowledge rather than patents that require substantial disclosure of the underlying knowledge. Then the market for knowledge requires transacting whole organizational units rather than disembodied technology. In contrast, when the market for technology is more transparent and efficient, it becomes easier to trade in patents. Thus, the appropriability regime is likely to influence actors' appropriation strategies, including their organizational choices of commercialization, or business models (cf. Gans and Stern, 2003; Teece, 2006).

Business models are even more complex where digital content is concerned. Digital communication technologies have rendered appropriation of returns to creative content extremely challenging and, therefore, inventors or creators are forced to try to capture returns to investments through informal organizational strategies rather than formal legal strategies. The case of the music industry illustrates the challenges faced. A pop musician used to work with record labels to develop their content and record it. Record labels used to take care of production, distribution, marketing, and contract management for the musician, in exchange for a lucrative share in the revenue streams (Fisher, 2004). The recording artist used to assign a sound-recording copyright to the record company and, if the artist was also the composer of the music, he or she would assign musical-work copyright to a music publisher who would grant a mechanical license to the record company. According to Fisher, before the Internet, record companies used to retain about 28 per cent of the retail price of a compact disc to develop repertoire, market the album, cover the administrative overhead, and generate a profit (Fisher, 2004: 19). However, as album sales have shrunk in the past ten years, in part because of digital piracy, but also because of increased competition from new types of digital entertainment for people's time and attention (Peitz and Waelbroeck, 2004), these revenue streams are in danger.

From most musicians' perspective, the Internet has not changed much. A large majority of musicians never actually made a living from albums but from touring, and in fact, the Internet may have made them more competitive because of the lower cost of advertising. Superstars are a different story, and their response has been more varied. Some of them have embraced free (even illegal) distribution on the Internet and use it to market their new albums and put an increased effort in touring. Others have signed new types of '360°' contracts with their record companies, which enable record companies not only to take a cut from the album sales, but also from touring and merchandise. While this does not seem to make sense for established stars, for beginner musicians, help from the experienced record company staff in developing, producing, and marketing tours and merchandise probably adds value. On the other hand, some musicians have left record labels altogether and started to experiment with self-recording and publishing. This is enabled by the digital transformation of recording and editing that has dramatically reduced the cost of music production, as well as distribution.

From the record companies' perspective, even with the new types of contracts, the situation seems dire. As the power of copyright has weakened, they have resorted to massive litigation campaigns against infringers, both individual users and intermediaries such as Napster, Grokster, and most recently Pirate Bay and Megaupload. Additionally, record companies have sought to limit the availability of their content on the Internet to a select few sources. Their concern is that, independent of the Digital Rights Management (DRM) technology used, any content files that are released to the Internet are likely to be 'cracked' and made available for millions of downloads. Although DRM technologies have rapidly evolved over the years, it has proven impossible to come up with secure digital formats. For these reasons, record companies have tended to rely on closed systems such as Apple's iTunes music store and its proprietary content format combined with iPod/iPhone/iPad terminals. Whereas in the short term this might seem reasonable, it is probably not a wise long-term strategy to concentrate all sales through one single proprietary system controlled by a large outside partner.

For years, record companies have shunned 'webcasters' such as Rhapsody or Pandora internet radio stations, perhaps because they have seen themselves as being in the business of music *sales*, not streaming and licensing. Before the 2008 Webcaster Act settlement in the United States, SoundExchange—the organization in charge of collecting webcaster royalties for the Recording Industry Association of America (RIAA)—attempted to collect double the amount of hourly royalties from such Internet radio stations compared to satellite radio stations. At the same time, terrestrial-only radio stations were paying no such royalties. This strategy is contraindicated by the observation that the patterns of consumption of popular music among younger people are changing, whereby music tastes are becoming more fragmented and shift more frequently (cf. Molteni and Ordanini, 2003). As a result, the superstar-obsessed recording-industry business model may not be coming back, even if copyright enforcement is made substantially stronger and easier than currently, as the industry attempted in the proposed Stop Online Piracy Act of 2012 in the United States.

Because of its central position in youth culture, the music industry was the first to experience the erosion of revenue due to the changing patterns of digital consumption and piracy. Later other digitizable content industries such as movies, books, newspapers, and television followed suit. Thus, a continuing experiment is seeking to figure out how to redesign business models to continue to capture value from digital content. The earlier approach that involved using aggressive legal tactics by suing individuals who downloaded illegal copies of music clearly failed. Hilary Rosen, the former CEO and Chairman of the Recording Industry Association of America, the main recording industry group in the United States, wrote in 2005 just before the Grokster decision of the United States Supreme Court:

> While the victory of whoever wins maybe important psychologically, it just won't really matter in the marketplace.…Each time there is a successful enforcement they [illegal downloading services] reinvent themselves.…Allow future innovation to stimulate negotiation and not just confrontation. Consumer is left with a few legitimate services that offer some great content and lots more illegal [peer-to-peer] choices that offer ALL the content plus spyware, bad files and unwanted risk. These are not legal decisions or trade association PR responsibilities. They are fundamentally business issues that must be addressed in the marketplace.
>
> (Rosen, 2005)

The search for a working business model and an organization structure that enables musicians and their promoters to make a living and be able to invest in their creative activities thus continues. It is not clear to what degree the large record companies of the past will be part of this future. The digital content case, however, vividly illustrates the strong connection between intellectual property rights regimes and firms' organizational strategies to profit from creative investments, and this applies across a range of industries.

Intellectual property litigation is by no means limited to the realm of the copyright. Since patents are defined as legal rights to exclude others from using an invention, their enforcement may involve suing parties seen to infringe on the intellectual property. There is evidence that patent litigation has been increasing in recent years, even relative to the increasing number of patents granted (Bessen and Meurer, 2005a). This is aligned with the view that the US patent regime has been stronger since the launch of the specialized Appellate Court in 1985, thus making US patents more valuable (Hall and Ziedonis, 2001; Henry and Turner, 2006). Due to the globalization of markets for high-technology products, however, the strengthening of the patent regime appears to have spilled over to the European market, and perhaps to Asia, too. Furthermore, in many ICT industries, patents have become more integrated in interoperability standards, magnifying the value of the key essential patents (Bekkers et al., 2002).

The decision to litigate patents has been analysed from both strategic and economic perspectives. Somaya (2003) examines firms' decisions not to settle patent disputes in computer and pharmaceutical industries. He finds that greater strategic stakes—the potentially asymmetric values of the underlying invention to the two parties—significantly increase the likelihood that the dispute goes all the way to judgment in Court. Moreover, more important patents (measured by forward citations to them in subsequent patents) are likely to be selected for litigation.

Litigation behaviour also varies with firm and portfolio size. Cohen, Nelson, and Walsh (2000) find that perceptions of the effectiveness of patents as a protection method increases with firm size and business unit size. This may partly derive from their legal resources, as Cohen, Nelson, and Walsh also find that among the reasons *not* to patent, the cost of defending patents was less important for larger firms. On the other hand, Lanjouw and Schankerman (2004) analyse patent suits and settlements and find that patents owned by individuals and firms with smaller portfolios are *more* likely to file infringement suits, presumably because their intellectual property is more at risk of infringement by larger rivals. These authors also find that the portfolio effect is significantly greater for firms smaller than the median in number of employees, suggesting that these firms cannot rely on repeated interaction with rivals to support cooperation. This idea is also reinforced by the finding that settlement is more likely in concentrated technology areas where the firms are likely to frequently deal with the same peers over time.

In summary, different forms of intellectual property rights support different types of innovations. Whereas copyrights have recently become a weaker form of protection due to the digital transformation of content industries, patents have become stronger in terms of being more likely to be validated by courts, particularly in the United States. Both main forms of IP are increasingly enforced through litigation. Indeed, these legal forms of protection for innovations may be worthless unless the innovator has sufficient funds and expertise to pursue such enforcement. Obtaining a high-quality patent is costly, but its litigation is many more times costlier. According to data cited by Bessen and Meurer (2005b), in patent lawsuits with $1–25 million at stake, the median total cost of litigation is about $2 million.

STANDARD SETTING IN INNOVATION STRATEGY

Technical interoperability standards are very common in systemic technologies. Technical systems consist of many components or products. For example, computers consist of components such as memory, central processing unit, microprocessor, keyboard, screen, and so on. Wireless telecommunication systems consist of terminals, base stations, routers, and cables that connect them to fixed-line telecommunications and Internet communication systems. When multiple complementary components are required to make a product or a service functional, competitive and intellectual property rights strategies can become extremely complex.

The complementary assets framework of Teece (1986) suggests that imitability and access to complementary product or service providers determine which parties in an industry or value chain are likely to profit from an innovation, and which strategies are likely to be the most successful. As acknowledged by Pisano and Teece twenty years later (2007), however, this view is rather static in that it ignores the opportunities of firms to actively influence the industry 'architecture', that is, the nature and constellation of relevant assets, their ownership and other rights, the number and characteristics of competing organizations in the different layers of the value chain, and the relationships among the assets and among the organizations.

Firms might, for example, be able to influence the appropriability conditions in their industry with selective investments in either weakening or tightening the IPR regime. Firms might also be in a position to strategically weaken the IPR regime by placing intellectual property in the public domain, or tighten the regime by aggressively enforcing their IPRs and by lobbying law-makers. In the software industry, there are many examples of commercial firms donating code to open-source software projects, usually in order to foment competition in an adjacent industry. On the other hand, efforts by the record and movie industry associations in the United States to lobby Congress to tighten copyright laws and enforcement mechanisms are strategies to tighten appropriability.

Firms are also potentially capable of influencing the constellation of complementary assets in their industry. Simply by vertically or horizontally integrating into complementary activities may substantially alter the competitive dynamic. Depending on their implementation, such integration decisions change the number of rivals in relevant industry layers and hence have an impact on the degree and nature of competition.

Slightly subtler ways to influence competitive conditions in these complementary markets include corporate venture capital investments into emerging or adjacent markets and efforts to develop compatibility standards.

Technical compatibility standards can emerge from 'ex post' market competition or 'ex ante' cooperative standardization efforts, although in practice the market and committee processes often take place (at least partly) simultaneously. Market-based standardization is usually called 'de facto' standardization whereas committee-based or government-mandated standards are called 'de jure', even though many committee-based standards are not legally mandated. Leiponen (2006) discusses the variety of organizational arrangements to develop compatibility standards in the wireless communication industry. The organizational forms for standardization in much of communication and computer technologies appear to have converged towards the middle ground of industry associations, alliances, and consortia, rather than the extremes of purely market-based or government-mandated standardization. This can be seen to follow from the reduction of government involvement in standard setting in most parts of the world (hence fewer government-imposed standards have been created), and from the realization that de facto standardization can be extremely costly when the competition is winner-take-all and when consumers are reluctant to engage without clear sign that their sunk investments are not going to be 'stranded' on the losing side of the standard battle. Although cooperative standardization may be a slow process (cf. Farrell and Saloner, 1988; Simcoe, 2012) it may subsequently accelerate the market take-off and reduce the investment risk for all parties.

The implications of de facto vs. de jure standardization, and ex post vs. ex ante standards, were highlighted in the drawn-out battle for the domination of high-definition DVD player and content format markets. Similarly, the costliness of multiple competing standards in the marketplace was exemplified by the slow take-off of the second-generation mobile telephony in the United States, as compared with the European market where a unified standard (GSM) was adopted ex ante. These cases have demonstrated to ICT companies the risks associated with standard battles and the appeal of the alternative route to market domination through cooperative (committee-based) standardization. This mode of standardization, however, is no less competitive or less marred with thorny intellectual property issues than is the market-based mode.

Overall, committee-based standard setting emphasizes consensus decisions and collaborative drafting of technical specifications (technical features of the standard) that tend to take time to complete. Formal standard-setting organizations such as IEEE (Institute of the Electrical and Electronics Engineers), IETF (Internet Engineering Task Force), or 3GPP (Third Generation Partnership Project) are open for all interested parties to attend and participate, but in practice, influencing the specification development process requires substantial expertise, negotiation skills, and long-term commitment to the process (Spring et al., 1995). Cooperative standardization takes a lot of time and effort; the development of the third-generation mobile communication system UMTS at 3GPP took about a ten-year investment on the part of the major players in this industry. Some of the companies centrally involved in this effort, such as Nokia and Ericsson, had dozens of engineers working on the standard full time and travelling to committee meetings in different parts of the world every few weeks. This represents substantial investments of human resources and money, but without a seat at the table, firms risk being left out of critical information sharing about the future technological and commercial basis of their industry.

In addition to the formal standard-setting investments described above, most firms involved in communication and computer industries participate in several informal technical or marketing consortia (Chiao et al., 2007). In these consortia, firms research, discuss, and test emerging technical solutions and the rapidly evolving end-user market. Some industry consortia are simply fora for information exchange and coordination of marketing, whereas others actually involve coordination of technical solutions and subsequent submission of standard specifications for formal standard-setting bodies such as 3GPP or IEEE. Industry consortia are not always open for all comers, and they often have tiered membership structures, whereby a greater ability to influence the technology recommendations is associated with significantly higher membership fees. Nevertheless, as suggested by Leiponen (2008) and Delcamp and Leiponen (2012), participation in informal consortia in the mobile communication industry substantially increased firms' abilities to influence formal standard specification in 3GPP and invent technologies that are cited by subsequent essential patents for the UMTS standard. Informal consortia were thus central in the process of innovating and collaboratively developing this standard. A third paper suggests that different consortium strategies yield different types of power in standard setting. Access to information appears to be related to technical power to draft new features whereas access to many peers appears to be related to administrative power to make changes to specifications in progress (Leiponen and Ter Wal, 2012).

Participation in formal and informal standardization organizations can thus greatly benefit the member firms. They learn earlier and in more depth about their rivals' innovation strategies regarding the technical system to be standardized, and can connect with complementary technology input providers (Bar and Leiponen, 2012). They can better align their own innovation, competitive, and organizational strategies with expected outcomes from the committee process. Firms also have opportunities for inserting their own intellectual property in standard specifications, thus potentially forcing rivals to pay royalties for building and selling products that refer to the standard. Rysman and Simcoe (2008) analyse the effects of disclosure of 'essential' patents to standard-setting organizations on subsequent citations by other patents and notice a significant increase after disclosure. This suggests that the value of patents is substantially enhanced by their declaration as potentially required for the implementation of the standard. This (mandatory) disclosure, however, does not necessarily mean that the patents in question actually are essential for the standard. The set of patents that are actually licensed and the basis of royalties from rivals are determined through bilateral negotiation and legal enforcement.

Finally, participants in standards committees, if they have the necessary skills and expertise, may position themselves to actively influence the standard by proposing and developing features highly desired by their own customers or aligned with their own complementary assets. There may also be indirect financial and competitive benefits from standardization activities. Waguespack and Fleming (2009) find that participation in formal standard setting may enhance startup innovators' opportunities for successful 'exit' events. As a whole value chain may participate in the standard-setting process, upstream firms might also find opportunities to advertise their expertise or products by working with their clients in committees that draft new features.

The intersection of standards and patents is most clearly highlighted through the licensing terms that are negotiated for so-called essential patents. Many standard-setting organizations adhere to a policy of requiring members to license under Fair, Reasonable, and Non-Discriminatory (FRAND or RAND) terms. Because actual licensing terms are private and contain highly strategic information about bilateral negotiations, however, there are no guarantees that these terms actually are fair, reasonable, or non-discriminatory. In fact, judging by the information from wireless communications that has leaked to the public domain over the years, it seems clear that terms are often unfair and discriminatory (i.e. different industry parties get treated very differently), and occasionally unreasonable, in which case patent holders have been sued in courts. For instance, in 2006, Nokia sued Qualcomm arguing that its licensing terms were unreasonable, and indeed requested the US Court of Chancery of the State of Delaware to define what FRAND means.[2] Unfortunately, this case was settled out of court, as most such patent disputes are, and the community still does not know what FRAND means! Nokia argued in its pre-trial brief for the continued dispute with Qualcomm in 2008, however, that 'Fair and Reasonable' should take into account the contribution of the disputed patents to the standard and the appropriate total royalties for the standard. This is also what Nokia proposed, and Qualcomm resisted, in the contemporaneous ETSI (European Telecommunication Standards Institute) review of IPR policies.

Uncertainty surrounding essential patent licensing terms and their legal implications is a major problem for many ICT industries. Bekkers and West (2009) have reviewed the discussion within ETSI to revise the treatment of IPRs in communication standards, and noted that each proposal to improve the transparency and non-discriminatory nature of licensing terms was blocked by some industry party or another, with the major conflict between firms who both invest in R&D thus generating IPRs and manufacture communication equipment (such as Nokia and Ericsson) and firms who only invest in R&D to license IPRs without building products (such as Qualcomm and InterDigital). As a result, the IPR policy related to ETSI communication standards was revised only marginally, and the major issues persist.

One of the proposals in the ETSI discussion and elsewhere has focused on requiring members of a standardization organization to disclose their IPR licensing terms ex ante, that is, before the selection of the standard (Updegrove, 2006; Bekkers and West, 2009). This idea was also supported by the United States Federal Trade Commission (2011), which interpreted it to enhance IPR price competition rather than to represent collusion on the part of the IPR suppliers or buyers.

No intellectual property rights reform is likely to be appealing to all the parties involved in the current ICT industry structure, and hence it will be impossible to change the system endogenously. Unless American or European regulators are willing to step in, the situation is therefore likely to continue. The fundamental root cause is the strengthening of patent policies over the past two decades, and such property rights are likely to be upheld no matter what rules or negotiations are imposed by standard-setting organizations themselves. Only very rarely have regulators interfered with the IPR system in place, such as in the case where Rambus was seen to intentionally mislead its fellow JEDEC (Solid State Technology Association) members about its IPR status and licensing terms. Even here, an administrative court and competition authorities disagreed about the necessity of action (Federal Trade Commission, 2006).

In summary, integration of innovation, IPR, and standardization strategies to optimize influence on the industry architecture can generate substantial profits for innovators, small and large. This does not necessarily mean maximizing IPR protection (appropriability), because sometimes proliferation of an invention can be very beneficial, particularly for firms holding market power in complementary markets. Working through cooperative standard-setting organizations to influence the industry architecture and associated technological trajectories has become very commonplace in ICT industries, because these coordinated processes reduce the uncertainty around the standardization outcome for both suppliers and end-users. As a result, the risk of being stranded with sunk investments is mitigated, and the probability and speed of market take-off is enhanced. Cooperation in standard setting, however, does not mean that firms are not fiercely competing to dominate either the upstream markets for technologies and licensing or the downstream markets for products and services. Indeed, the 'component competition' (Farrell et al., 1998) that results from widely adopted open standards tends to have leaner margins than that resulting from closed and controlled standards. Firms are competing not only on standardized products and price (their piece of the 'pie') but on innovation and on evolving the industry and its standards in a direction that is the most beneficial for themselves (the size and shape of the 'pie').

RECENT DEVELOPMENTS IN INTELLECTUAL PROPERTY AND STANDARDS

Globalization of technology competition has in part made patenting an essential part of innovation strategy in ICT industries and chemicals. This is reflected in the growth of patent applications over the years since the mid-1980s. According to the World Intellectual Property Office, patent filings have doubled between 1995 and 2010.[3] The most dramatic growth was experienced by patent offices in the United States, Japan, South Korea, and China, with the European Patent Office handling a relatively more modest rate of growth in applications. Filings in Korean and Chinese patent offices now clearly exceed those in the European Patent Office. Patent filings in individual European countries including France, the United Kingdom, and Germany have been flat in comparison. Technologically, growth is largely concentrated in digital communications, which now exceeds pharmaceuticals in terms of global patent filings. It is also the main area of patent and standardization disputes. Rapid growth is also observed in the smaller technology fields (in terms of international patent classes) of IT methods for management (business models) and micro-structural technology and nanotechnology. For the regional patent offices, this rapid growth of patent applications has presented tremendous challenges, compounded by the fact that new technology areas such as software and business models in which patent offices traditionally have no expertise have been included within the realm of patent protection. It is argued that, together, these reasons have led to a decline in patent quality (Wagner, 2008).

In the constant debate around patentability, gene patents have been an area of controversy for decades. It is estimated that there are about 47,000 patents on genetic inventions and between 3,000 and 5,000 human gene patents in the United States. Some people view patents on human genes as unethical because genes are viewed as our common heritage rather than something that can be owned. Another concern is that gene patents may hinder the progress of scientific research on genetics. The Myriad Genetics breast cancer gene patents, for example, were feared to prevent scientific laboratories from doing research in that space (Paradise, 2004). As a consequence, these patents are ignored in Canada and restrictions on their applicability have been created in some European countries (Cook-Deegan, 2012).

International differences in intellectual property protection have also given rise to heated debates and lengthy international negotiations through the World Trade Organization (Trade Related aspects of Intellectual Property Rights, TRIPS). It has been argued that, although patenting is on the rise in emerging economies, these economies tend not to enforce intellectual property as vigorously as technology companies from industrialized countries would like (IIPA, 2009). This will influence the innovation and competitive strategies of such companies entering emerging economy environments. Firms' appropriation strategies are likely to rely on secrecy and design complexity under conditions of weak patent enforceability (Zhao, 2006).

In ICT industries, patents and compatibility standards have become the centre of much litigation and uncertainty about the future of technological trajectories, which in the aggregate and in the long term cannot be beneficial for innovators, manufacturers, service providers, or end-users. Simcoe, Graham, and Feldman (2009) document the impact of essential patent disclosure on litigation by small and large innovators. They suggest that standard-setting processes influence the incentives for activating legal strategies for patent holders. This is to be expected as the value of patents increases after disclosure (as documented by Rysman and Simcoe, 2008), and therefore the benefits from litigation are also magnified. They find, however, that smaller firms' probability of filing a lawsuit increases after disclosure, whereas large firms' probability of doing so remains the same. The differential rates of litigation for small and large firms lend support to the argument that firms that primarily profit from building products behave quite differently from firms that primarily profit from developing and licensing IPRs. Whereas strong patents and markets for technology have spurred many important innovations, the ability of innovators to hold up other firms commercializing complementary technologies can be detrimental to social welfare. However, the net effect has not been properly modelled or empirically quantified in the scholarly literature.

Another relatively recent aspect of IPR strategies has to do with so-called non-practising entities. As is clear from the discussion of standards-related litigation above, firms with manufacturing activities in a technology area have significantly different strategic drivers than firms with only R&D and intellectual property rights commercialization activities in that area. Moreover, a third business model has emerged that only concerns intellectual property rights acquisition, management, and commercialization—most often through litigation. Such entities are called patent trolls, somewhat derogatorily, but also more recently, patent assertion entities (Federal Trade Commission, 2011).

Although upstream R&D providers can also be very active in litigation, Layne-Farrar (2012) argues that a clear difference can be identified between these R&D-focused non-practising entities and the patent-assertion entities. R&D specialists can be, and often need to be, aggressive in the market for technology, but they are nevertheless engaged in a repeated game with manufacturing ('practising') entities, which keeps them focused on long-term performance rather than (only) on short-term hold-up strategies. They should thus generally be cooperative with their manufacturing counterparts, in order to get their technologies proliferated in the marketplace. Patent assertion entities, in contrast, acquire and litigate patents, solely in order to maximize short-term returns on their monopoly power.

Bessen, Ford, and Meurer (2011) suggest that, in the aggregate, patent assertion entities represent a welfare loss to society. These authors' study of lawsuits with non-practising entities as plaintiffs indicates half a trillion dollars of lost wealth to defendants between 1990 and 2010. Moreover, Bessen, Ford, and Meurer argue that 'loss of incentives [to innovate] to the defendant firms is not matched by an increase in incentives to other inventors'. Opponents of this view, however, suggest that non-practising entities make markets for technology more liquid, thus enhancing economic welfare (e.g. McDonough, 2006). However, the sheer cost of litigation is a large part of the welfare transfer away from the defendants, and this is only increasing lawyers' welfare.

An unresolved issue related to these novel intellectual property strategies concerns patents that are essential to a standard. In 2011, Nokia sold a subsidiary (Core Wireless) that was holding hundreds of its patents essential for 3G and LTE wireless telecom standards to a patent assertion entity called Mosaid. The deal will apparently allow Mosaid to pursue litigation and licensing strategies of the Core Wireless patents, but the returns to such activities will be shared among Mosaid, Nokia, and its alliance partner Microsoft. However, it is not known how courts will deal with standards-related patents litigated by patent trolls. The Rambus case suggests that deceitfully holding up industries that commercialize technologies associated with open standards will not be tolerated, but perfectly legal patent acquisitions and reassignments have not yet been dealt with. Competition agencies have also taken interest in IP licensing related to cooperatively standardized technologies.[4] Moreover, the recent America Invents Act of the United States makes it more difficult for patent assertion entities to pursue multiple defendants in a single lawsuit, thus increasing their costs of litigation. Layne-Farrar (2012) suggests this Act could also curb damages awarded to patent assertion entities.

Many of the aforementioned debates, laws, and lawsuits are closely related to the licensing terms for standardized technologies. If there was a way to transition into ex ante commitments to maximum royalties, and scope for legally enforcing them, even after the associated IP has been acquired by a third party, many of these wasteful lawsuits could perhaps be avoided. There are very powerful new tactics at play in the IP litigation sphere, and regulators, law-makers, and courts are currently taking a close look at them to weigh their benefits and costs to society. Fortunately, these new developments also provide many fascinating opportunities for theoretical and empirical research on the strategic management of intellectual property rights and standard setting.

NOTES

1. Financial Times, 30 March 2011. Retrieved from <http://www.ft.com/cms/s/0/b0da8540-5 aea-11e0-a290-00144feab49a.html#axzz1uKgPiqsm> on 8 May 2012.
2. See A. Updegrove (2006): 'Delaware Court is Asked To Define "FRAND": Another Reason to Call for Ex Ante'. Retrieved from <http://www.consortiuminfo.org/standardsblog/ article.php?story=20060813123034215> on 28 March 2012. Pre-trial brief retrieved from <http://www.nokia.com/NOKIA_COM_1/Press/Legal_News_(IPR_news)/IPR_News/ Latest_News/Redacted-Public_Nokia_Opening_Pretrial_Brief.pdf> on 28 March 2012.
3. Data retrieved from <http://www.wipo.int/ipstats/en/statistics/patents/> on 9 May 2012.
4. Reuters: EU regulators investigate Samsung over mobile patents (31 January 2012). Retrieved from <http://www.reuters.com/article/2012/01/31/us-eu-samsung-idUSTRE80U0NU20120131> on 3 April 2012.

REFERENCES

Arora, A., Fosfuri, A., and Gambardella, A. (2001). *Markets for Technology*. Cambridge, MA: MIT Press.

Arundel, A., (2001). 'The Relative Effectiveness of Patents and Secrecy for Appropriation', *Research Policy*, 30: 611–24.

Arundel, A., and Kabla, I. (1998). 'What Percentage of Innovations are Patented? Empirical Estimates for European Firms', *Research Policy*, 27: 127–41.

Bar, T., and Leiponen, A. (2012). 'Committees and Networking in Standard Setting'. Unpublished manuscript.

Bekkers, R., and Martinelli, A. (2012). 'Knowledge Positions in High-tech Markets: Trajectories, Standards, Strategies and True Innovators', *Technological Forecasting & Social Change*, 79: 1192–1216.

Bekkers, R., and West, J. (2009). 'The Limits to IPR Standardization Policies as Evidenced by Strategic Patenting in UMTS', *Telecommunications Policy*, 33: 80–97.

Bekkers, R., Duysters, G., and Verspagen, B. (2002). 'Intellectual Property Rights, Strategic Technology Agreements and Market Structure: The Case of GSM', *Research Policy*, 31(7): 1141–61.

Bessen, J., and Meurer, M. J. (2005a). 'The Patent Litigation Explosion', Boston University School of Law, Working Paper No. 05-18.

Bessen, J., and Meurer, M. J. (2005b). 'Lessons for Patent Policy from Empirical Research on Patent Litigation', *Lewis & Clark Law Review*, 9(1): 1–27.

Bessen, J., Ford, J., and Meurer, M. (2011). 'The Private and Social Costs of Patent trolls', Boston University School of Law, Working Paper No. 11-45. Available at <http://ssrn.com/abstract=1930272> (accessed 10 March 2012).

Chiao, B., Lerner, J., and Tirole, J. (2007). 'The Rules of Standard-setting Organizations: An Empirical Analysis', *RAND Journal of Economics*, 38(4): 905–30.

Cohen, W. M., Nelson, R. R., and Walsh, J. P. (2000). 'Protecting Their Intellectual Assets: Appropriability Conditions and Why U.S. Manufacturing Firms Patent (Or Not)', NBER Working Paper 7552.

Cook-Deegan, R. (2012). 'Gene Patents. The Hastings Center Bioethics Briefing Book'. Available at <http://www.thehastingscenter.org/Publications/BriefingBook/Detail.aspx?id=2174> (accessed 11 May 2012).

Delcamp, H. and Leiponen, A. (2012). 'Innovating Standards Through Informal Consortia: The Case of Wireless Telecommunications'. NBER Working Paper.

Farrell, J. and Saloner, G. (1988). 'Coordination through Committees and Markets', *RAND Journal of Economics*, 19(2): 235–52.

Farrell, J., Monroe, H. K., and Saloner, G. (1998). 'The Vertical Organization of Industry: Systems Competition versus Component Competition', *Journal of Economics and Management Strategy*, 7(2): 143–82.

Federal Trade Commission (2006). 'FTC Finds Rambus Unlawfully Obtained Monopoly Power'. Press release available at <http://www.ftc.gov/opa/2006/08/rambus.shtm> (accessed 4 March 2012).

Federal Trade Commission (2011). 'The Evolving IP Marketplace, Aligning Patent Notice and Remedies with Competition'. Report issued in March 2011.

Fisher, W. W. III (2004). *Promises to Keep: Technology, Law, and the Future of Entertainment*. Palo Alto: Stanford University Press.

Gallini, N., and Scotchmer, S. (2002). 'Intellectual Property: When is it the Best Incentive System?' in A. B. Jaffe, J. Lerner, and S. Stern (eds), *Innovation Policy and the Economy*. Cambridge, MA: MIT Press, 51–77.

Gans, J., and Stern, S. (2003). 'The Product Market and the Market for "Ideas": Commercialization Strategies for Technology Entrepreneurs', *Research Policy*, 32: 333–50.

Gans, J. S., Hsu, D., and Stern, S. (2002). 'When does Start-up Innovation Spur the Gale of Creative Destruction?' RAND Journal of Economics, 33(4): 571–86.

Griffith, R., Redding, S., and Van Reenen, J. (2004). 'Mapping the Two Faces of R&D: Productivity Growth in a Panel of OECD Industries', *Review of Economics and Statistics*, 86(4): 883–95.

Hall, B., and MacGarvie, M. (2010). 'The Private Value of Software Patents', *Research Policy*, 39: 994–1009.

Hall, B., and Ziedonis, R. (2001). 'The Patent Paradox Revisited: An Empirical Study of Patenting in the U.S. Semiconductor Industry, 1979-1995', *RAND Journal of Economics*, 32(1): 101–28.

Hall, B. H., Graham, S. J. H., Harhoff, D., and Mowery, D. C., (2003). 'Prospects for Improving U.S. Patent Quality via Post-grant Opposition', *Innovation Policy and the Economy*, 4: 115–43.

Harabi, N. (1995). 'Appropriability of Technical Innovations: An Empirical Analysis', *Research Policy*, 24: 981–92.

Henry, M. D. and Turner, J. L. (2006). 'The Court of Appeals for the Federal Circuit's impact on patent litigation', *Journal of Legal Studies*, 35(1): 85–117.

IIPA (2009). 'International Intellectual Property Alliance Special 301 Reports on Copyright Protection and Enforcement: Priority Watch List including People's Republic of China, India, Argentina, Indonesia, Russian Federation, Thailand etc.' Available at <http://www.iipa.com/2009_SPEC301_TOC.htm> (accessed 12 May 2012).

Jaffe, A. B. and Lerner, J. (2004). *Innovation and Its Discontents: How Our Broken Patent System is Endangering Innovation and Progress, and What to Do About It*. Princeton, NJ: Princeton University Press.

Jones, C. I. and Williams, J. C. (1998). 'Measuring the Social Return to R & D', *Quarterly Journal of Economics*, 113(4): 1119–35.

Lanjouw, J. O. and Schankerman, M. (2004). 'Protecting Intellectual Property Rights: Are Small Firms Handicapped?' *Journal of Law and Economics*, 47(1), 45–74.

Layne-Farrar, A. (2012). 'The Brothers Grimm Book of Business Models: A Survey of Literature and Developments in Patent Acquisition and Litigation'. Paper presented at the George Mason University Law School conference, The Digital Inventor: How Entrepreneurs Compete on Platforms, 24 February 2012.

Leiponen, A. (2006). 'National Styles in the Setting of Global Standards: The Relationship between Firms' Standardization Strategies and National Origin', in A. Newman and J. Zysman (eds), *How Revolutionary was the Revolution? National Responses, Market Transitions, and Global Technology in the Digital Era*. Palo Alto, CA: Stanford University Press, 350–72.

Leiponen, A. (2008). 'Competing Through Cooperation: Standard-Setting in Wireless Telecommunications', *Management Science*, 54(11): 1904–19.

Leiponen, A., and Byma, J. (2009). 'If You Cannot Block, You Better Run: Small Firms, Cooperative Innovation, and Appropriation Strategies', *Research Policy*, 38(9): 1478–88.

Leiponen, A., and Ter Wal, A. (2012). 'Small Worlds and Affiliation Networks in Wireless Telecom Standardization'. Unpublished manuscript.

Lemley, M. A. (2009). 'The Surprising Virtues of Treating Trade Secrets as IP Rights', *Stanford Law Review*, September 4. Available at http://legalworkshop.org/2009/09/04/the-surprising-virtues-of-treating-trade-secrets-as-ip-rights on 2 April 2012.

Lerner, J., and Merges, R. P. (1998). 'The Control of Technology Alliances: An Empirical Analysis of the Biotechnology Industry', *The Journal of Industrial Economics*, 46(2): 125–56.

McDonough, J. F. III (2006). 'The Myth of the Patent Troll: An Alternative View of the Function of Patent Dealers in an Idea Economy', *Emory Law Journal*, 56: 189–228.

Molteni, L., and Ordanini, A. (2003). 'Consumption Patterns, Digital Technology and Music Downloading', *Long Range Planning*, 36: 389–406.

Mowery, D. C., Sampat, B., and Ziedonis, A. (2002). 'Learning to Patent: Institutional Experience, Learning, and the Characteristics of U.S. University Patents After the Bayh–Dole Act, 1981–1992', *Management Science*, 48(1): 73–89.

Paradise, J. (2004). 'European Opposition to Exclusive Control over Predictive Breast Cancer Testing and the Inherent Implications for U.S. Patent Law and Public policy: A Case Study of the Myriad Genetics BRCA Patent Controversy', *Food and Drug Law Journal*, 59(1): 133–54.

Peitz, M., and Waelbroeck, P. (2004). 'An Economist's Guide to Digital Music', CESifo Working Paper No. 1333.

Pisano, G. P. and Teece, D. J. (2007). 'How to Capture Value from Innovation: Shaping Intellectual Property and Industry Architecture', *California Management Review*, 50(1): 278–96.

Rose, M. (1993). *Authors and Owners: The Invention of Copyright*. Boston, MA: Harvard University Press.

Rosen, H. (2005). 'The Supreme Wisdom of Not Relying on the Court'. Huffington Post 26 June. Available at <http://www.huffingtonpost.com/hilary-rosen/the-supreme-wisdom-of-not_b_3221.html> (accessed 27 March 2012).

Rysman, M., and Simcoe, T. (2008). 'Patents and the Performance of Voluntary Standard Setting Organizations', *Management Science*, 54(11): 1920–34.

Scotchmer, S. (1991). 'Standing on the Shoulders of Giants: Cumulative Research and the Patent Law', *Journal of Economic Perspectives*, 5(1): 29–41.

Simcoe, T. (2012). Standard Setting Committees: Consensus Governance for Shared Technology Platforms', *American Economic Review*, 102(1): 305–36.

Simcoe, T., Graham, S. J., and Feldman, M. (2009), 'Competing on Standards? Entrepreneurship, Intellectual Property and Platform Technologies', *Journal of Economics and Management Strategy*, 18(3): 775–816.

Somaya, D. (2003). 'Strategic Determinants of Decisions not to Settle Patent Litigation', *Strategic Management Journal*, 24: 17–38.

Spring, M. B., Grisham, C., O'Donnell, J., Skogseid, I., Snow, A., Tarr, G., and Wang, P. (1995).'Improving the Standardization Process: Working with Bulldogs and Turtles', in B. Kahin and J. Abbate (eds), *Standards Policy for Information Infrastructure*. Cambridge MA: MIT Press, 220–50.

Stobbs, G. A. (2000). *Software Patents*. New York NY: Aspen Law & Business.

Teece, D. J. (1986). 'Profiting from Technological Innovation: Implications for Integration, Collaboration, Licensing and Public Policy', *Research Policy*, 15: 285–305.

Teece, D. J. (2006). 'Reflections on "Profiting from Innovation"', *Research Policy*, 35(8): 1131–46.

Tether, B., and Massini, S. (2007). 'Services and the Innovation Infrastructure', in *Innovation in Services*. DTI Occasional Paper 9, 135–92.

Updegrove, A. (2006). 'Delaware Court is Asked To Define "FRAND": Another Reason to Call for Ex Ante'., <http://www.consortiuminfo.org/standardsblog/article.php?story=20060813123034215> (accessed 28 March 2012).

Wagner, R. P. (2008). 'Understanding Patent-quality Mechanisms', *University of Pennsylvania Law Review*, 157: 2135–57.

Waguespack, D. M., and Fleming, L. (2009). 'Scanning the Commons? Evidence on the Benefits to Startups Participating in Open Standards Development', *Management Science*, 55(2): 210–23.

Waldfogel, J. (2011). 'Copyright Protection, Technological Change, and the Quality of New Products: Evidence from Recorded Music since Napster'. NBER Working Paper 17503.

Wright, B. D. (1983). 'The Economics of Invention Incentives: Patents, Prizes, and Research Contracts', *American Economic Review*, 73(4): 691–700.

Zhao, M. (2006). 'Conducting R&D in Countries with Weak Intellectual Property Rights Protection', *Management Science*, 56(7): 1185–99.

CHAPTER 29

MERGERS AND ACQUISITIONS AND INNOVATION

GAUTAM AHUJA AND
ELENA NOVELLI

INTRODUCTION

MERGERS and acquisitions (henceforth, M&A) are an important organizational phenomenon and are studied from different disciplinary perspectives (e.g. management,
industrial economics) and in the context of many different organizational or economic
implications (e.g. do they create value, their antitrust implications, etc.). This chapter
focuses on an overview of extant management research in the context of innovation.
It identifies the different theoretical perspectives and abstractions used to conceptualize the M&A–Innovation relationship, reviews the literature on antecedents and consequences of mergers as well as integration of M&A in the context of innovation, and
identifies notable gaps in the literature.

In the context of innovation, M&A can be conceived of as mechanisms to accomplish several distinct ends including but not limited to (a) filling out a product line;
(b) reducing time to market; (c) obtaining scale economies in research and bringing down the unit cost of R&D; (d) enhancing the rate of innovative output of an
organization; (e) enabling absorption of new technology to enhance manufacturing
or operations capability; (f) helping organizations enhance an existing innovation

capability (e.g. in an area in which the organization is already active); and (g) helping organizations build a new innovation capability (e.g. in an area in which the organization is currently not active or marginally active). In terms of academic research effort has focused on (d), (f), and (g) in an approximately decreasing order of frequency.

Broadly speaking, the key conclusions of this chapter suggest that several important areas in this topic remain understudied. Central amongst these understudied areas are (a) 'strategic' mergers that are intended to place a bet on the future by either acquiring a technology to preclude its usage by others or acquiring a technology as an 'option' in an uncertain technological setting; (b) systems effects of mergers including the impact of mergers on sector-wide diffusion of technologies; (c) 'consequential' effects of mergers, in that innovation-focused M&A are evaluated in terms of their eventual impact on more 'final' or 'consequential' measures such as firm productivity and profitability; (d) mergers as providing context or 'shocks' to an activity system or organization that enable researchers to evaluate and understand other organizational theories and issues; and (e) divestments as 'shocks' that could be used to infer the effects of mergers.

THE IMPORTANCE OF THE ISSUE

Understanding the innovation implications of M&A is important for both theoretical and practical reasons. Theoretically, M&A represents an important arena for advancing our understanding of several important organizational issues including but not limited to (a) organizational boundaries and their effects; (b) the building of organizational capabilities; (c) the effects of shocks on organizational routines; and (d) market domination through control of technology. This very illustrative and short list suggests that in addition to being relevant for developing a theoretical understanding of the phenomenon itself, M&A can also provide a useful context for developing and testing arguments for each of the major theoretical perspectives commonly used in strategy research (transactions costs economics, capabilities and resource-based views, evolutionary economics, and structure-conduct performance type industrial economics approaches). Existing management research has perhaps focused heavily on (b) and (c), though it has examined (a) too. However, relatively limited emphasis has been put on (d).

From a managerial standpoint, very significant corporate resources are commonly invested in M&A for innovation-oriented goals ranging from acquiring firms for building capabilities or filling out product lines to providing an exit option for venture capitalists to monetize their investments. Understanding the effectiveness of such investments is naturally a matter of interest from both managerial and policy perspectives.

Major Themes and Research Findings, and their Implications

The Conceptualization of M&A in the Context of Innovation

A simple abstraction can serve to represent the mechanisms underlying M&A in the innovation context. Innovations generated by an organization are fundamentally a product of the organization's knowledge. Recognizing this, in very elemental terms we can represent the organization's knowledge-base as a set, with the individual elements of knowledge or things known by the organization being the elements of this set (Ahuja and Katila, 2001). This abstraction makes it possible to recast several dimensions of M&A analysis into simple set-theoretic arguments (see Figures 29.1 and 29.2). For instance, the merger of two firms can be thought of as a union of two knowledge-bases; the relatedness of two organizations in terms of their technology can be recast as the overlap (i.e. intersection) between their knowledge-bases; even the absolute and relative magnitude of knowledge resources available to each organization can be captured using the cardinal numbers of the original knowledge-bases and the resultant merged knowledge-base. This approximation can then be used to yield testable predictions about various dimensions of mergers. If it becomes possible to identify these knowledge elements empirically (e.g. by using patents or individual scientists as elements of this set) validation and testing of the arguments is also possible, a strategy followed by several researchers (e.g. Ahuja and Katila, 2001; Puranam and Srikanth, 2007).

Research has analysed the phenomenon of M&A and innovation from different theoretical lenses. From a *Resource Based View Logic* (RBV), acquisitions can be seen as a way to acquire new resources and capabilities and enhance innovation. Innovation requires the creation of new technologies, products, or processes. An acquisition expands the

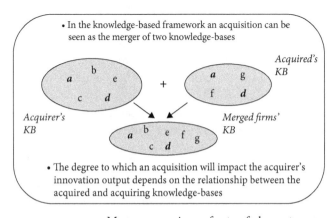

FIGURE 29.1 Mergers as unions of sets of elements

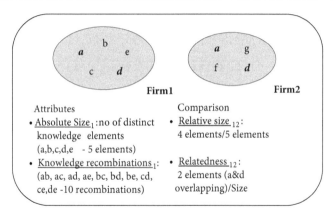

Firm1   **Firm2**

Attributes
- Absolute Size$_1$:no of distinct
  knowledge elements
  (a,b,c,d,e - 5 elements)
- Knowledge recombinations$_1$:
  (ab, ac, ad, ae, bc, bd, be, cd,
  ce,de -10 recombinations)

Comparison
- Relative size $_{12}$:
  4 elements/5 elements
- Relatedness $_{12}$:
  2 elements (a&d
  overlapping)/Size

FIGURE 29.2 Organizational knowledge-base

set of resources and capabilities available to the acquirer for recombination, and from a recombinant innovation logic (Fleming, 2001) may thus improve the innovation performance of the firm through two mechanisms. First, acquisitions may provide the firm with access to new knowledge elements, or technological resources or capabilities directly, and thus enhance the number of elements available for recombination. Second, since the range of resources and capabilities meaningful for innovating include not just technology and product development capabilities but also commercialization competences (Kaul, 2012) and complementary assets, acquisitions of such non-technical assets may also enable the firm to use its existing technologies resources and capabilities differently, and thus create innovations.

Further, acquisitions represent a way to transact for resources such as organizational routines or uncodifiable knowledge that are embedded in otherwise non-tradable assets (Capron et al., 1998; Ahuja and Katila, 2001). Internal development of resources and capabilities may represent an infeasible option for firms and—more importantly—may not be efficiently conducted in the short term (Dierickx and Cool, 1989). The speed with which acquisitions can provide access to resources becomes a clear benefit compared to other forms of inter-organizational knowledge-sourcing relationships (Hagedoorn, 2002; King et al., 2008). Acquisitions enable trade in competencies or otherwise non-tradable resources (Capron et al., 1998) by enabling trading in the unit itself that embeds the competency. These benefits are likely to be maximized in dynamic and fast-changing environments such as high-technology industries. In addition, acquisitions provide access to new resources, without the focal firm having to share the existing resources with other firms, as often happens in the case of alliances.

Building on these premises, literature in the RBV area has focused on investigating the issue of M&A in the context of innovation by focusing on the following broad research questions: What characteristics of the knowledge/resources of the acquired/acquiring firms increase the likelihood of engaging in acquisitions? Under what conditions do different characteristics of knowledge and resources of the acquired and acquirer firms lead to an improvement in innovation performance?

From a *transaction cost economics perspective* (TCE) acquisitions are viewed as a way to expand firm boundaries. Research in this area compares acquisitions to alternative governance modes to assess the benefits and implications of these different modes (Schilling and Steensma, 2002; Keil et al., 2008). In the innovation context this phenomenon is often investigated in connection with technology sourcing. Engaging in acquisitions has been presented as a response to the inefficiencies of technology markets (Hennart, 1991; Vanhaverbeke et al., 2002) as it corresponds to relying on the market 'for corporate control' (Hitt et al., 1996) as opposed to relying on buying the technology itself on the market (Arora et al., 2001). More generally, acquisitions of innovative firms provide more control compared to alliances (Hagedoorn, 1990, 1993), allow the firm to enjoy economies of scale and scope (Hoffmann and Schaper-Rinkel, 2001), and offer the benefit of speed compared to internal growth (Capron et al., 1998).

Compared to acquiring the technologies on the market, acquisitions tend to be more appropriate when the underlying knowledge is tacit (Ranft and Lord, 2002) and may require intense interaction between partners and a richer communication, since they reduce the conflict of interest and transaction costs that may be involved in other types of relationships (Khanna et al., 1998; Steensma and Corley, 2001; Keil et al., 2008). Since they often involve the transfer of different types of knowledge, acquisitions are found to be complementary to other types of external relationships (Arora and Gambardella, 1990). However, acquisitions also entail higher transaction costs to be incurred for the processes of target selection and negotiation (Singh and Montgomery, 1987).

Following these premises, research on M&A builds on TCE arguments to address the following broad questions: Under what conditions are acquisitions superior forms for successfully sourcing innovation? Under what conditions do the characteristics of the technology/knowledge to be sourced induce the firm to choose an acquisition as opposed to an alternative governance form?

From an *organizational learning* perspective acquisitions are seen as a way to enhance firm learning (Huber, 1991; Haleblian and Finkelstein, 1999). In this view, the objective of learning about a different market can motivate the decision to acquire another company (Ghoshal, 1987; Hitt et al., 1996; Ahuja and Katila, 2001; Kaul, 2012). One of the key mechanisms through which learning occurs in the context of acquisitions is through an expansion of the firm's absorptive capacity, which increases with the size of the target firm knowledge-base (Ahuja and Katila, 2001; Makri et al., 2010). Also, grafting the knowledge possessed by the target organization represents an additional way of pursuing learning through acquisitions (Huber, 1991). In such a context the past experience of the company and the direction of the desired learning affect firms' decision to acquire (Beckman and Haunschild, 2002). Research in the organizational learning area has therefore addressed the following broad research questions: What types of learning objectives motivate the choice to acquire? Under what conditions is learning enhanced in the context of an acquisition so that the subsequent innovation is increased?

The Relationship Between M&A and Innovation: Innovation and the Determinants of M&A

Three main factors tend to affect the extent of engagement or the probability of engaging in an acquisition in innovation intensive contexts: the magnitude and nature of the technological innovativeness of the parties, the performance of the parties, and the characteristics of the environment. Extant studies suggest that *innovative productivity* of the acquirer has an impact on the choice of engaging in an acquisition. Firms experiencing declines in internal productivity are more likely to engage in acquisitions in an effort to replenish their research pipelines (Higgins and Rodriguez, 2006). However, high innovative performance from the acquirer can also be associated with an increase in the likelihood of engaging in an acquisition (Kaul, 2012). This happens because acquisitions do not just represent opportunities to access new knowledge; rather they can also be efficient and effective means to acquire complementary resources for commercialization in response to a successful technological innovation. From a learning perspective, a technologically richer target can provide more opportunity to enhance focal firm knowledge (Ghoshal, 1987; Hitt et al., 1996). Consistent with a resource-based view perspective, the level of technological resources of the target firm justifies the choice of acquiring as opposed to allying with the target firm, to secure closer access to the relevant resources (Villalonga and McGahan, 2005).

The *nature of knowledge* of the parties can also influence the acquisition decision. Acquirers with less specialized knowledge see acquisitions as a mean of diversification and are more likely to engage in them (Miller, 2004). Moreover, when the technological knowledge capital of the focal firm is highly valuable, firms may choose to safeguard that capital and prefer more integrative forms of governance (such as acquisitions) as opposed to alliances (Villalonga and McGahan, 2005). In addition, characteristics of the knowledge to be sourced, such as its uniqueness, difficulty of imitation, and uncertainty, will also lead to a higher likelihood of acquisitions over other form of governance, due to the higher perceived threats of opportunism (Schilling and Steensma, 2002). Dyer et al. (2004) highlight that companies that use both strategies, acquisitions, and alliances grow faster than their rivals. However, more research is required to conclusively establish the direction of causality in this relationship.

A second determinant of the decision to engage in an acquisition in the context of innovation is the *performance of the firms involved*. For instance, Miller (2004) finds that firms diversifying through acquisitions have lower R&D intensity than matched firms that stay focused. Also, the probability of an acquisition increases when the performance of the focal firm falls below the aspiration levels and there is organizational slack available (Iyer and Miller, 2008). From the seller point of view, Graebner and Eisenhardt (2004) find that acquisitions occur when targets face strategic hurdles, such as a chief executive search or funding round, and strong personal motivations for sale, such as past failures and investments by friends.

Finally, the *characteristics of the technological environment* affect the likelihood of engaging in acquisitions. For instance, acquisitions of small high-tech firms tend to be a common option to acquire resources and capabilities in environments characterized by rapid technological change (Granstrand and Sjolander, 1990; Arora et al., 2001). Moreover, the characteristics of the knowledge in the target technical subfield, such as its complexity, specificity, and value, increase the likelihood of using acquisitions over other forms of governance (Carayannopoulos and Auster, 2010).

The Relationship Between M&A and Innovation: Innovation as the Outcome of M&A

A substantial set of studies has focused on investigating whether and how engaging in an acquisition may lead to higher innovation. This issue has attracted quite some interest because, despite the fact that many acquirers explicitly identify innovation as their ultimate goal and some studies do report a positive impact of engaging in acquisitions on the innovation outcome of firms (Capron and Mitchell, 1998; Desyllas and Hughes, 2010), other studies show the opposite effect. For instance, multiple studies report a negative relationship between acquisition intensity and the rate of internal innovation, due to the fact that acquisitions require time and attention for the extensive preparation, negotiations, and integration activities involved in the process (Hitt et al., 1990; Hitt et al., 1996). A negative impact on innovation can also be the result of managers overestimating their ability to manage an acquired business (Hitt et al., 1991). Chaudhuri and Tabrizi (1999) emphasize many common pitfalls that lead acquisitions to have disappointing results for acquiring firms. At the individual level, research has shown that acquisitions may have a negative impact on the productivity of inventors (Kapoor and Lim, 2007) and that often they result in key inventors leaving the acquired firm (Ernst and Vitt, 2000). Due to these conflicting results, a substantial body of research has focused on investigating the conditions under which acquisitions lead to an improvement on innovation.[1] The factors identified by extant literature in this area can be classified into three main categories: dyadic-level factors, firm-level factors, and factors related to acquisition implementation.

Dyadic-Level Factors

Many studies have recognized that the innovative outcome of an acquisition is dependent on the characteristics of the acquired firm in comparison to the acquirer. In particular the relative size and the relatedness of the firms involved in the acquisition have been identified as key factors.

Relative size of the acquired firm. Ahuja and Katila (2001) suggest that the relative size of the knowledge-base of the two firms involved in the acquisition affect the subsequent innovation output. Although a greater absolute size of the acquired knowledge-base may improve innovation, by enhancing the firm's absorptive capacity, scale, and scope economies and knowledge recombination possibilities, the authors find that the greater the relative size of the acquired knowledge-base, the less the subsequent innovation output of the acquiring firm. This effect originates from the fact that the understanding, assimilation, and application of new knowledge requires time and energy (Cohen and Levinthal, 1990), which eventually translates into a reduced innovative outcome when the target knowledge-base is disproportionally large compared to the acquirer. Moreover, the disruption of the existing organizational routines that is often involved in acquisitions (Haspeslagh and Jemison, 1991; Zollo and Singh, 2004) is going to be more dramatic in the face of a larger amount of knowledge to be incorporated and integrated within the organizational processes. Consistent results are reported by Clodt et al. (2006) who replicate the Ahuja and Katila study by extending it to a wider range of industries.

Relative size also influences intermediate outcomes in innovation-related acquisitions (Ranft and Lord, 2002). Acquirers tend to concede more autonomy to larger and better-performing targets, in an attempt to prevent the disruption of their successful routines and to reduce the adaptation required for its managers. However, while this leads to a greater level of autonomy and retention of key employees, it also negatively affects the communication between the two firms, reducing the likelihood of successfully transferring technologies and capabilities (Ranft and Lord, 2002).

Relatedness of the Acquired Knowledge-Base

Prior research on acquisitions in general has emphasized that the relatedness between the acquirer and the target significantly determines the performance of the entire operation (Singh and Montgomery, 1987). In the context of technological acquisitions, despite the fact that acquisitions are often considered a mode of diversification, moderate degrees of relatedness do appear to convey better innovation performance (Ahuja and Katila, 2001), due to the fact that relatedness leads to higher absorptive capacity (Cohen and Levinthal, 1990). Yet, we observe diminishing results when the similarity is such that no learning is involved in the process (Hitt et al., 1996; Capron and Mitchell, 1998; Cloodt et al., 2006). Multiple studies investigate this issue by looking at how this effect changes for different dimensions of relatedness. Cassiman et al. (2005) make a distinction between similarity and complementarity of technological and market knowledge and find that M&A partners that operate in the same technological fields tend to reduce their R&D effort and rationalize the R&D process after the M&A compared to the case of complementary technological fields. In the same study, they also investigate market relatedness to understand how competitive dynamics affect the investment in R&D. Their results suggest that former rivals that engage in a merger are less likely to expand their R&D effort compared to non-rival firms.

Makri et al. (2010) consider both scientific and technological knowledge and make a distinction between similarity and complementarity of knowledge. Their results suggest that a higher innovation performance is associated with complementarity in both scientific and technological knowledge. Moreover, the complementarity versus the similarity of knowledge tends to affect the nature of the innovation subsequently developed, with similarities in knowledge facilitating incremental renewal, while complementarities lead to more discontinuous outcomes. Complementarity between technology and marketing resources is also one of the main factors that are found to explain abnormal returns in the context of the acquisition of R&D-intensive firms (King et al., 2008). While acquiring firm marketing resources and target firm technology resources positively reinforce each other, acquirer and target firm technologies substitute for one another.

Firm-Level Factors

At the firm level, three main factors have been identified to determine the innovative outcome of an acquisition: the motivation, the characteristics of the underlying knowledge, and firm characteristics.

Motivation

The clarity and the nature of the objectives motivating the acquisition play a key role in determining the acquisition outcome. As Chaudhuri and Tabrizi (1999) emphasize, being too short-sighted regarding the acquisition's expected results and focusing only on the immediate benefits can lead firms to suboptimal performance. Ahuja and Katila (2001) suggest that the balance between the costs and benefits of acquisitions on the innovation performance of the firm is negative in the context of non-technological acquisitions, since they tend not to add much to the knowledge-base of the acquirer, while requiring a great deal of disruption in existing routines, overall reducing its productivity.

Characteristics of the Underlying Knowledge

Extant research has recognized the importance of the *absolute size* of the acquired knowledge-base in explaining the innovative performance of an acquisition (Ahuja and Katila, 2001; Cloodt et al., 2006). Using recombinant logic, innovation emerges from the recombination of existing knowledge elements in new ways (Henderson and Cockburn, 1996; Fleming, 2001). The larger the acquired knowledge-base, the higher the probability of accessing new knowledge elements and hence an increase in the number of new knowledge combinations available to the focal firm. Moreover, merging knowledge-bases can provide scale and scope benefits, leading to a rationalization of the R&D effort. The absolute size of the knowledge-base of the acquirer can also determine the success of the acquisition, due to the correspondingly higher absorptive capacity (Desyllas and Hughes, 2010). Superior absorptive capacity can help the buyer in better deploying the acquired resources and fully exploiting the innovation potential of the acquired knowledge-base. This effect is reduced in the context of unrelated acquisitions, where increases in knowledge-base concentration lead to impaired peripheral vision and core rigidities (Desyllas and Hughes, 2010).

The *nature of knowledge and technology* underlying the acquisition is an additional factor determining the outcome of such efforts. Acquisition activity tends to be positively related to the number of biotechnology-based products subsequently developed in the context of an emergent technological regime (Nicholls-Nixon and Woo, 2003).

Steensma and Corley (2001) suggest that in the context of technology sourcing partnerships, the tight coupling offered by an acquisition as opposed to other types of relationships provides beneficial results when the technology is unique, since tightly coupled sourcing partnerships will maintain its uniqueness by removing it from the marketplace. However, acquisitions are associated with better performance when the technology is more difficult to imitate and when the level of technological uncertainty and dynamism are lower. Ranft and Lord (2002) indicate that the tacitness and social complexity of the underlying knowledge makes it more difficult to comprehend and transfer; however, it is its effective integration that determines the success of the acquisition.

Firm Characteristics

In the context of technology acquisitions, the *age of the firm* affects the acquisition performance. Young targets tend to be associated with more flexible growth options and greater valuation uncertainty, implying respectively more opportunities for synergistic fit and lower prices. However, the negative effect of target age on acquirer value is partially mitigated if the target has recent patents or is privately held, due to the reduced amount of information disclosed by such firms (Ransbotham and Mitra, 2010).

The levels of *leverage and of leverage growth* are also associated with heterogeneous performance results (Desyllas and Hughes, 2010). The enhanced monitoring associated with higher leverage tends to induce efficiency, which eventually raises R&D productivity. However, the increased financial constraints and short-term vision induced by high-leverage growth eventually reduces R&D intensity.

Finally, *the experience of the acquirer* affects the technical output of the firm, as identified by Nicholls-Nixon and Woo (2003) in the context of emerging technological regimes, since prior experience affects the ability to select appropriate partners and learn from them. However, this effect is not significant in the context of foreign acquisitions (Belderbos, 2003).

Acquisition Implementation

A key area of research on this topic has concerned the implementation phase of an acquisition. Extant research has suggested that the implementation process determines the success or failure of the acquisition (Haspeslagh and Jemison, 1991). Schweizer (2005) implies that the complexities and multiple facets of an acquisition require going beyond a 'single integration' approach and designing more hybrid forms of integration, balancing short- and long-term motives and objectives. For instance, the R&D and non-R&D portions require distinct integration approaches that account for the specificities of these areas.

Engaging in the integration process between the acquirer and acquired firms implies facing a trade-off from a knowledge transfer perspective (Ranft and Lord, 2002; Puranam and Srikanth, 2007). As Puranam and Srikanth (2007) note, on the one hand, integration leads to improved coordination between the two parties, which start adhering to the same procedures, become aware of common goals, and coordinate their actions with actions of the other party. This has many beneficial effects including a more efficient transfer of knowledge across parties. On the down-side, however, integration leads to high costs in terms of loss of autonomy. Since the first effect tends to increase with the interdependence between the two parties, while the second is almost independent from interdependence, the authors conclude that structural integration is more likely in technology acquisitions when the acquisition is motivated by obtaining a component technology. Further, higher integration facil-itates leveraging of a given technology (what the acquired firm knows) but inhib-its ongoing development of new technologies through the acquired unit (what the acquired firm does). Interestingly, the existence of a pre-existing 'common ground' between the acquirer and acquired firms can act as a substitute for the integration process (Puranam et al., 2009).

The wrong integration approach can lead to detrimental results, neutralizing the ben-efits of the acquisition. At the individual level, it can impair the productivity of corporate scientists of acquired entities due to the loss in autonomy and the consequent sense of loss and failure associated with it (Paruchuri et al., 2006). This effect is conditional on the relative standing and social embeddedness losses: the more relative standing and social embeddedness an inventor loses on being acquired, the worse his productivity after the innovation. In addition, the inventor's divergence from the expertise of the acquirer is another factor that can affect negatively inventors' productivity. Puranam et al. (2006) suggest that one way to resolve the coordination autonomy dilemma is by recognizing that the effect of structural form on innovation outcomes depends on the developmental stage of acquired firms' innovation trajectories. The disruptive effects of integration are going to be worst if the acquired firm is still in an exploratory stage, since integration is more conducive to exploitation and disruption of organizational routines can under-mine exploratory activities. However, if integration occurs when the target has already launched its first products, the disruption is going to be limited. Moreover the disruptive effect of integration is going to be more salient for immediate innovations than for later ones, with the result that these effects tend to disappear with the evolution of the techno-logical trajectory after the acquisition. The acquisition speed can also affect the success of the integration process, with slow acquisition implementation leading to a more effec-tive preservation of knowledge compared to a faster one (Ranft and Lord, 2002).

Individuals can play a key role in solving the implementation dilemma. In her 2004 study, Graebner shows how key managers from the acquired firm tend to achieve both 'expected' and 'serendipitous' value, which in turn helps in maintaining the advan-tages of both integration and autonomy and managing exploration and exploitation simultaneously. Performance and incentives systems can also represent an important tool to address the implementation puzzle. At the inventor level, Kapoor and Lim (2007) find that the productivity of inventors from an acquired firm tends to be lower than that of similar firms that did not go through an acquisition process. They suggest that while a knowledge-based view perspective explains the temporary disruption of the first two years after the acquisition, it is only from an incentive-based perspective that we can explain why the productivity loss persists below the level for non-acquired inventors.

Future Development in the Area

As the foregoing review would indicate, the M&A context has been a fecund ground in the context of innovation. That notwithstanding, there remain many potential directions of further research that have both theoretical and practical implications. We identify a list of research areas and questions below.

The Conceptualization of M&A in the Context of Innovation

Relatively limited research has adopted a social network perspective to investigate M&A in connection with innovation. Yet, being acquired may represent entry into a new network for the target. Alternatively, the acquisition may provide the focal firm access to the network ties of the acquired firm. How does this dynamic affect the level and nature of the subsequent innovation activities of both firms? Under what conditions may such changes be beneficial or end up in an additional source of disruption? Similarly, limited research has adopted an institutional perspective for analysing M&A in the context of innovation. Institutional theory suggests that organizations imitate practices adopted by others in the attempt to acquire legitimacy (DiMaggio and Powell, 1983). How does such legitimacy-seeking behaviour affect the innovation outcomes of the merger? When acquired target companies face a new reference group, from an institutional theory perspective one might expect that this might induce the firms to modify the set of practices used in the acquired firm. However, this creates a puzzle since very often the incentive motivating the acquisition may be to acquire the target's competencies, but such modification of the target's set of innovative routines may undermine the value of the merger.

Another area of potential research that has been underemphasized is the realm of 'strategic' acquisitions. In technology contexts, companies are sometimes acquired to shut down their technologies and reduce competition. Examining post-merger technology trajectories to consider such possibilities would be interesting from both strategic and policy perspectives. Alternatively, in some cases firms acquire other firms, largely to preclude competitors from acquiring those firms' technologies. The logic and appropriateness of such strategic acquisitions is still not well explicated nor are they even documented in detail. A third type of strategic acquisition is the acquisition of small firms as 'options' in the face of an uncertain technology environment. Since technology can evolve along different trajectories the *ex ante* prospects of a particular technology trajectory may be unclear. To insure themselves firms may acquire other firms that are pursuing different technological paths. In such cases evaluating the success of the acquisitions through post-acquisition innovation output may be misleading. Understanding the contingencies under which such 'option' acquisitions are appropriate remains to be clarified.

Cross-Border Acquisitions

Cross-border acquisitions represent a wonderful study space from multiple perspectives. M&A in such contexts bring together not just different cultures, but also different institutions, costs structures, demand preferences, and management approaches (Kale et al., 2009). The cultural conflict possible in such mergers has been somewhat studied but relatively limited attention has been paid to other dimensions of difference in these mergers. For instance, many emerging market firms are acquiring Western firms with superior technology (e.g. Geely's acquisition of Volvo) with the intent of absorbing superior technical manufacturing capabilities. Yet, our understanding of whether such approaches to enhance technology-embedded manufacturing capabilities work remains nascent. These types of mergers, which are often increasingly commanding significant resources, have only been very limitedly examined. The topic of such emerging market acquirers also raises questions about the effectiveness of mergers between traditional innovators and pioneers of what is described as frugal innovation. Understanding the emergence of hybrid innovation models that combine the practices of frugal and regular innovation may be another fruitful direction of innovation research. How cross-border mergers navigate the differences in institutional contexts also remains an unaddressed issue for the most part.

Partners' Relatedness and Innovation Outcomes

In evaluating the returns from mergers scholars have often made a distinction between related and unrelated businesses in the context of a merger with the argument being that related businesses should be easier to absorb and integrate. The results from this research have not always been easy to interpret because in part it is unclear as to what relatedness really means. For instance, firms could be related by exposure to common buyers or common suppliers or common processes or common technologies or common inputs. Exposure to each of these categories may lead to different innovation outcomes, for instance in terms of product versus process innovations. Clubbing all of these into the same 'relatedness' category may hide many underlying differences. The literature on diversification has identified several different bases of relatedness, including skills, markets, technologies, and inputs (Ahuja et al., 2012), and these could be useful in this context.

More broadly, although much research has focused on the post-merger productivity of the acquirer, much of this research has used patents as measures of the innovative outcome. Yet, innovations are not created solely to create new patents, which are at best intermediate outcomes. Innovations are eventually intended to advance, develop, and upgrade manufacturing capabilities, and new products and processes. Evaluating the effects of M&A on these outcomes, and identifying the contingencies that are likely to underlie these relationships, is another natural path to extend the research.

Further, although extant research has shown that firms engage in acquisitions in order to improve their innovation performance, it has only limitedly been recognized that there is a high heterogeneity in terms of the specific innovation goals a firm may want to pursue. For instance, some firms aggressively target the pursuit of radical breakthrough inventions, while others focus on increasing the frequency of new product launches, even if this implies releasing incremental inventions. How do these different types of goals and incentive affect the success of the acquisition process? What does the final innovative outcome originating from the merger of such diverse firms look like? Future research might explore further these issues.

System Effects of M&A

Taking a broader perspective on the problem one could also consider the broader or 'systemic' effects of mergers. Relatively little research has examined these, even though their substantive consequences may be very significant. We identify two such broad areas for consideration, though clearly many other 'systemic effects' could also be studied. Research on M&A patterns shows that M&A often occur in waves, with a large number of mergers in a given sector being compressed into a short period of time. An interesting question that then arises is how such merger waves influence innovation outcomes.

Since a merger wave leads to significant consolidation in a sector of the economy, at least two possibly contrasting effects on subsequent innovation performance can be identified. First, increasing consolidation could enable economies of scale and thus foster innovation. Alternatively, the increased consolidation may lead to fewer independent lines of research being supported within the industry. Given that the best path to innovation is generally unclear *ex ante*, a reduction in the variety of approaches may lead to reduced innovation. Industry concentration may also result in a reduced incentive for firms to be innovative in the first place. Which of these two effects plays out and is stronger under which circumstance is not clear, and is a target for research efforts. For instance, it could be conceived that the scale benefits of size enhance innovative productivity, but the fewer lines of enquiry aspect diminish the likelihood of breakthrough innovations. Thus, the final effect of a merger wave might well be an increased innovative productivity but lowered innovation quality. The experience of the United States pharmaceutical industry which has seen massive consolidation in the search for scale but has also seen a subsequent dearth of breakthrough innovation provides at least anecdotal support for such a theory.

In the vein of systemic thinking another research opportunity lies in understanding the transformative effects of mergers on given sectors of the economy. In many sectors of the economy, increasingly final assembly of products is done by mega-assemblers (e.g. Boeing, GM, GE), that bring together the manufacturing and technology efforts of many component manufacturers. An acquisition event in the case of any of these mega-assemblers can then influence the fortunes of many of the other firms that are part of the focal assemblers' network. Prior research on innovation and productivity suggests that entry of foreign firms using advanced technologies into a local ecology provides knowledge spillovers for the local firms. In a quasi-reversal of the process we have increasingly seen that acquiring firms from emerging markets (e.g. Geely mentioned earlier) have been taking over established developed country targets that have been the repositories of advanced technologies and processes (e.g. Volvo). Given the strong links between the acquiring firms and their original home markets it is then conceivable to imagine knowledge spillover flows from the acquired company to the acquiring company and then to the acquiring company's network of suppliers, fostering their technological advancement and innovative outcomes. Thus, such mergers, beyond enhancing the innovative performance of individual firms, may well affect the technical capabilities of entire sectors and may have implications for international competitive advantage. Evaluating such knowledge-flows from foreign acquisitions of assemblers to domestic sector-wide productivity effects remains to be studied.

Empirical Considerations

In addition to opening many interesting paths from a theoretical standpoint, acquisitions may also further research from a methodological angle. Acquisitions represent a discrete shock in the life of an organization and thus could potentially be used as a mechanism to help identify other organizational effects and theories. Although the acquisition shock may not always be argued to be exogenous at the level of the organization, it is definitely a tenable assumption that it is an exogenous shock for individual sub-units within the organization (e.g. inventors, research groups, research projects). Using acquisitions as shocks to help us understand other organizational processes remains a generally underutilized research strategy. For instance, using acquisitions of

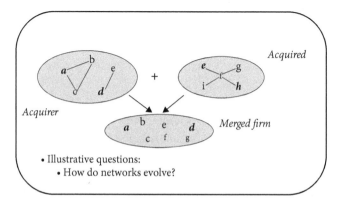

FIGURE 29.3 Broadening the abstraction: merger as a union of networks

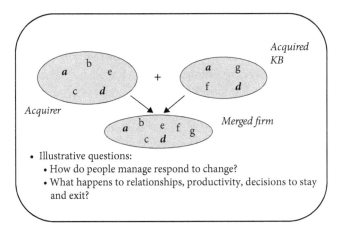

FIGURE 29.4 Broadening the abstraction: merger as a union of people

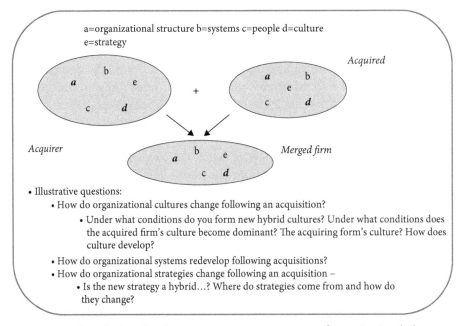

FIGURE 29.5 Broadening the abstraction: merger as a union of organizational elements

small firms by large firms as a focal shock, researchers may unwrap the effects of organizational size on innovative productivity. Returning to the abstraction of an acquisition as a union of two sets of elements we note that the abstraction can be broadened in its application by recognizing that the elements need not only represent knowledge, they could also represent people, organizational design elements, activities, or capabilities (see Figures 29.3 to 29.6). The common feature across these interpretations is that the

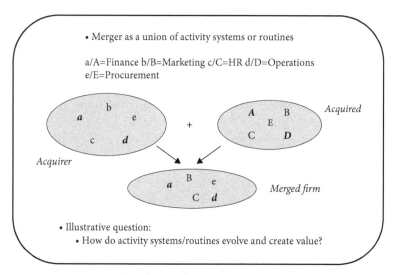

FIGURE 29.6 Broadening the abstraction: activity systems

acquisition is treated as a shock to an existing assembly of elements. Interpreting the elements as people we can think of acquisitions as a forum to study network evolution. Prior to the acquisition the individuals are bound to other individuals in the existing organization. That organization is subject to a shock when the merger occurs and existing network ties are now open to change. How new network ties form and how old ones are influenced following the merger can help us uncover much about how social networks evolve. The same principle can be used in studying the evolution of strategies, cultures, and activity systems and even the productivity and satisfaction of individuals (Figures 29.4, 29.5, and 29.6).

To conclude, while much work has been done in the context of mergers and acquisitions in the innovation context, much remains to be done in furthering our understanding of this important industrial phenomenon. It is our hope that the above review and the directions of thought identified in this chapter will help in pushing the literature further.

NOTES

1. In particular these studies have evaluated the innovative performance of firms on multiple dimensions, corresponding to different measures of performance such as the patent and citations counts (Ahuja and Katila, 2001; Hagedoorn and Duysters 2002; Kapoor and Lim, 2007; Keil et al., 2008; Zhao 2009); number of products and firm reputation (Nicholls-Nixon and Woo, 2003); R&D intensity and patent intensity (Hitt et al., 1991) and R&D productivity (Desyllas and Hughes, 2010). In addition different studies have looked at the nature of the innovation created through an innovation, such as exploratory and non-exploratory (Wagner 2011); incremental vs discontinuous (Makri et al., 2010).

References

Ahuja, G., and Katila, R. (2001). 'Technological Acquisitions and the Innovation Performance of Acquiring Firms: A Longitudinal Study', *Strategic Management Journal*, 22: 197–220.

Ahuja, G., Lampert, C., and Tandon, V. (2012). 'Paradigm-Changing Versus Paradigm-Deepening Innovation: How Firm Scope Influences Firm Technological Response to Shocks'. Working paper.

Arora, A., Fosfuri, A., and Gambardella, A. (2001). *Markets for Technology: The Economics of Innovation and Corporate Strategy*. Cambridge, MA: The MIT Press.

Arora, A., and Gambardella, A. (1990). 'Complementarity and External Linkages: The Strategies of the Large Firms in Biotechnology', *Journal of Industrial Economics*, 28(4): 361–79.

Beckman, C. M., and Haunschild, P. R. (2002). 'Network Learning: The Effects of Partners' Heterogeneity of Experience on Corporate Acquisitions', *Administrative Science Quarterly*, 47(1): 92–124.

Belderbos, R. (2003). 'Entry Mode, Organizational Learning, and R & D in Foreign Affiliates: Evidence from Japanese Firms', *Strategic Management Journal*, 24: 235–59.

Capron, L., Dussauge, P., and Mitchell, W. (1998). 'Resource Redeployment Following Horizontal Acquisitions in Europe and North America, 1988–1992', *Strategic Management Journal*, 19(7): 631–61.

Capron, L., and Mitchell, W. (1998). 'Bilateral Resource Redeployment Following Horizontal Acquisitions: A Multi-Dimensional Study of Business Reconfiguration', *Industry and Corporate Change*, 7: 453–84.

Carayannopoulos, S., and Auster, E. R. (2010). 'External Knowledge Sourcing in Biotechnology through Acquisition Versus Alliance: A KBV Approach', *Research Policy*, 39: 254–67.

Cassiman, B., Colombo, M. G., Garrone, P., and Veugelers, R. (2005). 'The Impact of M&A on the R&D Process: An Empirical Analysis of the Role of Technological- and Market-Relatedness', *Research Policy*, 34: 195–220.

Chaudhuri, S., and Tabrizi, B. (1999). 'Capturing the Real Value in High-Tech Acquisitions', *Harvard Business Review*, 77: 123–30.

Cloodt, M., Hagedoorn, J., and Van Kranenburg, H. (2006). 'Mergers and Acquisitions: Their Effect on the Innovative Performance of Companies in High-Tech Industries', *Research Policy*, 35: 642–54.

Cohen, W. M., and Levinthal, D. A. (1990). 'Absorptive Capacity: A New Perspective on Learning and Innovation', *Administrative Sciences Quarterly*, 35: 569–96.

Desyllas, P., and Hughes, A. (2010). 'Do High Technology Acquirers Become More Innovative?', *Research Policy*, 39: 1105–21.

Dierickx, I., and Cool, K. (1989). 'Asset Stock Accumulation and Sustainability of Competitive Advantage', *Management Science*, 35: 1504–11.

DiMaggio, P. J., and Powell, W. (1983). 'The Iron Cage Revisited': Institutional Isomorphism and Collective Rationality in Organizational Fields', *American Sociological Review*, 48: 147–60.

Dyer, J., Kale, P., and Singh, H. (2004). 'When to Ally and when to Acquire?', *Harvard Business Review*, 82: 102–15.

Ernst, H., and Vitt, J. (2000). 'The Influence of Corporate Acquisitions on the Behavior of Key Inventors', *R&D Management*, 30: 105–19.

Fleming, L. (2001). 'Recombinant Uncertainty in Technological Search', *Management Science*, 47: 117–32.

Ghoshal, S. (1987). 'Global Strategy: An Organizing Framework', *Strategic Management Journal*, 8: 425–40.

Graebner, M. E. (2004). 'Momentum and Serendipity: How Acquired Leaders Create Value in the Integration of Technology Firms', *Strategic Management Journal*, 25: 751–77.

Graebner, M. E., and Eisenhardt, K. M. (2004). 'The Seller's Side of the Story: Acquisition as Courtship and Governance as Syndicate in Entrepreneurial Firms', *Administrative Science Quarterly*, 49: 366–403.

Granstrand, O., and Sjolander, S. (1990). 'The Acquisition of Technology and Small Firms by Large Firms', *Journal of Economic Behavior & Organization*, 13(3): 367–86.

Hagedoorn, J. (1990). 'Organisational Modes of Inter-Firm Cooperation and Technology Transfer', *Technovation*, 10(1): 17–31.

Hagedoorn, J. (1993). 'Understanding the Rationale of Strategic Technology Partnering: In ter-Organizational Modes of Cooperation and Sectoral Differences', *Strategic Management Journal*, 14: 371–85.

Hagedoorn, J. (2002). 'Inter-Firm R&D Partnerships: An Overview of Major Trends and Patterns since 1960', *Research Policy*, 31: 477–92.

Haleblian, J., and Finkelstein, S. (1999). 'The Influence of Organizational Acquisition Experience on Acquisition Performance: A Behavioral Learning Perspective', *Administrative Science Quarterly*, 44: 29–56.

Haspeslagh, P. C., and Jemison, D. B. (1991). *Managing Acquisitions: Creating Value Through Corporate Renewal*, New York: The Free Press.

Henderson, R., and Cockburn, I. (1996). 'Scale, Scope and Spillovers: The Determinants of Research Productivity in Drug Discovery', *RAND Journal of Economics*, 27: 32–59.

Hennart, J. F. (1991). 'A Transaction Cost Theory of Joint Ventures: An Empirical Study of Japanese Subsidiaries in the United States', *Management Science*, 37: 483–97.

Higgins, M. J., and Rodriguez, D. (2006). 'The Outsourcing of R&D through Acquisition in the Pharmaceutical Industry', *Journal of Financial Economics*, 80: 351–83.

Hitt, M. A., Hoskisson, R. E., and Ireland, R. D. (1990). 'Mergers and Acquisitions and Managerial Commitment to Innovation in M-Form Firms', *Strategic Management Journal*, 11 (Special Issue): 29–47.

Hitt, M. A., Hoskisson, R. E., Ireland, R. D., and Harrison, J. S. (1991). 'Effects of Acquisitions on R&D Inputs and Outputs', *Academy of Management Journal*, 34: 693–706.

Hitt, M. A., Hoskisson, R. E., Johnson, R. A., and Moesel, D. D. (1996). 'The Market for Corporate Control and Firm Innovation', *Academy of Management Journal*, 39: 1084–119.

Hoffmann, W. H., and Schaper-Rinkel, W. (2001). 'Acquire or Ally? A Strategy Framework for Deciding between Acquisitions and Cooperation', *Management International Review*, 41: 131–59.

Huber, G. (1991). 'Organizational Learning: The Contributing Processes and the Literatures', *Organization Science*, 2(1): 88–115.

Iyer, D. N., and Miller, K. D. (2008). 'Performance Feedback, Slack, and the Timing of Acqui- sitions', *Academy of Management Journal*, 51(4): 808–22.

Kale, P., Singh, H., and Raman A. (2009). 'Don't Integrate your Acquisitions, Partner with them', *Harvard Business Review*, 87(12): 109–15.

Kapoor, R., and Lim, K. (2007). 'The Impact of Acquisitions on the Productivity of Inventors at Semiconductor-rms: A Synthesis of Knowledge-Based and Incentive-Based Perspectives', *Academy of Management Journal*, 50(5): 1133–55.

Kaul, A. (2012). 'Technology and Corporate Scope: Firm and Rival Innovation as Antecedents of Corporate Transactions', *Strategic Management Journal*, 33: 347–67.

Keil, T., Maula, M., Schildt, H., and Zahra, S. A. (2008). 'The Effect of Governance Modes and Relatedness of External Business Development Activities on Innovation Performance', *Strategic Management Journal*, 29(8): 895–907.

Khanna, T., Gulati, R., and Nohria, N. (1998). 'The Dynamics of Learning Alliances: Competition, Cooperation, and Relative Scope', *Strategic Management Journal*, 19(3): 193–210.

King, D. R., Slotegraaf, R. J., and Kesner, I. (2008). 'Performance Implications of Firm Resource Interactions in the Acquisition of R&D-Intensive Firms', *Organization Science*, 19: 327–40.

Makri, M., Hitt, M. A., and Lane, P. J. (2010). 'Complementary Technologies, Knowledge Relatedness and Innovation Outcomes in High-Technology Mergers and Acquisitions', *Strategic Management Journal*, 31: 602–28.

Miller, D. J. (2004). 'Firms' Technological Resources and the Performance Effects of Diversification: A Longitudinal Study', *Strategic Management Journal*, 25(11): 1097–119.

Nicholls-Nixon, C. L., and Woo, C. U. (2003). 'Technology Sourcing and Output of Established Firms in a Regime of Encompassing Technological Change', *Strategic Management Journal*, 24: 651–66.

Paruchuri, S., Nerkar, A., and Hambrick, D. C. (2006). 'Acquisition Integration and Productivity Losses in the Technical Core: Disruption of Inventors in Acquired Companies', *Organization Science*, 17(5): 545–62.

Puranam, P., Singh, H., and Chaudhuri, S. (2009). 'Integrating Acquired Capabilities: When Structural Integration is (Un)necessary', *Organization Science*, 20: 313–28.

Puranam, P., Singh, H., and Zollo, M. (2006). 'Organizing for Innovation: Managing the Coordination Autonomy Dilemma in Technology Acquisitions', *Academy of Management Journal*, 2: 263–80.

Puranam, P., and Srikanth, K. (2007). 'What they Know vs. what they Do: How Acquirers Leverage Technology Acquisitions', *Strategic Management Journal*, 28: 805–25.

Ranft, A., and Lord, M. (2002). 'Acquiring New Technologies and Capabilities: A Grounded Model of Acquisition Implementation', *Organization Science*, 13: 420–41.

Ransbotham, S., and Mitra, S. (2010). 'Target Age and the Acquisition of Innovation in High Technology Industries', *Management Science*, 56(11): 2076–93.

Schilling, M., and Steensma, H. (2002). 'Disentangling the Theories of Firms Boundaries: A Path Model and Empirical Test', *Organization Science*, 13(4): 387–401.

Schweizer, L. (2005). 'Organizational Integration of Acquired Biotechnology Companies in Pharmaceutical Companies: The Need for a Hybrid Approach', *Academy of Management Journal*, 6: 1051–74.

Singh, H., and Montgomery, C. (1987), 'Corporate Acquisition Strategies and Economic Performance', *Strategic Management Journal*, 8(4): 377–86.

Steensma, H. K., and Corley, K. (2001). 'Organizational Context as a Moderator of Theories on Firm Boundaries for Technology Sourcing', *Academy of Management Journal*, 44: 271–91.

Vanhaverbeke, W., Duysters, G., and Noorderhaven, N. (2002). 'External Technology Sourcing through Alliances or Acquisitions: An Analysis of the Application-Specific Integrated Circuits Industry', *Organization Science*, 13(6): 714–33.

Villalonga, B., and McGahan, A. (2005). 'The Choice among Acquisitions, Alliances and Divestitures', *Strategic Management Journal*, 26: 1183–208.

Wagner, M. 2011. To explore or to exploit? An empirical investigation of acquisitions by large incumbents. *Research Policy*, 40: 1217–1225.

Zhao, X. 2009. Technological innovation and acquisitions. *Management Science* 55(7): 1170–1183.

Zollo, M., and Singh, H. (2004). 'Deliberate Learning in Corporate Acquisitions: Post-Acquisition Strategies and Integration Capability in U.S. Bank Mergers', *Strategic Management Journal*, 25: 1233–56.

CHAPTER 30

SERVICES, INNOVATION, AND MANAGING SERVICE INNOVATION

BRUCE S. TETHER

INTRODUCTION

THIS chapter discusses services, innovation, and their interplay, particularly through 'service innovation' and the design of new and improved services. 'Services' is a very big space.[1] Innovation management comprises the strategies and practices that decision-makers use to achieve organizational benefits from innovation, which is here defined as the commercial exploitation of new ideas, and includes both the introduction of 'radically new services' and, more commonly, the incremental improvement of existing services (Avlonitis et al, 2001; Alam, 2006b). Services are a wide set of activities which share certain features in common (which we shall discuss). As such, services provide contexts for innovation of various types (Sirilli and Evangelista, 1998; den Hertog, 2000; Hipp et al, 2000; Hipp and Grupp, 2005; Tether and Howells, 2007; Tether and Tajar, 2008), and some service sectors, such as knowledge-intensive and professional services, and telecommunication and IT services, are considered much more innovative than others (e.g. transportation and public services) (Evangelista and Sirilli, 1998; Tether, 2003; Camacho and Rodriguez, 2005).[2]

In general, service innovation tends not to be organized through specific departments, and especially R&D,[3] but is instead typically more distributed, involving people from a variety of functions or departments, and frequently involves users and complementary service providers. To increase runway capacity at severely congested airports such as London's Heathrow and Frankfurt in Germany, for example, airline pilots, air traffic controllers, and the airport operators have developed practices to work very closely together to 'find' additional capacity (Tether and Metcalfe, 2003). Service innovation is also more likely to involve re-skilling and training than is typically the case with technological product and process innovation in manufacturing. Studies have also found that innovating service firms, whether large or small, tend to outperform their non-innovating counterparts in terms of productivity (income per head) and growth (Cainelli et al., 2004; Matear et al., 2004). Like manufacturers, service firms tend to emphasize 'product' innovation over process innovation (Damonpour and Gopalakrishnan, 2001), although organizational innovations have been found to be particularly important in services (van der Aa and Elfring, 2002; Tether and Tajar, 2008).

Whilst service firms and sectors innovate, 'service innovation' is also a type of innovation, distinct from physical product innovation (i.e. goods innovation), process innovation, or organizational innovation. Service innovations are not confined to service sector activities. The introduction of self-service check-in/outs at a hotel, for example, would be a service innovation within a service business. Manufacturers can also introduce service innovations (Berry et al., 2006). A manufacturer might introduce a service to install and maintain its equipment within a client's business, for example, ensuring it operates efficiently, and perhaps even being paid by performance, rather than in a one-off transaction. Some manufacturers, such as Rolls Royce, Xerox, and Alsthom have innovated in this way (Shelton, 2009; Gremyr et al., 2010), and increasingly compete by focusing on the utility of their products, which is what they do (i.e. the service they provide), not what they are.

Rolls Royce, for example, now derives around half of its total revenues from services (Neely et al., 2011), and most of its aircraft engines are sold with service agreements, such as 'Total Care'. This has required a shift in focus from technologies to customer needs. As one executive explained: 'We design these complex high value propulsion systems with turbine blades that see temperatures exceeding the melting point of the metal from which they are made, but our airline customers see them as a tube they stick on their aircraft to get them to where they want in a predictable and reliable way'. In other words, the customer is not interested in the technology per se, but rather in what it does, and that it does it efficiently. This change in focus has been accompanied by both a shift in mind-set and a stream of technical and organizational innovations, many of which do not involve the traditional hub of technological product innovation, the R&D department. Similarly, the market for mobile phones is now less focused on the core product as the service element of mobile phones is increasingly important. Nokia's challenge is not due to its production of inferior products, but rather because it has become part of an inferior ecosystem, which needs to provide more attractive, value adding services. Interestingly, larger firms do not appear to hold an advantage when it comes to the introduction of service innovation, whilst they do have the upper hand with product and process innovations (Pires et al., 2008).

Service businesses, meanwhile, can introduce 'product innovations'. McDonald's is a prime example, where a service has essentially been turned into a product (or suite of products), and with great success (Levitt, 1976). Many other service organizations, such as retail banks, essentially provide products, albeit intangible products, rather than true services (Hill, 1999; Avlonitis et al., 2001).

In manufacturing we use one word to describe the productive activity—'manufacturing'—and another to describe the principal output—the 'product'. In services the same word ('service') is used for both, which creates considerable confusion. Although service activities dominate advanced economies, our understanding of services, innovation in services, and service innovation, lags far behind our understanding of manufacturing, of innovation in manufacturing, and of (physical) product innovation (Page and Schirr, 2008). This can be seen by a search for articles on the Thompson-Reuters 'Web of Knowledge' database. A search using the term 'new product development' finds over 1,300 items with this term in the title: a very substantial literature. Meanwhile, a search using the term 'new service development' found only fifty-eight articles with this term in the title, the first having been published in 1985.[4] Similarly, a search using the terms 'product innovation' and 'innovation in manufacturing', found 700 articles, whereas a search using the terms 'service innovation' or 'innovation in service(s)' found half as many. But while 'service innovation' and 'new service development' received relatively little attention in the past, this situation is changing rapidly, and indeed most of the articles on these topics have been published in the last decade. In marketing especially, there has been a strong drive to focus on service, notably through the concept of 'service dominant logic' (Vargo and Lusch, 2004; Lusch et al., 2007). And a new literature is now emerging on service design (Moritz, 2005; Stickdorn and Schneider, 2010; de Meroni and Sangiorgi, 2011; see also Chapter 8 by Verganti and Dell'Era).

Here, we will (1) provide an overview of what services are, arguing that it is still useful to separate services from tangible and intangible products, for this distinction has implications for our second and third objectives, which are: (2) to understand different types of service innovations, and (3) to discuss some of the tools and techniques used to manage the development of service innovations.

WHAT ARE SERVICES?

Services are notoriously difficult to define. One jocular definition is that they are 'the fruits of economic activity that you cannot drop on your foot' (Pennant-Rea and Emmott, 1983). The act of dropping things on feet, or indeed on other things, could however be a service! This is a good first approximation, which highlights the intangible or immaterial aspects of services; it also highlights that services are acts or activities (Gallouj and Weinstein, 1997), rather than outputs. With services, the outcome and the activity are often inseparable (e.g. a 'dropping service'). In other words, the service 'product' is a process, act, or performance.

Services are often derided as being menial, but paradoxically they also contain the most knowledge intensive of all activities, including surgery, higher education, and consulting of various forms. The same word 'service' relates to both the 'best' and the 'worst' jobs in advanced economies, or both the 'top' and the 'bottom' of the knowledge economy.

One of the reasons why services are hard to define is that they have traditionally been a residual category. When the system of national accounts was developed in the 1930s and 1940s, 'industrial activities' were considered central, and economies were divided into three broad classes of activity: the primary sector (including agriculture, fishing, and mining), which extracts raw materials from the earth; the secondary, or industrial sector, which transforms those raw materials (and semi-manufacturing) into end products, and the tertiary, or 'service sector' which contains everything else. Various attempts have been made to update this, and in particular to differentiate amongst different types of services, but these have not found widespread acceptance.

A long-held view is that services are 'unproductive': at best necessary, and at worst a burden. This dates back to Adam Smith, who emphasized the distinction between goods and services. For Smith, manufacturing was particularly important because it 'fixes' value into devices (goods) and therefore builds the capital stock. The production of capital releases labour to engage in further productive activities, partly through the utilization of previously produced capital. Services, by contrast, which 'perish in the very instant of their performance, and seldom leave any trace or value behind them for which an equal quantity of services could afterwards be procured', do not 'fix value'.[5] This points to an important distinction, between 'true services', which are enacted or performed, and services which do indeed 'fix value' in such a way that the benefits from these activities endure. An unscripted speech, for example, may endure only in the memories of those present, but a written document or a video recording does 'fix value'.

A feature of services, and especially of 'true services', is that because they are enacted or performed, it is often difficult for them to achieve substantial and continuous productivity growth. For example, a nurse today probably takes a similar amount of time to dress a wound as it took Florence Nightingale over a century ago. And a live performance of Beethoven's Ninth requires the same time and number of players today as it did when it was written in 1824. The cost of bandages and musical instruments may have declined massively over time, and hospitals and auditoriums may be larger, but substantially and continuously enhancing productivity in the core service activity is impossible. This leads to the 'cost disease' that Baumol and Bowen (1966) recognized: essentially that the price of true services tends to rise relative to the price of goods, making them increasingly costly to provide. This is one of the reasons why many service providers seek to innovate by 'productizing' their service (e.g. by selling recordings of the orchestra playing—i.e. creating tangible products), by shifting activities from themselves to the customer (i.e. self-servicing), or by cutting back on the 'service elements' of their offer (e.g. less frequent mail deliveries; or employing fewer people to help guests at a hotel). The latter strategy can also involve introducing higher and lower classes of services. In general, as the quality (and price) of the service increases, we can expect that both the quality of both the service (S) and of the 'product' aspects (P) will improve. For example, an airline's First Class (FC) will be better than its Business Class (BC), which will be better than its Coach or Economy Class (EC), and therefore: $P_{FC} > P_{BC} > P_{EC}$, and $S_{FC} > S_{BC} > S_{EC}$, but less obvious is that the ratio of service to product is also likely to increase: $[(S_{FC})/(P_{FC})] > [(S_{BC})/(P_{BC})] > [(S_{EC})/(P_{EC})]$. The extent to which service providers can change the level of service provision, and/or transfer activities to their customers is partially dependent on the development of new technologies, partly dependent on changing customer preferences (e.g. self-service kiosks that are easily used, and ideally preferred by customers), and partially a matter of service design.

While services are often considered 'unproductive' and problematic, they also include the most generative and creative sectors of the economy. Beethoven was a service provider commissioned by his patron to write music. In the present day both R&D and market research are essentially service activities, without which most products would not exist. Indeed, one can argue that these knowledge-intensive services are now the primary economic sector: they are critical 'extractive' economic activities, upon which advanced economies are built.

A positive way to think about services is to consider what they are, what they act on, and how they 'add value'. Services can act on people, individually or in groups, and can act on our minds, bodies, or both. They can also act on things, from tangible property (e.g. maintenance and repair) to information (e.g. banking and insurance services). What is acted on has implications for the nature of services provided, for their design, and for the possible trajectories for innovation. Data, for example, are much more easily dispersed across space, and manipulated, than either people or physical objects, and one cannot, or at least should not, treat people like physical objects.

'True' or 'classic' services are considered to have four characteristics that distinguish them from physical products, or goods, sometimes captured by the acronym: IHIP (Sasser, Olson, and Wyckoff, 1978; Johne and Storey, 1998):

- First, *intangibility*, or immateriality. This means the service cannot be touched, seen, tasted, or heard. Legal advice is just one example; a wedding another. A service may of course involve things that can be touched, seen, tasted, and/or heard, and these physical aspects of a service are often very important because they are considered to represent the values of the service (e.g. the cleanliness of a restaurant gives a clue as to its general quality; a wedding needs a cake!, etc.).
- Second, *inseparability* between what is provided and who is providing it. Whereas a physical product (such as a book) exists independently of the entity (person or organization) that produced it, this is not the case with a 'true service'. A 'true service', such as a rock concert, cannot be separated from the unit of its provision. It is for this reason that 'true services' are inherently relational, rather than transactional, as is the case with physical products. Inseparability also relates to another of the defining characteristics of services, which is that their enactment does not involve a transfer of ownership (Hill, 1999). Because services do not have an independent existence, they are often consumed as they are produced, something which is referred to as simultaneity (or co-production). Sometimes simultaneity requires both the presence and/or active participation of the customer, user, or recipient (education is a good example: a teacher helps the student to learn, but cannot force the student to learn—learning is co-produced by the student and teacher), but not all services are 'done with' the client: many services are 'done for'

(or on behalf of) the client, with the service provider acting as the client's agent (e.g. lawyers in real estate conveyancing, or divorce settlements).

- Third, services are temporal and *perishable*: they exist in time. This property partially follows from their inseparability. Because outputs have no independent existence they cannot be stocked as physical products can be stocked. An unoccupied airline seat is a revenue opportunity lost forever once the aircraft leaves the departure gate. The temporal nature of services is particularly important if demand is unpredictable or stochastic over time, and leads to lots and varied attempts to manage demand (e.g. peak and off-peak ticket prices, etc.).

- Fourth, because services are enacted, they are often variable or *heterogeneous*. Whereas with classic manufacturing the aim is to first design and then produce in volume such that each individual product is identical (and therefore the client is indifferent to which particular item he buys). Such a high degree of conformity, or standardization, is very difficult to achieve in enacted services, especially where people are involved in their provision. This can be seen as advantageous in services: every night in the theatre should be a little different from any other night; it is not the same as watching a film.

It is, however, not difficult to think of many things that are commonly described as services which have few (and possibly none) of these characteristics. In part, this is because of forms of innovation in services which seek to relax the classic characteristics of services. Meanwhile, many 'products' are adopting some of the characteristics of services, becoming for example more customizable or configurable, rather than standardized (e.g. mobile phones). And product-related brands are used to forge enduring relationships with customers: we are asked not only to buy a product, but to buy into a lifestyle.

Forms of Service Innovation

Service innovations can involve the introduction of new services (with all, or most, of the IHIP characteristics outlined above) which can be provided *for the client* (i.e. without their active involvement), or which can be provided *with the client* being actively involved in the execution of the service. A grocery retailer might, for example, introduce a new service that not only delivers groceries but brings them into the home and places them in cupboards, fridges, and so on. Or a restaurant might offer a new service whereby a chef comes to the customer's home and cooks meals there. Some degree of participation by the clients is required, but fundamentally these services substitute for the activities that customers would have previously done for themselves. Other new services are done with the client as an active participant. For example, for $200,000 per person, Virgin Galactic will soon take you to the edge of space. The service involves two days of flight preparation, training, and other activities: you are not a 'mere passenger'.

Overall, and unlike new technologies, 'radical new services' are not easy to identify. Most service innovation involves relatively minor changes to existing services, and much has to do with finding ways to escape classic IHIP characteristics. Because of these characteristics, classic or true services are provided in close physical proximity to the user and are difficult to scale. Various approaches to innovation break these bonds, allowing services to be scaled, both in the volume of customers served and across geographical space. Approaches include:

Increasing tangibility: Although the essence of a service may remain intangible, one form of innovation involves increasing the tangible, or material aspects of the service experience. An example is retail banking. A generation ago it was common for customers to be known personally by bank staff, and loan judgements could usually be made by the local branch manager. Except for banks that have exceptionally affluent clients (e.g. Coutts & Co), this is no longer normal practice. Instead, banks have introduced a lot of tangible artefacts, including plastic cards, cheque books, pay-in slips, ATM machines, and so on, all of which increase the tangible aspects of the service. These tangible aspects are much easier to design and control, and allow a productization of the service, which enables it to be scaled relatively cheaply, both in the volume of customers served and in the geographical space covered: customers can use ATMs around the world, not just their local branch.

Addressing inseparability: Because a service is inherently tied to the entity that provides it, there are two basic ways to innovate with respect to this characteristic. The first is to reduce the inseparability. For example, traditional university education is campus-based, delivered primarily through lectures and seminars. The UK's Open University, established in 1969, broke that mould by carefully preparing textbooks and other study materials so that students could mainly self-study off-campus. This also relates to the second basic way to address inseparability, which is to transfer activities from the service provider to the service user/customer. The move to self-servicing, often accompanied by automation (increasing the tangibility of services and/or simplifying the service), is a widely used service innovation strategy, especially in countries where labour costs are high. Related to this is simultaneity, which implies that services are produced and consumed at the same time (e.g. a music concert). Historically, enjoyment of the concert meant being in the auditorium, but less so now. The concert can be broadcast using telecommunications, potentially to anywhere in the world, and can be recorded for later viewing. The experience is not exactly the same as 'being there', but enjoyable, distributable versions of the concert are attainable. Some services, such as surgery, have historically required the co-presence of the service provider and service consumer. But even here, robot technologies are being developed to allow surgeons to remotely operate on patients.

Addressing temporality: Services are typically demanded in time and space, and innovation can relate to breaking both these constraints. For example, 'out of hours', or 'extended hours' services make services available outside of normal working hours, often through developing different organizational arrangements. Examples include using telephone call centres located in different time zones (e.g. using call centres in India to serve UK customers), or altering how services are provided. In the UK, for example, large stores are only allowed to open for six hours on Sunday. This is one of the reasons why the major supermarket chains have expanded their number of small format, 'local' stores, the trading hours of which are less restricted. Ensuring quality service provision is particularly challenging if levels of demand are unpredictable, or stochastic. In manufacturing a degree of unpredictable demand is normally accommodated by holding stock (which is effective with all but the most perishable, or customized, of goods). Because services cannot generally be stocked, this buffering is not generally possible. The simplest and most common way of dealing with the problem is to stock demand, in the form of queues waiting to be served. Expecting customers to wait, especially for long periods, is often not acceptable, and there are a host of ways in which the service provider can deal with stochastic demand. One is to maintain excess capacity to serve, which implies additional costs. A second is to try to make demand more predictable by using discriminant pricing (and other techniques), charging high tariffs when capacity is scarce and low tariffs when it is abundant. Another is to vary the volume or speed of processing, although this implies different levels of service, and often a reduced quality of service at the busiest times. More innovative approaches include the introduction of highly flexible systems of provision, such that the capacity to serve can be expanded or contracted rapidly to accommodate shifting patterns of demand. Some call-centre operators, for example, have a 'reserve army' of home workers who can be added to the workforce, albeit working from home, if the level of demand means that callers are waiting too long to be served.

Addressing heterogeneity: At one extreme, each service event is unique, or bespoke, and designed to meet the specific needs of a particular client. But unique or bespoke services are typically expensive to provide. A common strategy is to 'industrialize' services, using productization strategies, of which McDonalds is a classic example (Levitt, 1976). This strategy involves de-skilling the workforce, and embodying the service not in the knowledge of people providing the service, but in the practices and routines to be followed (Metcalfe et al., 2005). At the extreme, this results in a highly standardized 'product' (a Big Mac is a Big Mac anywhere in the world), but arguably with very little service. More generally, service-level agreements are widely used to define the key metrics of a service (e.g. average waiting times, turn-around time, time taken to resolve, etc.), whilst scripting is often used to guide, if not define, what front-line staff say to customers. Many of these practices involve removing some of the tacit or discretionary elements of service, for which customers will pay premium prices. A more subtle approach is not to standardize the service, but rather to customize it, often on the basis of building service platforms or modules, which can be configured to the needs of particular customers. The modularization of services is quite commonly used in activities like education and health care, but difficulties can arise in: 1) standardizing the interfaces which is necessary for true modularity; and 2) maintaining a consistency in style where is a danger that the service becomes disjointed, with different elements increasingly dissimilar to one another.

Intra- and Inter-Organizational and Relational Forms of Innovation

Services are largely experienced, yet a complete service frequently depends on a number of organizations (or departments within an organization) working together to provide the service. A journey by air, for example, is likely to involve a number of organizations, only some of which the passenger will come into direct contact with (e.g. the passenger will not come into direct contact with air traffic control, nor with baggage handlers). The effective orchestration of the service requires that these organizations or departments work together. Very often, problems arise at the interface of organizations, and intra- and inter-organizational innovations play a significant role in services (van der Aa and Elfring, 2002; Tether and Tajar, 2008; Eisingerich et al., 2009). For example, in the UK a significant problem for the National Health Service (NHS) is 'bed blocking', which occurs where someone (usually an elderly patient) remains in hospital, thereby occupying a hospital bed, not because they require further medical treatment but because they cannot go home and there is nowhere else for them to go.[6] 'Bed blocking' is estimated to cost the NHS more than £500,000 a day. It places patients in an unpleasant and unsettling situation, and means that hospital beds are not available for other patients, which may extend their waiting times. The situation largely arises due to breakdowns between the health care system and social services, including those provided privately (care homes) and those provided by the state.

A very significant form of innovation in services therefore relates to inter- and intra-organizational cooperation. One example of effective inter-organizational cooperation that led to sustained innovation was that between the pilots, air traffic controllers, and the airport operators at London Heathrow, Frankfurt, and other highly congested airports (Tether and Metcalfe, 2003). At Heathrow in particular, by these organizations working closely together, learning by interacting, and discovering, then implementing a series of incremental innovations, a significant amount of additional runway capacity was 'discovered' despite the system being considered full. Many of the innovations were micro-changes to procedures, but their cumulative benefit in terms of the additional capacity realized was significant, putting back the need to build an additional runway by several years.

THE MANAGEMENT OF SERVICE INNOVATION

Barriers to Innovating Services

Most of the barriers to innovation that exist in manufacturing also apply to services, but arguably services also face some important additional barriers. One is that because services do not have an independent physical existence, but are instead provided, there is a much closer relationship between the service output (or product) and the means of provision (the process). The relationship between these often becomes embedded into working practices, and job definitions. For example, in the airline industry, pilots' unions have negotiated 'scope clauses' with the airlines, which limit the number and/ or size of aircraft that the airline can contract out to smaller, regional airlines. These clauses were intended to protect jobs with the major airlines by preventing regional airlines from operating large aircraft. But the existence of these clauses (and corresponding job definitions) effectively reinforced the predominant 'hub-and-spoke' model of the major airline's operations, and therefore added to the rigidities that 'the majors' faced when dealing with the threat of the 'no frills' airlines that primarily operate a point-to-point model using smaller aircraft. Similarly, conventional 'bricks and mortar' shops found it difficult to respond to new e-tailers like Amazon.com because their business model was based on a very different proposition and set of investments. The main point here is that major innovations in services often involve business model innovation and a fundamental change in the organization of service provision.

The Appropriability Regime for Innovating Services

A second important issue is the 'appropriability regime', which concerns how firms can protect their inventions and innovations against copying or imitation, and how they can extract commercial value from them (i.e. gain the returns to innovation). Generally, firms can benefit from two forms of protection: formal forms, and strategic forms (Tether and Massini, 2007). Formal forms include intellectual property rights, such as patents, copyrights, design rights, and trademarks.

Patents, which can be the strongest form of intellectual property protection for technological innovations (Cohen et al., 2000 and see Chapter 28 by Leiponen), are not generally applicable to services, although in some jurisdictions, including the United States and Japan, business method (and software) patents are increasingly allowed and are used to protect services.[7] In other jurisdictions, such as the EU and Australia, patents cannot be used to protect 'pure' business methods (i.e. those that do not have a technical aspect and inventive step).

The existence of effective business method patents changes the service innovation landscape fundamentally, as it encourages the deliberate invention and protection of service-related business methods. Potentially, this may become a specialist function in the firm, akin to the R&D department in technology-based manufacturing firms. It may also accelerate the development of new and highly innovative services. However, the patenting of business methods can also be used to anticipate future risks to an incumbent's business models, and may be used to block the emergence of new, rival business models.

Some services that have a physical presence can and do make use of both technology patents and registered designs (which protect the external form of products).

Virgin Atlantic, for example, used both to protect its lie-flat business class beds and the herringbone seating arrangement used for its 'upperclass' seating.[8] But more generally the weakness of effective intellectual property protection for service inventions is thought to encourage a highly incremental approach to innovation. This is because firms are reluctant to take the risk associated with developing fundamental service innovations because, as the first mover, they would bear the risk that the innovation fails in the market whilst not being able to stop 'fast followers' from benefitting from their pioneering efforts. Unless they have strong complementary assets which help protect the innovation from rapid imitation, incumbents therefore typically prefer to introduce incremental innovations (Alam, 2006b).

Brands, trademarks, and servicemarks are more relevant to services: a brand is a repository or sign of qualities that are difficult, if not impossible, to quantify. They are especially important in retailing and other consumer services, such as banking and insurance, but are also increasingly used by professional service organizations and even public-service organizations.[9] Brands are, or can be, complementary assets (Teece, 1986), which can help firms to appropriate the returns to their innovation activities.

Copyright is also relevant, although copyrights protect intangible products, rather than true services. Indeed, the invention of printing, essentially a manufacturing technology, spurred the introduction of copyrights. Copyrights are also used to protect websites and web content.

Strong IP protection also encourages a division of labour between inventors and producers, because it makes clear 'who owns what' and encourages a market for ideas (Gans and Stern, 2003, 2010). In these circumstances the outsourcing of the production of ideas (brands, designs, patents, etc.) to specialist service providers, or the outsourcing of production, becomes more commonplace. Strong specific capabilities and reputation effects also encourage the splitting of creative service activities from production activities, for example architectural practices are service businesses split from construction, whilst advertising agencies are separate from the producers of the products they promote.

In general and with the possible exception of trademarks (i.e. brands), formal forms of intellectual property protection can be used to protect the periphery rather than the core of the services. Strategic forms of protection (including secrecy, complementary assets and activities, social capital or connections, and reputation) are generally more significant than formal IP protection, with both less widely used than in manufacturing (Tether and Massini, 2007).[10]

Approaches to Managing Service Innovation

Because services lack an independent physical existence but are increasingly an important source of competitive advantage for modern organizations, Cooper and Edgett (1999) consider that service innovation is the most challenging and difficult endeavour facing modern corporations. Aside from being intangible, service innovation is more difficult than conventional product innovation because what is provided as a service is usually much more tightly coupled to the process of provision than is a product. In other words, whereas in manufacturing product, process, and organizational innovation are often loosely tied or coupled (a new product can often be innovated without making substantial changes to production processes or organizational arrangements), this is often not the case in services.

Success in innovation involves picking the right projects and doing projects right. Many service organizations fail on both counts: first, because there is no systematic process or rationale for the selection of projects; second, because projects are managed in an ad hoc and often informal manner (Sundbo, 1997; Kelly and Storey, 2000; Dolfsma, 2004), with formal approaches to project selection and development being much less widespread than amongst manufacturing firms. The prevailing approach is often to generate an idea (often in response to a competitor's actions), do a minimum of background work, and move directly to development.

It is important to consider the fit between the strategy of the business and its innovation strategy (Teece, 1986). These can easily become misaligned, such that innovation undermines rather than reinforces the strategy of the business. For example, the innovation strategy of a low-cost airline like Ryan Air is to automate as much as possible and to try to shift as much activity (or work) as possible to the consumer, charging high prices for 'extras' such as having a boarding card printed at the airport. This cost-minimization strategy would be wholly inappropriate for a high-service airline such as Singapore Airlines.

A weak development process is also associated with a high 'opportunity cost', as alternative projects are not considered, and is more likely to give rise to costly failures, which further reduce confidence that the organization can innovate successfully. A further problem is that organizations often pursue too many innovation projects simultaneously, which can lead to incoherence and can spread the available resources too thinly.

To be effective, service innovation requires alignment to a clear strategy, effective processes, and adequate resourcing (Cooper and Edgett, 1999). Often, service organizations try to innovate without adequately resourcing the activity—those involved are expected to do it in their 'free time' alongside their other duties. Instead, it is often better to set aside time and resource dedicated to innovation. Many service firms also lack institutional arrangements to support innovation activities. For example, whereas manufacturing firms typically have an accounting line for 'development', this is rare in service organizations. Getting an effective innovation process implemented in service organizations is often difficult, as this involves challenging deep-seated organizational practices.

Consideration should be given to the various spaces, or dimensions of service innovation. Den Hertog (2000) considers that there are four 'dimensions' of service innovation (which are not wholly independent of one another), including: 1) The Service Concept, business model, or value proposition (Edvardsson and Olsson, 1996). The most radical forms of service innovation involve reconceptualizing the service concept. Examples include Skype compared with conventional telephony. Clearly, engaging in this form of service innovation may be hazardous to an existing business model as what were considered core assets may become liabilities. 2) The Client Interface: that is, innovating the front-of-house activities, or touchpoints where the service provider and clients meet. This is an important space for innovation, and the service provider needs to be clear about how innovation will impact on the client experience (we discuss this further below). 3) The Service Delivery System: that is, considering how the service is provided or delivered to the user/customer. There is usually a back-stage and a front-stage aspect to this, the activities of each of which should be carefully considered. And 4) Technological Options: that is, the role of technologies in providing or supporting the service. Care should be given as to how technologies can be used to enhance the service, rather than just reduce costs, and how to align this with strategy.

Other important aspects to consider include the extent to which customers are involved in the service innovation process. Evidence suggests this tends to be beneficial, particularly to the identification of more radical innovations that create more value to customers (den Hertog, 2000; Abramovici and Bancel-Charensol, 2004; Alam, 2006a). However, involving customers can also raise expectations about innovations that are very difficult to implement (Magnusson et al., 2003). The use of cross-functional or departmental teams is also generally recommended (Avlonitis et al., 2001), as this brings together a variety of knowledge-bases and perspectives (Leiponen, 2005, 2006), and builds 'buy-in'. The use of cross-functional teams, together with an effective innovation management process, has been found to increase the speed, reduce the cost, and improve the quality of innovation (Hull, 2003; Fay et al., 2006). But care should be taken not to allow innovation teams to become too large, as inter-team communications are much easier and more effective in small teams. In this context, the involvement of front-line staff should not be overlooked or undervalued, as their involvement provides access to their 'coal face' insights and builds 'buy-in'. However, this practice often runs against hierarchical norms in organizations (de Brentani, 2001; Ramirez, 2004).

Service Design and Development

In recent years, a literature on service design has developed (e.g. Moritz, 2005; Stickdorn and Schneider, 2010; de Meroni and Sangiorgi, 2011). As with the design and development of new physical products, commentators have advocated that new services design and development be broken down into stages, or phases (e.g. Cooper and Edgett, 1999). The number of stages, and role of each, varies, and should be appropriate to the specific organization. Three broad stages are: identify (or scan), (select and) build, and implement. We outline these phases below, and some of the tools used in each of them:

Stage 1: Identify
The first stage of service design and innovation is the identification of the problem or opportunity to be addressed. Many organizations spend far too little time on these front-end activities, which include scanning the environment to look for both threats

and opportunities. They fail to actively look for problems or opportunities, only dealing with these as they arise (if indeed they really deal with them at all). They also often
take problems at face value, considering them entirely within existing frameworks or
constructs, thereby failing to 'play with the problem' and possibly reconceptualize it
(Hargadon, 2002). This, however, is a significant part of the design process. For example, when the architect Frank Gehry was commissioned to design the new School
of Management building at Case Western Reserve University, he started by asking
fundamental questions, such as 'what is education?' and 'what is faculty?'. These challenged those commissioning and designing the building to reassess some of their
basic assumptions about the structure of the building as a platform for providing
services.

Innovations can arise from observations. For example, the design and innovation
consultancy IDEO was challenged by Bank of America (BoA) to find new ways to entice
people to bank with them. After undertaking observations in three US cities, IDEO
noticed that people would often round up their financial transactions for speed and convenience. They also discovered that many people want to save, but fail to do so because
of a lack of resources or willpower. Based on these observations and subsequent brainstorming, IDEO proposed a 'piggy bank' solution based on existing habits. Called 'Keep
the Change', the service rounds up to the nearest dollar purchases made with a BoA
debit card, and transfers the difference from individuals' checking accounts into their
savings accounts, thereby helping individuals to save in a convenient way. Launched in
2005, 'Keep the Change' attracted 12 million customers to BoA, helping them to save
over $3bn.

Observations can lead to the identification of lots of opportunities for innovation. The
challenge is to understand both the customer's journey through the service (and interactions with it), and the structuring of service provision. Service design is both a form
giving challenge and an organizational structuring challenge. Service blueprinting, first
developed by Shostack (1982, 1984; Bitner et al., 2008) is a useful and potentially powerful tool for mapping out services in order to understand them, and to identify issues
requiring improvement. A Service blueprint includes:[11]

1. *Customer actions*: that is, all of the steps that the customer takes as part of his/
   her engagement with the service. This includes the 'moments of truth', which
   are the instances where the customer and service provider come into contact
   with one another in a manner that gives the customer an opportunity to either
   form or change an impression about the service provider. Favourable moments
   of truth, where the service provider has exceeded the customer's expectations,
   are sometimes called 'moments of magic'. By contrast, 'moments of misery' are
   negative customer interaction, where the customer feels poorly served and dissatisfied. Service providers need to be aware of issues such as confirmation bias.
   If a customer is very (un)happy with one service encounter, this is likely to carry
   over and influence her perception of subsequent encounters; service journeys
   are to a considerable extent subjective.

2. Onstage, or 'front of house' employee actions, that is, the face-to-face actions between physically present employees and customers. The activities of visible, front-of-house, customer-contact employees, are separated from customers' own actions by the 'line of interaction'.
3. Back-stage, or invisible contact employee actions. These are employees who engage with customers but who are not visible to them: for example, call centre operatives. A 'line of visibility' separates back-stage employees and their activities from customers.
4. Support processes: that is, the processes which support on-stage and/or back-stage contact employees. An 'internal line of interaction' separates the contact employees from these processes.
5. Physical evidence: the physical evidence or environment that customers come in contact with (e.g. the airline cabin, or hotel room).

Blueprints can be high level and 'macro', or focus on specific elements of a wider service, and the process starts by determining the bounds of the service to be blueprinted. But deciding when a service starts and ends is often an illuminating matter in itself, and there are often significant opportunities for innovation before the service is conventionally thought to start, or after it has 'ended'. The second step involves identifying customers' actions through the service (i.e. the customer journey), with the actions of on-stage and back-stage employees, the support processes, and physical evidence then added in. Blueprints can be very useful in developing a shared understanding as to how a service fits together, which is often missing in highly siloed service organizations.

Stage 2: Build

The identify stage should lead to a full, or at least fuller, understanding of the service and the possibilities for innovation. For example, what actions can be transferred to or from customers? Airlines, for example, have been encouraging passengers to print their own boarding cards rather than have these printed for them at the airport. What actions can be transferred from visible, front-stage employees to invisible, back-stage employees? In 1985, for example, Direct Line 'reinvented' retail motor insurance in the UK by moving this from being sold in shops to being sold over the telephone. With its distinctive red-telephone on wheels logo, and effective call centres, Direct Line rapidly became the UK's largest motor vehicle insurer. There are also opportunities to automate the service, or to encourage customers to self-serve through IT systems rather than engage with the service provider's employees. Booking online via websites is now commonplace. Very effective services can remove or minimize interactions between customers and the service providers' employees. The vast majority of transactions on Amazon do not involve any direct interactions between Amazon's employees and its customers.

Unlike physical products, services are difficult to prototype. With products, proto-types are widely used to both develop and refine ideas, and moreover prototypes are a powerful communications device and 'boundary object' that enables shared under-standing between departments, such as marketing, R&D, and others (Dougherty, 1992; Carlile, 2004). Because services do not have an independent physical existence, proto-typing, and developing a shared mental model of the services, and possible innovations to it, is more difficult. Indeed, because the tangible aspects of a service are those that are most easily prototyped, this can lead to attention being lavished on these 'product aspects', whilst the 'service aspects' are neglected. Various techniques have, however, been developed to prototype services. These include:

1. The service blueprints discussed above and developed during the observation phase. These can be used to 'see' and develop the service, first from the per-spective of the customer, and then from the perspective of the front-stage and back-stage employees. A useful technique within this is the use of personas. Personas are made-up people or characters that represent the different types of users who can be expected to engage with the service. They are intended to create empathy for different user groups and help service innovators to both see the service from perspectives other than their own, and 'put a human face' on both customers and service employees; they help make services less mechanistic.

2. Using storyboards or comic strips. This technique is borrowed from the film industry, whereby the service journey is depicted in a series of drawings or pic-tures, put together in a narrative sequence. Touchpoints and 'moments of truth' are usually central to this process, with these represented in individual pictures. The use of storyboarding allows the order in which touchpoints occur to be reconsidered and re-sequenced, which can innovate a service.

3. Role play and theatre. Borrowed from theatre and drama, this technique experi-ments with playing out different service scenarios. This highlights that services are enacted.

4. 'Faking the advert'. Developed by IDEO, the objective of this technique is to crystallize or encapsulate the new service proposition into a short fake advert. Using a severe time constraint forces the team to really focus on the essence of the service and what it offers to customers.

Before moving to implementation, it is recommended that reflection and evalua-tion of the proposed new service should take place. Cooper and Edgett (1999) advo-cate that new service development should be focused on what they call 'superior services'. Superior services are those that achieve one or more of the following tar-gets: 1) Deliver unique or superior benefits to customers; 2) Provide better value for money than previous offers; 3) Feature a better service outcome than rival services; 4) Be more reliable and have fewer fail points; and 5) Have a higher quality image. They find that new services that meet these 'superior service' criteria are more likely to be successful and meet their profit objectives than 'me too' services. The character-istics of 'superior services' can be used as a scale against which to assess a proposed new service offering.

Stage 3: Implement

Implementation is challenging for a number of reasons:

1. It can be difficult to fully specify a service in advance. Moreover, it may not be wholly desirable to do so, especially if some of the finer details of how the service is provided is to be left to the discretion of the individual service providers. Services are sometimes said to exist in a state of 'perpetual beta', constantly being tweaked, or subject to minor improvements.
2. It can be difficult to fully understand any existing service to be innovated, and therefore to anticipate all of the 'teething problems' that arise when introducing a new service. This can be mitigated by involving front-line staff with knowledge of 'field problems' and issues in the development of the service. A staged roll-out process is also advisable so that teething problems can be understood and addressed before mass implementation takes place.
3. Involving front-line staff in the service design/innovation process also helps to build 'buy in' amongst these service providers. This is important because these employees are central to the enactment of the new or changed service. Special care needs to be taken if the responsibilities of front-line staff are reduced without their consent but they are still integral to service provision. This can lead to disaster, as Flowers' (1996) examination of innovation at the London Ambulance Service demonstrates.
4. The new service may be so successful, or so desirable, that customers demand exceeds the capacity to provide. In public services especially this leads to accusations of unfairness, and 'postcode lotteries'.

For these reasons, it is often advisable to quietly introduce the new or innovative service and to address teething problems before any later formal launch.

Is Formalizing the Service Innovation Process Effective?

Evidence suggests that formalizing the service innovation is effective, as it improves the quality of execution of a number of service innovation activities and reduces the number of failed service innovations, increasing the quality of the services introduced, and improving the timeframe over which they are developed (Froehle et al., 2000; de Brentani, 2001). Storey and Kelly (2001), meanwhile, investigated how new service development activities are evaluated, and found that in the most innovative firms measures include the cost and speed of development, and the effectiveness of the process, whereas less innovative firms typically confine their evaluation to financial measures of performance.

CONCLUSIONS

Approximately three-quarters of value added in advanced economies is due to services, and product-based companies are increasingly providing and innovating services. Yet our understanding of services, and how to innovate services, is much less developed than our understanding of manufacturing, and how to innovate physical products, production processes, and the organizational arrangements for production. This situation is changing, as the literature on services, and service innovation has grown rapidly in recent years.

This chapter has provided an overview of services, and emphasized how they differ from physical products. In this context we should stress that many services are becoming more 'product-like', whilst physical products are often being blended with services (e.g. the Apple iPod/iPhone is not just an attractive physical product, but an attractive bundle of product and services, many of which the user can configure to meet his own specific needs. Apple also maintains its own chain of stores so that it can exert greater control over the customer experience).

New services are developed that remove the burden of activities from the customer ('done for' services), whilst other new services are 'done with' the customer, and still others effectively transfer work to the customer. Most innovation in services is less radical, involving shifts away from the fundamental characteristics of services: intangibility, heterogeneity, inseparability, and perishability. Much is also organizational rather than technological in nature. The evidence suggests that carefully managed approaches to service innovation are less widespread than in otherwise similar manufacturing organizations, but that the adoption of these practices is highly beneficial; indeed, possibly more beneficial than in manufacturing, precisely because they are less widespread.

The literature on services, service design, and service innovation has been growing rapidly in recent years, so many of the gaps in our understanding are being addressed. But there is still more to do, and here we make three suggestions for further research that would aid understanding of the management of service innovation. These are:

1. A more detailed understanding of how to formalize the innovation process for services in different contexts and the effectiveness of this. Some studies have found benefits from formalized innovation processes in services, but how substantial are these benefits, and does the scale of the benefits vary between different service contexts (e.g. large service retailers versus small bespoke consultancies)? Also, more detailed guidelines on what to include in the formal process would be very useful to managers of service design and innovation processes: for example, when and how to involve front line staff in new service development activities.

Work in this space would be of considerable practical value as well as academic interest.

2. A fuller understanding of how workforce and managerial skills, and human resource practices, impact on service innovation. The whole area of interconnections between skills and innovation is surprisingly underdeveloped (Tether et al., 2005). Generalizing somewhat, services are much more dependent on skills, and especially 'soft skills' than are manufacturers, so an understanding of skills and human resource issues is more critical. We know, for example, that service innovation often involves training and up-skilling, but we know little about their content and general effectiveness.

3. A fuller understanding of intellectual property protection strategies in services, and how firms use (or don't use) various forms of IP protection to develop innovations and appropriate the returns to these (Tether and Massini, 2007). This is especially interesting given differences between countries in whether or not service innovations, or key aspects of them, can be protected by intellectual property rights. The differences in business method patenting between the US and EU should alter the balance of incentives between incumbents and insurgents in the development of service innovations in these contexts. Much could be learnt by studying the same service systems in contrasting institutional contexts.

Notes

1. Various attempts have been made to distinguish different types of service. Schmenner (1986) develops a two-by-two matrix based on labour intensity (i.e. the capital to labour ratio) (low; high) and degree of customization (low; high). He defines as 'service factories' those low on both (e.g. airlines, trucking, retail banking); as 'professional services' those high on both; as 'mass services' those high on labour intensity but low on customization (e.g. retailing, schools); and as 'service shops' those that are low on labour intensity but high on customization (e.g. hospitals and auto repair). This matrix is static rather than dynamic, so it is interesting to consider how innovation has altered (or not) both capital intensity and customization in these activities.

2. Amongst services, telecoms, and IT, financial services (and especially retail banking), knowledge-based services, and public services (including health care) have received by far the most attention by researchers. Many services have received very little attention.

3. Although exceptions exist, especially in telecommunications, computer software, and R&D service companies.

4. Admittedly a few of the articles on 'new product development' relate to service organizations (and especially financial services) or contexts. But arguably, conceiving of services as products is problematic.

5. The view that services are unproductive has long been contested. See Hill, 1977 for a full discussion.

6. During the 15-month period from August 2010 to the end of October 2011, the NHS claims more than 900,000 <Compositor: thin space please>hospital bed days were lost to bed

blocking. The government estimates that each day lost to bed blocking costs the NHS about £260.

7. An infamous example is that of Amazon's 1-Click, or one-click-buying. In the United States the patenting of business methods (i.e. class 705) has been growing at 15 per cent per annum over the period from 1970 to 2010.

8. The airline has both licensed the design to others, thereby partially recovering the development costs, and has taken legal action against infringements, some of which it has lost.

9. Walmart, Vodafone, and Amazon are service businesses that have amongst the ten most valuable global brands. Google, arguably a service business, is another.

10. However, some technical services, including telecommunications, computer services, R&D services providers, and architectural and engineering consultancies do make greater use of formal forms of protection.

11. Similarly, Voss and Zomerdijk (2007) identify five aspects of experiential design that apply to services, and the service experience: the physical environment, the actions of visible service employees, the service delivery process, the actions or behaviours of fellow customers, and, behind the line of visibility, back-office.

REFERENCES

Abramovici, M., and Bancel-Charensol, L. (2004). 'How to Take Customers into Consideration in Service Innovation Projects', *Service Industries Journal*, 24(1): 56–78.

Alam, I. (2002). 'An Exploratory Investigation on User Involvement in New Service Development', *Journal of the Academy of Marketing Science*, 30(3): 250–61.

Alam, I. (2006a). 'Removing the Fuzziness from the Fuzzy Front-End of Service Innovations through Customer Interactions', *Industrial Marketing Management*, 35(4): 468–80.

Alam, I. (2006b). 'Service Innovation Strategy and Process: A Cross National Comparative Analysis', *International Marketing Review*, 23(3): 234–54.

Anand, N., Gardner, H. K., and Morris, T. (2007). 'Knowledge Based Innovation: Emergence and Embedding of New Practice Areas in Management Consulting Firms', *Academy of Management Journal*, 50(2): 406–28.

Avlonitis, G. J., Papastathopoulou, P. G., and Gounaris, S. P. (2001). 'An Empirically-Based Typology of Product Innovativeness for New Financial Services: Success and Failure Scenarios', *Journal of Product Innovation Management*, 18(5): 324–42.

Baumol, W. J., and Bowen W. G. (1966). *Performing Arts: The Economic Dilemma*. New York: The Twentieth Century Fund.

Berry, L. L., Shankar, V., Parish J. T., Cadwallader, S., and Dotzel, T. (2006). 'Creating New Markets through Service Innovation', *MIT Sloan Management Review*, 47(2): 56–63.

Blindenbach-Drissen, F., and van den Ende, J. (2006). 'Innovation in Project-Based Firms: The Context Dependency of Success Factors', *Research Policy*, 35: 545–61.

Bitner, M. J., Ostrom, A. L, and Morgan, F. N. (2008). 'Service Blueprinting: A Practical Technique for Service Innovation', *California Management Review*, 50(3): 66–94.

Cainelli, G., Evangelista, R., and Savona, M. (2004). 'The Impact of Innovation on Economic Performance in Services', *Service Industries Journal*, 24(1): 116–30.

Camacho, J. and Rodriguez, M. (2005). 'How Innovative are Services? An Empirical Analysis for Spain', *Service Industries Journal*, 25(2): 253–71.

Carbonell, P., Rodriguez-Escudero, A. I., and Pujari, D. (2009). 'Customer Involvement in New Service Development: An Examination of Antecedents and Outcomes', *Journal of Product Innovation Management*. 26(5): 536–50.

Carlile, P. R. (2004). 'A Pragmatic View of Knowledge and Boundaries: Boundary Objects in New Product Development', *Organization Science*, 13(4): 442–55.

Cohen, W. M., Nelson, R. R., and Walsh, J. P. (2000). *Protecting their Intellectual Assets: Appropriability Conditions and Why U.S. Manufacturing Firms Patent (or not)*, NBER Working paper 7552. Cambridge, MA: National Bureau of Economic Research.

Cooper, R. G., and Edgett, S. J. (1999). *Product Development for the Service Sector*. New York: Perseus Publishing.

Damonpour, F., and Gopalakrishnan, S. (2001). 'The Dynamics of the Adoption of Product and Process Innovation in Organizations', *Journal of Management Studies*, 38(1): 45–65.

de Brentani, U. (2001). 'Innovative Versus Incremental New Business Services: Different Keys for Achieving Success', *Journal of Product Innovation Management*, 18: 169–87.

de Jong, J. P. J., and Klemp, R. (2003). 'Determinants of Co-Workers' Innovative Behaviour: An Investigation into Knowledge Intensive Services'. *International Journal of Innovation Management*, 7(2): 189–212.

de Meroni, A., and Sangiorgi, D. (2011). *Design for Services*. Gower, Farnham, UK and Burlington, VT.

de Vries, E. J. (2006). 'Innovation in Services in Networks of Organizations and in the Distribution of Services', *Research Policy*, 35(7): 1037–51.

den Hertog, P. (2000). 'Knowledge Intensive Business Services as Co-Producers of Innovation', *International Journal of Innovation Management*, 4(4): 491–528.

den Hertog, P., van der Aa, W., and de Jong, M. (2010). 'Capabilities for Managing Service Innovation: Towards a Conceptual Framework', *Journal of Service Management*, 21(4): 490–514.

Dolfsma, W. (2004). 'The Process of New Service Development: Issues of Formalization and Appropriability', *International Journal of Innovation Management*, 8(3): 319–37.

Dougherty, D. (1992). 'Interpretive Barriers to Successful Product Innovation In Large Firms', *Organization Science*, 3: 179–203.

Droege, H., Hildebrand, D., and Heras Forcada, M. A. (2009). 'Innovation in Services: Present Findings and Future Pathways', *Journal of Service Management*, 20(2): 131–55.

Easingwood, C. J. (1986). 'New Product Development for Service Companies', *Journal of Product Innovation Management*, 3(4): 264–75.

Edvardsson B., and Olsson J. (1996). 'Key Concepts for New Service Development', *Service Industries Journal*, 162: 140–164.

Eisingerich, A. B., Rubera, G., and Seifert, M. (2009). 'Managing Service Innovation and Inter-Organizational Relationships for Firm Performance: To Commit or Diversify?' *Journal of Service Research*, 11(4): 344–56.

Evangelista, R., and Sirilli, G., (1998). 'Innovation in the Service Sector: Results from the Italian Statistical Survey', *Technological Forecasting and Social Change*, 58(3): 251–69.

Fay, D., Borrill, C., Amir, Z., Haward, R., and West, M. A. (2006). 'Getting the Most out of Multidisciplinary Teams: A Multi-Sample Study of Team Innovation in Health Care', *Journal of Occupational and Organizational Psychology*, 79: 553–67.

Flowers, S. (1996). *Software Failure: Management Failure*. Chichester, UK: John Wiley and Sons.

Froehle, C. M., Roth, A. V., Chase, R. B., and Voss, C. A. (2000). 'Antecedents of New Service Development Effectiveness: An Exploratory Examination of Strategic Operations Choices', *Journal of Service Research*, 3(1): 3–17.

Froehle, C. M., and Roth, A. V. (2007). 'A Resource-Process Framework of New Service Development', *Production and Operations Management*, 16(2): 169–88.

Gadrey, J., and Gallouj, F. (1998). 'The Provider-Customer Interface in Business and Professional Services', *Service Industries Journal*, 18(2): 1–15.

Gallouj, F., and Savona, M. (2009). 'Innovation in Services: A Review of the Debate and a Research Agenda', *Journal of Evolutionary Economics*, 19(2): 149–72.

Gallouj, F., and Weinstein O. (1997). 'Innovation in Services', *Research Policy*, 26(4–5): 537–56.

Gans, J. S., and Stern, S. (2003). 'The Product Market and the Market for "Ideas": Commercialization Strategies for Technology Entrepreneurs', *Research Policy*, 32(2): 333–50.

Gans, J. S., and Stern, S. (2010). 'Is There a Market for Ideas?', *Industrial and Corporate Change*, 19(3): 805–37.

Gremyr I., Lofberg N., and Witell, L. (2010). 'Service Innovations in Manufacturing Firms', *Managing Service Quality*, 20(2): 161–75.

Gustafsson, A., Ekdahl, F., and Edvardsson, B. (1999). 'Customer Focused Service Development in Practice: A Case Study at Scandinavian Airlines Systems (SAS)', *International Journal of Service Industry Management*, 10(4): 344–58.

Hargadon, A. B. (2002). 'Brokering Knowledge: Linking Learning and Innovation', *Research in Organizational Behavior*, 24: 41–85.

Hill, P. (1999). 'Tangibles, Intangibles, and Services: A New Taxonomy for the Classification of Output', *Canadian Journal of Economics Revue*, 32(2): 426–47.

Hill, T. P. (1977). 'On Goods and Services', *Review of Income and Wealth*, 23: 315–38.

Hipp, C., and Grupp, H. (2005). 'Innovation in the Service Sector: The Demand for Service Specific Innovation Measurement Concepts and Typologies', *Research Policy*, 34(4): 517–35.

Hipp, C., Tether, B. S., and Miles, I. D. (2000). 'The Incidence and Effects of Innovation in Services: Evidence from Germany', *International Journal of Innovation Management*, 4(4): 417–53.

Hull, F. M. (2003). 'Simultaneous Involvement in Service Product Development: A Strategic Contingency Approach', *International Journal of Innovation Management*, 7(3): 339–70.

Hull, F. M. (2004). 'Innovation Strategy and the Impact of a Composite Model of Service Product Development on Performance', *Journal of Service Research*, 7(2): 167–80.

Hurmelinna-Laukkanen P., and Ritala, P. (2010). 'Protection for Profiting from Collaborative Service Innovation', *Journal of Service Management*, 21(1): 6–24.

Jaw, C., Lo, J.-Y., and Lin, Y-H. (2010). 'The Determinants of New Service Development: Service Characteristics, Market Orientation and Actualizing Innovation Effort', *Technovation*, 30(4): 265–77.

Johne, A., and Storey, C. (1998). 'New Service Development: A Review of the Literature and Annotated Bibliography', *European Journal of Marketing*, 32(3–4): 184–251.

Karniouchina, E. V., Victorino, L., and Verma, R. (2006). 'Product and Service Innovation: Ideas for Future Cross-Disciplinary Research', *Journal of Product Innovation Management*, 23(3): 274–80.

Kelly, D., and Storey, C. (2000). 'New Service Development: Initiation Strategies', *International Journal of Service Industry Management*, 11(1): 45–62.

Kristensson, P., Gustafsson, A., and Archer, T. (2004). 'Harnessing the Creative Potential among Users', *Journal of Product Innovation Management*, 21: 4–14.

Lee, Y.-C., and Jih-Kuang, C. (2009). 'A New Service Development Integrated Model', *Service Industries Journal*, 29(12): 1669–86.

Leiponen, A. (2005). 'Organization of Knowledge and Innovation: The Case of Finnish Business Services', *Industry and Innovation*, 12(2): 185–203.

Leiponen, A. (2006). Managing Knowledge for Innovation: The Case of Business to Business Services', *Journal of Product Innovation Management*, 23: 238–58.

Levitt, T. (1976). 'Industrialization of Service', *Harvard Business Review*, 54(5): 63–74.

Lusch, R. F., Vargo, S. L., and O'Brien, M. (2007). 'Competing through Service: Insights from Service-Dominant Logic', *Journal of Retailing*, 83(1): 5–18.

Lyons, R. K., Chatman, J. A., and Joyce, C. K. (2007). 'Innovation in Services: Corporate Culture and Investment Banking', *California Management Review*, 50(1): 174–91.

Magnusson, P. R., Matthing, J., and Kristensson, P. (2003). 'Managing User Involvement in Service Innovation: Experimenting with Innovating End Users', *Journal of Service Research*, 6(2): 111–24.

Matear, S., Gray, B. J., and Garrett, T. (2004). 'Market Orientation, Brand Investment, New Service Development, Market Position and Performance for Service Organisations', *International Journal of Service Industry Management*, 15(3–4): 284–301.

Matthing, J., Sanden, B., and Edvardsson, B. (2004). 'New Service Development: Learning from and with Customers', *International Journal of Service Industry Management*, 15(5): 479–98.

Menor, L. J., and Roth A. V. (2008). 'New Service Development Competence and Performance: An Empirical Investigation in Retail Banking', *Production and Operations Management*, 17(3): 267–84.

Menor, L. J., Tatikonda, M. V., and Sampson, S. E. (2002). 'New Service Development: Areas of Exploitation and Exploration', *Journal of Operations Management*, 20(2), Special Issue: 135–57.

Metcalfe, J. S., James, A., and Mina, A. (2005). 'Emergent Innovation Systems and the Delivery of Clinical Services: The Case of Intra-Ocular Lenses', *Research Policy*, 34(9): 1283–1304.

Miles, I. (2008). 'Patterns of Innovation in Service Industries', *IBM Systems Journal*, 47(1): 115–28.

Moller, K., Rajala, R., and Westerlund, M. (2007). 'Service Innovation Myopia? A New Recipe for Value Creation', *California Management Review*, 50(3): 31–48.

Moritz, S. (2005). *Service Design: Practical Access to an Evolving Field*. Cologne, Germany: Köln International School of Design (KISD).

Neely, A., Benedettini, O., and Visnjic, I. (2011). 'The Servitization of Manufacturing: Further Evidence'. Working paper. Cambridge, UK: Cambridge Service Alliance, University of Cambridge.

Nijssen, E. J., Hillebrand, B., Vermeulen, P. A. M., and Kemp, R. G. M. (2006). 'Exploring Product and Service Innovation Similarities and Differences', *International Journal of Research in Marketing*, 23(3): 241–51.

Oke, A. (2007). 'Innovation Types and Innovation Management Practices in Service Companies', *International Journal of Operations and Production Management*, 27(6): 564–87.

Olsen, N. V., and Sallis J. (2006). 'Market Scanning for New Service Development', *European Journal of Marketing*, 40(5–6): 466–84.

Page, A. L., and Schirr, G. R. (2008). 'Growth and Development of a Body of Knowledge: 16 Years of New Product Development Research', *Journal of Product Innovation Management*, 25(3): 233–48.

Pennant-Rea, R., and Emmott, B. (1983). *Pocket Economist*, 2nd edn. London: The Economist Publications and Basil Blackwell.

Pires, C. P., Sarkar, S., and Carvalho, L. (2008). 'Innovation in Services: How Different from Manufacturing?', *Service Industries Journal*, 28(10): 1339–56.

Ramirez, M. (2004). 'Innovation, Network Services and the Restructuring of Work Organization in Customer Services', *Service Industries Journal*, 24(1): 99–115.

Sasser, W. E., Olsen, R. P., and Wyckoff, D. D. (1978). *Management of Service Operations: Text and Cases*. Boston: Allyn and Bacon.

Schleimer S. C., and Shulman, A. D. (2011). 'A Comparison of New Services Versus New Product Development: Configurations of Collaborative Intensity as Predictors of Performance', *Journal of Product Innovation Management*, 28(4): 521–35.

Schmenner, R. W. (1986). 'How Can Service Businesses Survive and Prosper?', *Sloan Management Review*, 27(3): 21–32.

Shelton, R. (2009). 'Integrating Product and Service Innovation', *Research-Technology Management*, 52(3): 38–44.

Shostack, G. L. (1982). 'How to Design a Service', *European Journal of Marketing*, 16(1): 49–63.

Shostack, G. L. (1984). 'Designing Services that Deliver', *Harvard Business Review*, 62(1): 133–9.

Sirilli, G., and Evangelista, R. (1998). 'Technological Innovation in Services and Manufacturing: Results from Italian Surveys', *Research Policy*, 27(9): 881–99.

Spohrer, J., and Maglio, P. P. (2008). 'The Emergence of Service Science: Toward Systematic Service Innovations to Accelerate Co-Creation of Value', *Product and Operations Management*, 17(3): 238–46.

Stickdorn, M., and Schneider, J. (2010). *This is Service Design Thinking*. Amsterdam: BIS Publishers,

Storey, C., and Kelly, D. (2001). 'Measuring the Performance of New Service Development Activities', *Service Industries Journal*, 21(2): 71–90.

Sundbo, J. (1997). 'Management of Innovation in Services', *Service Industries Journal*, 17(3): 432–55.

Teece, D. J. (1986). 'Profiting from Technological Innovation: Implications for Integration, Collaboration, Licensing and Public Policy', *Research Policy*, 15(6): 285–305.

Tether, B. S. (2003). 'The Sources and Aims of Innovation in Services: Variety between and within Sectors', *Economics of Innovation and New Technology*, 12(6): 481–505.

Tether, B. S. (2005). 'Do Services Innovate (Differently)? Insights from the European Innobarometer Survey', *Industry and Innovation*, 12(2): 153–84.

Tether, B. S., and Howells, J. (2007). 'Changing Understanding of Innovation in Services', in DTI Occasional Paper #9, *Innovation in Services*. London: Department of Trade and Industry.

Tether, B. S., Mina, A., Consoli, D., and Gagliardi, D. (2005). *Innovation Impact on the Demand for Skills and How Do Skills Drive Innovation?* CRIC Report for The Department of Trade and Industry. Manchester, UK: University of Manchester. Available at <http://www.berr.gov.uk/files/file11008.pdf> (accessed 23 July 2013).

Tether, B. S., and Massini, S. (2007). 'Services and the Innovation Infrastructure', in DTI Occasional Paper #9, *Innovation in Services*, London: Department of Trade and Industry.

Tether, B. S., and Metcalfe, J. S. (2003). 'Horndal at Heathrow? Capacity Creation through co-Operation and System Evolution', *Industrial and Corporate Change*, 12(3): 437–76.

Tether, B. S., and Tajar, A. (2008). 'The Organisational-Cooperation Mode of Innovation and its Prominence amongst European Service Firms', *Research Policy*, 37(4): 720–39.

van den Ende, J. (2003). 'Modes of Governance of New Service Development of Mobile Networks: A Life Cycle Perspective', *Research Policy*, 32: 1501–18.

van der Aa, W., and Elfring, T. (2002). 'Realizing Innovation in Services', *Scandinavian Journal of Management*, 18: 155–71.

van Riel, A. C. R., Lemmink, J., and Ouwsterloot, H. (2004). 'High Technology Service Innovation Success: A Decision Making Perspective', *Journal of Product Innovation Management*, 21: 348–59.

Vargo, S. L., and Lusch, R. F. (2004). 'Evolving to a New Dominant Logic for Marketing', *Journal of Marketing*, 68(1): 1–17.

Voss, C., and Zomerdijk, L. (2007). 'Innovation in Experiential Services: An Empirical View', In DTI Occasional Paper #9, *Innovation in Services*, London: Department of Trade and Industry.

Windrum P., and Garcia-Goni, M. (2008). 'A neo-Schumpeterian Model of Health Services Innovation', *Research Policy*, 37(4): 649–72.

CHAPTER 31

INNOVATION AND PROJECT MANAGEMENT

ANDREW DAVIES

INTRODUCTION

INNOVATION and projects are closely connected. A project is a temporary organization and process established to create a novel or unique outcome. Projects are used to sustain and grow an organization's existing activities, but they play a more fundamental role as the engine of innovation. Organizations use projects to create novel products, processes, and services, develop new technologies, launch entrepreneurial ventures, implement strategies, and produce complex infrastructure ranging from systems and iconic buildings to ecologically sustainable cities. But innovation creates project uncertainty (Freeman and Soete, 1997). Organizations cannot predict the future and therefore cannot be certain that a project will achieve its original goals. The uncertainties range from known and predictable circumstances to unforeseeable events and unknown interactions among different parts of a project (Loch et al., 2006). Project management processes, tools, and techniques were developed to help select, plan, manage, and reduce the uncertainties associated with the process and outcome of innovation.

The close bond between the management of projects and innovation was well understood in the 1950s when pioneering organizations created new structures, techniques, and processes to manage complex and highly uncertain research and development projects in technologically advanced weapons and defence-related industries (Morris, 1994; Hughes, 1998). Over subsequent decades, there have been some points of convergence when researchers investigated the nexus of innovation and project management, such as Japanese product development practices in the 1980s (Takeuchi and Nonaka, 1986; Clark and Wheelwright, 1992). But in the main, the literatures on project and innovation management followed distinct, largely self-contained trajectories of theoretical, professional, and practical development. In recent years, however, there has been a strong convergence and cross-fertilization of ideas between innovation and project management. A new wave of research is investigating the different ways in which organizations use projects to manage the uncertainties associated with innovation processes and outcomes (Pich et al., 2002; Davies and Hobday, 2005; Loch et al., 2006; Shenhar and Dvir, 2007; Lenfle, 2008; Lenfle and Loch, 2010; Brady and Hobday, 2011).

This chapter suggests that project and innovation management research developed along different paths as relatively autonomous bodies of knowledge because they were influenced by very different assumptions about how organizations cope with uncertainty. Building on a classic contribution by Klein and Meckling (1958) and more recent conceptual thinking, this chapter identifies two models—optimal and adaptive—which describe contrasting processes of project-based innovation.

The 'optimal model' is closely associated with the foundations of project management in the 1950s and its development as professional practitioner-oriented discipline. It assumes that the goal of innovation and path to achieving that goal can be identified up-front and then carefully managed and optimized as a well-understood, predictable, and rational process. While there is no single theoretical foundation which drives all conceptual thinking about project management, most textbooks tend to conform to the rational, optimizing model based on systems engineering, operations research, and deterministic planning and control processes (Winter et al., 2006; Söderlund, 2011). This model has been criticized for failing to deal with the uncertainties associated with front-end planning and execution (Morris, 1994), organizational design (e.g. Galbraith, 1973), and project complexity, uncertainty, and novelty (Shenhar and Dvir, 1996; Hobday, 1998; Shenhar, 2001; Loch et al., 2006). As a discipline, project management has promoted the idea that all projects are the same and can be managed using standardized and universally applicable processes, structures, and scheduling techniques (Kerzner, 2006; Project Management Institute, 2009).

The 'adaptive model' is more closely related to innovation management and organizational theory. It recognizes that there is no single best way of organizing innovative projects; the form of organization and process must be able to adaptable to the uncertainty and complexity of each project (e.g. Galbraith, 1973; Shenhar and Dvir, 2007). It assumes that the goal of an innovation process and path to achieving that goal are inherently uncertain (Klein and Meckling, 1958). Adaptation depends on trial-and-error learning, flexibility, intuition, and managerial judgement to cope with unexpected situations facing a project and our inability to predict the future. Research recognizes that while highly uncertain and novel projects require a fully adaptive, flexible, and real-time learning process, optimal, rational, and standardized processes may apply to predictable and routine projects (Eisenhardt and Tabrizi, 1995).

Drawing upon a variety of contributions, which are illustrated in Table 31.1, the chapter identifies the main features of an 'adaptive model' of project-based innovation. The recent convergence of innovation and project management research has opened up promising avenues of research on projects as temporary organizational designs and fast, flexible, and real-time adaptive processes for managing the uncertainties associated with innovation.

Table 31.1 The adaptive model of project-based innovation: constructs from innovation, project management, and organizational theory

Literature	Temporary Organization	Adaptive Process
Projects and Innovation: Early Formulations	Project structure for integrating knowledge and promoting innovation (Gaddis 1959; Middleton 1967)	Models of the innovation process in uncertain projects: Mr Optimiser vs Mr Skeptic (Klein and Meckling 1958) Projects as 'voyages of discovery' (Hirschman 1967)
Innovation Management	Contingency theory identifies organic-adaptive structure (Burns and Stalker 1961; Bennis and Slater 1968) associated with project producers (Woodward 1965; Mintzberg 1983; Hayes and Wheelwright 1984) Project and matrix structures (Galbraith 1973; Davis and Lawrence 1977) Innovation management: project production and stage in product life cycle Derivative, platform and breakthrough projects (Wheelwright and Clark 1992a and 1992b) Heavyweight projects support breakthrough innovation (Clark and Wheelwright 1992) Kanter (1990) newstream vs mainstream projects; O'Reilly and Tushman (2004) ambidextrous organization	'Fast and flexible process': sequential vs overlay process (Takeuchi and Nonaka 1986; Nonaka and Takeuchi 1995) 'Compression' vs 'experiential mode' of adaptation (Eisenhardt and Tabrizi 1995) Christensen (1997) sustaining and disruptive innovation Projects as experiments (Thomke 2003) 'Real-time' flexible development process to match the degree of market uncertainty (Bhattacharya et al. 1998).
Project Management	Critique of 'one-size-fits-all' approach of traditional project management and use of contingency theory to identify uncertainties, complexities, and dynamics of change in projects (Shenhar and Dvir 2007) Diamond model helps select temporary organization (Shenhar and Dvir 2007)	Measures of project success depend on project uncertainty and dynamics over time (Pinto and Kharbanda 1995; Shenhar and Dvir 2007) Limits of traditional phase-based approach and importance of multiple and parallel trials (Lenfle and Loch 2010) Managing project uncertainty depends on learning and selectionism (Loch et al. 2006) Diamond model helps select appropriate process (Shenhar and Dvir 2007)

(Continued)

Table 31.1 *(Continued)*

Literature	Temporary Organization	Adaptive Process
Converging Research: Recent Formulations	Projects as a form of temporary organization (Bechky 2006; Jones and Lichtenstein 2008; Bakker 2010) range from 'standalone projects' to 'fully embedded projects' (Schwab and Miner 2008) Project-based firms (Gann and Salter 2000; Whitley 2006) and project-based organizations (Hobday 2000) Projects embedded in history, context, networks, and ecologies (Engwall 2003; Grabher 2004; Sydow et al. 2004; Manning and Sydow 2011)	Learning in project-based firms (Prencipe and Tell 2001) depends on lessons encoded in project routines (Stinchcombe and Heimer 1985) and project capabilities (Davies and Brady 2000; Brady and Davies 2004; Söderlund and Tell 2009) Experiential mode of adaptation associated with emergent, novel, and improvised action in project teams (Gersick and Hackman 1990; Weick and Roberts 1993, Weick and Sutcliffe 2001) Improvisation can be accidental and deliberate (Miner et al. 2001; Vera and Crossan 2005) Teams engage in bricolage to tackle the unexpected (Bechky and Okhuysen 2011) Creative projects drive future and envisioned outcomes (Obstfeld 2012)

INNOVATION AND PROJECTS: EARLY FORMULATIONS

Early formulations about the relationship between innovation and project management can be traced back to the 1950s and 1960s when government-sponsored 'change-generating projects' were established to create complex weapons, defence, and aerospace systems, such as the Manhattan project, Atlas, Titan, and Polaris ballistic missiles, SAGE air defence system, and the Apollo moon landing programme (Morris, 1994; Johnson, 1997; Hughes, 1998). To solve these research and development problems and keep pace with rapid technological innovation, scientists, engineers, and managers developed radically new project structures, systems engineering, and operations research.

These advanced technology projects brought together interdisciplinary and cross-functional project teams of scientists and engineers. New forms of project organization were created to integrate and coordinate the specialized knowledge and resources required to achieve a desired innovative outcome on time, within budget, and to specification (Gaddis, 1959; Middleton, 1967). Matrix forms of organization (combining functional and project lines of authority) were established to manage multiple defence and aerospace projects (Galbraith, 1973; Davis and Lawrence, 1977). Systems engineering knowledge and techniques supported by 'systems integrator' organizations, such as the Ramo-Wooldridge, were created to coordinate the design, concurrent development, and integration of complex, multiple, and evolving technologies supplied by a large network of contractors (Sapolsky, 1972). Operations research emerged as a discipline to analyse military operational environments. Hughes (1998) argues that these systems engineering, operations research, and project management approaches to innovation created a managerial revolution less well known, but comparable in importance to the mass production system established by firms such as Ford and General Motors.

In the late 1950s, economists and social scientists at the RAND Corporation began to analyse the organizational processes associated with innovation in complex weapons systems projects such as fighter jets and inter-continental ballistic missiles. These projects were highly uncertain in terms of performance, cost, time, quality, and future operational environments (Klein et al., 1962). The RAND studies identified a variety of factors impinging on the innovation process, such as the discrepancies between estimated and actual project costs and the time spent on procurement (Freeman and Soete, 1997). The RAND research recognized that the highly uncertain process of innovation in complex systems must be distinguished from known and predictable processes such as mass production. Some eminent scholars at the time recognized that the RAND studies identified an adaptive process of change which was broadly applicable to organizations facing uncertainties involving innovation in processes and products (Hirschman and Lindblom, 1962). For example, in his study of infrastructure projects in developing countries Hirschman (1967) identified the 'supply uncertainties' related to the processes by which a project's output is produced (including a variety of technological, organizational, managerial, and financial conditions) and 'demand uncertainties' concerning a project's outcome and ongoing operational performance. The RAND research showed that certain types of projects had to be responsive, flexible, and adaptive to a complex, changing, and uncertain environment; ideas which have recently resurfaced in research on project management and innovation (Shenhar and Dvir, 2007; Lenfle and Loch, 2010).

Klein and Meckling (1958) identified two alternative models—'Mr Optimizer' and 'Mr Skeptic'—presented as ideal types for managing complex and uncertain projects, which we call the optimal and adaptive models. The optimal model—Mr Optimizer—relies on rational planning, formal processes, and analytical techniques at the start of a project to predict future conditions and reach a decision about the best end-product from a range of alternatives. This requires careful up-front planning to select the optimal technologies, an exact scheduling of project tasks, and pre-arranged integration of proven components in the final system. For example, the Special Projects Office developed Program Evaluation Review Technique (PERT) in 1957 to plan, schedule, and control the Polaris ballistic missile programme (Sapolsky, 1972). According to Klein and Meckling (1958), the optimizing approach fails to address uncertainties or emergent situations that may be encountered as the project proceeds, such as the introduction of novel technologies, new strategic factors, and changes in the operational environment, not originally foreseen. The cost of making revisions to an integrated system when Mr Optimizer's predictions turn out to be wrong can be very high.

The adaptive model—'Mr Skeptic'—recognizes that the goal of innovation and path to achieving that goal are uncertain. Rather than rely on up-front plans and formal processes, Mr Skeptic relies on intuitive judgement, informal processes, and learning gained from trial-and-error experience to guide his decision-making. He experiments, tests, and evaluates a range of alternatives before selecting the most desirable solution. Instead of attempting to set optimal performance targets, the original goal of the project is reviewed or modified when new information becomes available. This model recognizes that innovative projects are 'voyages of discovery' (Hirschman, 1967: 78). Organizations presiding over these ventures must gather real-time information and feedback gained by learning to reduce the risks and emergent problems encountered along the way. Efforts to establish rigid performance specifications of the desired product or system—or early 'design freeze'—should be avoided at the start of an innovative project or risk the possibility of excluding more advanced technologies or performance requirements while the project is underway. According to Klein and Meckling (1958), a successful outcome five or ten years in the future cannot be fully appreciated until the product or system has undergone some development and testing in its operational environment. Uncertainties encountered at an early stage can be reduced by engaging in multiple and parallel approaches to buy in valuable information before selecting the one 'best' approach (Hirschman, 1967: 82). The costs of experimental prototypes and repeated tests may be less than the cost of deciding on a single technology at the outset, which subsequently faces major difficulties not originally envisaged or becomes outdated when the product is launched.

The close bond between innovation and projects identified in these studies of complex systems was not restricted by disciplinary boundaries, communities of professional interests, or theoretical and practical differences between innovation and projects. During the subsequent decades, the innovation management and project management literatures have largely followed distinct and diverging intellectual and practical trajectories. But a common research question pursued in different ways in both sets of literature is: How can organizations manage the uncertainty associated with innovative projects?

How Innovation Management Thinks about Projects

Drawing upon diverse theoretical foundations, such as industrial economics, business strategy, and organizational theory, studies of innovation focus on the scientific, organizational, financial, and technological activities involved in the highly uncertain development and commercialization of a new product, process, or service (Dodgson et al., 2008). Under conditions of rapid technological and market change, innovation is considered vital to the survival and success of firms (Utterback, 1994). Although projects are identified as the core structure and process supporting rapid innovation—with few exceptions (e.g. Wheelwright and Clark, 1992a)—innovation research rarely refers to the mainstream project management literature (Lenfle, 2008). Unlike project management research, the innovation literature has emphasized theory development inspired by emerging new practices.

Early research on the 'management of innovation' largely builds on contingency theory and studies of organizational design. Burns and Stalker's (1961) highly influential study laid the foundations for the contingency theory of innovation by challenging prevailing assumption of a single best model of organizational efficiency. Whereas 'mechanistic' structures refer to hierarchical management and specialized tasks which work well under stable and predictable technological and market conditions, Burns and Stalker (1961) argued that 'organic' structures refer to flatter management, fluid structures, and continuously evolving tasks which are appropriate for promoting innovation and responding rapidly and flexibly to changing and highly uncertain situations. Although Burns and Stalker (1961) do not identify the project form, subsequent research influenced by contingency theory argued that this temporary, 'organic-adaptive' structure is ideally suited to promoting innovation (Bennis and Slater, 1968; Mintzberg and McHugh, 1985). Researchers identified a range of organizational designs—from functional through matrix to pure project organizations—and administrative structures for coping with the speed of change, complexity, and uncertainty associated with different technological and market environments. Lawrence and Lorsch (1967) treated an organization as a system that adapts to changes or contextual contingencies in its external environment. A project or matrix structure is a way of integrating cross-functional resources and knowledge to cope with higher degrees of uncertainty, complexity, and change in the environment (Galbraith, 1973; Davis and Lawrence, 1977).

Whereas contingency theory has had relatively little impact on mainstream project management—a notable exception is Shenhar (2001)—studies of innovation recognize that a project is a particular form of organization and process uniquely suited to performing complex tasks in a rapidly changing environment. The project form of organization and process is an extreme case of Woodward's (1965) 'unit producer'; an organization that produces a variety of customized, tailor made, or unique products and services to each customer's order, such as buildings, film making, engineering prototypes, and product development (Mintzberg, 1983: 270; Davies and Frederiksen, 2010). Project production is the first stage along a spectrum of organizations from one-off/unit production to high-volume mass and continuous flow processes (Hayes and Wheelwright, 1984: 176). Firms adopt project organizations to design and produce one-off or tailor-made products and services and perform product development. Repetitive, standardized, and sequential operational tasks used in volume production are rarely applicable to a project process. A project performs complex and interdependent tasks in a time-sequence using scheduling tools such as PERT to represent the phasing and network of relationships (Hayes and Wheelwright, 1984: 176–7).

Project or unit production is associated with the initial 'fluid' stage in the 'product life cycle' (Hayes and Wheelwright, 1984; Hobday, 1998); an influential model of innovation developed by Abernathy and Utterback (1975) showing how the interaction between product and process innovation is behind the emergence, growth, and maturity of products and industries (Utterback, 1994). Research has examined the dynamics of innovation in project-based firms and industries, focusing on the capabilities and structures required to design and produce complex products and systems, such as flight simulators, intelligent buildings, and mobile communications (Hobday, 1998; Hobday, 2000; Gann and Salter, 2000; Davies and Brady, 2000; Davies and Hobday, 2005).

Many new and important concepts, tools, and frameworks were developed in the 1980s and 1990s when innovation scholars attempted to understand and encourage the emulation of successful novel project-based practices associated with accelerated product development in the Japanese car and electronics industries led by firms such as Toyota, Honda, and Sony (Takeuchi and Nonaka, 1986; Wheelwright and Clark, 1992a, 1992b). The research recognized that the uncertainty associated with innovation requires specific forms of project organization and time-limited processes which are highly adaptive, flexible, and responsive to a rapidly changing technology, market, and competitive environment.

In formulating ways of emulating Japanese development practices, Wheelwright and Clark (1992a, 1992b) identified three types of innovation projects according to the degree of change or novelty in the product or process on a continuum from incremental to radical innovation. Derivative projects are based on incremental innovation and range from cost reductions in existing products or enhancements to an existing production process. At the other end of the spectrum, breakthrough projects are based on radical innovation because they introduce entirely novel processes or untried new products, such as the Sony Walkman in 1981 and Apple iPod in 2001. Breakthrough projects differ fundamentally from previous generations and can create a new market or an entirely new industry. Platform projects, which develop new products (sharing standardized, modular, and common components) for existing customers and known markets, lie in the middle of the incremental–radical innovation continuum. Firms achieve their strategic objectives by mapping the different types of innovative projects, sequencing them over time in waves, and compressing the time taken to perform them individually and collectively.

Specific organizational structures are required to manage different types of innovation projects. Platform and breakthrough innovations require a 'heavyweight project team' led by a senior manager within the organization with the authority to claim the resources required to meet the project goals (Clark and Wheelwright, 1992). Whereas matrix structures are suitable for developing incremental innovation in derivative projects, more radical innovations require an autonomous project team which is temporarily dedicated and physically co-located around the heavyweight manager who has full control of team members. The team has a 'clean sheet of paper', is entirely focused on achieving successful project outcomes, and is well suited to developing breakthrough innovations that can be the birthplace of new markets or industries.

Organizational designs must strike a balance between exploiting a firm's project capabilities in its existing technologies and markets and exploring novel and strategic breakthrough innovations. Kanter (1990) distinguishes between a firm's established mainstream projects and newstream projects. Whereas mainstream projects require certainty, newstream projects specialize in managing the uncertainties associated with breakthrough innovations:

> Multiple approaches, flexibility, and speed are required for innovation because of the advance of new ideas through random and often highly intuitive insights and because of the discovery of unanticipated problems. Project teams need to work unencumbered by formal plans, board approval, and other 'bureaucratic delays', that might act as constraints against the change of direction.
>
> (Kanter, 1990: 205)

Similarly, according to Christensen (1997) the mainstream organization can be highly effective at developing and implementing 'sustaining' or incremental innovations in existing markets for known customers with defined and predictable needs that are amenable to carefully prepared project planning and execution. However, mainstream organizations often fail when attempting to develop highly uncertain 'disruptive' (or breakthrough) innovations, which create entirely new sets of customers and markets. A potentially disruptive technology, such as an electric vehicle, cannot be developed by a firm's mainstream organization because it does not satisfy the performance requirements of existing customers. Whereas in less risky sustaining innovation projects plans are developed before action is taken, in 'discovery-driven' markets for highly uncertain disruptive innovations actions must be taken before careful project plans are produced (Christensen, 1997: 160). The successful commercialization of disruptive innovation within established firms depends on the creation of autonomous project organizations, such as 'skunk works' (Rich and Janos, 1994) or spin-off units, with a capacity for learning and gathering real-time information about new markets so that uncertainties are resolved before incurring expensive investments in time and money. A similar point is made by Thomke (2003), who advocates the use of 'projects as experiments' for testing, adapting to change, promoting organizational learning, and resolving the uncertainty associated with innovation.

O'Reilly and Tushman (2004) claim that successful firms often fail in their attempts to promote breakthrough innovations in markets beyond their existing base of customers or markets. For example, in 1976 Eastman Kodak accounted for 90 per cent of film and 85 per cent of camera sales in the USA. Despite developing one of the first digital cameras in 1975, Kodak continued to promote sustaining innovation in its mainstream camera business and failed to switch from film to digital technology. An 'ambidextrous' organization is identified as the most successful structure for continuing to expand an existing business, while supporting breakthrough innovation. These structures encompass two different types of business units. Exploitative business units are formal and mechanistic (functional or matrix) structures focused on honing, developing and extending a firm's existing capabilities and supporting its current mainstream projects. Exploratory business units are adaptive, project-based structures focused on exploring newstream projects and entrepreneurial ventures associated with breakthrough innovation. Because a firm has no experience base in the new market and there are no current customers, forecasts about user requirements are difficult to produce, project schedules are unrealistic, and costs are likely to overrun. Breakthrough project teams are established as autonomous organizations which are structurally independent from the mainstream organization, but supported by top management and tightly integrated into the wider corporate structure.

Research on breakthrough innovation projects in large Japanese and US firms conducted in the 1980s and 1990s argued that the traditional phased-based approach to product development was being superseded by a 'fast and flexible' process (Takeuchi and Nonaka, 1986; Nonaka and Takeuchi, 1995). The traditional 'sequential' approach—typified by NASA's 'phased program planning' (PPP) system—was analogous to passing the baton in a relay race. The product moved sequentially through highly structured phases, from one functionally specialized group to the next: concept development, feasibility testing, product design, development, pilot, and final production. In contrast, the 'overlay' approach, created by firms such as Fuji-Xerox, Canon, and Honda to accelerate product development, is more analogous to a game of rugby. The phases of development overlap, enabling members of the project team to absorb new information and engage in problem-solving before moving forward. The new product is the result of a constant interaction among members of a multidisciplinary team, autonomous project organizations, and iterative experimentation which can occur at late stages of development. Using the overlay approach, team members are highly responsive to changing market conditions, engaging in an iterative and dynamic process of trial and error learning to narrow down the number of alternatives they must consider. Takeuchi and Nonaka recognize that 'because projects do not proceed in a totally rational and consistent manner, adaptability is particularly important' (1986: 141). This adaptive approach to product design is now widely associated with 'concurrent engineering'; which ironically has its roots in older practices pioneered by the US military during the 1940s and 1950s to synchronize the simultaneous development of component technologies in complex weapons systems projects (Sapolsky, 1972; Johnson, 1997).

In a review of the product development literature, Eisenhardt and Tabrizi (1995) distinguished between two modes of adaptive processes under varying conditions, predictable, and uncertain. The 'compression' mode assumes a well-understood, rational, and predictable process and relies on overlapping development steps to compress or squeeze development times. The 'experiential' mode of adaptation assumes that the process is unpredictable and relies on improvisation, flexibility, trial-and-error experience, and access to real-time information. As we will see below, experiential adaptation builds on a different theoretical tradition based on intuition, improvisation, and choice.

The compression mode of adaptation—with its emphasis on highly structured, engineering-oriented processes—is prominent in the specialized field of new product development. The well-known 'stage-gate system', for example, was devised to optimize and compress the time taken to move from novel idea to product launch (Cooper and Kleinschmidt, 1987; Cooper, 1990). The innovation process is divided into predetermined stages of development work and gates (or checkpoints) which determine whether a product passes from one stage to the next (see Chapter 26 by Zedtwitz et al.). Product specifications should be defined at an early stage to maintain a predictable development process which is not subject to unnecessary changes in scope and expensive rework. The stage-gate system may be well suited to low-uncertainty projects, but may not be optimal in highly uncertain projects when an early design freeze results in an incorrect product definition. Bhattacharya et al. (1998) suggest that a 'real-time definition' model is required to 'adapt' the product definition process to the degree of market uncertainty (Bhattacharya et al., 1998). Uncertainty in product definition is resolved through repeated interactions with customers and using an iterative and flexible development process.

How Project Management Thinks about Innovation

In the project management literature, a project is defined as a temporary organization and process used to create novel and unique outcomes, such as new and improved products and services tailored to individual customer requirements (Kerzner, 2006; Pinto, 2007). Unlike permanent organizations (e.g. firms or government agencies), projects are time-limited organizations, which can be a standalone team or form part of a larger organization or a multi-firm alliance. Project management textbooks often distinguish between projects and operations. Whereas a project performs many novel, unpredictable, and complex tasks, operations refer to the repetitive, predictable, and routine tasks associated with volume production of standardized products or services. With some notable exceptions (Loch et al., 2006; Shenhar and Dvir, 2007), the terms 'novel' and 'unique' are often vague and poorly defined because project management textbooks do not refer to the more robust concepts and typologies developed by innovation management scholars.

Although a project is identified with novel and uncertain outcomes, project management as a discipline has emphasized the rational, formal, and predictable processes required to plan and manage projects, such as PERT in 1959 and Critical Path Analysis (CPA) developed in 1957–9 by Du Pont, the chemical manufacturer. According to Kerzner (2006), the theoretical foundations of project management can be found in systems theory (e.g. Boulder, 1956), which identifies the importance of integrating the many different component practices, disciplines, and knowledge into a system of project management. Traditional project management has largely evolved into a 'one size fits all' normative and applied form of systems management (Kerzner, 2006) as exemplified by the proliferation of project management text books and 'fairly thin' development of theory development over subsequent decades (Morris et al., 2011). The 'optimization school' is the first of several different theoretical traditions shaping the intellectual development of project management in all its aspects (Söderlund, 2011) and remains the most pervasive and influential approach advocated by professionals and used by practitioners to manage the uncertainty associated with innovation.

Since the 1960s, project management has extended into all other industries and across private and public organizations, ranging in size from small teams and start-up ventures to globally integrated companies, such as IBM and GE (Morris, 1994). Professional bodies, such as the Project Management Institute (PMI) founded in the USA in 1969 and Association of Project Management (APM) established in the UK in 1972, have been hugely influential in promoting and extending the discipline of project management as standardized set of practices. The PMI and numerous textbooks on the subject define project management as the application of a body of knowledge and techniques to manage the trade-off or 'triple constraints' between the time, cost, and quality specification. An important concept is the project life cycle which identifies a set of phased activities to successfully manage phases of a project, from project definition, through execution, to commissioning, start-up, and operations. While there are numerous project management textbooks, many of them continue to assume 'a project is a project, is a project'. Informed by the one best way thinking about organizational design and industrial efficiency first advocated by Frederick Taylor and Henry Gantt, project management processes and techniques are considered to be universally applicable all types of projects, large and small, simple and complex, routine and novel.

Projects have long been considered 'the lifeblood of innovation and today's project managers must create innovation in order to compete in a changing world' (Randolph and Posner, 1988). But these authors reveal the optimizing assumptions of traditional project management when they state that effective project managers 'plan, then manage the plan' and should concentrate on 'getting innovative projects done on time, within budget, and according to the desired quality standards' (Randolph and Posner, 1988). It is only in recent years that project management scholars—working within the discipline but influenced by innovation research and organizational theory—have begun to suggest ways of radically revising project management, focusing in particular on its inability to address the manifold uncertainties associated with innovation (Loch et al., 2006; Shenhar and Dvir, 2007; Pinto, 2007; Lenfle and Loch, 2010). This research questions the validity of three optimizing axioms of traditional project management: the 'triple constraints', 'phased stage-gate', and 'project risk management' approaches.

According to the triple constraints measure of project success, any deviation from time, cost, or quality has to be prevented or corrected to get the project back on track. Although the triple constraints model identifies the trade-offs that impact on the efficiency of a project in the short term, it fails to address an important 'fourth constraint'— whether a project meets a client's needs (Pinto and Kharbanda, 1995). It is essential to measure the short-term efficiency of a project and its impact on the customer, but other measures of success are required to assess the longer-term outcomes of highly uncertain innovative projects (Shenhar and Dvir, 1997, 2007). Project success depends on the degree of uncertainty. Low-risk routine projects must meet short-term measures of time, cost, and quality and impact on the customer. In high-uncertainty innovative projects, by contrast, time, cost, and quality overruns in the short term may be offset by longer-term benefits, such as creating new markets, launching new products, and developing new technologies. For example, Steve Jobs, Apple's former CEO, encouraged his project development teams to focus on launching breakthrough products and creating new markets, such as the digital hub created by the iPod and iTunes, and was notoriously impatient with 'those that made compromises in order to get the product out on time and on budget' (Isaacs, 2011: 123).

The optimizing model advocated by the PMI, APM, and other professional bodies assumes that all projects should follow carefully planned, defined and controlled phases (Slevin and Pinto, 1987). The completion of each phase or stage-gate in the project life cycle has an output—such as a scope statement, plan, baseline document, or milestone— which is reviewed before proceeding to the next stage. It is often assumed that the phased stage-gate model was pioneered by the Manhattan project in the 1940s, applied to Atlas, Titan, and Polaris ballistic missile projects in the 1950s, and subsequently adopted as the standard practice for managing projects. Lenfle and Loch (2010) challenge the received wisdom. They argue that the Manhattan and first ballistic missile projects did not remotely conform to or rigorously apply phased-based principles and tools. For example, in reality PERT was mainly used to sell the Polaris project to visiting politicians and to justify the need to secure additional resources, rather than improve the managerial efficiency of phased-based project control and execution (Sapolsky, 1972). These defence projects applied a combination of trial-and-error experiments and multiple and parallel trials 'in order to "push the envelope", that is, to achieve outcomes considered impossible at the outset' (Lenfle and Loch, 2010: 32). Lenfle and Loch (2010) argue that the phased stage-gate approach may apply to low-uncertainty projects, but is unable to support the problem-solving search, adaptation, and flexible learning required to tackle the numerous unforeseen events encountered in highly innovative, strategic, and uncertain projects.

Traditional project management assumes that uncertainties can be identified at the outset of a project and mitigated by the application of project risk management tools and techniques. Project risk is the probability of a risk occurring and the extent of its impact on the project if it occurs. For example, a firm designing and building a 4G mobile communications network can develop a risk register that identifies all the major difficulties likely to occur on the project. The organization can prepare contingency plans—such as reserve funding and time buffers—to tackle any foreseen eventualities. There may be some unaccounted for residual risk that has to be dealt with during the execution of the project, but this optimizing approach to risk management depends on up-front formal planning and problem solving before the project is under way.

Loch et al. (2006) argue that traditional risk management is applicable to highly routine and predictable projects, but is unable to cope with novel or innovative projects facing unforeseen uncertainties (Pich et al., 2002; De Meyer et al., 2002; Sommer and Loch, 2004; Loch et al., 2006). When a firm is developing a breakthrough project or radically new technology, traditional project risk management techniques encourage efforts to 'get back to the plan' rather than learning new things and changing direction to resolve the unexpected problems. Novel projects face major risks that are unknown and cannot be foreseen. Even if some events can be foreseen, the complex interactions of events and influences encountered during a project mean that it is difficult to develop detailed up-front plans for managing novel projects.

Rather than attempting to tackle unforeseen uncertainties by the more rigorous application of planning and formal risk management, Loch et al. (2006) argue that project managers should rely on a combination of two adaptive approaches for managing novel and uncertain projects: trial-and-error learning and selectionism. First, the project organization needs to adapt and adjust the project as one learns more about the project and its interaction with the environment. As the project moves towards its goal, the team must be prepared to adjust a project's original goal and ongoing processes as new information becomes available. Second, the project organization needs to pursue multiple approaches in parallel and independently of each other, before selecting the best approach. For example, Microsoft used selectionism in the 1980s when it simultaneously developed several operating systems—Dos, Windows, OS/2, and Unix—when it was unclear which one was the best solution. The twin processes of learning and selectionism help managers to assess the type of uncertainty associated with each project and rapidly adapt their management approach to cope with it (De Meyer et al., 2002).

Strongly influenced by studies of innovation management, the new project management research recognizes that the optimal model, founded upon rational planning, exact scheduling, and predictable processes, applies to some routine and predictable projects, but often breaks down when applied to novel and uncertain projects (Loch et al., 2006, Lenfle and Loch, 2010; Brady and Hobday, 2011). Informed by the contingency theory of innovation, Shenhar and Dvir (2007) develop a radical alternative and critique of the traditional one-size-fits-all and triple constraints assumptions of traditional project management (see also Shenhar, 2001). Bringing together new thinking about project management and innovation research, Shenhar and Dvir (2007) maintain that a new 'adaptive project management' model is required to drive innovation and manage projects which are uncertain, complex, and strongly affected by a dynamically changing technology, market, and business environment. The project organization and process must be flexible and adaptive to the goal, task, and environment of each specific project and unforeseen events, contingencies, and changes it encounters over time.

To address differences among projects, Shenhar and Dvir (2007) develop an integrated conceptual approach—the 'diamond framework'—which classifies each project according to four dimensions: technological uncertainty, market novelty, complexity, and pace. The novelty dimension refers to the uncertainty of a project's goal and how new the project's product is to customers, users, and organizations in a particular market. Adopting Wheelwright and Clark's classification (1992b), novelty includes derivative, platform, and breakthrough projects. Technology represents the degree of uncertainty associated with how much new technology is incorporated into a project. The more technology adopted or developed on a project, the greater the chance of cost and time and quality overruns, and risk of not achieving planned performance objectives. Complexity classifies a project's product according to a hierarchy of systems and subsystems. Complexity affects the tasks, the interdependencies among them, and type of organization required to design, integrate, and produce the product or system (Shenhar and Dvir, 1996; Hobday, 1998; Shenhar, 2001). For example, whereas relatively simple projects can be conducted in-house, the most complex 'system of systems' project (e.g. London Olympics 2012 construction programme) typically requires the creation of a dedicated 'umbrella organization', contracts tailored to each sub-project, and a bespoke process to coordinate multiple suppliers in a large programme. Building on prior research on time-based strategies (Brown and Eisenhardt, 1995, 1997; Eisenhardt and Tabrizi, 1995) the 'project pace' dimension refers to the urgency of the project and how much time there is complete the task in an uncertain and changing environment.

Shenhar and Dvir (2007) help to bridge project management and innovation research. Their formal and normative model of project management attempts to capture a series of momentary 'snapshots' of the project uncertainty, complexity, and changing conditions facing a project over time. It aims to help managers to select the right type of project organization and process at one moment such as the start or mid-point of a project, which can be adjusted to cope with the uncertainty, complexity, and contingent events encountered and 'get the project back on track' at a later stage. The framework pays less attention to the importance of 'real-time' experiential learning, emergent and unfolding processes of discovery identified by Loch et al. (2006).

CONVERGING THEMES OF RESEARCH: RECENT FORMULATIONS

The convergence of innovation and project management research over the past decade underlines the importance of the adaptive model. This final section illustrates how scholars from a variety of theoretical and empirical backgrounds have deepened our understanding of three different streams of research: projects as temporary organizations; project-based learning, capabilities, and routines; and a more diverse literature concerned with improvised, emergent, and creative responses to uncertainty.

The opportunity for project-based learning and adaptation depends on the form of temporary organization (Bechky, 2006; Jones and Lichtenstein, 2008; Bakker, 2010). Different types of projects can be located on a continuum of temporary organizations ranging from 'standalone projects' to 'fully-embedded projects' (Schwab and Miner, 2008). At one extreme, a standalone project is a temporary organization in which multiple individual participants and independent organizations join a new project or repeated collaboration, such as a network of participants recurrently involved in independent movie projects (DeFillippi and Arthur, 1998; Lampel and Shamsie, 2003; Bechky, 2006).

At the other end of the spectrum, fully embedded projects are centrally controlled at a higher level and entirely incorporated in a permanent organization or firm, such as internal R&D, ad hoc task force and product development projects (Wheelwright and Clark, 1992b). The middle ground between these two extremes consists of 'hybrid' project organizations where the formation and coordination of a project is influenced by a combination of participants and centrally controlled organizations (Schwab and Miner, 2008) collaborating in embedded networks (Manning and Sydow, 2011). A construction project involving a temporary coalition of firms led by a large client or prime contractor is an example of a hybrid organization oriented towards the standalone end of the spectrum (Davies et al., 2009). An example of a hybrid structure in a more centrally controlled setting is TV production company that combines in-house expertise with outside specialists in a series of temporary projects and long-term collaborations with clients (Manning and Sydow, 2011).

Although temporary organizations are widely associated with their adaptive capability (Bechky, 2006), they face difficulties in learning from their own experiences because the project disbands on completion of its goal (Hobday, 2000). Unless embedded in a permanent structure, members of a project cannot rely on a stable firm's capabilities and memory as the engine of adaptation (Schwab and Miner, 2008). Learning from projects is possible, however, because most projects are embedded in the historical context of a permanent organizational unit, stable inter-organizational network, and wider project ecology (Engwall, 2003; Sydow et al., 2004; Grabher, 2004). In standalone collaborative projects, learning is possible because participants form a network of enduring relationships established by working together on prior projects. If collaboration is repeated, the trust established among the participants involved can function as a repository of knowledge and shared learning, resources and capabilities which can be retrieved when participants work together on subsequent projects (Sydow et al., 2004; Manning and Sydow, 2011).

Projects embedded in stable organizations offer greater opportunities for learning and adaptation over the longer term. In project-based firms and organizations, most productive tasks are undertaken in projects, ranging from standalone multi-firm collaborations to embedded R&D and product development projects (DeFillippi and Arthur, 1998; Hobday, 2000; Gann and Salter, 2000; Keegan and Turner, 2002; Whitley, 2006; Gann et al., 2012). Research on innovation in project-based firms (Hobday, 2000; Davies and Brady, 2000; Gann and Salter, 2000; Brady and Davies, 2004; Davies and Hobday, 2005; Whitley, 2006) draws upon theoretical contributions identifying how a firm's knowledge, prior history, and learning encoded in stable routines and organizational capabilities shapes its adaptive action and future behaviour (Nelson and Winter, 1982; Levitt and March, 1988; March, 1991; Teece et al., 1997). Embedding the learning gained from temporary project experiences in stable, repetitive, and enduring 'project routines' provides an important source of efficiency improvements when applied repeatedly across multiple projects (Stinchcombe and Heimer, 1985). Stinchcombe distinguishes between predictable project routines and the uncertainties encountered in projects that must be resolved by innovation (Stinchcombe and Heimer, 1985: 26).

Project-based firms depend on their permanent structures (e.g. functional and corporate departments), routines, and capabilities to learn from individual projects, memorize, and adapt to an uncertain and rapidly changing environment (Prencipe and Tell, 2001). A process of learning, capability building, and organizational renewal can occur when project-based firms launch innovative projects to develop novel technologies and create new markets (Brady and Davies, 2004; Shamsie et al., 2009). It begins with a phase of exploratory learning when a firm establishes a new venture or vanguard project to explore strategic opportunities, experiment with new approaches, and resolve uncertainties (Yoo et al., 2006; Frederiksen and Davies, 2008). Such projects are often set up as independent units within an ambidextrous structure to provide the freedom and autonomy to adapt existing routines and invent new ones. Efforts to exploit the learning gained, transfer knowledge to subsequent projects, and replicate the new practices across the wider organization can help a firm institutionalize new project routines and build the 'project capabilities' required to perform a growing number of projects (Davies and Brady, 2000; Shamsie et al., 2009), often extending over years or even decades (Söderlund and Tell, 2009).

The capabilities research focuses on the strategies, deliberate learning, and purposeful actions undertaken by project-based firms to manage the uncertainties associated with planned innovation. A more eclectic body of research on experiential adaptation identifies the emergent, responsive, and unplanned innovative or improvised action that occurs when organizational units and project teams (e.g. product development teams, emergency rooms in hospitals, fire-fighting units, police SWAT teams, and film crews) tackle unexpected events in real time. Early research on project teams recognized that they may have to abandon their existing 'habitual routines' and respond innovatively by creating new routines when faced with novel and uncertain situations (Gersick and Hackman, 1990). When teams face extreme degrees of uncertainty they must remain 'mindful' of the complete situation, learn rapidly, and act swiftly to tackle unexpected events (Weick and Roberts, 1993; Weick and Sutcliffe, 2001). Research has identified how prior learning and existing routines shape improvisation (Brown and Eisenhardt, 1995) and recognizes that organizations often rely on managerial judgement and intuition to create entirely novel, creative, and improvised responses to the unexpected (Weick, 1998). Such improvised activity often occurs outside of pre-existing routines and formal plans, and can refer to the deliberate—as well as accidental—creation of novel activity (Miner et al., 2001; Vera and Crossan, 2005). Bechky and Okhuysen (2011) found that project teams engage in bricolage by drawing upon combinations of resources to hand, adapting their routines, and responding innovatively to unexpected surprises. Obstfeld (2012) distinguishes between routine projects based on repetitive tasks and past learning and 'creative projects' associated with novel tasks and forward-looking, envisioned future outcomes.

Prior research on temporary organizations has focused on stable project-based organizations, embedded organizational units, and standalone teams. Future research could deepen and extend our understanding of experiential adaptation in temporary organizations by studying how innovation occurs in very large, standalone, and complex inter-organizational projects, such as transport, energy, sustainable cities, and Olympics projects. Some important studies identify the complex organizational structures, ambiguities, and institutional contexts surrounding the management of risks and uncertainty in 'grand-scale projects' (Shapira and Berndt, 1997), 'large-engineering projects' (Miller and Lessard, 2000), 'mega-projects' (Flyvbjerg et al., 2003), 'unique projects' (Pitsis et al., 2003), and 'global projects' (Scott et al., 2011). However, this research is largely silent about the relationship between innovation and uncertainty in temporary organizations. Following Stinchcombe and Heimer's (1985) pioneering study of North Sea oil and gas projects, there is great opportunity for future research to investigate how innovation and routines are combined and adapted in real time to tackle the uncertainty, complexity, and dynamically changing conditions found in large-scale projects— a hybrid setting involving conditions ranging from routine, planned, and predictable to novel, emergent, and highly uncertain.

Conclusion

This chapter began with the observation that projects and innovation are closely connected. It identified two different models—optimal and adaptive—which highlight the contrasting processes involved in the management of innovative projects. Building on past and recent literature, we suggest that the adaptive model is emerging as a new paradigm for understanding the relationship between project-based innovation and uncertainty. Two core constructs emerge from the converging streams of research: temporary organization and adaptive process. Both are required to explain how organizations deal with the uncertainty associated with managing innovative projects.

Future research should deepen our understanding of the close and complementary relationship between innovation and projects. As illustrated in this Handbook, research on innovation continues to attract scholars informed by a variety of theoretical traditions interested in exploring different, novel, and emerging aspects of the innovation process. Research on project management has evolved from its foundations in the rational, systems-based analysis and normative orientation of the optimization school to embrace various theoretical bases such as inter-disciplinary research associated with the 'management of projects' paradigm (Morris, 1994, 2013), the rebirth of contingency theory (Miller and Lessard, 2000; Shenhar and Dvir, 2007), and projects as organizational entities endowed with capabilities, routines, and managerial judgement to support real-time learning, innovation, and creative improvisation (Davies and Hobday, 2005; Cattani et al., 2011). The recent confluence of innovation and project management thinking is an encouraging sign that more research is focusing on temporary organizations and the adaptive processes required to manage innovative and uncertain projects.

REFERENCES

Abernathy, W. J., and Utterback, J. M. (1975). 'A Dynamic Model of Process and Product Innovation', *Omega*, 3(6): 639–56.

Bakker, R. M. (2010). 'Taking Stock of Temporary Organizational Forms: A Systematic Review and Research Agenda', *International Journal of Management Reviews*, 12(4): 466–86.

Bechky, B. A. (2006). 'Gaffers, Gofers, and Grips: Role-Based Coordination in Temporary Organizations', *Organization Science*, 17(1): 3–21.

Bechky, B. A., and Okhuysen, G. O. (2011). 'Expecting the Unexpected? How SWAT Officers and Film Crews Handle Surprises', *Academy of Management Review*, 54(2): 239–61.

Bennis, W. G., and Slater, P. L. (1968). *The Temporary Society*. New York: Harper and Row.

Bhattacharya, S., Krishan, V., and Mahajan, V. (1998). 'Managing New Product Definition in Highly Dynamic Environments', *Management Science*, 44(11): 50–64.

Boulder, K. E. (1956). 'General Systems Theory: The Skeleton of Science', *Management Science*, 3(2): 197–208.

Brady, T., and Davies, A. (2004). 'Building Project Capabilities: From Exploratory to Exploitative Learning', *Organization Studies*, 25(9): 1601–21.

Brady, T., and Hobday, M. (2011). 'Projects and Innovation: Innovation and Projects', in P. W. G. Morris, J. K. Pinto, and J. Söderlund (eds), *The Oxford Handbook of Project Management*. Oxford: Oxford University Press, chapter 11, 273–94.

Brown, S. L., and Eisenhardt, K. M. (1995). 'Product Development: Past Research, Present Findings, and Future Directions', *Academy of Management Review*, 20(2): 343–78.

Brown, S. L., and Eisenhardt, K. M. (1997). 'The Art of Continuous Change: Linking Complexity Theory and Time-Paced Evolution in Relentlessly Shifting Organizations', *Administrative Science Quarterly*, 42: 1–34.

Burns, T., and Stalker, G. M. (1961). *The Management of Innovation*. Oxford: Oxford University Press.

Cattani, G., Ferriani, S., Frederiksen, L., and Täube, F. (2011). 'Project-Based Organizing and Strategic Management: A Long-Term Research Agenda on Temporary Organizational Forms', *Advances in Strategic Management*, 28: 3–26.

Christensen, C. M. (1997). *The Innovator's Dilemma: When New Technologies Cause Great Firms to Fail*. Boston: Harvard Business School Press.

Clark, K. B., and Wheelwright, S. C. (1992). 'Organizing and Leading "Heavyweight" Development Teams', *Californian Management Review*, 34(3): 9–28.

Cooper, R. G. (1993). *Winning at New Products*. Reading, MA: Addison-Wesley.

Cooper, R. G. (1990). 'Stage-Gate Systems: A New Tool for Managing New Products', *Business Horizons*, May–June: 44–54.

Cooper, R. G., and Kleinschmidt, E. (1987). 'New Products: What Separates Winners and Losers?', *Journal of Product Innovation Management*, 4: 169–84.

Davies, A., and Brady, T. (2000). 'Organisational Capabilities and Learning in Complex Product Systems: Towards Repeatable Solutions', *Research Policy*, 29: 931–53.

Davies, A., and Frederiksen, L. (2010). 'Project Modes of Innovation: The World after Woodward', *Research in the Sociology of Organizations*, 29: 177–215.

Davies, A., Gann, D., and Douglas, T. (2009). 'Innovation in Megaprojects: Systems Integration at London Heathrow Terminal 5', *California Management Review*, 51(2): 101–25.

Davies, A., and Hobday, M. (2005). *The Business of Projects: Managing Innovation in Complex Products and Systems*. Cambridge: Cambridge University Press.

Davis, S. M., and Lawrence, P. R. (1977). *Matrix*. Reading, MA: Addison-Wesley Publishing Company.

DeFillippi, R. J., and Arthur, M. B. (1998). 'Paradox in Project-Based Enterprise: The Case of Film Making', *California Management Review*, 40(2): 125–39.

De Meyer, A., Loch, C. H., and Pich, M. T. (2002). 'Managing Project Uncertainty', *Sloan Management Review*, 43(2): 6–67.

Dodgson, M., Gann, D., and Salter, A. (2008). *The Management of Technological Innovation: Strategy and Practice*. Oxford: Oxford University Press.

Eisenhardt, K. M., and Tabrizi, B. N. (1995). 'Accelerating Adaptive Processes: Product Innovation in the Global Computer Industry', *Administrative Science Quarterly*, 40: 84–110.

Engwall, M. (2003). 'No Project is an Island: Linking Projects to History and Context', *Research Policy*, 32: 789–808.

Flyvbjerg, B., Bruzelius, N., and Rothengatter, W. (2003). *Megaprojects and Risk: An Anatomy of Ambition*. Cambridge: Cambridge University Press.

Frederiksen, L., and Davies, A. (2008). 'Vanguards and Ventures: Projects as Vehicles for Corporate Entrepreneurship', *International Journal of Project Management*, 26: 487–96.

Freeman, C., and Soete, L. (1997). *The Economics of Industrial Innovation*, 3rd edn. London: Pinter.

Gaddis, P. O. (1959). 'The Project Manager', *Harvard Business Review* (May–June): 89–97.

Galbraith, J. R. (1973). *Designing Complex Organizations*. Reading, MA: Addison-Wesley.

Gann, D. M., and Salter, A. (2000). 'Innovation in Project-Based, Service-Enhanced Firms: The Construction of Complex Products and Systems', *Research Policy*, 29: 955–72.

Gann, D., Salter, A., Dodgson, M., and Phillips, N. (2012). 'Inside the World of the Project Baron', *MIT Sloan Management Review*, 53(3): 63–71.

Gersick, C. J. G., and Hackman, J. R. (1990). 'Habitual Routines in Task-Performing Groups', *Organizational Behaviour and Human Decision Processes*, 47: 65–97.

Grabher, G. (2004). 'Temporary Architectures of Learning: Knowledge Governance in Project Ecologies', *Organization Studies*, 25(9): 1491–514.

Hayes, R. H., and Wheelwright, S. C. (1984). *Restoring our Competitive Edge: Competing through Manufacturing*. New York: Wiley.

Hirschman, A. O. (1967). *Development Projects Observed*. Washington, DC: The Brookings Institution.

Hirschman, A. O., and Lindblom, C. E. (1962). 'Economic Development, Research and Development, Policy Making: Some Converging Views', *Behavioral Science*, 7(2): 211–22.

Hobday, M. (1998). 'Product Complexity, Innovation and Industrial Organisation', *Research Policy*, 26: 689–710.

Hobday, M. (2000). 'The Project-Based Organisation: An Ideal Form for Management of Complex Products and Systems', *Research Policy*, 29: 871–93.

Hughes, T. P. (1998). *Rescuing Prometheus*. New York: Pantheon Books.

Isaacs, W. (2011). *Steve Jobs*. St Ives: Little Brown.

Johnson, S. B. (1997). 'Three Approaches to Big Technology: Operations Research, Systems Engineering and Project Management', *Technology and Culture*, 38(4): 891–919.

Jones, C., and Lichtenstein, B. (2008). 'Temporary Inter-Organizational Projects: How Temporal and Social Embeddedness Enhance Coordination and Manage Uncertainty', in S. Cropper, M. Ebers, C. Huxman, and P. Smith Ring (eds), *The Oxford Handbook of Inter-Organizational Relations*. Oxford: Oxford University Press, 231–55.

Kanter, R. M. (1990). *When Elephants Learn to Dance: Mastering the Challenges of Strategy, Management, and Careers in the 1990s*. London: Unwin Paperbacks.

Keegan, A., and Turner, J. R. (2002). 'The Management of Innovation in Project-Based Firms', *Long Range Planning*, 35(4): 367–88.

Kerzner, H. (2006). *Project Management*. New York: Wiley.

Klein, B., and Meckling, W. (1958). 'Application of Operations Research to Development Decisions', *Operations Research*, 6: 352–63.

Klein, B. H., Marshak, T. A., Marshall, A. W., Meckling, W. H., and Nelson, R. R. (1962). *The Rate and Direction of Inventive Activity*. Princeton: Princeton University Press.

Lampel, J., and Shamsie, J. (2003). 'Capabilities in Motion: New Organizational Forms and the Reshaping of the Hollywood Movie Industry', *Journal of Management Studies*, 40(8): 2002–380.

Lawrence, P. R., and Lorsch, J. W. (1967). *Organization and Environment: Managing Differentiation and Integration*. Boston: Harvard Business School Press.

Lenfle, S. (2008). 'Exploration and Project Management', *International Journal of Project Management*, 26: 469–78.

Lenfle, S., and Loch, C. (2010). 'Lost Roots: How Project Management Came to Emphasize Control over Flexibility and Novelty', *Californian Management Review*, 53(1): 32–55.

Levitt, B., and March, J. G. (1988). 'Organizational Learning', *Annual Review of Sociology*, 14: 319–40.

Loch, C. H., De Meyer, A., and Pich, M. T. (2006). *Managing the Unknown: A New Approach to Managing High Uncertainty and Risk in Projects*. Hoboken, NJ: John Wiley and Sons.

Manning, S., and Sydow, J. (2011). 'Projects, Paths and Practices: Sustaining and Leveraging Project-Based Relationships', *Industrial and Corporate Change*, 20(5): 1369–402.

March, J. G. (1991). 'Exploration and Exploitation in Organizational Learning', *Organization Science*, 2(1): 71–87.

Middleton, C. J. (1967). 'How to Set up a Project Organization', *Harvard Business Review*, March–April: 73–82.

Miller, R., and Lessard, D. R. (2000). *The Strategic Management of Large Engineering Projects: Shaping Institutions, Risks, and Governance*. Cambridge, MA: The MIT Press.

Miner, A. S., Bassoff, P., and Moorman, C. (2001). 'Organizational Improvisation and Learning: A Field Study', *Administrative Science Quarterly*, 46: 304–37.

Mintzberg, H. (1983). *Structures in Fives: Designing Effective Organizations*. Englewood Cliffs, NJ: Prentice Hall.

Mintzberg, H., and McHugh, A. (1985). 'Strategy Formulation in an Adhocracy', *Administrative Science Quarterly*, 30: 160–97.

Morris, P. W. G. (1994). *The Management of Projects*. London: Thomas Telford.

Morris, P. W. G. (2013). *Reconstructing Project Management*. Chichester: Wiley-Blackwell.

Morris, P. W. G., Pinto, J. K., and Söderlund, J. (eds) (2011). *The Oxford Handbook of Project Management*. Oxford: Oxford University Press.

Nelson, R. N., and Winter, S. G. (1982). *An Evolutionary Theory of Economic Change*. Cambridge, MA: The Belknap Press of Harvard University Press.

Nonaka, I., and Takeuchi, H. (1995). *The Knowledge-Creating Company: How Japanese Companies Create the Dynamics of Innovation*. Oxford: Oxford University Press.

Obstfeld, D. (2012). 'Creative Projects: A Less Routine Approach toward Getting New Things Done', *Organization Science*, forthcoming.

O'Reilly, C. A., and Tushman, M. L. (2004). 'The Ambidextrous Organization', *Harvard Business Review*, April: 74–833.

Pich, M. T., Loch, C. H., and De Meyer, A. (2002). 'On Uncertainty, Ambiguity and Complexity in Project Management', *Management Science*, 48(8): 1008–23.

Pinto, J. K. (2007). *Project Management: Achieving Competitive Advantage*. Upper Saddle River, NJ: Pearson Prentice Hall.

Pinto, J. K., and Kharbanda, O. P. (1995). 'Lessons for an Accidental Profession', *Business Horizons*, 5: 41–50.

Pitsis, T., Clegg, S. R., Marosszeky, M., and Rura-Polley, T. (2003). 'Constructing the Olympic Dream: A Future Perfect Strategy of Project Management', *Organization Science*, 14(5): 574–90.

Prencipe, A., and Tell, F. (2001). 'Inter-Project Learning: Processes and Outcomes of Knowledge Codification in Project-Based Firms', *Research Policy*, 30: 1373–94.

Project Management Institute (2009). *A Guide to the Project Management Body of Knowledge: PMBOK Guide*, 4th edn. Project Management Institute.

Randolph, W. A., and Posner, B. Z. (1988). 'What Every Manager Needs to Know about Project Management', *Sloan Management Review*, 29(4): 65–73.

Rich, B. R., and Janos, L. (1994). *Skunk Works: A Personal Memoir of my Years at Lockheed*. London: Warner Books.

Sapolsky, H. M. (1972). *The Polaris System Development: Bureaucratic and Programmatic Success in Government*. Cambridge, MA: Harvard University Press.

Schwab, A., and Miner, A. S. (2008). 'Learning in Hybrid-Project Systems: The Effects of Project Performance on Repeated Collaboration', *Academy of Management Journal*, 51(6): 1117–49.

Scott, W. R., Levitt, R. E., and Orr, R. J. (2011). *Global Projects: Institutional and Political Challenges*. Cambridge: Cambridge University Press.

Shamsie, J., Martin, X., and Miller, D. (2009). 'In with the Old, in with the New: Capabilities, Strategies, and Performance among the Hollywood Studios', *Strategic Management Journal*, 30: 1440–52.

Shapira, Z., and Berndt, D. J. (1997). 'Managing Grand-Scale Construction Projects: A Risk-Taking Perspective', *Research in Organizational Behaviour*, 19: 303–60.

Shenhar, A. J. (2001). 'One Size does not Fit All Projects: Exploring Classical Contingency Domains', *Management Science*, 47(3): 394–414.

Shenhar, A. J., and Dvir, D. (1996). 'Toward a Typological Theory of Project Management', *Research Policy*, 25: 607–32.

Shenhar, A. J., and Dvir, D. (2007). *Reinventing Project Management: The Diamond Approach to Successful Growth and Innovation*. Boston: Harvard Business School Press.

Slevin, D. P., and Pinto, J. K. (1987). 'Balancing Strategy and Tactics in Project Implementation', *Sloan Management Review*, Fall: 33–41.

Söderlund, J. (2011). 'Theoretical Foundations of Project Management: Suggestions for Pluralistic Thinking', in P. W. G. Morris, J. K. Pinto, and J. Söderlund (eds), *The Oxford Handbook of Project Management*. Oxford: Oxford University Press, chapter 2, 37–64.

Söderlund, J., and Tell, F. (2009). 'The P-Form Organization and the Dynamics of Project Competence: Project Epochs in Asea/ABB, 1950–2000', *International Journal of Project Management*, 27: 101–12.

Sommer, S. C., and Loch, C. H. (2004). 'Selectionism and Learning in Projects with Complexity and Unforeseeable Uncertainty', *Management Science*, 50(10): 1334–47.

Stinchcombe, A. L., and Heimer, C. A. (1985). *Organization Theory and Project Management: Administering Uncertainty in Norwegian Offshore Oil*. Oslo: Norwegian University Press and Oxford University Press.

Sydow, J., Lindkvist, L., and DeFillippi, R. (2004). 'Project-Based Organizations, Embedded-ness and Repositories of Knowledge: Editorial', *Organization Studies*, 25(9): 1475–89.

Takeuchi, H., and Nonaka, I. (1986). 'The New Product Development Game', *Harvard Business Review*, January–February: 137–46.

Teece, D., Pisano, G., and Shuen, A. (1997). 'Dynamic Capabilities and Strategic Management', *Strategic Management Journal*, 18(7): 509–33.

Thomke, S. H. (2003). *Experimentation Matters*. Boston: Harvard Business School Press.

Utterback, J. M. (1994). *Mastering the Dynamics of Innovation: How Companies Can Seize Opportunities in the Face of Technological Change*. Boston: Harvard Business School Press.

Vera, D., and Crossan. M. (2005). 'Improvisation and Innovative Performance in Teams', *Organization Science*, 16(3): 203–24.

Weick, K. E. (1998). 'Improvisation as a Mindset for Organizational Analysis', *Organization Science*, 9/5: 543–55.

Weick, K. E., and Roberts, K. H. (1993). 'Collective Mind in Organizations: Heedful Interrelating on Flight Decks', *Administrative Science Quarterly*, 38: 357–81.

Weick, K. E., and Sutcliffe, K. M. (2001). *Managing the Unexpected: Resilient Performance in an Age of Uncertainty*. San Francisco: John Wiley and Sons.

Wheelwright, S. C., and Clark, K. B. (1992a). 'Creating Project Plans to Focus Product Development', *Harvard Business Review*, 5: 70–82.

Wheelwright, S. C., and Clark, K. B. (1992b). *Revolutionizing Product Development*. New York: Free Press.

Whitley, R. (2006). 'Project-Based Firms: New Organizational Form or Variations on a Theme?', *Industrial and Corporate Change*, 15(1): 77–99.

Winter, M., Smith, C., Morris, P., and Cicmil, S. (2006). 'Directions for Future Research in Project Management: The Main Findings of a UK Government-Funded Research Network', *International Journal of Project Management*, 24: 638–49.

Woodward, J. (1965). *Industrial Organization: Theory and Practice*. Oxford: Oxford University Press.

Yoo, Y., Boland, R. J., and Lyytinen, K. (2006). 'From Organization Design to Organization Designing', *Organization Science*, 17(2): 215–29.

CHAPTER 32

··

PLATFORMS AND INNOVATION

··

ANNABELLE GAWER AND
MICHAEL A. CUSUMANO

INTRODUCTION

THIS chapter is about the role that platforms can play in innovation and their implications for innovation management. First, we define the term 'platform' and consider why this concept is important. Second, we discuss the different types of platforms as well as basic economic and strategic concepts associated with them as identified by researchers working in the field. Third, we examine a few major cases of platform leadership and innovation challenges that companies face as markets, technologies, and competition evolve. Finally, we review some of major remaining issues for future research on platforms and innovation management.[1]

THE IMPORTANCE OF PLATFORMS

Platforms exist in a variety of industries, especially in high-tech businesses driven by information technology. Microsoft, Apple, Google, Intel, Cisco, ARM, Qualcomm, EMC, and hundreds if not thousands of other firms, small and large, build hardware and software products as well as applications, and provide a variety of services, for computers, cell phones, and consumer electronics devices. All these firms and their 'ecosystem partners' participate in platform-based innovation (see Chapter 11 by Autio and Thomas). Even non-technology products, such as the Barbie doll, can be considered platforms for innovation in the sense that hundreds of firms license the rights to make clothing, accessories, and toys, or publish books, that rely on the Barbie brand name and enable children to play with each other and trade these accessories. The more users who adopt the platform, the more valuable the platform becomes to the owner and to the users because there are increasing incentives for more firms to join the ecosystem and add their own complementary innovations.

As platforms and associated innovations have become increasingly pervasive, researchers have focused on the phenomenon, resulting in several distinct academic literatures all using the term 'platform'. We see this in the new product development and operations management field (e.g. Meyer and Lehnerd, 1997; Simpson et al., 2005). The term is widely used in technology strategy (e.g. Gawer and Cusumano, 2002, 2008; Eisenmann et al., 2006). It also appears in industrial economics (e.g. Rochet and Tirole, 2003; Evans, 2003; Armstrong, 2006). Usage and meaning of the term often differs, however, which suggests that there are different types of platforms and, therefore, different ways that platforms can impact the innovation process.

Major Themes and Research Findings

Internal Platforms

The first popular usage of the term platform seems to have been in the context of new product development and incremental innovation around reused components or technologies. We refer to these as *internal platforms* in that a firm, working alone or with suppliers, can build a family of related products or sets of new features by reusing the components. Wheelwright and Clark (1992), for example, describe how these kinds of internal 'product platforms' can meet the needs of different customers simply by modifying, adding, or subtracting different features. McGrath (1995), Meyer and Lehnerd (1997), Krishnan and Gupta (2001), and Muffatto and Roveda (2002), all define platforms as a set of subsystems and interfaces that form a common structure from which a company can efficiently develop and produce a stream of derivative products. Robertson and Ulrich (1998) propose an even broader definition, viewing platforms as the collection of assets (i.e. components, processes, knowledge, people, and relationships) that a set of products share. In the marketing literature, Sawhney (1998) even suggests that managers should move from 'portfolio thinking' to 'platform thinking', which he defines as aiming to understand the common strands that tie the firm's offerings, markets, and processes together, and exploit these commonalities to create leveraged growth and variety.

This literature has identified, with a large degree of consensus, several potential benefits from designing and using internal platforms as a basis for creating new products: savings in fixed costs, efficiency gains in product development through the reuse of common parts and, in particular, the ability to produce a large number of derivative products with limited resources, as well as flexibility gains in product design. One key objective of platform-based new product development seems to be the ability to increase product variety and meet diverse customer requirements, business needs, and technical advances while maintaining economies of scale and scope within manufacturing processes—an approach also associated with 'mass customisation' (Pine, 1993).

The empirical evidence indicates that, in practice, companies have successfully used product platforms to control high production and inventory costs, as well as reduce the time to market. Most of the research is about durable goods, whose production processes involve manufacturing, such as in the automotive, aircraft, equipment manufacturing, and consumer electronics sectors. Some of the companies most frequently associated with module-based product families include Sony, Hewlett-Packard, NDC (Nippon Denso), Boeing, Honda, Rolls Royce, and Black & Decker.

In particular, we can think of internal platforms as promoting both efficiency (economies of scale and scope in design, engineering, and manufacturing) as well as incremental or modular innovation (new types or varieties of products and features derived from existing components and technologies). Simpson et al. (2005) in particular pursue this idea when they distinguish between platforms used to develop *module-based* product families and platforms used to create *scale-based* product families. A common example is Sony, which built all of its Walkman® models around key modules and platforms. It used modular design and flexible manufacturing techniques to introduce more than 250 models in the United States alone in the 1980s (Sanderson and Uzumeri, 1997). In addition, Hewlett Packard successfully developed several inkjet and LaserJet printers around modular components that reused many of the same manufacturing and assembly processes (Feitzinger and Lee, 1997). NDC made an array of automotive components for a variety of automotive manufacturers using a combinatorial strategy that involved several different modules with standardized interfaces. For instance, it was able to make 288 different types of panel meters from seventeen standardized subassemblies (Whitney, 1993).

Scale-based product families are developed by scaling one or more variables to 'stretch' or 'shrink' the platform and create products whose performance vary accordingly to satisfy a variety of market niches. Platform scaling is a common strategy employed in many industries. For example, Lehnerd (1987) explains how Black & Decker developed a family of universal motors for its power tools in response to a new safety regulation: double insulation. Prior to that, Black & Decker used different motors in each of their 122 basic tools with hundreds of variations. Rothwell and Gardiner (1990) describe how Rolls Royce scaled its RB211 aircraft engine by a factor of 1.8 to realize a family of engines with up-rated, de-rated, and re-rated Shaft Horse Power and Thrust. Sabbagh (1996) relates how Boeing developed many of its commercial airplanes by 'stretching' the aircraft to accommodate more passengers, carry more cargo, and/or increase flight range.

Automobile manufacturers often use a common platform for different products, either across models with similar quality levels, or across multiple models with different quality levels. An automobile platform generally consists of the core framework of cars, including the floor plan, drive train, and axle, which can be stretched in both width and length. Examples of platform sharing across products with similar quality levels include Mitsubishi, which shares a common platform between its Endeavour and Galant models, and Honda which shares a common platform between its CR-V and Civic (Rechtin and Kranz, 2003). Automobile manufacturers can also share a common platform across multiple products with different quality levels. For example, Toyota uses a common platform for its Landcruiser and Lexus LX 470, and Honda uses a common platform for its CR-V and Acura RDX (Anonymous, 2006; Rechtin and Kranz, 2003). Naughton et al. (1997) also relate how Honda developed an automobile platform to make a 'world car' after failing to satisfy Japanese and US markets with a single platform.

Simpson et al. (2005) also report that in the automotive industry, platforms enable greater flexibility between plants and increased plant usage—sharing underbodies between models can yield a 50 per cent reduction in capital investment, especially in welding equipment—and can reduce product lead times by as much as 30 per cent (Muffatto, 1999). In the 1990s, firms in the automotive industry that used a platform-based product development approach gained a 5.1 per cent market share per year, while firms that did not lost 2.2 per cent (Cusumano and Nobeoka, 1998). In the late 1990s, Volkswagen saved $1.5 billion per year in development and capital costs using platforms, and they produced three of the six automotive platforms that successfully achieved production volumes over one million in 1999 (Bremmer, 1999, 2000). Its platform group consist of the floor group, drive system, and running gear, along with the unseen part of the cockpit—and is shared across nineteen models marketed under its four brands Volkswagen, Audi, Seat, and Skoda.

The literature also identifies a few fundamental design principles or 'design rules' that appear to operate in internal product platforms, in particular the stability of the system architecture, and the systematic or planned reuse of modular components (Baldwin and Clark, 2000; Baldwin and Woodward, 2009). It also recognizes a fundamental trade-off couched in terms of functionality and performance: the optimization of any particular subsystem may result in the sub-optimization of the overall system (Meyer and Lehnerd, 1997). In this sense, platforms may tend to promote only incremental innovation or have a constraining effect on innovation.

Supply-Chain Platforms

A supply-chain platform extends the internal platform concept to a set of firms that follow specific guidelines or rules to supply intermediate products or components to the platform leader or the final product assembler. A major potential benefit for innovation is that a firm with access to a platform supply chain can go outside its internal capabilities to find more innovative components or technologies. It may also find these components or technologies at lower cost than it could produce itself. At the same time, a firm may have less control over the components and technology, which can have its own negative consequences. Managers therefore need to balance the pluses and minuses of going outside the firm for components or new technologies versus developing them internally, much like common discussions of make versus buy in vertical integration strategy.

Supply-chain platforms are common in assembly industries, such as consumer electronics, computers, and automobiles. For example, Renault and Nissan (as members of the Renault-Nissan alliance) have developed a common platform for the Nissan Micra and the Renault Clio, and they reduced the number of platforms they used from thirty-four in 2000 to ten in 2010 (Tierney et al., 2000; Bremner et al., 2004). Szczesny (2003) reports platform sharing between Ford Motor and Mazda. Porsche and Volkswagen use a common platform for Porsche's Cayenne and Volkswagen's Touareg, where the former is more luxurious than the latter. Sako (2009) discusses supply-chain platforms in the context of automotive supplier parts in Brazil. Other studies on supply-chain platforms include Zirpoli and Becker (2008), Zirpoli and Caputo (2002)—both in the automotive industry—Brusoni (2005), and Brusoni and Prencipe (2006) in aerospace.

The objective of these platforms is similar to internal product platforms, that is, to improve efficiency and reduce cost. The most obvious benefits come from reducing the variety of parts to design, maintain, and manufacture. Other benefits can include increases in product variety at minimal cost as well as access to new sources of innovation.

The main design principles for supply-chain platforms are very similar to those for internal platforms: the systematic reuse of modular components, and the stability of the system architecture, or at least the interfaces to the different subsystems. However, supply-chain platforms can create a specific set of challenges for managers. In particular, divergent incentives between the members of a supply chain, or of an alliance, are not uncommon. Even internal product platforms can involve a trade-off between optimizing the performance of subsystems and optimizing the performance of the overall system. In a supply-chain platform, even though there are several actors that may have divergent objectives and incentives, there is usually a clear hierarchy, with bargaining power resting with the final assembler.

Important insights from recent research on this subject of supply-chain platforms have come from Sako (2003, 2009) on modularity and outsourcing; Doran (2004) on the transfer of value-added activities within modular supply chains; and Zirpoli and Becker (2008) on the negative consequences of knowledge erosion that may accompany extreme forms of outsourcing along a supply chain. There are links between these literatures as well to other research on inter-firm modularity (Staudenmayer et al., 2005), the limits of modularity as a design strategy (Brusoni and Prencipe, 2001), and industry architecture (Jacobides et al., 2006; Pisano and Teece, 2007).

Industry Platforms

Industry platforms are products, services, or technologies that are developed by one or more firms, and which serve as foundations upon which a larger number of firms can build complementary innovations, again, in the form of specific products, related services or component technologies. There is a similarity to internal platforms or supply-chain platforms in that industry platforms provide a foundation of common components or technologies, but they differ in that this foundation is 'open' to outside firms. The degree of openness can vary on a number of dimensions—such as degree of access to information on interfaces to link to the platform or utilize its capabilities, the type of rules governing use of the platform, or cost of access (as in the form of patent or licensing fees). In general, despite different degrees of openness, various products and technologies serve as industry platforms: the Microsoft Windows and Linux operating systems, Intel and ARM microprocessors, Apple's iPod, iPhone, and iPad along with the iOS operating system and iTunes and the Apple App Store, Google's Internet search engine and Android operating system for smartphones, social networking sites such as Facebook, LinkedIn, and Twitter, video-game consoles, and even the Internet itself. So too do credit and debit cards. But a key distinction between supply-chain platforms and industry platforms is that, in the case of industry platforms, the firms developing the complementary innovations—such as applications for Windows or the Apple App Store—do not necessarily buy or sell from each other. Nor are they usually part of the same supply chain or sharing patterns of cross-ownership, such as Toyota with its major component suppliers.

The first research on industry platforms and their innovation ecosystems generally focused on computing, telecommunications, and other information-technology intensive industries. For example, Bresnahan and Greenstein (1999), in their study of the emergence of computer platforms, analysed platforms as a bundle of standard components around which buyers and sellers coordinated their efforts. West (2003) defined a computer platform as an architecture of related standards that allowed modular substitution of complementary assets such as software and peripheral hardware. Iansiti and Levien (2004) called a 'keystone firm' the equivalent of what Gawer and Cusumano (2002, 2008) referred to as a platform leader, that is, a firm that drives industry-wide innovation for an evolving system of separately developed components. Gawer and Henderson (2007) described a product as a platform when it is one component or subsystem of an evolving technological system, when it is strongly functionally interdependent with most of the other components of this system, and when end-user demand is for the overall system, so that there is no demand for components when they are isolated from the overall system.

These studies suggest several generalizations with regard to how industry platforms affect competitive dynamics as well as innovation at industry level, with implications for managers. Positions of industrial leadership are often contested and lost when industry platforms emerge, as the balance of power between assemblers and component makers changes. And, at the same time, industry platforms tend to facilitate and increase the degree of innovation on complementary products and services; so called 'complements', produced by 'complementors'. The more innovation there is on complements, the more value it creates for the platform and its users via network effects, creating a cumulative advantage for existing platforms: as they grow, they become harder to dislodge by rivals or new entrants, the growing number of complements acting as a barrier to entry. The rise of industry platforms raises complex social welfare questions as well regarding the trade-offs between the social benefits of platform-compatible innovation, versus the potentially negative effects of preventing competition on overall systems.

As for internal and supply-chain platforms, industry platforms tend to be designed and managed strategically, to further the competitive advantage of the platform leader or owner. However, there are important differences: in particular, industry platform leaders aim to tap into the innovative capabilities of *external* firms, which are not necessarily part of their supply chain. While there are some obvious links with the work of Chesbrough (2003) on open innovation, recent research on platforms has highlighted the complex trade-offs between 'open' and 'closed' (Eisenmann, Parker, and Van Alstyne, 2009; Greenstein, 2009; Schilling, 2009; see also Chapter 22 by Alexy and Dahlander). And beyond the choices about opening or closing intellectual property, platform leaders tend to strategically facilitate and stimulate complementary third-party innovation through the careful and coherent management of their ecosystem relationships as well as decisions on design and intellectual property (Cusumano and Gawer, 2002; Iansiti and Levien, 2004).

Gawer and Cusumano (2002) highlight how managers of platform firms need to pay particular attention to the governance of platforms, which requires a coherent approach using four distinct levers. The first lever is *firm scope*: the choice of which activities to perform in-house vs. which to leave to other firms. This decision is about whether the platform leader should make at least some of its own complements in-house. The second lever is *technology design and intellectual property*: what functionality or features to include in the platform, whether the platform should be modular, and to what degree the platform interfaces should be open to outside complementors and at what price. The third lever covers *external relationships with complementors*: the process by which the platform leader manages complementors and encourages them to contribute to a vibrant ecosystem. The fourth lever is *internal organization*: how and to what extent platform leaders should use their organizational structure and internal processes to give assurances to external complementors that they are genuinely working for the overall good of the ecosystem. This last lever often requires the platform leader to create a neutral group inside the company, with no direct profit-and-loss responsibility, as well as a 'Chinese wall' between the platform developers and other groups that are potentially competing with their own complementary products or services. Taken together, and dealt with in a coherent manner, the four levers offer a management template for sustaining a position of platform leadership.

Again, however, we can see platform leaders both encouraging and constraining innovation. As described in Gawer and Cusumano (2002), Intel separated internal product or R&D groups that might have conflicting interests among themselves or clash with third-party complementors, such as chipset and motherboard producers. The latter relied on Intel's advance cooperation to make sure their products were compatible. When Intel decided that these chipset and motherboard producers were not making new versions of their products fast enough to help sell new versions of microprocessors, Intel started making some of these intermediate products itself to stimulate the end-user market. But it still kept its laboratories in a neutral position to work with ecosystem partners.

In contrast, Microsoft claimed not to have such a wall between its operating systems and applications groups despite potential conflicts. Microsoft also insisted that 'integration' of different applications, systems, and networking technologies (such as embedding its own Internet browser, media player, and instant messaging technology into Windows) was good for customers because it improved performance of the overall system. There is some truth to this, and it is one reason why it is often held that the user experience with the far more integrated Macintosh system is better than the Windows–Intel PC experience. But Microsoft leveraged the market power of Windows and its other platform, Office—which by the latter 1990s had evolved into another set of services and tools used by various companies to build their own desktop application products—to influence the direction of the software business and, in some ways, constrain innovation as well as competition.

It is not illegal in itself to have a monopolistic market share, often considered to be around 70 per cent. Apple has this position with the iPod and iTunes as well as the iPad in its product categories. Intel has surpassed this level with desktop microprocessors and Cisco with basic Internet routers. But it is illegal to utilize a monopoly to harm consumers and competitors, such as through predatory pricing or contracts that impede competition. It is also illegal to use a monopoly in one product market to enter an adjacent market and thereby restrain competition. Microsoft committed this violation when it bundled Internet Explorer with Windows and did not charge extra for it, as well as pressuring PC makers not to load Netscape Navigator on their machines—essentially destroying Netscape's browser business and reducing competition in this market. Microsoft argued that the browser was an integral part of Windows. But Microsoft also sold or distributed the browser as a separate product, as did Netscape and several other companies, so this argument made little sense. Antitrust enforcement in the United States, Europe, and Asia has frequently forced Microsoft to adjust its behaviour, though usually too late to make much difference in the current market (Cusumano and Yoffie, 1998; Cusumano, 2010).

The design principles or 'design rules' of industry platforms also overlap somewhat with those for internal and supply-chain platforms. In particular, the stability of the architecture of the platform is still essential. There are, however, important differences. In contrast to what happens for internal and supply-chain platforms, in industry platforms, the logic of design is inverted. Instead of a firm being a 'master designer' (and it was always the assembler in previous contexts, that conceived of and designed an end-product, to later modularize it and dispatch various modular tasks to other groups or firms), here we start with a core component that is part of an encompassing modular structure, and the final result of the assembly is either unknown *ex ante*, or is incomplete. In fact, in industry platforms, the end use of the end-product or service is not pre-determined. This creates unprecedented scope for innovation on complementary products, services, and technologies—and in general for innovation on the nature of complementary markets. The situation also poses the fundamental question of how incentives (for third parties) to innovate can be embedded in the design of the platforms. This leads to another design rule for industry platforms: the interfaces around the platform must allow outside firms to 'plug in' complements, as well as innovate on these complements and make money from their investments.

There are also specific strategic management questions that arise in the context of industry platforms. For example, Gawer and Cusumano (2008) argue that not all products, services, or technologies can become industry platforms. To perform this industry-wide role and convince other firms to adopt the platform as their own, the platform must (a) perform a *function* that is essential to a broader technological system, and (b) solve a *business* problem for many firms and users in the industry. Nor is it clear how firms can transform their products, technologies, or services into industry platforms, or how a platform leader can stimulate complementary innovations by other firms, including some competitors, while simultaneously taking advantage of owning the platform.

One particular challenge for innovation dynamics is that platform leaders or aspirants must navigate a complex strategic landscape where both competition and collaboration occur, sometimes among the same actors. For example, as a technology evolves, platform owners often possess the opportunity to extend the scope of their platform and integrate into complementary markets. This can create disincentives for complementors to invest in innovation in these complementary markets. For example, Farrell and Katz (2000) identified the difficulty for platform owners not to squeeze the profit margins of their complementors. Gawer and Henderson (2007) show how Intel's careful selection of which complementary markets to enter (the connectors) while giving away corresponding intellectual property allowed the firm to push forward the platform–applications interface, thereby retaining control of the architecture, while renewing incentives for complementors to innovate 'on top of' the newly extended platform. Another challenge is that, as technology is constantly evolving, the business decisions about the technology or design decisions have to be taken in a coherent manner. This is difficult to achieve since these decisions are often made by different teams within organizations. Hence, to make the whole greater than the sum of the parts, we can see the need in many complex systems industries for one firm or a small group of firms to act as a 'platform leader' (Gawer and Cusumano, 2002).

Platform Dynamics: Network Effects and Multi-Sided Markets

Perhaps the most critical distinguishing feature of industry platforms compared to internal company platforms or supply chains is the potential creation of network effects. These are positive feedback loops that can cause the value of the platform to grow at exponentially increasing rates as adoption of the platform and the number of complements rise. These network effects can be very powerful, especially when they are 'direct' (sometimes called 'same-side') between the platform and the user of the complementary innovation and reinforced by a technical compatibility or interface standard that makes using multiple platforms ('multi-homing') difficult or costly. For example, Windows applications or Apple iPhone applications only work on compatible devices. Facebook users can only view profiles of friends and family within their groups. The network effects can also be 'indirect' or 'cross-side', and sometimes these are very powerful as well. These occur when, for example, advertisers become attracted to the Google search engine because of the large number of users. Companies can also innovate in business models and find ways of charging different sides of the market to make money from their platform or from complements and different kinds of transactions or advertising (Eisenmann et al., 2006).

There may be some limits to the power of network effects, however. Boudreau (2012), in a study of ecosystems for mobile computing and communications platforms, has found that, while there is a positive feedback loop to the number of complementors, this positive impact does not perpetuate itself infinitely. Too many complementors at some point seem to discourage additional firms from making the investment to join the ecosystem.

In parallel with the strategy literature, some researchers in industrial organization economics have begun using the term 'platform' to denote markets with two or more sides, and potentially with network effects that cross different sides. Such a 'multi-sided market' provides goods or services to several distinct groups of customers, all of whom need each other in some way and rely on the platform to mediate their transactions (Evans, 2003; Rochet and Tirole, 2003, 2006). For example, advertisers need Google to access the select group of users whom they want to reach through their targeted ads positioned alongside the users' Google search results. While the concept of a multi-sided market can sometimes apply to supply-chain platforms as well as industry platforms, it does not entirely conform to either category.

There are important similarities between industry platforms and multi-sided markets. Among the similarities are the existence of indirect network effects that arise between two different sides of a market when customer groups must be affiliated with the platform in order to be able to interact or transact with one another (Caillaud and Jullien, 2003; Evans, 2003; Rochet and Tirole, 2003, 2006; Armstrong, 2006; Hagiu, 2006). At the same time, though, not all multi-sided markets are industry platforms as we describe them in this chapter. The literature on double-sided markets helps us understand the 'chicken-and-egg problem' of how to encourage access to a platform for distinct groups of buyers or sellers. Double-sided markets, where the role of the platform is purely to facilitate exchange or trade, without the possibility for other players to innovate on complementary markets, seem to belong to the supply-chain category. A multi-sided market that stimulates external innovation could be regarded as an industry platform. However, while all industry platforms function in this way, not all multi-sided markets do. For example, dating bars and websites, a common example used in the literature, can certainly be seen as double-sided markets since they facilitate transactions between two distinct groups of customers. But there need not be a market for complementary innovations facilitated by the existence of the platform.

The Challenges of Platform Evolution and Change

Platform leaders supported by a global ecosystem of complementors and strong network effects should be more difficult for competitors to dislodge than companies focused on standalone products more subject to competition based on fashion or price. Even the best firms, however, face a potential challenge perhaps best described by Clay Christensen as *The Innovator's Dilemma* (1997): success ties a firm to its existing customers, technology base, and business model, and this fact makes it difficult to change, even though the technology must evolve or it will likely become obsolete. As new platform wars emerge around mobile phones, video games, cloud computing, and social networking, this section highlights some practical lessons from four well-known cases of platform leadership where companies have experienced this type of innovation dilemma.

IBM: Successful Evolution to Software and Services

IBM created the first global platform in the modern computer era, based on the IBM 360 mainframe software and family of compatible computers, introduced in the mid-1960s. Antitrust initiatives pressured IBM to release information to independent maintenance providers. This eventually led to an ecosystem of hardware 'clone' makers led by Amdahl and Fujitsu as well as software product and service companies focused on IBM customers. But IBM had the deepest knowledge of its market. It had sold early electronic computers since the early 1950s and for decades before that dominated in electro-mechanical tabulating machines and other office equipment. In the 2000s, IBM still dominated the mainframe business and undertook pioneering work in high-performance computing. After the introduction of personal computers in the 1970s and 1980s, however, enterprise computing has evolved to a much more heterogeneous world of machines and software of different shapes and sizes.

By 1980, a few key IBM executives had realized that a platform shift was occurring and the company introduced its own personal computer design in 1981. The operating system and microprocessor turned out to be the two key components of this new PC platform, and IBM ceded control over these elements to its supply-chain partners, Microsoft and Intel. This is a case where a supply-chain platform evolved to become an industry platform but under the control of the key suppliers, not the original platform architect and leader. After absorbing billions of dollars in losses, IBM developed a new strategy under CEO Louis Gerstner, hired from RJR Nabisco in 1993, when it became the champion of 'open systems' (Linux, Java, the Internet, ubiquitous computing, and the cloud). Gerstner and his successors also sold off commodity hardware businesses and rebuilt the company around services and middleware that helped customers better utilize different platform technologies (Gerstner, 2002).

The lesson here for innovation management is that platforms will evolve and the leader of one generation may lose control over the next, especially if the platform evolves from a hardware-driven system to one more based around software and services. All is not lost, however, if the erstwhile leader has distinctive capabilities that keep it linked to customers and able to recreate a competitive advantage. In the IBM case, this involved decades of experience that helped the firm understand the data-processing needs of enterprise users and other large organizations. This is where the firm kept its focus. The shift in platforms away from the mainframe and the loss of control over the PC were very damaging financially, but these changes created a new beginning for a new IBM with a much greater focus on software and services.

JVC and Sony: a Struggle to Evolve Beyond Hardware Platforms

In the 1970s and 1980s, video-cassette recorders (VCRs) became the highest volume consumer electronics product as everyone with a television set became a potential customer. Although Sony won the race to create the first viable home device, Japan Victor Corporation (JVC) ended up as the market winner. Several Japanese firms had studied the technology of a US company, Ampex, in the late 1950s, which sold machines to broadcasters. Both JVC and Sony found ways to miniaturize and improve the technology for the mass market, beating their rivals in Japan, the United States, and Europe. Sony introduced the Betamax product in 1975 and JVC countered with the VHS in 1976. By 1978 VHS had passed Betamax in sales. It became a global platform as JVC licensed the VHS technology widely, allowing other companies such as RCA and GE to influence feature development (mainly recording time), and cultivating a large set of outside firms for video content licensing and distribution. Sony may have built a better product and was first to market, but it did not cultivate ecosystem partners as well as JVC did. In the 1980s and 1990s, JVC became a multi-billion-dollar company, based mainly on its VHS platform. It gradually diversified from audio and video to computer storage products, but never dominated another market. In 2008, JVC merged with another Japanese audio equipment producer, Kenwood.

Sony had broader technical skills and greater resources than JVC, but still lost this particular platform contest as JVC accumulated more licensing partners and distributors of pre-recorded tapes. The Betamax episode caused Sony managers to cooperate better with other firms in next-generation digital video standards as well as the PlayStation platform for video games, and its Blu-ray format for DVDs. Nevertheless, Sony failed to grasp how new software and networking technologies would change future platforms for consumer electronics. The success of the Walkman, for example, introduced in 1979, was not replicated. Nor did Sony evolve the Walkman into a networked device for digital content, such as Apple did with the iPod and then the iPhone and iPad. JVC also never repeated its VHS success once networking and digital technologies came to dominate the consumer electronics market.

The lesson here for innovation management is again that platform leaders need to evolve their platforms as technologies and markets change, even when they are highly successful with their present businesses. This means creating a flexible organization that can exploit a current platform advantage while learning how to move on to the next generation of technologies and business models. JVC could have done better after the VCR era had it evolved its skills more quickly from analogue to digital technology, and then to networked systems and hardware driven by software, rather than software driven by hardware. Sony faced the same challenges and, with much greater resources, fared only slightly better. By 2012, although it still made Walkman multimedia devices as well as PCs, smartphones, and video-game consoles, and owned its own music label and movie studio, Sony was still searching for hit hardware products. It always seemed to find itself trailing in the newer platform markets that were more focused on networking capabilities, software skills, and third-party content.

Google: Thinking Broadly about Platforms and Business Models

Google's platform was initially the Internet, with a product that improved on prior search engine technology. But Google also made its platform nearly ubiquitous on PC desktops as a downloadable and free toolbar that other companies could embed in their own websites. It then built an Internet portal, with email, maps, applications, storage, and other services, to complement and feed users into the search engine. Google monetizes its leadership position by selling targeted ads that accompany searches, but it continually looks for profitable activities from organic growth and acquisition. The company quickly appreciated the value of mobile devices, for example, so it bought and then refined the Android operating system (which is based on Linux) and created the Chrome browser to facilitate mobile computing (and mobile searches as well as advertising). Google is now the largest smartphone operating system provider.

At the same time, Google has not done everything right. It was late in mounting a challenge to Facebook, with Google Plus. Its coalition of partners aiming to gain access to more social networking and social media content has been slow to create cross-platform applications and sell more search and advertising.

The management lesson here is that platform leaders should force themselves to think broadly about how to evolve their platforms and business models while extending their technical and marketing capabilities. Platforms are often useful because they can evolve and attract different types of innovations that complement the initial core technology and functions. The platform owner as well as its ecosystem partners can introduce these innovations. Google has always focused on search, but computing has evolved from the desktop to mobile devices as well as applications and content that reside within open (such as the Internet) and closed (such as Facebook) networks. Google evolved far beyond the desktop. Moreover, Google has challenged the *modus operandi* of the computer industry, that is, proprietary technology and license fees. Its software platform for mobile phones and other devices such as Netbooks and tablets is both free and technically open. It will be difficult for companies that charge for their technology and do not have large advertising income—such as Apple, Microsoft, and Nokia—to beat free and open access.

Microsoft versus Apple: A Contrast in Vision and Capabilities

Some platform leaders, despite the awareness that change is occurring, still find it difficult to evolve much beyond their original platform technologies and business models. This is especially clear when we contrast Microsoft with Apple.

Steve Ballmer, Microsoft's CEO since Bill Gates became its chairman in 2000, was often criticized for not being able to move much beyond the PC platform. Indeed, at the time of writing, Windows desktop and server and the Office suite still account for nearly 80 per cent of Microsoft's revenues and almost all its profits. Ballmer was under particular pressure because Microsoft's share price had been stagnant for more than a decade (although this was also true of Intel, Cisco, Nokia, and a host of other high-tech firms). And arch-rival Apple, despite the small (but rising) global market share of the Macintosh personal computer, and despite its near bankruptcy only a few years ago, was growing at 50 per cent a year and surpassed Microsoft in market value during 2010. Apple grew so fast because it had become a major player in consumer electronics as well as mobile phones and the distribution of digital content and software products. On the strength of its high-margin digital service platforms (iTunes, App Store, and iCloud), Apple could aspire to match or surpass Microsoft in profitability, even though reproducing digital bits is much less costly than reproducing hardware boxes.

For most of its history, Microsoft has remained the most profitable of the high-tech giants, including Apple and Google. It survived radically disruptive technological transitions and daunting business-model challenges (character-based to graphical computing, the Internet, Software-as-a-Service and cloud computing, mobile computing, and social networking). It survived antitrust scrutiny and violation. Microsoft continues to rely on the enormously profitable gross margins of the packaged software business, but change has been occurring, albeit slowly. Billions of dollars in losses ('investment') from MSN and Bing over some fifteen years has prepared Microsoft for the online world and cloud computing funded by advertising revenue. The technical failure of the launch of its Vista operating system in the early 2000s has caused it to break up Windows into smaller, more manageable technological chunks, which can also help deliver new Internet and cloud-based services. The Windows Azure cloud offering and Software-as-a-Service versions of major products have had good receptions in the marketplace and appear to be competitive, though not dominant, offerings for the future. Microsoft's acquisition of Skype in early 2011 is intended to help it integrate new technologies into the old platform by gaining access to new customers as well as better

Internet voice and video technology. Its alliance with Nokia to take over its future smartphone software, and an earlier alliance with RIM to take over the search business on the Blackberry smartphones, were conceived in the same way.

The lesson here is that great success in platform leadership and associated revenue streams can be a mixed blessing. It can promote as well as constrain innovation. Bill Gates's major mistake (in the late 1990s) was probably to insist that Microsoft remain a Windows company, rather than become a broader platform company and move quickly into new technologies and new markets. Microsoft engineers tried to integrate Windows into the new platforms, the Internet, and then mobile phones, rather than create new optimized software systems that could link the new platforms back to Windows. Windows on the desktop has, of course, been exceptionally profitable. It is not hard to understand why Gates and Ballmer were reluctant to cannibalize this business. Personal computers has been a slow growth industry for several years, however, and Microsoft's slow growth in sales and stagnant market reflect the realities of its limited vision and capabilities.

Apple, by contrast, was never wedded to the original Macintosh platform, which never caught on at the industry level and failed as a business in the 1980s and 1990s. Apple later replaced the first Mac OS with NeXT software, which was based on UNIX. But Apple did remain wedded to its remarkable capabilities in user interface design and visionary product innovation. Those skills are the basis for Apple's business success with the iPod, iPhone, iTunes, and iPad and its remarkable transformation into a global platform leader that also leads in product design and innovation. Apple has moved far beyond the personal computer and now grows at the much faster rates of several high-growth industries—smartphones, digital content distribution (music, video, books), and software product distribution.

ISSUES FOR FUTURE RESEARCH

Although various authors have been writing about theoretical concepts and case examples related to industry platforms for the past two decades, there are at least three areas where we need more research: how platforms form, how ecosystems form, and the impact of platforms and ecosystems on competitive advantage and survival.

We still do not understand, for example, if there is a common pattern or process to how most industry platforms emerge and evolve. The economics literature has so far not tackled this question, as researchers tend to assume that the platform already exists (as well as its associated markets on each 'side' of the platform). The literature on technological change and competitive dynamics, and on organizational processes, could usefully address the question of platform emergence and ecosystem creation as well. The classification of platforms offered in this chapter may indicate that, under certain conditions, there could be an evolution from internal platforms to external platforms. This hypothesis needs to be refined and tested, and we need to generate alternative explanations.

A related question is how do business ecosystems associated with platforms emerge and evolve? How often is this a strategic process driven by managers or more of a random process driven by exogenous or chance events? The networks approach from the organizational literature (see Brass et al., 2004 for a review), by bringing its insights on network dynamics and field evolution (Powell et al., 2005) and strategic networks (Lorenzoni and Lipparini, 1999; Gulati et al., 2000), is well-positioned to make significant contributions in this area (see also Chapter 6 by Kastelle and Steen). In particular, recent work by Nambisan and Sawhney (2011), building on Dhanaraj and Parkhe (2006), develops explicitly the link between platform leadership and orchestration processes in network-centric innovation. The new institutional literature rooted in sociology offers concepts such as legitimacy, collective identity, and institutional work, which can be useful to determine whether and how platform leaders can successfully establish themselves as trustworthy brokers.

Finally, we need to improve our understanding of the impact of platforms on innovation and competition. In the economics, innovation, operations, and strategy literatures, technological platforms are associated with a positive impact on innovation. The positive effect stems from the fact that, by offering unified and easy ways to connect to common components and foundational technologies, platform leaders help reduce the cost of entry in complementary markets, and provide demand for complements, often fuelled by network effects. Platforms offer, therefore, a setting where it is in the interest of private firms to elicit and encourage innovation by others. Concern over the dominant positions that platform leaders such as IBM, Microsoft, Google, or Apple can achieve, however, has raised awareness that platforms may have a potentially negative effect on competition and possibly on innovation, especially non-incremental innovation.

Further theory development could examine the role of interfaces and architecture, and how platform design might focus the attention of innovators onto specific trajectories of technological change (Dosi, 1982). These might take the form of what Rosenberg (1969) called 'inducement mechanisms and focusing devices'. It is possible that platform leaders tend to successfully stimulate a certain kind of externally-developed innovation (that would complement the platform), while aiming to discourage another kind of innovation (that would diminish the appeal or the perceived value of the platform). This type of research would highlight the potential trade-offs between innovation on modules or discrete products versus innovation on systems.

NOTES

1. This chapter builds on Gawer (2009b) ['Platform Dynamics and Strategies: from Products to Services'] and Cusumano (2011) [The Platform Leader's Dilemma].

REFERENCES

Anonymous. (2006). 'Vehicle Profiles: A User's Guide', *Consumer Reports*, 71 (April): 39–78.

Armstrong, M. (2006). 'Competition in Two-Sided Markets', *RAND Journal of Economics*, 37: 668–91.

Baldwin, C. Y., and Clark, K. B. (2000). *Design Rules: The Power of Modularity*. Cambridge, MA: MIT Press.

Baldwin, C. Y., and Woodward, C. J. (2009). 'The Architecture of Platforms: A Unified View', in A. Gawer (ed.), *Platforms, Markets and Innovation*. Cheltenham, UK and Northampton, MA: Edward Elgar, 19–44.

Boudreau, K. (2012). 'Let a Thousand Flowers Bloom? An Early Look at Large Numbers of Software app Developers and Patterns of Innovation', *Organization Science*, 23(5): 1409–27.

Brass, D. J., Galaskiewicz, J., Greve, H. R., and Tsai, W. (2004). 'Taking Stock of Networks and Organizations: A Multilevel Perspective', *Academy of Management Journal*, 47(6): 795–817.

Bremmer, R. (1999). 'Cutting-edge Platforms', *Financial Times Automotive World*, September: 30–8.

Bremmer, R. (2000). 'Big, Bigger, Biggest', *Automotive World*, June: 36–44.

Bremner, B., Edmondson, G., and Dawson, C. (2004). 'Nissan's boss', *Business Week*, 4 October: 50.

Bresnahan, T., and Greenstein, S. (1999). 'Technological Competition and the Structure of the Computer Industry', *Journal of Industrial Economics*, 47: 1–40.

Brusoni, S. (2005). 'The Limits to Specialization: Problem-solving and Coordination in Modular Networks', *Organization Studies*, 26(12): 1885–907.

Brusoni, S., and Prencipe, A. (2001). 'Unpacking the Black Box of Modularity: Technologies, Products and Organizations', *Industrial and Corporate Change*, 10(1): 179–204.

Brusoni, S., and Prencipe, A. (2006). 'Making Design Rules: A Multi-domain Perspective', *Organization Science*, 17(2): 179–89.

Caillaud, B., and Jullien, B., (2003). 'Chicken and Egg: Competition among Intermediation Service Providers', *RAND Journal of Economics*, 34: 309–28.

Chesbrough, H. W. (2003). *Open Innovation: The New Imperative for Creating and Profiting from Technology*. Boston, MA: Harvard Business School Press.

Christensen, C. M. (1997), *The Innovator's Dilemma*. Boston, MA: Harvard Business School Press.

Cusumano, M. A. (2010). 'The Evolution of Platform Thinking', *Communications of the ACM*, 53(1): 32–4.

Cusumano, M. A. (2011). 'The Platform Leader's Dilemma', *Communications of the ACM*, 54 (10): 21.

Cusumano, M. A., and Gawer, A. (2002). 'The Elements of Platform Leadership', *MIT Sloan Management Review*, 43(3): 51–8.

Cusumano, M. A., and Nobeoka, K. (1998). *Thinking Beyond Lean*. New York: The Free Press.

Cusumano, M. A., and Yoffie, D. B. (1998). *Competing on Internet Time: Lessons from Netscape and its Battle with Microsoft*. New York: The Free Press.

Dhanaraj, C., and Parkhe, A. (2006). 'Orchestrating Innovation Networks', *Academy of Management Review*, 31(3): 659–69.

Doran, D. (2004). 'Rethinking the Supply Chain: An Automotive Perspective', *Supply Chain Management: An International Journal*, 9(1): 102–9.

Dosi, G. (1982). 'Technological Paradigms and Technological Trajectories: A Suggested Interpretation of the Determinants and Directions of Technical Change', *Research Policy*, 11 (3): 147–62.

Eisenmann, T., Parker, G., and Van Alstyne, M. (2006). 'Strategies for Two-sided Markets', *Harvard Business Review*, 84(10): 92–101.

Eisenmann, T., Parker, G., and Van Alstyne, M. (2009). 'Opening Platforms: How, When and Why?', in A. Gawer (ed.), *Platforms, Markets and Innovation*. Cheltenham and Northampton, MA: Edward Elgar, 131–62.

Evans, D. S. (2003). 'The Antitrust Economics of Multi-sided Platform Markets', *Yale Journal on Regulation*, 20: 325–82.

Farrell, J., and Katz, M. L. (2000). 'Innovation, Rent Extraction, and Integration in Systems Markets', *Journal of Industrial Economics*, 97(4): 413–32.

Feitzinger, E., and Lee, H. L. (1997). 'Mass Customization at Hewlett-Packard: The Power of Postponement', *Harvard Business Review*, 75(1): 116–21.

Gawer, A. (2009a), 'Platforms, Markets and Innovation: An Introduction', in A. Gawer (ed.), *Platforms, Markets and Innovation*. Cheltenham and Northampton, MA: Edward Elgar, 1–16.

Gawer, A. (2009b), 'Platform Dynamics and Strategies: From Products to Service', in A. Gawer (ed.), *Platforms, Markets and Innovation*. Cheltenham and Northampton, MA: Edward Elgar, 45–76.

Gawer, A., and Cusumano, M. A. (2002). *Platform Leadership: How Intel, Microsoft, and Cisco Drive Industry Innovation*. Boston, MA: Harvard Business School Press.

Gawer, A., and Cusumano, M. A. (2008). 'How Companies become Platform Leaders', *MIT Sloan Management Review*, 49(2): 28–35.

Gawer, A., and Henderson, R. (2007). 'Platform Owner Entry and Innovation in Complementary Markets: Evidence from Intel', *Journal of Economics and Management Strategy*, 16(1): 1–34.

Gerstner, L. V., (2002) *Who Says Elephants Can't Dance? Leading a Great Enterprise through Dramatic Change*. New York: HarperBusiness.

Greenstein, S. (2009). 'Open Platform Development and the Commercial Internet', in A. Gawer (ed.), *Platforms, Markets and Innovation*. Cheltenham and Northampton, MA: Edward Elgar, 219–48.

Gulati, R., Nohria, N., and Zaheer, A. (2000). 'Strategic Networks', *Strategic Management Journal*, 21: 203–16.

Hagiu, A. (2006). 'Pricing and Commitment by Two-sided Platforms', *RAND Journal of Economics*, 37(3): 720–37.

Iansiti, M., and Levien R. (2004). *The Keystone Advantage: What the New Dynamics of Business Ecosystems Mean for Strategy, Innovation, and Sustainability*. Boston, MA: Harvard University Press.

Jacobides, M. G., Knudsen, T., and Augier, M. (2006). 'Benefiting from Innovation: Value Creation, Value Appropriation and the Role of Industry Architectures', *Research Policy*, 35(6): 1200–21.

Krishnan, V., and Gupta, G. (2001). 'Appropriateness and Impact of Platform-based Product Development', *Management Science*, 47: 52–68.

Lehnerd, A. P. (1987). 'Revitalizing the Manufacture and Design of Mature Global Products', in B. R. Guile and H. Brooks (eds), *Technology and Global Industry: Companies and Nations in the World Economy*. Washington, DC: National Academy Press, 49–64.

Lorenzoni, G., and Lipparini, A. (1999). 'The Leveraging of Interfirm Relationships as a Distinctive Organizational Capability: A Longitudinal Study', *Strategic Management Journal*, 20(4): 317–39.

McGrath, M. E. (1995). *Product Strategy for High-Technology Companies*. New York: Irwin Professional Publishing.

Meyer, M. H., and Lehnerd, A. P. (1997). *The Power of Product Platforms: Building Value and Cost Leadership*. New York: The Free Press.

Muffatto, M. (1999). 'Platform Strategies in International New Product Development', *International Journal of Operations and Production Management*, 19(5/6): 449–59.

Muffatto, M., and Roveda, M. (2002). 'Product Architecture and Platforms: A Conceptual Framework', *International Journal of Technology Management*, 24(1): 1–16.

Nambisan, S., and Sawhney, M. (2011). 'Orchestration Processes in Network-centric Innovation: Evidence from the Field', *Academy of Management Perspectives*, 25(3): 40–57.

Naughton, K., Thornton, E., Kerwin, K., and Dawley, H. (1997). 'Can Honda Build a World Car?' *Business Week*, 100(7).

Pine, B. J. (1993). *Mass Customization: The New Frontier in Business Competition*. Boston, MA: Harvard Business School Press.

Pisano, G. P., and Teece, D. J. (2007). 'How to Capture Value from Innovation: Shaping Intellectual Property and Industry Architecture', *California Management Review*, 50(1): 278–96.

Powell, W. W., White, D. R., Koput, K. W., and Owen-Smith, J. (2005). 'Network Dynamics and Field Evolution: The Growth of Interorganizational Collaboration in the Life Sciences', *American Journal of Sociology* 110(4): 1132–205.

Rechtin, M., and Kranz, R. (2003). 'Japanese Step Up Product Charge', *Automotive News*, 77 (August 18): 26–30.

Robertson, D., and Ulrich, K. (1998). 'Planning for Product Platforms', *MIT Sloan Management Review*, 39(4): 19–31.

Rochet, J.-C., and Tirole, J. (2003). 'Platform Competition in Two-sided Markets', *Journal of the European Economic Association*, 1(4): 990–1029.

Rochet, J.-C., and Tirole, J. (2006). 'Two-sided Markets: A Progress Report', *RAND Journal of Economics*, 35: 645–67.

Rosenberg, N. (1969). 'The Direction of Technological Change: Inducement Mechanisms and Focusing Devices', *Economic Development and Cultural Change*, 18(1): 1–24.

Rothwell, R., and Gardiner, P. (1990). 'Robustness and Product Design Families', in M. Oackley (ed.), *Design Management: A Handbook of Issues and Methods*. Cambridge, MA: Basil Blackwell Inc., 279–92.

Sabbagh, K. (1996). *Twenty-First Century Jet: The Making and Marketing of the Boeing 777*. New York: Scribner.

Sako, M. (2003). 'Modularity and Outsourcing: The nature of Co-evolution of Product Architecture in the Global Automotive Industry', in A. Prencipe, A. Davies, and M. Hobday (eds), *The Business of Systems Integration*. Oxford: Oxford University Press, 229–53.

Sako, M. (2009). 'Outsourcing of Tasks and Outsourcing of Assets: Evidence from Automotive Suppliers Parks in Brazil', in A. Gawer (ed.), *Platforms, Markets and Innovation*. Cheltenham and Northampton, MA: Edward Elgar, 251–72.

Sanderson, S. W., and Uzumeri, M. (1997). *Managing Product Families*. Chicago, IL: Irwin.

Sawhney, M. S. (1998). 'Leveraged High-variety Strategies: From Portfolio Thinking to Platform Thinking', *Journal of the Academy of Marketing Science*, 26(1): 54–61.

Schilling, M. A. (2009). 'Protecting or Diffusing a Technology Platform: Trade-offs in Appropriability, Network Externalities, and Architectural Control', in A. Gawer (ed.), *Platforms, Markets and Innovation*. Cheltenham and Northampton: Edward Elgar, 192–218.

Simpson, T. W., Siddique, Z., and Jiao, J. (2005). 'Platform-based Product Family Development: Introduction and Overview', in T. W. Simpson, Z. Siddique, and J. Jiao (eds), *Product Platforms and Product Family Design: Methods and Applications*. New York: Springer, 1–16.

Staudenmayer, N., Tripsas, M., and Tucci, C. L. (2005). 'Interfirm Modularity and its Implications for Product Development', *Journal of Product Innovation Management*, 22: 303–21.

Szczesny, J. (2003). 'Mazda Ushers in New Ford Era: Platform Sharing Across global brands is Ford's New Way of Doing Business'. Available at http://www.thecarconnection.com/index.asp?article=6574pf=1 (accessed 14 February 2006).

Tierney, C., Bawden, A., and Kunii, M. (2000). 'Dynamic Duo', *Business Week* (23 October): 26.

West, J. (2003). 'How Open is Open Enough? Melding Proprietary and Open Source Platform Strategies', *Research Policy*, 32: 1259–85.

Wheelwright, S. C., and Clark, K. B. (1992). 'Creating Project Plans to Focus Product Development', *Harvard Business Review*, 70(2): 67–83.

Whitney, D. E. (1993). 'Nippondenso Co. Ltd: A Case Study of Strategic Product Design', *Research in Engineering Design*, 5(1): 1–20.

Zirpoli, F., and Becker, M. C. (2008). 'The Limits of Design and Engineering Outsourcing: Performance Integration and the Unfulfilled Promises of Modularity', Mimeo, presented at the International Workshop on Collaborative Innovation and Design Capabilities, Mines ParisTech, Paris.

Zirpoli, F., and Caputo, M. (2002). 'The Nature of Buyer–Supplier Relationships in Co-design Activities: The Italian Auto Industry Case', *International Journal of Operations and Production Management*, 22(12): 1389–410.

Index

CPSIA information can be obtained
at www.ICGtesting.com
Printed in the USA
LVHW102129301218
602226LV00010B/384/P

9 780198 746492